Advances in Condition Monitoring, Optimization and Control for Complex Industrial Processes

Advances in Condition Monitoring, Optimization and Control for Complex Industrial Processes

Editors

Zhiwei Gao
Michael Z. Q. Chen
Dapeng Zhang

MDPI • Basel • Beijing • Wuhan • Barcelona • Belgrade • Manchester • Tokyo • Cluj • Tianjin

Editors
Zhiwei Gao
Northumbria University
UK

Michael Z. Q. Chen
Nanjing University of Science
and Technology
China

Dapeng Zhang
Tianjin University
China

Editorial Office
MDPI
St. Alban-Anlage 66
4052 Basel, Switzerland

This is a reprint of articles from the Special Issue published online in the open access journal *Processes* (ISSN 2227-9717) (available at: https://www.mdpi.com/journal/processes/special_issues/monitor_industrial_processes).

For citation purposes, cite each article independently as indicated on the article page online and as indicated below:

LastName, A.A.; LastName, B.B.; LastName, C.C. Article Title. *Journal Name* **Year**, *Volume Number*, Page Range.

ISBN 978-3-0365-0688-3 (Hbk)
ISBN 978-3-0365-0689-0 (PDF)

Contents

About the Editors

Zhiwei Gao is a Reader at Northumbria University and Head of Electrical Power and Control System Research Group. His research interests include complex industrial systems, offshore energy systems, power electronics, control engineering, and artificial intelligence techniques. He has led over 10 Special Issues in premier journals.

Michael Z. Q. Chen is Professor in the School of Automation, Nanjing University of Science and Technology. He is a Senior Member of IEEE. His research interests include passive network synthesis, inerter systems, mechanical control systems, and complex networks.

Dapeng Zhang is a Lecturer at Tianjin University and a member of automation research group. His research interests include condition monitoring, industrial control, process optimization, and artificial intelligence techniques. He has led numerous industrial projects.

Preface to "Advances in Condition Monitoring, Optimization and Control for Complex Industrial Processes"

Complex industrial automation systems/processes place high demands on system operation performance and reliability. Recent advances in monitoring, optimization, and control techniques have enabled an improved understanding of systems dynamics and an enhanced achievement of reliable monitoring and control in complex industrial processes. This Special Issue, composed of state-of-the-art review paper and intensive research papers, has highlighted advances in this field. This collection demonstrates the versatility of the area, ranging from theoretical algorithms to experimental implementation of complex industrial processes.

Zhiwei Gao, Michael Z. Q. Chen, Dapeng Zhang
Editors

Editorial

Special Issue on "Advances in Condition Monitoring, Optimization and Control for Complex Industrial Processes"

Zhiwei Gao [1,*], **Michael Z. Q. Chen** [2] **and Dapeng Zhang** [3]

[1]   Faculty of Engineering and Environment, University of Northumbria at Newcastle, Newcastle upon Tyne NE1 8ST, UK

[2]   School of Automation, Nanjing University of Science and Technology, Nanjing 210094, China; mzqchen@njust.edu.cn

[3]   School of Electrical Engineering and Automation, Tianjin University, Tianjin 300072, China; zdp@tju.edu.cn

[*]   Correspondence: zhiwei.gao@northumbria.ac.uk

Complex industrial automation systems and processes, such as chemical processes, manufacturing systems, wireless network systems, power and energy systems, smart grids and so forth, have greatly contributed to our daily life. Complex engineering systems are rather expensive, with a high requirement for system reliability and control and production performance [1,2]. As a result, there has been increasing demand in the complex industries to develop reliable condition-monitoring techniques to monitor real-time system status and promote advanced optimization algorithms and resilient control methods to ensure the desired control and operation performance. Recently, artificial intelligence, data-driven techniques, cyber-physical systems, and cloud and cognitive computation have further stimulated research and applications of monitoring, optimization, and control techniques.

This Special Issue aims to provide a platform for researchers and engineers to report their recent results, exchange research ideas, and have an overview of emerging research and application directions in condition monitoring, optimization, and advanced control for complex industrial processes. There are 25 papers included in this Special Issue after a rigorous review process, which are presented in Table 1 according to their categories.

Table 1. Categories of the papers included in the Special Issue.

Categories	Condition Monitoring & Resilient Strategies	Control Applications	Optimization
Papers	[3–11]	[12–20]	[21–27]

Citation: Gao, Z.; Chen, M.Z.Q.; Zhang, D. Special Issue on "Advances in Condition Monitoring, Optimization and Control for Complex Industrial Processes". *Processes* **2021**, *9*, 664. https://doi.org/10.3390/pr9040664

Received: 8 April 2021
Accepted: 8 April 2021
Published: 9 April 2021

Publisher's Note: MDPI stays neutral with regard to jurisdictional claims in published maps and institutional affiliations.

Condition Monitoring and Resilient Strategies for Complex Industrial Processes

Condition monitoring is the process of monitoring system parameters in an industrial process to identify a significant change that is indicative of a developing fault. Condition monitoring can be regarded as fault detection, which can tell whether the system is healthy and when a fault occurs. Along with condition monitoring and fault detection, fault isolation is used to find out which component is faulty, and fault identification aims to determine the size and type of the faults, which is important evaluation on the severity degree of a fault. Prognosis aims to predict the remaining useful life of a system or component. Condition monitoring, fault diagnosis and prognosis play a key role for predictive maintenance. Resilient methods are used to accommodate the faults so that the system can operate with a tolerant performance degradation. As a result, condition monitoring and resilience strategies are paramount to improve the reliability, safety, availability, and productiveness of an industrial automation process. Gao and Liu in [3] provided a comprehensive overview on condition monitoring, fault diagnosis, and prognosis and resilient design on wind turbine energy systems. Condition monitoring and fault diagnosis approaches were reviewed following the categories of model-based approaches, signal-based methods, knowledge-based techniques, and the hybrid of the approaches above. Prognosis methods were surveyed following the classes of model-based and data-based methods, and their combination. Resilience strategies were overviewed based on the sets of passive and active techniques and their combinations. The comparisons on different techniques were discussed, and their advantages and disadvantages were commented on. Both gear-box coupled generator-based (i.e., doubly fed induction generators) wind turbine systems and direct-drive generator-based (e.g., permanent-magnet synchronous generators) wind turbine systems were reviewed in an unified framework. An overview of further research directions in this field was also provided. The comprehensive survey paper with 106 references will much benefit the researchers and engineers so that they can get insight into this field conveniently.

Fu et al., in [4] integrated fast Fourier transform (FFT) and uncorrelated multilinear principal component analysis (UMPCA) techniques for fault detection and classification of a 4.8 MW wind turbine benchmark system under five faulty scenarios. By using the detailed comparison studies, the effectiveness of the proposed algorithm was illustrated and demonstrated. It concluded that, among all the used algorithms in this paper, the FFT showed a significantly positive impact on the improvement of the performance of the fault diagnosis and classification. Moreover, it showed that the proposed FFT plus UMPCA algorithm can also recognize the differences between the data within the same class.

A control chart plays an important role in production processes for monitoring the quality of the products and preventing defects. Aslam et al., in [5] addressed a control-chart algorithm to monitor the mean time between two events using a belief estimator under the neutron-sophic gamma distribution. The proposed control chart approach was demonstrated to be more effective to detect the causes of the variations in the process than the conventional chart using classical statistics under uncertain environments.

In [6], Xie et al., determined the key parameters of the Holmquist–Johnson–Cook (HJC) constitutive model for coal by using a series of experimental tests, which were important to understand the occurrence mechanism and predict coal-rock dynamic disasters. By implementing split Hopkinson pressure bar (SHPB) experiments and simulation studies on the impact damage of the coal using ANSYS/LS-DYNA software, the reliability of the HJC constitutive model parameters for briquette were analyzed and validated. The HJC constitutive model parameters were used to analyze impact damage of the tunnel face in simulations; it was shown that the failure process of the coal seam in the roadway was visually present. The results of this paper would benefit understanding of the mechanism of coal-rock dynamic disasters better.

Operation conditions of the flotation process can be reflected by the features of the froth image in a zinc flotation process, where the bubble size is the most evident feature. In the paper authored by Tang et al. [7], the bubble size cumulative distribution function was adopted as a new feature of the froth image, and the estimation approach for the cumulative distribution function of the bubble sizes was addressed. The froth image features would change continuously with time, caused by the change of the reagent dosage. The relationship between the time series of the bubble size cumulative distribution function and the reagent dosage was analyzed, and a nonlinear relationship model between the dynamic feature vectors and the dosage of reagent in the flotation process was established. The industrial experiment results demonstrated the effectiveness of the proposed approach for operation condition monitoring.

It is important but challenging to find out the load state of a wet ball mill during the grinding process. In the work by Cai et al. [8], a novel approach for mill load identification was proposed by synthesizing empirical wavelet transform, multiscale fuzzy entropy, and adaptive evolution particle swarm optimization probabilistic neural network classification. The feasibility of the presented method was verified using grinding experiments, which showed that overall recognition rate was as high as 97.3%, outperforming the existing methods in the literature. It concluded that the addressed method can accurately identify and monitor different load states of a ball mill.

Taking the Xiashanmao coal mine as an engineering object, the stress distribution coal seam mining process was investigated by Sun et al. [9]. Using the new technology of gob-side entry retaining by roof cutting without pillar, the mechanical model of the roof structure was established. Implementing numerical simulation, the distributions of strike and lateral abutment pressure of the working faces were obtained. Mine pressure monitoring data were used to verify the proposed methods, showing the consistency between the simulated results and monitoring results.

Modern power systems are usually complex and high-nonlinear, which can be modelled as directed graphs. The graphs can be divided into communities, and it is thus of interest to find out an optimal community to alleviate cascading failure propagation. In [10], Hu and Lee proposed three low-degree-node-based link-addition strategies to optimize the original topology, by considering islanding characteristics and node vulnerability. An evaluation index was addressed to measure impacts from sequential attacks on the network. The results of this paper would benefit design of an optimal power network to mitigate power system cascading failures.

Modern industrial systems are prone to cyber-attacks due to potential vulnerabilities of the underlying entities in systems. It is important for a system to have resilience so that it can recover to a normal state under attacks. In the paper authored by Ibrahim and Alsheikh [11], a hybrid attack graph was introduced to deal with system resilience under various attacks. By using the hybrid attack graph, the evolution of both logical and real values of system parameters can be captured under attack and recovery actions, where the hybrid attack graph was generated automatically and visualized using Java-based tools. The effectiveness of the results was illustrated by a communication network example.

Control Applications for Complex Industrial Systems

Thin coal seam mining is significant but challenging as it is subjected to small working space, low-level automation and drilling deviation. In the work authored by Ji and Liu [12], an integration of nonlinear adaptive backstepping controller and disturbance observer was addressed for position tracking control to achieve directional drilling on a coal auger. A stability analysis of the deviation control system was proved using Lyapunov stability theory. An electro-hydraulic servo displacement control experiment was set up to validate the proposed control strategy. The addressed control strategy would benefit technical breakthrough on horizontal directional drilling for thin coal seam mining.

Boost control for a variable geometry turbocharger-equipped diesel engine is a challenging task due to its strong coupling with the exhaust gas recirculation system and large delays. In the paper authored by Hu et al. [13], as one of the powerful model-free deep reinforcement learning algorithms, the deep deterministic policy gradient algorithm was used to track the target boost pressure under transient driving cycles. It showed that the proposed algorithm can achieve a satisfactory transient control performance from scratch by autonomously learning the interaction with environments. It is worthy to point out the proposed strategy can adapt to varying environment and hardware aging over time owing to its capability of self-learning on-line.

There is a growing demand for temperature control of thermal processing systems. In the paper authored by Xu et al. [14], a novel slow-mode-based control approach was presented for multipoint temperature control systems, where the temperature differences and the transient responses can be regulated, ensuring the outputs of the fast modes to follow that of the slow mode. The experiments were implemented under a DSP control platform. The proposed control algorithms were demonstrated to be effective compared with the conventional PI control methods.

In the paper authored by Xu et al. [15], a pole-zero cancellation method was proposed for temperature control in heating process systems, where the temperature differences and transient properties of all points can be adjusted by implementing dead time difference compensation and pole-zero cancellation with the feedforward reference model. Compared with conventional control approaches, the control efficiency of the proposed approach was well-demonstrated.

Voltage source converters can regulate active and reactive power rapidly, which has many applications such as in renewable energy systems and so on. In the paper by Jiang et al. [16], a novel optimal nonlinear adaptive control scheme was proposed to control voltage source converters. An extended state observer was used to estimate uncertain perturbations, and the perturbation compensation was implemented through state feedback. The proposed control strategy can ensure a consistent control performance even when operation conditions are varied. A hardware experiment was implemented to demonstrate the effectiveness of the proposed control design.

Steam/water loop is an important part of a steam power plant, which operates in harsh environments. As a result, it is challenging to design an effective controller to deliver satisfactory control performance for steam/water loop systems. Motivated by the above, Zhao et al., in [17] studied the feasibility of a distributed model predictive control strategy for steam/water loop systems. A multiple objective model predictive control approach was addressed to improve computing speed. The stability and convergence of the system under distributed model predictive control was discussed. Simulation tests were carried out on a steam/water loop system with five different sub-loops, demonstrating the effectiveness of the proposed control strategy.

The raceway reactor is the most common reactor on an industrial scale as it has advantages such as simplicity of operation and low cost for maintenance. In the work by Rodríguez-Miranda et al. [18], an event-based control architecture for PID controllers was presented, which aims to tune classical time-driven PI parameters for pH control, and then to build on event-based abilities while keeping the initial PI control design. The proposed event-based PI controller can achieve better performance by reducing the actuator effort and saving costs relevant to gas consumption, compared with the traditional on–off controllers.

In the paper authored by Ohrem et al. [19], control structure analysis and controller design were addressed for a novel multipipe separator, in order to enhance efficient production of hydrocarbons on the seabed in waters. PI controllers and model reference adaptive controllers were designed for different control loops. The proposed control methods were implemented and tested on a prototype of the separator concept, showing the effectiveness and drawing resultant conclusions.

In the paper authored by Zhang and Gao [20], an online data-driven approach was presented to improve the conversion efficiency of a refrigeration system under varying load conditions. A reinforcement learning approach was used to find out optimal actions using online data in the process level, and a coarse model was developed to evaluate action values. The actions were achieved as preset variables by implementing a single loop control. The effectiveness of the proposed approach was demonstrated by simulation studies on a test bed.

Optimization for Complex Industrial Processes

The paper authored by Tang et al., addressed a case study on optimization of the support design for a tunnel boring machine—an excavated coal mine roadway in Zhangji, China in [21]. An improved rock constitutive model of the roadway surrounding rocks was derived, and an updated failure criterion was presented based on laboratory rock tests. The proposed model and the failure criterion were used in simulation studies, and an optimal roadway support design was proposed based on simulation analyses. The feasibility and effectiveness of the support design was verified by in situ monitoring results.

Distributed generation systems play an important role in modern power networks. It is noticed that if the placement and sizing of the distributed generation systems were not selected properly, it would cause power system safety hazards. As a result, it is of significance to have a proper design for the placement and sizing of the distributed generators. In the work by Liu et al., in [22], an improved genetic algorithm was proposed to optimize the siting and sizing of the distributed generation units. The proposed optimization algorithm was demonstrated to be effective via various simulation experiments.

Han et al., proposed a novel optimization algorithm, by combining a simulated annealing algorithm-based Hopfield neural network algorithm and local scheduling rules, to solve the flexible flow shop scheduling problem with a public buffer [23]. The addressed local scheduling rules were used to control the moving process of the workpieces, reduce the production blockage, and improve the efficiency of the workpiece transfer. Based on a simulation using actual production data, the proposed method was proved to outperform the conventional methods in searching efficiency and optimization target.

In the work by Han et al. [24], a scheduling problem in a flexible flow shop with setup times was investigated, where practical constraints of the multiqueue limited buffer were taken into account in the addressed model. An improved compact genetic algorithm with local dispatching rules was presented, which was verified by using the real data from a bus manufacture production line, showing satisfactory performance.

In the work by Li et al., Aspen Plus software was used to simulate and optimize the separation of the aqueous acetonitrile solution by pressure swing distillation [25]. The total annual cost was used as the objective function, and the tray number, reflux ratio, and feeding position were the design variables to be optimized. Moreover, pressure swing distillation optimizations were compared with and without full-heat integration process. It concluded that it was more economical to separate the acetonitrile and water mixture by pressure swing distillation with full-heat integration.

In the paper authored by Cao et al. [26], reactive power optimization was investigated for large-scale power systems. A novel transfer bees optimizer was used, where Q-learning was employed to construct the learning mode of bees in order to improve the intelligence of bees through task division and cooperation. The simulation results demonstrated the proposed optimization algorithm possessed better convergent performance compared with the traditional artificial intelligence algorithms.

In the paper authored by Gao et al. [27], grouping semiconductor wire bonding equipment was investigated using processing task matching. Ahead of establishing the associated relationship between devices, a clustering by fast search and find of density peaks (CFSFDP) algorithm was addressed to cluster device attribute information and achieve the maximum number of groups of device sets, so that the resulting device groups can be obtained. The experimental results demonstrated that the improved equipment grouping method with CFSFDP algorithm outperformed the conventional methods. The equipment utilization of the bonding process segment was improved, and a good dynamic grouping was achieved so that the efficiency of the entire semiconductor production line was improved correspondingly.

Author Contributions: Conceptualization, Z.G.; writing—original draft preparation, Z.G.; writing—revision, Z.G., M.Z.Q.C., D.Z.; project administration, Z.G., M.Z.Q.C., D.Z. All authors have read and agreed to the published version of the manuscript.

Funding: There are no funding supports.

Acknowledgments: The guest editors would like to express great appreciation to all authors who contributed this Special Issue. Moreover, the guest editor team is deeply indebted to all reviewers for their constructive comments that helped shape this Special Issue.

Conflicts of Interest: The authors declare no conflict of interest.

References

1. Gao, Z.; Kong, D.; Gao, C. Modelling and control of complex dynamic systems: Applied mathematical aspects. *J. Appl. Math.* **2012**. [CrossRef]
2. Gao, Z.; Nguang, S.; Kong, D. Advances in modelling, monitoring, and control for complex industrial systems. *Complexity* **2019**. [CrossRef]

3.	Gao, Z.; Liu, X. An overview on fault diagnosis, prognosis and resilient control for wind turbine systems. *Processes* **2021**, *9*, 300. [CrossRef]
4.	Fu, Y.; Gao, Z.; Liu, Y.; Zhang, A.; Yin, X. Actuator and sensor fault classification for wind turbine systems based on fast Fourier transform and uncorrelated multi-linear principal component analysis techniques. *Processes* **2020**, *8*, 1066. [CrossRef]
5.	Aslam, M.; Bantan, R.; Khan, N. Monitoring the process based on belief statistic for neutrosophic gamma distributed product. *Processes* **2019**, *7*, 209. [CrossRef]
6.	Xie, B.; Yan, Z.; Du, Y.; Zhao, Z.; Zhang, X. Determination of Holmquist–Johnson–Cook constitutive parameters of coal: Laboratory study and numerical simulation. *Processes* **2019**, *7*, 386. [CrossRef]
7.	Tang, Z.; Tang, L.; Zhang, G.; Xie, Y.; Liu, J. Intelligent setting method of reagent dosage based on time series froth image in zinc flotation process. *Processes* **2020**, *8*, 536. [CrossRef]
8.	Cai, G.; Liu, X.; Dai, C.; Luo, X. Load state identification method for ball mills based on improved EWT, multiscale fuzzy entropy and AEPSO_PNN classification. *Processes* **2019**, *7*, 725. [CrossRef]
9.	Sun, X.; Liu, Y.; Wang, J.; Li, J.; Sun, S.; Cui, X. Study on three-dimensional stress field of gob-side entry retaining by roof cutting without pillar under near-group coal seam mining. *Processes* **2019**, *7*, 552. [CrossRef]
10.	Hu, P.; Lee, L. Community-based link-addition strategies for mitigating cascading failures in modern power systems. *Processes* **2020**, *8*, 126. [CrossRef]
11.	Ibrahim, M.; Alsheikh, A. Automatic hybrid attack graph (AHAG) generation for complex engineering systems. *Processes* **2019**, *7*, 787. [CrossRef]
12.	Ji, H.; Liu, S. Position deviation control of drilling machine using a nonlinear adaptive backstepping controller based on a disturbance observer. *Processes* **2021**, *9*, 237. [CrossRef]
13.	Hu, B.; Yang, J.; Li, J.; Li, S.; Bai, H. Intelligent control strategy for transient response of a variable geometry turbocharger system based on deep reinforcement learning. *Processes* **2019**, *7*, 601. [CrossRef]
14.	Xu, S.; Hashimoto, S.; Jiang, W.; Jiang, Y.; Izaki, K.; Kihara, T.; Ikeda, R. Slow mode-based control method for multi-point temperature control system. *Processes* **2019**, *7*, 533. [CrossRef]
15.	Xu, S.; Hashimoto, S.; Jiang, W. Pole-zero cancellation method for multi input multi output (mimo) temperature control in heating process system. *Processes* **2019**, *7*, 497. [CrossRef]
16.	Jiang, Y.; Jin, X.; Wang, H.; Fu, Y.; Ge, W.; Yang, B.; Yu, T. Optimal nonlinear adaptive control for voltage source converters via memetic salp swarm algorithm: Design and hardware implementation. *Processes* **2019**, *7*, 490. [CrossRef]
17.	Zhao, S.; Maxim, A.; Liu, S.; Keyser, R.; Ionescu, C. Distributed model predictive control of steam/water loop in large scale ships. *Processes* **2019**, *7*, 442. [CrossRef]
18.	Rodríguez-Miranda, E.; Beschi, M.; Guzmán, J.; Berenguel, M.; Visioli, A. Daytime/nighttime event-based PI control for the pH of a microalgae raceway reactor. *Processes* **2019**, *7*, 247. [CrossRef]
19.	Ohrem, S.; Skjefstad, H.; Stanko, M.; Holden, C. Controller design and control structure analysis for a novel oil–water multi-pipe separator. *Processes* **2019**, *7*, 190. [CrossRef]
20.	Zhang, D.; Gao, Z. Improvement of refrigeration efficiency by combining reinforcement learning with a coarse model. *Processes* **2019**, *7*, 967. [CrossRef]
21.	Tang, B.; Cheng, H.; Tang, Y.; Zheng, T.; Yao, Z.; Wang, C.; Rong, C. Supporting design optimization of tunnel boring machines-excavated coal mine roadways: A case study in Zhangji, China. *Processes* **2020**, *8*, 46. [CrossRef]
22.	Liu, W.; Luo, F.; Liu, Y.; Ding, W. Optimal siting and sizing of distributed generation based on improved nondominated sorting genetic algorithm II. *Processes* **2019**, *7*, 955. [CrossRef]
23.	Han, Z.; Han, C.; Lin, S.; Dong, X.; Shi, H. Flexible flow shop scheduling method with public buffer. *Processes* **2019**, *7*, 681. [CrossRef]
24.	Han, Z.; Zhang, Q.; Shi, H.; Zhang, J. An Improved compact genetic algorithm for scheduling problems in a flexible flow shop with a multi-queue buffer. *Processes* **2019**, *7*, 302. [CrossRef]
25.	Li, J.; Wang, K.; Lian, M.; Li, Z.; Du, T. Process simulation of the separation of aqueous acetonitrile solution by pressure swing distillation. *Processes* **2019**, *7*, 409. [CrossRef]
26.	Cao, H.; Yu, T.; Zhang, X.; Yang, B.; Wu, Y. Reactive power optimization of large-scale power systems: A transfer bees optimizer application. *Processes* **2019**, *7*, 321. [CrossRef]
27.	Gao, Z.-J.; Si, W.; Han, Z.; Peng, J.; Qiao, F. Grouping method of semiconductor bonding equipment based on clustering by fast search and find of density peaks for dynamic matching according to processing tasks. *Processes* **2019**, *7*, 566. [CrossRef]

processes

MDPI

Review

An Overview on Fault Diagnosis, Prognosis and Resilient Control for Wind Turbine Systems

Zhiwei Gao [1],* and Xiaoxu Liu [2]

[1] Faculty of Engineering and Environment, University of Northumbria at Newcastle, Newcastle upon Tyne NE1 8ST, UK

[2] Sino-German College of Intelligent Manufacturing, Shenzhen Technology University, Shenzhen 518118, China; liuixiaoxu@sztu.edu.cn

* Correspondence: zhiwei.gao@northumbria.ac.uk

Abstract: Wind energy is contributing to more and more portions in the world energy market. However, one deterrent to even greater investment in wind energy is the considerable failure rate of turbines. In particular, large wind turbines are expensive, with less tolerance for system performance degradations, unscheduled system shut downs, and even system damages caused by various malfunctions or faults occurring in system components such as rotor blades, hydraulic systems, generator, electronic control units, electric systems, sensors, and so forth. As a result, there is a high demand to improve the operation reliability, availability, and productivity of wind turbine systems. It is thus paramount to detect and identify any kinds of abnormalities as early as possible, predict potential faults and the remaining useful life of the components, and implement resilient control and management for minimizing performance degradation and economic cost, and avoiding dangerous situations. During the last 20 years, interesting and intensive research results were reported on fault diagnosis, prognosis, and resilient control techniques for wind turbine systems. This paper aims to provide a state-of-the-art overview on the existing fault diagnosis, prognosis, and resilient control methods and techniques for wind turbine systems, with particular attention on the results reported during the last decade. Finally, an overlook on the future development of the fault diagnosis, prognosis, and resilient control techniques for wind turbine systems is presented.

Keywords: wind turbine; energy conversion systems; condition monitoring; fault diagnosis; fault prognosis; resilient control

Citation: Gao, Z.; Liu, X. An Overview on Fault Diagnosis, Prognosis and Resilient Control for Wind Turbine Systems. *Processes* **2021**, *9*, 300. https://doi.org/10.3390/pr9020300

Academic Editor: Krzysztof Rogowski

Received: 31 December 2020

Accepted: 2 February 2021

Published: 5 February 2021

Publisher's Note: MDPI stays neutral with regard to jurisdictional claims in published maps and institutional affiliations.

1. Introduction

In order to enhance the capability of harvesting wind energy, wind turbines have become larger, but more complex and expensive. It would cost 3.3 million pounds per megawatt for installing offshore wind turbines, and spend 1.25 million pounds per megawatt for installing on-shore wind turbines. Working under harsh environments and varying load conditions, wind turbine systems are unavoidably subjected to a variety of anomalies and faults. As is known, the operation and maintenance cost for a wind turbine is relatively high, especially for one built offshore. The operation and maintenance costs for onshore and offshore wind turbines, respectively, make up 10–15% and 20–35% of the total life costs in wind turbine systems [1,2]. Therefore, there is a high demand in wind energy industries to improve the reliability, safety, availability, and productiveness of the wind turbine systems. One of the important techniques is condition monitoring and fault diagnosis, which is to monitor whether a system is healthy, detect any faults or malfunctions in their early stages, determine where the faults occur, and assess the severity of the faults so that appropriate actions can be taken in order to avoid further damages and even dangerous situations in wind turbine systems. Prognosis is a technique to predict potential faults, and estimate the remaining useful life of wind turbine systems so that timely predictive maintenances and repairing can be scheduled. Resilient control is a technique to minimize the effects from the faulty components or unexpected disruptions so that the wind turbine system can work with tolerant performance degradation under some abnormal conditions. During the past two decades, essential studies were carried out in the area of monitoring, fault diagnosis, prognosis, and resilient control for wind energy systems, which were well documented in the survey papers [3–22]. Table 1 presents some existing survey papers categorized by years and topics. Specifically, non-destructive testing methods of wind turbines at manufacture and in-service were reviewed in [3], aiming to inspect potential flaws in wind turbines. In [4], a brief review was provided for condition monitoring and fault diagnosis for various subsystems and components, such as doubly-fed induction generators, blades, and driven trains. Condition and performance monitoring techniques were surveyed in [5], with focus on blades, rotors, generators, and braking systems. In [6], a variety of condition monitoring and fault diagnosis algorithms for wind turbine systems were overviewed. [7] provided a brief discussion of advantages and limitations of different monitoring and diagnosis methods specified in each subsystem. In [8], the review focused on condition monitoring and diagnosis for wind turbine components using signal processing methods. In [9], condition monitoring techniques were reviewed, respectively, from off-line and on-line viewpoints. In [10], condition monitoring techniques for wind turbines were overviewed with discussions on trends and challenges in wind turbine maintenance. In [11], typical faults in wind turbines were discussed and structure condition monitoring and fault diagnosis approaches were inspected. Off-shore wind turbines have received much more attention in wind turbine industries recently, owing to high capability for power generations. In [12], health monitoring and safety evaluation for off-shore wind turbines were surveyed with a focus on blades, tower, and foundation in off-shore wind turbine systems. In [13], wind turbine main bearings were reviewed from the angles of design, operation, modeling, damage mechanism, and corresponding fault diagnosis methods. In the two-part survey papers [14,15], wind turbine components and their potential faults, and fault diagnosis algorithms were reviewed from the viewpoint of signal processing. In [16], machine learning based condition monitoring and diagnosis approaches were surveyed. In [17], an overview was presented for condition monitoring, fault diagnosis, and operation control (including online maintenance and fault tolerant control) on electric power conversion systems in direct-drive wind turbines. In [18], major failures in off-shore wind turbines such as grid failure, yaw system failure, electrical control failure, hydraulic failure, blade failure, and gearbox failure, were discussed and possible prognosis approaches for the failures were commented. For low-speed bearings and multistage gearbox faults in wind turbines, diagnosis and prognosis approaches were overviewed in [19] according to the applicability to wind turbine farm-level health management. In [20], a concise and specific review was carried out on prognosis and remaining useful life estimation methods for critical components in wind turbines. In [21], a state-of-the-art review for fault prognosis and predictive maintenance was documented. In [22], a brief review for fault tolerant control approaches in wind turbine systems was provided. In [23], a concise review was given on diagnosis, prognosis, and resilient control for wind turbine systems. From Table 1, one can see the majority of the review papers was focused on condition monitoring and fault diagnosis, and there were very few overview papers dealing with prognosis and resilient control. Moreover, the survey papers mainly concentrated on single topic, either diagnosis or prognosis or resilient control. It is noted that [23] was a unique review paper covering all the monitoring and diagnosis, prognosis, and resilient control. However, [23] is actually editorial review for 23 papers included in a special issue. Consequently, this motivates us to provide a comprehensive review on fault diagnosis, prognosis, and resilient control for wind turbine systems in a single paper, which would benefit the readers to appreciate the current state of the art of heath monitoring and management and control in wind turbine conversion systems.

Table 1. Review papers on monitoring and diagnosis, prognosis, and resilient control for wind energy systems.

Years	Monitoring and Diagnosis	Prognosis	Resilient Control
2000s	[3,4,6,7]		
2010s	[5,8–17,23]	[18–21,23]	[17,22,23]

The rest of this paper is organized as follows: In Section 2, the structure and typical faults of wind turbines are introduced, and the schematic methodologies of fault diagnosis, prognosis and resilient control are illustrated briefly. Section 3 presents a comprehensive survey of condition monitoring and fault diagnosis methods of wind turbines in terms of four categorizations from the viewpoint of different types of information redundancy, namely, model-based techniques, signal-based techniques, knowledge-based techniques, and hybrid techniques. Recent development of wind turbine prognosis and resilient control approaches will be reviewed in Sections 4 and 5, respectively. Finally, Section 6 concludes the work and proposes an overlook about future research about maintenance operation of wind energy systems.

2. System Overview and Fault Modes of Wind Turbines

A wind turbine system is a complex electromechanical system that converts wind energy to electrical energy. A wind turbine constitutes various subsystems and components such as blades, rotor, gearbox, generator, yaw, tower, controller, anemometer, break, and so forth. A typical structure of a wind turbine system is shown in Figure 1. Wind flows force the blades and rotor to run, which rotates the main shaft and speeds up through the gearbox to drive the generator, converting wind energy into mechanical energy, and further to electrical energy. Pitch angles can be regulated to adapt the change of wind speed, while the yaw system can align turbine with the direction of the wind identified by the anemometer. The controller is used to ensure to generate desired electricity, and the housing (or "nacelle") is mounted at the top of a tower to cover most of these components.

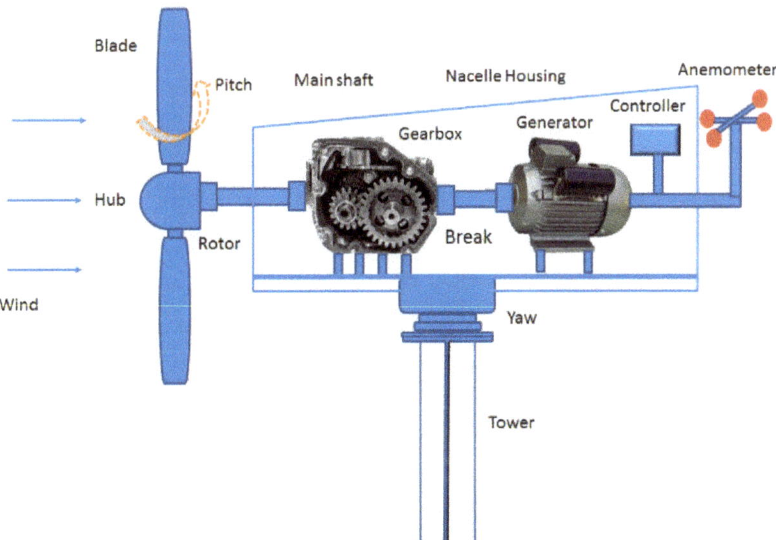

Figure 1. Wind turbine systems.

In practice, wind turbine components are prone to malfunctions or faults due to either ephemeral events or aging degradation, leading to system interruptions and economic losses. Unexpected abnormal behaviors of wind turbines can be categorized into faults and failures. A fault is recognized as an unacceptable deviation of the system structure or the system parameters from the nominal situation, whereas a failure is defined as inability of a system or a component to fulfil its function [24,25]. The pie chart in Figure 2 shows the percentages of the typical faults in wind turbines, and the main causes of the typical faults are summarized in Table 2. If an unexpected fault is not detected in the early stage and a timely action is not taken, it may cause consequent failures.

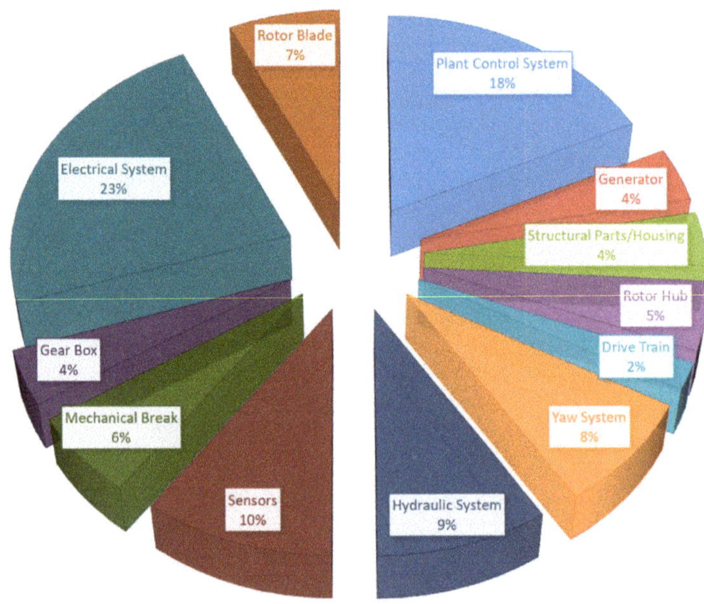

Figure 2. Percentages of typical faults in wind turbines [26].

Table 2. Typical faults in wind turbines [1,14,18].

Types of Faults	Causes of Faults
Faults on blades and rotors	Corrosion of blades and hub; crack; reduced stiffness; increased surface roughness; deformation of the blades; errors of pitch angle; and imbalance of rotors, etc.
Faults on gearbox	Imbalance and misalignment of shaft; damage of shaft, bearing and gear; broken shaft; high oil temperature; leaking oil; and poor lubrication, etc.
Faults on generator	Excessive vibrations of generator; overheating of generator and bearing; abnormal noises; and insulation damage, etc.
Faults on bearing	Overheating; and premature wear caused by unpredictable stress, etc.
Faults on main shaft	Misalignment; crack; corrosion; and coupling failure, etc.
Hydraulic faults	Sliding valve blockage; oil leakage, etc.
Faults on mechanical braking system	Hydraulic failures; and wind speed exceeding the limit, etc.
Faults on tower	Poor quality control during the manufacturing process; improper installation and loading; harsh environment, etc.
Faults on electrical systems/devices	Broken buried metal lines; corrosion or crack of traces; board delamination; component misalignment; electrical leaks; and cold-solder joints, etc.
Faults on sensors	Malfunction or physical failure of a sensor; malfunction of hardware or the communication link; and error of data processing or communication software, etc.

It is noticed that wind turbine generators can be classified into gear-box coupled wind turbine generators (e.g., doubly-fed induction generators) and direct-drive wind turbine generators (e.g., permanent-magnet synchronous generators). Although Figures 1 and 2 and Table 2 include gearbox, this survey paper will review fault diagnosis, prognosis, and resilient control approaches for both geared and gearless wind turbine systems.

Condition monitoring is defined as a process to monitor operation parameters of machinery in order to identify significant changes as an indication of a developing fault. Fault diagnosis aspires to detect the occurrence of faults, locate the faulty components, and identify the types, magnitudes, and patterns of the faults at an early stage, and the three tasks aforementioned are named as fault detection, fault isolation, and fault identification, respectively. Prognosis aims at predicting remaining operation time before faults result in failures, while resilient control is to design control laws such that the adverse influences from faults can be mitigated, ensuring the system to work normally even under faulty conditions, which may not necessarily induce an immediate component replacement or repairing for non-vital faults. The schematic diagram of the three issues is illustrated in Figure 3, where f_a, f_c, and f_s denote actuator faults, process or component faults, and sensor faults, respectively; v, u, and y are, respectively, the reference inputs, control inputs and measurement outputs. Based on recorded input and output data, fault diagnosis can be implemented to detect and locate the faulty components. The recorded data can be further used for fault prognosis and remaining useful life prediction. Based on fault diagnosis and prognosis information, resilient controls and decisions can be carried out to mitigate the adverse influences from the faults. By implementing appropriate fault diagnosis, prognosis, and resilient control strategies, the reliability and safety of wind turbine systems can be enhanced, and the maintenance cost and downtime can be reduced, which are of significant importance to achieve economic operation and increase productivity.

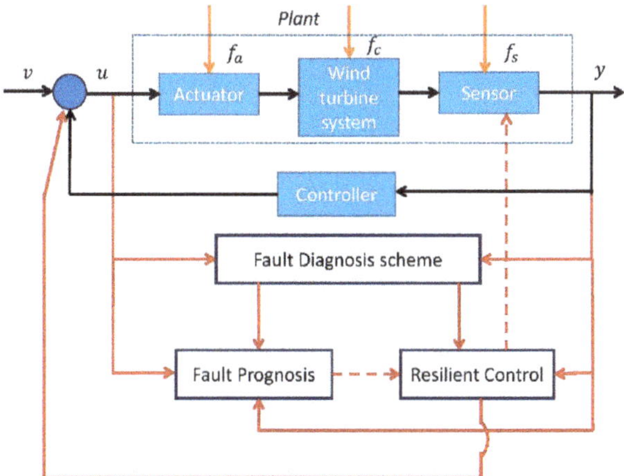

Figure 3. Schematic diagram of fault diagnosis, prognosis and resilient control.

3. Fault Diagnosis of Wind Turbines

Condition monitoring aims to check operation parameters of wind turbines to provide an early indication of faults, and fault diagnosis is conducted to detect, locate, and identify occurring faults, which allows us to plan system repair strategies prior to complete failures. Condition monitoring is actually kind of fault detection, therefore in this paper we will survey condition monitoring and fault diagnosis within a framework. From the viewpoint of different types of the information redundancy, fault diagnosis can be categorized into mode-based methods, signal-based methods, knowledge-based methods, and hybrid methods by combining the three above-mentioned methods.

3.1. Model-Based Fault Diagnosis for Wind Turbine Systems

Model-based fault diagnosis is suitable for non-stationary operation for wind turbines. This method requires models of wind turbine systems established by using either physical principles or systems identification techniques. In [27], a versatile wind velocity model was established, delivering a capability of simulating a wide range of wind variations and usual disturbances. In [28], a dynamic model was derived to simulate a doubly fed induction generation (DFIG) wind turbine with a single-cage and double-cage description of the generator rotor, and a characterization of its control and protection circuits. In [29], an industrial standard simulation tool, namely PSCAD/ EMTDC, was used to address dynamic modeling and simulation of a grid connected variable speed wind turbine. A 4.8 MW wind turbine benchmark model was originally addressed in [30] for a generic three-blade horizontal variable speed wind turbine with a full converter coupling, and the model was described with more detail in [31]. Based on wind turbine modeling software FAST, a 5MW enhanced wind turbine benchmark model was built in [32] by considering more realistic wind inputs and nonlinear behavior of aerodynamics. The wind turbine models have facilitated the development and applications of model-based fault diagnosis for wind turbine systems.

A schematic diagram of model-based fault diagnosis is depicted by Figure 4. The basic idea for model-based fault diagnosis is to provide the same inputs to the real-time wind turbine and wind turbine model, and monitor the differences between the real-time wind turbine outputs and model outputs. If the difference or called residual is zero or less than a preset threshold, the wind turbine is under healthy conditions. Otherwise, the real-time wind turbine outputs are inconsistent with the model outputs, which indicates a fault occurs in wind turbine systems. It is noticed that wind turbine models usually suffer modeling errors, and real-time turbine systems are subjected to external disturbances and varying loads, the aforementioned simple detection strategy may cause considerable false alarm rate. In order to reduce false alarm rate and improve fault detection and diagnosis performance, great efforts were paid to developing effective diagnosis algorithms so that the residual was sensitive to faults but robust against modeling errors and external disturbances.

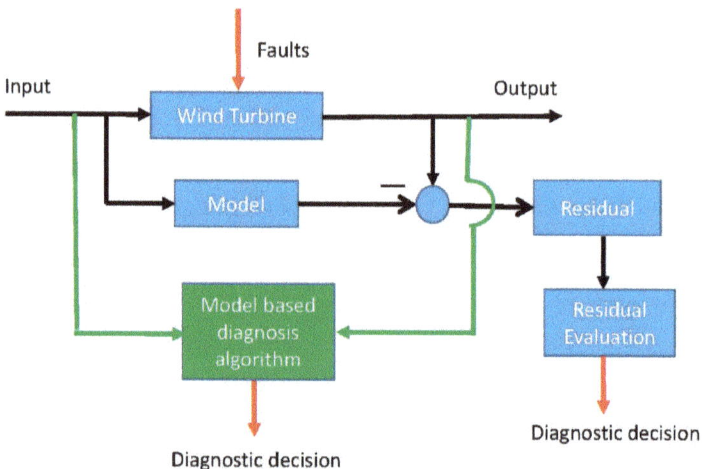

Figure 4. Schematic diagram of model-based fault diagnosis.

One of the most popular approaches is observer based fault detection approach. The key idea is to design an observer to estimate the model output, and monitor the residuals between the wind turbine outputs and the estimated model outputs. Optimization approaches are used to find a suitable observer gain to enhance the fault effects on the residuals but attenuate the influences from uncertainties to the residuals. In [33], a Luenberger observer based fault detection algorithm was addressed to detect actuator and sensor faults for a linearized 3MW wind turbine system, where parameter eigenvalue assignment approach and evolutionary algorithm were amalgamated to search an optimal observer gain so that the residual was sensitive to faults but robust against noises and perturbations. In [34], a fault diagnosis method was presented for multiple open-circuit faults in back-to-back converters of a permanent magnet synchronous generator (PMSG) drive for wind turbine systems where a Luenberger observer and adaptive threshold were used to ensure a reliable diagnosis independent of drive operation conditions. Motivated by the challenges to handle nonlinearities and partially known properties, which are difficult for mathematical modeling of wind turbines, Takagi-Sugeno (T-S) fuzzy models have drawn much attention by approximating nonlinear dynamics by using weighted aggression of a set of linear models valid around selected operating points, such that the complexity of the nonlinear problems can be decreased to linear range. In [35], a T-S fuzzy model was established for a 4.8 MW benchmark wind turbine, and the corresponding residual based fault diagnosis methodology was developed based on developed fuzzy representation. In [36], a T-S fuzzy model was built for a 4.8 MW wind turbine system, and a fault estimator was addressed to estimate actuator and sensor faults, where an augmented system approach, robust observer technique, and linear matrix inequality optimization method were integrated to ensure a robust fault estimation for generator torque actuator fault and rotor speed sensor fault against modeling errors and noises. Time-varying model has been a powerful alternative to describe wind turbine system dynamics. In [37], a time-varying model was created for a 4.8 MW wind turbine system where the blade pitch angle, tip-speed ratio, and rotor speed were the scheduling parameters to be updated real-time. Based on the time-varying model, an augmented time-varying observer was addressed to estimate parameter fault and actuator fault in wind turbine systems. In [38], internal model was used to describe wind turbine dynamics where uncertainties were located in the parameters bounding their values by intervals. Internal observers were then designed for a 5 MW wind turbine system, and fault detection was achieved by checking if the real-time measurements fall inside the estimated output interval.

In parallel with the observer, Kalman filter also plays an important role in fault detection and diagnosis for wind turbine systems where process and measurement noises are assumed to be random since wind turbine dynamics are more or less subjected to random noises in either wind speed or measurements. Kalman filter has a similar structure with observers, associated with various statistic tools (e.g., generalized likelihood ratio test), the nature of the faults can be extracted by testing on whiteness, mean, and covariance of the residuals [39]. In [40], for a three-blade horizontal axis wind turbine, a system identification algorithm was used to establish a state-space linearized model, and a Kalman filter based diagnosis algorithm was then addressed to detect additive and multiplicative sensor faults. Cascaded Kalman filters were addressed in [41] to detect faults in wind turbines which can achieve fast detection but may fail under low fault-to-noise signal ratio scenario. In order to capture more accurate means and covariance of faults, unscented Kalman filter [42] was employed to identify three fault modes, namely, gearbox faults, lubrication oil leakage, and pitch damages.

The parameter estimation approach was based on the assumption that the faults were reflected in the physical parameters, such as oil temperatures and electrical voltages of the systems. In this approach, residuals are computed as the parameter estimation errors, which are used to check the consistency of the estimated parameters with real process parameters. [43] presented an adaptive parameter estimation based fault diagnosis method to detect and isolate faults in wind turbine hydraulic pitching systems. In [44], a nonlinear parameter estimation approach was addressed for wind turbine generators by monitoring temperature trend. This method is straightforward if the model parameters have an explicit mapping with the physical coefficients. However, the diagnostic performance strongly depends on the accuracy of the measured parameters, which would be a constraint of this approach.

3.2. Signal-Based Fault Diagnosis for Wind Turbine Systems

Signal-based methods rely on appropriate sensors installed in wind turbines, rather than explicit input–output models. Sensors measure wind turbine signals such as electrical signals, vibration, and sound signals. Signal processing techniques are used to extract symptoms which are highly reflected by corresponding faults. The symptoms of real-time signals are checked with the symptoms of healthy signals from prior knowledge and experiences so that a diagnostic decision can be made. A schematic diagram is shown in Figure 5 to describe signal-based fault diagnosis approach. In general, signal-based fault diagnosis can be classified into time-domain method, frequency-domain approach, and time-frequency technique.

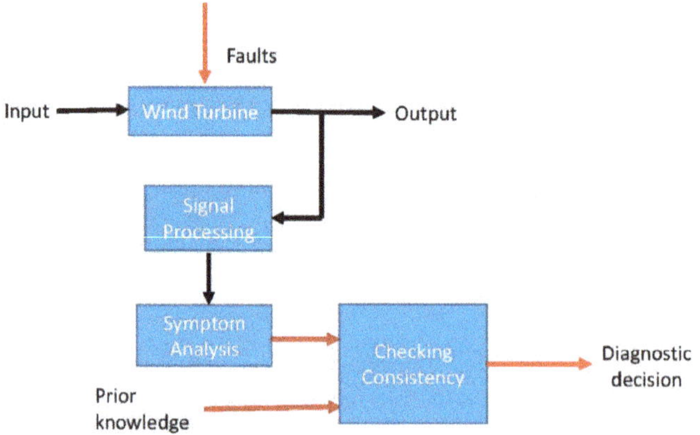

Figure 5. Schematic diagram of signal-based fault diagnosis.

Time-domain signal-based fault diagnosis utilizes time-domain parameters reflecting component malfunctions or failures such as root mean square, peak value, and kurtosis straightforwardly to monitor wind turbine dynamics. In [45], a fault diagnosis approach for multiple open-circuit faults in two converters of permanent-magnet synchronous generator drives for wind turbine application was presented by using the absolute value of the derivative of the Park's vector phase angle as a fault indicator.

Frequency-domain signal-based fault diagnosis approaches use a variety of spectrum analysis techniques, such as discrete Fourier transformation (DFT) which can be calculated by using fast Fourier transformation (FFT) [46,47], to transform a time-domain waveform into its frequency-domain equivalence, consequently used for monitoring and fault diagnosis. In [46], a two-stage fault diagnosis algorithm for wind turbine gearbox was addressed, where FFT was utilized to convert raw time-domain vibration signals to frequency spectrum, and kurtosis values were used to compute severity factors and levels by comparing with desired frequencies of the non-fault conditions. In [47], gear tooth damages were detected by checking gear vibration spectra.

In order to improve the processing ability for signals, time-frequency analysis approaches, by combining both time-domain waveform and corresponding frequency spectrum, have received much attention. In condition monitoring and fault diagnosis of wind turbines, commonly used time-frequency analysis techniques include wavelet transforms [48–50], Hilbert transform [51,52], Wigner–Ville distribution (WVD) [53], and short-time Fourier transform (STFT) [54], and so forth. Wavelet transform is applicable to non-stationary signals to enhance signal-to-noise ratio (SNR). Continuous wavelet transform was used for faulty symptom extraction, while discrete wavelet transform was employed to achieve noise cancellation in [48,50]. Hilbert transform is usually combined with other tools, such as Empirical Mode Decomposition (EMD) and time synchronous average (TSA) to reduce the influences from noises and uncertainties. Specifically, Hilbert–Huang transforms (HHT), which is an improved scheme by employing Hilbert transform and EMD, was utilized in [51] to detect gear-pitting faults. [52] utilized a time synchronous average (TSA) to extract periodic waveform from noisy signals in vibration signals when faults were detected by Hilbert transform. STFT and TSA were combined in [54] and spectral kurtosis was used to detect tooth crack faults. In practice, the aforementioned methods are often jointly used to achieve better diagnosis performances. For example, [53] proposed Morlet continuous wavelet transforms to handle extra noises and Smoothed Pseudo Wigner–Ville distribution (SPWVD) spectrum to cope with cross terms. Taking both advantages of wavelet transform and EMD, empirical wavelet transform (EWT) was adopted in [55] for generator bearing fault diagnosis.

Signal-based monitoring methods do not need to establish an explicit of mathematical model for wind turbine system. It is easily implemented by using various signal processing techniques. In general, it is suitable for monitoring and diagnosing rotating components of wind turbines, such as wheels and bearings of gearbox, bearings of generator and main bearing.

3.3. Knowledge-Based Fault Diagnosis for Wind Turbine Systems

Different from model based and signal based diagnosis methods which reply either prior mathematical model or known signal pattern, knowledge-based approach relies on a large volume of historical data available and symbolic and computational intelligence techniques to extract knowledge base, representing dependency of system variables explicitly. A diagnostic decision is made by checking the consistency between the knowledge base and the real-time operation behavior with the help from a classifier. The schematic diagram of knowledge-based fault diagnosis is shown in Figure 6. From the perspective of extraction process of historical knowledge, knowledge-based fault diagnosis methods can be classified into qualitative approaches and quantitative approaches.

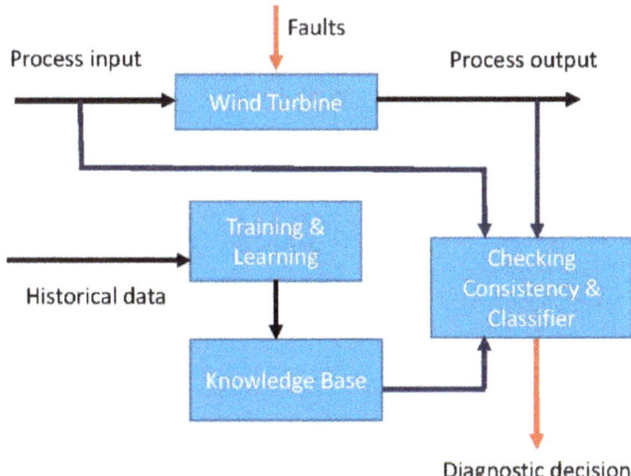

Figure 6. Schematic diagram of knowledge-based fault diagnosis.

A Root cause and fault tree analysis approach and expert system-based method can be generally regarded as qualitative knowledge approaches for condition monitoring and fault diagnosis [56,57], which have been successfully applied to wind turbines systems [58–61]. Specifically, [58] utilized fault tree analysis to describe a set of potential system failures, and cost-priority-number values were calculated to evaluate the severity of faults. In [59], a fuzzy fault tree analysis approach was addressed for risk and failure mode analysis in offshore wind turbine systems, where expert knowledge was expressed using fuzzy linguistics terms, and grey theory analysis was then integrated to determine the risk priority of the failure modes. In [60], fault tree analysis was employed to identify possible causes of top event, and expert system was then designed to implement diagnosis for gear box in a wind turbine. In [61], a fuzzy expert system was introduced by setting rules to determine the levels of faults of wind turbine gear box.

Quantitative, knowledge-based methods can be either statistical-analysis-based or non-statistical-analysis-based. Due to using a large amount of historical data, knowledge-based approaches here are often called data-driven approaches. Commonly used statistical data-driven fault diagnosis techniques include principal component analysis (PCA) [62], independent component analysis (ICA) [63], subspace aided approach (SAP) [64], fisher discriminant analysis (FDA) [65], and support vector machine (SVM) [66,67], and so on. The basic idea for PCA, ICA, SAP, and FDA is to use a variety of dimensionality reduction approaches to preserve significant trends of original data set in order to achieve promising results in fault extraction. The SVM is a nonparametric statistical method which can be used to capture faulty response of wind turbines owing to its excellence capability for classification. In [66], least squares SVM was used to train function of weather and turbine response variables; and distinguish faulty conditions from healthy conditions. Associated with appropriate nonlinear kernels tested on dataset, statistical-analysis-based approaches can attain more accurate and reliable identifications. For instance, in [67], a comparative study was carried out to demonstrate advantages of a kernel-based SVM diagnosis approach in wind turbines compared with traditional methods. In addition to statistical data-driven diagnostic techniques, non-statistical approaches, such as neural network (NN) [68,69] and fuzzy logic (FL) [70], are widely used to carry out fault diagnosis and condition monitoring for wind turbines. FL is an approach of partitioning a feature space into fuzzy sets and utilizing fuzzy rules for reasoning, which can essentially provide approximate human reasoning. Cluster center fuzzy logic approach was used in [70] to estimate wind turbine power curve. A well-trained NN has an ability of making intelligent decisions even when noises, system disturbances, and corrupted data are present. In [71], a deep neural network based fault detection approach was presented for direct-drive wind turbine systems. In [72], a fault diagnosis algorithm was proposed by using multiple extreme learning machines (ELM) layers to achieve feature learning and fault classification.

The aforementioned statistic and non-statistic data-driven fault diagnosis techniques have their own advantages and disadvantages. In practice, these methodologies are often used jointly. For instance, FL-based methods require extensive expert knowledge of the system, which may be difficult to be derived. NN is ideal for situations like this, where the knowledge, describing the behavior of the system, is stored in a large volume of quantitative datasets. However, the output is difficult to back-track due to the black-box data processing structure which causes slow convergence speed. Recent developments have shown an interest in adaptive Neuro-Fuzzy Inference System (ANFIS) to integrate these two methods, so that better fault diagnosis performances can be achieved. The joint method, addressed in [73], proved to be faster than NN in monitoring abnormal behaviors in wind turbines. Additionally, several data-driven algorithms, including PCA, k-nearest neighbor algorithms, and evolutionary strategy were combined to monitor operation of a wind farm in [74].

3.4. Hybrid Fault Diagnosis for Wind Turbine Systems

Model-based fault diagnosis method has excellent fault detection, fault isolation, and fault identification capability from a system level when a system model is available. Due to the nature of off-line design and on-line implementation, model-based fault diagnosis approach has excellent real-time performance. Signal-based approaches are independent of explicit mathematical models, which mainly focus on measurement outputs and require less knowledge about input signals. Nevertheless, it is not realistic to install sensors on all components of wind turbines from the viewpoint of economic cost, and space and weight consideration. Knowledge-based (data-driven) approaches relay on historical data and symbolic and computational intelligence, and supervisory control and data acquisition (SCADA) and smart sensors equipped in wind farms make data-driven approach feasible and attractive. It is noted that it is time consuming for training and learning. Recently, hybrid approaches by adopting more than one of the approaches are usually used to enhance fault diagnosis performance for wind turbines.

In [75], FFT was used to spot the main frequency of disturbances, and evolutionary algorithm was employed to seek an optimal observer gain to minimize the effects in the estimation error from dominant disturbances as well as low-frequency faults so that a robust fault estimation algorithm was developed for a 5 MW wind turbine system. The work in [75] is a combination of signal-based and model-based methods.

In [76], artificial neural network was used to estimate nonlinear term in wind turbine model, and linear matrix inequality optimization was addressed to find an optimal observer gain so that a robust actuator fault estimation was achieved for 4.8 MW wind turbine Benchmark system. In [77], fault detection and isolation for wind turbines were addressed by using a mixed Bayesian/Set-membership, where modeling errors were described as unknown but bounded perturbations from the viewpoint of set membership method, while measurement noises were characterized as bounded noises following a statistical distribution. The approaches in [76,77] are actually a hybrid of model-based and data-driven approaches.

Vibration signals in rotational parts of wind turbines are nonstationary and no-Gaussian, and fault samples are usually limited. In order to solve the issue above, [78] proposed a fault diagnosis algorithm based on diagonal spectrum and SVM where diagonal spectrum can be used to extract fault features from the vibration signals, and SVM can realize fault classification effectively. In [79], FFT method and uncorrelated multi-linear principal component analysis technique were integrated to achieve an effective three-dimensional space visualization for fault diagnosis and classification under five actuator and sensor faulty scenarios in a 4.8 MW wind turbine benchmark system. The methods of [78,79] are essentially a hybrid of signal-based and data-driven based approaches.

4. Prognosis for Wind Turbine Systems

Prognostics is a process to predict the progression of a deviation of a system from expected normal operating condition into a failure and estimate the remaining useful life of wind turbines. Based on various fault diagnosis and condition monitoring strategies, health status of wind turbines can be assessed and degradation patterns can be indicated, which allow a prognostic scheme to be introduced to predict when machine will fail. A schematic diagram of fault prognosis is depicted in Figure 7. When a system performance degradation (i.e., the performance departs from the normal performance) is recognized, remaining useful life (RUL) estimation will be implemented. Prognosis approaches can be generally classified into model-based, data-driven, and hybrid approaches. Model-based prognosis methods use physical and mathematical expressions to describe the degradation trend, and identify the performance degradation from real-time monitoring, and estimate the RUL of wind turbines. Data-driven approaches use historical data and machine learning techniques to train and learn system performance dynamics, identify current performance degradation from rea-time data, and predict the RUL of wind turbines. Hybrid approach is a combination of the two approaches in order to obtain a better prognosis.

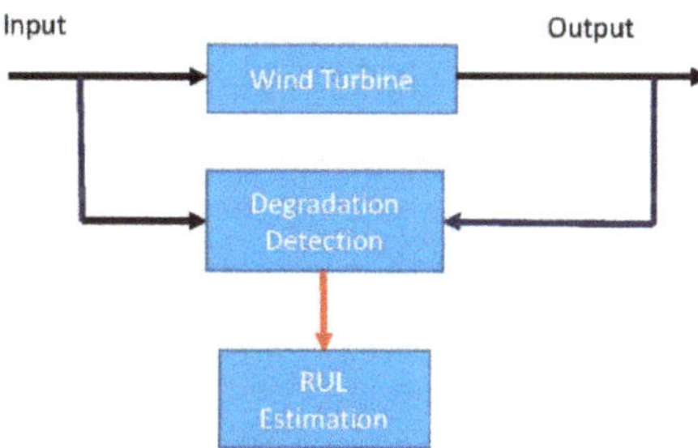

Figure 7. Schematic diagram of fault prognosis.

4.1. Model-Based Prognosis for Wind Turbine Systems

As wind turbines are subject to strongly varying loads which make the components fatigue during operation, predicting their reliable lifetime is of significance. Different fatigue models for lifetime prediction were proposed, including fatigue life models [80,81] and progressive damage models [82,83]. Fatigue life models are based on a well-known S–N curve to describe allowable cycles of failures. Prediction of fatigue life from the random load spectrum was addressed in [80,81] for medium-scale and small-scale wind turbine blades, respectively, using S–N curves. In progressive damage models, variables that describe the deterioration of the composite component were selected to assess damages. [82] presented a reduced order model to predict the occurrence and progression of damages in blades by integrating Thine-wall beam and progressive failure analysis. A progressive damage model, based on the cohesive zone concept with mixed-mode bilinear constitutive law, was addressed in [83] for analyzing fatigue of the adhesive joint root of the wind turbine blades. In [84], a probabilistic damage-growth model was utilized to characterize performance degradation of individual wind turbine, and failure prognosis informed decision-making tool was developed.

4.2. Data-Driven Prognosis for Wind Turbine Systems

One promising data-driving technique for prognosis of wind turbines is adaptive neuro-fuzzy inference system (ANFIS), which is a hybrid learning algorithm by integrating the best features of fuzzy systems and artificial neural networks. [85,86] developed ANFIS-based pitch faults prognosis for a wind farm composed of 26 wind turbines, incorporated by a priori knowledge of six known faults to train the system. Artificial intelligence systems, by integrating fuzzy logic, neural networks, and expert systems, were utilized for predictive maintenance of the wind turbine gearbox in [61]. Another popular methodology for prognosis successfully implemented in wind conversion systems is genetic algorithm (GA), inspired by Darwinian evolutionary models, which was used in [87], such that blade pitch faults can be predicted 5–60 min in advance. Moreover, a number of data mining approaches, consisting of NN, ensemble NN, standard classification, and regression tree, boosting tree algorithm, and SVM were cooperated in [88] to achieve fault prediction at three levels, namely, identification of existence of a fault, prediction of the severity of the fault, and prediction of specific fault.

4.3. Hybrid Prognosis for Wind Turbine Systems

Since prognosis of a problem is more challenging than diagnosis, it is common to integrate different types of approaches. For instance, in [89], a model was derived to describe mathematical relationship between lubrication, oil degradation, and particle contamination level, and a particle filtering technique-based RUL prediction tool was used to achieve lubrication oil prognosis by means of predicting state values in terms of probability density function (PDF). This method is a hybrid of model-based and data-driven method.

5. Resilient Control of Wind Turbine Systems

Resilient control strategies aim to mitigate the influences from unexpected faults (rather than failures) or unexpected dynamics (e.g., unknown delays) such that the overall function of the wind turbines can be maintained, although the operation and production performance may be reduced but tolerated. There are two types of resilient control approaches: passive resilient control and active resilient control.

A schematic diagram of resilient control for wind turbines is depicted by Figure 8, where f_a, f_c, and f_s denote the actuator faults, parameter (or process) faults, and sensor faults, respectively. In passive resilient control, a fixed controller is designed that tolerates all considered faulty conditions (or abnormalities) of the plant. This approach requires no on-line detection of the abnormalities, and is therefore more attractive computationally. However, a passive resilient control would be invalid if an unexpected abnormality occurred that was not considered in the design. In active resilient control, the controller is reconfigured by control laws that react to abnormalities based on information extracted by real-time monitoring and diagnosis scheme. Once a fault or abnormality is identified, effective configuration strategies can be conducted to attenuate the impact of the abnormalities on the wind turbine systems.

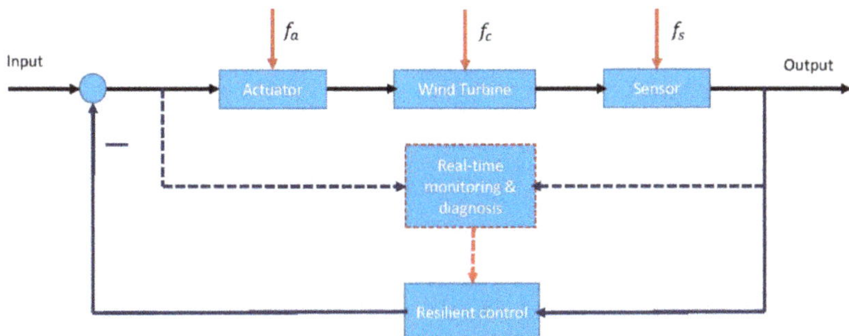

Figure 8. Schematic diagram of resilient control.

5.1. Passive Resilient Control

In [90], a passive resilient control strategy was addressed to avoid saturation caused by potential faults in 5MW wind turbine, and the key idea used was to manipulate the reference power and generator speed set-points hysterically. In [91], wind turbine system was described by linear parameter varying (LPV) system, and a robust control strategy was developed so that the system was resilient against a fault in a pitch system without need of the information from monitoring and fault diagnosis. In [92], a passive, fault tolerant cooperative control scheme was presented for a wind farm under power generation faults where fuzzy model reference control was used in a cooperative framework. In [93], a robust super-twisting algorithm-based control scheme was designed for a large floating offshore wind turbine disrupted by wind turbulence and pitch actuator faults, so that a tolerant operation was procured.

5.2. Active Resilient Control

Fault estimation and compensation have proven a powerful tool for resilient control design and implementation. In [36], a 4.8MW wind turbine system was approximated by a Takagi-Sugeno fuzzy model. By using an augmented unknown input observer, actuator and sensor faults were estimated, and signal compensation techniques were used to mitigate the effects from the actuator faults on the system dynamics and the influences of the sensor faults on the system outputs. It was proved and demonstrated that the existing controllers with compensation can ensure a tolerant operation of the wind turbine under predefined low-frequency actuator and sensor faults. The approach in [36] can deliver both real-time fault diagnosis (fault estimation) and resilient control, but there is no need for an on-line controlupdate. A hydraulic press drive unit may cause unknown delays of the pitch dynamics, which has adverse effects on wind turbine operation performance. In [94], an augmented observer was proposed to estimate a perturbed term caused by unknown delays of the pitch system, and a sensor compensation technique was addressed to mitigate the adverse effect of the unknown delay on the pitch output dynamics in a 4.8 MW wind turbine system. In [95], a disturbance observer was addressed to estimate pitch actuator fault, and a fault tolerant control with actuator compensation was designed to achieve tolerant operation of a 5MW wind turbine, under a pitch actuator fault. In [96], an adaptive sliding mode observer was addressed to estimate a pitch actuator fault in a wind turbine, and the estimated fault signal was used to compensate the effects from the actuator fault. In [97], a perturbation observer was used to estimate time-varying external disturbances including grid faults, voltage dips, and intermittent wind power inputs, and a nonlinear adaptive control with compensation was used to enhance the fault ride-through capability for a full-rated converter wind turbine. In [98], an adaptive sliding mode tolerant controller with compensation was addressed to alleviate the fluctuations in rotor speed, generator speed, and generator power under faulty conditions in a 5MW wind turbine system. In [99], an adaptive tolerant control algorithm, with the aid of fault estimation, was presented for wind turbines subjected to effectiveness loss faults in pitch actuators. In [100], a tolerant control strategy was proposed for wind turbine systems under bias faults of converter actuators in a 2 MW wind turbine system, in which fault detection and estimation were achieved by using residual filter and fault estimator, and receding horizon control technique was used to reconfigure control parameters so that the turbine health such as maximum power and less fatigue reduction was attained under faults.

In [101], a resilient configuration of doubly fed induction generator (DFIG) in 1.5 MW wind turbine was addressed to achieve tolerant operation under various kinds of grid faults, where nine-switch converter was used to replace conventional six-switch converter, and appropriate control algorithm was designed to ensure a seamless fault-ride through under grid faults. Power converter is recognized as one of the most fragile parts in wind turbine conversion systems, which contributes about 14% of the downtime of a wind turbine. In [102], a fault-tolerant operation strategy against switch faults was addressed where an additional power switch leg was used to replace a faulty leg using fault diagnosis information and corresponding control algorithms. In [103], a fault tolerant control method was addressed for direct-drive wind turbine systems under open circuit faults in machine side converters by regulating SVPWM switching patterns. It is evident that the aforementioned techniques are model-based active resilient control techniques.

In [104], a data-driven resilient control approach was addressed for a wind turbine benchmark system. Specifically, a residual generator was constructed directly identified from the input and output data, which should be sensitive to faults. The residual was embedded into the control loop to mitigate the effects from the faults and achieve tolerant operation performance under faulty conditions. In [105], a data-driven fault tolerant control scheme was presented for wind turbine systems in which the residual generator was included in the control loop so that the key performance indicator (e.g., the quality of produced power) was maintained in the admissible range under faulty conditions. In [106], a data-driven fault tolerant control approach was developed for 10 MW off-shore wind turbines, where a subspace algorithm was employed to identify a linearized-dynamics of the wind turbine, and an adaptive repetitive control law was formulated to mitigate faulty induced loads.

6. Conclusions and Overlook

The presented paper has provided a comprehensive survey covering three crucial topics, namely fault diagnosis, prognosis, and resilient control, of wind turbines, which are beneficial to maintain operation, improve energy productivity, prolong the life of usage and enhance system safety.

For fault diagnosis, it has been reviewed following the categories of model-based, signal-based, knowledge based, and hybrid approaches. Model-based monitoring and diagnosis approaches need a mathematical model to describe explicit relationships between system inputs and outputs in wind turbine systems, which are effective and powerful to carry out real-time monitoring and fault diagnosis from a system level. How to develop an accurate mathematical model and how to enhance the robustness of the model-based fault diagnosis algorithms against modeling errors and external disturbances, and sensitivity to the faults monitored are the key factors for model-based fault diagnosis approaches. Owing to off-line design, and on-board implementation, model-based monitoring, and fault diagnosis algorithms have excellent real-time performance. Signal-based monitoring and diagnosis approaches do not need system models, but rely on measurement signals from sensors, which are convenient for implementation. The measured signals are mainly dependent on system outputs, but with less attention on inputs, signal-based approaches would be sensitive to external disturbances and load changes. Knowledge-based approaches do not need to establish an explicit mathematical model, but use historical data to train and search in order to represent an implicit relationship among the variables. Knowledge-based approaches are effective for monitoring and diagnosis for both system-level faults and structural faults in wind turbines. A knowledge-based approach is highly dependent on the quality of the recorded data, and is time-consuming for training and searching. The three approaches above have own advantages and disadvantages, it would be a better solution to integrate them to lead a hybrid design and implementation to achieve a reliable and effective monitoring and diagnosis for wind turbines.

For prognosis and remaining useful life prediction, it has been reviewed following model-based, data-based, and hybrid approaches. Model-based method needs to derive an explicit physical or mathematical expression to describe the performance degradation trend, and the remaining useful life is estimated once upon the performance degradation status is identified by real-time monitoring. A model-based method needs a thorough understanding on how a physical parameter or symptom relates the performance degradation. Data-driven methods reply on historical run-to-failure data, but do not need a mathematical model. It would be difficult to obtain sufficient and reliable run-to-failure data in practice, particularly for wind turbines as they are expensive, and the machines generally stop before a collapse happens. It would be a better solution to integrate model-based and data-driven based prognosis approaches for an effective and reliable fault prediction. Compared with condition monitoring and fault diagnosis approaches, prognosis and remaining useful life estimation need much more research and development due to the complexity of wind turbine systems.

For resilient control, it has been surveyed following the categories of passive resilient control method and active resilient control method. Passive resilient control approaches do not need the information of healthy status in wind turbines, but design a robust controller so that the stability and operation of wind turbines are robust against both disturbances and faults. Resilient passive control is simple to implement, but generally has limited tolerant capabilities to accommodate faults. Active resilient control approaches need the information from real-time monitoring and fault diagnosis, and the controllers are reconfigured to mitigate the adverse effects from the faults, and achieve a tolerant operation performance. Active resilient control approaches are more attractive as they are integrated with fault diagnosis, which can effectively adapt to faulty conditions by appropriate control configurations in terms of monitored faults. It is noticed that the majority of resilient control approaches for wind turbines systems are model-based, and only a few works use data-driven approaches. It is encouraged to develop data-driven based resilient control approaches for wind turbine systems with the aid of large amount of data available and machine learning techniques.

Recently, offshore wind turbines have received much more attention, owing to their capabilities for capturing larger wind power compared with on-shore wind turbines. Offshore wind turbines are classified into fixed-foundation offshore wind turbines and floating offshore wind turbines. Floating offshore wind turbines can be installed in deep water over 50 m, which can harvest more and steadier wind power, and have less environment effect. As a result, floating offshore wind turbines will be being invested more and more, and would dominate wind turbine industries in the future. Due to the limited accessibility and a more complex structure integrated with wind turbine machine, mooring lines and floating platform, it is challenging but promising to further stimulate the research and development of real-time monitoring and fault diagnosis, prognosis and remaining useful life prediction, and resilient control for floating off-shore wind turbines to improve the reliability, availability, and productiveness.

In addition, wireless sensory and distributed networked wind farms would bring new opportunities and challenges for reliability and safety of wind turbine systems. Diagnosis and resilient control against cyber-attacks in wind turbine systems would be a promising research topic in the near future.

We have tried to comprise as many up-to-date references for fault diagnosis, prognosis, and resilient control for wind turbines as possible. Woefully, it is impossible to include all the existing publications due to the limit of space. We hope this review paper can bring a light to the researchers and engineers so that they can get insight into this field conveniently.

Author Contributions: Conceptualization, Z.G.; writing—original draft preparation, X.L., Z.G.; writing—revision, Z.G.; supervision, Z.G.; project administration, Z.G.; All authors have read and agreed to the published version of the manuscript.

Funding: This research was funded by the National Nature Science Foundation of China, grant number 61673074; Guangdong Basic and Applied Basic Research Foundation, grant number 1515110234; and Nature Science Foundation of Top Talent of SZTU, grant number 2020106.

Acknowledgments: The authors would like to acknowledge the research support from Northumbria University (UK), and the National Nature Science Foundation of China under the grant 61673074.

Conflicts of Interest: The authors declare no conflict of interest.

References

1. Verbruggen, T. Wind turbine operation & maintenance based on condition monitoring. In *ECN Wind Energy*; Technical Report ECN-C-03-047; ECN: Petten, The Netherlands, 2003.
2. McMillan, D.; Ault, G. Quantification of condition monitoring benefit for offshore wind turbines. *Wind Energy* **2007**, *31*, 267–285. [CrossRef]
3. Drewry, M.; Georgiou, G. A review of NDT techniques for wind turbines. *Insight* **2007**, *49*, 137–141. [CrossRef]
4. Amirat, Y.; Benbouzid, M.; Al-Ahmar, E.; Bensaker, B.; Turri, S. A brief status on condition monitoring and fault diagnosis in wind energy conversion systems. *Renew. Sustain. Energy Rev.* **2009**, *13*, 2629–2636. [CrossRef]
5. Aval, S.; Ahadi, A. Wind turbine fault diagnosis techniques and related algorithms. *Int. J. Renew. Energy Res.* **2016**, *6*, 80–89.
6. Hameed, Z.; Hong, Y.; Cho, Y.; Ahn, S.; Song, C. Condition monitoring and fault detection of wind turbines and related algorithms: A review. *Renew. Sustain. Energy Rev.* **2009**, *13*, 1–39. [CrossRef]
7. Lu, B.; Li, Y.; Wu, X.; Yang, Z. A review of recent advances in wind turbine condition monitoring and fault diagnosis. In Proceedings of the IEEE Power Electronics and Machines in Wind Applications, Lincoln, NE, USA, 24–26 June 2009; pp. 1–7.
8. Márquez, F.; Tobias, A.; Pérez, J.; Papaelias, M. Condition monitoring of wind turbines: Techniques and methods. *Renew. Energy* **2012**, *46*, 169–178. [CrossRef]
9. Sharma, S.; Mahto, D. Condition monitoring of wind turbines: A review. *Int. J. Sci. Eng. Res.* **2013**, *4*, 35–50.
10. Tchakoua, P.; Wamkeue, R.; Ouebec, M.; Slaoui-Hasnaoui, F.; Tameghe, T.; Ekemb, G. Wind turbine condition monitoring: State-of-the-art review, new trends, and future challenges. *Energies* **2014**, *7*, 2595–2630. [CrossRef]
11. Liu, W.; Tang, B.; Han, J.; Lu, X.; He, Z. The structure healthy condition monitoring and fault diagnosis methods in wind turbines: A review. *Renew. Sustain. Energy Rev.* **2015**, *44*, 466–472. [CrossRef]
12. Lian, J.; Cai, O.; Dong, X.; Jiang, Q.; Zhao, Y. Health monitoring and safety evaluation of the offshore wind turbine structure: A review and discussion of future development. *Sustainability* **2019**, *11*, 494. [CrossRef]
13. Hart, E.; Clarker, B.; Nicholas, G.; Amiri, A.; Stirling, J.; Carroll, J.; Dwyer-Joyce, R.; McDonald, A.; Long, H. A review of wind turbine main bearings: Design, operation, modelling, damage mechanisms and fault detection. *Wind Energy Sci.* **2020**, *5*, 105–204. [CrossRef]
14. Qiao, W.; Lu, D. A survey on wind turbine condition monitoring and fault diagnosis-part I: Components and subsystems. *IEEE Trans. Ind. Electron.* **2015**, *62*, 6536–6545. [CrossRef]
15. Qiao, W.; Lu, D. A survey on wind turbine condition monitoring and fault diagnosis-part II: Signals and signal processing methods. *IEEE Trans. Ind. Electron.* **2015**, *62*, 6546–6557. [CrossRef]
16. Stetco, A.; Dinmohammadi, F.; Zhao, X.; Robu, V.; Flynn, D.; Barnes, M.; Keane, J.; Nenadic, G. Machine learning methods for wind turbine condition monitoring: A review. *Renew. Energy* **2019**, *133*, 620–635. [CrossRef]
17. Huang, S.; Wu, X.; Liu, X.; Gao, J.; He, Y. Overview condition monitoring and operation control of electric power conversion systems in direct-drive wind turbines under faults. *Front Mech. Eng.* **2017**, *12*, 281–302. [CrossRef]
18. Lau, B.; Ma, E.; Pecht, M. Review of offshore wind turbine failures and fault prognostic methods. In Proceedings of the IEEE Conference on Prognostics and System Health Management (PHM), Beijing, China, 23–25 May 2012; pp. 1–5.
19. Kandukuri, S.; Klausen, A.; Karimi, H.; Robbersmyr, K. A review of diagnostics and prognostics of low-speed machinery towards wind turbine farm-level health management. *Renew. Sustain. Energy Rev.* **2016**, *53*, 697–708. [CrossRef]
20. Leite, G.; Araujo, A.; Rosas, P. Prognostic techniques applied to maintenance of wind turbines: A concise and specific review. *Renew. Sustain. Energy Rev.* **2018**, *81*, 1917–1925. [CrossRef]

21. Abid, K.; Mouchaweh, M.; Cornez, L. Fault prognosis for the predictive maintenance of wind turbines: State of the art. In Proceedings of the Joint European Conference on Machine Learning and Knowledge Discovery in Databases, Dublin, Ireland, 10–14 September 2018; pp. 113–125.
22. Pourmohammad, S.; Fekih, A. Fault-tolerant control of wind turbine systems-a review. In Proceedings of the IEEE Green Technologies Conference, Baton Rouge, LA, USA, 21 April 2011; pp. 1–6.
23. Gao, Z.; Sheng, S. Real-time monitoring, prognosis, and resilient control for wind turbine systems. *Renew. Energy* **2018**, *116*, 1–4. [CrossRef]
24. Schrick, D. Remarks on terminology in the field of supervision, fault detection and diagnosis. In Proceedings of the IFAC Symposium on Fault Detection, Supervision Safety for Technical Processes, Hull, UK, 26–28 August 1997; pp. 959–964.
25. Gao, Z.; Cecati, C.; Ding, S. A Survey of fault diagnosis and fault-tolerant techniques part I: Fault diagnosis with model- and signal-based approaches. *IEEE Trans. Ind. Electron.* **2015**, *62*, 3575–3767. [CrossRef]
26. Hahn, B.; Durstewitz, M.; Rohrig, K. Reliability of Wind Turbines. *Wind Energy Eng.* **2007**, 329–332.
27. Anderson, P.; Bose, A. Stability simulation of wind turbine systems. *IEEE Trans. Power Syst.* **1983**, *102*, 3791–3795. [CrossRef]
28. Ekanayake, J.; Holdsworth, L.; Wu, X.; Jenkins, N. Dynamic modelling of doubly fed induction generator wind turbines. *IEEE Trans. Power Syst.* **2003**, *18*, 803–809. [CrossRef]
29. Kim, S.; Kim, E. PSCAD/EMTDC-based modelling and analysis of a gearless variable speed wind turbine. *IEEE Trans. Energy Convers.* **2007**, *22*, 421–430. [CrossRef]
30. Odgaard, P.; Stoustrup, J.; Kinnaert, M. Fault tolerant control of wind turbines—A benchmark model. In Proceedings of the 7th IFAC Symposium on Fault Detection, Supervision and Safety of Technical Processes, Barcelona, Spain, 30 June–3 July 2009; pp. 155–160.
31. Odgaard, P.; Stoustrup, J.; Kinnaert, M. Fault-tolerant control of wind turbines: A benchmark model. *IEEE Trans. Control Syst. Technol.* **2013**, *21*, 1168–1182. [CrossRef]
32. Odgaard, P.; Johnson, K. Wind turbine fault detection and fault tolerant control-an enhanced benchmark challenge. In Proceedings of the American Control Conference, Washington, DC, USA, 17–19 June 2013; pp. 4447–4452.
33. Zhu, Y.; Gao, Z. Robust observer-based fault detection via evolutionary optimization with applications to wind turbine systems. In Proceedings of the IEEE 9th Conference on Industrial Electronics and Applications (ICIEA), Hangzhou, China, 9–11 June 2014; pp. 1627–1632.
34. Jlassi, I.; Estima, J.; Khil, S.; Bellaaj, N.; Cardoso, A. Multiple open-circuit faults diagnosis in back-to-back converters of PMSG drives for wind turbine systems. *IEEE Trans. Power Electron.* **2015**, *30*, 2689–2702. [CrossRef]
35. Simani, S.; Farsoni, S.; Castaldi, P. Fault diagnosis of a wind turbine benchmark via identified fuzzy models. *IEEE Trans. Ind. Electron.* **2015**, *62*, 3775–3782. [CrossRef]
36. Liu, X.; Gao, Z.; Chen, M. Takagi-Sugeno fuzzy model based fault estimation and signal compensation with application to wind turbines. *IEEE Trans. Ind. Electron.* **2017**, *64*, 5678–5689. [CrossRef]
37. Shao, H.; Gao, Z.; Liu, X.; Busawon, K. Parameter-varying modelling and fault reconstruction for wind turbine systems. *Renew. Energy* **2018**, *116*, 145–152. [CrossRef]
38. Sanchez, H.; Escobet, T.; Puig, V.; Odgaard, P. Fault diagnosis of an advanced wind turbine benchmark using interval-based ARRs and observers. *IEEE Trans. Ind. Electron.* **2015**, *62*, 3783–3793. [CrossRef]
39. Kalman, R. A new approach to linear filtering and prediction problems. *J. Basic Eng.* **1960**, *82*, 35–45. [CrossRef]
40. Wei, X.; Verhaegen, M.; Engelen, T. Sensor fault detection and isolation for wind turbines based on subspace identification and Kalman filter techniques. *Int. J. Adapt. Control Signal Process.* **2010**, *24*, 687–707. [CrossRef]
41. Dey, S.; Pisu, P.; Ayalew, B. A comparitive study of three fault diagnosis schemes for wind turbines. *IEEE Trans. Control Syst. Technol.* **2015**, *23*, 1853–1868. [CrossRef]
42. Cao, M.; Qiu, Y.; Feng, Y.; Wang, H.; Li, D. Study of wind turbine fault diagnosis based on unscented Kalman filter and SCADA data. *Energies* **2016**, *9*, 847. [CrossRef]
43. Wu, X.; Li, Y.; Li, F.; Yang, Z.; Teng, W. Adaptive estimation-based leakage detection for a wind turbine hydraulic pitching system. *IEEE Trans. Mechatron.* **2012**, *17*, 907–914. [CrossRef]
44. Guo, P.; Infield, D.; Yang, X. Wind turbine generator condition monitoring using temperature trend analysis. *IEEE Trans. Sustain. Energy* **2012**, *3*, 124–133. [CrossRef]
45. Freire, N.; Estima, J.; Cardoso, A. Open-circuit fault diagnosis in PMSG drives for wind turbine applications. *IEEE Trans. Ind. Electron.* **2013**, *60*, 3957–3967. [CrossRef]
46. Tamilselvan, P.; Wang, P.; Sheng, S.; Twomey, J. A two-stage diagnosis frame work for wind turbine gearbox condition monitoring. *Int. J. Progn. Health Manag.* **2013**, *4*, 21–31.
47. Zappalá, D.; Tavner, P.; Crabtree, C.; Sheng, S. Side-band algorithm for automatic wind turbine gearbox fault detection and diagnosis. *IET Renew. Power Gener.* **2014**, *8*, 380–389. [CrossRef]
48. Yang, W.; Tavner, P.; Wilkinson, M. Condition monitoring and fault diagnosis of a wind turbine synchronous generator drive train. *IET Renew. Power Gener.* **2009**, *3*, 1–11. [CrossRef]
49. Watson, S.; Xiang, B.; Yang, W.; Tavner, P.; Crabtree, C. Condition monitoring of the power output of wind turbine generators using wavelets. *IEEE Trans. Energy Convers.* **2010**, *25*, 715–721. [CrossRef]

50. Yang, W.; Tavner, P.; Crabtree, C.; Wilkinson, M. Cost-effective condition monitoring for wind turbines. *IEEE Trans. Ind. Electron.* **2010**, *57*, 263–271. [CrossRef]

51. Teng, W.; Wang, F.; Zhang, K.; Liu, Y.; Ding, X. Pitting fault detection of a wind turbine gearbox using empirical mode decomposition. *J. Mech. Eng.* **2014**, *60*, 12–20. [CrossRef]

52. Yoon, J.; He, D.; Hecke, B. On the use of a single piezoelectric strain sensor for wind turbine planetary gearbox fault diagnosis. *IEEE Trans. Ind. Electron.* **2015**, *62*, 6585–6593. [CrossRef]

53. Tang, B.; Liu, W.; Song, T. Wind turbine fault diagnosis based on Morlet wavelet transformation and Wigner-Ville distribution. *Renew. Energy* **2010**, *35*, 2822–2866. [CrossRef]

54. Barszcz, T.; Randall, R. Application of spectral kurtosis for detection of a tooth crack in the planetary gear of a wind turbine. *Mech. Syst. Signal Process* **2009**, *23*, 1352–1365. [CrossRef]

55. Chen, J.; Pan, J.; Li, Z.; Zi, Y.; Chen, X. Generator bearing fault diagnosis via empirical wavelet transform using measured vibration signals. *Renew. Energy* **2016**, *89*, 80–92. [CrossRef]

56. Kum, S.; Sahin, B. A root cause analysis for arctic marine accidents from 1993 to 2011. *Saf. Sci.* **2015**, *74*, 206–220. [CrossRef]

57. Unver, B.; Gurgen, S.; Sahin, B.; Altin, I. Crankcase explosion for two-stroke marine diesel engine by using fault tree analysis method in fuzzy environment. *Eng. Fail. Anal.* **2019**, *97*, 288–299. [CrossRef]

58. Shafiee, M.; Dinmohammadi, F. An FMEA-based risk assessment approach for wind turbine systems: A comparative study of onshore and offshore. *Energies* **2014**, *7*, 619–642. [CrossRef]

59. Dinmohammadi, F.; Shafiee, M. A fuzzy FEMA-based risk assessment approach for offshore wind turbines. *Int. J. Progn. Health Manag.* **2013**, *4*, 1–10.

60. Yang, Z.; Wang, B.; Dong, X.; Liu, H. Expert system of fault diagnosis for gear box in wind turbine. *Syst. Eng. Procedia* **2012**, *4*, 189–195.

61. Garcia, M.; Sanz-Bobi, M.; Pico, J. SIMAP: Intelligent system for predictive maintenance: Application to the health condition monitoring of a wind turbine gearbox. *Comput. Ind.* **2006**, *57*, 552–568. [CrossRef]

62. Pozo, F.; Vidal, Y. Wind turbine fault detection through principal component analysis and statistical hypothesis testing. *Energies* **2016**, *9*, 3. [CrossRef]

63. Wang, J.; Gao, R.; Yan, R. Integration of EEMD and ICA for wind turbine gearbox diagnosis. *Wind Energy* **2014**, *7*, 757–773. [CrossRef]

64. Ding, S.; Zhang, P.; Naik, A.; Ding, E.; Huang, B. Subspace method aided data-driven design of fault detection and isolation systems. *J. Process Control* **2009**, *19*, 1496–1510. [CrossRef]

65. Lou, J.; Lu, H.; Xu, J.; Qu, Z. A data-mining approach for wind turbine power generation performance monitoring based on power curve. *Int. J. Smart Home* **2016**, *10*, 137–152. [CrossRef]

66. Yampikulsakul, N.; Byon, E.; Huang, S.; Sheng, S.; You, M. Condition monitoring of wind power system with nonparametric regression analysis. *IEEE Trans. Energy Convers.* **2014**, *29*, 288–299.

67. Santos, P.; Villa, L.; Reñones, A.; Bustillo, A.; Maudes, J. An SVM-based solution for fault detection in wind turbines. *Sensors* **2015**, *15*, 5627–5648. [CrossRef]

68. Schlechtingen, M.; Santos, I. Comparative analysis of neural network and regression based condition monitoring approaches for wind turbine fault detection. *Mech. Syst. Signal Process.* **2011**, *25*, 1849–1875. [CrossRef]

69. Xiang, J.; Watson, S.; Liu, Y. Smart monitoring of wind turbines using neural networks. In *Sustainability in Energy and Buildings*; Springer: Berlin/Heidelberg, Germany, 2009; pp. 1–8.

70. Üstüntaş, T.; Şahin, A. Wind turbine power curve estimation based on cluster center fuzzy logic modeling. *J. Wind Eng. Ind. Aerodyn.* **2008**, *96*, 611–620. [CrossRef]

71. Teng, W.; Cheng, H.; Ding, X.; Liu, Y.; Ma, Z.; Mu, H. DNN based method for fault detection in a direct drive wind turbine. *IET Renew. Energy Gener.* **2018**, *12*, 1164–1171. [CrossRef]

72. Yang, Z.; Wang, X.; Zhong, J. Representational learning for fault diagnosis of wind turbine equipment: A multi-layered extreme learning machines approach. *Energies* **2016**, *9*, 379. [CrossRef]

73. Schlechtingen, M.; Santos, I.; Achiche, S. Wind turbine condition monitoring based on SCADA data using normal behaviour models. Part 1: System description. *Appl. Soft Comput.* **2013**, *13*, 259–270. [CrossRef]

74. Kusiak, A.; Zheng, H.; Song, Z. Models for monitoring wind farm power. *Renew. Energy* **2009**, *34*, 583–590. [CrossRef]

75. Odofin, S.; Gao, Z.; Sun, K. Robust fault estimation in wind turbine systems using GA optimization. In Proceedings of the 13rd IEEE Conferences on Industrial Informatics, Cambridge, UK, 22–24 June 2015; pp. 580–585.

76. Rahimilarki, R.; Gao, Z.; Zhang, A.; Binns, R. Robust neural network fault estimation approach for nonlinear dynamic systems with applications to wind turbine systems. *IEEE Trans. Ind. Inform.* **2019**, *15*, 6302–6312. [CrossRef]

77. Fernandes-Canti, R.; Blesa, J.; Tornil-Sin, S.; Puig, V. Fault detection and isolation for a wind turbine benchmark using a mixed Bayesian/set-membership approach. *Annu. Rev. Control* **2015**, *40*, 59–69. [CrossRef]

78. Liu, W.; Wang, Z.; Han, J.; Wang, G. Wind turbine fault diagnosis method based on diagonal spectrum and clustering binary tree SVM. *Renew. Energy* **2013**, *50*, 1–6.

79. Fu, Y.; Gao, Z.; Liu, Y.; Zhang, A.; Yin, X. Actuator and sensor fault classification for wind turbine systems based on fast Fourier transform and uncorrelated multi-linear principal component analysis techniques. *Processes* **2020**, *8*, 1066. [CrossRef]

80. Konga, C.; Kima, T.; Hanb, D.; Sugiyamac, Y. Investigation of fatigue life for a medium scale composite wind turbine blade. *Int. J. Fatigue* **2006**, *28*, 1382–1388. [CrossRef]
81. Epaarachchi, J.; Clausen, P. The development of a fatigue loading spectrum for small wind turbine blades. *J. Wind Eng. Ind. Aerodyn.* **2006**, *94*, 207–223. [CrossRef]
82. Cárdenas, D.; Elizalde, H.; Marzocca, P.; Gallegos, S.; Probst, O. A coupled aeroelastic damage progression model for wind turbine blades. *Compos. Struct.* **2012**, *94*, 3072–3081. [CrossRef]
83. Hosseini-Toudeshky, H.; Jahanmardi, M.; Goodarzi, M. Progressive debonding analysis of composite blade root joint of wind turbines under fatigue loading. *Compos. Struct.* **2015**, *120*, 417–427. [CrossRef]
84. Wang, P.; Tamilselvan, P.; Twomey, J.; Youn, B. Prognosis-informed wind farm operation and maintenance for concurrent economic and environmental benefits. *Int. J. Precis. Eng. Manuf.* **2013**, *14*, 1049–1056. [CrossRef]
85. Chen, B.; Matthews, P.; Tavner, P. Wind turbine pitch faults prognosis using a-priori knowledge-based ANFIS. *Expert Syst. Appl.* **2013**, *40*, 6863–6876. [CrossRef]
86. Chen, B.; Matthews, P.; Tavner, P. Automated on-line fault prognosis for wind turbine pitch systems using supervisory control and data acquisition. *IET Renew. Power Gener.* **2015**, *9*, 503–513. [CrossRef]
87. Kusiak, A.; Verma, A. A data-driven approach for monitoring blade pitch faults in wind turbines. *IEEE Trans. Sustain. Energy* **2011**, *2*, 87–96. [CrossRef]
88. Kusiak, A.; Li, W. The prediction and diagnosis of wind turbine faults. *Renew. Energy* **2011**, *36*, 16–23. [CrossRef]
89. Zhu, J.; Yoon, J.; He, D.; Bechhoefer, E. Online particle contaminated lubrication oil condition monitoring and remaining useful life prediction for wind turbines. *Wind Energy* **2015**, *18*, 1131–1149. [CrossRef]
90. Acho, L.; Rodellar, J.; Tutiven, C.; Vidal, Y. Passive fault tolerant control strategy in controlled wind turbines. In Proceedings of the Conference on Control and Fault-Tolerant Systems, Barcelona, Spain, 7–9 September 2016; pp. 636–641.
91. Sloth, C.; Esbensen, T.; Stoustrup, T. Active and passive fault-tolerant LPV control of wind turbines. In Proceedings of the American Control Conference, Baltimore, MD, USA, 30 June–2 July 2010; pp. 4640–4646.
92. Badihi, H.; Zhang, Y. Passive fault-tolerant cooperative control in an offshore wind farm. *Energy Procedia* **2017**, *105*, 2959–2964. [CrossRef]
93. Rodellar, J.; Tutiven, C.; Acho, L.; Vidal, Y. Fault tolerant control design of floating offshore wind turbines. In Proceedings of the European Conference on Structural Control, Sheffield, UK, 11–13 July 2016; p. 160.
94. Gao, R.; Gao, Z. Pitch control for wind turbine systems using optimization, estimation and compensation. *Renew. Energy* **2016**, *91*, 501–515. [CrossRef]
95. Vidal, Y.; Tutivén, C.; Rodellar, J.; Acho, L. Fault diagnosis and fault tolerant control of wind turbines via a discrete time controller with disturbance compensator. *Energies* **2015**, *8*, 4300–4316. [CrossRef]
96. Lan, J.; Patton, R.; Zhu, X. Fault-tolerant wind turbine pitch control using adaptive sliding mode estimation. *Renew. Energy* **2018**, *116*, 219–231. [CrossRef]
97. Chen, J.; Jiang, L.; Yao, W.; Wu, Q. Perturbation estimation based nonlinear adaptive control of a full-rated converter wind turbine for fault ride-through capability enhancement. *IEEE Trans. Power Syst.* **2014**, *29*, 2733–2743. [CrossRef]
98. Azizi, A.; Nourisolaa, S.; Shoja-Majidabad, S. Fault tolerant control of wind turbines with an adaptive output feedback sliding mode controller. *Renew. Energy* **2019**, *135*, 55–56. [CrossRef]
99. Maati, Y.; Bahir, L. Optimal fault tolerant control of large-scale wind turbines in the case of the pitch actuator partial faults. *Complexity* **2020**. [CrossRef]
100. Jain, T.; Yame, J. Health-aware fault-tolerant receding horizon control of wind turbines. *Control Eng. Pract.* **2020**, *95*. [CrossRef]
101. Kanjiya, P.; Ambati, B.; Khadkikar, V. A novel fault-tolerant DFIG-based wind energy conversion system for seamless operation during grid faults. *IEEE Trans. Power Syst.* **2014**, *29*, 1296–1305. [CrossRef]
102. Shahbazi, M.; Poure, P.; Saadate, S. Real-time power switch fault diagnosis and fault-tolerant operation in a DFIG-based wind energy system. *Renew. Energy* **2018**, *116*, 209–218. [CrossRef]
103. Bolbolnia, R.; Heydari, E.; Abbaszadeh, K. Fault tolerant control in direct-drive PMSG wind turbine systems under open circuit faults. In Proceedings of the 11th Power Electronics, Drive Systems, and Technologies Conference, Tehran, Iran, 4–6 February 2020. [CrossRef]
104. Wang, G.; Huang, Z. Data-driven fault-tolerant control design for wind turbines with robust residual generator. *IET Control Theory Appl.* **2015**, *9*, 1173–1179. [CrossRef]
105. Luo, H.; Ding, S.; Haghani, A.; Hao, H.; Yin, S.; Jeinsch, T. Data-driven design of KPI-related fault-tolerant control system for wind turbines. In Proceedings of the American Control Conference, Washington, DC, USA, 17–19 June 2013; pp. 4465–4470.
106. Liu, Y.; Frederik, J.; Ferrari, R.; Wu, P.; Li, S.; Wingerden, J. Adaptive fault accommodation of pitch actuator stuck type of fault in floating offshore wind turbines: A subspace predictive repetitive control approach. In Proceedings of the American Control Conference, Denver, CO, USA, 1–3 July 2020; pp. 4077–4082.

 *processes*

Article

Actuator and Sensor Fault Classification for Wind Turbine Systems Based on Fast Fourier Transform and Uncorrelated Multi-Linear Principal Component Analysis Techniques

Yichuan Fu [1], Zhiwei Gao [1,*], Yuanhong Liu [2], Aihua Zhang [3] and Xiuxia Yin [4]

[1] Department of Mathematics, Physics and Electrical Engineering, Faculty of Engineering and Environment, University of Northumbria, Newcastle upon Tyne NE1 8ST, UK; yichuan.fu@northumbria.ac.uk

[2] School of Electrical Engineering and Information, Northeast Petroleum University, Daqing 163318, China; liuyuanhong@nepu.edu.cn

[3] College of Engineering, Bohai University, Jinzhou 121000, China; jsxinxi_zah@163.com

[4] Department of Mathematics, School of Science, Nanchang University, Nanchang 330000, China; yinxiuxiaa0635@163.com

* Correspondence: zhiwei.gao@northumbria.ac.uk

Received: 4 May 2020; Accepted: 25 August 2020; Published: 1 September 2020

Abstract: In response to the high demand of the operation reliability and predictive maintenance, health monitoring and fault diagnosis and classification have been paramount for complex industrial systems (e.g., wind turbine energy systems). In this study, data-driven fault diagnosis and fault classification strategies are addressed for wind turbine energy systems under various faulty scenarios. A novel algorithm is addressed by integrating fast Fourier transform and uncorrelated multi-linear principal component analysis techniques in order to achieve effective three-dimensional space visualization for fault diagnosis and classification under a variety of actuator and sensor faulty scenarios in 4.8 MW wind turbine benchmark systems. Moreover, comparison studies are implemented by using multi-linear principal component analysis with and without fast Fourier transform, and uncorrelated multi-linear principal component analysis with and without fast Fourier transformation data pre-processing, respectively. The effectiveness of the proposed algorithm is demonstrated and validated via the wind turbine benchmark.

Keywords: fault diagnosis; fault classification; fast Fourier transform (FFT); multi-linear principal component analysis (MPCA); uncorrelated multi-linear principal component analysis (UMPCA); additive white Gaussian noises (AWGN); wind turbine systems

---

1. Introduction

With the development of advanced technologies to increase production, modern industrial systems become more complex and expensive. The components of industrial systems are prone to malfunction, which could bring unanticipated economic costs due to unscheduled shutdown and repair/maintenance. Therefore, it is of particular interest to design effective fault diagnosis and fault classification approaches to automatically monitor the behaviour of industrial systems and prevent damage caused by unexpected faults. Motivated by environmental considerations and the shortage of fossil fuels, wind turbines, as one of renewable energy sources, have contributed to a large portion of the world's power production [1,2]. As a clean energy, wind energy has been significantly exploited via the onshore and offshore wind turbines. By the end of 2019, the overall installation capacity of all wind turbines worldwide reached 651 GW, and European countries contributed to 205 GW.

Moreover, wind power contributed 15% electricity generation in Europe and 20% electricity production in the UK in 2019 [3].

Wind farms consisting of hundreds of wind turbine units are being established in many different locations around the country, for instance, in offshore, arctic, and desert regions. In recent years, some different topologies of generators, such as doubly fed induction generators (DFIGs) and permanent magnet synchronous generators (PMSGs), are widely utilized in wind turbine systems. However, like any other industrial systems, wind turbines are sophisticated and prone to faults. It is observed that the operation and maintenance costs for onshore and offshore wind turbines make up 10~15% and 20~35%, respectively, of the total life costs of wind energy conversion systems. Furthermore, wind turbine systems are complex and expensive; therefore, there is a high demand for improving the reliability and availability, and reducing unscheduled down time in wind turbine industries [4]. Motivated by the above, monitoring and fault diagnosis for wind turbine systems have received wide attention in wind turbine industries [5–9].

Fault diagnosis approaches can be classified into model-based, signal-based, and knowledge-based methods. The model-based fault diagnosis approach requires a well-established model of practical processes developed by either physical principles or systems identification techniques. By checking the residual between the model output and the real-time process output, the decision for fault diagnosis can be made [10,11]. Signal-based fault diagnosis is relying on appropriate sensors, whose locations are normally installed in plant components. The faults in the process are reflected in the measured signals, and the time-domain, frequency-domain, or time-frequency-domain techniques are used to extract features. By checking the consistency between the features of the real-time process and the prior knowledge on the symptoms of the healthy system, a diagnostic decision can be made [12]. Knowledge-based approaches utilize a large volume of historic data available to train universal estimations or approximations on behalf of implementing to recognize faulty conditions [13]. It is worthy to point out that the knowledge-based approach more depends on the data processing and data-based learning, including processing historical data and real-time data. Therefore, the knowledge-based fault diagnosis approach is often called the data-driven approach [14,15].

Machine learning techniques play an important role for data-driven fault diagnosis. Generally speaking, machine learning techniques can mainly be classified into three categories, which are unsupervised, semi-supervised, and supervised learning algorithms, respectively [16]. Unsupervised machine learning aims to learn structure in the data, such as sparse or low-dimensional feature representation [17–20]. According to the tasks of the supervised machine learning, such as prediction and classification, the aim is to learn a knowledge base, on the basis of the known or labelled examples of the target pattern [21,22]. Semi-supervised machine learning represents a class of algorithms that include both supervised and unsupervised tasks [23–25].

It is noted that the dataset generally has a great volume of data with existence in high-dimensional space. Feature extractions thus play an important role in data-driven fault diagnosis [26–29] as well as dimensionality reduction for the samples/datasets. The geometric distribution of the datasets in high-dimensional space can be analyzed in order to effectively extract significant features. There are several methods to solve this problem and one of the most popular techniques is the principal component analysis (PCA) algorithm [30–34]. The PCA, as an unsupervised learning technique, is a statistical procedure that utilizes an orthogonal transformation to convert a set of correlated variables into linearly uncorrelated variables, namely principal components [35]. The number of principal components should be generally less than the number of the original variables [36–38]. The transformation in the PCA is carried out in a way so that the first principal component has the largest possible variance, and each succeeding component in turns has the highest variance possible under the constraint that it is orthogonal to the preceding components [39]. As a result, the PCA has become a popular tool for fault detection and fault classification on the basis of a large volume of high-dimensional experimental samples/datasets [40–42].

A wind turbine system is a complex industrial system, and the operation condition is harsh. Therefore, the conventional PCA technique may become invalid for fault diagnosis and fault classification in wind turbine systems subjected to multiple faults. As a result, there is a strong motivation to develop advanced PCA-based fault diagnosis and classification techniques for wind turbine systems. In this study, uncorrelated multi-linear principal component analysis (UMPCA) is integrated with FFT data preprocessing to form an algorithm, which is applied to a 4.8 MW wind turbine benchmark system for fault diagnosis and classification. Furthermore, comparison studies are carried out to demonstrate the effectiveness of the proposed algorithm by comparing with the known algorithms.

The rest of this paper is organized as follows: In Section 2, the fundamentals of the 4.8 MW wind turbine benchmark model are introduced, and actuator and sensor faults of wind turbines are explained. In Section 3, An algorithm integrated with FFT and UMPCA techniques is addressed for dimensionality reduction and feature extraction. Experimentation designs are proposed in terms of different topologies of the actuator and sensor faults of wind turbines in Section 4. Simulation studies are illustrated in Section 5. In order to demonstrate the effectiveness of the addressed FFT plus UMPCA method, the simulated studies of the fault diagnosis and classification for wind turbines respectively by using MPCA, FFT plus MPCA, and UMPCA are also discussed. Finally, this paper is ended by summarizing the conclusions in Section 6.

2. Wind Turbine Benchmark Systems

A wind turbine is a complex electro-mechanical system that converts wind energy to electrical energy. Most wind turbines are horizontal three–bladed unites, which are composed of blades, low-speed and high-speed shafts, gearbox, generator, yaw, tower, brake, and controller, and so forth. A typical structure of the wind turbine is depicted by Figure 1. The wind flow in the nature drives the blades and rotor to rotate, converting wind energy to mechanical energy. The rotor drives the generator via the high-speed shaft so that the mechanical energy is converted into electric energy. The pitch angle is controlled to adapt to the varying wind speed to achieve the desired output power. The functionality of the yaw system contributes to align the turbine with the direction of the wind detected by the anemometer.

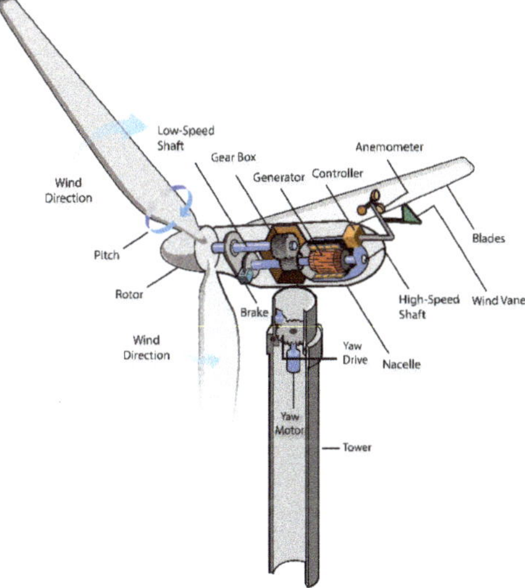

Figure 1. A schematic diagram of the wind turbine system.

A benchmark model of a 4.8 MW wind turbine system was developed in [43,44], which has been widely used for the algorithm validation in control and fault diagnosis. The definitions of the parameters of the benchmark model are shown in Table 1.

Table 1. Parameters of the 4.8 MW wind turbine benchmark system [43,44].

Symbol	Definition	Symbol	Definition
β_r	Pitch angle Reference	θ_Δ	Torsion Angle
$\tau_{g,r}$	Generator Torque Reference	ζ	Damping Ratio
β	Pitch Angle	B_{dt}	Torsion Damping Coefficient
ω_g	Generator Rotating Speed	B_g	Generator External Damping
ω_r	Rotor Angular Speed	B_r	Rotor External Damping
τ_g	Generator Torque	C_q	Torque Coefficient
α_{gc}	Generator and Converter Parameter	J_g	Generator Moment of Inertia
η_{dt}	Efficiency of Drive Train	J_r	Rotor Moment of Inertia
λ	Tip-Speed-Ratio	K_{dt}	Torsion Stiffness
ω_n	Natural Frequency	N_g	Gear Ratio
ρ	Air Density	R	Rotor Radius

The diagram of the 4.8 MW wind turbine benchmark system is shown by Figure 2, which is composed of the blade and pitch subsystem, drive train subsystem, generator and convertor subsystem, and controller, respectively.

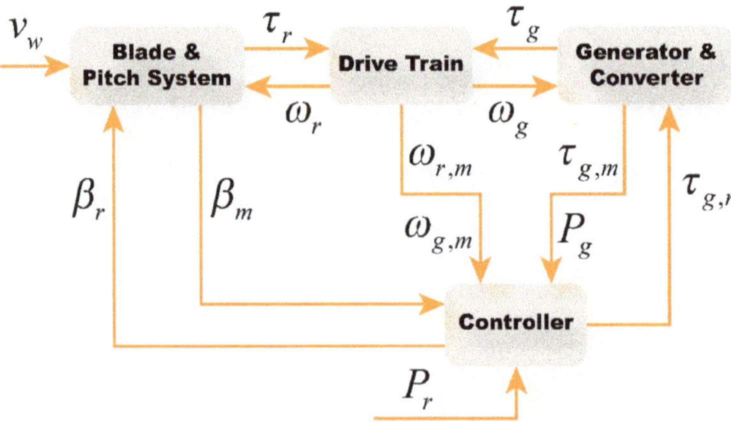

Figure 2. Block diagram of the 4.8 MW benchmark wind turbine model.

The wind turbine benchmark system has an external input (e.g., varying wind speed), two control reference inputs composed of the reference pitch angle (β_r) and generator torque reference ($\tau_{g,r}$).

The wind speed is shown in Figure 3, from which one can see the wind speed ranges from 5 to 20 m/s, with the peak spike over 25 m/s, showing a good coverage of the operation conditions under a healthy situation.

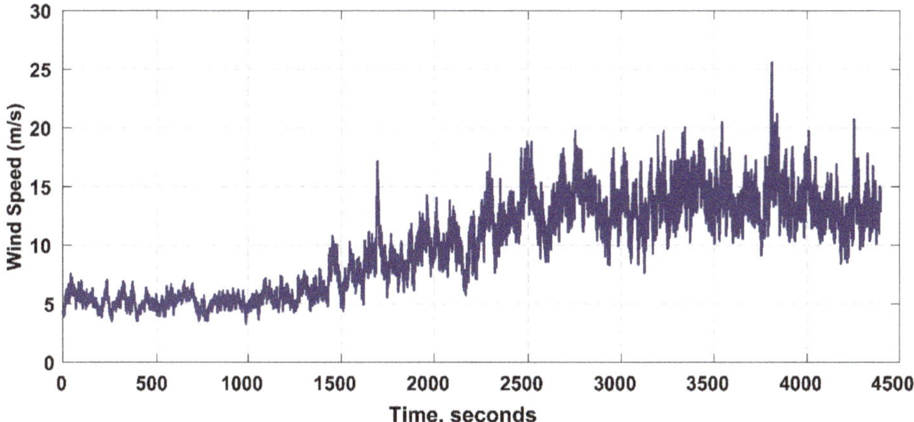

Figure 3. Wind speed sequence used in the benchmark wind turbine under fault-free condition.

In this study, we focus on the actuator faults and sensor faults of the wind turbines. Suppose that $u(t)$ is the control input, $f_A(t)$ is the actuator fault, and $u_R(t)$ is the actuation signal applied to the system; $y(t)$ is the measured output, $f_S(t)$ is the sensor fault, and $y_R(t)$ is the output from the system. It is clear that $u_R(t) = u(t) + f_A(t)$, and $y(t) = y_R(t) + f_S(t)$. As a result, the faults $f_A(t)$ and $f_S(t)$ will divert the performance of the system states and outputs from the normal. The topologies of the actuator faults and sensor faults are depicted by Figure 4.

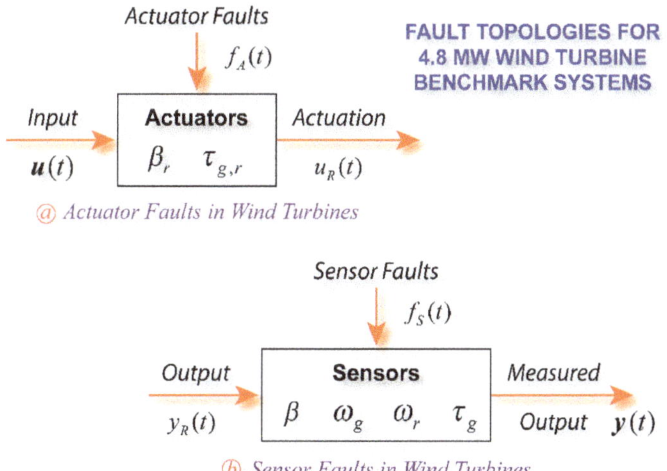

Figure 4. Topologies of the faults in the 4.8 MW wind turbine benchmark system: (**a**). Actuator faults, and (**b**). Sensor faults, respectively.

3. Methodology

3.1. Data Set Construction

The 4.8 MW wind turbine benchmark system has four measurement outputs, namely the pitch angle β, the generator rotating speed ω_g, the rotor angular speed ω_r, and generator torque τ_g. By using the measurement outputs above, the data set recorded from each measurement, denoted by β_s, ω_{gs}, ω_{rs}, and τ_{gs}, can be obtained as follows:

$$
\beta_s = \begin{bmatrix} \beta_{s11} & \beta_{s12} & \cdots & \beta_{s1\gamma} \\ \beta_{s21} & \beta_{s22} & \cdots & \beta_{s2\gamma} \\ \vdots & \vdots & \ddots & \vdots \\ \beta_{sN1} & \beta_{sN2} & \cdots & \beta_{sN\gamma} \end{bmatrix} \in R^{N\times\gamma}, \quad
\omega_{gs} = \begin{bmatrix} \omega_{gs11} & \omega_{gs12} & \cdots & \omega_{gs1\gamma} \\ \omega_{gs21} & \omega_{gs22} & \cdots & \omega_{gs2\gamma} \\ \vdots & \vdots & \ddots & \vdots \\ \omega_{gsN1} & \omega_{gsN2} & \cdots & \omega_{gsN\gamma} \end{bmatrix} \in R^{N\times\gamma}
$$

$$
\omega_{rs} = \begin{bmatrix} \omega_{rs11} & \omega_{rs12} & \cdots & \omega_{rs1\gamma} \\ \omega_{rs21} & \omega_{rs22} & \cdots & \omega_{rs2\gamma} \\ \vdots & \vdots & \ddots & \vdots \\ \omega_{rsN1} & \omega_{rsN2} & \cdots & \omega_{rsN\gamma} \end{bmatrix} \in R^{N\times\gamma}, \quad
\tau_{gs} = \begin{bmatrix} \tau_{gs11} & \tau_{gs12} & \cdots & \tau_{gs1\gamma} \\ \tau_{gs21} & \tau_{gs22} & \cdots & \tau_{gs2\gamma} \\ \vdots & \vdots & \ddots & \vdots \\ \tau_{gsN1} & \tau_{gsN2} & \cdots & \tau_{gsN\gamma} \end{bmatrix} \in R^{N\times\gamma} \tag{1}
$$

where N is the number of the measurement points recorded, and γ is the number of the measurement scenarios. Specifically, for each measurement output, the dataset is recorded under γ operation scenarios (including the fault-free condition, and various faulty conditions), and N measurement points are documented at each scenario. As a result, the original data set can be described by:

$$
X = \begin{bmatrix} \beta_s \\ \omega_{gs} \\ \omega_{rs} \\ \tau_{gs} \end{bmatrix} \in R^{\overline{N}\times\gamma}, \tag{2}
$$

where $\overline{N} = 4N$.

3.2. Data Set Pre-Processing

In order to enhance the feature extraction capability, the time-domain data is pre-proceeded to generate frequency-domain data with a reshaping expression.

According to the original data-set model X defined in (2), we can rewrite it as:

$$
X = \begin{bmatrix} X_1 & X_2 & \cdots & X_\gamma \end{bmatrix} = \begin{bmatrix} x_{11} & x_{12} & \cdots & x_{1\gamma} \\ x_{21} & x_{22} & \cdots & x_{2\gamma} \\ \vdots & \vdots & \ddots & \vdots \\ x_{\overline{N}1} & x_{\overline{N}2} & \cdots & x_{\overline{N}\gamma} \end{bmatrix}, \tag{3}
$$

where $X_i = \begin{bmatrix} x_{1i} & x_{2i} & \cdots & x_{\overline{N}i} \end{bmatrix}^T$, $i = 1, 2, \cdots, \gamma$, and $[\cdot]^T$ represents the transpose of the vector $[\cdot]$. The Fourier transform of X_i can be calculated as follows:

$$
X_i(k) = \sum_{t=1}^{\overline{N}} x_{ti} e^{-\frac{j2\pi}{\overline{N}}k(t-1)}, \tag{4}
$$

where $k = 0, 1, 2, \cdots, \overline{N} - 1$.

In terms of (4), the discrete-time Fourier transform can transform a sequence of $\overline{N}$ numbers $x_{1i} \quad x_{2i} \quad \cdots \quad x_{\overline{N}i}$ into a sequence of complex numbers $X_i(0)$, $X_i(1)$, $\cdots$, $X_i(\overline{N} - 1)$, which can also be denoted by the symbols $f_i^{(1)}, f_i^{(2)}, \cdots, f_i^{(\overline{N})}$. By arranging the sequence of the complex numbers as a vector, we have:

$$
\begin{bmatrix} X_i(0) \\ X_i(1) \\ \vdots \\ X_i(\overline{N}-1) \end{bmatrix} = \Omega \begin{bmatrix} x_{1i} \\ x_{2i} \\ x_{3i} \\ \vdots \\ x_{\overline{N}i} \end{bmatrix} := \begin{bmatrix} f_i^{(1)} \\ f_i^{(2)} \\ \vdots \\ f_i^{(\overline{N})} \end{bmatrix},
\tag{5}
$$

where:

$$
\Omega = \begin{bmatrix} 1 & 1 & 1 & \cdots & 1 \\ 1 & e^{\frac{-j2\pi}{\overline{N}}} & e^{\frac{-j4\pi}{\overline{N}}} & \cdots & e^{\frac{-j2(\overline{N}-1)\pi}{\overline{N}}} \\ \vdots & \vdots & \vdots & \ddots & \vdots \\ 1 & e^{\frac{-j2(\overline{N}-1)\pi}{\overline{N}}} & e^{\frac{-j4(\overline{N}-1)\pi}{\overline{N}}} & \cdots & e^{\frac{-j2(\overline{N}-1)^2\pi}{\overline{N}}} \end{bmatrix},
\tag{6}
$$

and Ω is called the Fourier transform base. It is clear that $i = 1, 2, \cdots, \gamma$ in (5).

The Fourier transform above can be calculated by using the fast Fourier transform algorithm [45,46]. The fast Fourier transform algorithm treats the columns of a matrix as vectors and returns the Fourier transform vector for each column, leading to a Fourier transform matrix.

Taking magnitude and reshaping the vector in (5), one can obtain the matrix expression as follows:

$$
F_i = \begin{bmatrix} \left|f_i^{(1)}\right| & \left|f_i^{(r+1)}\right| & \cdots & \left|f_i^{((l-1)r+1)}\right| \\ \left|f_i^{(2)}\right| & \left|f_i^{(r+2)}\right| & \cdots & \left|f_i^{((l-1)r+2)}\right| \\ \vdots & \vdots & \ddots & \vdots \\ \left|f_i^{(r)}\right| & \left|f_i^{(2r)}\right| & \cdots & \left|f_i^{(\overline{N})}\right| \end{bmatrix} \in R^{r \times l}, \; i = 1, 2, \cdots, \gamma,
\tag{7}
$$

where r indicates the number of rows, l stands for the number of columns, $\left|(\cdot)\right|$ represents the absolute value or magnitude of the complex number $(\cdot)$, and $\overline{N} = lr$. By determining two parameters r and l, the frequency-domain data of the wind turbine can be described as follows:

$$
\left\{ \mathcal{F} \mid \mathcal{F} = [F_1, F_2, \dots, F_i, \dots, F_\gamma] \right\} \in R^{r \times l \times \gamma}.
\tag{8}
$$

Therefore, the dataset has been reformatted as a tensor data expression. From (8), one can see the dataset has γ samples, and the size of each sample is $r \times l$.

The reshaping process of the obtained data set above can be described by the flowchart in Figure 5. From this figure, one can see the data vector (e.g., $X_i = \begin{bmatrix} x_{1i} & x_{2i} & \cdots & x_{\overline{N}i} \end{bmatrix}^T, i = 1, 2, \cdots, \gamma$) is projected into a frequency-domain space relying on the Fourier transform base, and the tensor representation is further generated in terms of (7) and (8).

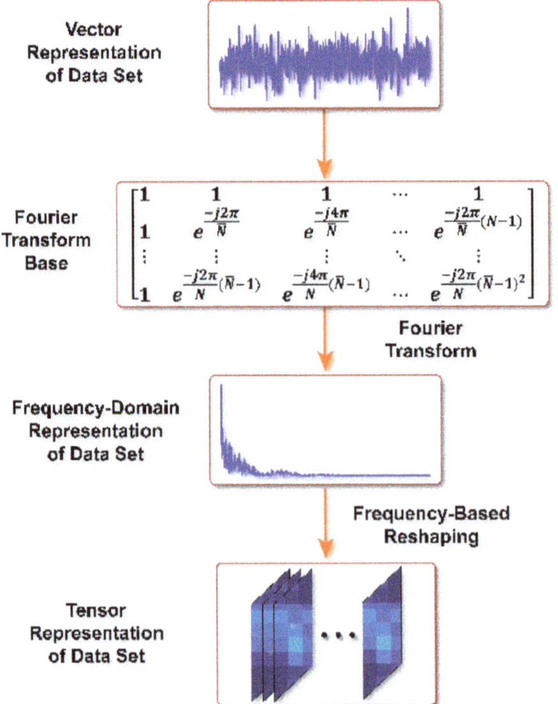

Figure 5. Reprocessing and reshaping of the experimental data.

3.3. Dimensionality Reduction and Feature Extraction for Wind Turbines by Using the Uncorrelated Multi-Linear Principal Component Analysis Method

The multi-linear principal component analysis (MPCA) technique [47], which belongs to one of the unsupervised machine learning algorithms, is usually to cope with the tensor expression dataset. However, some of the correlations of the principal components amongst the projected directions are neglected to some extent, which means the final features obtained by MPCA would be redundant. In contrast to other multilinear PCA techniques, such as MPCA, two-dimensional PCA, and so forth, UMPCA seeks a tensor-to-vector projection, which can capture the maximum number of the uncorrelated multilinear features [39,48]. In this paper, UMPCA is thus used to extract the significant features of the 4.8 MW benchmark wind turbines.

The n-mode product of a tensor $\mathcal{F}$ by a matrix U is denoted by $\mathcal{F} \times_n U$ [39,48].

Suppose the dataset $\{z_i(p), i = 1, 2, \ldots, \gamma\}$ represents the pth principal components (e.g., low-dimensional features), where $z_i(p)$ is the projection of the ith data sample F_i by the p-th elementary multi-linear projection (EMP) $U_p = \left\{ \left(u_p^{(n)} \right)^T, n = 1, 2, \ldots, Q \right\}$, where Q represents the number of projection directions. As a result, the formula of $z_i(p)$ can be described as follows [39,48]:

$$z_i(p) = F_i \times_{n=1}^{Q} \left\{ \left(u_p^{(n)} \right)^T, n = 1, 2, \ldots, Q \right\}, i = 1, 2, \ldots, \gamma. \tag{9}$$

The objective of the UMPCA methodology is to seek U_p that projects F_i into a feature subspace to determine a tensor-to-vector projection, whose functionality will guarantee the implementation for the maximum direction of the original data sets, and the significant features extracted are uncorrelated. Based on the above, the variance can be calculated by [39,48]:

$$S^z_{T_p} = \sum_{i=1}^{\gamma} \left[z_i(p) - \bar{z}_p \right]^2, \tag{10}$$

where $\bar{z}_p = \sum_{i=1}^{\gamma} \frac{z_i(p)}{\gamma}$. Let h_p denote the pth coordinate vector, describing the training sample in the pth EMP space. The ith component of h_p equals the p-th component of z_i, that is, $h_p(i) = z_i(p)$.

In order to determine a set of projection directions $U_p = \left\{ \left(u_p^{(n)} \right)^T, n = 1, 2, \ldots, Q \right\}$ to maximize the variance and generate uncorrelated features, the cost function can be given as follows [39,48]:

$$\left\{ \left(u_p^{(n)} \right)^T, n = 1, 2, \ldots, Q \right\} = argmax S^z_{T_p}$$

$$s.t. \left\{ u_p^{(n)} \right\}^T \cdot u_p^{(n)} = 1, \text{ and } \frac{(h_p)^T h_q}{\|h_p\| \cdot \|h_q\|} = \delta_{pq}, \; (p, q = 1, 2, \ldots, P), \tag{11}$$

where P is the dimensionality of the projected space, and:

$$\delta_{pq} = \left\{ \begin{array}{ll} 1, & if \; p = q \\ 0, & otherwise. \end{array} \right. \tag{12}$$

In terms of the background of the benchmark wind turbine in Section 2 and the fundamental principle of the UMPCA [39,48] mentioned above, the specific procedures of the significant feature extraction for wind turbines can be illustrated as follows.

Step 1: Explore the first projection direction $U_1 = \left\{ \left(u_1^{(n)} \right)^T, n = 1, 2, \ldots, Q \right\}$ by maximizing $S^z_{T_1}$.

Step 2: Compute the second project direction $u_2 = \left\{ \left(u_2^{(n)} \right)^T, n = 1, 2, \ldots, Q \right\}$ by maximizing $S^z_{T_2}$ subjected to $h_2^T h_1 = 0$.

Step 3: Determine the third project direction $u_3 = \left\{ \left(u_3^{(n)} \right)^T, n = 1, 2, \ldots, Q \right\}$ by maximizing $S^z_{T_3}$ subjected to $h_3^T h_2 = 0$.

Step 4: Calculate the p-th project direction $u_p = \left\{ \left(u_p^{(n)} \right)^T, n = 1, 2, \ldots, Q \right\}, p = 4, \cdots, P$, by maximizing $S^z_{T_p}$ subjected to $h_p^T h_q = 0$, when $q = 1, 2, \cdots, p - 1$.

Step 5: Based on all the obtained project directions from the steps above, the final features can be obtained by:

$$z_i = F_i \times_{n=1}^Q \left\{ \left(u_p^{(n)} \right)^T, n = 1, 2, \ldots, Q \right\}_{p=1}^P, \; i = 1, 2, \ldots, \gamma. \tag{13}$$

3.4. FFT Plus UMPCA Algorithm

The specific procedures of the dimensionality reduction and feature extraction based on FFT plus the UMPCA technique for wind turbines can be described as follows:

Algorithm 1

Input: Date set $\left\{ \mathcal{F} \,\middle|\, \mathcal{F} = \left[F_1, F_2, \ldots, F_i, \ldots, F_\gamma \right] \right\}$.
Output: Significant features

$$z_i = F_i \times_{n=1}^{Q} \left\{ \left(u_p^{(n)} \right)^T, n = 1, 2, \ldots, Q \right\}_{p=1}^{P}, \; i = 1, 2, \ldots, \gamma.$$

- (i) **Step 1:** Collect the original data set X by (2)
- (ii) **Step 2:** Pre-process the data set by using Fourier transform base to construct the tensor dataset by (7) and (8).
- (iii) **Step 3:** Calculate the projection directions $U_1, U_2, \cdots, U_P$;
- (iv) **Step 4:** Project the FFT data space into a vector subspace by using

$$z_i = F_i \times_{n=1}^{Q} \left\{ \left(u_p^{(n)} \right)^T, n = 1, 2, \ldots, Q \right\}_{p=1}^{P}, \; i = 1, 2, \ldots, \gamma. \text{ As a result, for the tensor dataset } \mathcal{F},$$

the resultant UMPCA feature vector z can be given as $z = \mathcal{F} \times_{n=1}^{Q} \left\{ \left(u_p^{(n)} \right)^T, n = 1, 2, \ldots, Q \right\}_{p=1}^{P}$.

---

4. Experimentation Designs

4.1. Brief Description and Definition

In this section, in order to validate the applicability of the proposed methodology for fault diagnosis and fault classification in wind turbine systems, five different topologies of experimentation are addressed subsequently. Furthermore, actuator and sensor faults are simultaneously considered in each group of experiment. The size of each data set is $1000 \times 440{,}001$.

For the simplicity of the description for the subsequent experimentations, we define some abbreviations for different types of faulty conditions in two actuators and four sensors. 'A1' represents the first actuator relevant to the pitch angle reference (β_r); 'A2' stands for the second actuator relevant to the generator torque reference ($\tau_{g,r}$); 'S1' is the first sensor to measure the pitch angle (β), 'S2' indicates the second sensor to measure the generator rotating speed (ω_g), 'S3' stands for the third sensor to measure the rotor angular speed (ω_r), and 'S4' defines as the fourth sensor to measure the generator torque (τ_g). The detailed information is shown in Table 2.

Table 2. Symbols and acronyms of the actuator and sensor for 4.8 MW wind turbines.

	Actuator			Sensor		
Symbol	β_r	$\tau_{g,r}$	β	ω_g	ω_r	τ_g
Acronym	A1	A2	S1	S2	S3	S4

In addition, 'FF' indicates fault free. 'EL', 'SWD', and 'RN' represent effectiveness losses, sinusoidal wave disturbances, and random numbers, respectively. Their combination, including 'EL + SWD', 'EL + RN', 'SWD + RN', and 'EL + SWD + RN', are also taken into consideration.

The other abbreviations of the parameters for faulty signals are defined as follows, whose specifications are explained in Table 3:

- (1) 'EL': Percentage (P);
- (2) 'SWD': Amplitude (A) and Bias (B), namely A/B
- (3) 'RN': Mean (M), and Variance (V), namely M/V;

(4) 'EL + SWD': Percentage (P), Amplitude (A), and Bias (B), namely P/A/B;

(5) 'EL + RN': Percentage (P), Mean (M), and Variance (V), namely P/M/V;

(6) 'SWD + RN': Amplitude (A), Bias (B), Mean (M), and Variance (V), namely A/B/M/V;

(7) 'EL + SWD + RN': Percentage (P), Amplitude (A), Bias (B), Mean (M), and Variance (V), namely P/A/B/M/V.

Table 3. Operation Conditions, Parameters, and Acronyms for 4.8 MW Wind Turbine Systems.

Operation Conditions	Abbreviations	Parameters	Acronyms
Fault Free	FF	-	-
Effectiveness Losses	EL	Percentage	P
Sinusoidal Wave Disturbances	SWD	Amplitude & Bias	A/B
Random Numbers	RN	Mean & Variance	M/V
Effectiveness Losses + Sinusoidal Wave Disturbances	EL + SWD	Percentage + Amplitude + Bias	P/A/B
Effectiveness Losses + Random Numbers	EL + RN	Percentage + Mean + Variance	P/M/V
Sinusoidal Wave Disturbances + Random Numbers	SWD + RN	Amplitude + Bias + Mean + Variance	A/B/M/V
Effectiveness Losses + Sinusoidal Wave Disturbances + Random Numbers	EL + SWD + RN	Amplitude + Bias + Mean + Variance	P/A/B/M/V

4.2. Experimental Statement

In the experiment, the fault signals are shown in Table 4. For instance, the effective loss (EL) of every single actuator or sensor is selected as 1%, 2%, 3%, ... , 19% and 20% of the normal value, respectively, which means there are 20 faulty cases for the typical fault EL. More detailed information on other faults can refer to Table 4.

Table 4. Actuator and sensor fault signals: Experimentation design.

Faulty Conditions	Name of Parameters	Actuator		Sensor			
		β_r	$\tau_{g,r}$	β	ω_g	ω_r	τ_g
EL	P			1.00–20.00%			
SWD	A	0.01–0.20	5.20–9.00	0.01–0.20	5.20–9.00	0.01–0.20	5.20–9.00
	B	0.10–2.00	501–520	0.10–2.00	50.10–52.00	0.01–0.20	501–520
RN	M	0.10–2.00	1.00–20.00	0.10–2.00	1.00–20.00	0.01–0.20	1.00–20.00
	V	0.20–2.10	91.00–110.00	0.20–2.10	1.10–2.05	0.01–0.20	91.00–110.00
EL + SWD	P/A/B	β_r: P/A/B—From 1.00%/0.01/0.10 to 20.00%/0.20/2.00; $\tau_{g,r}$: P/A/B—From 1.00%/5.20/501 to 20.00%/9.00/520; β: P/A/B—From 1.00%/0.01/0.10 to 20.00%/0.20/2.00; ω_g: P/A/B—From 1.00%/5.20/50.10 to 20.00%/9.00/52.00; ω_r: P/A/B—From 1.00%/0.01/0.01 to 20.00%/0.20/0.20; τ_g: P/A/B—From 1.00%/5.20/501 to 20.00%/9.00/520.					
EL + RN	P/M/V	β_r: P/M/V—From 1.00%/0.10/0.20 to 20.00%/2.00/2.10; $\tau_{g,r}$: P/M/V—From 1.00%/1.00/91 to 20.00%/20.00/110; β: P/M/V—From 1.00%/0.10/0.20 to 20.00%/2.00/2.10; ω_g: P/M/V—From 1.00%/1.00/1.10 to 20.00%/20.00/2.05; ω_r: P/M/V—From 1.00%/0.01/0.01 to 20.00%/0.20/0.20; τ_g: P/M/V—From 1.00%/1.00/91 to 20.00%/20.00/110.					

Table 4. *Cont.*

Faulty Conditions	Name of Parameters	Actuator and Sensor Faults					
		Actuator		Sensor			
		β_r	$\tau_{g,r}$	β	ω_g	ω_r	τ_g
SWD + RN	A/B/M/V	β_r: A/B/M/V—From 0.01/0.10/0.10/0.20 to 0.20/2.00/2.00/2.10; $\tau_{g,r}$: A/B/M/V—From 5.20/501/1.00/91 to 9.00/520/20.00/110; β: A/B/M/V—From 0.01/0.10/0.10/0.20 to 0.20/2.00/2.00/2.10; ω_g: A/B/M/V—From 5.20/50.10/1.00/1.10 to 9.00/52.00/20.00/2.05; ω_r: A/B/M/V—From 0.01/0.01/0.01/0.01 to 0.20/0.20/0.20/0.20; τ_g: A/B/M/V—From 5.20/501/1.00/91 to 9.00/520/20.00/110.					
EL + SWD + RN	P/A/B/M/V	β_r: P/A/B/M/V—From 1.00%/0.01/0.10/0.10/0.20 to 20.00%/0.20/2.00/2.00/2.10; $\tau_{g,r}$: P/A/B/M/V—From 1.00%/5.20/501/1.00/91 to 20.00%/9.00/520/20.00/110; β: P/A/B/M/V—From 1.00%/0.01/0.10/0.10/0.20 to 20.00%/0.20/2.00/2.00/2.10; ω_g: P/A/B/M/V—From 1.00%/5.20/501/1.00/1.10 to 20.00%/9.00/52.00/20.00/2.05; ω_r: P/A/B/M/V—From 1.00%/0.01/0.01/0.01/0.01 to 20.00%/0.20/0.20/0.20/0.20; τ_g: P/A/B/M/V—From 1.00%/5.20/501/1.00/91 to 20.00%/9.00/520/20.00/110.					

Supplementary Explanations: (i). AWGN signals are introduced to each faulty condition, and the number of AWGN signals is equal to 50; (ii). For β_r, the EL is increased from 1.00 to 20.00% with an increase of 1.00%, and the Amplitude of the SWD increases from 0.01 to 0.20 with an increase by 0.01, and the Bias varies between 0.10 and 2.00 with the interval of 0.10, gradually, as well as the Mean of RN increases from 0.01 to 0.20 with an increase by 0.01, and the Variance increases between 0.20 and 2.10 with the interval of 0.10.

In this section, five groups of experiments of multiple actuator and sensor faults are discussed:

(i)   Scenario I: single actuator and three sensor faults, '1AF + 3SFs'; Types of fault: $C_2^1 \cdot C_4^3 = 8$;

(ii)   Scenario II: single actuator and four sensor faults, '1AF + 4SFs'; Types of fault: $C_2^1 \cdot C_4^4 = 2$;

(iii) Scenario III: two actuators and two sensor faults, '2AFs + 2SFs'; Types of fault: $C_2^2 \cdot C_4^2 = 6$;

(iv)   Scenario IV: two actuators and three sensor faults, '2AFs + 3SFs'; Types of fault: $C_2^2 \cdot C_4^3 = 4$;

(v)   Scenario V: two actuators and four sensor faults, '2AFs + 4SFs'; Types of fault: $C_2^2 \cdot C_4^4 = 1$.

These scenarios are further illustrated by Figures 6–8. From Figure 6, one can see there are eight combinations of actuator and sensor faults under Scenario I, and two combinations in Scenario II. Figure 7 describes Scenario III and Figure 8 explains Scenarios IV and V, respectively.

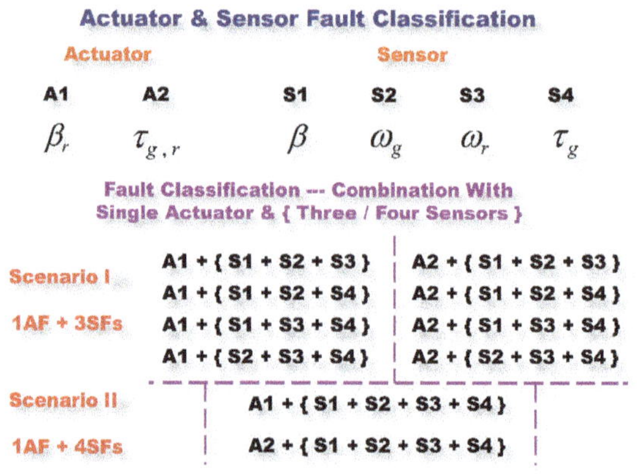

Figure 6. Experimentation design for actuator and sensor fault classification, under Scenario I (1AF + 3SFs) and Scenario II (1AF + 4SFs).

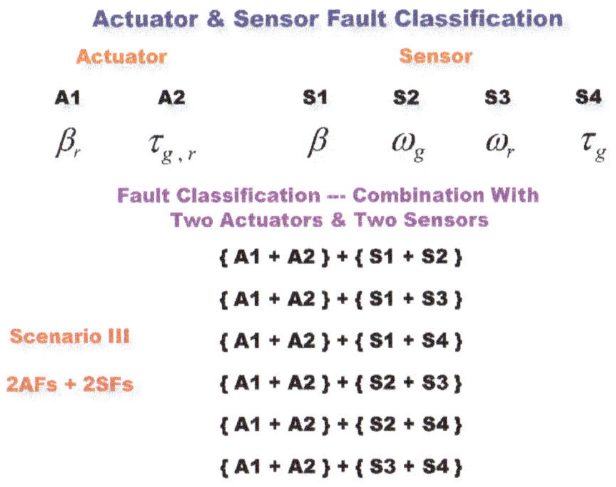

Actuator & Sensor Fault Classification

Actuator		Sensor			
A1	A2	S1	S2	S3	S4
β_r	$\tau_{g,r}$	β	ω_g	ω_r	τ_g

**Fault Classification --- Combination With
Two Actuators & Two Sensors**

	{ A1 + A2 } + { S1 + S2 }
	{ A1 + A2 } + { S1 + S3 }
Scenario III	{ A1 + A2 } + { S1 + S4 }
	{ A1 + A2 } + { S2 + S3 }
2AFs + 2SFs	{ A1 + A2 } + { S2 + S4 }
	{ A1 + A2 } + { S3 + S4 }

Figure 7. Experimentation design for actuator and sensor fault classification, under Scenario III (2AFs + 2SFs).

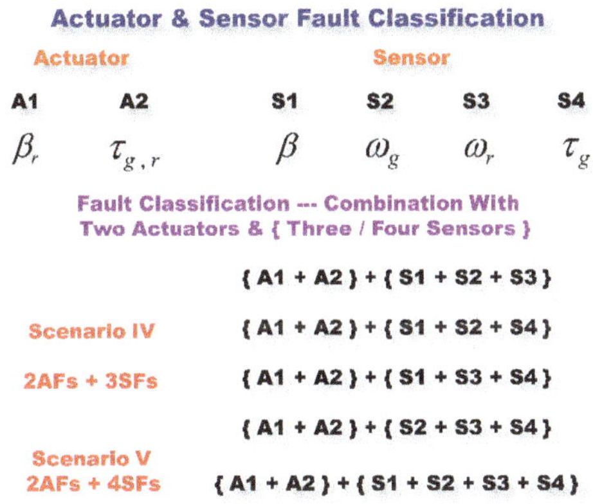

Actuator & Sensor Fault Classification

Actuator		Sensor			
A1	A2	S1	S2	S3	S4
β_r	$\tau_{g,r}$	β	ω_g	ω_r	τ_g

**Fault Classification --- Combination With
Two Actuators & { Three / Four Sensors }**

	{ A1 + A2 } + { S1 + S2 + S3 }
Scenario IV	{ A1 + A2 } + { S1 + S2 + S4 }
2AFs + 3SFs	{ A1 + A2 } + { S1 + S3 + S4 }
	{ A1 + A2 } + { S2 + S3 + S4 }
Scenario V **2AFs + 4SFs**	{ A1 + A2 } + { S1 + S2 + S3 + S4 }

Figure 8. Experimentation design for actuator and sensor fault classification, under Scenario IV (2AFs + 3SFs) and Scenario V (2AFs + 4SFs).

In order to evaluate the feasibility and capability of the proposed FFT + UMPCA algorithm, the MPCA, UMPCA, and FFT + MPCA techniques are also discussed and analyzed. The datasets of the experiments using the algorithms MPCA, UMPCA, FFT + MPCA, and FFT + UMPCA, respectively, are shown in Tables 5 and 6. In Table 5, X_{I}^{MPCA}, X_{II}^{MPCA}, X_{III}^{MPCA}, X_{IV}^{MPCA}, and X_{V}^{MPCA} are the tensor datasets for the MPCA algorithm under scenarios I, II, III, IV and V, respectively. X_{I}^{UMPCA}, X_{II}^{UMPCA}, X_{III}^{UMPCA}, X_{IV}^{UMPCA}, X_{V}^{UMPCA} denote the tensor datasets for the UMPCA algorithm under scenarios I, II, III, IV, and V, respectively. In Table 6, $X_{\mathrm{I}}^{FFT+MPCA}$, $X_{\mathrm{II}}^{FFT+MPCA}$, $X_{\mathrm{III}}^{FFT+MPCA}$, $X_{\mathrm{IV}}^{FFT+MPCA}$, and $X_{\mathrm{V}}^{FFT+MPCA}$ represent the tensor datasets for the FFT + MPCA algorithm under scenarios I, II, III, IV, and V, respectively. $X_{\mathrm{I}}^{FFT+UMPCA}$, $X_{\mathrm{II}}^{FFT+UMPCA}$, $X_{\mathrm{III}}^{FFT+UMPCA}$, $X_{\mathrm{IV}}^{FFT+UMPCA}$, and $X_{\mathrm{V}}^{FFT+UMPCA}$ are the tensor datasets for the FFT + UMPCA algorithm under scenarios I, II, III, IV, and V, respectively.

Table 5. Datasets of experimentations with AWGN noises based on different topologies of the data-driven methodologies: MPCA and UMPCA.

Name of Experimentation	Types	Data Sets with AWGN Noises Based on Different Topologies of Data-Driven Methodologies	
		MPCA	**UMPCA**
FF + 1AF + 3SFs	9	$X_{\mathrm{I}}^{MPCA} \in R^{\{440,000\times4\times9000\}}$	$X_{\mathrm{I}}^{UMPCA} \in R^{\{22,000\times80\times9000\}}$
FF + 1AF + 4SFs	3	$X_{\mathrm{II}}^{MPCA} \in R^{\{440,000\times4\times3000\}}$	$X_{\mathrm{II}}^{UMPCA} \in R^{\{22,000\times80\times3000\}}$
FF + 2AFs + 2SFs	7	$X_{\mathrm{III}}^{MPCA} \in R^{\{440,000\times4\times7000\}}$	$X_{\mathrm{III}}^{UMPCA} \in R^{\{22,000\times80\times7000\}}$
FF + 2AFs + 3SFs	5	$X_{\mathrm{IV}}^{MPCA} \in R^{\{440,000\times4\times5000\}}$	$X_{\mathrm{IV}}^{UMPCA} \in R^{\{22,000\times80\times5000\}}$
FF + 2AFs + 4SFs	2	$X_{\mathrm{V}}^{MPCA} \in R^{\{440,000\times4\times2000\}}$	$X_{\mathrm{V}}^{UMPCA} \in R^{\{22,000\times80\times2000\}}$

Table 6. Datasets of experimentations with AWGN noises based on different topologies of the data-driven methodologies: FFT + MPCA and FFT + UMPCA.

Name of Experimentation	Types	Data Sets with AWGN Noises Based on Different Topologies of Data-Driven Methodologies	
		FFT + MPCA	**FFT + UMPCA**
FF + 1AF + 3SFs	9	$X_{\mathrm{I}}^{FFT+MPCA} \in R^{\{550\times800\times4\times9000\}}$	$X_{\mathrm{I}}^{FFT+UMPCA} \in R^{\{100\times220\times80\times9000\}}$
FF + 1AF + 4SFs	3	$X_{\mathrm{II}}^{FFT+MPCA} \in R^{\{550\times800\times4\times3000\}}$	$X_{\mathrm{II}}^{FFT+UMPCA} \in R^{\{100\times220\times80\times3000\}}$
FF + 2AFs + 2SFs	7	$X_{\mathrm{III}}^{FFT+MPCA} \in R^{\{550\times800\times4\times7000\}}$	$X_{\mathrm{III}}^{FFT+UMPCA} \in R^{\{100\times220\times80\times7000\}}$
FF + 2AFs + 3SFs	5	$X_{\mathrm{IV}}^{FFT+MPCA} \in R^{\{550\times800\times4\times5000\}}$	$X_{\mathrm{IV}}^{FFT+UMPCA} \in R^{\{100\times220\times80\times5000\}}$
FF + 2AFs + 4SFs	2	$X_{\mathrm{V}}^{FFT+MPCA} \in R^{\{550\times800\times4\times2000\}}$	$X_{\mathrm{V}}^{FFT+UMPCA} \in R^{\{100\times220\times80\times2000\}}$

Simulations were operated under the environment of Windows Server 2016 Technical Preview 5 OS and software MathWorks MATLAB R2018a, and run on a server with DELL PowerEdge C6100 4 Nodes Server Dual Intel Xeon 5670, Hex-Core, 2.93 GHz CPU, 384 GB memory, and 3 TB storage (Overall: 48-Core CPU, 1.50 TB Memory, and 36 TB Storage).

5. Simulation Results

5.1. Time-Domain Space Characteristics of Wind Turbine Benchmark Systems

The curves displayed in Figure 9a–d show the time-domain responses of the four measurement outputs β, ω_g, ω_r, and τ_g under fault-free, and various faulty conditions of the actuator and sensor faults, including 'EL', 'SWD', 'RN', 'EL + SWD', 'EL + RN', 'SWD + RN', and 'EL + SWD + RN', respectively.

From Figure 9a–c, one can see that the curves are difficult to distinguish between fault-free and faulty situations. In Figure 9d, from 0–2300 s, it is impossible to find differences among the fault-free and faulty cases. From 2300–4400 s, one can see the fault-free curve is distinguishable from the faulty curves; however, it is hard to see the differences between the faulty curves. As a result, fault classification and diagnosis techniques are needed. It is noted that the overall simulated time of the 4.8 MW wind turbine benchmark system is 4400 s with the interval of 0.01 s. Consequently, the dimension of each experimental sample is 440,001.

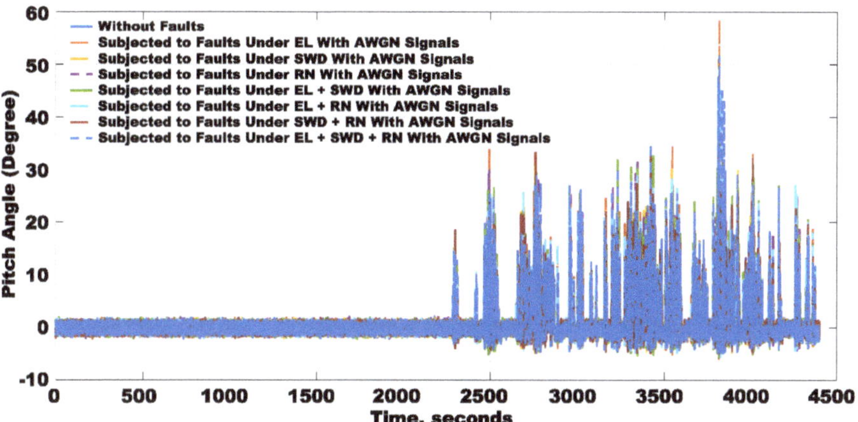

(**a**) Output response of the pitch angle β under healthy and faulty conditions, respectively.

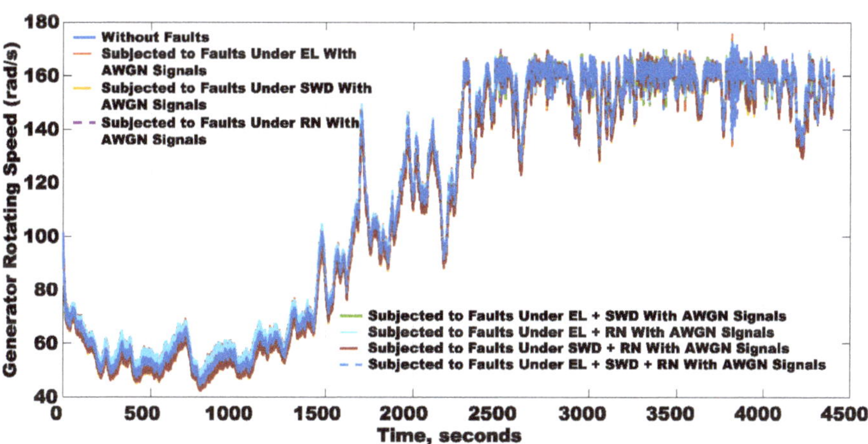

(**b**) Output response of the generator speed ω_g under healthy and faulty conditions, respectively.

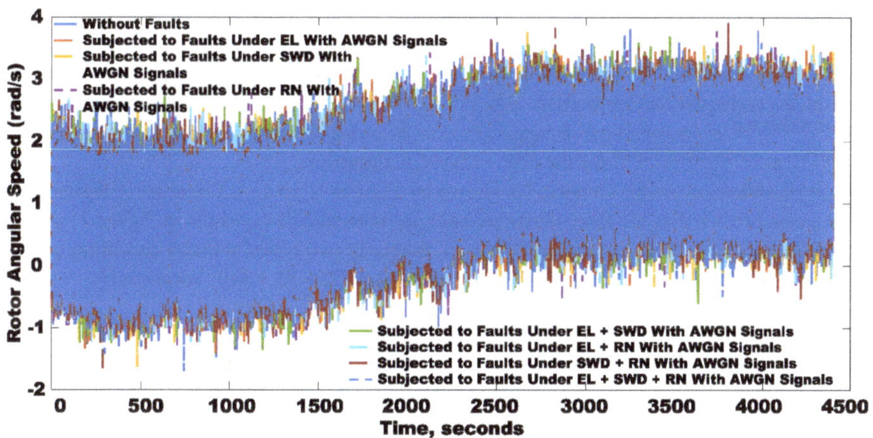

(**c**) Output response of the rotor speed ω_r under healthy and faulty conditions, respectively.

Figure 9. *Cont.*

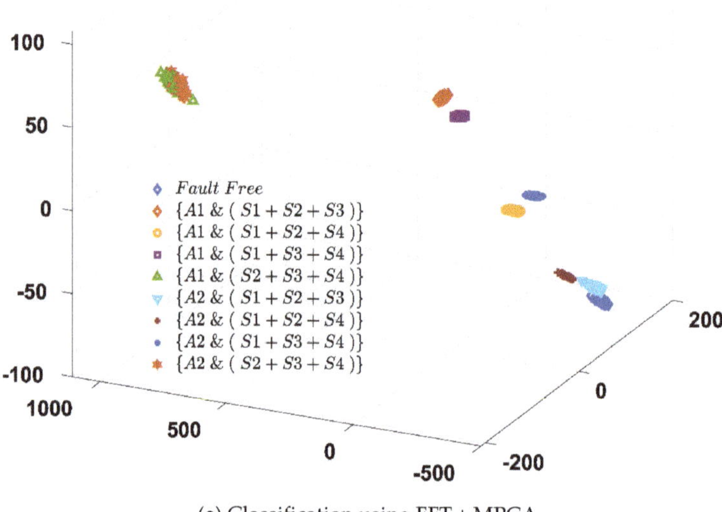

(**a**) Classification using FFT + MPCA

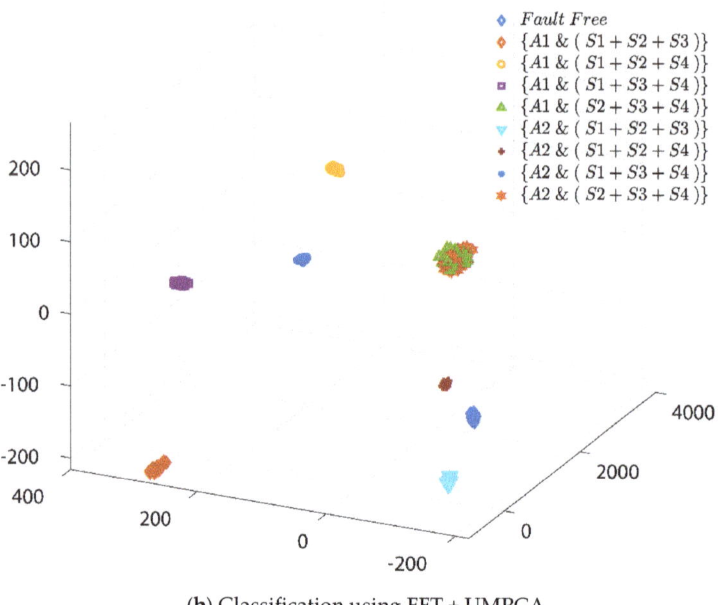

(**b**) Classification using FFT + UMPCA

Figure 11. Three-dimensional space visualization performance for fault classification for wind turbines subjected to single actuator fault and three sensor faults under AWGN noises, using (**a**) FFT + MPCA and (**b**) FFT + UMPCA, respectively.

Specifically, from Figure 10a, the data generally cluster in three large groups, by using the MPCA algorithm, indicating a poor classification performance. To see the details, one can see one of the overlapping occurs between 'Fault Free' and '{A1 & (S1 + S3 + S4)}', and the other exists between '{A1 & (S2 + S3 + S4)}' and '{A2 & (S2 + S3 + S4)}'. Moreover, the rest of the five classes of faulty situations indistinguishably cluster together. From Figure 10b based on the UMPCA, the visualized results cluster around more groups, but are still unsatisfactory for classification.

From Figure 11a,b, one can see that both methods, that is, FFT + MPCA and FFT + UMPCA, can successfully classify seven classes of faulty/heathy conditions. It is noticed that the spatial distance amongst these generated features in Figure 11a is closer than that in Figure 11b in the corresponding three-dimensional space. In other words, there are larger distances between different faulty data in Figure 11b comparing with Figure 11a, indicating a better classification performance of the FFT + UMPCA algorithm. As a result, it is evident that the proposed FFT + UMPCA algorithm outperforms the MPCA, UMPCA, and FFT + MPCA for fault classification under scenario I.

5.3. Feature Extractions and Fault Classifications Based under Scenario II

Data Set for Scenario II: In this data set, it is composed of 'FF' samples and two types of '1AF + 4SFs' samples. The detailed information is shown in Figure 6—Scenario II. In order to evaluate the effectiveness of the proposed algorithm by comparison, four types of datasets are constructed, which is $X_{II}^{MPCA} \in R^{\{440,000 \times 4 \times 3000\}}$, $X_{II}^{UMPCA} \in R^{\{22,000 \times 80 \times 3000\}}$, $X_{II}^{FFT+MPCA} \in R^{\{550 \times 800 \times 4 \times 3000\}}$, and $X_{II}^{FFT+UMPCA} \in R^{\{100 \times 220 \times 80 \times 3000\}}$, respectively. All the detailed information can be found in Tables 5 and 6.

For $X_{II}^{MPCA} \in R^{\{440,000 \times 4 \times 3000\}}$: '440,000' represents the dimensionality of the feature subspace, '4' stands for the dimensionality of parameter subspace, and '3000' illustrates the dimensionality of the sample subspace. For $X_{II}^{UMPCA} \in R^{\{22,000 \times 80 \times 3000\}}$: '22,000' represents the dimensionality of the feature subspace, '80' stands for the dimensionality of the parameter subspace, and '3000' illustrates the dimensionality of the sample subspace.

For $X_{II}^{FFT+MPCA} \in R^{\{550 \times 800 \times 4 \times 3000\}}$: The original data set $X_{II} \in R^{\{440,000 \times 12,000\}}$ is projected into a frequency-domain subspace and reshaped into a tensor dataset $X_{II}^{FFT+MPCA} \in R^{\{550 \times 800 \times 4 \times 3000\}}$ for the FFT + MPCA algorithm. For $X_{II}^{FFT+UMPCA} \in R^{\{100 \times 220 \times 80 \times 3000\}}$: The original data set $X_{II} \in R^{\{440,000 \times 12,000\}}$ is projected into a frequency-domain subspace and reshaped into a tensor dataset $X_{II}^{FFT+UMPCA} \in R^{\{100 \times 220 \times 80 \times 3000\}}$ for the FFT + UMPCA algorithm..

In this section, Figures 12 and 13 illustrate the three-dimensional space visualization performance for fault classification for wind turbine systems subjected to an actuator fault and four sensors faults simultaneously under AWGN noises, respectively using MPCA, UMPCA, FFT + MPCA, and FFT + UMPCA. From the simulated result observation, all types of faulty condition can only be successfully classified by using the FFT + MPCA and FFT + UMPCA methodologies. Therefore, FFT has a positive impact on the improvement of the performance of the dimensionality reduction and feature extraction.

Specifically, from Figure 12a,b, the data sets cluster in a distributive way, although the UMPCA performs a bit better in classification. Encouragingly, from Figure 13a,b, the data sets cluster in three clear groups, indicating a clear fault classification and diagnosis for the three faulty/healthy conditions concerned. From Figure 13b, it is interesting to observe the data in the same group shape distinguishably. It is noted that faulty data in this study includes seven types of faults, such as effectiveness loss, sinusoidal faults, and random number disturbances and so forth, and the fault-free data are subjected to stochastic noises. Therefore, the classification by using the FFT + MPCA can recognize the difference between the data in the same large group. In other words, Figure 13b can also reflect the intrinsic properties of the original samples of the 4.8 MW wind turbine benchmark system.

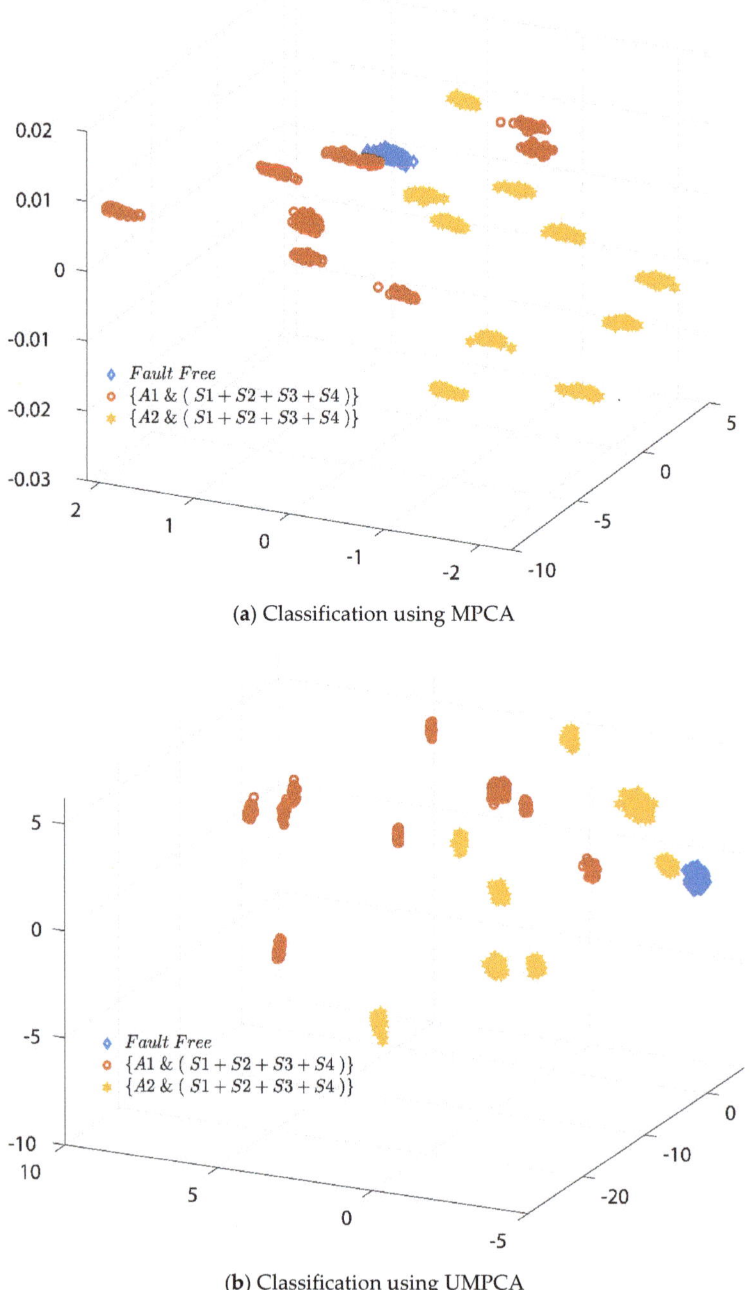

(a) Classification using MPCA

(b) Classification using UMPCA

Figure 12. Three-dimensional space visualization performance for fault classification for wind turbines subjected to single actuator and four sensor faults under AWGN noises, using (a) MPCA and (b) UMPCA, respectively.

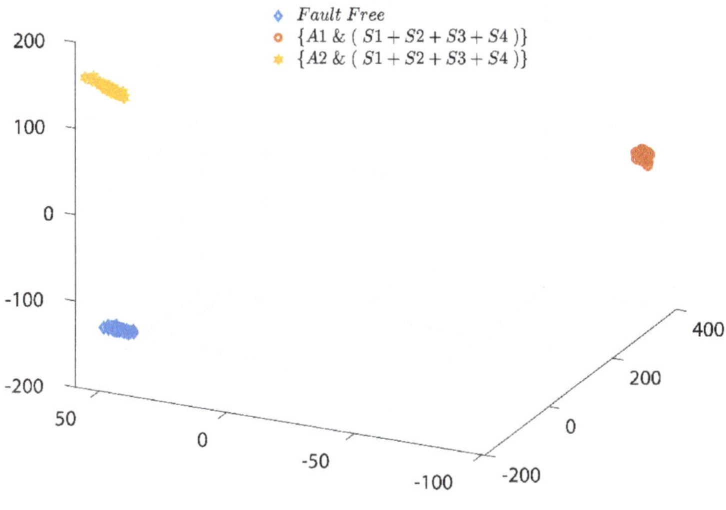

(**a**) Classification using FFT + MPCA

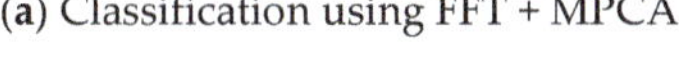

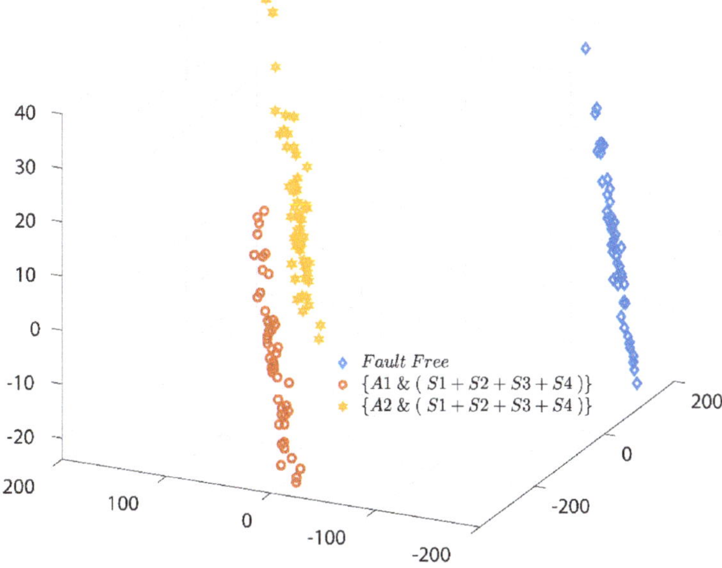

(**b**) Classification using FFT + UMPCA

Figure 13. Three-dimensional space visualization performance for fault classification for wind turbines subjected to the single actuator and four sensor faults under AWGN noises, using (**a**) FFT + MPCA and (**b**) FFT + UMPCA, respectively.

5.4. Feature Extractions and Fault Classifications under Scenario III

Data Set for Scenario III: In this data set, it includes 'FF' samples and six types of '2AFs + 2SFs' samples. The detailed information is shown in Figure 7—Scenario III. In order to validate the effectiveness of the proposed algorithm by comparison, four types of datasets are addressed, which is $X_{III}^{MPCA} \in R^{\{440,000\times4\times7000\}}$, $X_{III}^{UMPCA} \in R^{\{22,000\times80\times7000\}}$, $X_{III}^{FFT+MPCA} \in R^{\{550\times800\times4\times7000\}}$, and $X_{III}^{FFT+UMPCA} \in R^{\{100\times220\times80\times7000\}}$, respectively. All the detailed information can be found in Tables 5 and 6.

For $X_{III}^{MPCA} \in R^{\{440,000\times4\times7000\}}$: '440,000' represents the dimensionality of the feature subspace, '4' stands for the dimensionality of the parameter subspace, and '7000' illustrates the dimensionality of the sample subspace. For $X_{III}^{UMPCA} \in R^{\{22,000\times80\times7000\}}$: '22,000' represents the dimensionality of the feature subspace, '80' stands for the dimensionality of parameter subspace, and '7000' illustrates the dimensionality of the sample subspace.

For $X_{III}^{FFT+MPCA} \in R^{\{550\times800\times4\times7000\}}$: The original data set $X_{III} \in R^{\{440,000\times28,000\}}$ is projected into a frequency-domain subspace and reshaped into a tensor dataset $X_{III}^{FFT+MPCA} \in R^{\{550\times800\times4\times7000\}}$ for using the FFT + MPCA algorithm. For $X_{III}^{FFT+UMPCA} \in R^{\{100\times220\times80\times7000\}}$: The original data set $X_{III} \in R^{\{440,000\times28,000\}}$ is projected into a frequency-domain subspace and reshaped into a tensor dataset $X_{III}^{FFT+UMPCA} \in R^{\{100\times220\times80\times7000\}}$ for the FFT + UMPCA algorithm.

In this section, Figures 14 and 15 exhibit the three-dimensional space visualization performance for fault classification for wind turbine systems subjected to two actuator faults and two sensor faults simultaneously under AWGN noise corruption, respectively by using MPCA, UMPCA, FFT + MPCA, and FFT + UMPCA.

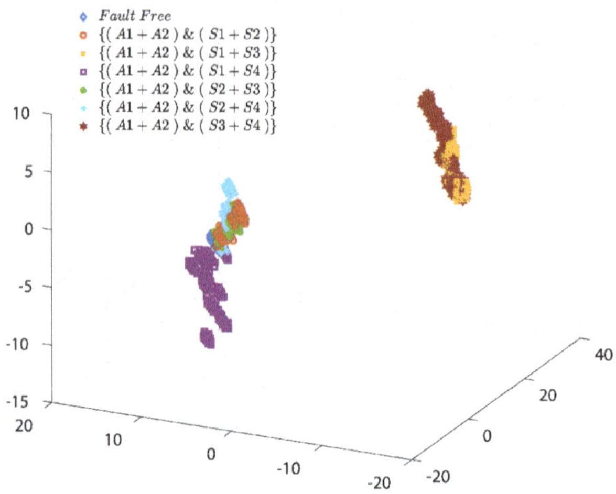

(**a**) Classification using MPCA

Figure 14. *Cont.*

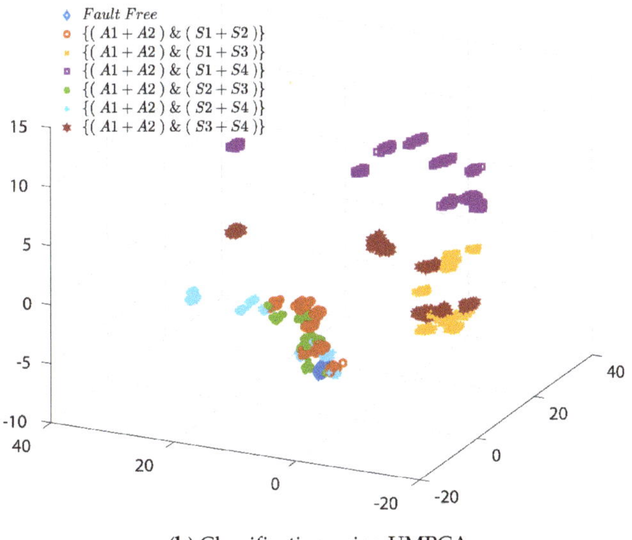

(**b**) Classification using UMPCA

Figure 14. Three-dimensional space visualization performance for fault classification for wind turbines subjected to two actuator and two sensor faults under AWGN noises, using (**a**) MPCA and (**b**) UMPCA, respectively.

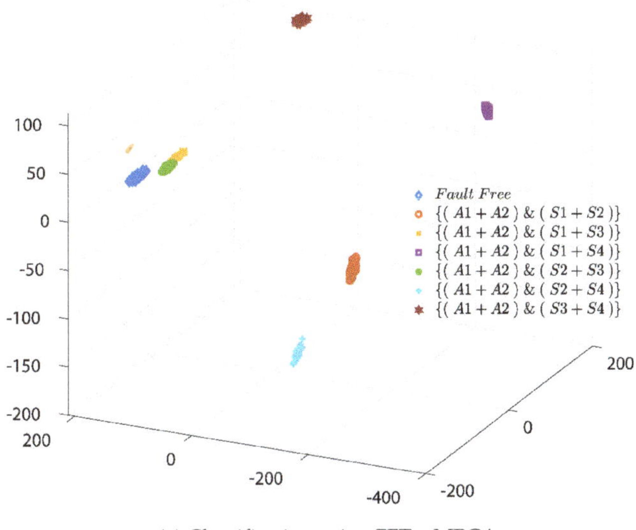

(**a**) Classification using FFT + MPCA

Figure 15. *Cont.*

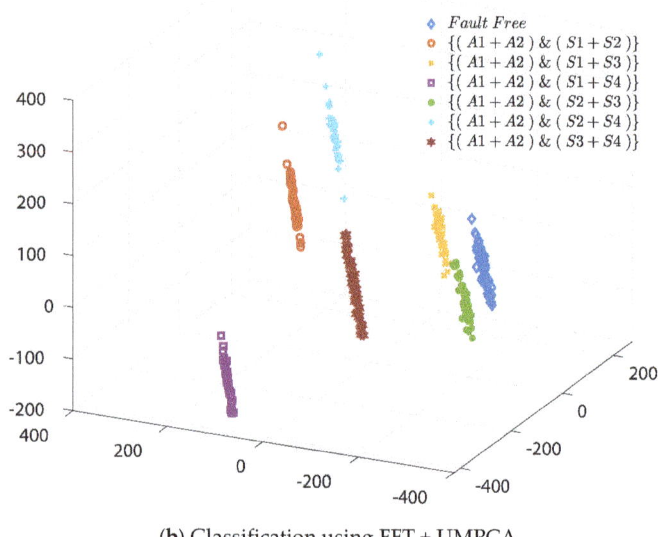

(**b**) Classification using FFT + UMPCA

Figure 15. Three-dimensional space visualization performance for fault classification for wind turbines subjected to two actuator and two sensor faults under AWGN noises, using (**a**) FFT + MPCA and (**b**) FFT + UMPCA, respectively.

From Figure 14a, it is shown that two large groups are formed in the corresponding three-dimensional space based on the MPCA method. It is observed that the overlapping occurs between '{(A1 + A2) & (S1 + S3)}' and '{(A1 + A2) & (S3 + S4)}', and another overlapping happens among 'Fault Free', '{(A1 + A2) & (S1 + S2)}', '{(A1 + A2) & (S1 + S4)}', '{(A1 + A2) & (S2 + S3)}', and '{(A1 + A2) & (S2 + S4)}'. From Figure 14b based on the UMPCA technique, the visualization performance, with more formed data groups, is slightly better than that using the MPCA but is far from acceptable for classification.

Seven classes of faulty/healthy situations were successfully classified respectively by using the FFT + MPCA shown in Figure 15a and FFT + UMPCA exhibited by Figure 15b. More interesting, Figure 15b can clearly reflect the intrinsic properties of the original samples of the wind turbines, which indicates the FFT + UMPCA approach can also sense different types of faults in every single faulty situation.

5.5. Feature Extractions and Fault Classifications Based under Scenario IV

Data Set for Scenario IV: In this data set, it consists in 'FF' samples and four types of '2AFs + 3SFs' samples. The detailed information is shown in Figure 8—Scenario IV. In order to evaluate the effectiveness of the proposed algorithm by comparison, four types of datasets are determined: $X_{IV}^{MPCA} \in R^{\{440,000\times4\times5000\}}$, $X_{IV}^{UMPCA} \in R^{\{22,000\times80\times5000\}}$, $X_{IV}^{FFT+MPCA} \in R^{\{550\times800\times4\times5000\}}$, and $X_{IV}^{FFT+UMPCA} \in R^{\{100\times220\times80\times5000\}}$, respectively. All the detailed information can be found in Tables 5 and 6.

For $X_{IV}^{MPCA} \in R^{\{440,000\times4\times5000\}}$: '440,000' represents the dimensionality of the feature subspace, '4' stands for the dimensionality of the parameter subspace, and '5000' illustrates the dimensionality of the sample subspace. For $X_{IV}^{UMPCA} \in R^{\{22,000\times80\times5000\}}$: '22,000' represents the dimensionality of the feature subspace, '80' stands for the dimensionality of the parameter subspace, and '5000' illustrates the dimensionality of the sample subspace.

For $X_{IV}^{FFT+MPCA} \in R^{\{550 \times 800 \times 4 \times 5000\}}$: The original data set $X_{IV} \in R^{\{440,000 \times 20,000\}}$ is projected into a frequency-domain subspace and reshaped into a tensor representation $X_{IV}^{FFT+MPCA} \in R^{\{550 \times 800 \times 4 \times 5000\}}$ for the use of the FFT + MPCA algorithm. For $X_{IV}^{FFT+UMPCA} \in R^{\{100 \times 220 \times 80 \times 5000\}}$: The original data set $X_{IV} \in R^{\{440,000 \times 20,000\}}$ is projected into a frequency-domain subspace and reshaped into a tensor representation $X_{IV}^{FFT+UMPCA} \in R^{\{100 \times 220 \times 80 \times 5000\}}$ for the implementation of the FFT + UMPCA technique.

In this subsection, Figures 16 and 17 exhibit the three-dimensional space visualization performance for fault classification for wind turbine systems subjected to two simultaneous actuator faults and three simultaneous sensors under AWGN noise corruptions, by using the MPCA, UMPCA, FFT + MPCA, and FFT + UMPCA algorithms, respectively.

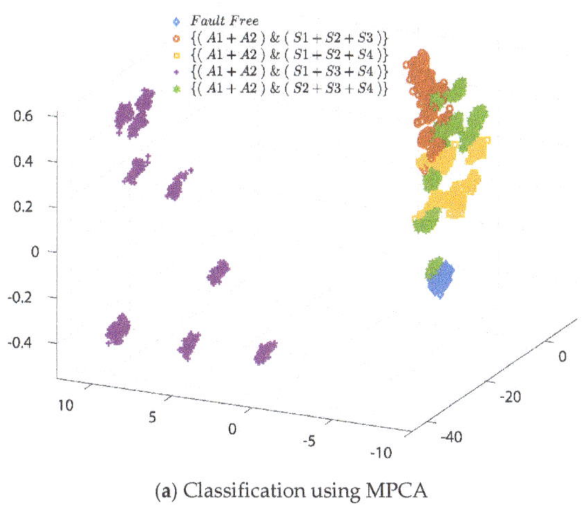

(a) Classification using MPCA

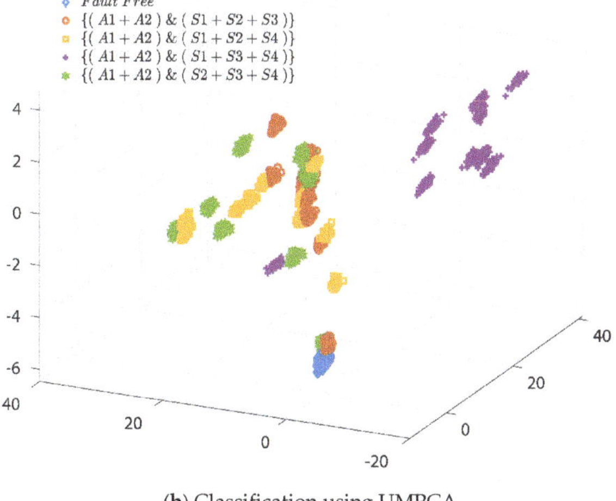

(b) Classification using UMPCA

Figure 16. Three-dimensional space visualization performance for fault classification for wind turbines subjected to two actuator and three sensor faults under AWGN noises, by using (a) MPCA and (b) UMPCA, respectively.

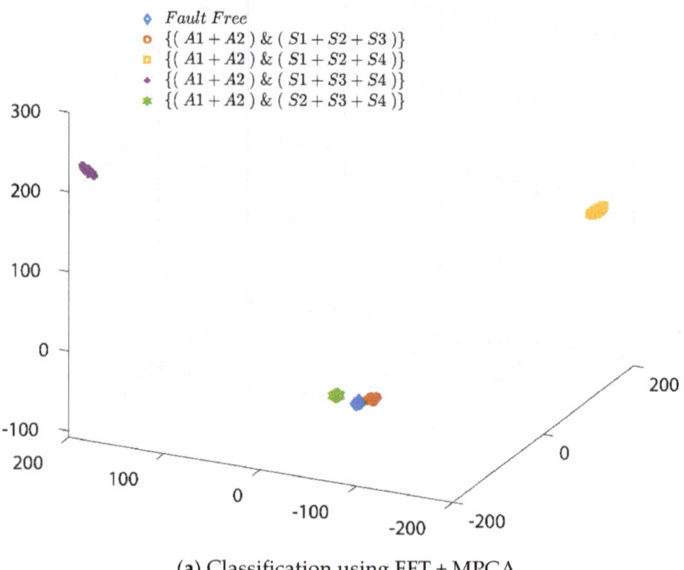

(**a**) Classification using FFT + MPCA

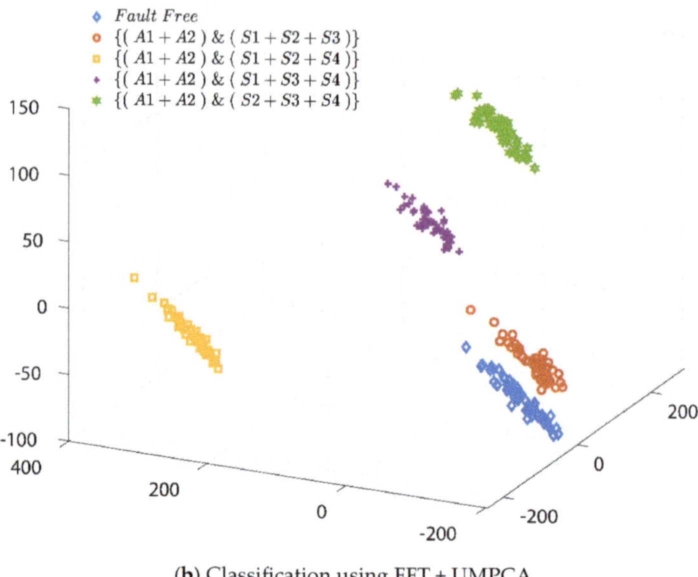

(**b**) Classification using FFT + UMPCA

Figure 17. Three-dimensional space visualization performance for fault classification for wind turbines subjected to two actuator and three sensor faults under AWGN noises, (**a**) FFT + MPCA and (**b**) FFT + UMPCA, respectively.

From Figure 16a based on the MPCA, the data are clustering around three large sets, while in Figure 16b using the UMPCA, the data are clustering in a more distributed way. Both of the visualized results in Figure 16a,b fail to classify and diagnose the faults.

From Figure 17a,b, faulty conditions can be successfully classified by both FFT + MPCA and FFT + UMPCA algorithms. Specifically, it is worthy to point out that the corresponding three-dimensional space visualization behaviours in Figure 17a,b shape differently in comparison with Figure 16a,b, respectively. From what is exhibited in Figure 17a, one can see that the FFT + MPCA methods outperform the MPCA. The reason behind this is that the intrinsic structures of the obtained experimental data sets were reconstructed by using Fourier transform bases. Additionally, these samples are mapped into the multi-dimensional frequency-domain subspace, which means the visualized performance/behaviour are different from the MPCA-based circumstances. Furthermore, the performance of the FFT + UMPCA in Figure 17b is much better than that based on the UMPCA in Figure 16b. Consequently, the FFT has a positive impact on the improvement of the performance of the fault classification and diagnosis. From Figure 17b, one can see that the FFT + UMPCA approach can also recognize the differences of data in the same group.

5.6. Feature Extractions and Fault Classifications Based under Scenario V

Data Set for Scenario V: In this data set, it is combined by 'FF' and '2AFs + 4SFs' samples. The detailed information is shown in Figure 8—Scenario V. In order to validate the effectiveness of the proposed algorithm by comparison, four types of datasets are built: $X_V^{MPCA} \in R^{\{440,000 \times 4 \times 2000\}}$, $X_V^{UMPCA} \in R^{\{22,000 \times 80 \times 2000\}}$, $X_V^{FFT+MPCA} \in R^{\{550 \times 800 \times 4 \times 2000\}}$, and $X_V^{FFT+UMPCA} \in R^{\{100 \times 220 \times 80 \times 2000\}}$, respectively. All the detailed information can be found in Tables 5 and 6.

For $X_V^{MPCA} \in R^{\{440,000 \times 4 \times 2000\}}$: '440,000' represents the dimensionality of the feature subspace, '4' stands for the dimensionality of the parameter subspace, and '2000' indicates the dimensionality of the sample subspace. For $X_V^{UMPCA} \in R^{\{22,000 \times 80 \times 2000\}}$: '22,000' represents the dimensionality of the feature subspace, '80' stands for the dimensionality of the parameter subspace, and '2000' represents the dimensionality of the sample subspace.

For $X_V^{FFT+MPCA} \in R^{\{550 \times 800 \times 4 \times 2000\}}$: The original data set $X_V \in R^{\{440,000 \times 8000\}}$ is projected into a frequency-domain subspace and reshaped into a tensor dataset $X_V^{FFT+MPCA} \in R^{\{550 \times 800 \times 4 \times 2000\}}$ for the use of the FFT + MPCA algorithm. The original data set $X_V \in R^{\{440,000 \times 8000\}}$ is projected into a frequency-domain subspace and reshaped into a tensor representation $X_V^{FFT+UMPCA} \in R^{\{100 \times 220 \times 80 \times 2000\}}$ for the FFT + UMPCA technique.

Figures 18 and 19 show the three-dimensional space visualization performance for fault classification for wind turbine systems subjected to two simultaneous actuators faults and four simultaneous sensors faults corrupted by AWGN noisy signals, respectively using different algorithms, such as MPCA, UMPCA, FFT + MPCA, and FFT + UMPCA.

From Figure 18a based on the MPCA, the faulty-data cluster in 10 groups, and one of them is overlapped with the fault-free data, indicating unsuccessful fault classification. From Figure 18b based on the UMPCA technique, the visualization performance is relatively better than that using the MPCA, as there is no overlapping between the faulty-data and fault-free data. However, the performance in Figure 18b is still not satisfactory, since the distances between the faulty-data are too large and the distances between the fault-free data and some of the faulty-data are quite close.

From Figure 19a, one can see that FFT + MPCA method has a much better classification performance than the MPCA, shown in Figure 18a, as the faulty data and fault-free data are clearly classified into two separated groups by using FFT + MPCA. Comparing Figure 19b by using FFT + UMPCA to Figure 18b via the UMPCA, the faulty data and fault-free data are separated into two large groups in Figure 19b, showing a clear classification between the faulty data and fault free data. As a result, it is evident that the FFT has a positive impact on the improvement of the performance of the fault classification and diagnosis. In addition, from Figure 19b, the data in the same group are not clustered so close compared with Figure 19a. As the faulty data is a combination of the data subjected to different types, such as effectiveness loss, sinusoidal fault signal, random number disturbances, and so forth, this means the FFT + UMPCA can sense the difference between these data.

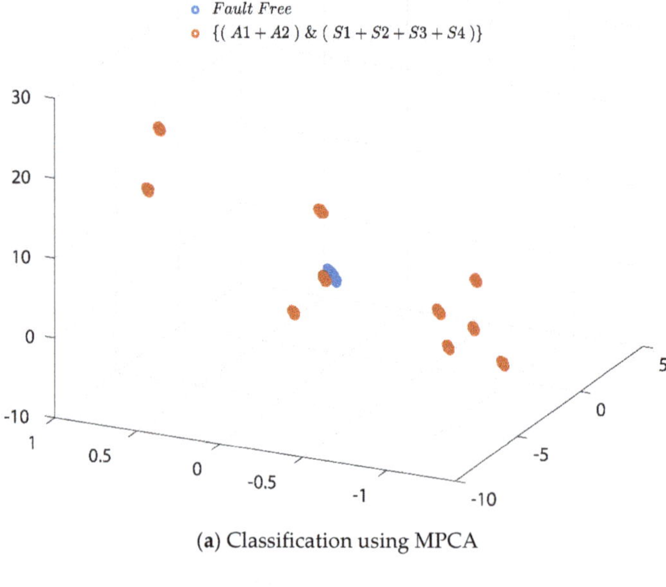

(**a**) Classification using MPCA

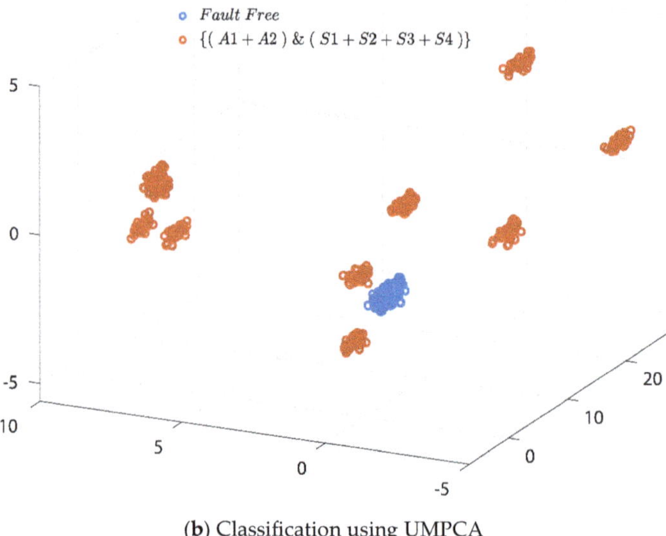

(**b**) Classification using UMPCA

Figure 18. Three-dimensional space visualization performance for fault classification for wind turbines subjected to two actuator and four sensor faults under AWGN noises, using (**a**) MPCA and (**b**) UMPCA, respectively.

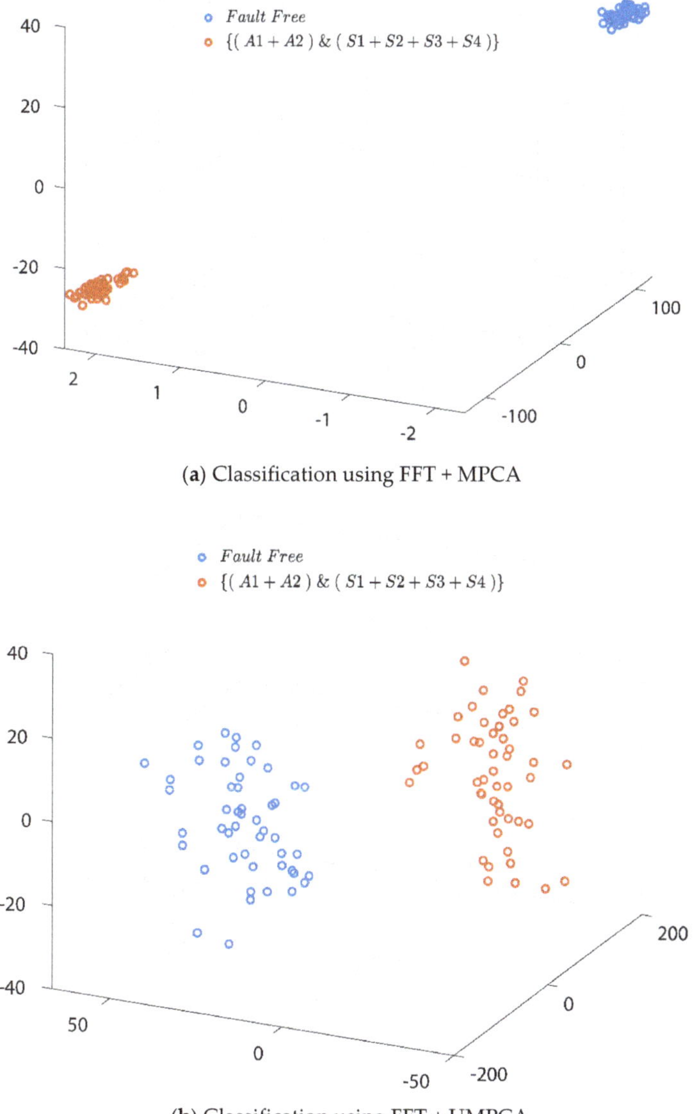

(**a**) Classification using FFT + MPCA

(**b**) Classification using FFT + UMPCA

Figure 19. Three-dimensional space visualization performance for fault classification for wind turbines subjected to two actuator and four sensor faults under AWGN noises, using (**a**) FFT + MPCA and (**b**) FFT + UMPCA, respectively.

It is noted that the MPCA approach determines a tensor-to-tensor projection that captures most of the signal variation present in the original tensor representation, whereas, the UMPCA method uses the tensor-to-vector projection. For the MPCA technique, some of the correlations of the principal components among the projected directions are neglected to some extent. Compared with MPCA, UMPCA can exclude the possibilities of getting significant features with similar geometric structures, depending on the methodology of tensor-to-vector projection. The reason behind is that the UMPCA algorithm concentrates on extracting and determining the uncorrelated principal components rather than the conventional principal components in the MPCA technique. Moreover, the FFT preprocessing technique can enhance the data classification capability of the UMPCA. As a result, this is why the proposed FFT + UMPCA can effectively classify the fault under all five scenarios above.

6. Conclusions

In this paper, fast Fourier transform (FFT) and uncorrelated multi-linear principal component analysis (UMPCA) techniques were integrated for fault classification of the 4.8 MW benchmark wind turbine systems subjected to multiple actuator and sensor faults under five scenarios of actuator and sensor faults. The detailed comparison studies were carried out, and the effectiveness of the proposed algorithm was well demonstrated. It is worthy to point out, among all the used algorithms, the FFT has a positive impact on the improvement of the performance of the fault diagnosis and classification. The proposed FFT plus UMPCA algorithm can not only classify the various classes of faulty conditions but can also recognize the differences between the data within the same class.

In the future, it is of interest to investigate data-driven fault prognosis and remaining useful life prediction for wind turbine systems. It is also promising to enhance fault diagnosis and prognosis performance by using hybrid methods (by integrating various data-driven fault diagnosis/prognosis methods or even by combining model-based approaches and data-driven based methods).

Author Contributions: Conceptualization, Y.F., Z.G. and Y.L.; methodology, Y.F., Z.G. and Y.L.; software, Y.F. and Y.L.; validation, Y.F.; formal analysis, Y.F., Y.L., A.Z., X.Y. and Z.G.; writing—original draft preparation, Y.F.; writing—review and editing, Z.G.; supervision, Z.G., Y.L.; funding acquisition, Z.G. and A.Z. All authors have read and agreed to the published version of the manuscript.

Funding: This research was funded by the National Nature Science Foundation of China (NNSFC) under grant 61673074, and the Alexander von Humboldt Foundation under grant GRO/1117303 STP.

Acknowledgments: The authors would like to thank the research support from the E & E faculty at University of Northumbria (UK), the National Nature Science Foundation of China (NNSFC) under grant 61673074, and the Alexander von Humboldt Foundation under grant GRO/1117303 STP.

Conflicts of Interest: The authors declare no conflict of interest.

References

1. Rashid, M.H. *Electric Renewable Energy Systems*; Elsevier, Academic Press: San Diego, CA, USA, 2016.
2. Gao, R.; Gao, Z. Pitch control for wind turbine systems using optimization, estimation and compensation. *Renew. Energy* **2016**, *91*, 501–515. [CrossRef]
3. Wind Europe. Wind Energy in Europe in 2019: Trends and Statistics. 2020. Available online: https://windeurope.org/wp-content/uploads/files/about-wind/statistics/WindEurope-Annual-Statistics-2019.pdf (accessed on 10 August 2020).
4. Gao, Z.; Sheng, S. Real-time monitoring, prognosis, and resilient control for wind turbine systems. *Renew. Energy* **2018**, *116*, 1–4. [CrossRef]
5. Hameed, Z.; Hong, Y.; Cho, Y.; Ahn, S.-H.; Song, C. Condition monitoring and fault detection of wind turbines and related algorithms: A review. *Renew. Sustain. Energy Rev.* **2009**, *13*, 1–39. [CrossRef]
6. Márquez, F.; Tobias, A.M.; Pérez, J.M.P.; Papaelias, M. Condition monitoring of wind turbines: Techniques and methods. *Renew. Energy* **2012**, *46*, 169–178. [CrossRef]
7. Liu, X.; Gao, Z.; Chen, M.Z.Q. Takagi–Sugeno Fuzzy Model Based Fault Estimation and Signal Compensation with Application to Wind Turbines. *IEEE Trans. Ind. Electron.* **2017**, *64*, 5678–5689. [CrossRef]
8. Simani, S.; Farsoni, S.; Castaldi, P. Fault Diagnosis of a Wind Turbine Benchmark via Identified Fuzzy Models. *IEEE Trans. Ind. Electron.* **2014**, *62*, 3775–3782. [CrossRef]

9. Rahimilarki, R.; Gao, Z.; Zhang, A.; Binns, R. Robust Neural Network Fault Estimation Approach for Nonlinear Dynamic Systems with Applications to Wind Turbine Systems. *IEEE Trans. Ind. Inform.* **2019**, *15*, 6302–6312. [CrossRef]

10. Chen, J.; Patton, R.J. *Robust Model-Based Fault Diagnosis for Dynamic Systems*; Kluwer: Boston, MA, USA, 1999.

11. Venkatasubramanian, V.; Rengaswamy, R.; Yin, K.; Kavuri, S.N. A review of process fault detection and diagnosis. *Comput. Chem. Eng.* **2003**, *27*, 293–311. [CrossRef]

12. Gao, Z.; Cecati, C.; Ding, S.X. A Survey of Fault Diagnosis and Fault-Tolerant Techniques—Part I: Fault Diagnosis with Model-Based and Signal-Based Approaches. *IEEE Trans. Ind. Electron.* **2015**, *62*, 3757–3767. [CrossRef]

13. Gao, Z.; Cecati, C.; Ding, S.X. A Survey of Fault Diagnosis and Fault-Tolerant Techniques Part II: Fault Diagnosis with Knowledge-Based and Hybrid/Active Approaches. *IEEE Trans. Ind. Electron.* **2015**, *62*, 3768–3774. [CrossRef]

14. Gao, Z.; Ding, S.X.; Cecati, C. Real-time fault diagnosis and fault-tolerant control. *IEEE Trans. Ind. Electron.* **2015**, *62*, 3752–3756. [CrossRef]

15. Zhang, D.; Gao, Z. Improvement of Refrigeration Efficiency by Combining Reinforcement Learning with a Coarse Model. *Processes* **2019**, *7*, 967. [CrossRef]

16. Bergen, K.J.; Johnson, P.A.; De Hoop, M.V.; Beroza, G.C. Machine learning for data-driven discovery in solid Earth geoscience. *Science* **2019**, *363*, eaau0323. [CrossRef] [PubMed]

17. Oh, H.; Jung, J.H.; Jeon, B.C.; Youn, B.D. Scalable and Unsupervised Feature Engineering Using Vibration-Imaging and Deep Learning for Rotor System Diagnosis. *IEEE Trans. Ind. Electron.* **2018**, *65*, 3539–3549. [CrossRef]

18. Song, W.; Wen, L.; Gao, L.; Li, X. Unsupervised fault diagnosis method based on iterative multi-manifold spectral clustering. *IET Collab. Intell. Manuf.* **2019**, *1*, 48–55. [CrossRef]

19. Holtzman, B.K.; Paté, A.; Paisley, J.; Waldhauser, F.; Repetto, D. Machine learning reveals cyclic changes in seismic source spectra in Geysers geothermal field. *Sci. Adv.* **2018**, *4*, 5. [CrossRef]

20. Phan, H.T.H.; Kumar, A.; Feng, D.; Fulham, M.J.; Kim, J. Unsupervised Two-Path Neural Network for Cell Event Detection and Classification Using Spatiotemporal Patterns. *IEEE Trans. Med. Imaging* **2018**, *38*, 1477–1487. [CrossRef]

21. Zhao, R.; Wang, D.; Yan, R.; Mao, K.; Shen, F.; Wang, J. Machine Health Monitoring Using Local Feature-Based Gated Recurrent Unit Networks. *IEEE Trans. Ind. Electron.* **2018**, *65*, 1539–1548. [CrossRef]

22. Elforjani, M.; Shanbr, S. Prognosis of Bearing Acoustic Emission Signals Using Supervised Machine Learning. *IEEE Trans. Ind. Electron.* **2018**, *65*, 5864–5871. [CrossRef]

23. Adeli, E.; Thung, K.-H.; An, L.; Wu, G.; Shi, F.; Wang, P.; Shen, D. Semi-Supervised Discriminative Classification Robust to Sample-Outliers and Feature-Noises. *IEEE Trans. Pattern Anal. Mach. Intell.* **2019**, *41*, 515–522. [CrossRef]

24. Razavi-Far, R.; Hallaji, E.; Farajzadeh-Zanjani, M.; Saif, M.; Kia, S.H.; Henao, H.; Capolino, G.-A. Information Fusion and Semi-Supervised Deep Learning Scheme for Diagnosing Gear Faults in Induction Machine Systems. *IEEE Trans. Ind. Electron.* **2019**, *66*, 6331–6342. [CrossRef]

25. Zhao, M.; Kang, M.; Tang, B.; Pecht, M. Multiple Wavelet Coefficients Fusion in Deep Residual Networks for Fault Diagnosis. *IEEE Trans. Ind. Electron.* **2019**, *66*, 4696–4706. [CrossRef]

26. Ge, Z.; Song, Z.; Ding, S.X.; Huang, B. Data Mining and Analytics in the Process Industry: The Role of Machine Learning. *IEEE Access* **2017**, *5*, 20590–20616. [CrossRef]

27. Guo, M.-F.; Yang, N.-C.; Chen, W.-F. Deep-Learning-Based Fault Classification Using Hilbert–Huang Transform and Convolutional Neural Network in Power Distribution Systems. *IEEE Sens. J.* **2019**, *19*, 6905–6913. [CrossRef]

28. Pan, T.; Chen, J.; Zhou, Z.; Wang, C.; He, S. A Novel Deep Learning Network via Multiscale Inner Product With Locally Connected Feature Extraction for Intelligent Fault Detection. *IEEE Trans. Ind. Inform.* **2019**, *15*, 5119–5128. [CrossRef]

29. Abid, A.; Khan, M.T.; Khan, M.S. Multidomain Features-Based GA Optimized Artificial Immune System for Bearing Fault Detection. *IEEE Trans. Syst. Man Cybern. Syst.* **2020**, *50*, 348–359. [CrossRef]

30. Isermann, R. Fault detection with Principal Component Analysis (PCA). In *Fault-Diagnosis Systems*; Springer: Berlin/Heidelberg, Germany, 2006; pp. 267–278.

31. Luu, K.; Savvides, M.; Bui, T.D.; Suen, C.Y. Compressed Submanifold Multifactor Analysis. *IEEE Trans. Pattern Anal. Mach. Intell.* **2016**, *39*, 444–456. [CrossRef]

32. He, H.; Tan, Y. Pattern Clustering of Hysteresis Time Series with Multivalued Mapping Using Tensor Decomposition. *IEEE Trans. Syst. Man Cybern. Syst.* **2017**, *48*, 993–1004. [CrossRef]

33. Yan, S.; Xu, N.; Yang, Q.; Zhang, L.; Tang, X.; Zhang, H.-J. Multilinear discriminant analysis for face recognition. *IEEE Trans. Image Process.* **2007**, *16*, 212–220. [CrossRef]

34. Liu, J.; Rao, B.D. Robust PCA via $l_0 - l_1$ Regularization. *IEEE Trans. Signal Process.* **2018**, *67*, 535–549. [CrossRef]

35. Lenz, M.; Mueller, F.-J.; Zenke, M.; Schuppert, A.A. Principal components analysis and the reported low intrinsic dimensionality of gene expression microarray data. *Sci. Rep.* **2016**, *6*, 1–11. [CrossRef] [PubMed]

36. Zhou, F.; Park, J.H.; Liu, Y. Differential feature based hierarchical PCA fault detection method for dynamic fault. *Neurocomputing* **2016**, *202*, 27–35. [CrossRef]

37. Vaswani, N.; Bouwmans, T.; Javed, S.; Narayanamurthy, P. Robust Subspace Learning: Robust PCA, Robust Subspace Tracking, and Robust Subspace Recovery. *IEEE Signal Process. Mag.* **2018**, *35*, 32–55. [CrossRef]

38. Shi, Q.; Cheung, Y.-M.; Zhao, Q.; Lu, H. Feature Extraction for Incomplete Data Via Low-Rank Tensor Decomposition With Feature Regularization. *IEEE Trans. Neural Netw. Learn. Syst.* **2018**, *30*, 1803–1817. [CrossRef] [PubMed]

39. Lu, H.; Plataniotis, K.N.; Venetsanopoulos, A. *Multilinear Subspace Learning*; Dimensionality Reduction of Multidimensional Data; Chapman and Hall/CRC: New York, NY, USA, 2013.

40. Zhou, C.; Wang, L.; Zhang, Q.; Wei, X. Face recognition based on PCA and logistic regression analysis. *Optik* **2014**, *125*, 5916–5919. [CrossRef]

41. Huang, Y.; Guan, Y. On the linear discriminant analysis for large number of classes. *Eng. Appl. Artif. Intell.* **2015**, *43*, 15–26. [CrossRef]

42. Fu, Y.; Liu, Y.; Gao, Z. Fault Classification in Wind Turbines Using Principal Component Analysis Technique. In Proceedings of the 2019 IEEE 17th International Conference on Industrial Informatics (INDIN); Institute of Electrical and Electronics Engineers (IEEE), Helsinki-Espoo, Finland, 22–25 July 2019; Volume 1, pp. 1303–1308.

43. Odgaard, P.F.; Stoustrup, J.; Kinnaert, M. Fault Tolerant Control of Wind Turbines—A benchmark model. In Proceedings of the 7th IFAC Symposium on Fault Detection, Supervision and Safety of Technical Processes, Barcelona, Spain, 30 June–3 July 2009; Volume 1, pp. 155–160.

44. Odgaard, P.; Kinnaert, M.; Stoustrup, J. Fault-Tolerant Control of Wind Turbines: A Benchmark Model. *IEEE Trans. Control Syst. Technol.* **2013**, *21*, 1168–1182. [CrossRef]

45. Vaibhav, V. Higher Order Convergent Fast Nonlinear Fourier Transform. *IEEE Photonics Technol. Lett.* **2018**, *30*, 700–703. [CrossRef]

46. Wang, S.; Patel, V.M.; Petropulu, A.P. Multidimensional Sparse Fourier Transform Based on the Fourier Projection-Slice Theorem. *IEEE Trans. Signal Process.* **2019**, *67*, 54–69. [CrossRef]

47. Fu, Y.; Liu, Y.; Zhang, A.; Gao, Z. Multiple Actuator Fault Classification for Wind Turbine Systems by Integrating Fast Fourier Transform (FFT) and Multi-linear Principal Component Analysis (MPCA). In Proceedings of the IECON 2019—45th Annual Conference of the IEEE Industrial Electronics Society, Lisbon, Portugal, 14–17 October 2019; Volume 1, pp. 3761–3766.

48. Lu, H.; Plataniotis, K.N.K.; Venetsanopoulos, A.N. Uncorrelated multilinear principal component analysis for unsupervised multilinear subspace learning. *IEEE Trans. Neural Netw.* **2009**, *20*, 1820–1836. [CrossRef]

Article

Monitoring the Process Based on Belief Statistic for Neutrosophic Gamma Distributed Product

Muhammad Aslam [1,*], Rashad A. R. Bantan [2] and Nasrullah Khan [3]

[1] Department of Statistics, Faculty of Science, King Abdulaziz University, Jeddah 21551, Saudi Arabia

[2] Department of Marine Geology, Faculty of Marine Science, King Abdulaziz University, Jeddah 21551, Saudi Arabia; rbantan@kau.edu.sa

[3] Department of Statistics, University of Veterinary and Animal Sciences (Jhang campus), Lahore 54000, Pakistan; nas_shan1@hotmail.com

* Correspondence: aslam_ravian@hotmail.com or magmuhammad@kau.edu.sa; Tel.: +96-59-3329841

Received: 12 March 2019; Accepted: 8 April 2019; Published: 12 April 2019

Abstract: In this paper, we developed a control chart methodology for the monitoring the mean time between two events using the belief estimator under the neutrosophic gamma distribution. The proposed control chart coefficients and the neutrosophic average run length (NARL) have been determined using different process settings. The performance of the proposed chart is compared with the control chart under classical statistics in terms of NARL using the simulation data and real example. From comparisons, it is concluded that the proposed chart is efficient, effective and adequate to be used under uncertainty environment than the chart under classical statistics.

Keywords: control chart; fuzzy logic; neutrosophic statistic; incomplete data; belief statistic; gamma distribution

1. Introduction

The control chart is an important tool of Statistical Process Control (SPC) used in production processes for monitoring the quality of the products and effective in defect prevention. The diagnosis and correction of many production problems which often cause huge loss to the production unit can substantially be improved with the utilization of the effective control chart technique [1]. Once the control chart is established from the initial observations of the interested quality characteristic normally known as the Phase-I control limits and revised according to monitoring parameters with the prime objective of the maintaining the repute of the product in the market and the profit maximization. In the quick monitoring of the quality of the product in this era of fast technology, being used in the production units, a minute delay can result in the production of a huge amount of defective items or items which need reworking. To maintain the production process at the required quality level, the continuous improvement of the production process and the identification of sources of the variation are the prime objectives of any process monitoring scheme [2]. Vigilant monitoring is demanded by the production process to identify the root cause and preventing it from reoccurring of unwanted situation. The Shewhart control charts are technically the most sophisticated tool for monitoring such unusual changes in the processes Montgomery [3]. The tool of a control chart is used for the need of conducting timely corrective action if any abnormality creeps into the production process [4]. Fuzzy transformation methods to determine the tightness of the inspection for the linguist data are a valuable development [5]. The belief estimator has been thoroughly defined and discussed by Fallah Nezhad and Akhavan Niaki [6]. Variable control charts are developed when the characteristic under study is continuous in nature. The best-fitted distribution is the normal distribution for the continuous data when collected in groups or when the form of the distribution is known. In general, there are many instances when the data are not collected in the form of the groups or the shape of the distribution is skewed or unknown. When this occurs then the use of normal distribution may lead to erroneous results. The most commonly suggested distribution for this type of data is the gamma distribution. The study of control charts using the gamma distribution has been explored by many authors, for instance, see references [7–13]. The Shewhart control charts also known as the classic control charts are used to analyze variations in the quality characteristic when the collected data are quite exact and precise. However, the collected data may not always be so clear and exact in practice. We observe that uncertainty is a natural phenomenon which is associated with the human world, for which researches are continuously struggling for devising chart to address the uncertainty through probability theory or the fuzzy set theory. Also, the data collected from the human subjectivity cannot be treated as the exact numeric data and the construction of the control limits based upon such data will lead to erroneous conclusions. The construction of the control chart for such vague data can be best represented by using the most common logic of fuzzy control charts [14]. The fuzzy logic deals with data of the situation, which is, not clear, ambiguous or not well defined. The fuzzy logic is a special case of the neutrosophic statistic (see Smarandache [15]). In the literature of quality control, the fuzzy control charts are developed when the data are vague, incomplete, ambiguous and not well defined [16]. The notion of fuzzy sets was exposed by Zadeh [17]. The use of the fuzzy concept in the control chart literature started when Wang and Raz [18] published their paper. They applied two approaches to the construction of control charts for linguistic data. The linguistic data can provide thorough investigation than the binary classification used in attribute control charts. Raz and Wang [19] provided more results from the findings of Wang and Raz [18]. Taleb and Limam [20] proposed three sets of membership functions with different degrees of fuzziness for the fuzzy and probabilistic models using the average run lengths for comparisons. Kanagawa and Tamaki [21] developed control charts for the process average and process variability based on linguistic or imprecise data. Erginel and Sentürk [22] developed the fuzzy control chart for monitoring the food industry using fuzzy $\widetilde{\overline{X}}$ and $\widetilde{S}$. El-Shal and Morris [23] suggested a fuzzy rule-based algorithm for quality improvement monitoring through control charts. Rowlands and Wang [24] studied the fuzzy logic operation and its functioning in the control chart literature. Gülbay and Kahraman [1] used the α-cut control charts for the tightened inspection of observation for the fuzzy sets. Aslam [25] developed a sampling plan for the Neutrosophic statistics under the process loss index. Senturk and Erginel [26] constructed the fuzzy $\widetilde{\overline{X}}$–$\widetilde{R}$ and $\widetilde{\overline{X}}$–$\widetilde{S}$ charts with the α-cuts. Sentürk [27] developed the fuzzy regression control chart for the α-cut fuzzy numbers. Kaya and Kahraman [28] proposed a fuzzy control chart for process capability analysis based upon fuzzy measurements. Broumi and Smarandache [29] studied the correlation coefficient of the interval Neutrosophic set. Şentürk and Erginel [16] developed a fuzzy exponentially weighted moving average chart for univariate data with a real case application.

As mentioned by Smarandache [30] that the neutrosophic logic which considered the measure of indeterminacy is an extension of the fuzzy logic. The neutrosophic statistics which is based on neutrosophic numbers is the generalization of classical statistics and has been used under uncertainty, see Smarandache [31] and Smarandache [32]. The Neutrosophic statistics is the extension of the classic statistics used to analyze the data of vague, undefined, imprecise, incomplete, and indeterminate nature in opposition to the clear, certain, and crisp observations or parameters in which classical statistics is suitable, (see [32] and [25]). Due to wide application of the neutrosophic statistics, several authors applied it in various fields. Neutrosophic Statistical Numbers were introduced in [32], page 11. Chen et al. [33] and Chen et al. [34] introduced the neutrosophic statistical numbers to measure the rock roughness. Aslam [25] introduced the neutrosophic statistical quality control. Aslam [35] proposed a neutrosophic reliability plan. Aslam [36] designed the plan for the exponential distribution using neutrosophic statistics. Aslam [37] presented a neutrosophic attribute sampling plan. Aslam et al. [38] proposed the attribute control chart using neutrosophic statistics. Aslam et al. [39] worked on neutrosophic variance chart. Aslam et al. [40] proposed the chart for the gamma distribution under the neutrosophic statistics. Aslam [41] proposed the plan with measurement error in uncertainty. Aslam and Arif [42] worked on sudden death tests under the uncertainty. Aslam and Raza [43] designed the neutrosophic plan for multiple manufacturing lines. Peng and Dai [44] has given the bibliographic review of the Neutrosophic statistics for the last two decades. Peng and Dai [45], initiated a new axiomatic definition of single-valued neutrosophic distance measure and proposed a novel measure. More literature on Neutrosophic can be seen in [46–51].

In this paper, a control chart scheme has been developed for monitoring the mean time between two events under the neutrosophic statistics using the belief estimator for the NGD which according to the best knowledge of the authors has not been explored by any researcher. It is mentioned here that the proposed chart is reduced to the classical chart when no parameter is obtained as vague, imprecise, indeterminate or incomplete. Gao, Cecati, and Ding [52] provided a sophisticated state-of-the-art overview on data-driven and machine learning based fault detection and diagnosis approaches. The proposed control chart will indicate the change in the process mean and will be helpful to correct the fault during the process. The rest of the paper is organized as: the design of the proposed chart is explained in Section 2. In Section 3 the simulation study of the proposed scheme has been discussed. In Section 4 the application of the proposed chart has been explained by using a real-world example. In the last section, the conclusion of the proposed chart has been described.

2. The Neutrosophic Gamma Distribution

The neutrosophic gamma distribution (NGD) is introduced by Aslam et al. [40]. Let $T_N \epsilon [T_L, T_U]$ be the neutrosophic random variable (NRV) of size $n_N \epsilon [n_L, n_U]$ of the quality of interest that follows the NGD, where T_L and T_U are the lower and upper failure times of the indeterminacy interval. The neutrosophic cumulative distribution function (ncdf) of the NGD with two neutrosophic parameters $a_N \epsilon [a_L, a_U]$ and $b_N \epsilon [b_L, b_U]$, where $a_N \epsilon [a_L, a_U]$ is the shape parameter and $b_N \epsilon [b_L, b_U]$ is the scale parameter is given by

$$P(T_N \leq t_N) = 1 - \sum_{j=0}^{a_N-1} \frac{e^{\frac{-t_N}{b_N}} \left(\frac{t_N}{b_N}\right)^j}{j!}; \ T_N \epsilon [T_L, T_U], \ a_N \epsilon [a_L, a_U], \ b_N \epsilon [b_L, b_U] \tag{1}$$

The NGD is the generalization of the several distributions. The NGD reduces to neutrosophic exponential distribution when $a_N \epsilon [1, 1]$, see [36]. The NGD reduced to exponential distribution under classical distribution when $a_L = a_U$ and $b_L = b_U$. According to Wilson and Hilferty [53] using the transformation of $T_N^* = T_N^{1/3}; T_N^* \epsilon [T_L^*, T_U^*]$ in the NGD tends to form the approximate neutrosophic normal distribution, see Smarandache [32], with the mean and variance can be described as:

$$\mu_{T_N^*} = E\left(T_N^*\right) = \frac{b_N^{1/3} \Gamma(a_N + 1/3)}{\Gamma(a_N)}; \ T_N^* \epsilon [T_L^*, T_U^*], \ a_N \epsilon [a_L, a_U], \ b_N \epsilon [b_L, b_U] \tag{2}$$

$$\sigma_{T_N^*} = b_N^{1/3} \sqrt{\frac{\Gamma(a_N + 2/3)}{\Gamma(a_N)} - \left(\frac{\Gamma(a_N + 1/3)}{\Gamma(a_N)}\right)^2}; \ a_N \epsilon [a_L, a_U], \ b_N \epsilon [b_L, b_U] \tag{3}$$

So, the approximate neutrosophic normal distribution of $T_N^*; T_N^* \epsilon [T_L^*, T_U^*]$ is given as:

$$T_N^* \sim N\left(\frac{b_N^{1/3} \Gamma(a_N + 1/3)}{\Gamma(a_N)}, b_N^{2/3}\left[\frac{\Gamma(a_N + 2/3)}{\Gamma(a_N)} - \left(\frac{\Gamma(a_N + 1/3)}{\Gamma(a_N)}\right)^2\right]\right); \ T_N^* \epsilon [T_L^*, T_U^*],$$
$$a_N \epsilon [a_L, a_U], \ b_N \epsilon [b_L, b_U] \tag{4}$$

3. Designing of the Proposed Chart

Suppose that a single observation of the quality of interest is collected at every iteration or subgroup with $n_N \epsilon [1, 1]$ then the *kth* observation of T_{K_N} and $O_{K_N} = T_{N1}, T_{N2}, T_{N3} \ldots \ldots, T_{NK}$ be defined as the *kth* iteration. We define the posterior belief and the prior belief as $B\left(O_{K_N}\right)$ and $B\left(O_{K_N-1}\right)$ respectively with

$O_{K_N} = (T_{NK}, O_{K_N-1})$. The new observed variable $T_N^* = T_N^{1/3}$; $T_N^* \epsilon [T_L^*, T_U^*]$ is updated using $B(O_{K_N})$ and $B(O_{K_N-1})$ for the posterior belief using the following equation

$$B(O_{k_N}) = B(T_{K_N}, O_{K_N-1}) = \frac{B(O_{K_N-1})e^{\frac{T_N^* - \mu_{T_N^*}}{\sigma_{T_N^*}}}}{B(O_{K_N-1})e^{\frac{T_N^* - \mu_{T_N^*}}{\sigma_{T_N^*}}} + (1 - B(O_{K_N-1}))} \; ; \; T_N^* \epsilon [T_L^*, T_U^*] \tag{5}$$

The variable T_N^* is given without the subscript K_N just for the purpose of simplicity. Using a new statistic based upon $B(O_{K_N})$ and $B(O_{K_N-1})$ suggested by Fallah Nezhad and Akhavan Niaki [6] given as:

$$Z_{k_N} = \frac{B(O_{K_N})}{1 - B(O_{K_N})} \tag{6}$$

The recursion relation is given as

$$Z_{k_N} = Z_{k_N-1}e^{\frac{T_N^* - \mu_{T_N^*}}{\sigma_{T_N^*}}} \tag{7}$$

Let the initial value of $B(O_0) = 0.5$ and $Z_0 = 1$ then according to Fallah Nezhad and Akhavan Niaki [6] the statistic proposed below follows the normal distribution with mean 0 and variance $k_N \epsilon [k_L, k_U]$. Thus the neutrosophic lower and upper control limits of the proposed chart can be written as:

$$UCL_N = L_N \sqrt{k_N} \tag{8}$$

$$LCL_N = -L_N \sqrt{k_N} \tag{9}$$

Using the Equations (8) and (9) the control coefficient $L_N \epsilon [L_L, L_U]$ is computed for the specific level of type-I error and the predefined in-control average run length values.

3.1. The Proposed Chart

The procedure of the proposed control chart can be summarized using the following steps:

Step 1: Measure the quality characteristic T_{K_N} of the *kth* subgroup selected at random. Find $T_{K_N}^* = T_{K_N}^{1/3}$ and then calculate

$$ln(Z_{K_N}) = ln(Z_{K_N-1}) + \frac{T_{K_N}^* - \mu_{T_{K_N}^*}}{\sigma_{T_{K_N}^*}} \tag{10}$$

Step 2: If $LCL_N \leq ln(Z_{k_N}) \leq UCL_N$, Declare the process in-control and if $ln(Z_{k_N}) > UCL_N$ or $ln(Z_{k_N}) < LCL_N$ then the process is declared as out-of-control.

For the purpose of calculating different measures of the in-control and out-of-control processes we suppose that the scale parameter of the gamma distribution is not constant while other parameters, i.e., shape parameter, is declared as a constant parameter. Let b_{0N} and b_{1N} be the in-control and the out-of-control values of the shape parameter respectively then the probability of declaring the process as out-of-control when the process is in-control may be defined as

$$P_{outN}^0 = P\{ln(Z_{k_N}) < LCL_N | b_N = b_{0N}\} + P\{ln(Z_{k_N}) > UCL_N | b_N = b_{0N}\}$$
$$= 1 - \Phi_N\left(\frac{UCL_N}{\sqrt{k_N}}\right) + \Phi\left(\frac{LCL_N}{\sqrt{k_N}}\right) \tag{11}$$

where $\Phi_N(x)$ denotes the cumulative distribution function of neutrosophic standard normal distribution, see Smarandache [31].

Finally, Equation (11) reduces to

$$P^0_{out} = 1 - \Phi(L_N) + \Phi(-L_N) \tag{12}$$

It is to be noted that P^0_{out} is independent of K_N.

3.2. Neutrosophic Average Run Length for In-Control Process

As mentioned above the neutrosophic average run length (NARL) is defined as the average number of samples before the process is indicated as the out-of-control. It is used very commonly in the literature of the control chart for the evaluation and the comparison of the proposed control chart [3]. Two types of ARLs has been used by the quality control researchers as the in-control ARL is denoted by ARL_0 and the out-of-control is denoted by ARL_1 [54]. More literature on ARL can be seen in references [55–62].

The $NARL_0$ of the proposed chart can be calculated as

$$NARL_0 = \frac{1}{P^0_{out,N}} \tag{13}$$

3.3. Neutrosophic Average Run Length for Shifted Process

Now we develop the procedure for the shifted process. Here we suppose that a shift in the scale parameter of the gamma distribution is introduced to observe the efficiency of the proposed scheme in quick detection of this shift. It is to be noted that the NARL of the in-control process ($NARL_0$) is some predefined value of 200, 300, and 370 based upon the false alarm rate which is larger values while NARL of the shifted process ($NARL_1$) should be smaller values for the more efficient proposed chart. Now let $b_1 = sb_0$ is the shift in the scale parameter of the gamma distribution with a shift constant is s. Then the mean and variance of the $T^*_N \epsilon \left[T^*_L, T^*_U \right]$ fort the shifted process can be determined as:

$$E\left(T^*_N \middle| b_{1N}\right) = s^{1/3}\, b_{0N}{}^{1/3}\, \frac{\Gamma(a_N + 1/3)}{\Gamma(a_N)}; \, T^*_N \epsilon \left[T^*_L, T^*_U \right] \tag{14}$$

$$Var\left(T^*_N \middle| b_{1N}\right) = s^{2/3}\, b_{0N}{}^{2/3}\left[\frac{\Gamma(a_N + 2/3)}{\Gamma(a_N)} - \left(\frac{\Gamma(a_N + 1/3)}{\Gamma(a_N)} \right)^2 \right]; \, T^*_N \epsilon \left[T^*_L, T^*_U \right] \tag{15}$$

Then the mean and variance of $ln\left(Z_{k_N}\right)$ at b_{1N} can be calculated as:

$$E\left[ln\left(Z_{k_N}\right)\middle| b_{1N}\right] = k_N \frac{\frac{\Gamma(a_N + 1/3)}{\Gamma(a)}\left(s^{1/3} - 1\right)}{\sqrt{\frac{\Gamma(a_N + 2/3)}{\Gamma(a_N)} - \left(\frac{\Gamma(a_N + 1/3)}{\Gamma(a_N)}\right)^2}}; \, k_N \epsilon [k_L, k_U] \tag{16}$$

$$Var\left(ln\left(Z_{k_N}\right)\middle| b_{1N}\right) = k_N s^{2/3}; \, k_N \epsilon [k_L, k_U] \tag{17}$$

The probability of shifted process to declare as an out-of-control process at kth subgroup is calculated as:

$$\begin{aligned}
P^1_{out,k_N} &= P\left\{ln\left(Z_{k_N}\right) < LCL_N \middle| b_N = b_{1N}\right\} + P\left\{ln\left(Z_{k_N}\right) > UCL_N \middle| b_N = b_{1N}\right\}; k_N \epsilon [k_L, k_U] \\
&= P\left\{ln\left(Z_{k_N}\right) < -L_N\,\sqrt{k_N}\,\middle| b_N = b_{1N}\right\} + P\left\{ln\left(Z_{k_N}\right) > L_N\,\sqrt{k_N}\,\middle| b_N = b_{1N}\right\}; k_N \epsilon [k_L, k_U]
\end{aligned} \tag{18}$$

Finally, we have

$$
P^1_{out,k_N} = 1 - \Phi\left(\frac{L_N\sqrt{k_N} - \dfrac{k_N \cdot \frac{\Gamma(a_N + 1/3)}{\Gamma(a_N)}\left(s^{1/3}-1\right)}{\sqrt{\frac{\Gamma(a_N + 2/3)}{\Gamma(a_N)} - \left(\frac{\Gamma(a_N + 1/3)}{\Gamma(a_N)}\right)^2}}}{\sqrt{k_N s^{2/3}}} \right) + \Phi\left(\frac{-L_N\sqrt{k_N} - \dfrac{k_N \cdot \frac{\Gamma(a_N + 1/3)}{\Gamma(a_N)}\left(s^{1/3}-1\right)}{\sqrt{\frac{\Gamma(a_N + 2/3)}{\Gamma(a_N)} - \left(\frac{\Gamma(a_N + 1/3)}{\Gamma(a_N)}\right)^2}}}{\sqrt{k_N s^{2/3}}} \right);
$$
$$k_N \epsilon [k_L, k_U]$$
(19)

The probability of declaring out-of-control at $(k + j)$th subgroup when the process shift occurs at k is expressed as

$$P\{RL = j\} = \left(1 - P^1_{out,k_N+1}\right)\left(1 - P^1_{out,k_N+2}\right)\cdots\left(1 - P^1_{out,k_N+j-1}\right)P^1_{out,k_N+j} \tag{20}$$

where RL is a random variable representing $NARL_1$.

Therefore, ARL_1 under the proposed control chart is given as:

$$NARL_{1N} = P^1_{out,k_N+1} + 2\left(1 - P^1_{out,k_N+1}\right)P^1_{out,k_N+2} + 3\left(1 - P^1_{out,k_N+1}\right)\left(1 - P^1_{out,k_N+2}\right)P^1_{out,k_N+3} + \cdots \tag{21}$$

Note here that the formulae in Equation (1) to Equation (21) under the neutrosophic statistics is the generalization of the formulae in Aslam et al. [63].

To determine the control coefficient L_N, $NARL_0$ denoted by (r_{N0}) and the $NARL_1$ using the above-mentioned methodology, the following stepwise algorithm can be described as:

Step 1: Choose a range of control coefficient L_N

Step 2: Calculate L_N such that $ARL_{0N} \geq r_{N0}$

Step 3: For a fixed level of k_N and various shift constants s calculate P^1_{out,k_N} using Equation (19).

Step 4: Calculate the values of ARL_{1N} for a fixed k_N for various shift constants s.

Using the above mentioned methodology, an R-language code program was written and run for different parameters such that $a_N \epsilon[1, 1]$, $a_N \epsilon[5, 5]$ and $a_N \epsilon[10, 10]$ and using different in-control NARL values as $r_{0N} \epsilon[200, 200]$, $r_{0N} \epsilon[300, 300]$ and $r_{0N} \epsilon[370, 370]$ and different shift levels as $s = 4.00, 3.00, 2.80, 2.50, 2.25, 2.00, 1.90, 1.80, 1.70, 1.60, 1.50, 1.40, 1.30, 1.20, 1.10, 1.00, 0.80, 0.75, 0.70, 0.60, 0.50, 0.40, 0.30, 0.25, 0.15, 0.10,$ and 0.05. The $NARL_1$ values are determined and given in Tables 1–4.

Table 1. The values of neutrosophic average run length (NARL) when $a_N \epsilon[1.95, 2.05]$ and $k_N \epsilon[3, 5]$.

k_N	[2.8071,2.8141]	[2.9354,2.9416]	[3.0003,3.0012]
s	NARL		
4.00	[1.28,1.06]	[1.32,1.07]	[1.34,1.07]
3.00	[1.76,1.24]	[1.88,1.28]	[1.95,1.31]
2.80	[1.98,1.34]	[2.13,1.40]	[2.22,1.43]
2.50	[2.50,1.58]	[2.75,1.68]	[2.89,1.73]
2.25	[3.27,1.97]	[3.67,2.13]	[3.91,2.22]
2.00	[4.75,2.74]	[5.48,3.05]	[5.92,3.22]
1.90	[5.74,3.28]	[6.71,3.70]	[7.29,3.93]
1.80	[7.13,4.05]	[8.47,4.66]	[9.28,4.98]
1.70	[9.18,5.24]	[11.1,6.12]	[12.26,6.61]
1.60	[12.32,7.12]	[15.19,8.51]	[16.96,9.28]
1.50	[17.4,10.34]	[21.93,12.67]	[24.76,13.98]
1.40	[26.08,16.27]	[33.74,20.51]	[38.61,22.94]
1.30	[41.89,28.25]	[55.90,36.86]	[64.99,41.90]
1.20	[72.12,54.92]	[99.88,74.68]	[118.35,86.58]
1.10	[127.96,115.71]	[185.0,165.6]	[224.15,196.69]
1.00	[200.02,204.41]	[300.17,306.26]	[370.82,371.83]

Table 1. *Cont.*

k_N	[2.8071,2.8141]	[2.9354,2.9416]	[3.0003,3.0012]
s	NARL		
0.80	[152.58,98.67]	[227.95,143.72]	[281.15,172.31]
0.75	[119.34,67.59]	[177.29,97.32]	[218.11,116.09]
0.70	[90.73,45.44]	[134.07,64.61]	[164.56,76.64]
0.60	[49.26,19.71]	[71.88,27.17]	[87.73,31.78]
0.50	[24.63,8.22]	[35.26,10.86]	[42.65,12.45]
0.40	[11.21,3.44]	[15.56,4.26]	[18.55,4.75]
0.30	[4.64,1.60]	[6.10,1.82]	[7.09,1.95]
0.25	[2.91,1.22]	[3.68,1.32]	[4.19,1.37]
0.15	[1.27,1.00]	[1.4,1.01]	[1.49,1.01]
0.10	[1.03,1.00]	[1.05,1.00]	[1.07,1.00]
0.05	[1.00,1.00]	[1.00,1.00]	[1.00,1.00]

Table 2. The values of NARL when $a_N \epsilon [1.95, 2.05]$ and $k_N \epsilon [8, 10]$.

k_N	[2.8071,2.8164]	[2.9354,2.9436]	[2.9997,3.007]
s	NARL		
4.00	[1.01,1]	[1.01,1]	[1.01,1]
3.00	[1.07,1.02]	[1.08,1.03]	[1.09,1.03]
2.80	[1.11,1.04]	[1.13,1.05]	[1.14,1.05]
2.50	[1.23,1.1]	[1.27,1.12]	[1.29,1.13]
2.25	[1.43,1.22]	[1.5,1.26]	[1.54,1.28]
2.00	[1.86,1.5]	[2.01,1.59]	[2.1,1.64]
1.90	[2.17,1.71]	[2.38,1.83]	[2.5,1.9]
1.80	[2.63,2.02]	[2.93,2.21]	[3.1,2.31]
1.70	[3.34,2.52]	[3.8,2.81]	[4.07,2.97]
1.60	[4.51,3.36]	[5.25,3.83]	[5.69,4.1]
1.50	[6.58,4.88]	[7.87,5.72]	[8.64,6.22]
1.40	[10.58,7.92]	[13.06,9.6]	[14.57,10.61]
1.30	[19.27,14.85]	[24.68,18.73]	[28.06,21.11]
1.20	[40.86,33.47]	[54.79,44.29]	[63.76,51.16]
1.10	[99.7,91.45]	[141.65,128.72]	[169.75,153.39]
1.00	[200.01,205.94]	[300.2,308.26]	[370.01,379.03]
0.80	[63.82,47.69]	[91.14,66.87]	[109.64,79.63]
0.75	[39.6,27.68]	[55.57,37.96]	[66.28,44.72]
0.70	[24.4,16.13]	[33.57,21.58]	[39.66,25.12]
0.60	[9.26,5.74]	[12.14,7.23]	[14.01,8.17]
0.50	[3.69,2.34]	[4.54,2.74]	[5.07,2.98]
0.40	[1.71,1.27]	[1.94,1.36]	[2.08,1.42]
0.30	[1.09,1.01]	[1.14,1.02]	[1.16,1.03]
0.25	[1.02,1.00]	[1.03,1.00]	[1.03,1.00]
0.15	[1.00,1.00]	[1.00,1.00]	[1.00,1.00]
0.10	[1.00,1.00]	[1.00,1.00]	[1.00,1.00]
0.05	[1.00,1.00]	[1.00,1.00]	[1.00,1.00]

Table 3. The values of NARL when $a_N \epsilon [0.95, 1.05]$ and $k_N \epsilon [3, 5]$.

k_N	[2.8071,2.8142]	[2.9352,2.9392]	[2.9997,3.0056]
s	NARL		
4.00	[2.09,1.4]	[2.24,1.45]	[2.32,1.49]
3.00	[3.34,2.02]	[3.71,2.17]	[3.92,2.26]
2.80	[3.86,2.29]	[4.33,2.49]	[4.61,2.61]
2.50	[5.04,2.93]	[5.78,3.25]	[6.21,3.43]
2.25	[6.74,3.88]	[7.88,4.38]	[8.56,4.69]
2.00	[9.78,5.66]	[11.74,6.58]	[12.92,7.15]

Table 3. *Cont.*

k_N	[2.8071,2.8142]	[2.9352,2.9392]	[2.9997,3.0056]
s	NARL		
1.90	[11.71,6.84]	[14.24,8.05]	[15.76,8.81]
1.80	[14.33,8.5]	[17.66,10.15]	[19.69,11.19]
1.70	[17.99,10.91]	[22.52,13.24]	[25.31,14.73]
1.60	[23.25,14.56]	[29.63,18]	[33.6,20.23]
1.50	[31.1,20.36]	[40.43,25.71]	[46.32,29.23]
1.40	[43.21,30.06]	[57.47,38.92]	[66.63,44.84]
1.30	[62.46,47.23]	[85.32,62.95]	[100.27,73.68]
1.20	[93.37,78.73]	[131.5,108.66]	[156.99,129.57]
1.10	[140.64,134.1]	[204.74,192.56]	[248.65,234.62]
1.00	[200.02,204.48]	[300.01,303.91]	[370.04,377.32]
0.80	[238.37,175.18]	[365.34,260.87]	[455.85,324.45]
0.75	[220.33,142.4]	[338.02,211.22]	[422.1,262.24]
0.70	[197.09,112.31]	[302.88,166.02]	[378.67,205.82]
0.60	[147.18,65.42]	[227.51,95.98]	[285.52,118.6]
0.50	[101.86,34.98]	[158.66,50.71]	[200.09,62.32]
0.40	[64.33,16.86]	[100.95,23.93]	[127.95,29.12]
0.30	[35.38,7.15]	[55.76,9.77]	[70.97,11.66]
0.25	[24.24,4.43]	[38.16,5.86]	[48.62,6.88]
0.15	[8.68,1.64]	[13.36,1.93]	[16.9,2.13]
0.10	[4.15,1.12]	[6.10,1.20]	[7.59,1.26]
0.05	[1.60,1.00]	[2.05,1.00]	[2.39,1.00]

Table 4. The values of NARL when $a_N \epsilon [0.95, 1.05]$ and $k_N \epsilon [8, 10]$.

k_N	[2.8071,2.8145]	[2.9354,2.9399]	[2.9998,3.0019]
s	NARL		
4.00	[1.18,1.07]	[1.21,1.08]	[1.22,1.08]
3.00	[1.55,1.27]	[1.64,1.32]	[1.68,1.34]
2.80	[1.72,1.38]	[1.83,1.44]	[1.90,1.47]
2.50	[2.13,1.64]	[2.32,1.74]	[2.42,1.80]
2.25	[2.76,2.05]	[3.06,2.22]	[3.24,2.32]
2.00	[3.97,2.87]	[4.53,3.20]	[4.86,3.39]
1.90	[4.78,3.43]	[5.54,3.88]	[5.98,4.14]
1.80	[5.95,4.25]	[7.00,4.89]	[7.62,5.25]
1.70	[7.68,5.50]	[9.20,6.43]	[10.10,6.97]
1.60	[10.37,7.48]	[12.67,8.94]	[14.06,9.79]
1.50	[14.79,10.84]	[18.49,13.27]	[20.77,14.71]
1.40	[22.54,17]	[28.94,21.4]	[32.96,24.07]
1.30	[37.13,29.36]	[49.22,38.19]	[56.97,43.67]
1.20	[66.38,56.51]	[91.46,76.54]	[107.95,89.33]
1.10	[123.82,117.26]	[178.57,166.92]	[215.76,199.71]
1.00	[200.02,204.72]	[300.15,304.57]	[370.2,372.66]
0.80	[133.37,103.02]	[197.81,149.39]	[242.69,180.6]
0.75	[100.03,71.35]	[147.26,102.35]	[180.06,123.11]
0.70	[73.11,48.45]	[106.83,68.69]	[130.19,82.17]
0.60	[36.85,21.38]	[52.88,29.45]	[63.9,34.75]
0.50	[17.23,9.02]	[24.08,11.93]	[28.73,13.8]
0.40	[7.48,3.76]	[10.03,4.69]	[11.73,5.28]
0.30	[3.10,1.71]	[3.89,1.96]	[4.40,2.11]
0.25	[2.02,1.28]	[2.42,1.39]	[2.67,1.46]
0.15	[1.09,1.00]	[1.14,1.01]	[1.18,1.01]
0.10	[1.00,1.00]	[1.01,1.00]	[1.01,1.00]
0.05	[1.00,1.00]	[1.00,1.00]	[1.00,1.00]

From Tables 1–4, we note that the indeterminacy interval of NARL decreases when $k_N \epsilon [k_L, k_U]$ decreases. We also note the increases trend in indeterminacy interval of NARL when $a_N \epsilon [a_L, a_U]$ decreases.

4. Advantages of the Proposed Chart

For the data having some ambiguous observations, Chen et al. [33] mentioned that the method which provides the parameters in an indeterminacy interval is said to be more efficient and effective to be applied than the method which provides the determined value of the parameters. Related to the control chart theory, a control chart which provided the smaller values of NARL is called the efficient control chart, see Aslam et al. [38]. Now we discuss the advantages of the proposed control chart over the control chart proposed by Aslam et al. [63] under classical statistics.

4.1. By NARL

To compare both control chart in terms of NARL, the values of NARL are reported for the same values of control chart parameters. Let $NARL = ARL_1 + IARL_1$, where ARL_1 denotes the values of ARL of the chart under classical statistics and $I \epsilon [infI, infU]$ be the indeterminacy interval. Table 5 is presented for both control chart when $a_N \epsilon [0.95, 0.95]$ and $k_N \epsilon [8, 8]$. From Table 5, it can be noted that the proposed control provides the values of NARL in the indeterminacy interval while the existing control chart proposed by Aslam et al. [63] provides the determined values of ARL. For example, when $s = 1.50$, the indeterminacy interval from the proposed chart is $NARL = 14.77 + I14.77; \; I\epsilon [0, 0.4122]$. The value of ARL from [63] chart is 20.76. It means that when $s = 1.50$, the control chart will be out-of-control between 14th and 20th sample. By comparing both the control chart, it is concluded that the proposed control chart under uncertain situations is more effective than the control chart proposed by Aslam et al. [63].

Table 5. Comparison of average run length values at different levels of shift when $a = 0.95$ and $k = 8$.

s	Control Chart [63]			The Proposed Chart		
	ARLs			NARL		
4.00	1.178764	1.206864	1.2224	[1.18,1.07]	[1.21,1.08]	[1.22,1.08]
3.00	1.550192	1.635354	1.68279	[1.55,1.27]	[1.64,1.32]	[1.68,1.34]
2.80	1.720708	1.833704	1.896919	[1.72,1.38]	[1.83,1.44]	[1.90,1.47]
2.50	2.133142	2.317927	2.422306	[2.13,1.64]	[2.32,1.74]	[2.42,1.80]
2.25	2.757243	3.061253	3.23504	[2.76,2.05]	[3.06,2.22]	[3.24,2.32]
2.00	3.967114	4.530703	4.858373	[3.97,2.87]	[4.53,3.20]	[4.86,3.39]
1.90	4.783924	5.539304	5.982269	[4.78,3.43]	[5.54,3.88]	[5.98,4.14]
1.80	5.950011	6.99735	7.617622	[5.95,4.25]	[7.00,4.89]	[7.62,5.25]
1.70	7.680794	9.193211	10.09918	[7.68,5.50]	[9.20,6.43]	[10.10,6.97]
1.60	10.37155	12.66559	14.058	[10.37,7.48]	[12.67,8.94]	[14.06,9.79]
1.50	14.79182	18.4856	20.76225	[14.79,10.84]	[18.49,13.27]	[20.77,14.71]
1.40	22.53872	28.93451	32.94814	[22.54,17]	[28.94,21.4]	[32.96,24.07]
1.30	37.12749	49.20506	56.94885	[37.13,29.36]	[49.22,38.19]	[56.97,43.67]
1.20	66.37585	91.41983	107.9055	[66.38,56.51]	[91.46,76.54]	[107.95,89.33]
1.10	123.8084	178.4871	215.6592	[123.82,117.26]	[178.57,166.92]	[215.76,199.71]
1.00	200	300	370.0001	[200.02,204.72]	[300.15,304.57]	[370.2,372.66]
0.80	133.3607	197.7073	242.5684	[133.37,103.02]	[197.81,149.39]	[242.69,180.6]
0.75	100.0185	147.1854	179.9717	[100.03,71.35]	[147.26,102.35]	[180.06,123.11]
0.70	73.10437	106.7805	130.1201	[73.11,48.45]	[106.83,68.69]	[130.19,82.17]
0.60	36.84876	52.85517	63.86527	[36.85,21.38]	[52.88,29.45]	[63.9,34.75]
0.50	17.23191	24.06818	28.71763	[17.23,9.02]	[24.08,11.93]	[28.73,13.8]
0.40	7.479311	10.02559	11.72831	[7.48,3.76]	[10.03,4.69]	[11.73,5.28]
0.30	3.097514	3.884251	4.398122	[3.10,1.71]	[3.89,1.96]	[4.40,2.11]
0.25	2.023985	2.417816	2.672091	[2.02,1.28]	[2.42,1.39]	[2.67,1.46]
0.15	1.090389	1.144401	1.180756	[1.09,1.00]	[1.14,1.01]	[1.18,1.01]
0.10	1.003689	1.008166	1.011881	[1.00,1.00]	[1.01,1.00]	[1.01,1.00]
0.05	1.000	1.000	1.000001	[1.00,1.00]	[1.00,1.00]	[1.00,1.00]

4.2. By Simulation

Control charts are used for monitoring the process for unusual changes in the process. The proposed scheme has been developed for the efficient monitoring of the mean time of the process under uncertainty environment. The proposed control chart has been examined using the simulation data of the mean time between two events. The first 20 neutrosophic observations are generated from the neutrosophic gamma distribution with neutrosophic parameters $K_N = [3,5]$, $a_N = [1.95, 2.05]$ and $b_N = [2, 2.2]$. The next 20 observations are generated from the same distribution with $s = 1.50$. The gamma distributed data is transformed into a neutrosophic normal distribution using the transformation $T_N^* = T_N^{1/3}$. At these parameters, the tabulated value of $\text{ARLN}_1 \epsilon [24.76, 13.98]$. It is expected that the process will be out-of-control between 14th sample and 24th sample. We plotted the values of statistic $ln(Z_{k_N})$ on the control chart in Figure 1. From Figure 1, it is clear that the first shift is at the 38th sample. The values of $ln(Z_{k_N})$ are also calculated for Aslam et al. [63] and plotted in Figure 2. From Figure 2, we note no shift indication in the process. By comparing both control charts, it is concluded that the proposed control chart provides the values of NARL in indeterminacy interval and gives the quick indication about the shift in the process as compared to the existing sampling plan. A quick indication in the shift in the process helps industrial engineers to identify the cause of variation which resulted in minimizing the non-conforming items.

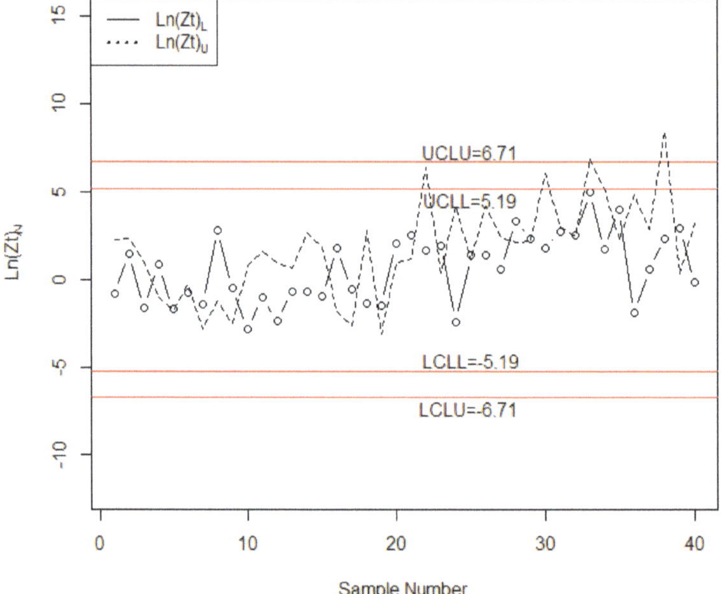

Figure 1. The proposed chart for the simulated data.

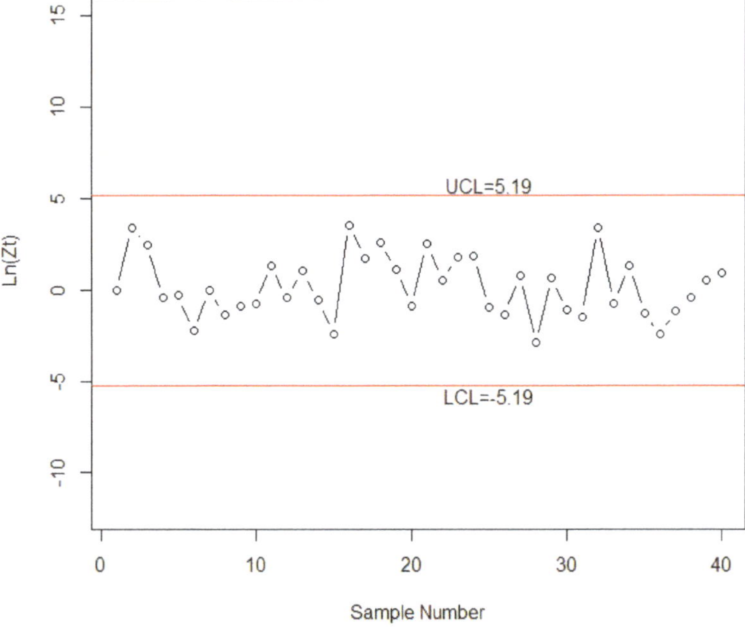

Figure 2. The existing chart for the simulated data.

5. Real Example

The application of the proposed control will be given in the healthcare department. Healthcare practitioners are interested in applying the proposed control chart for the monitoring of urinary tract infections (UTI) among male patients in a large hospital. A similar study was done by Santiago and Smith [64] and Aslam et al. [63] using classical statistics. The UTIs data is measured with the help of measurement devices. Therefore, there is a chance that observations are more fuzzy or imprecise. Under this uncertainty situation, the application of the existing control chart under classical statistics may mislead healthcare practitioners in monitoring the UTIs. Therefore, the proposed control chart is quite reasonable to apply for the monitoring of the UTIs infection. The neutrosophic data, which follows the gamma distribution with $a_N\epsilon[1.95,\ 2.05]$ and $b_N\epsilon[2,\ 2.2]$ is reported in Table 6.

The control limits are calculated as follows: $UCL_N = L_N\sqrt{k_N}$; $UCL_N\epsilon[-5.19,-6.71]$ and $LCL_N = -L_N\sqrt{k_N}$; $LCL_N\epsilon[5.19,6.71]$. We set the initial value of $Z_0 = 1$ and $B(O_0) = 0.5$. As, $Z_{k_N} = Z_{k_N-1}exp\left(\frac{T_N^*-\mu_{T_N^*}}{\sigma_{T_N^*}}\right)$ which yield $Z_{k_1}\epsilon[0.985,\ 54.568]$. The values of neutrosophic statistic $ln\left(Z_{k_N}\right)$ are plotted in Figure 3. From Figure 3, we note although the process is an in-control state, the 16th sample and 33rd sample are near the control limits which indicate some issue in the process. The values of $ln\left(Z_{k_N}\right)$ under classical statistics are also plotted in Figure 4. From Figure 4, it can be seen that only one point is near to control limit. By comparing both charts, it is concluded that the proposed control chart indicates that there is some issue in the process which should be identified. Therefore, the proposed control is more beneficial to be applied for the monitoring of UTIs inspection among male patients.

Table 6. The data for a real example.

Sr. #	B(k)	z(k)	ln(zk)
1	[0.496,0.982]	[0.985,54.568]	[−0.015,3.999]
2	[0.968,0.261]	[30.555,0.353]	[3.42,−1.04]
3	[0.922,0.252]	[11.788,0.338]	[2.467,−1.086]
4	[0.403,0.290]	[0.675,0.408]	[−0.393,−0.897]
5	[0.432,0.654]	[0.761,1.891]	[−0.274,0.637]
6	[0.096,0.652]	[0.106,1.872]	[−2.247,0.627]
7	[0.490,0.988]	[0.962,83.351]	[−0.039,4.423]
8	[0.204,0.264]	[0.256,0.358]	[−1.363,−1.028]
9	[0.287,0.820]	[0.403,4.546]	[−0.908,1.514]
10	[0.325,0.519]	[0.481,1.078]	[−0.732,0.075]
11	[0.795,0.904]	[3.885,9.443]	[1.357,2.245]
12	[0.396,0.941]	[0.656,15.825]	[−0.421,2.762]
13	[0.740,0.049]	[2.843,0.051]	[1.045,−2.974]
14	[0.361,0.017]	[0.564,0.017]	[−0.572,−4.048]
15	[0.084,0.331]	[0.091,0.496]	[−2.393,−0.701]
16	[0.972,0.278]	[34.857,0.385]	[3.551,−0.956]
17	[0.849,0.109]	[5.611,0.122]	[1.725,−2.101]
18	[0.932,0.06]	[13.683,0.064]	[2.616,−2.75]
19	[0.752,0.086]	[3.028,0.094]	[1.108,−2.364]
20	[0.293,0.621]	[0.414,1.64]	[−0.883,0.495]
21	[0.924,0.304]	[12.2,0.436]	[2.501,−0.829]
22	[0.631,0.636]	[1.709,1.748]	[0.536,0.558]
23	[0.859,0.611]	[6.108,1.572]	[1.810,0.452]
24	[0.863,0.812]	[6.296,4.306]	[1.840,1.460]
25	[0.279,0.983]	[0.388,59.465]	[−0.947,4.085]
26	[0.208,0.191]	[0.263,0.236]	[−1.337,−1.445]
27	[0.691,0.232]	[2.236,0.303]	[0.805,−1.195]
28	[0.052,0.896]	[0.055,8.661]	[−2.908,2.159]
29	[0.659,0.608]	[1.936,1.551]	[0.660,0.439]
30	[0.252,0.981]	[0.337,50.315]	[−1.087,3.918]
31	[0.186,0.221]	[0.229,0.283]	[−1.475,−1.262]
32	[0.968,0.286]	[29.788,0.4]	[3.394,−0.916]
33	[0.324,0.279]	[0.479,0.387]	[−0.736,−0.95]
34	[0.791,0.157]	[3.78,0.186]	[1.33,−1.684]
35	[0.217,0.812]	[0.277,4.321]	[−1.284,1.464]
36	[0.08,0.942]	[0.086,16.316]	[−2.449,2.792]
37	[0.246,0.273]	[0.327,0.376]	[−1.119,−0.979]
38	[0.398,0.914]	[0.66,10.644]	[−0.415,2.365]
39	[0.625,0.358]	[1.665,0.557]	[0.51,−0.586]
40	[0.720,0.286]	[2.569,0.400]	[0.944,−0.916]

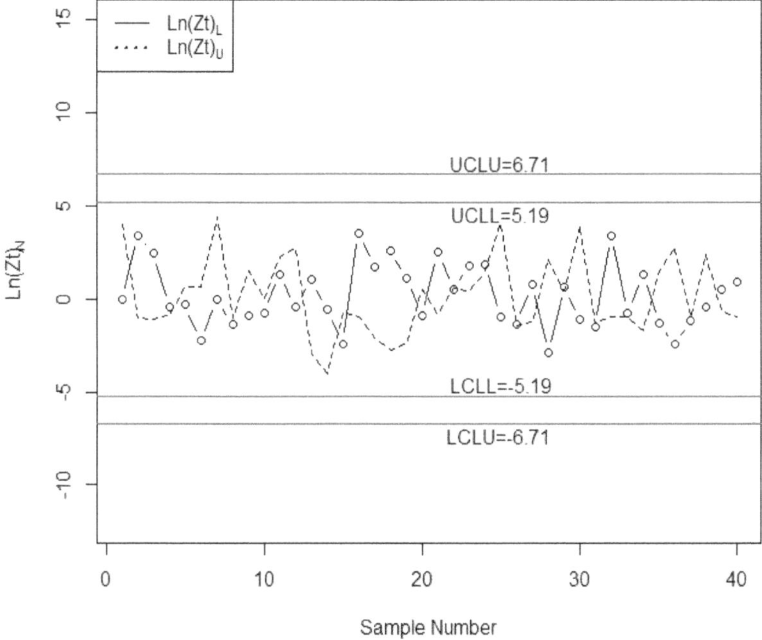

Figure 3. The proposed chart for the real data.

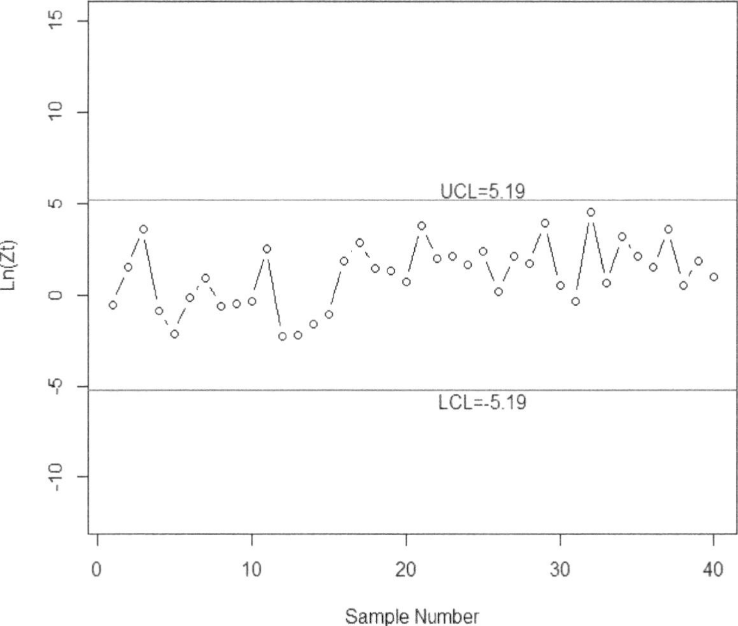

Figure 4. The existing chart for the real data.

6. Conclusions

In this article, a control chart for the belief statistic under the gamma distribution has been presented when the interested quality characteristics of the process are imprecise, incomplete and vague. Although there are numerous techniques available in the literature like the fuzzy logic the proposed scheme is effective in dealing with vague information. The control limits have been determined for different settings of the parameters and different levels of process shifts. In this paper, the average run lengths for many settings of the proposed technique have been tabulated. The comparison of the proposed chart with the existing chart for different process shifts have been tested and it has been observed that the proposed chart is an effective addition in the toolkit of the quality control experts. We conclude that the proposed control is more robustness in detecting cause of variation in the process than the chart under classical statistics in uncertainty. The proposed chart can further be extended for other probability distributions, see, for example, references [65] and [66], particularly for the multivariate case.

Author Contributions: M.A.; R.A.R.B., N.K., and M.A. conceived and designed the experiments; M.A. and N.K. performed the experiments; M.A. and N.K. analyzed the data; M.A. contributed reagents/materials/analysis tools; M.A. wrote the paper.

Funding: This work was supported by the Deanship of Scientific Research (DSR), King Abdulaziz University, Jeddah, under grant No. (130-107-D1440). The author, therefore, gratefully acknowledge the DSR technical and financial support.

Acknowledgments: The authors are deeply thankful to the editor and reviewers for their valuable suggestions to improve the quality of this manuscript.

Conflicts of Interest: The authors declare no conflict of interest regarding this paper.

References

1. Gülbay, M.; Kahraman, C.; Ruan, D. α-Cut fuzzy control charts for linguistic data. *Int. J. Intell. Syst.* **2004**, *19*, 1173–1195. [CrossRef]
2. Demirli, K.; Vijayakumar, S. Fuzzy logic based assignable cause diagnosis using control chart patterns. *Inf. Sci.* **2010**, *180*, 3258–3272. [CrossRef]
3. Montgomery, C.D. *Introduction to Statistical Quality Control*, 6th ed.; John Wiley & Sons, Inc.: New York, NY, USA, 2009.
4. Cheng, C.-B. Fuzzy process control: Construction of control charts with fuzzy numbers. *Fuzzy Sets Syst.* **2005**, *154*, 287–303. [CrossRef]
5. Gülbay, M.; Kahraman, C. An alternative approach to fuzzy control charts: Direct fuzzy approach. *Inf. Sci.* **2007**, *177*, 1463–1480. [CrossRef]
6. Fallah Nezhad, M.S.; Akhavan Niaki, S.T. A new monitoring design for uni-variate statistical quality control charts. *Inf. Sci.* **2010**, *180*, 1051–1059. [CrossRef]
7. Aksoy, H. Use of gamma distribution in hydrological analysis. *Turk. J. Eng. Environ. Sci.* **2000**, *24*, 419–428.
8. Zhang, C.; Zhang, C.W.; Xie, M.; Liu, J.Y.; Goh, T.N. A control chart for the Gamma distribution as a model of time between events. *Int. J. Prod. Res.* **2007**, *45*, 5649–5666. [CrossRef]
9. Aslam, M.; Arif, O.-H.; Jun, C.-H. A Control Chart for Gamma Distribution using Multiple Dependent State Sampling. *Ind. Eng. Manag. Syst.* **2017**, *16*, 109–117. [CrossRef]
10. Chen, F.; Yeh, C.-H. Economic statistical design of non-uniform sampling scheme X bar control charts under non-normality and Gamma shock using genetic algorithm. *Expert Syst. Appl.* **2009**, *36*, 9488–9497. [CrossRef]
11. Al-Oraini, H.A.; Rahim, M. Economic statistical design of X control charts for systems with Gamma (λ, 2) in-control times. *Comput. Ind. Eng.* **2002**, *43*, 645–654. [CrossRef]
12. Bhaumik, D.K.; Gibbons, R.D. One-sided approximate prediction intervals for at least p of m observations from a gamma population at each of r locations. *Technometrics* **2006**, *48*, 112–119. [CrossRef]
13. Khan, N.; Aslam, M.; Ahmad, L.; Jun, C.H. A Control Chart for Gamma Distributed Variables Using Repetitive Sampling Scheme. *Pak. J. Stat. Oper. Res.* **2017**, *13*, 47–61. [CrossRef]
14. Kaya, I.; Erdoğan, M.; Yıldız, C. Analysis and control of variability by using fuzzy individual control charts. *Appl. Soft Comput.* **2017**, *51*, 370–381. [CrossRef]

15. Smarandache, F. Neutrosophic logic-generalization of the intuitionistic fuzzy logic. *arXiv*, 2003; arXiv:math/0303009.
16. Şentürk, S.; Erginel, N.; Kaya, İ.; Kahraman, C. Fuzzy exponentially weighted moving average control chart for univariate data with a real case application. *Appl. Soft Comput.* **2014**, *22*, 1–10. [CrossRef]
17. Zadeh, L.A. Fuzzy sets. *Inf. Control* **1965**, *8*, 338–353. [CrossRef]
18. Wang, J.-H.; Raz, T. On the construction of control charts using linguistic variables. *Int. J. Prod. Res.* **1990**, *28*, 477–487. [CrossRef]
19. Raz, T.; Wang, J.-H. Probabilistic and membership approaches in the construction of control charts for linguistic data. *Prod. Plan. Control* **1990**, *1*, 147–157. [CrossRef]
20. Taleb, H.; Limam, M. On fuzzy and probabilistic control charts. *Int. J. Prod. Res.* **2002**, *40*, 2849–2863. [CrossRef]
21. Kanagawa, A.; Tamaki, F.; Ohta, H. Control charts for process average and variability based on linguistic data. *Int. J. Prod. Res.* **1993**, *31*, 913–922. [CrossRef]
22. Erginel, N.; Sentürk, S.; Kahraman, C.; Kaya, I. Evaluating the packing process in food industry using fuzzy and [stilde] control charts. *Int. J. Comput. Intell. Syst.* **2011**, *4*, 509–520. [CrossRef]
23. El-Shal, S.M.; Morris, A.S. A fuzzy rule-based algorithm to improve the performance of statistical process control in quality systems. *J. Intell. Fuzzy Syst.* **2000**, *9*, 207–223.
24. Rowlands, H.; Wang, L.R. An approach of fuzzy logic evaluation and control in SPC. *Qual. Reliab. Eng. Int.* **2000**, *16*, 91–98. [CrossRef]
25. Aslam, M. A New Sampling Plan Using Neutrosophic Process Loss Consideration. *Symmetry* **2018**, *10*, 132. [CrossRef]
26. Senturk, S.; Erginel, N. Development of fuzzy $\widetilde{\overline{x}} - \widetilde{R}$ and $\widetilde{\overline{x}} - \widetilde{S}$ control charts using α-cuts. *Inf. Sci.* **2009**, *179*, 1542–1551. [CrossRef]
27. Sentürk, S. Fuzzy regression control chart based on α-cut approximation. *Int. J. Comput. Intell. Syst.* **2010**, *3*, 123–140. [CrossRef]
28. Kaya, İ.; Kahraman, C. Process capability analyses based on fuzzy measurements and fuzzy control charts. *Expert Syst. Appl.* **2011**, *38*, 3172–3184. [CrossRef]
29. Broumi, S.; Smarandache, F. Correlation coefficient of interval neutrosophic set. In *Applied Mechanics and Materials*; Trans Tech Publication: Stafa-Zurich, Switzerland, 2013; pp. 511–517.
30. Smarandache, F. Neutrosophic Logic-A Generalization of the Intuitionistic Fuzzy Logic. *Multispace Multistruct. Neutrosophic Transdiscipl.* **2010**, *4*, 396. [CrossRef]
31. Smarandache, F. *Neutrosophy: Neutrosophic Probability, Set, and Logic: Analytic Synthesis & Synthetic Analysis*; American Research Press: Rehoboth, DE, USA, 1998; p. 105.
32. Smarandache, F. *Introduction to Neutrosophic Statistics*; Infinite Study: Hollywood, FL, USA, 2014.
33. Chen, J.; Ye, J.; Du, S. Scale effect and anisotropy analyzed for neutrosophic numbers of rock joint roughness coefficient based on neutrosophic statistics. *Symmetry* **2017**, *9*, 208. [CrossRef]
34. Chen, J.; Ye, J.; Du, S.; Yong, R. Expressions of rock joint roughness coefficient using neutrosophic interval statistical numbers. *Symmetry* **2017**, *9*, 123. [CrossRef]
35. Aslam, M. A New Failure-Censored Reliability Test Using Neutrosophic Statistical Interval Method. *Int. J. Fuzzy Syst.* **2019**, 1–7. [CrossRef]
36. Aslam, M. Design of Sampling Plan for Exponential Distribution under Neutrosophic Statistical Interval Method. *IEEE Access* **2018**, *6*, 64153–64158. [CrossRef]
37. Aslam, M. A new attribute sampling plan using neutrosophic statistical interval method. *Complex Intell. Syst.* **2019**, 1–6. [CrossRef]
38. Aslam, M.; Bantan, R.A.; Khan, N. Design of a New Attribute Control Chart Under Neutrosophic Statistics. *Int. J. Fuzzy Syst.* **2019**, *21*, 433–440. [CrossRef]
39. Aslam, M.; Khan, N.; Khan, M. Monitoring the Variability in the Process Using Neutrosophic Statistical Interval Method. *Symmetry* **2018**, *10*, 562. [CrossRef]
40. Aslam, M.; Bantan, R.A.; Khan, N. Design of a Control Chart for Gamma Distributed Variables under the Indeterminate Environment. *IEEE Access* **2019**, *7*, 8858–8864. [CrossRef]
41. Aslam, M. Product Acceptance Determination with Measurement Error Using the Neutrosophic Statistics. *Adv. Fuzzy Syst.* **2019**, *2019*, 8953051. [CrossRef]

42. Aslam, M.; Arif, O. Testing of Grouped Product for the Weibull Distribution Using Neutrosophic Statistics. *Symmetry* **2018**, *10*, 403. [CrossRef]
43. Aslam, M.; Raza, M.A. Design of New Sampling Plans for Multiple Manufacturing Lines Under Uncertainty. *Int. J. Fuzzy Syst.* **2019**, *21*, 978–992. [CrossRef]
44. Peng, X.; Dai, J. A bibliometric analysis of neutrosophic set: Two decades review from 1998 to 2017. *Artif. Intell. Rev.* **2018**, 1–57. [CrossRef]
45. Peng, X.; Dai, J. Approaches to single-valued neutrosophic MADM based on MABAC, TOPSIS and new similarity measure with score function. *Neural Comput. Appl.* **2018**, *29*, 939–954. [CrossRef]
46. Smarandache, F. Neutrosophic set-a generalization of the intuitionistic fuzzy set. *J. Def. Resour. Manag.* **2010**, *1*, 107.
47. Smarandache, F. A geometric interpretation of the neutrosophic set-A generalization of the intuitionistic fuzzy set. *arXiv*, 2004; arXiv:math/0404520.
48. Smarandache, F. n-Valued Refined Neutrosophic Logic and Its Applications to Physics. Available online: https://arxiv.org/pdf/1407.104 (accessed on 12 March 2019).
49. Rivieccio, U. Neutrosophic logics: Prospects and problems. *Fuzzy Sets Syst.* **2008**, *159*, 1860–1868. [CrossRef]
50. Wang, H.; Smarandache, F.; Zhang, Y.; Sunderraman, R. Single Valued Neutrosophic Sets. Available online: fs.unm.edu/SingleValuedNeutrosophicSets.pdf (accessed on 12 March 2019).
51. Wang, H.; Smarandache, F.; Sunderraman, R.; Zhang, Y.Q. Interval Neutrosophic Sets and Logic: Theory and Applications in Computing: Theory and Applications in Computing. Available online: https://arxiv.org/abs/cs/0505014 (accessed on 12 March 2019).
52. Gao, Z.; Cecati, C.; Ding, S.X. A survey of fault diagnosis and fault-tolerant techniques—Part II: Fault diagnosis with knowledge-based and hybrid/active approaches. *IEEE Trans. Ind. Electron.* **2015**, *62*, 3768–3774. [CrossRef]
53. Wilson, E.B.; Hilferty, M.M. The distribution of chi-square. *Proc. Natl. Acad. Sci. USA* **1931**, *17*, 684–688. [CrossRef]
54. Ahmad, L.; Aslam, M.; Jun, C.-H. Designing of X-bar control charts based on process capability index using repetitive sampling. *Trans. Inst. Meas. Control* **2014**, *36*, 367–374. [CrossRef]
55. Knoth, S. Accurate ARL Calculation for EWMA Control Charts Monitoring Normal Mean and Variance Simultaneously. *Seq. Anal.* **2007**, *26*, 251–263. [CrossRef]
56. Li, Z.H.; Zou, C.; Gong, Z.; Wang, Z. The computation of average run length and average time to signal: An overview. *J. Stat. Comput. Simul.* **2014**, *84*, 1779–1802. [CrossRef]
57. Lee, M.; Khoo, M.B. Optimal statistical design of a multivariate EWMA chart based on ARL and MRL. *Commun. Stat. Simul. Comput.* **2006**, *35*, 831–847. [CrossRef]
58. Phanyaem, S.; Areepong, Y.; Sukparungsee, S. Numerical Integration of Average Run Length of CUSUM Control Chart for ARMA Process. *Int. J. Appl. Phys. Math.* **2014**, *4*, 232–235. [CrossRef]
59. Busaba, J.; Sukparungsee, S.; Areepong, Y. Numerical approximations of average run length for AR (1) on exponential CUSUM. *Comput. Sci. Telecommun.* **2012**, *19*, 23.
60. Aslam, M.; Khan, N.; Ahmad, L.; Jun, C.H.; Hussain, J. A mixed control chart using process capability index. *Seq. Anal.* **2017**, *36*, 278–289. [CrossRef]
61. Ahmad, L.; Aslam, M.; Khan, N.; Jun, C.H. Double moving average control chart for exponential distributed life using EWMA. In *AIP Conference Proceedings*; AIP Publishing: New York, NY, USA, 2017.
62. Ahmad, L.; Aslam, M.; Jun, C.-H. Coal Quality Monitoring With Improved Control Charts. *Eur. J. Sci. Res.* **2014**, *125*, 427–434.
63. Aslam, M.; Khan, N.; Jun, C.-H. A control chart using belief information for a gamma distribution. *Oper. Res. Decis.* **2016**, *26*, 5–19.
64. Santiago, E.; Smith, J. Control charts based on the exponential distribution: Adapting runs rules for the t chart. *Qual. Eng.* **2013**, *25*, 85–96. [CrossRef]

65. Smarandache, F. Introduction to Neutrosophic Measure, Neutrosophic Integral, and Neutrosophic Probability. Available online: https://arxiv.org/abs/1311.7139 (accessed on 12 March 2019).
66. Smarandache, F. Neutrosophic Precalculus and Neutrosophic Calculus: Neutrosophic Applications. Available online: https://arxiv.org/pdf/1509.07723 (accessed on 12 March 2019).

Article

Determination of Holmquist–Johnson–Cook Constitutive Parameters of Coal: Laboratory Study and Numerical Simulation

Beijing Xie *, Zheng Yan *, Yujing Du, Zeming Zhao and Xiaoqian Zhang

School of Emergency Management and Safety Engineering, China University of Mining and Technology (Beijing), Beijing 100083, China; duyj223@163.com (Y.D.); zhaozeming1994@163.com (Z.Z.); sdslcszxq@163.com (X.Z.)
* Correspondence: bjxie1984@cumtb.edu.cn (B.X.); yanzh0329@163.com (Z.Y.)

Received: 22 April 2019; Accepted: 17 June 2019; Published: 21 June 2019

Abstract: The main sensitivity parameters of the Holmquist–Johnson–Cook constitutive model for coal were obtained from a variety of tests such as uniaxial compression, uniaxial cyclic loading, splitting and triaxial compression tests, as well as the indirect derivation equation of a briquette. The mechanical properties of briquettes under dynamic impact were investigated using a split Hopkinson pressure bar experiment. Based on the experimental measurement of the Holmquist–Johnson–Cook constitutive model, the numerical simulation of briquette was performed using ANSYS/LS-DYNA software. A comparison between experimental and simulation results verified the correctness of simulation parameters. This research concluded that the failure of briquette at different impact velocities started from an axial crack in the middle of the coal body, and the sample was swollen to some extent. By the increase of impact velocity, the severity of damage in the coal body was increased, while the size of the coal block was decreased. Moreover, there was good compliance between experimental and simulated stress wave curves in terms of coal sample failure and fracture morphology at different speeds. Finally, the parameters of the validated Holmquist–Johnson–Cook constitutive model were applied to the numerical simulation model of the impact damage of heading face and the process of coal seam damage in the roadway was visually displayed. The obtained results showed that the Holmquist–Johnson–Cook constitutive model parameters suitable for the prominent coal body were of great significance for the improvement and exploration of the occurrence mechanism of coal and rock dynamic disasters.

Keywords: Holmquist–Johnson–Cook constitutive model of briquette; parameter acquisition; split Hopkinson pressure bar experiment; numerical simulation

---

1. Introduction

China is the largest energy consumer in the world and its main energy resource is coal. With the increase in the depth and intensity of coal mining activities, a variety of coal-rock dynamic disasters such as coal and gas outbursts, rock bursts, and large-scale pressure on stope, have become more serious and safety production problems have also arisen [1,2]. Coal and gas outbursts and rock bursts are common dynamic disasters in coal mine rocks. Due to their sudden and transient vibration and great destructive characteristics, these phenomena often cause serious casualties and resource waste, which seriously restrict the national economic development of China [3].

In order to understand the mechanism of coal rock dynamic disasters such as rock bursts, researchers have recently carried out a large number of experimental studies on the dynamic and static mechanical properties of coal. Xue [4] carried out orthogonal experiments of triaxial stress with CH_4 seepage, and a complete stress–strain relationship and the corresponding evolution of volumetric strain and permeability were obtained. Cai [5] carried out multistage uniaxial compression creep tests on lean and raw coal samples and found that in the multistage creep process, the coal samples were first hardened, then weakened, and finally failed due to crack growth. Li's [6] tests on gas seepage in raw coal under three paths were carried out with a seepage tester under triaxial stress conditions. It was found that the permeability was subjected to the dual influence of stress and damage accumulation.

Currently, split Hopkinson pressure bar (SHPB) technology is widely applied in the investigation of the dynamic mechanical properties of materials [7–9]. Zhao [7] used an SHPB system to measure the semi-circular bending of incision and investigated the crack propagation fractal characteristics of coal seams under impact loads. His results showed that the existence of bedding had a significant impact on crack propagation. Feng [8] used SHPB to carry out dynamic load tests on coal samples and analyzed their dynamic and energy consumption characteristics. Li [9] used SHPB to study the impact failure of coal at impact velocities of 4.174–17.652 m/s and investigated the variation of different mechanical parameters such as stress, strain, incident energy, and dissipated energy.

Due to the limitations of experimental methods and equipment, the existing experimental findings on coal impact damage are seriously insufficient, and it is impossible to accurately determine stress and strain changes inside coal rock samples at the impact moment [4–10]. However, through numerical simulations, the deformation and stress changes of coal during coal and gas outbursts could be visualized [11–14], and dynamic load tests carried out under limited laboratory conditions were supplemented and improved. The Holmquist–Johnson–Cook (HJC) constitutive model is a concrete constitutive model based on large strain, a high strain rate, and high pressure that was proposed by Holmquist [15]. This constitutive model has been successfully applied to the numerical simulation of dynamic impact damage of concrete, rock, and other materials [16–19]. Due to the similarity of the mechanical properties and dynamic failure processes of coal and rock, the HJC constitutive model could be used to numerically simulate the impact damage of coal.

Considering the similar mechanical properties of coal rock, ordinary rock, and concrete materials, Xie [20] used the HJC constitutive model to test coal samples. He also used finite element software LS-DYNA (one of the most commonly used explicit simulation software for the numerical simulation of explosion and shock) numerical simulation to demonstrate the SHPB process of coal impact failure at different impact velocities and found that simulation results were consistent with experimental measurements. Li and Wang [21,22] employed the HJC constitutive model to numerically simulate a SHPB experiment and passive confining pressure test of coal using LS-DYNA. The stress waveform of coal rock samples during the impact test, the oscillation of stress waves, and the damage of the test specimen were reproduced, and it was found that simulation results complied well with experimental findings.

Many researchers have provided HJC constitutive model parameters for various concretes, but to the best of our knowledge, no parameter values have been proposed for coal. In the numerical simulation of coal, basic parameters can be directly obtained, and the remaining parameters are generally considered to be the same as concrete model parameters, which decreases the accuracy of numerical simulation results. Therefore, it is essential to understand the mechanism of coal/rock dynamic disasters by studying the dynamic mechanical parameters of coal to propose a systematic method to determine HJC constitutive model parameters for coal outbursts.

In this paper, the parameters of the HJC constitutive model for briquette are studied using experimental and numerical simulation methods. HJC constitutive model parameters for briquette were obtained through a series of experiments. SHPB experiments were carried out and the impact damage of coal was numerically simulated using ANSYS/LS-DYNA software. Numerical simulation results and experimental findings were analyzed to verify the reliability of the HJC constitutive model parameters for briquette. The validated HJC constitutive model parameters were applied to the numerical simulation of the impact damage of tunnel face, and the failure process of coal seam in the roadway was visually displayed. The research findings are of great significance for improving and exploring the mechanism of coal-rock dynamic disasters.

2. Parameters of the HJC Constitutive Model

The HJC constitutive model contains 19 parameters and two additional parameters exist in LS-DYNA software, for a total of 21 parameters [15]. These parameters were divided into five categories: basic parameters, strength parameters, pressure parameters, damage parameters, and software parameters (Table 1).

Table 1. Holmquist–Johnson–Cook (HJC) constitutive model material parameter classification.

Category	Parameter Name	Symbol
Basic material parameters	Density	ρ_0
	Shear modulus	G
	Quasi-static uniaxial compressive strength	f_c
	Maximum stretching hydrostatic pressure	T
Material strength parameters	Normalized cohesive strength	A
	Normalized pressure hardening coefficient	B
	Pressure hardening index	N
	Strain rate coefficient	C
	Maximum normalized intensity	S_{max}
Material pressure parameters	Volume pressure at crushing point	p_C
	Volumetric strain at crushing point	μ_C
	Pressure at compaction point	p_l
	Volumetric strain at compaction point	μ_l
	Pressure constant	k_1
	Pressure constant	k_2
	Pressure constant	k_3
Material damage parameters	Minimum plastic strain at material failure point	$\varepsilon_{f\,min}$
	Damage parameter	D_1
	Damage parameter	D_2
Software parameter	Reference strain rate	$\dot{\varepsilon}_0$
	Failure type	f_S

In order to obtain specific values for the 21 parameters of the coal HJC constitutive model shown in Table 1, uniaxial compression, uniaxial cyclic loading, and splitting and triaxial compression tests were carried out according to previously reported methods [23,24]. Some other parameters could not be obtained because of the limitations of experimental conditions and lack of experimental data. Due to the low sensitivity of some parameters, coal HJC constitutive model parameters could be used for estimating the values [20].

3. Coal Sample Preparation

The coal samples used in this research were obtained from Yongcheng Cheji Coal Mine, Henan Province, China, which is an outburst coal seam. The coal samples were high-quality anthracite with ultra-low sulfur, ultra-low phosphorus, medium ash, and high calorific values. The coal moisture M_{ad} was 0.81%, ash content A_d was 11.6%, and total volatile V_{daf} was 9.07%. A large-volume coal mass was crushed using a heavy hammer, and the obtained pieces were placed in a ball mill to be pulverized. The pulverized coal sample with particle diameter of 0.25 mm or below was screened out [25]. To achieve the strength required for the experiment, 1000 g pulverized coal and 200 g coal tar were uniformly mixed to prepare coal samples. Standard cylindrical coal samples with a diameter of 50 mm and length of 100 mm as well as disc-shaped coal samples with a diameter of 50 mm and thickness of 25 mm were obtained by pouring the prepared pulverized coal into separate molds and pressing them under a pressure of 250 KN on a WAW-type electro-hydraulic servo press device (Jinan Tianchen Testing Machine Manufacturing Co., Ltd., Licheng District, Jinan City, Shandong Province, China). A total of 22 cylindrical and 4 disc-shaped coal samples were prepared (Figure 1), and the basic parameters of the briquettes were assumed to be as summarized in Tables 2 and 3.

(**a**) Preparation of briquette coal sample.   (**b**) Preparation of disctype coal samples

Figure 1. Preparation of briquette coal sample.

Table 2. Parameters of cylindrical briquette coal samples.

Sample No	Quality/g	Length/mm	Diameter/mm	Density/g.cm^{-1}
XM-01	258.4	100.50	51.70	1.225
XM-02	256.7	100.46	51.60	1.222
XM-03	255.1	100.52	51.60	1.214
XM-04	257.3	100.20	51.70	1.223
XM-05	255.6	100.27	51.70	1.214
XM-06	250.4	95.40	51.70	1.250
XM-07	251.5	96.25	51.58	1.251
XM-08	255.0	99.40	51.62	1.226
XM-09	256.3	97.37	51.60	1.259
XM-10	254.5	96.60	51.60	1.260
XM-11	252.0	99.78	51.70	1.203
XM-12	251.9	100.25	51.65	1.199
XM-13	254.4	100.06	51.60	1.216
XM-14	254.0	99.60	51.57	1.221
XM-15	253.9	100.34	51.70	1.205
XM-16	252.7	95.70	51.67	1.259
XM-17	255.1	100.40	51.60	1.215
XM-18	252.6	98.15	51.57	1.232
XM-19	253.7	98.32	51.70	1.229
XM-20	253.9	98.28	51.60	1.236
XM-21	254.1	99.95	51.60	1.216
XM-22	250.8	96.07	51.70	1.244

Table 3. Parameters of disc-type briquette samples.

Sample No	Quality/g	Length/mm	Diameter/mm
YP-01	68.3	25.40	52.30
YP-02	68.7	26.24	52.36
YP-03	68.2	25.18	52.00
YP-04	68.0	24.64	52.40

As shown in Figure 2, a compression-mirror analysis was conducted on press-formed briquette samples using a KYKY-EM6200 tungsten filament scanning electron microscope.

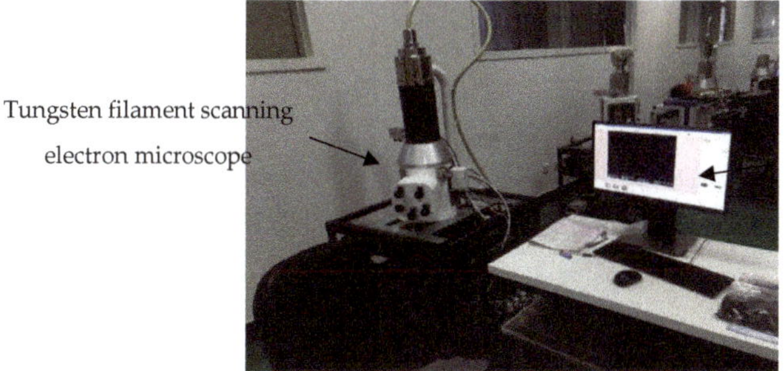

Figure 2. KYKY-EM6200 tungsten filament scanning electron microscope.

The strength of a briquette sample is closely related to its microstructure. If the gel content of the briquette sample is high, crystal development is enhanced, crystal distribution is more uniform, and briquette sample strength is higher [26]. The microstructure of the prepared briquette can be visually evaluated using scanning electron microscopy, as shown in Figure 3. It can be seen that the microstructure of coal samples had changed from granular to lamellar, and irregularly shaped crystals and gels were evenly distributed throughout the layered structure, indicating that the prepared briquette samples were well cemented, with high uniformity and strength. A large number of on-site experiments have shown that most prominent dangerous coal seams contain soft delamination with severe structural damage [27]. The structural strength of briquette samples conformed to soft stratification properties with high uniformity.

(a) Scanning electron microscope images of coal samples at 200 times magnification

(b) Scanning electron microscope images of coal samples at 2000 times magnification

Figure 3. Scanning electron microscope images of the prepared coal samples.

4. Determination of Coal HJC Constitutive Model Parameters

The parameter values of the HJC constitutive model for coal were obtained from experimental tests, equation derivations, and literature references. Uniaxial compression and cyclic loading experiments as well as splitting and triaxial compression tests were performed using a multifunctional three-axis testing machine in the State Key Laboratory of Geotechnical Mechanics and Underground Engineering at the China University of Mining and Technology (Beijing) (Figure 4).

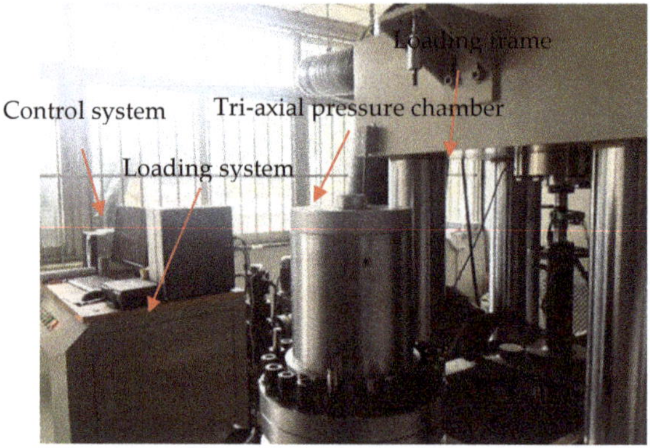

Figure 4. Multifunctional triaxial testing machine.

4.1. Determination of the Value of Parameter f_C

The value of the uniaxial compressive strength parameter f_C was obtained through a uniaxial compression test. Five uniaxial compression tests were carried out, and three sets of effective data were obtained. Based on the obtained data, the stress–strain curves of briquette samples under uniaxial compression were plotted. The curves are shown in Figure 5.

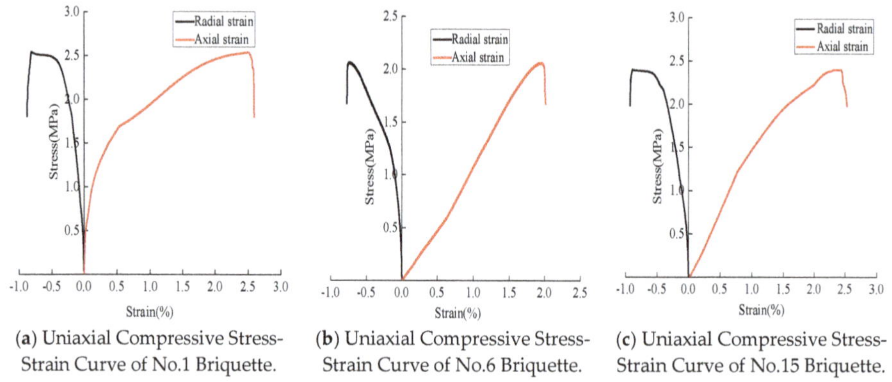

(**a**) Uniaxial Compressive Stress-Strain Curve of No.1 Briquette.

(**b**) Uniaxial Compressive Stress-Strain Curve of No.6 Briquette.

(**c**) Uniaxial Compressive Stress-Strain Curve of No.15 Briquette.

Figure 5. Uniaxial compression stress–strain curve.

As shown in Table 4, the value of the uniaxial compressive strength parameter f_C of the sample was obtained from the peak point of the stress–strain curve.

Table 4. Uniaxial compression test results.

Sample No.	Compressive Strength/MPa	Peak Axial Strain/10^{-2}	Peak Radial Strain/10^{-2}	Elastic Modulus/MPa	Poisson's Ratio
XM-01	2.54	2.51	−0.81	101.19	0.32
XM-06	2.05	1.95	−0.75	105.13	0.38
XM-15	2.39	2.43	−0.89	98.35	0.37

According to the experimental results presented in Table 4, the mean value of the three samples was considered as the value of uniaxial compressive strength parameter f_C for the briquette sample, which was 2.33 MPa.

4.2. Determination of the Value of Parameter T

A splitting test was required to obtain the value of the uniaxial tensile strength parameter T. The test equipment used for the splitting of coal samples was the same as that used for the uniaxial compression test. Samples were disc-shaped with a diameter of 50 mm and length of 25 mm. The axial load was applied by displacement control at a rate of 0.05 mm/min. When a crack appeared on the surface of the coal sample, the experiment was finished. Sample damage resulting from the four splitting experiments is shown in Figure 6.

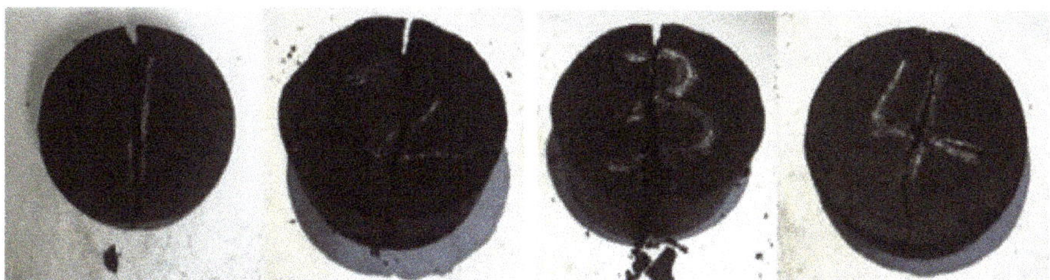

Figure 6. Splitting test specimen failure morphologies.

The load P at the failure point of the sample was obtained from the splitting test, and the tensile strength T of the sample was calculated using the equation $T = 2P/\pi dh$, where d and h are the diameter and length of the sample, respectively. The values of load P and tensile strength T at the fracture points of different samples are summarized in Table 5.

Table 5. Sample tensile strength values.

Sample No	Failure Load/KN	Length/mm	Diameter/mm	Tensile Strength/MPa
YP-01	1.133	25.40	52.30	0.54
YP-02	1.136	26.24	52.36	0.53
YP-03	1.243	25.18	52.00	0.60
YP-04	1.080	24.64	52.40	0.53

According to experimental results presented in Table 5, the mean value of the four samples was considered as the tensile strength of briquette sample, which was 0.55 MPa.

4.3. Determination of the Values of Parameters $\varepsilon_{f\min}$, D_1 and D_2

Damage constant $\varepsilon_{f\min}$ is the plastic strain at the moment when the minimum strength fracture of the material is achieved. The acquisition method is given in the literature [15] (Figure 7). Determination of the value of parameter $\varepsilon_{f\min}$ required a uniaxial compression cycle loading experiment, and a hypothetical failure interface was defined based on the curve drawn by the experimental results. The failure interface revealed that when the axial strain reached the intersection of the interface and the strain axis, the sample lost its strength completely, and the strain value was equal to the value of $\varepsilon_{f\min}$. When the equivalent fracture strain was achieved, $P^* = 1/6$ and T^* could be calculated as $T^* = T/f_C$. If $\varepsilon_{f\min} \leq D_1(P^* + T^*)^{D_2}$, then $D_1 = \varepsilon_{f\min}/(1/6 + T^*)$. Due to the lack of real data, we assumed that $D_2 = 1.0$.

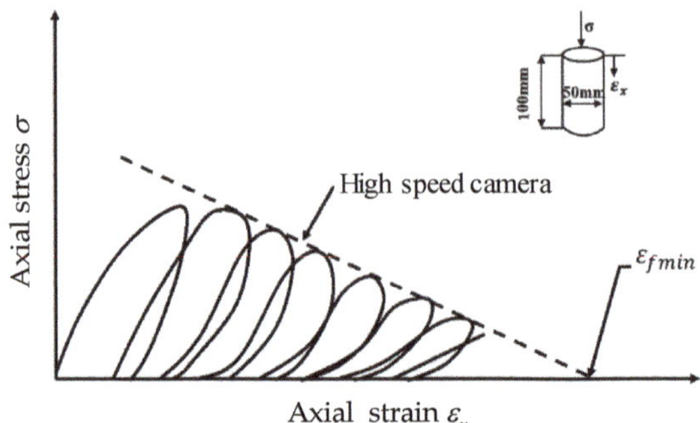

Figure 7. The acquisition method for parameter $\varepsilon_{f\min}$.

In this test, cylindrical samples with a diameter of 50 mm and length of 100 mm were used. During the loading process, the briquette samples were first loaded to 70% of their uniaxial compressive strength and then unloaded to zero at the same rate. Then, loading was repeated and its intensity was decreased by 10% at each cycle until the sample was destroyed. Uniaxial cyclic loading experiments were repeated for 5 test specimens. Due to operational error, only two sets of effective data were obtained. According to the experimental data, the stress–strain curve of the uniaxial compression cycle loading experiment was plotted, as shown in Figure 8.

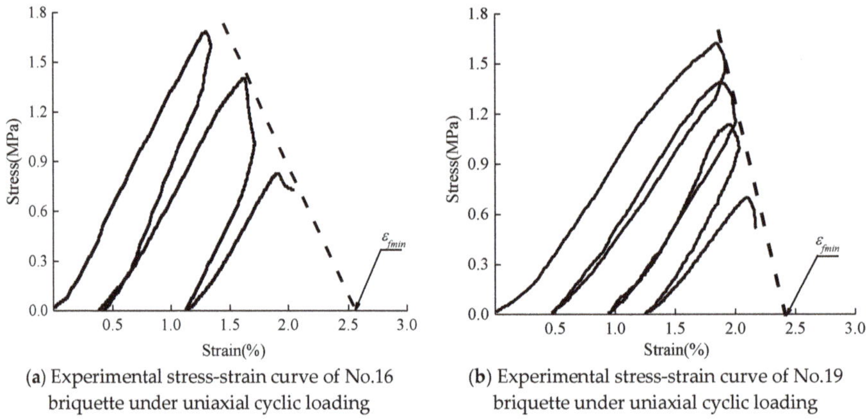

(**a**) Experimental stress-strain curve of No.16 briquette under uniaxial cyclic loading

(**b**) Experimental stress-strain curve of No.19 briquette under uniaxial cyclic loading

Figure 8. Experimental stress–strain curve of uniaxial cyclic loading.

A hypothetical failure interface was obtained on the stress–strain curve, and its intersection with the strain axis was considered to be the value of $\varepsilon_{f\min}$. The parameter values obtained by the above method are summarized in Table 6.

Table 6. Value of Parameter $\varepsilon_{f\min}$.

Sample No	$\varepsilon_{f\min}$	Average Value
XM-16	0.026	0.025
XM-19	0.024	

According to the results presented in Table 6, the damage parameter $\varepsilon_{f\min}$ takes the average of the values obtained for the two sets of samples, which is 0.025. $D_1 = 0.0131$ and $D_2 = 1.0$ were obtained according to the method described above.

4.4. Determination of the Values of Parameters B and N

Normalized pressure hardening coefficient B and pressure hardening index N were obtained by triaxial compression experiments. In these experiments, the confining pressure was set at $\sigma_2 = \sigma_3$. First, the three axes coordinately load to the values of the confining pressure, and then the confining pressure was kept constant at $\sigma_2 = \sigma_3$. In the next stage, a load was applied along the axial direction until the sample failed, and the peak value of axial stress σ_1 was recorded. The calculation of the normalized pressure hardening coefficient B and pressure hardening index N required hydrostatic pressure P and principal stress difference $\Delta\sigma$, respectively. The hydrostatic pressure P and principal stress difference $\Delta\sigma$ were calculated as $P = (\sigma + 2\sigma_3)/3$ and $\Delta\sigma = \sigma_1 - \sigma_3$, respectively. A series of values (P^*, σ^*) were obtained by normalizing the values of $(P, \Delta\sigma)$ according to the equations $P^* = P/f_c$ and $\sigma^* = \Delta\sigma/f_C$. The obtained values were fitted by equation $\sigma^* = BP^{*N}$ to obtain the values of B and N.

A triaxial compression test was performed using confining pressure at the rate of 0.02 MPa/s. After a stabilization period, the axial load was applied at the rate of 0.1 mm/min until the sample failed. Since the strength of the briquette sample was not high, excessively high confining pressures could break the sample. Therefore, confining pressure gradients were set at 1, 2, 3, and 4 MPa. The principal stress difference–axial strain curves of the samples using the experimental data obtained under different confining pressures are shown in Figure 9.

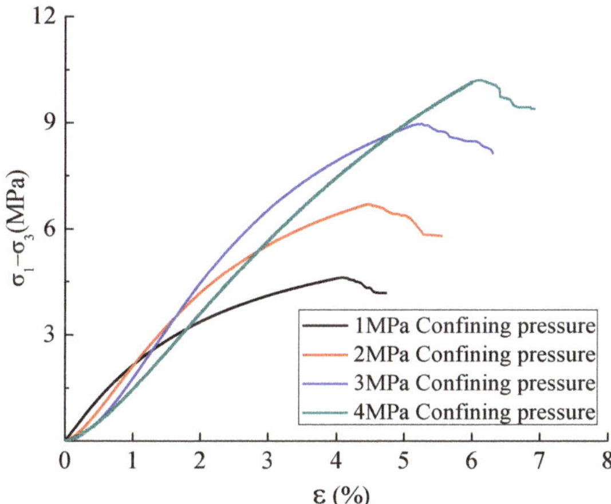

Figure 9. Curves of different principal stresses at different confining pressures.

The value of hydrostatic pressure P was calculated by equation $P = (\sigma + 2\sigma_3)/3$, and a series of $(P, \Delta\sigma)$ values were obtained. The obtained main stress difference $\Delta\sigma$ and hydrostatic pressure P were normalized to obtain a series of values (P^*, σ^*) (Table 7).

Table 7. Normalized principal stress difference σ^* and hydrostatic pressure P^*.

Principal Stress Difference/MPa	Hydrostatic Pressure/MPa
1.979	1.090
2.876	1.815
3.830	2.570
4.382	3.176

Using the data presented in Table 7, fitting was performed by equation $\sigma^* = BP^{*N}$, and the fitting curve was drawn as shown in Figure 10. The values of B and N were 1.86 and 0.75, respectively.

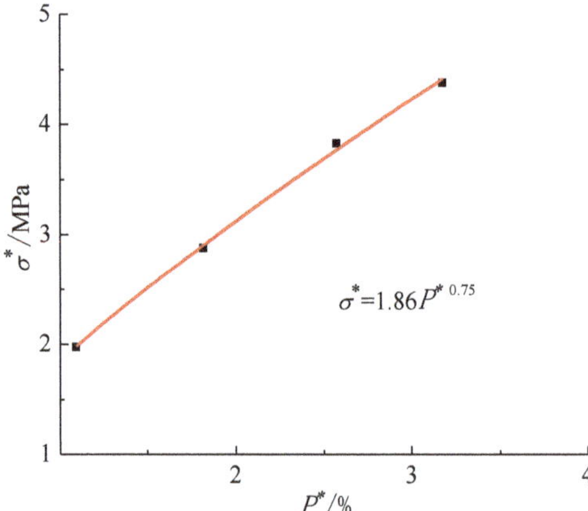

Figure 10. Curve of B and N fitting values.

4.5. Determination of the Values of the Remaining Parameters

The average density of the coal samples ρ_0, the elastic modulus E, Poisson's ratio v, volume pressure at crushing point P_C, and volumetric strain at crushing point μ_C were mainly derived indirectly using different equations. Based on the data shown in Table 2, the average density ρ_0 of the briquette samples was 1.228 g/cm^3. According to the data summarized in Table 4, the mean value of the three samples was taken as the elastic modulus and Poisson's ratio of the briquette samples, which were 101.56 MPa and 0.36, respectively. Shear modulus G was calculated using the equation $G = E/2(1 + v)$ to be 37.34 MPa. Based on previous literature, the value of P was calculated to be 0.78 MPa using the equation $P_C = f_C/3$ [14], where uniaxial compressive strength f_c was considered to be 2.33 MPa. Volumetric strain at crushing point μ_C was obtained to be 0.0064 using the equation $\mu_C = P_C/K$, where the bulk modulus K was calculated according to the equation $K = E/3(1 - 2v)$, where the elastic modulus E and Poisson's ratio v were assumed to be 101.56 MPa and 0.36, respectively.

The 11 parameters in the HJC constitutive model for coal were determined by experimental measurements and different equations. Pressure at compaction point P_1, volume strain at compaction point μ_1, pressure constants k_1, k_2, and k_3, normalized cohesive strength A, strain rate coefficient C, maximum normalized intensity S_{max}, and some other parameters were determined using a flying impact test [28]. The remaining parameters were not accessible due to their low sensitivity and limited experimental conditions. Therefore, their values were considered to be similar to those provided in [20]. Thus, the values of all parameters in the HJC constitutive model were determined, and the results are shown in Table 8.

Table 8. Coal HJC constitutive model parameter values.

ρ_0/g.cm^{-3}	G/Pa	f_C/Pa	A	B	C	N	S_{max}	D_1	D_2	$\varepsilon_{f\,min}$
1.228	37.34 × 10^6	2.33 × 10^6	0.4	1.86	0.005	0.75	7.0	0.0131	1.0	0.025
T/Pa	p_C/Pa	μ_C	p_l/Pa	μ_l	k_1/Pa	k_2/Pa	k_3/Pa	$\dot{\varepsilon}_0$	fs	
5.5 × 10^5	7.8 × 10^5	0.0064	1 × 10^9	0.12	8.5 × 10^{10}	1.7 × 10^{11}	2.08× 10^{11}	60	0.04	

5. SHPB Experiment and Numerical Simulation Analysis

5.1. SHPB Experimental Device

Dynamic impact tests on coal samples were carried out using an SHPB test device at China University of Mining and Technology (Beijing). The SHPB device consisted of a striking rod (bullet), an incident rod, and a projection rod. As shown in Figure 11, the bullet was a 540-mm long heavy double hammer-spun cone with a cone ratio of 310:100:130 [29]. As shown in Figure 12, the experimental and projection rods had a diameter of 75 mm and length of 2000 mm, and the tested coal sample was sandwiched between incident and projection rods. The SHPB experiment used coal briquettes with a diameter of 50 mm and length of 25 mm.

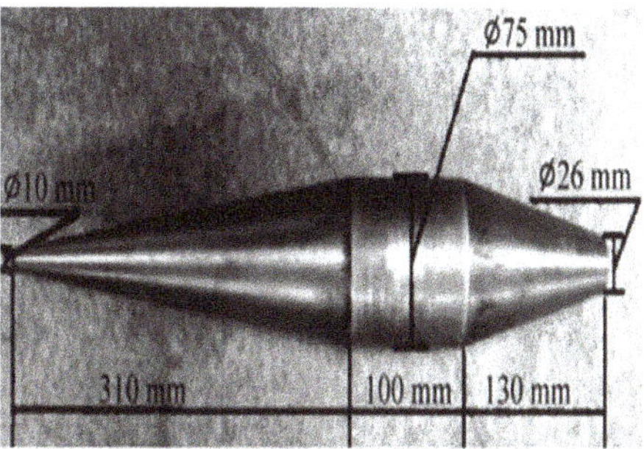

Figure 11. Spinning cone bullet.

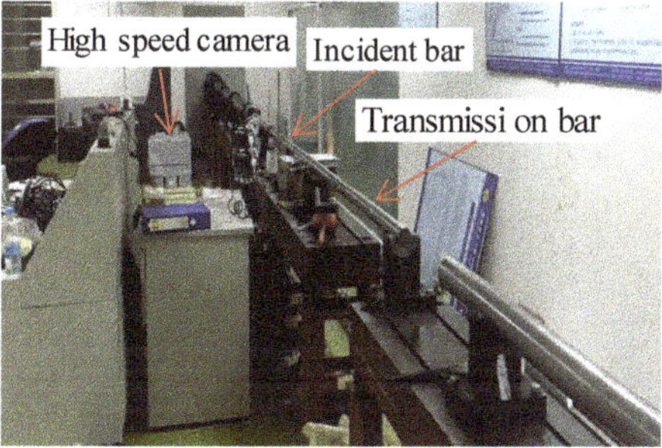

Figure 12. 50 mm split Hopkinson pressure bar (SHPB) experimental device.

5.2. Establishment of Finite Element Model

The finite element model in this work was built according to actual sizes and the components were all three-dimensional solid elements (Solid164). The finite element model was meshed, and the bullet, incident and transmission rods were divided into 15 parts along the radial direction. The bullet was divided into 40 parts along the axial direction, and incident and transmission rods were divided into 200 parts along the axial direction. To more intuitively reflect the impact damage of the test piece, it was finely meshed and was divided into 30 and 20 parts along radial and axial directions, respectively (Figure 13).

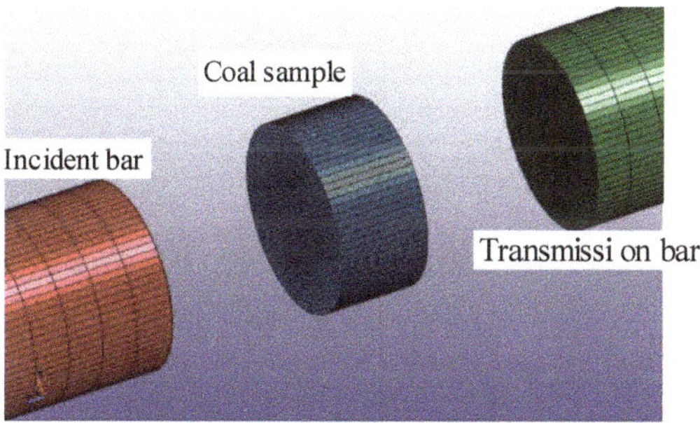

Figure 13. Establishment of the finite element model.

The material models of the bullet and compression bars were selected from linear elastic material models. The main parameters and their values were as follows: density 7800 kg/m³, elastic modulus 2.06 GPa, and Poisson's ratio 0.36. The coal sample was input into the HJC constitutive model and its parameter values were determined according to specific values listed in Table 8. The bullet was attached to the incident rod through an automatic contact between the two faces and the pressure bar was attached through an erosive contact with the test piece. During numerical simulation, the friction between contact surfaces was ignored. The value of 2 was taken as the contact stiffness penalty function factor f.

5.3. Waveform of the Stress Wave

In order to allow comparison with the experimental results, impact velocities in the numerical simulation were set at 4.732 and 7.267 m/s. Here, taking the stress waveform diagram of the incident rod and the middle of the projection rod at the impact velocity of 7.267 m/s as examples, experimental data and numerical simulation results were compared. The obtained results are shown in Figures 14 and 15.

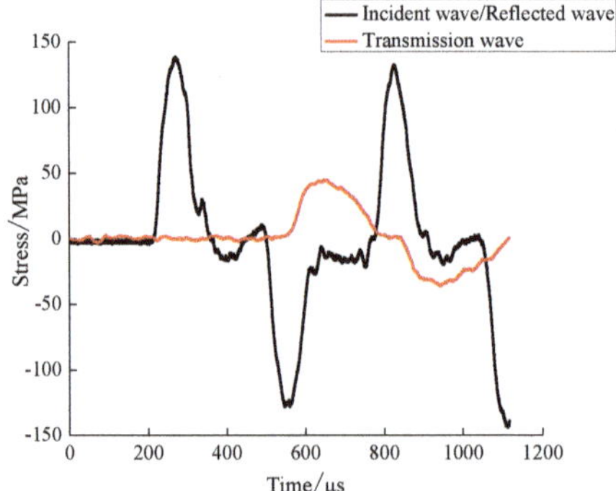

Figure 14. Experimental results (v = 7.267 m/s).

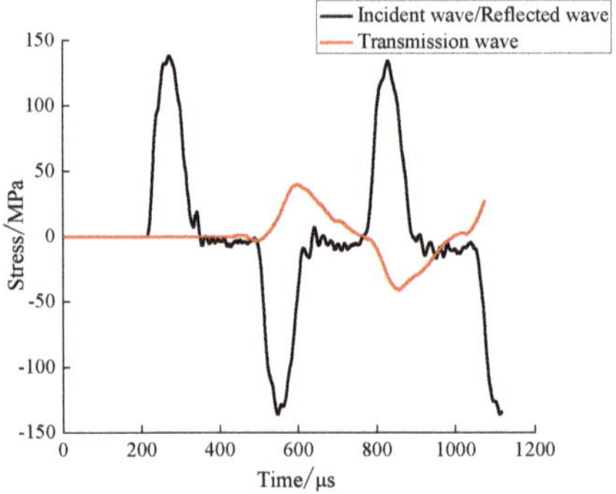

Figure 15. Simulation results (v = 7.267 m/s).

It can be seen from Figures 14 and 15 that the experimentally measured stress wave curve complied well with that drawn using simulation results, but there was a slight difference in curve volatilities. This was because the stress wave decayed with time during the propagation of the rod, and the components used in the experiment had inevitable little defects, which caused the stress wave to be weakened during propagation. Numerical simulation was carried out under ideal conditions. The end face of the rod was flat, and frictional force was neglected. Therefore, the stress wave obtained by numerical simulation was not attenuated, and the obtained stress wave curve was smoother; however, the peak value of the stress wave obtained by numerical simulation was slightly greater than that measured in real time. Although there were some differences between the simulation and experimental results obtained for the stress wave, they were generally consistent. Therefore, simulation results were considered to be accurate and reasonable.

5.4. Analysis of Coal Rock Damage

The damage process of the coal-rock SHPB experiment was captured with high-speed photography and compared to the simulated damage process, as shown in Figure 16.

Figure 16 shows that at the impact velocity of 4.732 m/s, the stress wave started to contact the coal sample at t = 578.93 μs. When t = 596.97 μs, a part of the coal body failed and obvious axial cracks appeared in the middle of the sample. When t = 623.97 μs, the number of surface cracks in the coal sample, and therefore the damage intensity, increased. When t = 725.93 μs, the coal sample broke into small pieces. When the impact velocity was 7.267 m/s, the stress wave started to contact the coal sample at t = 572.99 μs, and the coal body partially failed due to the propagation of the compressive stress wave. Radial cracks occurred at t = 584.96 μs, and when t = 583.96 μs, the internal stress of the sample gradually changed from compressive to tensile stress. At t = 617.98 μs, the coal sample was severely damaged and fell into pieces. At different impact velocities, axial cracks began to appear in the coal samples, which ultimately resulted in the failure of samples.

It can be seen from the high-speed images that when the impact velocity was 4.732 m/s, the sample was compressed by pressure, the coal was laterally deformed, and many transverse cracks were created parallel to the direction of stress wave propagation (indicated with a red circle in the Figure 16). With the development of cracks, the whole sample appeared to expand, and eventually, damage occurred under the joint action of upper crack expansion and the lower slip shear of the coal body. When the impact velocity was 7.267 m/s, first a through crack was created in the sample (shown by a red circle in Figure 16). Secondly, due to the large incident energy and high velocity of the bullet, the deformation of the coal continued to increase, and the crack expanded rapidly. Finally, the sample underwent compression expansion under impact loading, which caused tensile damage. As the speed continued to increase, the coal sample underwent a "comminuted" rupture, producing a large amount of fine granular coal dust. The severity of coal body damage positively correlated with the impact velocity, while the size of the coal block was decreased.

Comparing the experimental and simulated failure processes at impact velocities of 4.732 and 7.267 m/s revealed their high consistency. In general, numerical simulation using measured briquette HJC constitutive model parameters had a strong similarity to experimental findings, which verified the applicability of the HJC constitutive model parameters of briquette samples to simulate the failure process of coal samples under the impact of a dynamic load.

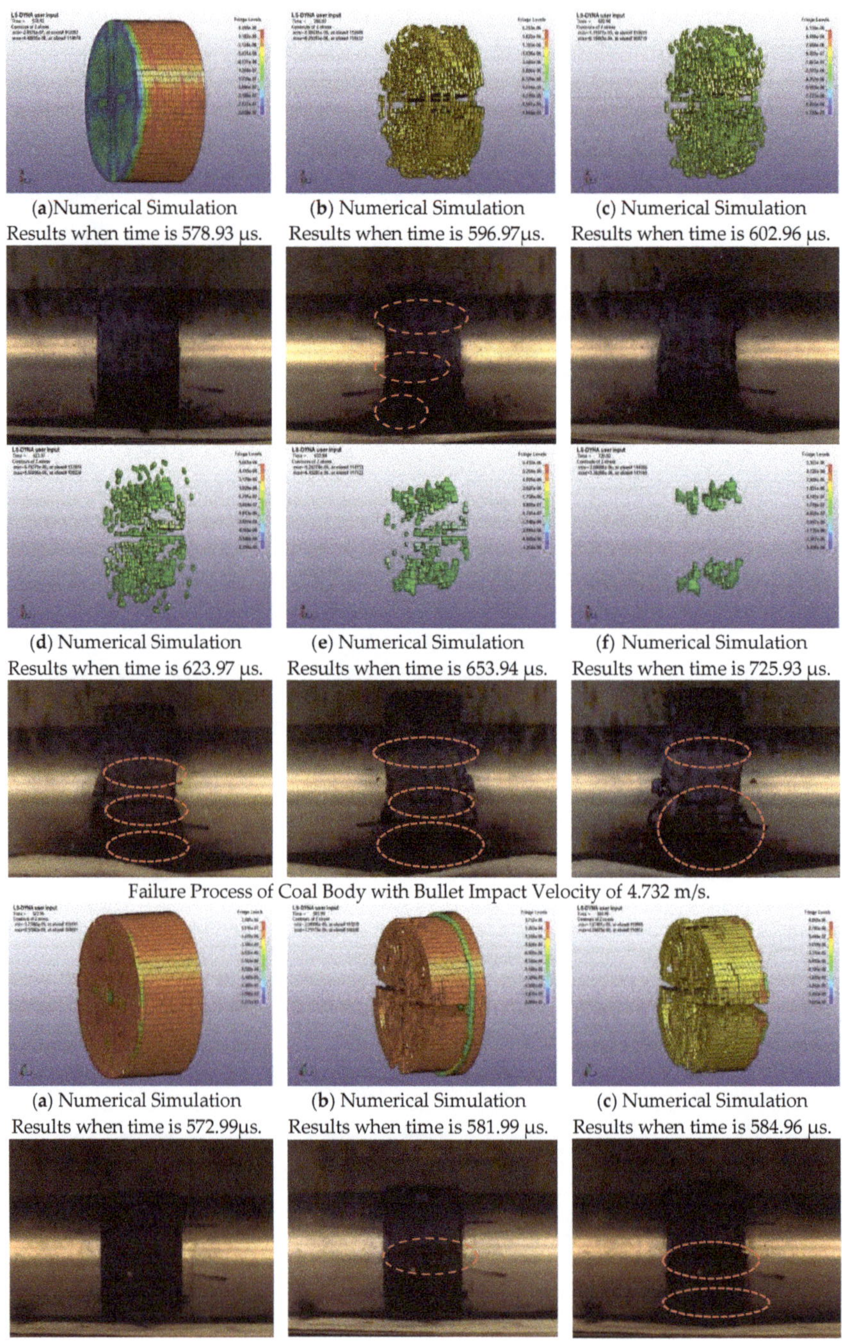

(**a**)Numerical Simulation Results when time is 578.93 μs.

(**b**) Numerical Simulation Results when time is 596.97μs.

(**c**) Numerical Simulation Results when time is 602.96 μs.

(**d**) Numerical Simulation Results when time is 623.97 μs.

(**e**) Numerical Simulation Results when time is 653.94 μs.

(**f**) Numerical Simulation Results when time is 725.93 μs.

Failure Process of Coal Body with Bullet Impact Velocity of 4.732 m/s.

(**a**) Numerical Simulation Results when time is 572.99μs.

(**b**) Numerical Simulation Results when time is 581.99 μs.

(**c**) Numerical Simulation Results when time is 584.96 μs.

Figure 16. *Cont.*

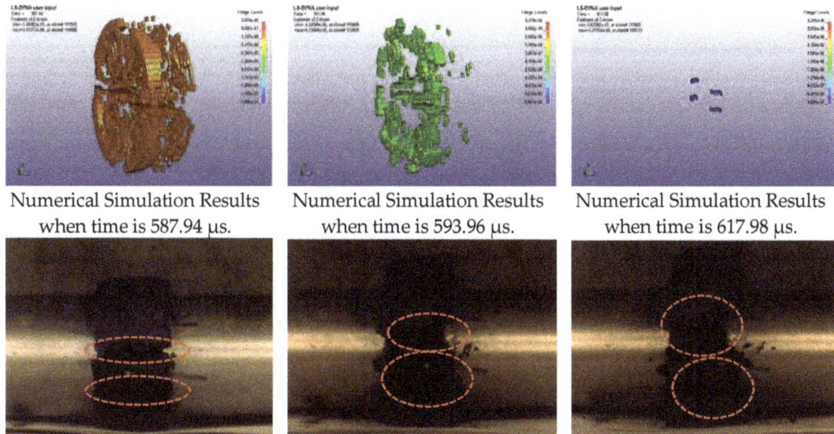

| Numerical Simulation Results when time is 587.94 μs. | Numerical Simulation Results when time is 593.96 μs. | Numerical Simulation Results when time is 617.98 μs. |

Failure Process of Coal Body with Bullet Impact Velocity of 7.267 m/s.

Figure 16. Dynamic change process of the impact failure of a coal body.

6. Numerical Simulation of Impact Damage on Heading Face

A coal-rock dynamic disaster refers to a strong dynamic phenomenon in which the surrounding coal and rock masses in an underground mining space are rapidly destroyed and a large amount of energy is suddenly released [30]. The HJC constitutive model parameters for briquette samples shown in Table 8 were applied in the numerical simulation of coal-rock dynamic hazards caused by rock bursts or blasting impact disturbances. Taking Yongcheng Cheji Coal Mine as an example, the impact damage surface of the excavation face was established. A numerical simulation model was used to verify the applicability of the HJC constitutive model parameters for briquette.

6.1. Model Establishment

The coal-rock layer was too thick or the size was too large, which increased the difficulty of meshing and the solution time. Therefore, the model was simplified. In the impact damage model developed for the tunnel face, the length and height of the roof and floor rock layers were considered to be 20 and 4 m, respectively. The length and height of the coal seam were 20 and 3 m, respectively (Figure 17), and the roadway size was 8 m. In order to simplify calculations, the rock layers of the top and bottom plates adopt an elastoplastic model with a density of 2500 kg/m^3, an elastic modulus of 18 Gpa, and a Poisson's ratio of 0.3 [31]. Coal seam was set based on the HJC constitutive model parameters for briquette shown in Table 8.

Considering the influence of ground stress on underground roadways, the maximum vertical ground stress of the roadway at 450 m depth for Yongcheng Cheji Coal Mine was measured to be 15.22 MPa. Therefore, during the simulation, a vertical static load was applied to the coal seam. The stress wave disturbance generated by excavation blasting was simplified into a semi-sinusoidal pulse, and based on the dynamic stress–strain curve obtained from the briquette SHPB experiment, 20 MPa was taken as the peak value (p_{max}) for a half-sinusoidal pulse.

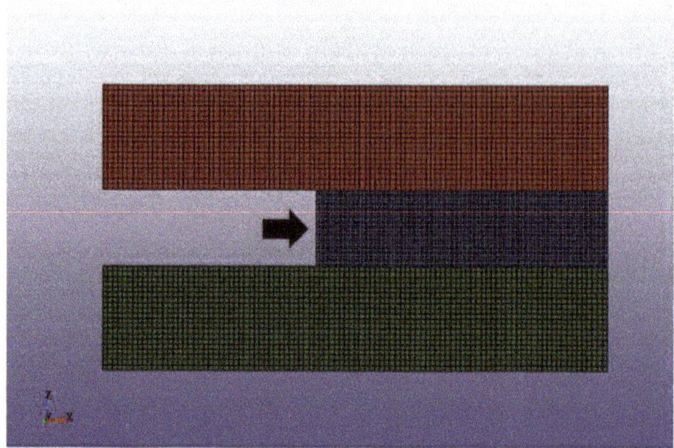

Figure 17. Establishment of a two-dimensional coal and rock mass model including a roadway.

6.2. Numerical Simulation of Impact Damage on Heading Face

In underground blasting operations, the shockwave generated by blasting releases huge amounts of energy to the surrounding area, which may cause the deformation, cracking, and even destruction of roadways and other structures. Explosion energy can loosen or destroy coal rock masses in the process of shockwave release and propagation. The numerical simulation of the impact damage of the tunneling face predicts the power of the disaster caused by the head. Numerical simulation results are shown in Figure 18.

Figure 18 clearly shows the destruction process of the coal seam in the roadway. As can be seen, at t = 158.61 μs, the spreading stress wave contacted the coal seam and triggered its failure. When t = 227.31 μs, the coal seam was cracked by the action of the compressive stress wave, which continued to propagate deep into the coal seam. At t = 278.52 μs, the crack gradually expanded and the coal seam was notched. When t = 416.76 μs, the size of the gap generated in the coal seam was increased with by the propagation of stress wave. Finally, when t = 536.62 μs, a large cavity was formed deep in the coal seam as the gap was increased.

The stress time-history curve of the coal seam in a two-dimensional model is shown in Figure 19. As can be seen, first the coal seam was cracked and destroyed by the propagation of the stress wave. The curve first rose sharply to the yield point and then began to fall. Then, as the stress wave propagated deep into the coal seam, the curve began to rise again and the coal seam was destroyed.

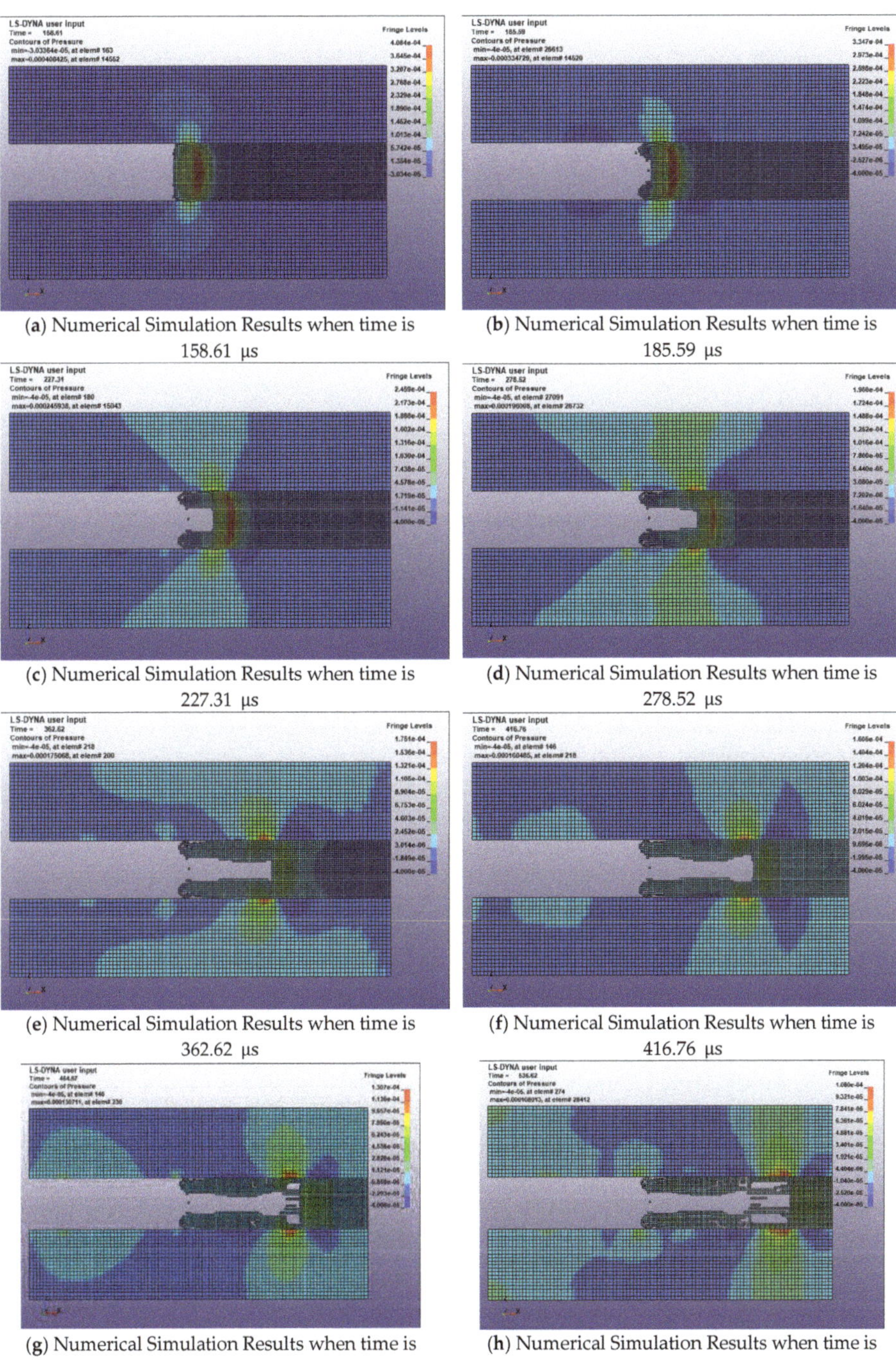

(**a**) Numerical Simulation Results when time is 158.61 µs

(**b**) Numerical Simulation Results when time is 185.59 µs

(**c**) Numerical Simulation Results when time is 227.31 µs

(**d**) Numerical Simulation Results when time is 278.52 µs

(**e**) Numerical Simulation Results when time is 362.62 µs

(**f**) Numerical Simulation Results when time is 416.76 µs

(**g**) Numerical Simulation Results when time is 464.57 µs

(**h**) Numerical Simulation Results when time is 536.62 µs

Figure 18. Excavation surface impact damage process (p_{max} = 20 MPa).

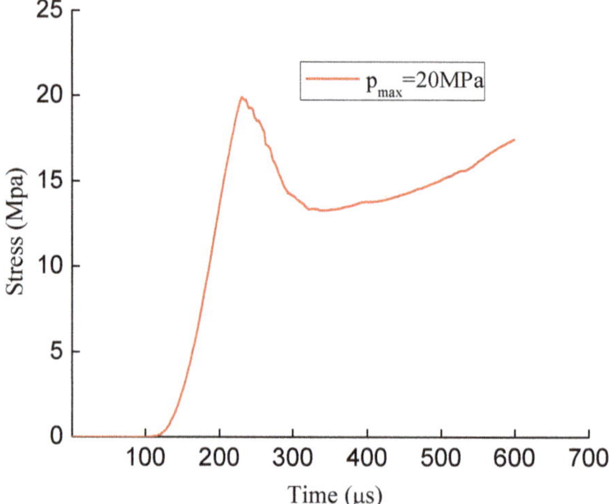

Figure 19. Time-history curve of coal seam stress in a 2D model.

7. Conclusions

Through SHPB experiments of briquette combined with experimentally measured HJC constitutive parameters and ANSYS/LS-DYNA software results, the following conclusions can be drawn:

(1) Through direct experimental measurements and indirect equation derivations, the main sensitivity parameters of the coal HJC constitutive model were obtained, which were of great importance in understanding the occurrence mechanism and predicting coal-rock dynamic disasters.

(2) The failure of coal briquettes at different impact velocities started with the creation of axial cracks in the middle of the coal body. The destructed coal body included slip failure, tensile failure caused by compression expansion, and crack expansion and slip shear joint failure phenomena.

(3) The fracture morphology of the briquette samples was different at different impact velocities. As the impact velocity increased, the severity of coal body damage increased and the size of the coal block decreased. At lower speeds, coal samples were intact and cracks appeared only on the surface. At higher impact velocities, however, coal samples underwent "comminuted" rupture, producing a large amount of fine granular coal dust.

(4) The numerical simulation of the SHPB experiment was carried out using the experimentally measured parameters of the HJC constitutive model for briquette. The simulated and experimentally measured stress wave curves complied well. Comparing the experimental and numerical simulation results obtained for coal sample failure and fracture morphology at different impact velocities (4.732 and 7.267 m/s) revealed that all coal samples produced axial cracks starting from the middle of the coal body and then failed. The experimental and numerical simulation results for the coal sample crushing process and fracture results were similar.

(5) The validated parameter values of the HJC constitutive model were applied to the numerical model of tunneling impact damage, which visually showed the process of coal seam damage in a roadway. It was observed that the simulation of the failure of coal samples under the impact of dynamic loads was possible using HJC constitutive model parameter values.

Author Contributions: B.X. conceived and designed the research; Z.Y. performed the experiment and wrote the manuscript. Y.D. undertook data curation; Z.Z. was responsible for software; X.Z. reviewed and edited the manuscript.

Funding: This research was supported by the National Natural Science Foundation of China (51404277, 51274206) and the State Key Laboratory Cultivation Base for Gas Geology and Gas Control (Henan Polytechnic University) (WS2018B08). This support is greatly acknowledged and appreciated.

Acknowledgments: Y.Y., J.J.Z., and S.Z.F. are acknowledged for their valuable technical support.

Conflicts of Interest: The authors declare no conflict of interest.

References

1. Li, X.J.; Liu, B.Q. Status of research and analysis on coal and gas outburst mechanism. *Coal Geol. Explor.* **2010**, *38*, 7–13.
2. Gong, F.Q.; Ye, H.; Luo, Y. The Effect of high loading rate on the behaviour and mechanical properties of coal-rock combined body. *Shock Vib.* **2018**, *2018*. [CrossRef]
3. He, X.Q.; Dou, L.M.; Mu, Z.L. Continuous monitoring and warning theory and technology of rock burst dynamic disaster of coal. *J. China Coal Soc.* **2014**, *39*, 1485–1491.
4. Xue, D.; Zhou, J.; Liu, Y.; Zhang, S. A strain-based percolation model and triaxial tests to investigate the evolution of permeability and critical dilatancy behavior of coal. *Processes* **2018**, *6*, 127–149. [CrossRef]
5. Cai, T.T.; Feng, Z.C.; Jiang, Y.L. An improved hardening-damage creep model of lean coal: A theoretical and experimental study. *Arab. J. Geosci.* **2018**, *11*, 645. [CrossRef]
6. Li, Q.; Liang, Y.; Zou, Q. Seepage and damage evolution characteristics of gas-bearing coal under different cyclic loading–unloading stress paths. *Processes* **2018**, *6*, 190–210. [CrossRef]
7. Zhao, Y.X.; Gong, S.; Zhang, C.G.; Zhang, C.G.; Zhang, Z.N.; Jiang, Y.D. Fractal characteristics of crack propagation in coal under impact loading. *Fractals-Complex Geom. Patterns Scaling Nat. Soc.* **2018**, *26*. [CrossRef]
8. Feng, J.J.; Wang, E.Y.; Shen, R.X.; Chen, L.; Li, X.L.; Xu, Z.Y. Investigation on energy dissipation and its mechanism of coal under dynamic loads. *Geomech. Eng.* **2016**, *11*, 657–670. [CrossRef]
9. Li, C.W.; Wang, Q.F.; Lv, P.Y. Study on electromagnetic radiation and mechanical characteristics of coal during an SHPB test. *J. Geophys. Eng.* **2016**, *13*, 391–398.
10. Gao, Z.; Ding, S.X.; Cecati, C. Real-time fault diagnosis and fault-tolerant control. *IEEE Trans. Ind. Electron.* **2015**, *62*, 3752–3756. [CrossRef]
11. Zhang, C.; Zhang, L.; Tu, S.H. Experimental and Numerical Study of the Influence of Gas Pressure on Gas Permeability in Pressure Relief Gas Drainage. *Transp. Porous Media* **2018**, *134*, 995–1015. [CrossRef]
12. Xie, J.; Xu, J.; Wang, F. Mining-induced stress distribution of the working face in a kilometer-deep coal mine-a case study in Tangshan coal mine-a case study in Tangshan coal mine. *J. Geophys. Eng.* **2018**, *15*, 2060–2070. [CrossRef]
13. Liu, Q.L.; Wang, E.Y.; Kong, X.G.; Li, Q.; Hu, S.B.; Li, D.X. Numerical simulation on the coupling law of stress and gas pressure in the uncovering tectonic coal by cross-cut. *Int. J. Rock Mech. Min. Sci.* **2018**, *103*, 33–42. [CrossRef]
14. Li, C.W.; Ai, D.H.; Sun, X.Y.; Xie, B.J. Crack identification and evolution law in the vibration failure process of loaded coal. *J. Geophys. Eng.* **2017**, *14*, 975–986. [CrossRef]
15. Holmquist, T.J.; Johnson, G.R. A computational constitutive model for glass subjected to large strains, high strain rates and high pressures. *J. Appl. Mech.* **2011**, *78*. [CrossRef]
16. Iqbal, M.A.; Kumar, V.; Mittal, A.K. Experimental and numerical studies on the drop impact resistance of prestressed concrete plates. *Int. J. Impact Eng.* **2016**, *11*, 657–670. [CrossRef]
17. Kong, X.Z.; Fang, Q.; Wu, H.; Peng, Y. Numerical predictions of cratering and scabbing in concrete slabs subjected to projectile impact using a modified version of HJC material model. *Int. J. Impact Eng.* **2016**, *95*, 61–71. [CrossRef]
18. Islam, M.J.; Swaddiwudhipong, S.; Liu, Z.S. Penetration of concrete targets using a modified Holmquist-Johnson-cook material model. *Int. J. Comput. Methods* **2012**, *9*. [CrossRef]
19. Yuan, P.; Ma, Q.; Ma, D. Stress uniformity analyses on nonparallel end-surface rock specimen during loading process in SHPB tests. *Adv. Civ. Eng.* **2018**, *2018*. [CrossRef]
20. Xie, B.J. Experimental research on characteristics of coal impact damage. *China Univ. Min. Technol. Beijing* **2013**, *3*, 56–60.
21. Li, C.W.; Wang, J.G.; Xie, B.J. Numerical simulation of SHPB tests for coal by using HJC model. *J. Min. Saf. Eng.* **2016**, *33*, 158–164.
22. Li, C.W.; Wang, J.G.; Xie, B.J. Numerical analysis of split hopkinson pressure bar test with passive confined pressure for coal. *J. Min. Saf. Eng.* **2014**, *31*, 957–962.

23. Chen, J.L.; Li, X.D.; Liu, K.X. Experimental research on parameters of constitutive model for a cement mortar. *Acta Sci. Nat. Univ. Pekin.* **2008**, *5*, 689–694.

24. Ren, G.M.; Wu, H.; Fang, Q.; Zhou, J.W. Determinations of HJC constitutive model parameters for normal strength concrete. *J. Vib. Shock* **2016**, *35*, 9–16.

25. Wang, K.; Jiang, Y.F.; Xu, C. Mechanical properties and statistical damage model of coal with different moisture contents under uniaxial compression. *Chin. J. Rock Mech. Eng.* **2018**, *37*, 1070–1079.

26. Chen, L.J.; Chai, Y.Y.; Zhu, C.H. Study of micro-structure of briquette. *J. China Coal Soc.* **1997**, *3*, 82–84.

27. Chen, Y.G.; Liu, C.D.; Zhong, S. Research on the influences of soft layer on coal thickness & coal and gas outburst. *Ind. Saf. Environ. Prot.* **2013**, *39*, 42–45.

28. Li, Y.; Li, H.P.; Wu, X.T. Research on the HJC dynamic constitutive model for concrete. *J. Hefei Univ. Technol. (Nat. Sci.)* **2009**, *32*, 1244–1248.

29. Xie, B.J.; Wang, X.Y.; Lu, P.Y. Dynamic properties of bedding coal and rock and the SHPB testing for its impact damage. *J. Vib. Shock.* **2017**, *36*, 117–124.

30. Lu, J.G.; Wang, T.; Ding, W.B. Induction mechanisms of coal bumps caused by thrust faults during deep mining. *J. China Coal Soc.* **2018**, *43*, 405–416.

31. Qin, H. Study on mechanism of roadway surrounding rock instability and rock burst. *China Univ. Min. Technol (Beijing).* **2008**, *5*, 77–81.

Article

Intelligent Setting Method of Reagent Dosage Based on Time Series Froth Image in Zinc Flotation Process

Zhaohui Tang [1], Liyong Tang [1], Guoyong Zhang [1,*], Yongfang Xie [1] and Jinping Liu [2]

[1] School of Automation, Central South University, Changsha 410083, China; zhtang@csu.edu.cn (Z.T.); tangliyong@csu.edu.cn (L.T.); yfxie@csu.edu.cn (Y.X.)

[2] School of Information Science and Engineering, Hunan Normal University, Changsha 410081, China; ljp@hunnu.edu.cn

* Correspondence: csu_zgy@csu.edu.cn

Received: 29 March 2020; Accepted: 30 April 2020; Published: 3 May 2020

Abstract: It is well known that the change of the reagent dosage during the flotation process will cause the froth image to change continuously with time. Therefore, an intelligent setting method based on the time series froth image in the zinc flotation process is proposed. Firstly, the sigmoid kernel function is used to estimate the cumulative distribution function of bubble size, and the cumulative distribution function shape is characterized by sigmoid kernel function parameters. Since the reagent will affect the froth image over a period of time, the time series of bubble size cumulative distribution function is processed by the ELMo model and the dynamic feature vectors are output. Finally, XGBoost is used to establish the nonlinear relationship modeling between reagent dosage and dynamic feature vectors. Industrial experiments have proved the effectiveness of the proposed method.

Keywords: flotation process; reagent dosage; time series froth image; cumulative distribution function

1. Introduction

Froth flotation is a physical and chemical process that occurs at the three-phase interface of solids, liquids, and gases, and is used to separate valuable minerals from gangue. Froth flotation is an industrial process of processing minerals. The flotation process is affected by a variety of operating variables, such as slurry level, pH value, and reagent dosage [1]. Of all operating variables, the reagent dosage is the most critical [2]. The zinc flotation process is a typical complex industrial system, and it is difficult to recognize a clear mechanism to establish a first principle model [3]. At present, the method of controlling the reagent dosage in the actual flotation process is mainly achieved by the operator observing the surface characteristics of the froth image, such as bubble size, texture, and gray value [4].

With the development of advanced technologies such as industrial automation, cloud computing, and artificial intelligence, the modeling, monitoring, and control of complex industrial systems have new technical support [5]. In reference [6], deep reinforcement learning is applied to the boost control of diesel engines. The results show that the performance is better than the traditional proportion integral derivative (PID) controller. In the reference [7], the intelligent algorithm is applied to the optimal sitting and sizing of a distributed generation system, which improves the economic benefits and safety of the distributed generation system.

In recent years, many scholars have studied the method of controlling the reagent dosage in the flotation process [8–10]. In reference [11], a control strategy based on the sensitive features of froth images is proposed. The strategy proposed to adjust the reagent dosage based on the feed grade, and established a model of feed grade estimation based on the sensitive features of the froth image. Then, based on the operation mode method, the reagent dosage is preset according to the feed grade. However, it only analyzes the froth image of a single moment, and it is difficult to avoid the noise effect of single frame image. In reference [12], the authors analyzed the effect of reagent dosage on performance indicators, and then proposed a collaborative optimization method for reagent dosage based on key feature changes and case-based reasoning. The proposed method achieved a certain effect on the antimony flotation process, but it was difficult for the method to overcome the time lag problem on the flotation process. In reference [13], the case-based reasoning method was used to optimize the control of the reagent dosage in the magnetite flotation process, the results show that the method reduces the fluctuation of tailings grade, but it was difficult to deal with the disturbance in the flotation process.

In reference [14], it is pointed out that the visual features of the froth image play an important role in the flotation process. In reference [15], the Wasserstein Distance-Based CycleGAN method was used to measure the color features of the froth image, and the color features were applied to guide the flotation process of bauxite. In reference [16], the author introduced the biologically inspired Gabor wavelet transform to extract the texture features of froth images, and applied the texture features to the online state recognition of the flotation process.

Among all visual features of the froth image, the bubble size is the most critical [17]. In reference [18], the authors modeled the relationship between flotation process performance and bubble size of froth through neural networks. The results prove that the average bubble size of the froth image is very important to guide the stability of the flotation process. In reference [19], a control strategy is proposed to optimize the recovery of valuable minerals by tracking the expected bubble size distribution. In reference [20], a control strategy based on bubble size distribution is proposed to control the reagent dosage by minimizing the difference between the output bubble size distribution and the optimal bubble size distribution. In reference [21], the feedback controller maintains the probability density function of bubble size at the setpoint through the reagent dosage. In reference [22], the method used the bubble size probability density function to characterize the froth state. The reagent dosage is optimally controlled by minimizing the difference between the current bubble size probability density function and the optimal bubble size probability density function.

The features of the froth image in the zinc flotation process can reflect the working conditions of the flotation process, and the bubble size is the most obvious among all the features. In this paper, the bubble size cumulative distribution function (CDF) is used as a new feature of the froth image, and the estimation method of bubble size CDF is proposed. Because the change of reagent dosage will cause the froth image features to change continuously with time, this paper analyzes the relationship between the time series of bubble size CDF and the reagent dosage. Compared with the single-frame bubble image feature at a single moment, the time series froth image feature can reduce the influence of noise. The proposed method does not require manual participation, avoids the disadvantages of unstable performance indicators, and large reagent consumption due to differences in manual experience in the zinc flotation process, which is conducive to improving the economic benefits of the flotation plant.

The rest of the paper is structured as follows. A description of the process is presented in Section 2. The method for the estimation of bubble size CDF is shown in Section 3. The reagent dosage intelligent setting based on time series of bubble size CDF is discussed in Section 4. The industrial experiment and discussion are contained in Section 5. Section 6 draws the conclusions of this article.

2. Process Description

In the actual production process, the zinc flotation process is often accompanied by lead flotation, and this work focuses on the zinc flotation process. The simplified process of zinc flotation process is shown in Figure 1.

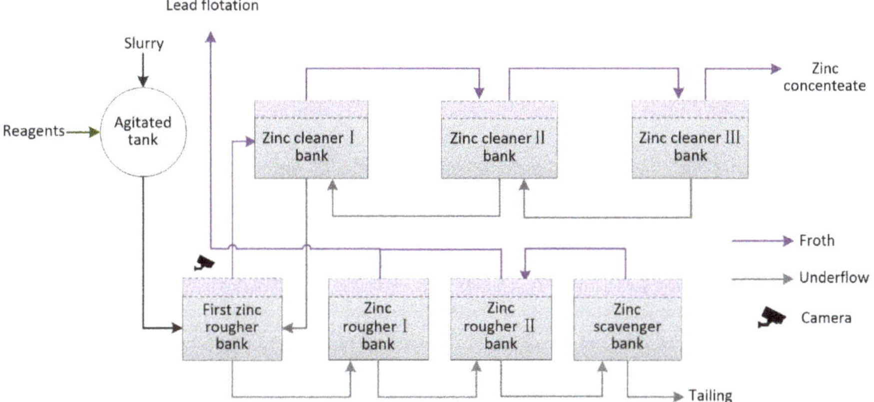

Figure 1. Flow diagram of the zinc flotation process.

The purpose of the zinc flotation process is to extract zinc elements from sphalerite. The zinc element in the slurry mainly exists in the form of zinc sulfide (ZnS). After the chemical reaction between the slurry and the reagent, the froth layer and underflow are formed in the flotation bank. In the zinc flotation process, the reagent is made up of foaming reagent (ROH), activating reagent ($CuSO_4$), capture reagent ($C_4H_9OCSSNa$) and PH adjustment reagent (H_2SO_4 and $Ca(OH)_2$), according to a certain proportion.

The capacity of the flotation bank is about 2.8 m^3, with a flow rate of 1.5 m^3/min–3 m^3/min. The zinc flotation process includes the following stages: first zinc rougher, zinc rougher I, zinc rougher II, zinc cleaner I and zinc cleaner II, zinc cleaner III, and zinc scavenger. The slurry and reagent react chemically in the agitated tank, and a large amount of bubbles are generated under the action of the foaming agent. The zinc minerals adhere to the bubble surface to form mineralized bubbles, overflow from the first zinc rougher bank, and enter the zinc cleaner I. At the same time, the water-soluble material forms an underflow in the first zinc rougher bank and flows into the zinc rougher I. The same principle is followed in the subsequent flotation process. Finally, the zinc concentrate product is obtained from the zinc cleaner III bank, and the underflow in the zinc scavenger bank forms tailings.

The reagent acting in the first zinc rougher bank has an important influence on the final performance index (concentrate grade, tailing grade) in the zinc flotation process. The zinc flotation process has a long flow, severe time lag, many operating variables, and strong coupling, resulting in a very complicated zinc flotation process mechanism, making it difficult to establish an accurate dynamic model. Currently, the flotation plant controls the dosage of reagent manually. The method for controlling the reagent dosage in the zinc flotation process is as follows: the operator repeatedly inspects the froth state in the flotation process, and judges the reagent dosage according to production experience. In the process, the operator's experience becomes extremely important. However, the experience level of different operators is very different, and it is difficult to ensure that the operating variables of the complex flotation process remain stable. A zinc flotation process monitoring system was established by installing a camera at the first zinc rougher bank [23]. Collecting and analyzing froth images through a monitoring system to reduce manual participation is a prerequisite for achieving automatic reagent control.

3. Estimation Method of Bubble Size CDF

In this paper, bubble size CDF is used to characterize the bubble size distribution. When determining a series of random variables x, if there is an integrable probability density function $p(x)$ on the real axis, the CDF is expressed as:

$$c(x) = \int_{-\infty}^{x} p(t)dt \tag{1}$$

In the Equation (1), x represents the bubble size, and $p(x)$ is the probability density function of the bubble size. We use the flotation process monitoring system to obtain 2-D froth images under different working conditions in the zinc flotation process, as shown in Figure 2. Class 1 appears in the case of under-dosing, class 2 appears in the case of normal dosing, and class 3 appears in the case of over-dosing.

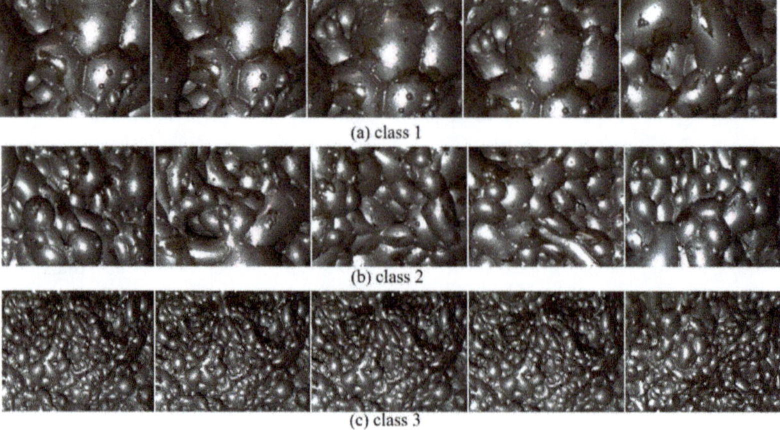

Figure 2. The typical froth images in zinc flotation process.

The resolution of the froth image captured by the camera in the flotation monitoring system is 692×518. With the original watershed algorithm it is difficult to accurately segment images such as the froth image without background, and this paper uses an improved watershed algorithm to segment the froth image to get the number of pixels contained in a single bubble [24]. The bubble size CDF of the different classes of the froth image is shown in Figure 3.

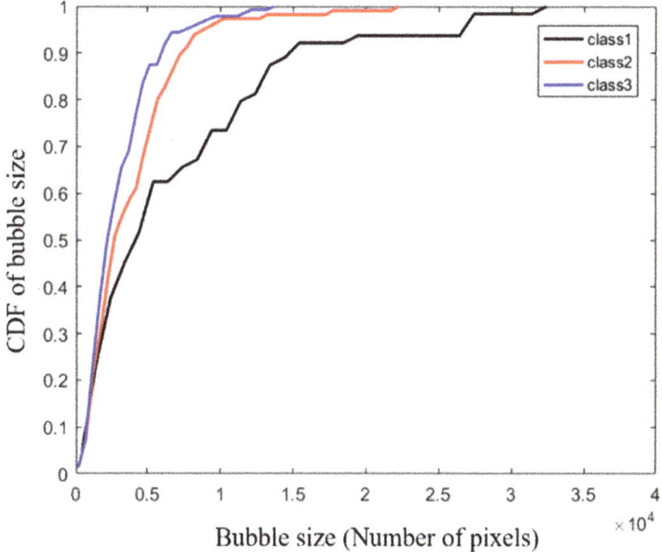

Figure 3. Bubble size cumulative distribution function (CDF) of different class froth image.

The standard sigmoid function expression:

$$s(x) = \frac{1}{1 + e^{-x}} \tag{2}$$

The shape of the sigmoid function in the first quadrant of the coordinate system is highly similar to the shape of the bubble size CDF, so it will be feasible to estimate the bubble size CDF with the sigmoid kernel function. Here, we use a sigmoid kernel function to estimate the CDF of bubble size. Expressed as:

$$\hat{c}(x) \approx c(x) = \frac{a}{1 + e^{-bx+c}} - d \tag{3}$$

The parameters a, b, c, d determine the shape of the CDF. The parameters a, b, c, d were obtained using the least-squares algorithm [25].

$$\min_{a,b,c,d} \sum_{j=1}^{N} \left(c(x_j) - \hat{c}(x_j) \right)^2 \tag{4}$$

where N is the estimated sample size, $c(x_j)$ is the CDF value when the bubble size is x_j, and $\hat{c}(\cdot)$ is the sigmoid kernel estimated value.

Figure 4 shows the experimental results of estimating the bubble size CDF using the sigmoid kernel function.

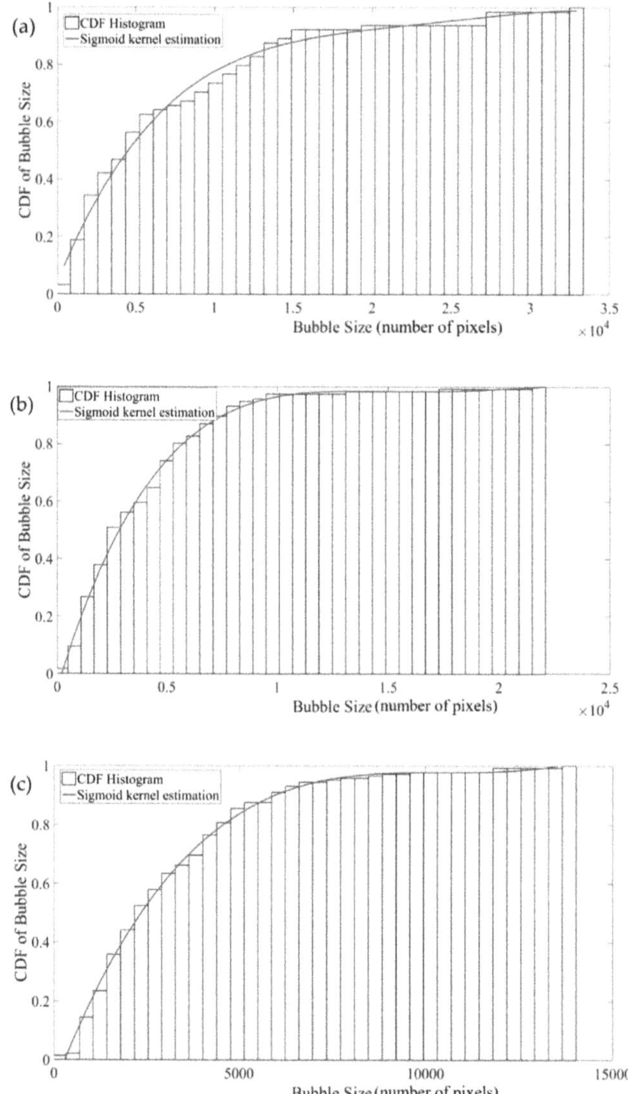

Figure 4. The estimation of bubble size CDF for different froth images: (**a**) class 1, (**b**) class 2, (**c**) class 3.

4. Reagent Dosage Intelligent Setting Based on Time Series of Bubble Size CDF

During the zinc flotation process, the bubble size CDF has an important influence on the grade of concentrate and tailing. The bubble size CDF under different working conditions shows different shapes. In this paper, the structure diagram of the intelligent setting of the reagent dosage based on time series froth image is shown in Figure 5.

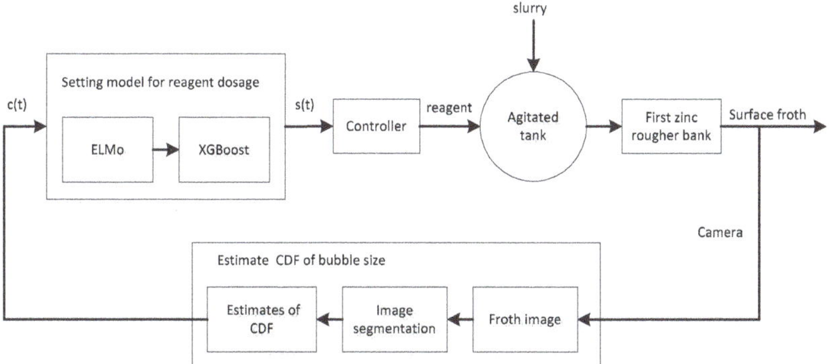

Figure 5. The structure diagram of the reagent dosage intelligent setting based on time series froth image.

Froth images of the first zinc rougher were captured by the camera. After the bubble image was segmented, the sigmoid kernel function was used to estimate the bubble size CDF, and the sigmoid kernel function parameters were used to characterize the CDF shape of bubble size. In the reagent dosage setting method, ELMo was used to process the time series features to obtain dynamic feature vectors, and XGBoost was used to establish the nonlinear relationship model between the dynamic feature vectors and the reagent dosage. In this way, the intelligent setting of the reagent dosage was realized.

4.1. Time Series Processing by Elmo

The ELMo model was proposed by Peter for natural language processing in 2018 [26]. Due to its efficient performance, ELMo has always been regarded as one of the excellent word embedding models. ELMo can recognize polysemy after training, that is, different word expresses a different meaning in different sentences. In the zinc flotation process, the current froth image is affected by the previous froth image, the time-series froth image after ELMo processing can express a deeper meaning.

ELMo model training is mainly divided into two steps. The first step is to build a bidirectional language model (biLM) based on LSTM, and adjust the model parameters through large-scale corpus training. The second step is to input the text into the ELMo model, and the weighted combination biLM multilayer output state is output as a dynamic feature vector. Its model structure is shown in Figure 6.

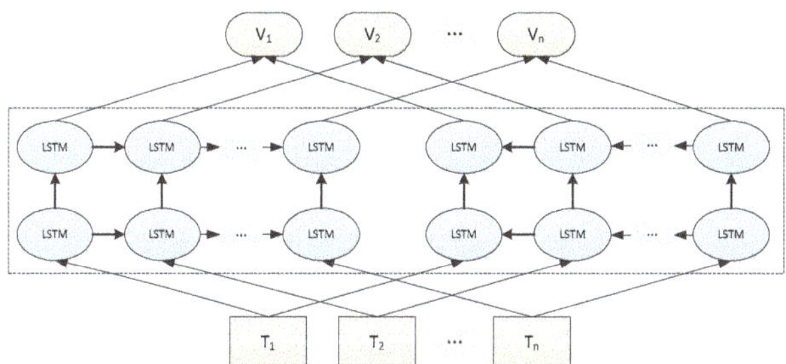

Figure 6. The framework of ELMo.

It is assumed that a continuous sequence $t_1, t_2, \ldots, t_N$ containing N tokens is given. The forward expression and backward expression are used to represent the relationship between the current token and the time series tokens before and after it.

The forward expression is:

$$p(t_1, t_2, \ldots, t_N) = \prod_{k=1}^{N} p(t_k | t_1, t_2, \ldots, t_{k-1}) \tag{5}$$

The backward expression is:

$$p(t_1, t_2, \ldots, t_N) = \prod_{k=1}^{N} p(t_k | t_{k+1}, t_{k+2}, \ldots, t_N) \tag{6}$$

The forward expression means predicting the current token through the previous $N-1$ tokens, and the backward expression means predicting the current token through the following $N-1$ tokens.

The objective function combines forward expression and backward expression. The meaning of the current token in the time series is determined by maximizing the likelihood probability of the objective function. This expression is shown in Equation (7).

$$\sum_{K=1}^{N} \left(\log p(t_k | t_1, t_2, \ldots, t_{k-1}, \Theta_x, \overrightarrow{\Theta}_{LSTM}, \Theta_s) + \log p(t_k | t_{k+1}, t_{k+2}, \ldots, t_N, \Theta_x, \overleftarrow{\Theta}_{LSTM}, \Theta_s) \right) \tag{7}$$

Among them, Θ_x represents the initial word embedding vector of the input biLM model, Θ_s represents the output of the LSTM layer input to softmax, $\overrightarrow{\Theta}_{LSTM}$ represent the forward LSTM layer, $\overleftarrow{\Theta}_{LSTM}$ represent the backward LSTM layer.

The dynamic feature vector is equal to the weighted combination of the output vectors of each LSTM layer, as shown in Equation (8).

$$ELMo = \gamma \sum_{j=0}^{2} \alpha_j . h_{k,j}^{LM} \tag{8}$$

where α_j is the weight of word embeddings in different layers, and it is a scalar parameter. $h_{k,j}^{LM}$ is the vector output from the jth layer LSTM.

The dynamic feature vector of the froth image features obtained by ELMo can characterize the state of the froth image over a period of time. In this paper, the time series bubble size CDF $c(t_1), c(t_2), \ldots, c(t_n)$ used to train ELMo. The output of the top LSTM in biLM is selected as the dynamic feature vector of ELMo.

4.2. Nonlinear Relationship Modeling by Xgboost

Chen et al. proposed the XGBoost algorithm in 2016 [27]. XGBoost is an efficient integrated learning method, and a few other integrated learning methods can be better than the configured XGBoost. XGBoost algorithm performs a second-order Taylor expansion of the loss function, and uses the complexity of the tree structure as a regular term, which can effectively avoid overfitting. In addition, XGBoost supports multi-threaded and can make full use of the machine's CPU core, thereby improving the speed and performance of the algorithm.

When there are n samples in a given dataset $D = \{(X_i, y_i)\}(|D| = n, X_i \in \mathbb{R}^m, y_i \in \mathbb{R})$, each sample contains m-dimensional features. XGBoost is used as a regression model to determine the estimated value $\hat{y}_i$ through the input feature X_i. XGBoost is trained through dataset D, and K trees are constructed. The accumulated value of these K trees is the output value. Expressed as:

$$\hat{y}_i = \phi(X_i) = \sum_{k=1}^{K} f_k(X_i), f_k \in F \tag{9}$$

where $F = \{f(x) = \omega_q(x)\}(q : \mathbb{R}^m \to T, \omega \epsilon \mathbb{R}^T)$ is the feature space of decision trees. T is the number of leaves in a tree. XGBoost is trained by minimizing the objective function. The objective function is shown in Equation (10).

$$L = \sum_i l(\hat{y}_i, y_i) + \sum_k \Omega(f_k) \tag{10}$$

where $\Omega(f_k) = \gamma T + \frac{1}{2}\lambda\|\omega\|^2$ represents the penalty term, which can reduce the risk of overfitting. γ represents the regularization parameter, λ means L2 regularization. l is a differentiable convex loss function. The second-order Taylor expansion of the loss function in the objective function is shown in Equation (11).

$$\widetilde{L}^{(t)} \simeq \sum_{i=1}^{n}\left[g_i f_t(x_i) + \frac{1}{2}h_i f_t^2(x_i)\right] + \Omega(f_t) \tag{11}$$

where $g_i = \partial_{\hat{y}^{(t-1)}} l(\hat{y}_i^{(t-1)}, y_i)$ and $h_i = \partial_{\hat{y}^{(t-1)}}^2 l(\hat{y}_i^{(t-1)}, y_i)$ are the first and the second order gradient statistic on the loss function.

By training XGBoost, a nonlinear relationship model of bubble size CDF time series and reagent dosage is established. The input is the dynamic feature vector generated by ELMo processing of the bubble size CDF time series, and the output is the reagent dosage.

5. Experiment and Discussion

The purpose of this work is to make a reasonable set value of the reagent dosage by analyzing the time series of the froth image of the first zinc rougher bank.

After the reagent works, the froth image of the first zinc rougher bank will respond first, and the froth of the first zinc rougher bank will affect the concentrate grade and tailing grade. When the dosage of the reagent is set low, the bubble size in the froth image is larger, indicating that there are fewer minerals attached to the surface of the bubble, which ultimately leads to a lower concentrate grade and a higher tailing grade. When the dosage of reagent is set higher, the result is the opposite. Figure 7 shows the different concentrate grades and tailing grades corresponding to different froth images.

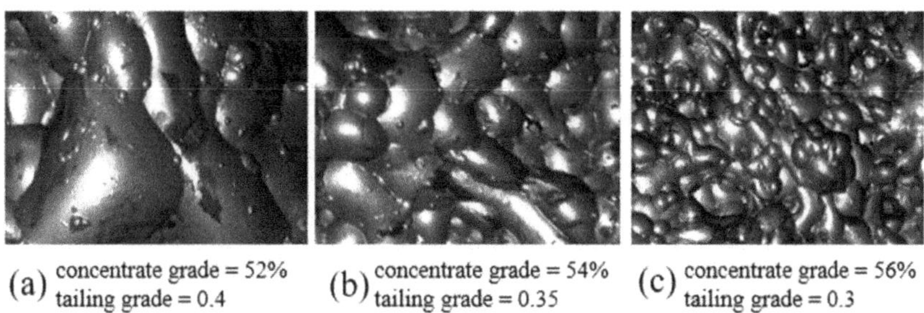

Figure 7. Typical froth image corresponds to concentrate grade and tailing grade.

In order to verify the intelligent setting method of reagent dosage, we collected 7 consecutive days of industrial data from a zinc flotation plant in China to train the relevant models in the proposed method. During the zinc flotation process, the operator recorded operating variables, including reagent dosage, concentrate grade, and tailing grade. The flotation monitoring system collects the froth image of the first zinc rougher bank in real time. The froth images at the first zinc rougher bank are captured at a rate of 1 frame/5 min; the concentrate grade and tailings grade are obtained using X-ray analysis instruments.

The design of ELMo's biLM consists of two bidirectional LSTMs. Each LSTM network layer consists of 512 cells, followed by a softmax layer. The initialization XGBoost parameters are as follows: boost tree depth max_depth = 5, learning rate learning_rate = 0.1, number of iterations n_estimators = 160.

Figure 8 shows the reagent dosage obtained by simulation using the proposed method, and the real reagent dosage.

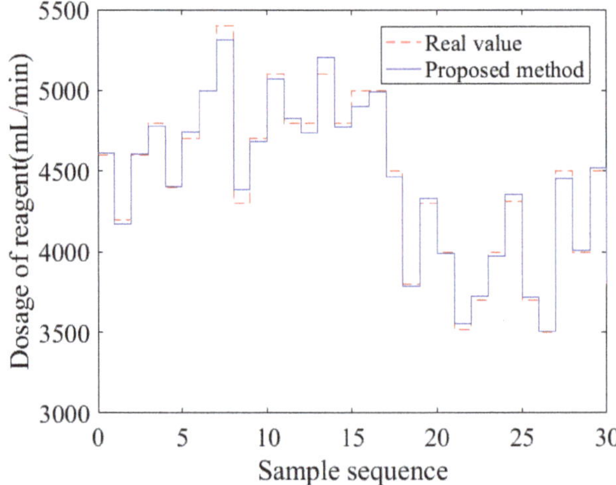

Figure 8. The proposed method is used to simulate the set value and real value of the reagent dosage.

To verify the feasibility of the proposed method, we applied it to a zinc flotation plant in China and recorded data in real time. According to the duty arrangement of the flotation plant, it is divided into three shifts of the morning shift, noon shift and night shift, each shift is 8 h. During each shift we recorded the relevant data and statistics. Figure 9 shows the set value of the reagent dosage for three shifts.

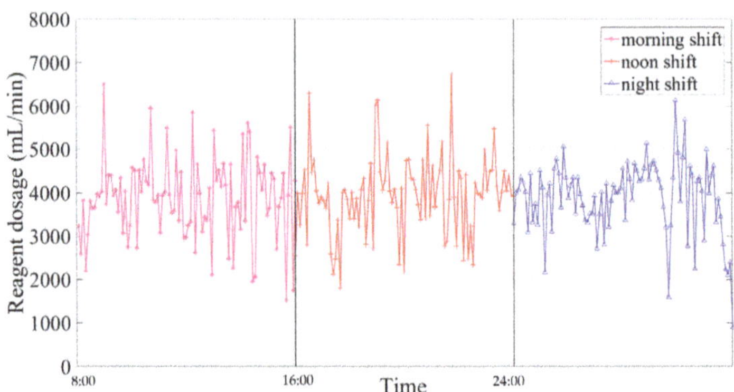

Figure 9. Using the proposed method for the dosage of reagent during the experiment.

According to further statistics, the total consumption of reagents on that day was 5620.32 L. Figures 10 and 11 show the test results of concentrate grade and tailing grade for three shifts.

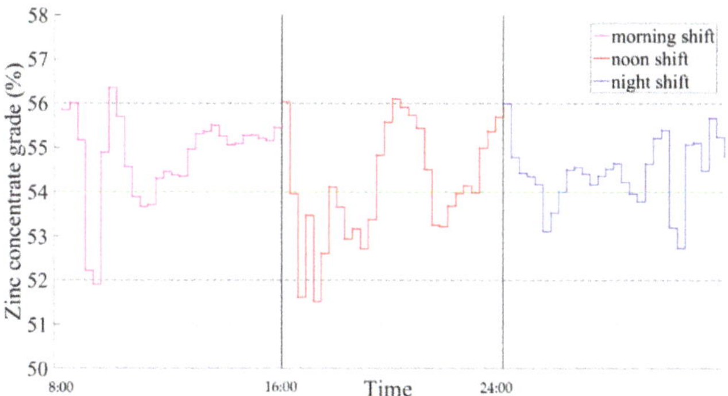

Figure 10. Using the proposed method for the concentrate grade during the experiment.

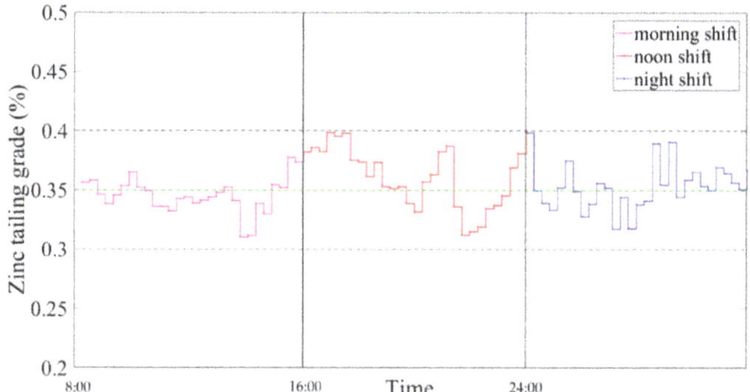

Figure 11. Using the proposed method for the tailing grade during the experiment.

According to the requirements of China's zinc flotation plant, it is necessary to stabilize the concentrate grade at about 54% and the tailings grade at about 0.35%. The results show that the proposed method can meet the production requirement.

The traditional dosage of reagents in the flotation process is determined by manual experience. It is worth noting that manual experience is different due to individual differences. Because the flotation plant is divided into three shifts in the morning, middle and evening, different people work. Figures 12 and 13 show the changes of concentrate grade and tailings grade from 21 to 23 March 2020 using the proposed method and manual method.

Table 1 summarizes the performance index evaluation parameters of the proposed method and manual method, where *Mean* represents the average and σ^2 represents the variance.

Table 1. Performance index evaluation parameters of the proposed method and manual method.

	Proposed Method		Manual Method	
	Mean (%)	σ^2 (%)	*Mean* (%)	σ^2 (%)
Concentrate grade	54.37	1.14×10^{-2}	54.39	1.14×10^{-2}
Tailing grade	0.36	7.35×10^{-6}	0.37	1.98×10^{-5}

The method proposed in this paper does not require manual participation, which reduces the employment cost of the plant. In addition, the experimental results show that the proposed method has a better control effect on the concentrate grade and tailings grade than the manual method.

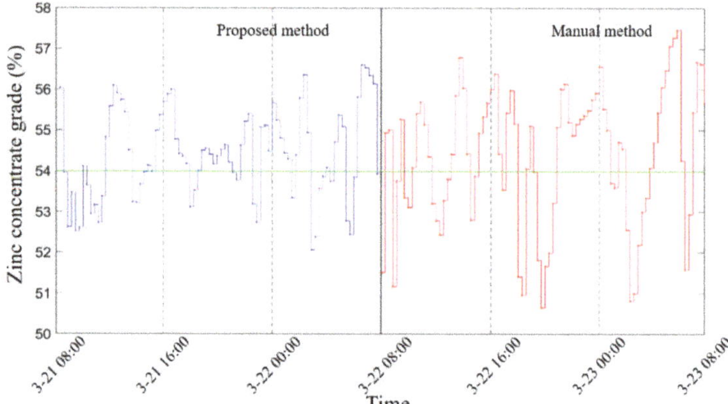

Figure 12. Comparison of change in concentrate grade by proposed method and manual method.

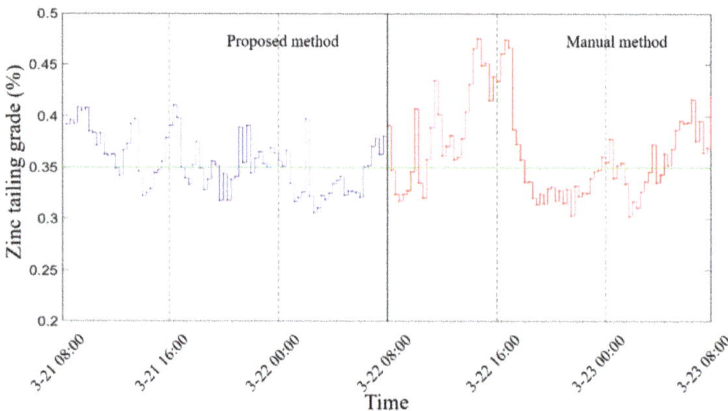

Figure 13. Comparison of change in concentrate grade by proposed method and manual method.

6. Conclusions

In this work, a method for intelligently setting the reagent dosage based on the time series froth image in the zinc flotation process is proposed. In this paper, the bubble size CDF is used as a new feature of the froth image, and the estimation method of bubble size CDF is proposed. Because the change of reagent dosage will cause the froth image features to change continuously with time, this paper uses the ELMo model to process the bubble size CDF time series, and generates dynamic feature vectors based on the bubble size CDF time series. Compared with the single frame froth image feature at a single moment, the dynamic feature vector can reduce the influence of noise in the flotation process. Finally, the efficient XGBoost algorithm was used to establish the nonlinear relationship model between the dynamic feature vectors and the dosage of reagent in the flotation process. The industrial experiment results show that the proposed method can stabilize the concentrate grade and tailing grade more than the traditional manual method in the zinc flotation process. In addition, the proposed method does not require manual participation, avoids inconsistent product quality due to differences in manual experience, and improves the economic benefits of the flotation plant.

Author Contributions: Z.T. performed data curation and provided the funding; L.T. conceived the methodology and wrote the original draft; G.Z. performed the review; Y.X. performed the formal analysis; J.L. provided research materials. All authors have read and agreed to the published version of the manuscript.

Funding: This research was funded by NSFC—Guangdong joint fund of key projects (No. U1701261), National Natural Science Foundation of China (No. 61771492).

Acknowledgments: The authors would like to acknowledge the research support from the NSFC—Guangdong joint fund of key projects (No. U1701261) and National Natural Science Foundation of China (No. 61771492).

Conflicts of Interest: The authors declare that the article will be reported without any conflict of interest.

References

1. Shean, B.J.; Cilliers, J.J. A review of froth flotation control. *Int. J. Miner. Process.* **2011**, *100*, 57–71.
2. Jovanovic, I.; Miljanovic, I.; Jovanovic, T. Soft computing-based modeling of flotation processes—A review. *Miner. Eng.* **2015**, *84*, 34–63.
3. Gao, Z.; Saxen, H.; Gao, C. Guest Editorial: Special section on data-driven approaches for complex industrial systems. *IEEE Trans. Ind. Inform.* **2013**, *9*, 2210–2212. [CrossRef]
4. Popli, K.; Afacan, A.; Liu, Q.; Prasad, V. Real-time monitoring of entrainment using fundamental models and froth images. *Miner. Eng.* **2018**, *124*. [CrossRef]
5. Gao, Z.; Nguang, S.K.; Kong, D. Advances in Modelling, Monitoring, and Control for Complex Industrial Systems. *Complexity* **2019**. [CrossRef]
6. Hu, B.; Yang, J.; Li, J. Intelligent Control Strategy for Transient Response of a Variable Geometry Turbocharger System Based on Deep Reinforcement Learning. *Processes* **2019**, *7*, 601. [CrossRef]
7. Liu, W.; Luo, F.; Liu, Y.; Ding, W. Optimal Siting and Sizing of Distributed Generation Based on Improved Nondominated Sorting Genetic Algorithm II. *Processes* **2019**, *7*, 955. [CrossRef]
8. Li, Z.M.; Gui, W.H. The Method of Reagent Control Based on Time Series Distribution of Bubble Size in a Gold-Antimony Flotation Process. *Asian J. Control* **2018**, *20*, 2223–2236.
9. Jiang, Y.; Fan, J.L.; Chai, T.Y.; Li, J.; Lewis, F.L. Data-Driven Flotation Industrial Process Operational Optimal Control Based on Reinforcement Learning. *IEEE Trans. Ind. Inform.* **2018**, *14*, 1974–1989. [CrossRef]
10. Zou, Y.L.; Yang, C.H.; Wang, X.L.; Zhao, L.; Xie, Y. Reagent Feed Intelligent Setting Based on Belief Rule Base for Antimony Roughing Flotation. In Proceedings of the 28th Chinese Control and Decision Conference, Yinchuan, China, 28–30 May 2016; pp. 848–853.
11. Xie, Y.F.; Wu, J.; Xu, D.G.; Yang, C.; Gui, W. Reagent Addition Control for Stibium Rougher Flotation Based on Sensitive Froth Image Features. *IEEE Trans. Ind. Electron.* **2017**, *64*, 4199–4206. [CrossRef]
12. Cao, B.F.; Xie, Y.F.; Gui, W.H.; Yang, C.; Li, J. Coordinated optimization setting of reagent dosages in roughing-scavenging process of antimony flotation. *J. Cent. South Univ.* **2018**, *25*, 95–106. [CrossRef]
13. Li, H.; Chai, T.; Zhang, L. Hybrid Intelligent Optimal Control for flotation processes. In Proceedings of the 2012 American Control Conference, Montreal, QC, Canada, 27–29 June 2012; pp. 4891–4896.
14. Aldrich, C.; Marais, C.; Shean, B.J.; Cilliers, J.J. Online monitoring and control of froth flotation systems with machine vision: A review. *Int. J. Miner. Process.* **2010**, *96*, 1–13. [CrossRef]
15. Liu, J.; He, J.; Xie, Y.; Gui, W.; Tang, Z.; Ma, T.; He, J.; Niyoyita, J.P. Illumination-Invariant Flotation Froth Color Measuring via Wasserstein Distance-Based CycleGAN With Structure-Preserving Constraint. *IEEE Trans. Cybern.* **2020**, *99*, 1–14. [CrossRef] [PubMed]
16. Liu, J.; Zhou, J.; Tang, Z.; Gui, W.; Xie, Y.; He, J.; Ma, T.; Niyoyita, J.P. Toward Flotation Process Operation-State Identification via Statistical Modeling of Biologically Inspired Gabor Filtering Responses. *IEEE Trans. Cybern.* **2019**, *99*, 1–14. [CrossRef] [PubMed]
17. Bournival, G.; Ata, S.; Jameson, G.J. Bubble and Froth Stabilizing Agents in Froth Flotation. *Miner. Process. Extr. Met. Rev.* **2017**, *38*, 366–387. [CrossRef]
18. Hosseini, M.R.; Shirazi, H.H.A.; Massinaei, M.; Mehrshad, N. Modeling the Relationship between Froth Bubble Size and Flotation Performance Using Image Analysis and Neural Networks. *Chem. Eng. Commun.* **2015**, *202*, 911–919. [CrossRef]
19. Riquelme, A.; Desbiens, A.; Del Villar, R.; Maldonado, M. Predictive control of the bubble size distribution in a two-phase pilot flotation column. *Miner. Eng.* **2016**, *89*, 71–76. [CrossRef]

20. Xie, Y.F.; Cao, B.F.; He, Y.P.; Gui, W.; Yang, C. Reagent dosages control based on bubble size characteristics for flotation process. *IET Control Theory Appl.* **2016**, *10*, 1404–1411. [CrossRef]

21. Jinping, L.; Jiezhou, H.; Wuxia, Z.; Xu, P.; Tang, X. TCvBsISM: Texture Classification via B-splines-based Image Statistical Modeling. *IEEE Access* **2018**, *6*, 44876–44893.

22. Zhu, J.Y.; Gui, W.H.; Yang, C.H.; Liu, J.; Tang, Y. Probability Density Function of Bubble Size Based Reagent Dosage Control for Flotation Process. *Asian J. Control* **2014**, *16*, 765–777. [CrossRef]

23. Zhaohui, T.; Heng, L.; Jinping, L.; Gui, W. Component based software design and development of froth flotation visual monitoring system. In Proceedings of the 2013 25th Chinese Control and Decision Conference (CCDC), Guiyang, China, 25–27 May 2013.

24. Jahedsaravani, A.; Massinaei, M.; Marhaban, M.H. An image segmentation algorithm for measurement of flotation froth bubble size distributions. *Measurement* **2017**, *111*, 29–37. [CrossRef]

25. Tripathi, Y.M.; Kayal, T.; Dey, S. Estimation of the PDF and the CDF of exponentiated moment exponential distribution. *Int. J. Syst. Assur. Eng. Manag.* **2017**, *8*, 1282–1296. [CrossRef]

26. Peters, M.; Neumann, M.; Iyyer, M.; Gardner, M.; Clark, C.; Lee, K.; Zettlemoyer, L. Deep Contextualized Word Representations. In Proceedings of the 2018 Conference of the North American Chapter of the Association for Computational Linguistics: Human Language Technologie, New Orleans, LA, USA, 1–6 June 2018.

27. Chen, T.; Guestrin, C.; Assoc Comp, M. XGBoost: A Scalable Tree Boosting System. In Proceedings of the 22nd ACM SIGKDD International Conference on Knowledge Discovery and Data Mining, San Francisco, CA, USA, 13–17 August 2016.

Article

Load State Identification Method for Ball Mills Based on Improved EWT, Multiscale Fuzzy Entropy and AEPSO_PNN Classification

Gaipin Cai *, Xin Liu, Congcong Dai and Xiaoyan Luo

School of Mechanical and Electrical Engineering, Jiangxi University of Science and Technology, Ganzhou 341000, China; zhuningyuan@jxust.edu.cn (X.L.); Daicongc@163.com (C.D.); LXY9416@163.com (X.L.)
* Correspondence: cgp4821@163.com; Tel.: +86-1397-970-8339

Received: 3 September 2019; Accepted: 8 October 2019; Published: 11 October 2019

Abstract: To overcome the difficulty of accurately determining the load state of a wet ball mill during the grinding process, a method of mill load identification based on improved empirical wavelet transform (EWT), multiscale fuzzy entropy (MFE), and adaptive evolution particle swarm optimization probabilistic neural network (AEPSO_PNN) classification is proposed. First, the concept of a sliding frequency window is introduced based on EWT, and the adaptive frequency window EWT algorithm, which is used to decompose the vibration signals recorded under different load states to obtain the intrinsic mode components, is proposed. Second, a correlation coefficient threshold is used to select the sensitive mode components that characterize the state of the original signal for signal reconstruction. Finally, the MFE of the reconstructed signal is used as the characteristic vector to characterize the load state of the mill, and the partial mean value of MFE is calculated. The results show that the mean value of MFE under different load states varies. To further identify the load state, a characteristic mill load vector is constructed from the MFE values of the reconstructed signal under different load conditions and is used as the input of the AEPSO_PNN model, which then outputs the predicted ball mill load state. Thus, a load state identification model is established. The feasibility of the method is verified based on grinding experiments. The results show that the overall recognition rate of the proposed method is as high as 97.3%. Compared with the back propagation (BP) neural network, Bayes discriminant method, and PNN classification, AEPSO_PNN classification increases the overall recognition rate by 8%, 5.3%, and 3.3%, respectively, which indicates that this method can be used to accurately identify the different load states of a ball mill.

Keywords: load identification; EWT; multiscale fuzzy entropy; PNN

1. Introduction

As the main type of mechanical equipment used for ore grinding, ball mills are widely used in the beneficiation process in mining operations [1]. It is imperative but challenging to develop effective modeling, monitoring, and control techniques for complex industrial systems [2–4]. Due to their complexity, it is difficult to investigate the internal charge dynamics of ball mills. Energy consumption is obviously related to rotational speed and mill load, and scholars have examined the influence of rotational speed on the energy consumption of mills and achieved good results [5]. For the mill load, it is important to be able to quickly and accurately identify the internal load of a ball mill to ensure that the mill is operating under the best possible working conditions, not only to reduce energy consumption during mineral processing, but also to ensure high grinding efficiency and output [6,7]. Therefore, a method of increasing the load recognition rate for ball mills would have great application value for improving the stability and economic benefits of the grinding process, and efforts to develop such methods have attracted the attention of many scholars at home and abroad [8,9]. To this end, studies have shown that the vibration signal generated by a ball mill during the grinding process is correlated with the load [10].

The vibration signal of a ball mill is nonlinear and nonstationary. Currently, the most widely used methods for processing such signals include the wavelet packet algorithm, empirical mode decomposition (EMD), variable mode decomposition (VMD), local mean decomposition (LMD), and the complete integrated empirical decomposition algorithm (CEEMDAN) [11–14]. Liu et al. [15] combined the EMD algorithm with principal component analysis (PCA) to extract the vibration signal from the cylinder of a wet ball mill. The results showed that this method can distinguish among different load states, but that the recognition rate requires improvement. Tang et al. [16,17] reported a method of extracting the vibration signal characteristics of a ball mill based on ensemble empirical mode decomposition (EEMD) and interval partial least squares (iPLS) modeling and extended this method to the study of ball mill sound signals. Although the signal features were successfully extracted, there was residual noise in the intrinsic mode functions (IMFs) after decomposition, and white noise with a different amplitude was added each time. Although the above methods can be used to successfully extract signal features, they face problems related to noise residuals and computational burden. Therefore, the key to mill load identification is to find an effective method of extracting the characteristic information of the vibration signal of the ball mill cylinder. The proposed empirical wavelet transform (EWT) algorithm effectively compensates for the above shortcomings. This algorithm not only suppresses the modal aliasing problem and reduces residual noise, but also improves the completeness of decomposition. In reference [18], the EMD, EEMD, and EWT algorithms were compared and analyzed. The EWT algorithm was found to have the best processing effect. Specifically, the EWT algorithm had a better processing speed and better ability to extract modal component signals than the other algorithms. However, in practical engineering, especially under the harsh working conditions of a ball mill, the Fourier spectrum of the EWT segmentation signal easily encounters interference from background noise and must be further improved. In this paper, the adaptive frequency window is used to improve EWT. Compared with traditional EWT and other signal processing algorithms, the denoising effect is more significant.

In recent years, many nonlinear dynamic methods, such as multiscale entropy (MSE), singular value entropy (SVE), permutation entropy (PE), and fuzzy entropy (FE), have been widely used for fault diagnosis, classification, and recognition because of their good performance in terms of feature extraction [19–21]. Miao Y et al. [22] applied SVE to the identification of the optimal frequency band. Zhao L et al. [23] completed the fault diagnosis of a gearbox using PE optimization and modified the modal decomposition algorithm. Chang J L et al. [24] applied MSE for load recognition in machine tools. Liu H et al. [25] reported an example of MSE applied for the fault diagnosis of rolling bearings, but the recognition accuracy required further improvement. To diagnose the problem of rolling bearing faults, Zheng H D et al. [26] adopted the method of multiscale fuzzy entropy (MFE), which effectively overcame the defect in the MSE mutation, and the diagnosis result was improved. Compared with the above methods, MFE has some advantages for feature extraction because of its unique performance and ability to accurately reflect the feature information of the original signal.

As a tool for recognition and classification, an artificial neural network is a model abstracted based on neural network theory that originates from the field of physiology. Such models can be used for arbitrary data clustering and pattern classification and are widely used for tasks such as pattern recognition [27–29]. Specifically, a probabilistic neural network (PNN) is an artificial neural network with the advantages of a fast training speed, simple parameter adjustment, and good classification performance [30]. However, the classification effect of a probabilistic neural network is greatly influenced by the smoothing parameter σ, and if the selection of σ is not appropriate, then inaccurate results may be obtained. To solve this problem, an adaptive evolutionary particle swarm optimization (AEPSO) algorithm is proposed in this paper to optimize the smoothing parameters in a probabilistic neural network (PNN) so that the optimized network can identify the load state of a ball mill. In this paper, the AEPSO algorithm is used to improve the PNN clustering method; compared with the traditional PNN clustering method and other clustering methods, it has the advantages of high speed and high accuracy.

Considering the nonstationary and nonlinear characteristics of the vibration signal from the cylinder of a ball mill, a load identification method for ball mills is proposed in this study based on improved EWT, MFE, and AEPSO_PNN classification. First, the vibration signals are decomposed using improved EWT, and the mode components of the reconstructed signals are selected using a correlation coefficient threshold. Then, the load state of the ball mill is determined based on the magnitude of the calculated MFE. Finally, AEPSO_PNN is used for learning and classification to enable the recognition of a different load state.

2. Principles of the Load State Identification Method

2.1. Principles of Improved EWT

2.1.1. Principles of EWT

EWT is a widely used method for the adaptive segmentation of signals [31]. The segmentation principle involves adaptively segmenting the Fourier spectrum by marking maximum points in the frequency domain, and a set of bandpass filters suitable for processing signals is constructed in the frequency domain to extract amplitude modulation and frequency modulation (AM-FM) components from the Fourier spectrum.

The Fourier axis $[0, \pi]$ is divided into n consecutive parts, that is, $\Lambda_n = [\omega_{n-1}, \omega_n] (\omega_0 = 0, \omega_n = \pi)$, where ω_n is the boundary point between two parts and the corresponding value is the minimum between the two adjacent maximum values in the Fourier spectrum of the signal. Figure 1 [32] shows the division diagram of the Fourier axis. In the figure, ω_n is defined as the center point of Λ_n. Then, a transition region with a width of $T_n = 2\tau_n$ is obtained.

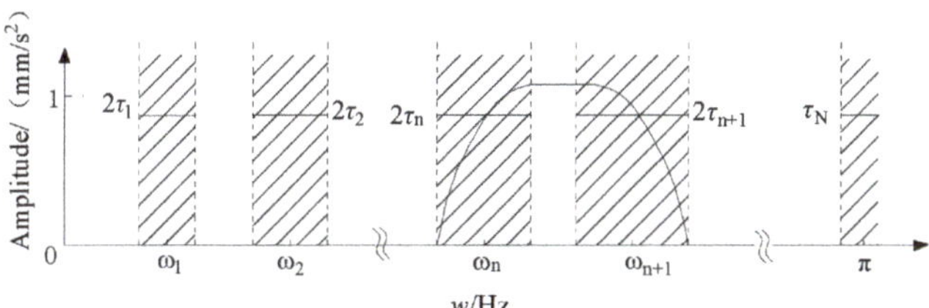

Figure 1. Region segmentation diagram of the Fourier axis.

Referring to the wavelet construction method of Littlewood–Paley and Meyer, the empirical wavelet function is constructed. After Λ_n is determined, the empirical wavelet is used as a bandpass filter. The formulas of the empirical wavelet function $\hat{\psi}_n(\omega)$ and the empirical scale function $\hat{\phi}_n(\omega)$ are as follows [33]:

$$\hat{\psi}_n(\omega) = \begin{cases} 1, & (|\omega| \leq (1-\gamma)\omega_n) \\ \cos\left\{\frac{\pi}{2}\beta\left[\frac{1}{2\gamma\omega_n}(|\omega| - (1-\gamma)\omega_n)\right]\right\}, & (1-\gamma)\omega_n \leq \omega \leq (1+\gamma)\omega_n \\ 0, & (\text{others}) \end{cases} \tag{1}$$

$$\hat{\phi}_n(\omega) = \begin{cases} 1, & (|\omega| \leq (1-\gamma)\omega_n) \\ \cos\left\{\frac{\pi}{2}\beta\left[\frac{1}{2\gamma\omega_n}(|\omega| - (1-\gamma)\omega_n)\right]\right\}, & (1-\gamma)\omega_n \leq \omega \leq (1+\gamma)\omega_n \\ 0, & (\text{others}) \end{cases} \tag{2}$$

where

$$\begin{aligned} \beta(x) &= x^4(35 - 84x + 70x^2 - 20x^3) \, ; \\ \tau_n &= \gamma\omega_n \qquad \gamma < \min_n\left(\frac{\omega_{n+1} - \omega_n}{\omega_{n+1} + \omega_n}\right) \end{aligned} \tag{3}$$

After the EWT, the approximation coefficient $W_{f1}(0,t)$ and the detail coefficient $W_{f2}(n,t)$ can be expressed as follows.

$$\begin{aligned} W_{f1}(0,t) \quad &=< x, \phi_1 >= \int x(\tau)\,\overline{\phi_1(\tau-t)}\,\mathrm{d}\tau \\ &= F^{-1}[x(\omega)\hat{\phi}_1(\omega)] \end{aligned} \tag{4}$$

$$\begin{aligned} W_{f2}(n,t) \quad &=< x, \psi_n >= \int x(\tau)\,\overline{\psi_n(\tau-t)}\,\mathrm{d}\tau \\ &= F^{-1}[x(\omega)\hat{\psi}_n(\omega)] \end{aligned} \tag{5}$$

Then, the functional expression of the reconstructed original signal a is as follows:

$$\begin{aligned} x(t) &= W_{f1}(0,t) * \varphi_1(t) + \sum_{n=1}^{N} W_{f2}(n,t) * \psi_n(t) \\ &= F^{-1}\left[\hat{W}_f(n,\omega)\varphi_1(\omega) + \sum_{n=1}^{N} \hat{W}_f(n,\omega)\hat{\psi}_n(\omega)\right] \end{aligned} \tag{6}$$

where "$*$" is a convolution operation and $\hat{W}_{f1}(0,\omega)$ and $\hat{W}_{f2}(n,\omega)$ are the Fourier transforms of the approximate coefficient $W_{f1}(0,t)$ and the detail coefficient $W_{f2}(n,t)$, respectively. Finally, the signal f is decomposed into the sum of several single component signals.

$$x(t) = \sum_{k=0}^{N-1} x_k(t) \tag{7}$$

2.1.2. Principle of the Adaptive Frequency Window EWT Algorithm

The division rules of the spectral boundaries of the traditional EWT algorithm are determined by the frequency domain extreme points, but a ball mill is vulnerable to strong noise, resulting in the disorderly arrangement of frequency domain extreme points. Considering these deficiencies, this paper uses the adaptive frequency window EWT to divide the spectral boundaries, as shown in Figure 2.

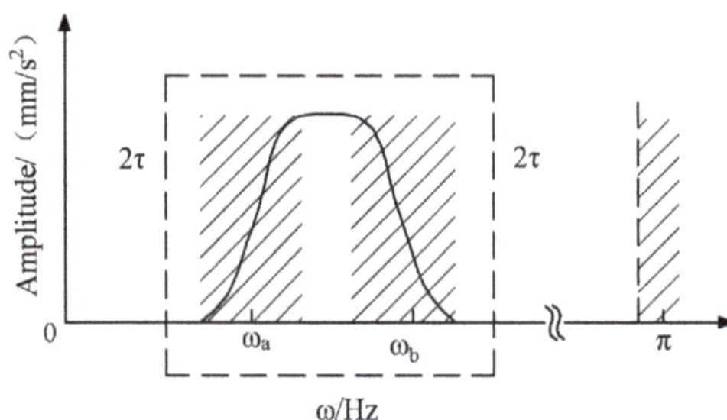

Figure 2. Diagram of empirical wavelet transform (EWT) spectral boundary division with an adaptive frequency window.

In Figure 2, the frequency window is represented as $[\omega_a, \omega_b]$, where ω_a, ω_b is the central frequency of the lower cutoff band of the window. The shaded area represents the transition region of the segmented portion of the spectrum with width 2τ. The range of the d support interval is $[0, \pi]$. The frequency window can slide freely in the interval, and the width range is adaptively variable.

After the frequency window segmentation is improved, Equations (1) and (5) are modified as follows.

$$\hat{\psi}_n(\omega) = \begin{cases} 1, & (\omega_a + \tau \le |\omega| \le \omega_b - \tau) \\ \cos\left\{\frac{\pi}{2}\beta\left[\frac{1}{2\tau}(|\omega| - \omega_b + \tau)\right]\right\}, & (\omega_b - \tau \le |\omega| \le \omega_b + \tau) \\ \sin\left\{\frac{\pi}{2}\beta\left[\frac{1}{2\tau}(|\omega| - \omega_a + \tau)\right]\right\}, & (\omega_a - \tau \le |\omega| \le \omega_a + \tau) \\ 0, & (\text{others}) \end{cases} \tag{8}$$

$$\begin{aligned} W'(t) &= < x, \psi >= \int x(\tau)\, \overline{\psi(\tau - t)}\, d\tau \\ &= F^{-1}[x(\omega)\hat{\psi}(\omega)] \end{aligned} \tag{9}$$

Additionally, Equation (8) must be satisfied as follows.

$$\begin{cases} \beta(x) = x^4(35 - 84x + 70x^2 - 20x^3) \\ \tau = \gamma\omega_a \qquad \gamma < (\omega_b - \omega_a)/(\omega_b + \omega_a) \end{cases} \tag{10}$$

Therefore, the modal component signal can be reconstructed as follows:

$$x^*(t) = W'(t) * \psi(\omega) = F^{-1}\left[\hat{W}(\omega)\hat{\psi}(\omega)\right] \tag{11}$$

where $\hat{\psi}(\omega)$ is a Fourier transform of $\psi(\omega)$ and $x^*(t)$ is an AM–FM component signal for improving EWT extraction.

2.1.3. Simulation and Comparative Analysis of Improved EWT

To verify the ability of the improved EWT method to extract the feature components of the signal, a simulation with the improved EWT approach is performed, and the results are compared with those of the traditional EWT. The simulation signal $x(t)$ is constructed as follows in Equation (12):

$$\begin{cases} x_1(t) = 2t^2 \\ x_2(t) = 1.1\sin(34\pi t) \\ x_3(t) = \begin{cases} 0.7\cos(56\pi t) & 0 < t < 0.5 \\ 0.8\cos(64\pi t) & t \ge 0.5 \end{cases} \\ x(t) = x_1(t) + x_2(t) + x_3(t) \end{cases} \tag{12}$$

where $x(t)$ is white noise, the signal-to-noise ratio (SNR) is set to 3, and $t \in [0, 1]$. Figure 3 shows the improved EWT and traditional EWT decomposition results for the simulation signal $x(t)$.

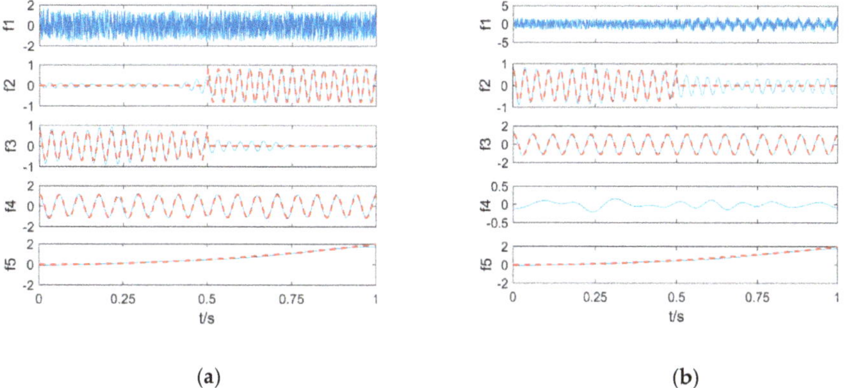

(a)

(b)

Figure 3. The improved EWT and traditional EWT decomposition results (red dotted lines represent the original signal; blue solid lines represent the decomposition results). (**a**) Improved EWT; (**b**) EWT.

In Figure 3, the components f2–f5 correspond to the signals $x_3(t)\sim x_1(t)$, respectively. Figure 3a shows that the noise contained in the signal is well decomposed by the improved EWT and that the degree of coincidence of each component is close to 90%. The two modes that originally belonged to the same component are decomposed because the two modal components obviously have distinct energy signals and can be regarded as two independent modes. Figure 3b shows that traditional EWT can decompose noise, but the components $x_1(t)$, $x_2(t)$, and $x_3(t)$ are deformed because the traditional EWT segmentation method is too simple. When analyzing local noise or nonstationary signals, some local maxima generated by noise and nonstationary components may appear and erroneously remain in the peak sequence, and some useful maxima may not be kept in the peak sequence, resulting in improper segmentation. The improved EWT uses the adaptive frequency window for spectrum segmentation, which can reduce the effects of noise and nonstationary components and greatly increase the reliability of spectrum segmentation.

This comparative study of simulated signals indicates that the improved EWT method can effectively detect the modal components in power spectra, extract components similar to the original signal components, and suppress modal aliasing. Thus, the decomposition effect of the improved EWT method is better than that of the traditional EWT method.

2.2. Principle of Multiscale Fuzzy Entropy

2.2.1. Principle of Fuzzy Entropy

FE is the probability of identifying a new pattern in a time series when the dimension changes, which reflects the complexity and irregularity of the time series. The larger the probability of the time series, the greater the FE value [34]. During the operation of a ball mill, the change in the load state will cause the characteristics of the vibration signal of a cylinder to change in an obvious manner, and FE can effectively characterize the state characteristics of the signal in each frequency band during the sampling time. Therefore, it is feasible to introduce FE as the characteristic parameter of the vibration signal of a ball mill cylinder. The algorithm steps are as follows.

1. The m-dimensional vector is obtained by processing the time series:

$$X_i^m = \{u(i), u(i+1), \ldots, u(i+m+1)\} - u_0(i)$$
$$u_0(i) = \frac{1}{m}\sum_{j=0}^{m-1} u(i+j) \qquad i = 1, 2, \ldots, i+m+1 \tag{13}$$

where X_i^m is the result of removing the mean $u_0(i)$ of the time series.

2. Calculate the maximum distance between X_i^m and X_j^m:

$$d_{ij}^m = d[X_i^m, X_j^m] = \max_{k \in (0, m-1)} \{|u(i+k) - u0(i) - (u(j+k) - u0(j))|\} \tag{14}$$

where $i, j = 1, 2, \cdots, N - m,\ i \neq j$.

3. The similarity between X_i^m and X_j^m is defined by a fuzzy function as follows:

$$D_{ij}^m = u(d_{ij}^m, n, r) = e^{-(d_{ij}^m/t)^n} \tag{15}$$

where $u(d_{ij}^m, n, r)$ is an exponential fuzzy membership function and n and r are the boundary gradients and widths of the fuzzy membership functions, respectively.

4. Define the functions as follows:

$$\phi^m(n, r) = \frac{1}{N-m}\sum_{i=1}^{N-m}\left(\frac{1}{N-m-1}\sum_{\substack{j=1 \\ j \neq i}}^{N-m} D_{ij}^m\right) \tag{16}$$

where $i, j = 1, 2, \cdots, N - m, \ i \neq j$.

5. The $m+1$ vector is constructed based on the above four steps.

$$\phi^{m+1}(n, r) = \frac{1}{N-m} \sum_{i=1}^{N-m} \left(\frac{1}{N-m-1} \sum_{\substack{j=1 \\ j \neq i}}^{N-m} D_{ij}^{m+1} \right) \tag{17}$$

6. The calculation formula of the FE value can be summarized as follows:

$$FuzzyEn(m, n, r) = \lim_{N \to \infty} \left[\ln \phi^m(n, r) - \ln \phi^{m+1}(n, r) \right] \tag{18}$$

where $i, j = 1, 2, \cdots, N - m, \ i \neq j$.

7. When N is limited, Equation (18) is transformed into the following formula.

$$FuzzyEn(m, n, r, N) = \ln \phi^m(n, r) - \ln \phi^{m+1}(n, r) \tag{19}$$

2.2.2. Principle of Multiscale Fuzzy Entropy

The characteristic frequency band and complexity of the vibration signal of a cylinder under different load conditions in a ball mill are different at different scales. Considering the FE of the vibration signal at different scales can improve the recognition accuracy, therefore the concept of multiple scales is introduced based on FE. The steps in the MFE algorithm are as follows.

1. Construct a new coarse granularity vector for the original time series $X_i = \{x_1, x_2, \cdots, x_n\}$ as follows:

$$y_j(\tau) = \frac{1}{\tau} \sum_{i=(j-1)\tau+1}^{j\tau} x_i \quad 1 \leq j \leq \frac{N}{\tau} \tag{20}$$

where $\tau = 1, 2, \cdots, n$ represents the scale factor. When $\tau = 1$, the coarse-grain vector is the original sequence. For a given τ, the original sequence is divided into coarse granularity vectors of length N/τ, and Figure 4 shows the coarse granularity process for $\tau = 2$ and $\tau = 3$.

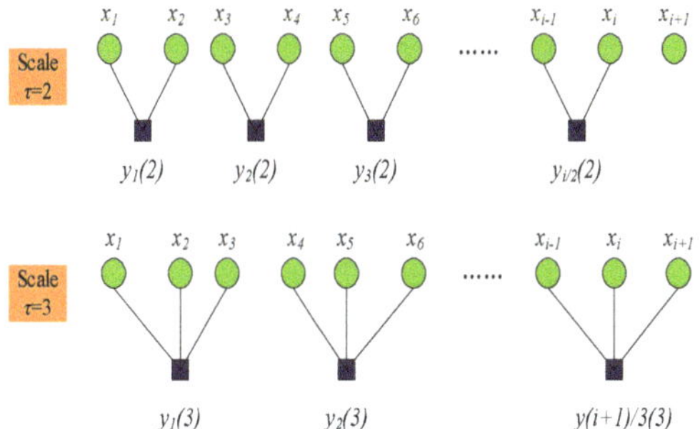

Figure 4. The coarse granularity process of multiscale fuzzy entropy (MFE).

2. The FuzzyEn of each coarse-grained sequence is determined by the standard deviation of the original sequence. The FuzzyEn value can be expressed as a function of the scale factor in MFE analysis.

2.2.3. Parameter Selection for MFE

According to the definition of MFE, the calculation of MFE is related to the embedding dimension m, similarity tolerance r, exponential function gradient n, and data length N. The selection rules are as follows.

1. A large embedding dimension m produces more information when the time series is dynamically reconstructed, and the data sequence $N = 10^m \sim 30^m$; thus, m is set to 2.

2. The similarity tolerance r represents the width of the boundary of the exponential function. If r is too large, then a large amount of statistical information will be lost, and if r is too small, then the sensitivity to noise will be high. r is usually set from 0.1 SD to 0.25 SD (SD denotes the standard deviation of the original time series). Considering the working characteristics of the ball mill, r is set to 0.15 SD.

3. n is a weighting factor in the calculation of FE vector similarity. A large n will result in a large gradient, but an overly small n will lead to the loss of detail. To obtain as much detailed information as possible, a small integer is usually used, and n is set to 2 in this case.

4. To obtain an accurate MFE calculation result, the data length N should be greater than $100\tau_{max}$. In addition, the maximum scale factor τ_{max} should also be considered when calculating the MFE, and the value of τ_{max} is usually between 10 and 20; thus, $a = 20$ is used in this study.

2.3. Principle of the AEPSO_PNN

2.3.1. PNN Principle

A PNN is a type of radial basis network that was first proposed by Dr. D.F. Speeht in 1989. The PNN is a supervised network classifier based on the Bayes minimum risk criterion [35]. As a feed-forward network, a PNN has the advantages of a fast training speed and simple parameter adjustment. Currently, PNNs are widely used in pattern classification [36]. Compared with other network classifiers, a PNN can not only guarantee real-time performance, but also produce classification and recognition results that are minimally influenced by complex parameter settings.

The signal sample vector can be represented as $X = [x_1, x_2, \ldots, x_i, \ldots, x_n]$ with states $Y = [y_1, y_2, \ldots, y_i, \ldots, y_n]$. Then, the prior probability, posterior probability, and class-specific probability density functions for each state can be represented by $P(y_i)$, $P(y_i/X)$, and $P(X/y_i)$, respectively. For a given identification target, $P(y_i)$ is a known parameter, and $P(X/y_i)$ can be estimated using the Parzen function. The corresponding formula is as follows:

$$P(X/y_i) = \frac{\sum_{j=1}^{N_i} \exp(-\frac{\|X-x_{ij}\|^2}{2\sigma^2})}{N_i(2\pi)^{\frac{d}{2}}\sigma^d} \tag{21}$$

where N_i is the total number of samples of the ith class, d is the dimensionality of the feature vector, x_{ij} is the jth sample of the ith class, and σ is the width of the Parzen function window, that is, the smoothing parameter.

The following formula is obtained from probabilistic and statistical theory.

$$P(y_i/X) = P(X/y_i)P(y_i)/P(X) \tag{22}$$

If the possibility of misjudgment is not considered, the Bayes rule can be expressed as follows.

$$\vee\, j \neq i = 1,2,3,\cdots,m, if\, P(y_i/X) > P(y_j/X), X \in y_j \tag{23}$$

However, because misjudgment can readily occur in real-world situations, it is necessary to introduce the risk coefficient λ_{ij}, yielding the following risk function R for the decision conditions.

$$R(y_i/X) = \sum_{j=1}^{m} \lambda_{ij}P(y_j/X) \tag{24}$$

In summary, the Bayes minimum risk criterion can be expressed as follows.

$$if\ R(y_i/X) > R(y_j/X), X \in y_j \tag{25}$$

In this paper, the minimum risk criterion is used as the basis of the feed-forward network that serves as the mill load state recognition model. By setting reasonable smoothing parameters, the network is trained on a set of sample feature vectors to estimate the probability densities of three distinct load states and enable the recognition of the mill load state.

2.3.2. Principle of AEPSO

The optimization speed and position updating formulas of the traditional particle swarm optimization algorithm [37] are as follows:

$$V_p^{k+1} = \omega V_p^k + c_1 r_1 (W_p - X_p^k) + c_2 r_2 (W_g - X_p^k) \tag{26}$$

$$X_p^{k+1} = X_p^k + V_p^{k+1} \tag{27}$$

where k is the number of iterations; ω is the inertial weight of the particle; c_1, c_2 are the learning factors of the particle, of which the former is the individual factor and the latter is the global factor; and r_1, r_2 are random numbers in the interval $[0, 1]$, which make the particles independent and diverse.

To address the nonlinear problem of ball mill load identification, the AEPSO algorithm introduces a nonlinear adaptive time-varying inertial weight.

$$\omega_t = \omega_{start} - (\omega_{start} - \omega_{end}) \times \exp(-\frac{1}{1 + 2t/t_{max}}) \tag{28}$$

For the learning factors c_1, c_2 of particles, the traditional particle colony algorithm usually sets $c_1 = c_2 = 2$, but this approach ignores the phase difference of the algorithm during training. The AEPSO algorithm adopts the strategy of managing the learning factor in segment, and the formula is as follows.

$$\begin{cases} c_1 = 2.5, c_2 = 1.5 & t < t_{max}/2 \\ c_1 = 1.5, c_2 = 2.5 & t \geq t_{max}/2 \end{cases} \tag{29}$$

To enhance the adaptability of particle swarm optimization after iteration, the AEPSO algorithm introduces the local search operator β in Equation (13). The revised formula is as follows:

$$X_p^{k+1} = X_p^k + \beta \times V_p^{k+1} \tag{30}$$

where $\beta = rand(\ )[rand(\ ) + 0.5]$ and $rand(\ )$ is a random number in $[0, 1]$.

2.3.3. Optimization of the PNN by AEPSO

The smoothing parameter σ in the PNN has a considerable influence on the training effect. The improper selection of the σ value makes it easy to misjudge the recognition of the mill load state. Therefore, this paper uses an AEPSO algorithm to optimize the smoothing parameters of the PNN so that the optimized network can effectively identify the state of the mill load. The specific steps in the algorithm are as follows.

1. The parameters of the PSO algorithm are initialized, the smoothing parameters σ of the PNN are used as the population particles, the number of iterations is set to 500, and a set of data (σ) is randomly generated as an initial parameter vector.
2. The training set samples are input, and the fitness function is used to calculate the fitness value. Then, the optimal individual fitness value and the global optimal fitness value of the group are traversed by comparing each particle (σ). Finally, the particles are adjusted.

3. After calculating each particle in the population, the termination condition is determined to be satisfied or not. If not, the state is updated according to the speed and position updating formula; then, the algorithm returns to step 2. Otherwise, the algorithm iterates until termination and outputs the search results.
4. The PNN model trained by the optimal parameter combination (σ) is used to classify the test sample set and output the target category.

The network structure of the AEPSO_PNN includes four parts: the input layer, the mode layer, the summation layer, and the output layer, as shown in Figure 4.

As Figure 5 shows, the training step of the load state identification model of a ball mill based on the AEPSO_PNN is as follows.

1. The input layer multiplies the received feature vector of the training sample by the weighting coefficient *Wj* and transmits the result to the mode layer for training. The number of neurons in this layer is the dimension of the feature vector.
2. The mode layer first uses the exponential function gj as the activation function. Then, the probability density of each neuron is determined, and finally, the result is transmitted to the summation layer.
3. The probability density is the weighted average of the summation, and the resulting estimated probability density is transmitted to the output layer.
4. Based on the Bayes minimum risk criterion, the output layer selects the category with the largest posterior probability as the final classification result of the sample.

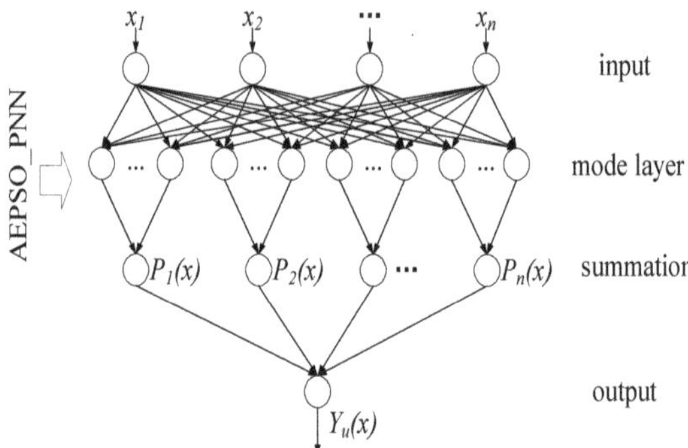

Figure 5. Adaptive evolution particle swarm optimization probabilistic neural network (AEPSO_PNN) network structure diagram.

3. Design of the Load State Identification Method for a Ball Mill

Based on the research on the EWT algorithm, MFE theory, and the PNN clustering algorithm combined with the characteristics of ball mill vibration signals, a feature extraction algorithm for vibration signals is proposed based on modified EWT, MFE, and AEPSO_PNN classification. The specific steps in the algorithm are as follows.

1. Decompose the recorded vibration signal from the cylinder of the ball mill via the adaptive frequency window EWT algorithm to obtain $AM - FM_i (i = 1, 2, \cdots, n)$.
2. Calculate the correlation coefficients for all $AM - FM_i$ components and the original signal in accordance with Equation (31), and select the sensitive $AM - FM$ components based on the threshold given in Equation (32).

$$\rho_{xy} = \frac{\sum\limits_{i=1}^{N} (x_i - \overline{x})(y_i - \overline{y})}{\sqrt{\sum\limits_{i=1}^{N} (x_i - \overline{x})^2} \sqrt{\sum\limits_{i=1}^{N} (y_i - \overline{y})^2}} \tag{31}$$

The correlation coefficient threshold is calculated as

$$\mu_h = \frac{\max(\mu_i)}{10 \times \max(\mu_i) - 3} \tag{32}$$

where μ_h is the threshold, μ_i is the correlation coefficient between the ith AM-FM component and the original signal, and max is the maximum correlation coefficient value. Each AM-FM component for which the value of the correlation coefficient with the original signal is greater than the threshold μ_h is retained as a sensitive AM-FM component. Each AM-FM component for which the correlation coefficient is smaller than the threshold μ_h is removed as a spurious component.

3. The sensitive AM-FM components are used to obtain the reconstructed vibration signals of different loads in the ball mill.
4. The MFE of the reconstructed vibration signal is calculated, and the result is used as the characteristic vector for the load classification of the ball mill.
5. The characteristic vector matrix is used as the input of AEPSO_PNN, and the load state is used as the output. Then, the load state of the mill is identified.

Thus, the overall flow of the ball mill load identification model that is proposed in this paper based on the modified EWT, MFE, and AEPSO_PNN classification methods can be summarized as shown in Figure 6.

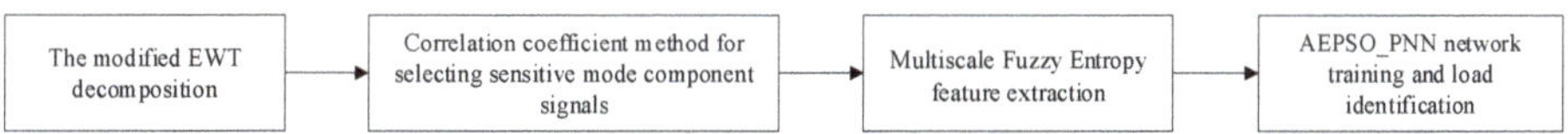

Figure 6. Algorithm flow based on the modified EWT, MFE, and AEPSO_PNN classification methods.

4. Experimental Analysis of Mill Load State Recognition

4.1. Data Collection

To verify the method proposed in this paper, a grinding experiment was performed using a $\Phi305 \times 305$ mm Bond index experimental ball mill. The experimental device is shown in Figure 7. The material used in the experiment was tungsten ore from a mine in Jiangxi, China, with a Protodyakonov scale of hardness of 14–18, a density of 1.8 t/m^3, and five grades of particle sizes: 1–3 mm, 3–6 mm, 6–9 mm, 9–11 mm, and >11 mm. The experimental parameters considered were the fill rate, powder-to-ball ratio, and grinding concentration. The vibration signal acquisition system of the mill cylinder consisted of a DH5922N dynamic data acquisition instrument and a DH131 acceleration sensor, which were used to collect the signals of various load parameters under three different load conditions. According to the literature, the mill load was divided into the following states: the underloaded state, corresponding to a fill rate of 10–20%; the normal load state, corresponding to a fill rate of 20–40%; and the overloaded state, corresponding to a fill rate of 40–60% [38].

Figure 7. Experimental device.

4.2. Decomposition of the Cylinder Vibration Signal

First, we present the typical working conditions corresponding to the three load conditions considered in this analysis: working condition 1 (a fill rate of 10%, a powder-to-ball ratio of 0.4, and a grinding concentration of 0.5), working condition 2 (a fill rate of 30%, a powder-to-ball ratio of 0.6, and a grinding concentration of 0.5), and working condition 3 (a fill rate of 50%, a powder-to-ball ratio of 0.8, and a grinding concentration of 0.5). The waveforms of the cylinder vibration signals recorded under these three working conditions are shown in Figure 8.

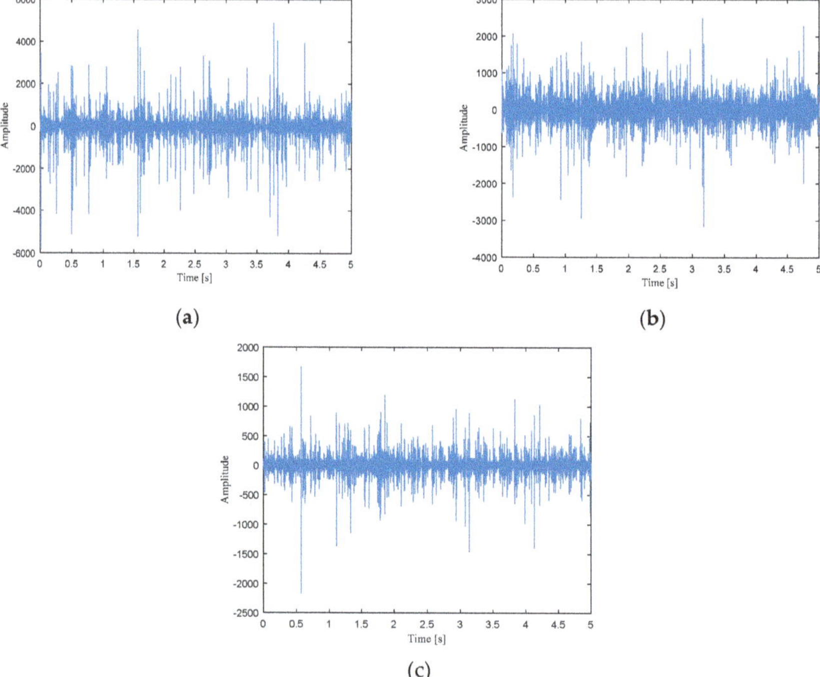

Figure 8. Waveforms of the original cylinder vibration signals: (**a**) working condition 1; (**b**) working condition 2; (**c**) working condition 3.

As Figure 8 shows, there is a large amount of noise in the vibration signal from the mill cylinder in all three load states, which makes it difficult to effectively extract feature information. To extract the characteristics of the vibration signal of the cylinder, the original signal must be preprocessed. The preprocessing steps are as follows.

1. The improved EWT algorithm was used to adaptively decompose the original signals under three typical working conditions. Then, 10 AM-FM components were obtained.
2. The correlation coefficients between the AM-FM components and the original cylinder vibration signal were calculated using Equation (29), and the threshold values were then calculated in accordance with Equation (30), yielding the following results: 0.21437 for working condition 1, 0.23872 for working condition 2, and 0.19905 for working condition 3. The correlation coefficient values and the threshold values of the vibration signals from the cylinder body under the three working conditions are shown in Figure 9.

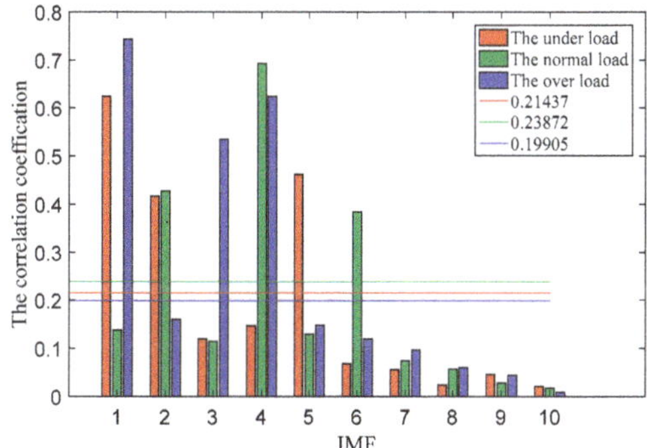

Figure 9. Relationship between the correlation coefficient and the sequence number of the amplitude modulation–frequency modulation (AM-FM) component.

As shown in Figure 9, the correlation coefficients between the AM-FM1, AM-FM2, and AM-FM5 components and the original signal for working condition 1 were greater than the threshold value of 0.21437. Thus, these components were identified as sensitive AM-FM components that characterize the vibration signal of the cylinder. For working condition 2, the AM-FM2, AM-FM4, and AM-FM6 components, with correlation coefficients greater than the threshold value of 0.23872, were selected as the sensitive mode components. For working condition 3, the AM-FM1, AM-FM3, and AM-FM4 components, with correlation coefficients greater than the threshold value of 0.23872, were selected as the sensitive mode components. All AM-FM components with correlation coefficients smaller than the corresponding threshold were removed.

3. The selected sensitive modal components are reconstructed, and the results are shown in Figure 10.

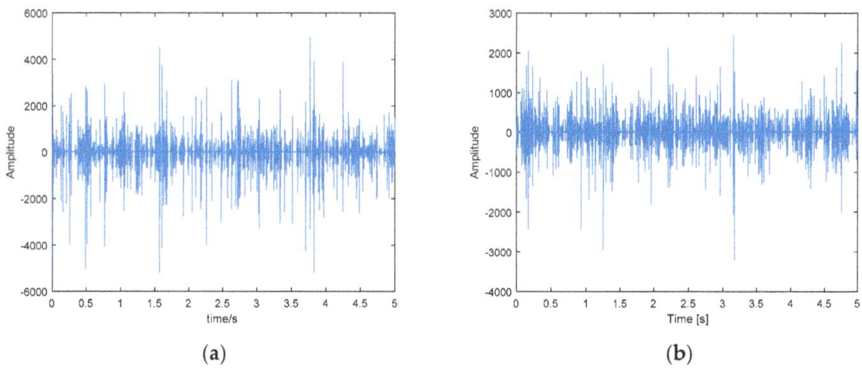

(a)   (b)

Figure 10. *Cont.*

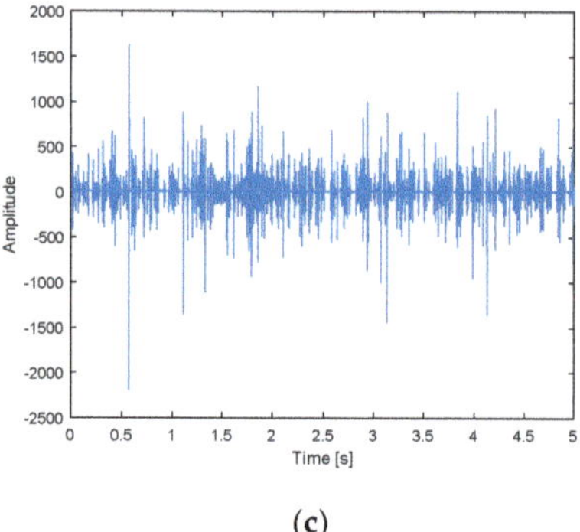

(c)

Figure 10. Waveforms of the reconstructed cylinder vibration signals: (**a**) working condition 1; (**b**) working condition 2; (**c**) working condition 3.

Based on a comparison of Figures 8 and 10, the trend of the reconstructed signal waveform is basically the same as that of the original signal. Compared with the original signal, the impact profile of the reconstructed signal curve is obviously distinct, but it preserves the characteristic information of the original signal while effectively denoising the signal. To further quantitatively highlight the preprocessing effect in this paper, the EMD algorithm, EWT algorithm, and improved EWT algorithm are used to decompose the original signals of the three working conditions, and the sensitive components are reconstructed by the correlation coefficient method. Additionally, the SNR is introduced into the comparative analysis before and after processing to qualitatively analyze the comparison results, and the results are shown in Table 1.

Table 1. Signal-to-noise ratio (SNR) before and after signal processing.

Working Conditions	The Original Signal (SNR/db)	Reconstructed Signals of Three Algorithms (SNR/db)		
		EMD	**EWT**	**Improved EWT**
1	7.91	13.97	17.22	21.23
2	9.58	15.35	18.94	22.36
3	7.02	14.61	19.07	24.54

As Table 1 shows, compared with the original signal, the SNR of the reconstructed signal processed by the improved EWT algorithm increases by 13.32 dB, 12.78 dB, and 17.52 dB under three typical working conditions, which indicates that the noise is considerably reduced after applying the improved EWT algorithm. Compared with those of the EMD algorithm and the EWT algorithm, the SNR of the reconstructed signal processed by the improved EWT algorithm increases the most. Therefore, the preprocessing effect of the improved EWT algorithm is best.

4.3. Decomposition of the Cylinder Vibration Signal

The FE of the reconstructed signal is calculated, and five groups of samples are assessed for each type of ball mill load state. The average value of the FE of the three-state data is calculated, as shown in Table 2.

Table 2. Fuzzy entropy values of three types of load state vibration signals.

Sample	Underloaded	Normal Load	Overloaded
1	1.19	1.01	0.45
2	1.31	0.88	0.59
3	1.03	0.92	0.45
4	1.42	0.73	0.38
5	1.30	1.11	0.57
Mean	1.25	0.93	0.48

Table 2 shows that the FE value of the vibration signal varies by load state and that the FE value of the vibration signal under the same load state fluctuates back and forth near the average value. By comparing the FE values of three different load vibration signals, the FE values of the underloaded state are found to be relatively large, which is due to the relatively small amount of steel ball and mineral material in the cylinder under this condition. Additionally, the collision frequency between the mineral material and the steel ball in grinding production increases with the movement of the cylinder body to a certain height under the action of friction, and the collision frequency with other steel balls, minerals, and the cylinder walls is high in the process of falling. Energy is mainly consumed in the collisions between the steel ball and the tube wall and between the steel ball and other steel balls; thus, the vibration signal is complex, and the signal is highly random. However, the FE value under overloaded conditions is relatively small because there are more steel balls and minerals in the cylinder under these conditions, causing the steel ball and minerals to undergo peristalsis in the grinding process. In this case, the randomness of the signal is small. Under a normal load, energy is mainly used for grinding the quantity of minerals, and so the complexity of generating a signal is relatively moderate. For underloaded conditions and a normal load state, the sample entropy values are similar, and individual overlap occurs, which results in a discriminating effect. Therefore, MSE is introduced into the analysis of the mill vibration signal. The MFE of the reconstructed signal that can characterize the characteristic information of the vibration signal under three different load conditions is calculated. To highlight the superiority of the feature extraction method used in this paper, four combination methods (EWT-MSE, EWT-MFE, improved EWT-MSE, and improved EWT-MFE) are used to analyze the vibration signals of the cylinder of the ball mill under three load conditions. The mean value and standard deviation curve of the three states (20 samples per group) are shown in Figure 11. The parameter selection process of the algorithm is as described above.

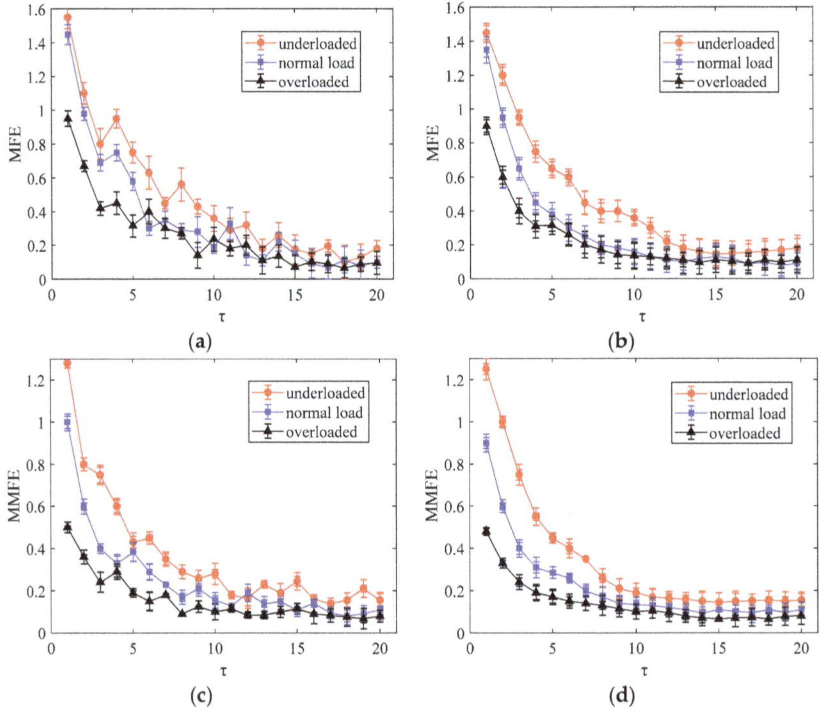

Figure 11. Reconstructed signal waveform under three working conditions: (**a**) EWT-MSE, (**b**) improved EWT-MSE, (**c**) EWT-MFE, and (**d**) improved EWT-MFE.

In Figure 11, it is evident that the order of the mean value of FE of the vibration signal of the ball mill cylinder under three working conditions displays the following order: underloaded > normal load > overloaded. Specifically, as the ball mill load increases, the amplitude of each component of the vibration signal in the spectrum obviously increases, which leads to a decrease in entropy. Although the variation trend of the FE of the cylinder vibration signal with the scale factor is the same in different load states, the fluctuation range of the entropy value varies, which indicates that FE can be effectively used to identify the load state. By comparing the four graphs, we see that there are obvious fluctuations and interval intersections between the EWT-MSE method and the EWT-MFE method. Although the entropy curve of the improved EWT-MSE method is smooth and the three states are distinguished to a certain extent, there are still overlap and intersection issues at small scales, which may lead to judgment errors. However, the MSE curves of the three load states obtained with the improved EWT-MFE method are smooth, and the fluctuation intervals have obvious limits. This finding indicates that the improved EWT-MFE method can effectively distinguish among the three load states of the ball mill.

4.4. Training and Testing

To verify the effectiveness of the proposed load identification model for a ball mill, 3×100 samples were randomly selected from each of the three classes of vibration signals, including 150 as training samples and 150 as test samples. The selected samples were first decomposed via the improved EWT method. The sensitive mode component signal with load state information was screened by the correlation coefficient method and reconstructed. Then, the MFE of the reconstructed signal was normalized as the input of the load state identification model of the ball mill based on AEPSO_PNN, and the load state of the ball mill was output. To highlight the superiority of AEPSO_PNN classification and identification, three clustering methods, namely, PNN classification, back propagation (BP) neural network, and Bayes identification methods, were trained and tested with the abovementioned samples. Then, the identification results were compared with the AEPSO_PNN identification results. For simplicity of description, the underloaded, normal load, and overloaded conditions are indicated by working condition numbers 1, 2 and 3, respectively. The identification effects of various classification methods are shown in Figure 12 and Table 3.

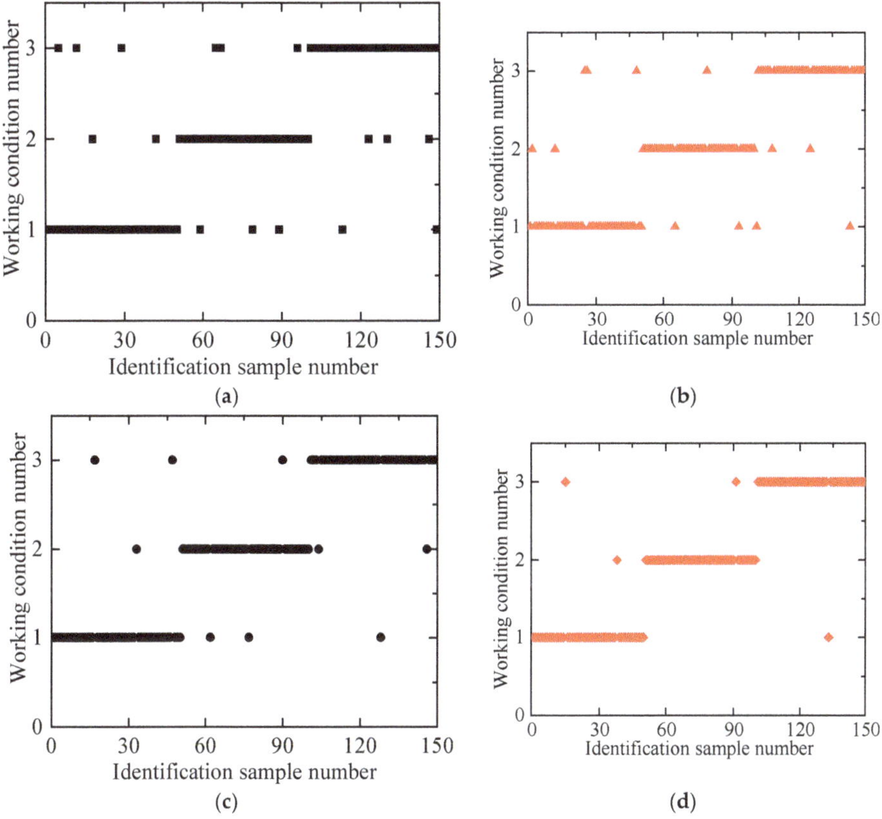

Figure 12. Recognition results of test samples for each classifier. (**a**) BP neural network; (**b**) Bayes identification method; (**c**) PNN; (**d**) AEPSO_PNN.

Table 3. Singular value entropy results for the three working conditions. BP: back propagation. probabilistic neural network (PNN): probabilistic neural network.

Classification Method	Correct Identifications	Load Recognition Accuracy
BP neural network	134	89.3%
Bayes identification method	138	92.0%
P NN classification	141	94.0%
AEPSO_PNN classification	146	97.3%

Figure 12 and Table 3 show that the predicted load state of the AEPSO_PNN model of ball mill load state recognition is largely consistent with the real state. Only four samples are misdiagnosed, and the overall recognition accuracy is 97.3%. Specifically, the recognition accuracy of AEPSO_PNN classification for three different load states is 96%, 98%, and 98%, all of which are high recognition levels. The BP neural network, Bayes discriminant method, and PNN classification can also achieve effective load identification. The highest accuracies of these methods are 89.3%, 92.0%, and 94.0%. Compared with the back propagation (BP) neural network, Bayes discriminant method, and PNN classification, AEPSO_PNN classification increases the overall recognition rate by 8%, 5.3%, and 3.3%. The results show that the mill load identification method based on the improved EWT-MFE method and AEPSO_PNN classification is effective, and the identification effect is excellent. Thus, this method provides a new approach for ball mill load identification.

5. Conclusions

By combining the improved EWT algorithm, MFE feature extraction, and AEPSO_PNN clustering, a load identification model of a ball mill is constructed. The main contributions to this work are as follows:

(1) The strong background noise, nonlinearity, and nonstationarity of the vibration signal of a ball mill cylinder hinder the recognition accuracy. The improved EWT algorithm proposed in this paper can effectively denoise the original signal and retain the feature information.
(2) The MFE algorithm has obvious advantages in terms of feature extraction. Notably, the MFE difference between underloaded, normal load, and overloaded conditions is large, and the proposed method can distinguish among the load states of the mill.
(3) The AEPSO_PNN classifier is introduced into the load recognition model of the ball mill to improve the recognition effect. Compared with the BP neural network, the Bayes discriminant method, and PNN classification, AEPSO_PNN classification provides a better recognition effect and the highest load recognition accuracy.
(4) The effectiveness of the method is verified based on a grinding experiment performed with a Bond work index ball mill in the laboratory.

In future research, the algorithm, structure, and parameter setting process of the proposed model will be optimized and improved to enhance the ability of the model to identify the ball mill load state.

Author Contributions: Conceptualization, G.C., X.L. (Xin Liu) and C.D.; methodology, G.C. and X.L. (Xin Liu); software, X.L. (Xin Liu) and C.D.; validation, G.C., X.L. (Xin Liu), and X.L. (Xiaoyan Luo); formal analysis, G.C. and X.L. (Xin Liu); investigation, G.C. and X.L. (Xiaoyan Luo); resources, X.L. (Xin Liu) and C.D.; data curation, G.C. and X.L. (Xin Liu); writing—original draft preparation, G.C. and X.L. (Xin Liu); writing—review and editing, G.C., X.L. (Xin Liu), and X.L. (Xiaoyan Luo); visualization, X.L. (Xin Liu) and C.D.; supervision, G.C.; project administration, G.C.; funding acquisition, G.C.

Funding: This research was supported by the National Natural Science Foundation of China (No. 51464017) and by a project of the Jiangxi Key Research and Development Plan, China (20181ACE50034).

Conflicts of Interest: The authors declare no conflicts of interest.

References

1. Sha, Y.; Chang, T.; Chang, J. Measure methods of ball mill's load. *Mod. Electr. Power* **2006**, *4*. (In Chinese) [CrossRef]
2. Gao, Z.W.; Sing, K.N.; Kong, D.X. Advances in Modelling, Monitoring, and Control for Complex Industrial Systems. *Complexity* **2019**, *3*. [CrossRef]
3. Gao, Z.W.; Kong, D.X.; Gao, C.H. Modeling and Control of Complex Dynamic Systems: Applied Mathematical Aspects. *J. Appl. Math.* **2012**, *5*. [CrossRef]
4. Gao, Z.W.; Saxen, H.; Gao, C. Special Section on Data-Driven Approaches for Complex Industrial Systems. *IEEE Trans. Ind. Inform.* **2013**, *9*, 2210–2212. [CrossRef]
5. Yin, Z.X.; Peng, Y.X.; Zhu, Z.C.; Ma, C.B.; Yu, Z.F.; Wu, G.Y. Effect of mill speed and slurry filling on the charge dynamics by an instrumented ball. *Adv. Powder Technol.* **2019**, *30*, 1611–1616. [CrossRef]
6. Das, S.P.; Das, D.P.; Behera, S.K.; Mishra, B.K. Interpretation of mill vibration signal via wireless sensing. *Miner. Eng.* **2011**, *24*, 245–251. [CrossRef]
7. Tang, J.; Zhao, L.; Zhou, J.; Yue, H.; Chai, T. Experimental analysis of wet mill load based on vibration signals of laboratory-scale ball mill shell. *Miner. Eng.* **2010**, *23*, 720–730. [CrossRef]

8. Jian, T.; Chai, T.; Wen, Y.; Zhao, L. Engineering modeling load parameters of ball mill in grinding process based on selective ensemble multisensor information. *IEEE Trans. Autom. Sci.* **2013**, *10*, 726–740. [CrossRef]

9. Zhou, P.; Chai, T.; Wang, H. Intelligent optimal-setting control for grinding circuits of mineral processing process. *IEEE Trans. Autom. Sci. Eng.* **2009**, *6*, 730–743. [CrossRef]

10. Jian, T.; Wen, Y.; Chai, T.; Zhuo, L.; Zhou, X. Selective ensemble modeling load parameters of ball mill based on multi-scale frequency spectral features and sphere criterion. *Mech. Syst. Signal Proc.* **2015**, *66–67*, 485–504. [CrossRef]

11. Lei, Z.; Su, W. Mold Level Predict of Continuous Casting Using Hybrid EMD-SVR-GA Algorithm. *Processes* **2019**, *7*, 177. [CrossRef]

12. Su, Z.G.; Wang, P.H.; Yu, X.J.; Lv, Z.Z. Experimental investigation of vibration signal of an industrial tubular ball mill: Monitoring and diagnosing. *Miner. Eng.* **2008**, *21*, 699–710. [CrossRef]

13. Zhang, J.; He, J.; Long, J.; Yao, M.; Zhou, W. A new denoising method for UHF PD signals using adaptive VMD and SSA-based shrinkage method. *Sensors* **2019**, *19*, 1594. [CrossRef] [PubMed]

14. Li, Y.; Chen, X.; Yu, J. A Hybrid Energy Feature Extraction Approach for Ship-Radiated Noise Based on CEEMDAN Combined with Energy Difference and Energy Entropy. *Processes* **2019**, *7*, 69. [CrossRef]

15. Zhuo, L.; Chai, T.; Wen, Y.; Jian, T. Multi-frequency signal modeling using empirical mode decomposition and PCA with application to mill load estimation. *Neurocomputing* **2015**, *169*, 392–402. [CrossRef]

16. Jian, T.; Wang, D.; Chai, T. Predicting mill load using partial least squares and extreme learning machines. *Soft Comput.* **2012**, *16*, 1585–1594. [CrossRef]

17. Jian, T.; Zhao, L.; Wen, Y.; Yue, H.; Chai, T. *Soft Sensor Modeling of Ball Mill Load via Principal Component Analysis and Support Vector Machines*; Springer: Berlin/Heidelberg, Germany, 2010; pp. 803–810.

18. Kedadouche, M.; Thomas, M.; Tahan, A. A comparative study between Empirical Wavelet Transforms and Empirical Mode Decomposition Methods:Application to bearing defect diagnosis. *Mech. Syst. Signal Process.* **2016**, *81*, 88–107. [CrossRef]

19. Costa, M.; Goldberger, A.L.; Peng, C.K. Multiscale Entropy Analysis of Complex Physiologic Time Series. *Phys. Rev. Lett.* **2007**, *89*, 705–708. [CrossRef]

20. Li, S.Y.; Yang, M.; Li, C.C.; Cai, P. Analysis of heart rate variability based on singular value decomposition entropy. *J. Shanghai Univ.* **2008**, *12*, 433–437. [CrossRef]

21. Han, L.; Li, C.; Zhan, L.; Li, X.L. Rolling Bearing Fault Diagnosis Method Based on EEMD Permutation Entropy and Fuzzy Clustering. In Proceedings of the Fifth International Conference on Instrumentation & Measurement, Qinhuangdao, China, 18–20 September 2015. [CrossRef]

22. Miao, Y.; Zhao, M.; Lin, J. Periodicity-impulsiveness spectrum based on singular value negentropy and its application for identification of optimal frequency band. *IEEE Trans. Ind. Electron.* **2018**. [CrossRef]

23. Zhao, L.; Yu, W.; Yan, R. Gearbox Fault Diagnosis Using Complementary Ensemble Empirical Mode Decomposition and Permutation Entropy. *Shock Vib.* **2016**, *2016*, 3891429. [CrossRef]

24. Chang, J.L.; Chao, J.A.; Huang, Y.C.; Chen, J.S. Prognostic Experiment for Ball Screw Preload Loss of Machine Tool through the Hilbert-Huang Transform and Multiscale Entropy Method. In Proceedings of the IEEE International Conference on Information & Automation, Harbin, China, 20–23 June 2010. [CrossRef]

25. Liu, H.; Han, M. A fault diagnosis method based on local mean decomposition and multi-scale entropy for roller bearings. *Mech. Mach. Theory* **2014**, *75*, 67–78. [CrossRef]

26. Zheng, J.; Pan, H.; Cheng, J. Rolling bearing fault detection and diagnosis based on composite multiscale fuzzy entropy and ensemble support vector machines. *Mech. Syst. Signal Process.* **2017**, *85*, 746–759. [CrossRef]

27. Alić, B.; Sejdinović, D.; Gurbeta, L.; Badnjevic, A. Classification of Stress Recognition using Artificial Neural Network. In Proceedings of the 2016 5th Mediterranean Conference on Embedded Computing (MECO), Bar, Montenegro, 12–16 June 2016. [CrossRef]

28. Pchelintseva, S.V.; Runnova, A.E.; Musatov, V.Y.; Hramov, A.E. Recognition and Classification of Oscillatory Patterns of Electric Brain Activity Using Artificial Neural Network Approach. In Proceedings of the Society of Photo-Optical Instrumentation Engineers, San Francisco, CA, USA, 23–27 January 2017. [CrossRef]

29. Petrosanu, D.M. Designing, Developing and Validating a Forecasting Method for the Month Ahead Hourly Electricity Consumption in the Case of Medium Industrial Consumers. *Processes* **2019**, *7*, 310. [CrossRef]

30. Rajagopal, R.; Ranganathan, V. Control evaluation of effect of unsupervised dimensionality reduction techniques on automated arrhythmia classification. *Biomed. Signal Process. Control* **2017**, *34*, 1–8. [CrossRef]

31. Li, Q.; Sun, Y.; Yu, Y.; Wang, C.; Ma, T. Short-term photovoltaic power forecasting for photovoltaic power station based on EWT-KMPMR. *Trans. Chin. Soc. Agric. Eng.* **2017**, *33*, 265–273. [CrossRef]
32. Gilles, J. Empirical Wavelet Transform. *IEEE Trans. Signal Process.* **2013**, *61*, 3999–4010. [CrossRef]
33. Cannone, M.; Meyer, Y. Littlewood-Paley decomposition and Navier-Stokes equations. *Methods Appl. Anal.* **1995**, *2*, 307–319. [CrossRef]
34. Morente-Molinera, J.A.; Mezei, J.; Carlsson, C.; Herrera-Viedma, E. Improving Supervised Learning Classification Methods Using Multigranular Linguistic Modeling and Fuzzy Entropy. *IEEE Trans. Fuzzy Syst.* **2017**, *25*, 1078–1089. [CrossRef]
35. Specht, D.F. Probabilistic neural networks. *Neural Netw.* **1990**, *3*, 109–118. [CrossRef]
36. Ouhibi, R.; Bouslama, S.; Laabidi, K. Faults Classification of Asynchronous Machine Based on the Probabilistic Neural Network (PNN). In Proceedings of the 2016 4th International Conference on Control Engineering & Information Technology (CEIT), Hammamet, Tunisia, 16–18 December 2017; IEEE: Hammamet, Tunisia, 2017. [CrossRef]
37. Chen, S.M.; Jian, W.S. Fuzzy forecasting based on two-factors second-order fuzzy-trend logical relationship groups, similarity measures and PSO techniques. *Inf. Sci.* **2016**, *391–392*, 65–79. [CrossRef]
38. Lu, X. Research on Ball Mill Load Forecasting Method Based on Multi-Source Signal Fusion Technology. Master's Thesis, Jiangxi University of Science and Technology, Ganzhou, China, 2017.

Article

Study on Three-Dimensional Stress Field of Gob-Side Entry Retaining by Roof Cutting without Pillar under Near-Group Coal Seam Mining

Xiaoming Sun [1,2,*,†], Yangyang Liu [1,2,*,†], Junwei Wang [1,2], Jiangbing Li [1,2], Shijie Sun [1,2] and Xuebin Cui [1,2]

1 State Key Laboratory for Geomechanics & Deep Underground Engineering, China University of Mining & Technology, Beijing 100083, China
2 School of Mechanics and Civil Engineering, China University of Mining & Technology, Beijing 100083, China
* Correspondence: zqt1700620120g@student.cumtb.edu.cn (X.S.);
  zqt1700620112g@student.cumtb.edu.cn (Y.L.); Tel.: +86-180-3922-5611 (X.S.)
† These authors contributed equally to this work.

Received: 13 June 2019; Accepted: 16 August 2019; Published: 22 August 2019

Abstract: In order to explore the distribution law of stress field under the mining mode of gob-side entry retaining by roof cutting without pillar (GERRCP) under goaf, based on the engineering background of 8102 and 9101 working faces in Xiashanmao coal mine, the stress field distribution of GERRCP and traditional remaining pillar was studied by means of theoretical analysis and numerical simulation. The simulation results showed that: (1) in the front of the working face, the vertical peak stress of non-pillar mining was smaller than that of the remaining pillar mining, and it could effectively control stress concentration in surrounding rock of the mining roadway; the trend of horizontal stress distribution of the two was the same, and the area, span and peak stress of stress the rise zone were the largest in large pillar mining and the minimum in non-pillar mining. (2) On the left side of the working face, the vertical stress presented increasing-decreasing characteristics under non-pillar mining mode and saddle-shaped distribution characteristics under the remaining pillar mining mode respectively. Among them, the peak stress was the smallest under non-pillar mining, and compared with the mining of the large pillar and small pillar, non-pillar mining decreased by 12–21% and 3–10% respectively. The position of peak stress of the former was closer to the mining roadway, indicating that the width of the plastic zone of the surrounding rock of the non-pillar mining was smaller and bearing capacity was higher. In the mining of the large and small pillar, the horizontal stress formed a high stress concentration in the pillar and 9102 working face respectively. In non-pillar mining, the horizontal stress concentration appeared in solid coal, but the concentration area was small.

Keywords: non-pillar; gob-side entry retaining by roof cutting; close distance coal seams; goaf; stress distribution

1. Introduction

In the 1960s and 1970s, longwall mining technology developed rapidly, and the "masonry beam theory" was put forward, forming the "121" construction method of longwall mining [1]. This technology requires two roadways to be tunneled for each working face, and a large pillar is set up to balance underground pressure. The "transfer rock beam theory" was put forward by analyzing the existence of the internal and external stress field in a high-stress area, forming the "121" small pillar construction method of longwall mining [2,3]. However, the traditional mining method of "121" will form a hanging roof with insufficient collapse at the side of the goaf, and the roof subsidence and rotary deformation are large, which greatly affects the stability of the pillar and support system on roadway side. In order to reduce the development ratio, increase the coal mining rate and improve the periodic pressure of roof, the "cutting cantilever beam theory" was born [4], and based on this theory, the mining technology of gob-side entry retaining by roof cutting without pillar (GERRCP) was proposed. In the new mining technology of GERRCP, only one roadway needs to be tunneled for each working face, and the other roadway is formed automatically by roof cutting and pressure relief. Moreover, there is no need to leave pillars, which reduces the waste of coal resources and avoids roof accidents, rock burst, gas outburst and other safety hazards caused by remaining pillars [5]. The phenomenon of stress concentration caused by a pillar is eliminated, and the pressure distribution of a stope is optimized, which makes coal mining more safe and efficient. Many scholars have carried out a lot of related research work on GERRCP using theoretical analysis, numerical simulation, laboratory experiments and field experiments. As one of the powerful methods, the numerical simulation has the advantage of low cost, high efficiency and good repeatability. It has been widely used in the related research of this technology and achieved good application results. With the introduction of GERRCP, its design principle and key technologies have been extensively investigated [6–8]. Guo et al. [9] studied the relationship between roof fracturing angle and stability of gob-side entry subjected to dynamic loading through establishing a numerical calculation model. Zhen et al. [10] investigated the influence of two methods of non-pillar-mining techniques by roof cutting and by filling artificial materials on the results of the entry retained via industrial case and numerical simulation. Guo et al. [11] studied the roof pre-fracturing and energy-absorbing support systems to evaluate the stress distribution and deformation control of gob-side entry by numerical simulation. Hu et al. [12] investigated the key parameters affecting GERRCP by theoretical analysis and numerical simulation. Combined with the above research, it can be found that the above researches on GERRCP were carried out under the condition of single coal-seam mining, and few researches on this technology when mining close distance coal seam. Therefore, it is of great significance to carry out relevant researches on GERRCP under the condition of near-group coal-seam mining.

The near-group coal-seam mining is very characteristic. When mining close distance coal-seam, the roof caving of the upper coal seam will cause various degrees of damage to the roof of the lower coal seam. As a result, the upper overburden structure and temporal and spatial distribution characteristics of the stress field during the mining of the lower coal seam are significantly different from those of a single coal seam. In particular, the mining direction of the lower coal seam is perpendicular to that of the upper coal seam, forming the vertical cross mining. Therefore, in order to ensure the safety of coal mining, it is of significance to analyze the distribution law of the stress field in lower coal seam when mining close distance coal seam. At present, domestic and foreign experts and scholars have conducted a lot of studies on the distribution law of a stress field in lower coal seam when mining close distance coal seam, and achieved fruitful results. Singh [13] established a numerical model and combined it with a double-yield model to assess its effectiveness in simulating the recovery of stress in goaf. Through theoretical analyses and physical modelling studies, the interaction between vertical stress distribution within goaf and surrounding rock mass in these systems was studied [14]. Zhang et al. [15] investigated the stress distribution, fracture development, and strata movement above a protective coal seam in longwall mining through numerical calculation. Liu et al. [16] analysed the stress distribution and roadway position of lower seams in the close distance coal seams by using numerical simulation. Zhang et al. [17] studied the floor failure depth of upper coal seam during close coal seams mining by building the mechanical model of floor failure of upper coal seam. Xu et al. [18] studied the stress propagation and distribution of a roadway by Kirsch equations and analyzed the changes of stress, displacement, and plastic zones around roadways during the mining of the upper coal seams by means of numerical simulation. Wang et al. [19] analyzed some key issues about abutment pressure and stress concentration shell by numerical simulations to study the distribution and evolution characteristics of the macroscopic stress field of surrounding rocks. Ma et al. [20] studied the stress distribution and deformation law of surrounding rocks for the water-dripping roadway below a contiguous seam goaf.

In-depth studies on the movement and instability characteristics and mining stress evolution law of the secondary mining structure of roof under goaf, were carried out by means of theoretical analysis, similarity simulation experiment, numerical simulation and field measurement [21–24].

The above experts and scholars have achieved fruitful results in research on the distribution law of a stress field when mining near-group coal seam. However, most previous studies have focused on conventional mining methods; few scholars have carried out relevant research on the stress distribution of the stope in GERRCP in the near-group coal-seam mining. In order to explore the distribution law of three-dimensional stress field of GERRCP in the near-group coal seam mining, this paper takes Xiashanmao coal mine as an engineering background, establishes the mechanical model of the roof structure of GERRCP through theoretical analysis, and establishes a three-dimensional numerical calculation model based on the finite difference program FLAC-3D (ITASCA, US) to study the distribution law of stress in the stope. Finally, the numerical simulation results are validated by field experiments.

2. Gob-Side Entry Retaining by Roof Cutting without Pillar (GERRCP)

2.1. Principle of GERRCP

As shown in Figure 1, the GERRCP adopts energy-gathered blasting technology to carry out advanced pre-splitting on the roof. The roof is cut off along the pre-splitting damaged structural surface through periodic weighting of the stope, and the fractured roof collapses naturally with the help of underground pressure. The side of the roadway is formed by the broken expanded characteristic of the collapsed gangue, and the flexible support body in a roadway is formed by the sliding and yielding gangue support structure and constant pressure retractable support equipment, which separates the goaf. At the same time, the high strength support of a roadway roof is formed by the constant-resistance large-deformation anchor cable (CRLDAC) with structural characteristics of negative Poisson's ratio, thus realizing the non-pillar mining of a single working face and single roadway [25,26]. This technology realizes the transformation of a roadway roof structure from a long-wall beam to short-wall beam by roof cutting, which ensures the stability of the roadway, can weaken the concentrated stress on the upper part of the coal body adjacent to working face, and can also avoid the roof collapse, rock burst and gas outburst caused by remaining pillars.

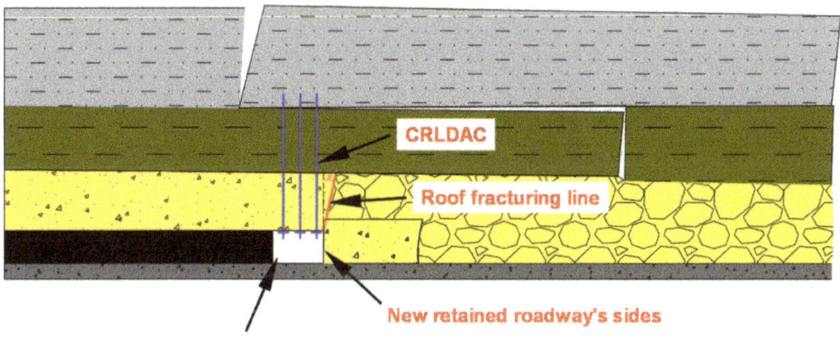

Figure 1. Section diagram of roof structure of gob-side entry retaining by roof cutting without pillar (GERRCP).

The mechanical model of the roof structure of GERRCP is introduced below (as shown in Figure 2). Under the action of periodic pressure, the immediate roof and main roof fracture and rotate. The main roof is fractured to form rock blocks A, B and C, and the interaction between the rock blocks forms a hinge structure. Rock block A is still supported by the immediate roof, which is relatively stable. Rock block C is supported by the gangue on the side of the goaf, and its stability is poor. Both ends of rock block B are supported by the immediate roof and gangue in the goaf, respectively, and rotate towards the goaf around the elastic-plastic boundary of solid coal. The following assumptions are made for the mechanical model of surrounding rock: (1) there is no interaction between rock block B and C and the gangue at the side of the goaf; (2) the shear force between rock strata such as immediate roof and main roof is ignored; (3) the supporting capacity of coal body in the lateral plastic zone of the retaining roadway is not considered; (4) the supporting force at the side of the roadway is neglected. The structural mechanical model established according to the assumed conditions is shown in Figure 2.

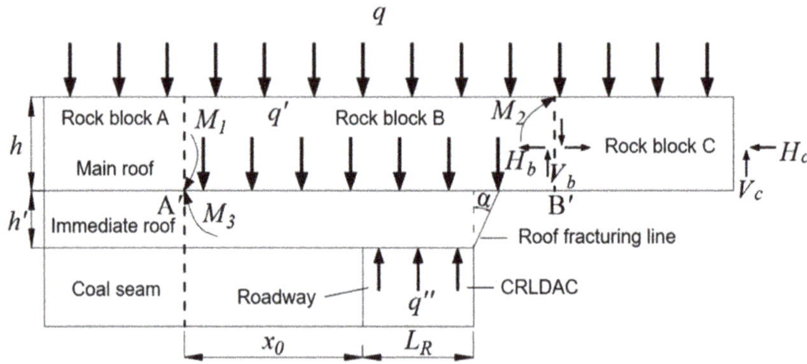

Figure 2. Mechanical model of roof structure of GERRCP.

The above picture is explained as follows. The key parameters of the system are as follows. After the rock strata fracture, the fracture length of key block B formed is [27]:

$$L = L_s\left(\sqrt{\frac{L_s^2}{L_f^2} + \frac{3}{2}} - \frac{L_s}{L_f}\right) \tag{1}$$

where, L: length of rock block; L_s: weighting interval of the immediate roof; L_f: working face length.

The horizontal force on rock block B is

$$H_b = \frac{qL}{2(h - S_b)} \tag{2}$$

where, H_b: horizontal force on rock block B; q: uniform load acting on the main roof; h: the thickness of basic roof; and S_b: the subsidence of rock block B.

The research on the mechanism of arch effect and its boundary conditions was studied, and the calculation of the plastic zone was referenced, and the width of the stress limit equilibrium zone in coal body was obtained [27,28].

$$x_0 = \frac{h_r A}{2tan\varphi_0} ln\left(\frac{k\gamma H + \frac{c_0}{tan\varphi_0}}{\frac{c_0}{tan\varphi_0} + \frac{p_x}{A}}\right) \tag{3}$$

where, x_0: width of stress limit equilibrium zone in coal body; h_r: roadway height; A: lateral pressure coefficient; C_0: cohesion of the interface between coal and rock; φ_0: internal friction angle; K: stress concentration factor; γ: average bulk density of overburden; H: roadway burial depth; P_x: support resistance of coal sides.

After mining, the main roof structure breaks under periodic pressure, forming a hinge structure, and the structure is in equilibrium. Through static analysis of hinge structure under this equilibrium condition, the static equilibrium equations of rock block C and B are established, such as Formulas 4 and 5. Among them, the support force provided by the support body in the roadway is simplified as the load collection degree of support, and the support load in the roadway is solved based on the above analysis.

(1) Rock block C:

$$\Sigma X = 0, H_b - H_c = 0 \tag{4}$$

$$\Sigma Y = 0, qL + V_b - V_c = 0 \quad \Sigma M_{B'} = 0, -M_2 + H_c(h/2 - S_c) + V_c L - H_b(h/2 - S_c) - qL^2/2 = 0$$

where, H_c: horizontal force on rock block C; V_c: the vertical force on rock block C; V_b: the vertical force on rock block B; M_2: moment of rock block C at section B; and S_c: the subsidence of rock block C.

(2) Rock block B:

$$\Sigma M_{A'} = 0, -M_1 - M_3 + q'' L_R(x_0 + L_R/2) + H_b(h/2 - S_b) - qL^2/2 + M_2 + V_b L + q'(x_0 + L_R + h' \tan \alpha)$$
$$\left[(L_R + x_0)^2 + (x_0 + L_R + h' \tan \alpha)^2 + (L_R + x_0)(x_0 + L_R + h' \tan \alpha)\right]/3(2x_0 + 2L_R + h' \tan \alpha) = 0 \tag{5}$$

where, M_1: moment of rock block B at section A; M_3: moment of immediate roof to basic roof; q'': load collection degree of support in the roadway; L_R: roadway width; q': the uniform load acting on the immediate roof; h': the thickness of immediate roof; and α: pre-cracking roof cutting angle.

(3) The load collection degree of support in the roadway can be obtained simultaneously.

$$q'' = \left\{ \begin{array}{c} M_1 + M_3 - 2M_2 + qL^2 - qL(h - 2S_b)/4(h - S_b) - q'(x_0 + L_R + h' \tan \alpha) \\ \left[(2x_0 + 2L_R + h' \tan \alpha)^2 - (x_0 + L_R)(x_0 + L_R + h' \tan \alpha)\right]/3(2x_0 + 2L_R + h' \tan \alpha) \end{array} \right\} / L_R(x_0 + L_R/2) \tag{6}$$

Based on the new technology of GERRCP, the mechanical model of the roof structure is established. Through the mechanical analysis of the model, the key parameters such as fracture length of key blocks in upper strata and horizontal force acting on it are introduced, and the extension depth of the plastic zone in the solid coal side of the roadway under this technical condition is obtained, which provides certain theoretical support for the support design of the solid coal side of the mining roadway. In addition, through the static analysis of the balanced hinge structure under the condition of GERRCP, the corresponding static equilibrium equation is established, and the support load in roadway under this condition is obtained, which provides corresponding theoretical support for the support problem of mining roadway.

2.2. Technical Process of GERRCP

The mining mode layout of the traditional "121" construction method is shown in Figure 3a. When the mining system is used for coal mining, pillars are left, which belongs to the mining method of "one working face and two roadways". Different from the traditional mining method, the GERRCP is shown in Figure 3b, which is a typical single working face and single roadway without a pillar mining method. That is to say, the up and down drifts on the first face should be excavated firstly, and then at the same time in the working face of the mining, the retaining roadway, as the transport roadway on the next working face, is formed through the reinforcement of the advance anchor cable, pre-cracking roof cutting and gangue support at the side of the goaf. Therefore, the mining ratio is reduced and non-pillar mining is realized.

The technological process of GERRCP is shown in Figure 4. Its core can be summarized as four steps: strengthening, cutting, protecting and closing, that is: (1) adopt the CRLDAC to actively strengthen the supporting roadway roof according to the designed supporting parameters (Figure 4a); (2) the energy-gathered pre-cracking blasting hole shall be constructed at a certain distance in advance of the working face, and the bidirectional energy-gathered pre-cracking blasting shall be carried out according to the parameters determined by the blasting test to form a slit on the roof at the side of the goaf (Figure 4b); (3) after mining at the working face, the sliding and yielding gangue support structure and the constant pressure retractable support equipment are adopted behind the working face to strengthen the support in time. Under the action of the underground pressure, the roof at the side of the goaf collapses along the structural surface of the roof cutting slit to form the side of a new roadway (Figure 4c); (4) the caving roof is gradually compacted as the working face advances, and the side of the roadway formed by caving is shotcreted to close the goaf. After the roadway is stabilized, the temporary support equipment is removed to realize the retaining new roadway (Figure 4d).

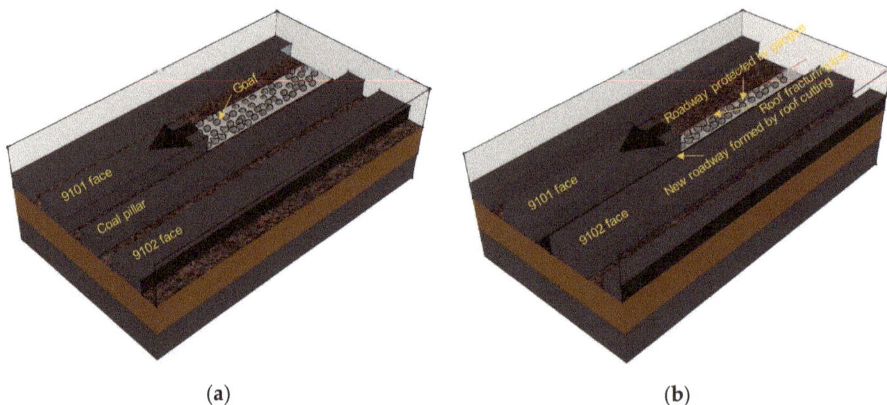

(**a**) (**b**)

Figure 3. Layout of mining mode. (**a**) Layout of traditional mining mode; (**b**) Layout of non-pillar mining mode.

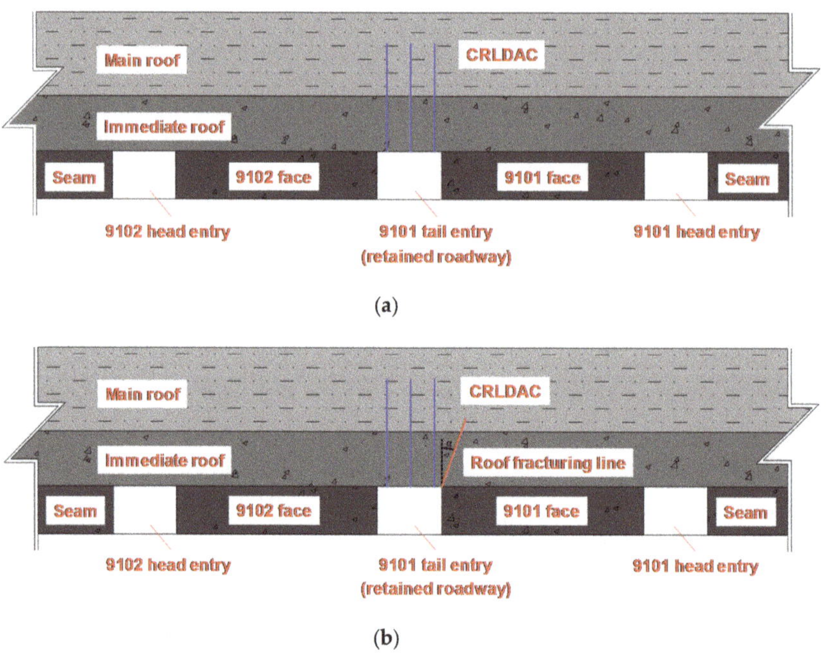

(**a**)

(**b**)

Figure 4. *Cont.*

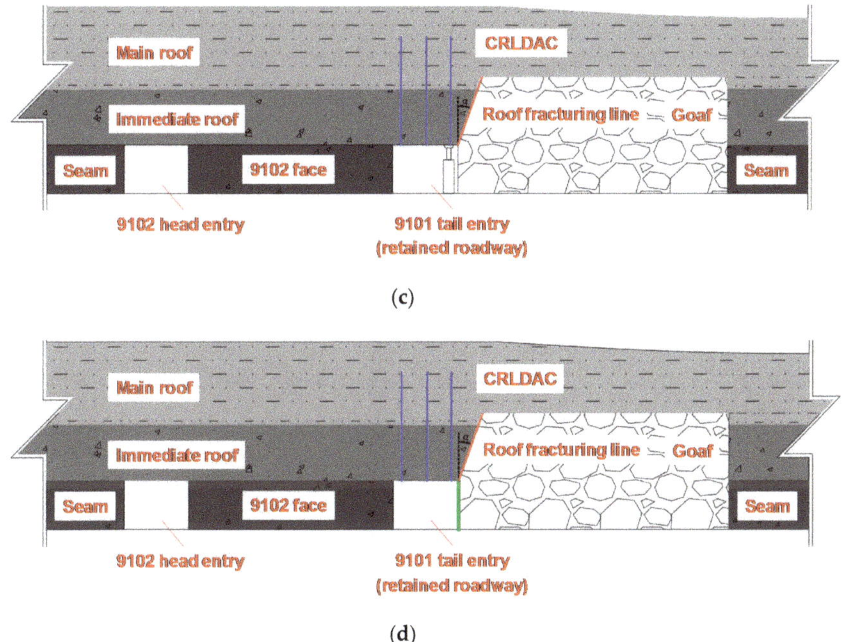

(c)

(d)

Figure 4. Technological process of GERRCP. (**a**) Strengthening roadway roof by constant-resistance large-deformation anchor cable (CRLDAC); (**b**) Pre-cracking roof cutting by energy-gathered blasting; (**c**) Gangue support; (**d**) Closing the goaf by shotcreting.

3. Engineering Background

The coal seam in the 9101 working face of Xiashanmao coal mine is located in the lower part of Taiyuan Formation. The thickness of the coal seam is 1.55–3.5 m, and the designed mining height is 3 m, which belongs to the medium-thick coal seam mining face. The dip angle of coal seam is 2–8° and the buried depth is 180–260 m. The distance between 8 # coal seam and 9 # coal seam is 11.20–19.60 m, with an average of 15.24 m. The immediate roof is mudstone, with an average thickness of 4.1 m; the basic roof is mainly sandy mudstone with an average thickness of 8.0 m, according to the drilling measurement on the roof and floor of the working face. The immediate and basic bottoms are mudstone and fine sandstone, respectively, as shown in Table 1.

Table 1. Lithological characteristics of roof and floor of coal seam.

Name of Roof and Floor	Lithology	Thickness/m	Feature Description
Main roof	Sandy mudstone	4.6–9.1	Grey, block structure
Immediate roof	Mudstone	3.6–4.2	Black, block structure
9 # coal seam	Coal seam	2.8–3.1	Black, vitreous luster, occurrence stability
Immediate bottom	Mudstone	4.8–8.5	Grey, block structure
Basic bottom	Fine sandstone	13.2–26.6	Grey, block structure, horizontal joint

The test face is the first working face of the first mining area of 9 # coal seams, with a strike length of 480 m and an inclination of 150 m (as showed in Figure 5). The roof is managed by all caving method, which adopts full-seam, comprehensive mechanized, and retreating mining methods. The average 12 m above the 9101 working face is the 8102 working face, which serves as the mining protective layer of 9101 working face. Among them, the 8102 working face adopts the "121" construction method of longwall mining, and the 9101 working face adopts the self-formed roadway without a pillar-mining system, and the mining direction is vertical intersection. The adjacent face is the 9102 working face, which is located at the south of the 9101 working face. The test roadway is the ventilation roadway of the 9101 working face. The roadway section is rectangular, with a width of 4000 mm and a height of 3200 mm.

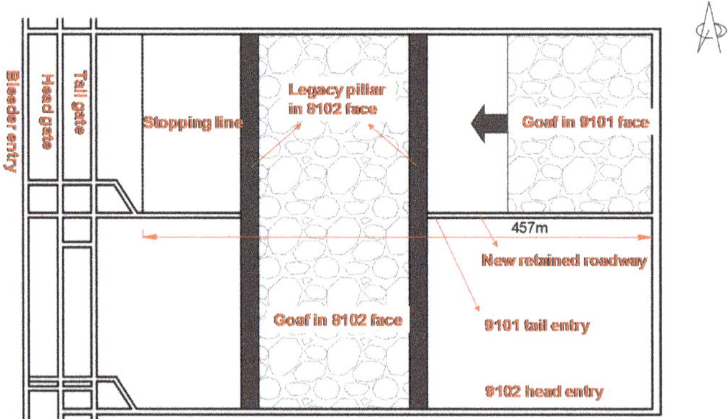

Figure 5. Layout of test working face of GERRCP in Xiashanmao coal mine.

4. Numerical Calculation and Analysis

4.1. Construction of Numerical Model

According to the specific engineering geological conditions of the 8102 and 9101 working faces of Xiashanmao coal mine, and combined with the existing underground pressure monitoring results, the finite difference software FLAC3D was used to establish a three-dimensional solid model. The distribution characteristics of mining stress in the mining process of the 9101 working face were studied. The numerical calculation model is shown in Figure 6. The calculation range was 330 m × 230 m × 100 m (length × width × height). The model simulated 12 layers of strata, including 8 # coal seam, 9 # coal seam and roof and floor strata, and truly reflected their occurrence conditions. Because the coal seam under actual working conditions could be regarded as a near-horizontal coal seam, the coal seam in the model was designed as a horizontal coal seam.

Figure 6. Three-dimensional numerical calculation model.

4.2. Determining Model Parameters

For rock mass materials, the elastic modulus of rock mass greatly influences the accuracy of simulation results. Therefore, in the process of numerical simulation, the rock elastic modulus should be corrected and verified repeatedly to reduce the error with the actual value to ensure the reliability of simulation results. The physical and mechanical parameters of strata were obtained according to the test of rock core samples from geological drilling in the testing field, and the elastic modulus of rock was taken to be 1/10 of the elastic modulus of rock block by comparing with the physical parameters of rock strata in the adjacent working face. In this study, the mole-coulomb model was selected as the constitutive model, and the effective physical and mechanical parameters of rock mass were finally determined based on the physical and mechanical parameters of the rock mass involved, as shown in Table 2.

Table 2. The physical and mechanical parameters of rock.

Lithology	Density (kg·m^{-3})	Bulk Modulus (GPa)	Shear Modulus (GPa)	Cohesion (MPa)	Internal Friction Angle (°)	Tensile Strength (MPa)
Upper strata	2620	6.47	4.09	1.61	35	0.82
Sandy mudstone	2512	12.36	7.21	2.04	33	0.74
Medium Sandstone	2670	23.46	15.20	4.45	40	5.14
Fine sandstone	2870	21.04	13.52	3.20	42	1.29
Sandy mudstone	2503	10.63	5.59	2.04	33	0.74
Limestone	2910	29.26	18.27	5.14	42	7.31
8 # coal seam	1380	4.91	2.01	1.25	32	0.15
Sandy mudstone	2531	14.13	9. 18	4.35	33	0.81
Mudstone	2488	9.97	7.35	1.20	32	0.58
9 # coal seam	1450	4.91	2.01	1.25	32	0.15
Mudstone	2460	5.12	4.73	1.20	32	0.58
Fine sandstone	2870	21.04	13.52	3.75	38	1.84

4.3. Simulation Scheme

According to the actual working conditions, the corresponding simulation scheme was formulated. The geometry and boundary conditions of the model are shown in Figures 7 and 8 respectively. The model was fixed around to limit the horizontal movement, and fixed at the bottom to limit the vertical movement, and 5 MPa uniform load was applied at the top to simulate the self-weight boundary of the overlying strata. The mining scheme of coal seam was as follows. For the large pillar mining, (a) the 8102 working face was firstly mined step by step, the excavation footage of each step was set as 10 m. After each step was balanced, the next step of excavation was calculated to balance, and the calculation was carried out step by step. (b) Three mining roadways of 9101 and 9102 working faces were excavated at one time, and a pillar with a width of 15 m was set, and the calculation was made to the model balance. (c) The 9101 working face was excavated step by step, and the excavation footage was set as 10 m. The mining direction was perpendicular to that of the 8102 working face. For the small pillar mining, (a) ditto; (b) three mining roadways of the 9101 and 9102 working faces were excavated at one time, a pillar with a width of 5 m was set, and the calculation was made to the model balance; (c) ditto. For the non-pillar mining, (a) ditto; (b) two mining roadways of the 9101 working face were excavated at one time, and the calculation was made to the model balance; (c) pre-cracking roof cutting was conducted firstly. Then the 9101 working face was excavated step by step, and the excavation footage was set to 10 m. The mining direction was perpendicular to that of the 8102 working face.

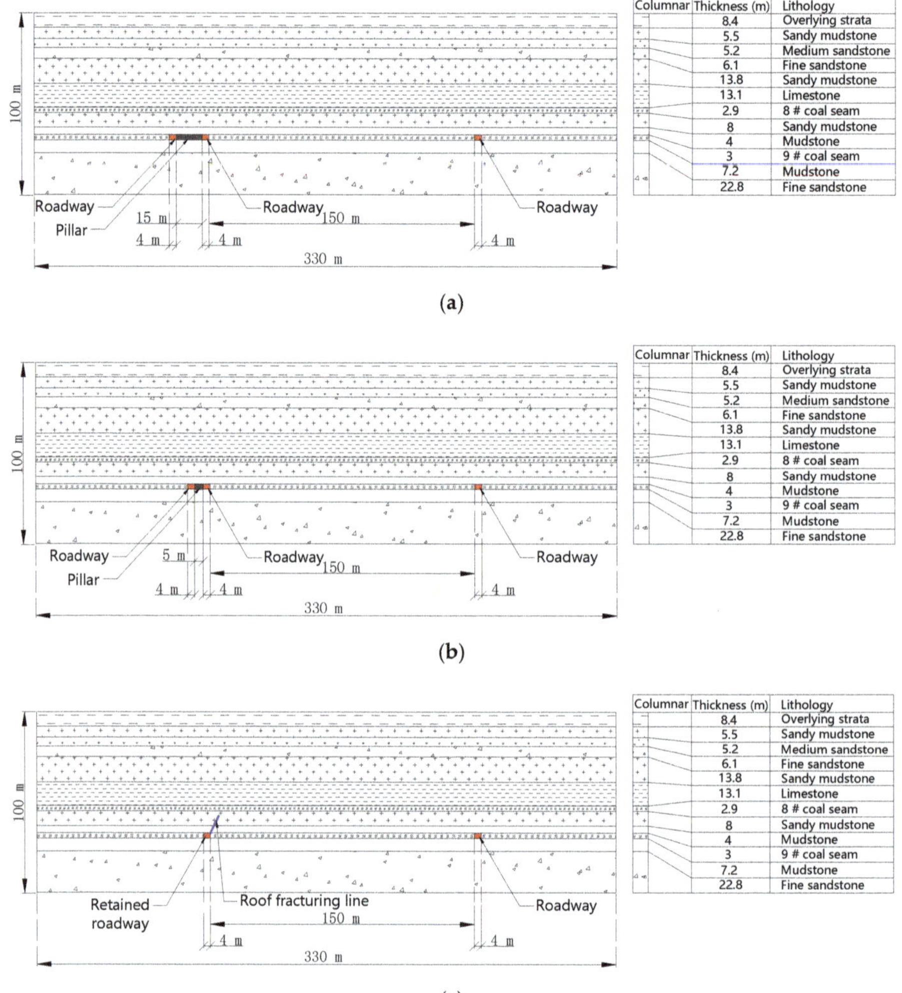

Figure 7. Geometry of the model. (**a**) Geometry of model of large pillar mining; (**b**) geometry of model of small pillar mining; (**c**) geometry of model of non-pillar mining.

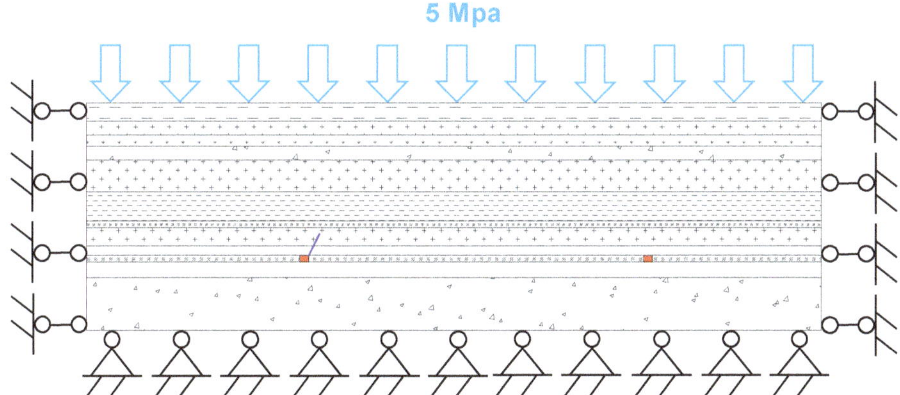

Figure 8. Boundary conditions of the model.

5. Distribution Law of Three-Dimensional Stress Field

5.1. Distribution Law of Vertical Stress

In order to analyze the evolution law of stress distribution in the surrounding rock of stope, when the working face advances to 120 m, monitoring lines are arranged at different positions of the model, and vertical stress is monitored by monitoring lines. The location of monitoring lines is shown in Figure 9. Among them, one, two and three monitoring lines are located in the inner 10 m of the 9101 ventilation roadway, the middle of the 9101 working face, and the inner 10 m of the 9101 haulage roadway; four, five and six monitoring lines are located 5 m, 10 m and 20 m in front of the working face, respectively; seven, eight and nine monitoring lines are located on the left side of the working face, 5 m, 10 m and 15 m in front of the working face respectively; 10, 11, 12 and 13 monitoring lines are located on the left side of working the face, 5 m, 10 m, 30 m, and 80 m behind the working face, respectively.

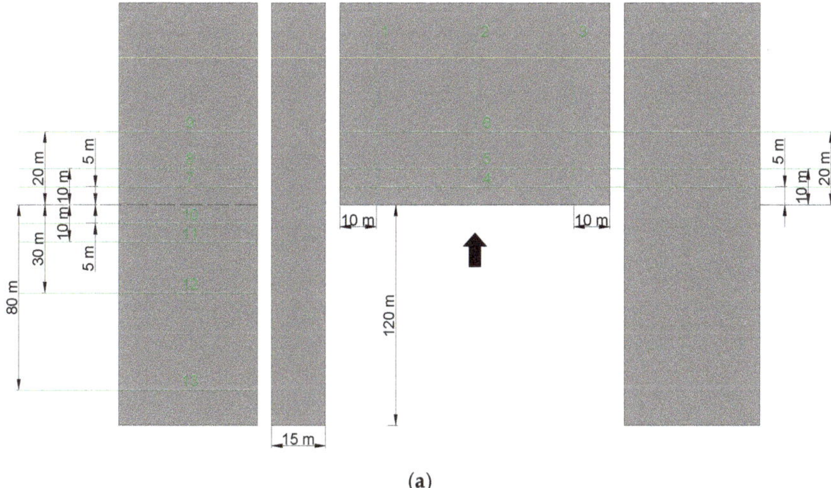

(a)

Figure 9. *Cont.*

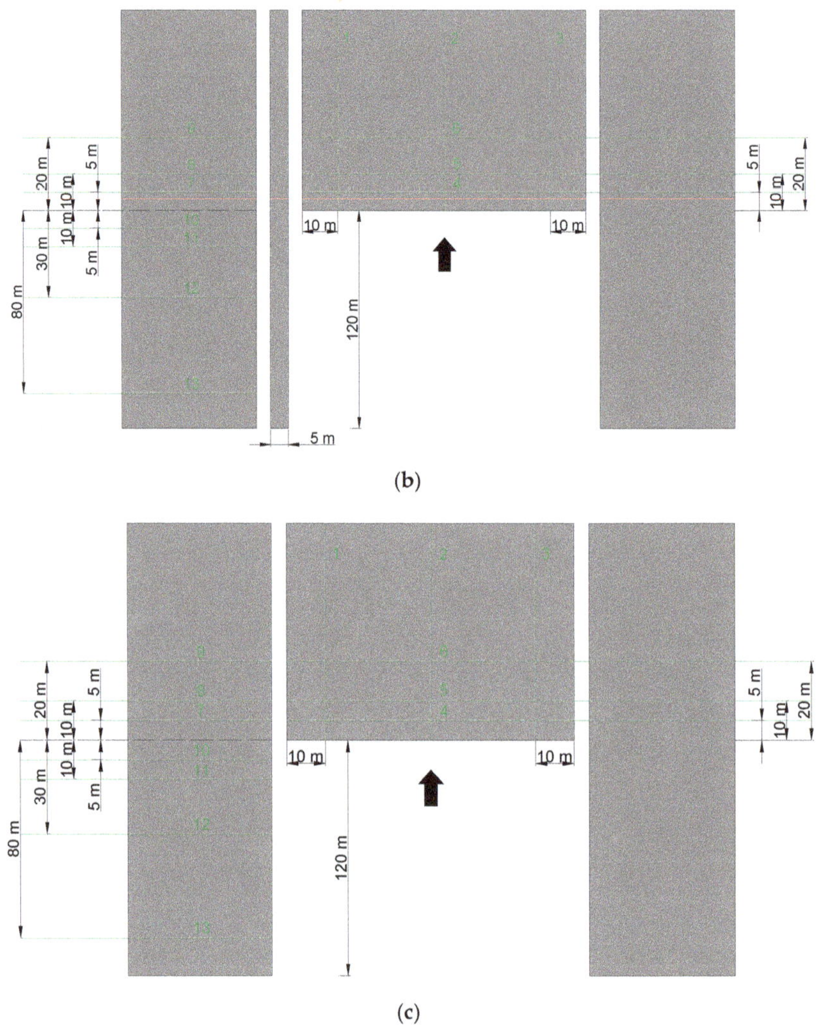

(b)

(c)

Figure 9. Location of monitoring lines. (**a**) Location of monitoring lines of large pillar mining; (**b**) Location of monitoring lines of small pillar mining; (**c**) Location of monitoring lines of non-pillar mining.

5.1.1. Stress Distribution in Front of Working Face

(1) Stress distribution along working face strike

The vertical stress distribution nephograms and stress distribution curves at monitoring line 1 are shown in Figure 10. Figure 10 shows that the vertical stress distribution along the working face strike was similar under the three mining methods, and the advanced mining stress concentration area of the remaining pillars mining was larger than that of non-pillar mining, and the stress value was slightly higher. The peak of advanced mining stress was located 10 m in front of the working face, which was about 3.3 times that of the mining height of the working face. Among them, the peak stress of non-pillar mining was 14% lower than that of large pillar mining and 10% lower than that of small pillar mining. Within 3 m from the face, was the pressure-released zone of the stope; the vertical stress value was lower than the original rock stress, and the coal body mainly underwent plastic deformation. Under the coupling effect of the overlying goaf and mining abutment pressure of this coal seam, a pressure boost belt was formed within the range of 3–33 m from the working face. The elastic deformation of coal body in this area caused the accumulation of elastic deformation energy, and the bearing capacity was higher. Within the range of 33–90 m from the working face, under the influence of pressure relief in the goaf of 8 # coal seam, the stress value was low. From 90–110 m away from working face, the residual pillar of 8 # coal seam formed a stress concentration zone here, and the stress value rose again. The stress distribution law at monitoring lines 2 and 3 was similar to that at monitoring line 1. The peak of advance mining stress was 10 m ahead of the working face. The peak stress of non-pillar mining was 23.88 MPa and 19.92 MPa respectively, which were 8%, 7% and −1%, 1% lower than that of traditional mining respectively (as shown in Table 3).

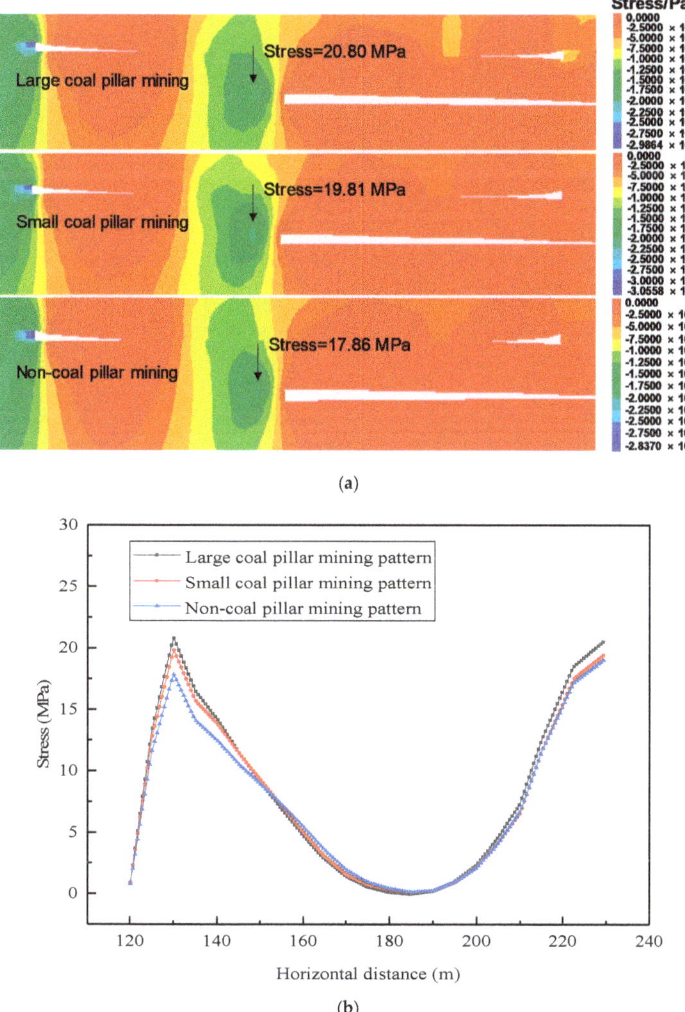

(a)

(b)

Figure 10. Vertical stress nephogram and stress distribution curve at the monitoring line 1 along the working face strike. (**a**) Vertical stress nephogram; (**b**) Stress distribution curve.

Table 3. Key parameters of the stress distribution characteristics along the working face strike.

Monitoring Line	Mining Mode	Stress Trend	Peak Position (m)	Peak Size (MPa)	Peak Reduction of (%)
1	Large	Similar	10	20.80	14
	Small		10	19.81	10
	GERRCP		10	17.86	Reference quantity
2	Large	Similar	10	26.05	8
	Small		10	25.63	7
	GERRCP		10	23.88	Reference quantity
3	Large	Similar	10	19.64	−1
	Small		10	20.05	1
	GERRCP		10	19.92	Reference quantity

Based on the above analysis, the key parameters of the stress distribution characteristics along the working face strike in the simulation results were summarized, and a matrix-type chart with resulting values and illustrations was made, so they can be visualized and compared in a single view (as shown in Table 3).

(2) Stress distribution along the inclination of working face

The vertical stress distribution nephograms and stress distribution curves at monitoring line 4 are shown in Figure 11. Figure 11 showed that, the distribution law of vertical stress along the inclination of the working face was similar under the three mining modes, that was, the stress data showed that the vertical stress increased first and then decreased. The stress curves of large and small pillars basically coincided, and the stress value was slightly higher than that of the non-pillar mining method. At the edge of the ventilation roadway, the three stresses were 10.8 MPa, 10.2 MPa and 7.6 MPa, respectively. The stress of non-pillar mining at the edge of the ventilation roadway was 30% and 25% lower than that of the large pillar and small pillar, respectively. At the distance of 10 m from the ventilation roadway, the three stresses reached 12.8 MPa, 12.3 MPa and 11.1 MPa. The stress of the non-coal pillar mining was 13% and 10% lower than that of the large pillar and small pillar mining, respectively. It could be seen that within the distance of 10 m from the ventilation roadway, the vertical stress increased greatly, while the non-coal pillar mining showed the characteristics of low stress compared with the traditional mining method. The main reason was that the roof rock beam was cut off by pre-cracking the roof cutting, which transformed it from a long-wall beam to short-wall beam and cut off the stress transfer between roofs, which improved the stress condition of surrounding rock obviously. The stress distribution law at monitoring lines 5 and 6 was similar to that at monitoring line 4. The stress increased rapidly within 10 m from the ventilation roadway and then slowed down. The stress distribution of traditional mining was axisymmetrical about the middle of the working face, while the stress of the side of the roof cutting of non-pillar mining was significantly lower than that of the non-roof cutting. Among them, the peak stress of non-pillar mining was 23.98 MPa and 19.41 MPa respectively, which were 7%, 7% and 10%, 8% lower than that of traditional mining respectively (as shown in Table 4).

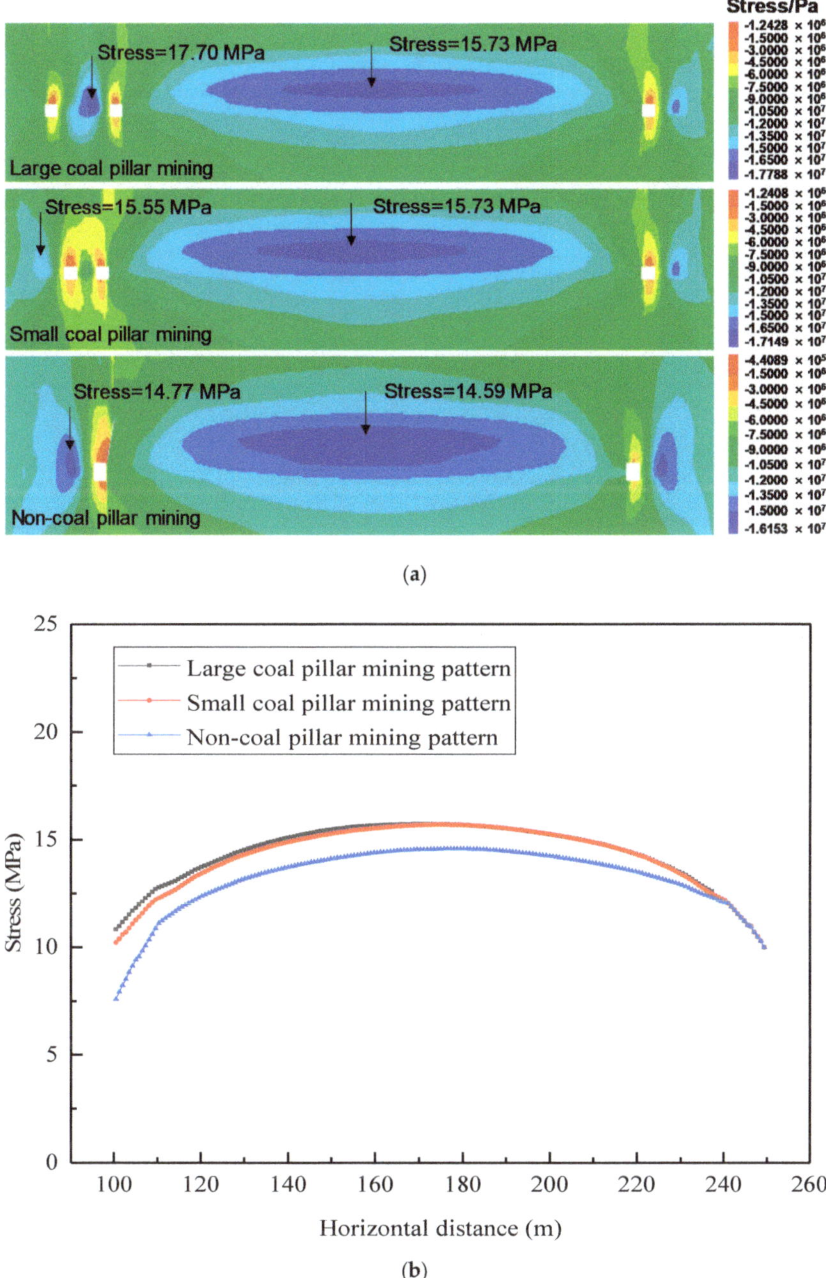

(**a**)

(**b**)

Figure 11. Vertical stress nephogram and stress distribution curve at the monitoring line 4 along the inclination of the working face. (**a**) Vertical stress nephogram; (**b**) Vertical stress distribution curve.

Table 4. Key parameters of the stress distribution characteristics along the inclination of the working face.

Monitoring Line	Mining Mode	Stress Concentration around Roadway (MPa)	Peak Reduction (%)	Stress Concentration of Working Face (MPa)	Peak Reduction (%)
4	Large	17.70	17	15.73	7
	Small	15.55	5	15.73	7
	GERRCP	14.77	Reference quantity	14.59	Reference quantity
5	Large	15.34	12	25.87	7
	Small	14.60	8	25.79	7
	GERRCP	13.50	Reference quantity	23.98	Reference quantity
6	Large	14.45	13	21.60	10
	Small	13.40	7	21.18	8
	GERRCP	12.50	Reference quantity	19.41	Reference quantity

Through the above analysis, the key parameters of the stress distribution characteristics along the inclination of the working face were summarized, and a matrix-type chart with resulting values and illustrations was made (as shown in Table 4).

5.1.2. Stress Distribution in Lateral Direction of Working Face

(1)   Stress distribution in left front of working face

The vertical stress distribution curves at monitoring line 7 are shown in Figure 12. Figure 12 showed that there were two peaks along the working face inclination under the condition of large and small pillars mining. The locations of peak stress were 6 m, 23 m and 2 m, 15 m away from the edge of ventilation roadway respectively, with sizes of 17.7 MPa, 12.3 MPa and 10.8 MPa, 15.6 MPa. The stress concentration factors were 2.6, 1.8 and 1.6, 2.3. There was a wave peak along the inclination of the working face under non-pillar mining, which was located 5 m outside ventilation roadway, with a size of 14.8 MPa, and the stress concentration factor was 2.2. Compared with the peak stress, it could be seen that the non-pillar mining decreased by 16% and 5% for large and small pillar mining respectively, indicating that the non-pillar mining had less influence on the advanced mining stress of the roadway surrounding rock, that was, the technology of roof cutting and pressure relief cut off the stress transfer between the working face and roadway roof, and the stress control effect was better. By analyzing the location of peak stress, it could be seen that the peak stress of the retaining pillar mining mode was 6 m outside the roadway, while that of the non-pillar mining mode was 5 m outside the roadway, which indicated that the plastic zone of surrounding rock in the advanced position of a roadway under non-pillar mining mode was smaller. In addition, the advantages of this mining method on the stress control of surrounding rock were further reflected. The stress distribution law at monitoring lines 8 and 9 was similar to that at monitoring line 7. The advance abutment pressure in lateral direction of the working face was lower than that of monitoring line 7. The peak stress of traditional mining was 6 m outside the ventilation roadway, while the peak stress of non-pillar mining was 5 m from the ventilation roadway. The peak stresses of non-pillar mining were 13.5 MPa and 12.5 MPa respectively, which were 12%, 8% and 13%, 7% lower than that of traditional mining respectively (as shown in Table 5).

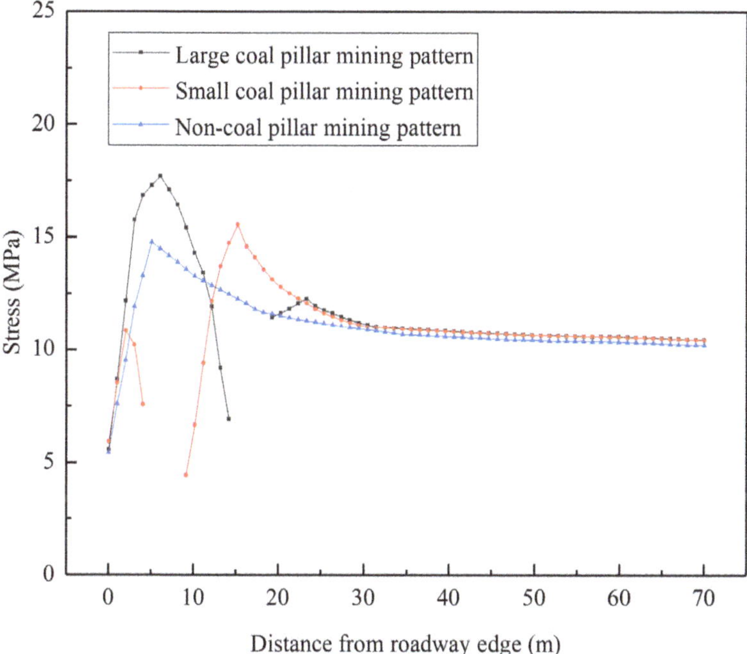

Figure 12. Vertical stress distribution curve at monitoring line 7 in lateral direction of the working face.

Table 5. Key parameters of the stress distribution characteristics in the left front of the working face.

Monitoring Line	Mining Mode	Number of Peaks	Peak Position (m)	Peak Size (MPa)	Peak Reduction (%)
7	Large	2	6	17.7	16
	Small	2	6	15.6	5
	GERRCP	1	5	14.8	Reference quantity
8	Large	2	6	15.3	12
	Small	2	6	14.6	8
	GERRCP	1	5	13.5	Reference quantity
9	Large	2	6	14.4	13
	Small	2	6	13.4	7
	GERRCP	1	5	12.5	Reference quantity

Combined with the analysis results, the key parameters of the stress distribution characteristics in the left front of the working face in the simulation results were summarized, and a matrix-type chart with resulting values and illustrations was made (as shown in Table 5).

(2) Stress distribution in left rear of working face

The vertical stress distribution curves at monitoring line 10 are shown in Figure 13. Figure 13 showed that, there were obvious differences in the stress distribution characteristics along the inclination of the working face under the three mining methods, which were mainly reflected in the stress concentration area and the degree of stress concentration. Figure 13b intuitively showed that, in the lateral direction of the working face, the stress of the large pillar mining method was the largest, followed by the small pillar mining and the minimum of the non-pillar mining method. Under the traditional mining method, the vertical stress showed that the distribution characteristics were saddle-shaped, in which the large pillar mining first formed the stress concentration in the large pillar, that was, the first wave peak appeared. Then a small degree of concentration appeared at the roadway edge of the 9102 working face, forming a second wave peak, indicating that the stope underground pressure was mainly borne by the large pillar behind the working face. The small pillar mining was obviously different from large pillar mining. The larger stress concentration was located at the 9102 working face, and a smaller concentration occurred in the small pillar, indicating that the small pillar had a low bearing capacity behind the working face due to the limitation of coal pillar width, and the stress bearing area transferred to the deep part of the working face. Because of the elimination of pillars for the non-pillar mining method, the weight of overlying strata was borne by the solid coal on the working face, and a stress bearing area was formed in the 9102 working face. Therefore, the stress concentration position was transferred from the traditional pillar area to the deep part of the working face, and the peak of non-pillar mining was the smallest, which was 21% and 10% lower than the traditional mining method of large pillar and small pillar respectively. It could be seen that the technique of gob-side roof cutting effectively reduced the stress concentration in stope and optimized the stress distribution in the lateral direction of the working face. The stress distribution law at monitoring lines 11, 12, and 13 was similar to that at monitoring line 10. Traditional mining methods still showed the characteristics of double-stress wave peak distribution, in which the high stress concentration in large pillar, small pillar and no pillar mining methods occurred in the pillar, 9102 working face and deep part of the solid coal, respectively. The peak stresses were the smallest of non-pillar mining, which were 18%, 10%; 16%, 6% and 13%, 7% lower than that of traditional mining respectively (as shown in Table 6).

Table 6. Key parameters of the stress distribution characteristics in the left rear of the working face.

Monitoring Line	Mining Mode	Number of Peaks	Peak Position (m)	Peak Size (MPa)	Peak Reduction (%)
10	Large	2	6	21.52	21
	Small	2	6	18.80	10
	GERRCP	1	5	16.99	Reference quantity
11	Large	2	6	21.68	18
	Small	2	6	19.84	10
	GERRCP	1	5	17.83	Reference quantity
12	Large	2	6	23.70	16
	Small	2	6	21.17	6
	GERRCP	1	5	20.00	Reference quantity
13	Large	2	6	27.17	21
	Small	2	6	22.16	3
	GERRCP	1	5	21.58	Reference quantity

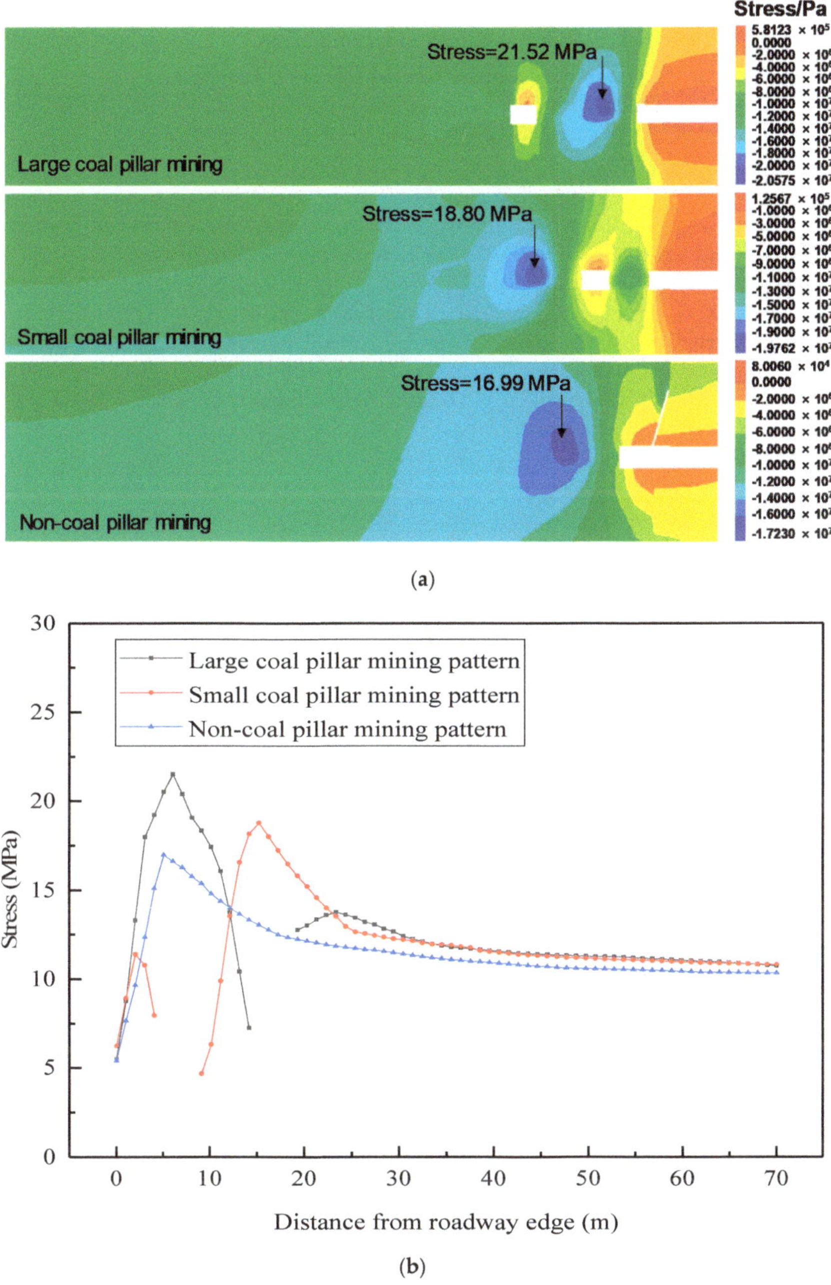

(a)

(b)

Figure 13. Vertical stress nephogram and stress distribution curve at monitoring line 10 in lateral direction of working face. (**a**) Vertical stress nephogram; (**b**) Vertical stress distribution curve.

With the help of the above analysis, the key parameters of the stress distribution characteristics in the left rear of the working face were summarized, and a matrix-type chart with resulting values and illustrations was made (as shown in Table 6).

5.2. Distribution Law of Horizontal Stress

In order to analyze the distribution of horizontal stress, the model was sliced horizontally. As this study focused on the analysis of the distribution characteristics of the mining stress field in 9 # coal seam, therefore, in the middle of the height range of the face, horizontal slices were made to analyze the evolution law of horizontal stress in the front and side of the working face.

5.2.1. Stress Distribution in Front of Working Face

When the working face advanced to 120 m, the horizontal stress distribution in the front and side of the working face under the three mining modes is shown in Figure 14. In front of the working face, the horizontal stress increased first and then decreased. Under the large pillar mining mode, the stress rising area was large with a span of about 31 m, and the peak stress was 20.5 MPa. The peak stress was located 10 m in front of the working face. Under the small pillar mining mode, the stress rising area was slightly smaller, the span was about 23 m, the peak stress was 16.8 MPa, and the stress wave peak was located 10 m ahead of the working face. The area of stress rise was the smallest under non-pillar mining, with a span of about 9 m, a peak stress of 15.2 MPa and the stress wave peak was located 10 m ahead of the working face.

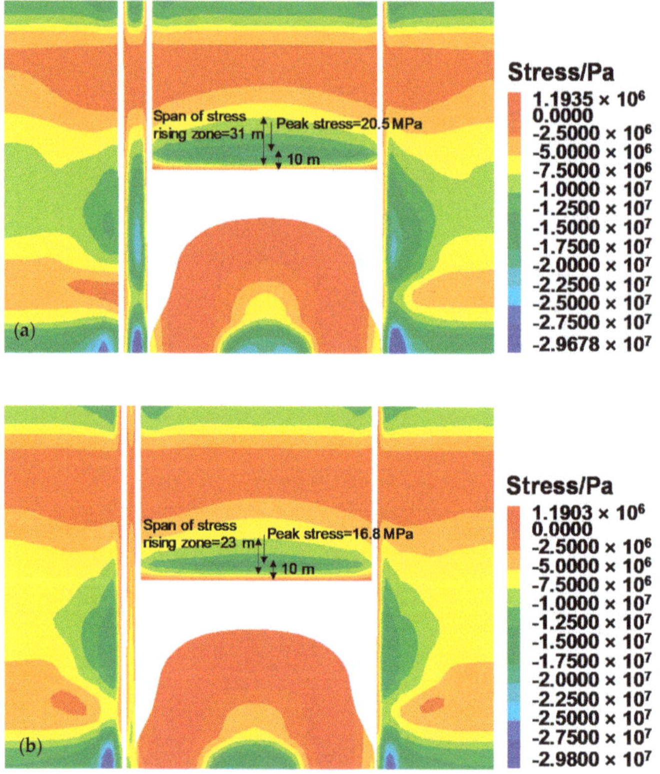

Figure 14. *Cont.*

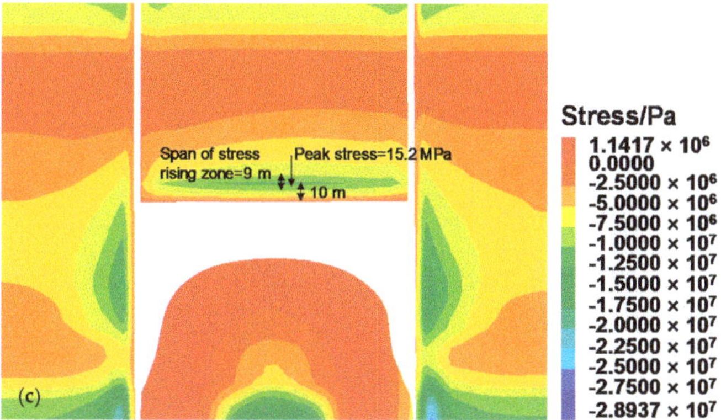

Figure 14. Horizontal stress distribution in front and side of the working face under the three mining modes. (**a**) Large pillar mining; (**b**) Small pillar mining; (**c**) Non-pillar mining.

The stress data showed that the non-pillar mining had the characteristics of a small stress rising area and small stress value. Therefore, it could be concluded that non-pillar mining technology could indeed improve the distribution of the surrounding rock stress field in front of the working face, which was of great significance to reduce the surrounding rock stress in the stope.

5.2.2. Stress Distribution in Lateral Direction of Working Face

Figure 14 showed that, in the lateral direction of the working face, high stress concentration was formed in the pillar under large pillar mining, and the stress concentration zone was also formed in the coal body of the 9102 working face; the stress concentration in the pillar under the small pillar mining was not obvious, but the phenomenon of a large stress concentration zone was formed in the 9102 working face; the phenomenon of stress rise occurred in the 9102 working face under non-pillar mining, but the area of the concentrated area was lower than that of the remaining pillars mining mode. This was because the retaining roadway by roof-cutting cut off the stress transfer between the working face and its lateral direction, so the value of the stress in the lateral direction of the working face was reduced and had the characteristics of optimizing the stress distribution in the lateral direction of the working face.

6. Mine Pressure Monitoring

6.1. Stress Monitoring

In former sections, the distributions of advance stress and lateral abutment pressure in 9101 working face were studied by numerical simulation. On-site measurements of the strike and lateral abutment pressure of 9101 working face were carried out in this section. With the help of monitoring data, the distribution law of mine pressure was analyzed to verify the numerical simulation results.

6.1.1. Monitoring System

At present, the KJ550 on-line stress monitoring system was used in Xiashanmao coal mine. The system consists of three main components (as shown in Figure 15): (1) The monitoring host and data processing and analysis system on the ground, adopting a high-performance integrated server workstation and high-performance computer, which can realize the functions of data storage and analysis. (2) The underground monitoring substation (including a power supply), whose main components are industrial high-performance electronic accessories. (3) The pressure sensing system arranged along the entry on both sides of the working face is composed of the borehole stress meter with hydraulic oil as the pressure-sensing medium and high-precision pressure sensors.

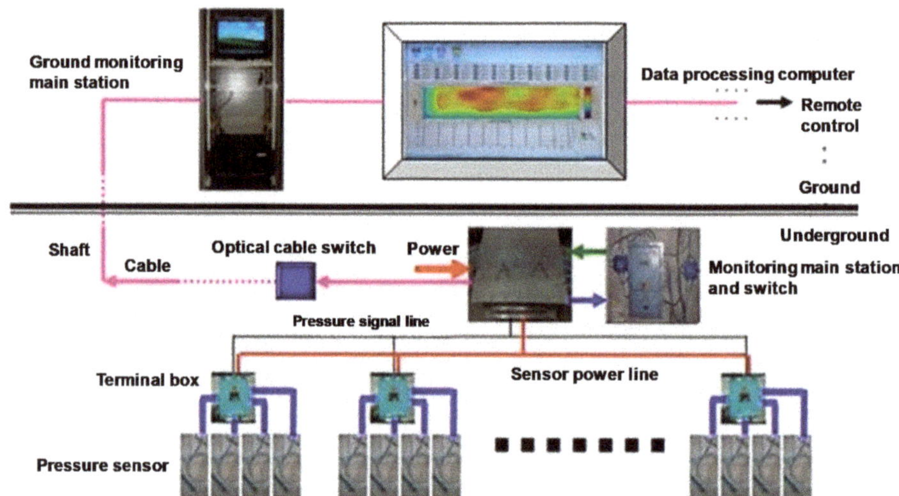

Figure 15. Schematic diagram of remote monitoring system.

The structure and working principle of the system are shown in Figure 15. Under the influence of mining, the pressure of coal and rock mass around the pressure sensor installed in the survey area changes. The sensor receives the pressure fluctuation signal and transmits it to the terminal box, which is transmitted through the pressure signal line to the monitoring main station and switch. The monitoring main station and switch convert the electrical signal into an optical signal, then transmit the optical signal to the ground-monitoring main station through the optical cable switch, and then transmit it to the data-processing computer for stress data processing, so as to realize remote control.

Real-time monitoring technology aims to detect and identify various potential abnormalities and faults, realizing real-time monitoring and early warning of dangerous situations, so as to take necessary measures for minimizing performance degradation and economic costs and avoid catastrophic situations [29–31]. Similarly, the KJ550 monitoring system can monitor the stress of coal body and rock mass in front of the working face and around the roadway in real time, and monitor and display the dynamic stress nephogram in front of the working face in real time, so as to realize the real-time monitoring and early warning of the hazardous area of rock burst. At the same time, it has the functions of remote control, data analysis and remote maintenance. Through the remote data processing and early warning center, the monitoring data can be analyzed and processed in real time.

6.1.2. Monitoring Programme

Figure 16 shows the layout of the 9101 working face and real-time monitoring system in Xiashanmao coal mine. It can be seen from the figure that two stations are arranged in the 9101 tail entry, and the measuring points of stations 1 and 2 are arranged in the solid coal side of the roadway to monitor the change of lateral abutment pressure of the coal side. There are four measuring points in stations 1 and 2, with buried depth of 3, 6, 9 and 12 m respectively; the distance between measuring points is 0.5–1 m, and the distance between measuring station 2 and station 1 is 100 m; Stations 3–22 are arranged in the 9101 working face to monitor the change of strike abutment pressure in working face, and each station is equipped with two measuring points with buried depths of 5 and 10 m. The distance between stations is 2 m and distance between measuring points is 0.5 m.

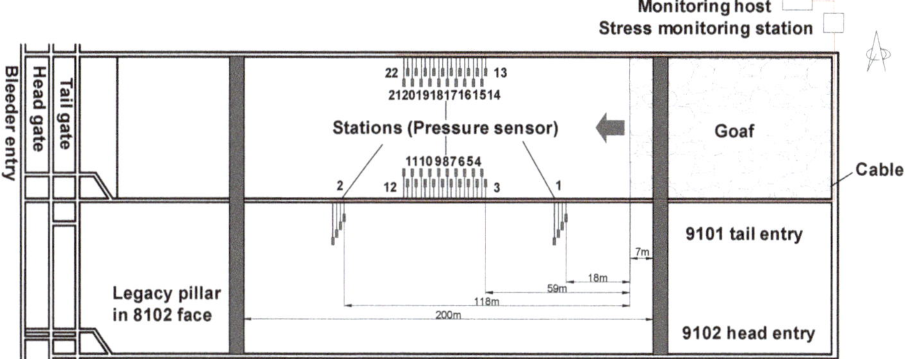

Figure 16. Layout of the 9101 working face and real-time monitoring system.

6.2. Analysis of Main Monitoring Results

6.2.1. Distribution Characteristics of Strike Abutment Pressure of Working Face

There are many stations for coal body layout in the inner side of drift in the 9101 working face. The monitoring results of two stations are selected for analysis below. Figure 17 is the relative vertical stress variation curve of each measuring point at station 3 of the lower drift and station 20 of the upper drift of the 9101 working face respectively.

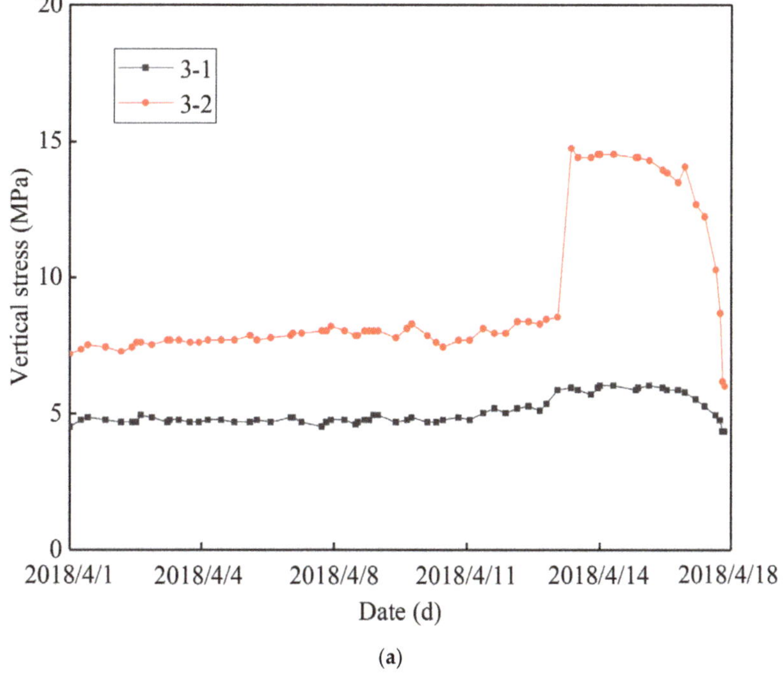

(**a**)

Figure 17. *Cont.*

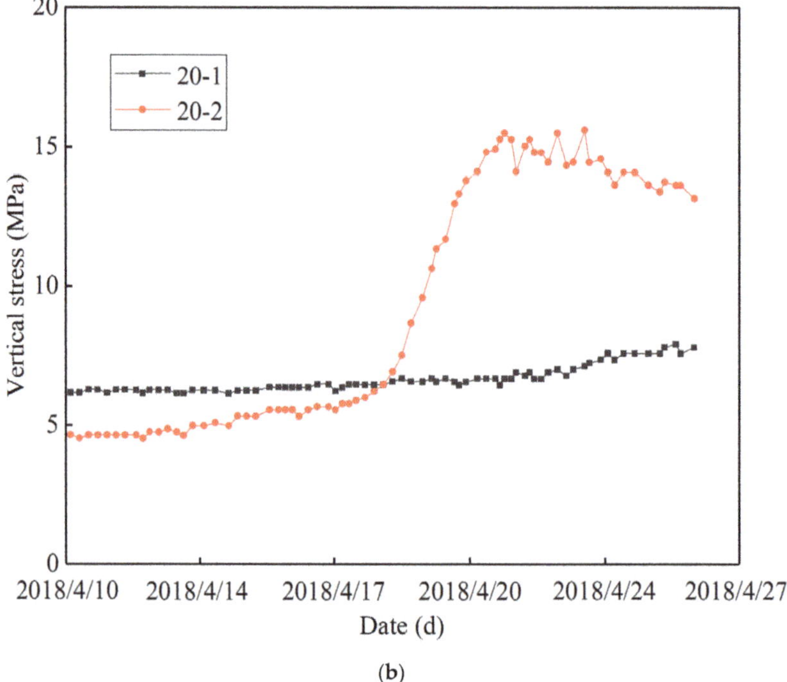

(b)

Figure 17. Relative vertical stress variation curve. (**a**) Relative vertical stress variation curve of station 3; (**b**) Relative vertical stress variation curve of station 20.

Figure 17a showed that, the vertical stress of the measuring point of the station increased significantly on 4/13, indicating that the measuring point began to enter the influence range of strike abutment pressure on the working face. At this time, the measuring point was 42.6 m away from the working face, that is, the influence range of strike abutment pressure on working face was 42.6 m. The vertical stress of the measuring point began to decrease on 4/17, indicating that the measuring point began to enter the plastic zone, which was 9 m away from the working face. It could be seen that the peak position of strike abutment pressure in the working face was 9 m away from the coal wall, and the continuous influence distance of abutment pressure was 33.6 m.

Figure 17b showed that the vertical stress of the 20-1 measuring point began to reach the peak abutment pressure on 4/20, when the measuring point was 54.2 m away from the working face, that is, the influence range of advance abutment pressure of the working face was 54.2 m. The vertical stress of the measuring point began to decrease significantly on 4/24, indicating that the measuring point has entered the plastic zone, which was 12 m away from working face, i.e., the peak position of the strike abutment pressure of the working face was 12 m away from the coal wall, and the continuous influence distance of the abutment pressure was 42.2 m.

6.2.2. Distribution Characteristics of Lateral Abutment Pressure of Working Face

Station 1 and Station 2 were continuously monitored for 45 days (Signal cable interruption in goaf at later stage). During this period, the 9101 working face pushed forward from 18 m in front of station 1 to 35 m after station 2, with a total of 153 m, and the distance of the signal cable entering the goaf was 60 m. Figure 18 showed the relative vertical stress variation curves of stations 1 and 2 respectively.

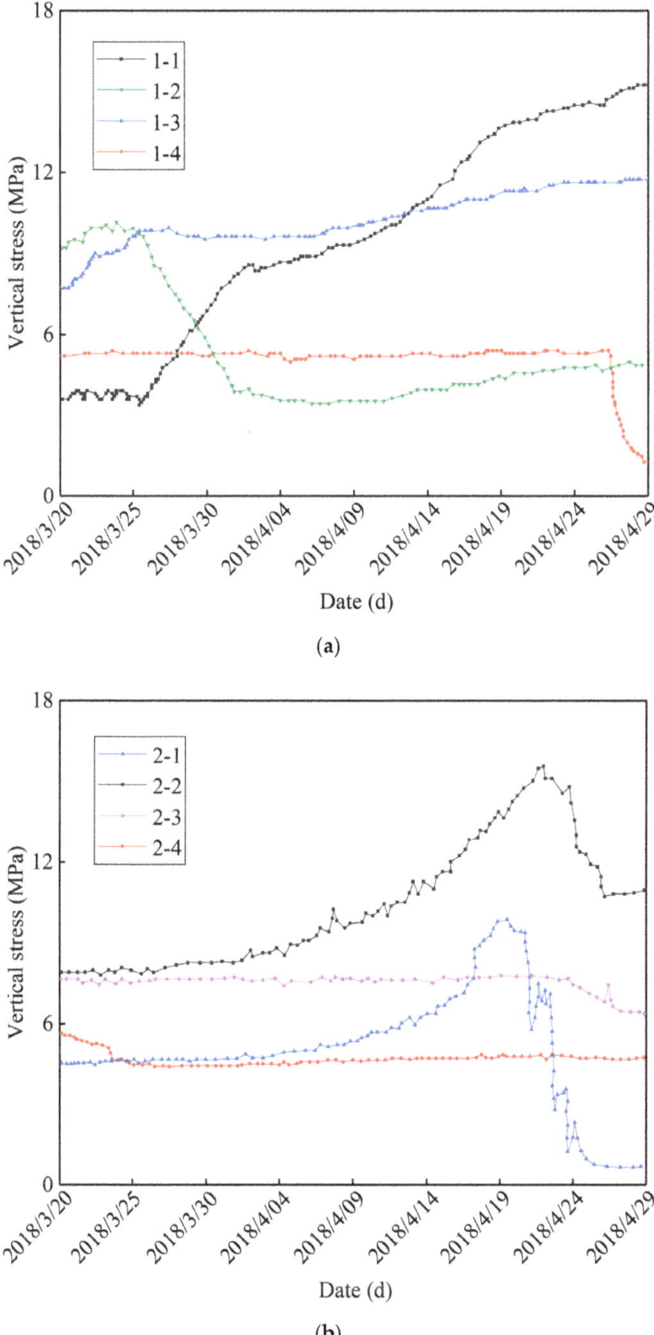

Figure 18. Relative vertical stress variation curve. (**a**) Relative vertical stress variation curve of station 1; (**b**) Relative vertical stress variation curve of station 2.

Figure 18a showed that, the vertical stress of the 1-2 measuring point reached its peak on 3/23, when the working face was 20 m ahead of the measuring point. The vertical stress of 1-3 measuring point reached the first peak on 3/26, when the working face position was 3.7 m behind the measuring point, and began to decrease slightly on 4/2, when the working face position was 40 m behind the measuring point. With the advancement of the working face, the vertical stress of each measuring point increased periodically. On 4/27, the vertical stresses of measuring points 1-1, 1-2 and 1-3 tended to be stable, and the vertical stress of measuring point 1-4 dropped sharply. At this time, the working face was located 60 m behind station 1.

Figure 18b showed that, the vertical stress of the measuring points 2-1, 2-2 and 2-3 began to rise on 3/26, when the working face was 96.3 m in front of station 2, and the vertical stress of the measuring point 2-1 reached its peak on 4/19, when the working face was 19 m in front of the measuring point. The vertical stress of the measuring point 2-2 reached its peak on 4/22, when the working face was 3 m behind the measuring point. The vertical stress of measuring point 2-3 increased slightly from 4/17 to 4/23 and decreased considerably on 4/24, when the working face advanced to 15 m behind the measuring point. From 4/18 to 4/25, the vertical stress of the 2-4 measuring point increased slightly, then decreased slightly. At this time, the working face was located 18 m behind the measuring point.

6.2.3. Comparative Analysis of Abutment Pressure

In order to verify the numerical simulation results, the field monitoring data were compared with simulation results. When the working face advanced to station 3, the abutment pressure in front of the working face was shown in Figure 19. The field monitoring results showed that the abutment pressure in front of the working face increased first and then decreased, reaching a peak at 9 m ahead of the working face, and the stress distribution tends to be stable as it was farther away from the working face. The numerical simulation results also showed the distribution law of first increasing and then decreasing, reaching the stress peak at 10 m ahead of the working face, and the distribution law of abutment pressure was consistent with the monitoring results.

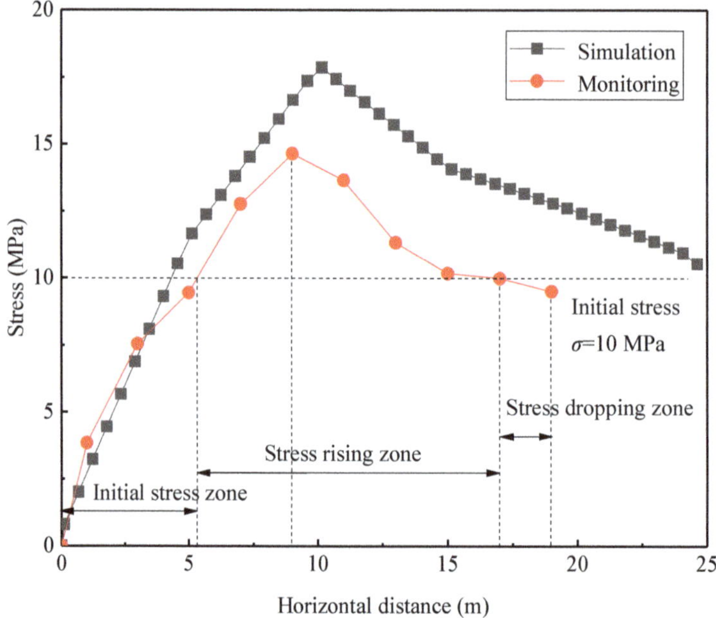

Figure 19. Distribution law of abutment pressure in front of the working face.

When the working face advanced to 146 m, station 2 lagged behind the working face by 28 m. The stress data extracted from the four measuring points could reflect the distribution characteristics of lateral abutment pressure, as shown in Figure 20. Field monitoring results showed that the lateral abutment pressure of the working face continued to increase within a range of 0–6 m from the roadway edge, reaching a peak value of 13.7 MPa at 6 m, and then gradually decreased. The numerical simulation results also showed a trend of first increasing and then decreasing, and reached the peak at 5 m away from the roadway edge. The stress distribution law was consistent with the monitoring results.

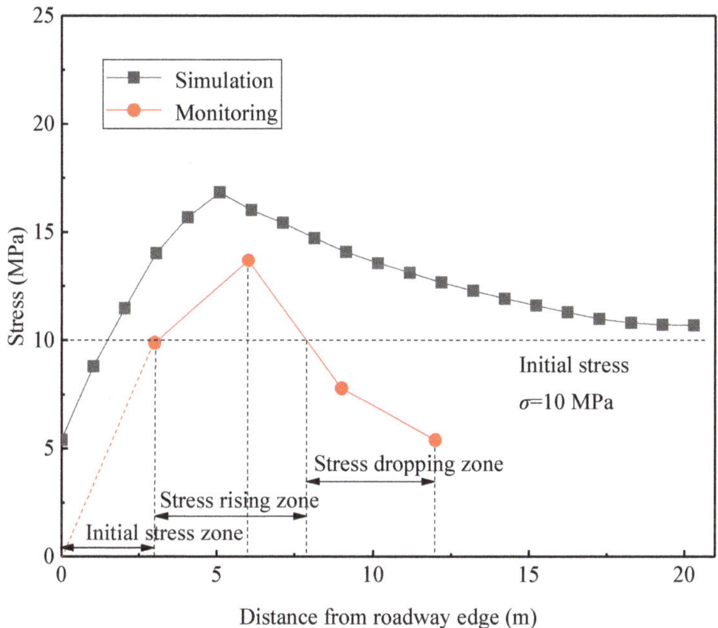

Figure 20. Distribution law of lateral abutment pressure in working face.

7. Conclusions

Taking Xiashanmao coal mine in Shanxi Province as the engineering background, the stress distribution in the process of coal seam mining was analyzed by establishing a numerical model, and the following conclusions were drawn:

(1) Based on the new technology of gob-side entry retaining by roof cutting without pillar, the mechanical model of the roof structure was established. Through the mechanical analysis of the model, the extension depth of the plastic zone on the solid coal side and the supporting load in the mining roadway were obtained under the condition of this technology, which provided certain theoretical support for the design of roadway support in the field.

(2) Through numerical simulation, the distributions of strike and lateral abutment pressure of the 9101 working face were obtained. Among them, the vertical stress distribution was as follows. In front of the working face: along the working face strike, the stress increased first, then decreased and then increased. The peak value of advance stress was formed at 10 m in front of the working face, and the peak value of non-pillar mining was reduced by 8–10% and 8–14% respectively compared with traditional mining. Along the inclination of the working face, the stress distribution of non-pillar mining and traditional mining was asymmetrical and symmetrical respectively, and the stress increased first and then decreased, while the stress at the side of roof cutting was significantly lower than that of the non-cutting. In the left side of the working face: along the inclination of the working face, the stress

of the non-pillar mining increased at first and then decreased, while the traditional mining showed the saddle-shaped distribution characteristics, and there were one and two peak stresses respectively. The peak stress of non-pillar mining was the smallest, which was 12–21% and 3–10% lower than that in the mining with large pillar and small pillar, respectively. The peak stress position in the former was closer to the mining roadway.

The horizontal stress distribution was as follows. In front of the working face: The stresses of the three increased first and then decreased. The peak position was 10 m ahead of the working face, in which the area, span and peak value of the stress rising area under large pillar mining were the largest, while those of non-pillar mining were the smallest. In the left side of the working face: high stress concentration was formed in the large pillar, not obvious in the small pillar, but a large area of stress concentration was formed in the 9102 working face. Because of the technology of roof cutting and retaining roadway, the phenomenon of stress increase appeared in the 9102 working face for non-pillar mining, but the area of the concentrated area was lower than that of traditional mining.

The results show that, in front of the working face: The stress increase area and peak stress of non-pillar mining were smaller than that of traditional mining. Around the retaining roadway, the stress transfer between roof rock beams was cut off by the roof cutting and pressure relief, which effectively weakens the stress concentration in deep surrounding rock. In the left side of the working face: The number of stress peaks was small and the peak stress was small for non-pillar mining. The stress-bearing areas of three mining methods were different, which were big coal pillar of large pillar mining, 9102 working face of small pillar mining and 9102 working face of non-pillar mining. The width of the plastic zone of surrounding rock of non-pillar mining was smaller and the bearing capacity was higher.

(3) The mine pressure monitoring data showed that the influence range of strike abutment pressure of the working face was 42.6–54.2 m. The distance between the peak position of abutment pressure and coal wall was 9–12 m. The sustained influence distance of abutment pressure was 33.6–42.2 m. The peak value of vertical stress at the deep-buried measuring point lagged behind that at the shallow-buried one. In the lateral side of the working face, the influence distance of mining in front of the 9101 working face was 48 m. With the advancement of the working face, the influence on it became more and more serious. The breaking and rotation of the hard strata overlying the working face caused the vertical stress of the shallow buried measuring point to rise. When 86 m ahead of the station, the stress at the measuring point was basically not affected by mining and reached a stable state. By comparing the field mine pressure monitoring results with the numerical simulation results, it could be found that the simulation results were consistent with the monitoring results.

Author Contributions: X.S. and Y.L. conceived and designed the research. Y.L. performed the numerical simulation. S.S. was involved in the construction of numerical model. J.W., X.C. and J.L. were involved in the numerical simulation. Y.L. analyzed the data and wrote the paper.

Funding: This work was supported by the National Key Research and Development Plan of China (2016YFC0600901), the National Natural Science Foundation of China (Grant No. 51874311), the Special Fund of Basic Research and Operating of China University of Mining & Technology, Beijing (Grant No. 2009QL03), which are gratefully acknowledged.

Conflicts of Interest: The authors declare no conflict of interest.

References

1. Qian, M.G. The equilibrium condition of overlying strata in stope. *J. China Min. Univ.* **1981**, *2*, 34–43.
2. Song, Z.Q. The basic law of overlying strata movement in stope. *J. Shandong Min. Univ.* **1979**, *1*, 64–77.
3. Zhu, Z.Q.; Song, Y.; Liu, Y.X.; Chen, M.B.; Hua, Y.F. The behavior regularity law and application of abutment pressure in stope. *J. Shandong Min. Univ.* **1982**, *1*, 1–25.
4. He, M.C.; Zhu, G.L.; Guo, Z.B. Longwall mining "cutting cantilever beam theory" and 110 mining method in China—The third mining science innovation. *J. Rock Mech. Geotech. Eng.* **2015**, *5*, 483–492. [CrossRef]

5. Zhang, G.F.; He, M.C.; Yu, X.P.; Huang, Z.G. Research on the technique of non-pillar mining with gob-side entry formed by advanced roof caving in the protective seam in Baijiao coal mine. *J. Min. Saf. Eng.* **2011**, *28*, 511–516.

6. Yang, J.; He, M.C.; Cao, C. Design principles and key technologies of gob side entry retaining by roof pre-fracturing. *Tunn. Space Technol.* **2019**, *90*, 309–318. [CrossRef]

7. Gao, Y.B.; Wang, Y.J.; Yang, J.; Zhang, X.Y.; He, M.C. Meso-and macroeffects of roof split blasting on the stability of gateroad surroundings in an innovative nonpillar mining method. *Tunn. Space Technol.* **2019**, *90*, 99–118. [CrossRef]

8. Wang, Q.; He, M.C.; Yang, J.; Gao, H.K.; Jiang, B.; Yu, H.C. Study of a no-pillar mining technique with automatically formed gob-side entry retaining for longwall mining in coal mines. *Int. J. Rock Mech. Min. Sci.* **2018**, *110*, 1–8. [CrossRef]

9. Guo, Z.B.; Zhang, L.; Ma, Z.B.; Zhong, F.X.; Yu, J.C.; Wang, S.M. Numerical investigation of the influence of roof fracturing angle on the stability of gob-side entry subjected to dynamic loading. *Shock Vib.* **2019**, *2019*, 1434135. [CrossRef]

10. Zhen, E.Z.; Gao, Y.B.; Wang, Y.J.; Wang, S.M. Comparative study on two types of nonpillar mining techniques by roof cutting and by filling artificial materials. *Adv. Civ. Eng.* **2019**, *2019*, 5267240. [CrossRef]

11. Guo, Z.B.; Wang, Q.; Li, Z.H.; He, M.C.; Ma, Z.B.; Zhong, F.X.; Hu, J. Surrounding rock control of an innovative gob-side entry retaining with energy-absorbing supporting in deep mining. *Int. J. Low Carbon Technol.* **2019**, *14*, 23–35. [CrossRef]

12. Hu, J.Z.; He, M.C.; Wang, J.; Ma, Z.M.; Wang, Y.J.; Zhang, X.Y. Key parameters of roof cutting of gob-side entry retaining in a deep inclined thick coal seam with hard roof. *Energies* **2019**, *12*, 934. [CrossRef]

13. Singh, G. Assessment of goaf characteristics and compaction in longwall caving. *Min. Technol.* **2011**, *120*, 222–232. [CrossRef]

14. Zhu, Z.; Zhu, C.; Yuan, H. Distribution and evolution characteristics of macroscopic stress field in gob-Side entry retaining by roof cutting. *Geotech. Geol. Eng.* **2019**, *37*, 2963–2976. [CrossRef]

15. Zhang, C.L.; Yu, L.; Feng, R.M.; Zhang, Y.; Zhang, G.J. A numerical study of stress distribution and fracture development above a protective coal seam in longwall mining. *Processes* **2018**, *6*, 146. [CrossRef]

16. Liu, X.J.; Li, X.M.; Pan, W.D. Analysis on the floor stress distribution and roadway position in the close distance coal seams. *Arab. J. Geosci.* **2016**, *9*, 83–91.

17. Zhang, W.; Zhang, D.S.; Qi, D.H.; Hu, W.M.; He, Z.M.; Zhang, W.S. Floor failure depth of upper coal seam during close coal seams mining and its novel detection method. *Energy Explor. Exploit.* **2018**, *36*, 1265–1278. [CrossRef]

18. Xu, Y.L.; Pan, K.R.; Zhang, H. Investigation of key techniques on floor roadway support under the impacts of superimposed mining: Theoretical analysis and field study. *Environ. Earth Sci.* **2019**, *78*, 436–450. [CrossRef]

19. Wang, P.; Zhao, J.; Feng, G.; Wang, Z. Interaction between vertical stress distribution within the goaf and surrounding rock mass in longwall panel systems. *J. S. Afr. Inst. Min. Metall.* **2018**, *118*, 745–756. [CrossRef]

20. Ma, C.Q.; Wang, P.; Jiang, L.S.; Wang, C.S. Deformation and control countermeasure of surrounding rocks for water-dripping roadway below a contiguous seam goaf. *Processes* **2018**, *6*, 77. [CrossRef]

21. Liu, C.Y.; Yang, J.X.; Yu, B.; Yang, P.J. Destabilization regularity of hard thick roof group under the multi gob. *J. China Coal Soc.* **2014**, *39*, 395–403.

22. Huang, Q.X.; Cao, J.; He, Y.P. Analysis of roof structure and support load of mining face under ultra-close goaf in shallow multiple seams. *Chin. J. Rock Mech. Eng.* **2018**, *37*, 3153–3159.

23. Yang, K.; He, X.; Liu, S.; Lu, W. Rib spalling mechanism and control with fully mechanized longwall mining in large inclination "three-soft" thick coal seam under closed distance mined gob. *J. Min. Saf. Eng.* **2016**, *33*, 611–617.

24. Gao, Z.W.; Saxen, H.; Gao, C.H. Special section on data-driven approaches for complex industrial systems. *IEEE Trans. Ind. Inform.* **2013**, *9*, 2210–2212. [CrossRef]

25. He, M.C.; Gong, W.L.; Wang, J.; Qi, P.; Tao, Z.G.; Du, S.; Peng, Y.Y. Development of a novel energy-absorbing bolt with extraordinarily large elongation and constant resistance. *Int. J. Rock Mech. Min. Sci.* **2014**, *67*, 29–42. [CrossRef]

26. He, M.C.; Li, C.; Gong, W.L.; Sousa, L.R.; Li, S.L. Dynamic tests for a Constant-Resistance-Large-Deformation bolt using a modified SHTB system. *Tunn. Space Technol.* **2017**, *64*, 103–116. [CrossRef]

27. Li, Y.F.; Hua, X.Z.; Cai, R.C. Mechanics analysis on the stability of key block in the gob-side entry retaining and engineering application. *J. Min. Saf. Eng.* **2012**, *29*, 357–364.
28. Lee, C.J.; Wu, B.R.; Chen, H.T.; Chiang, K.H. Tunnel stability and arching effects during tunneling in soft clayey soil. *Tunn. Space Technol.* **2006**, *21*, 119–132. [CrossRef]
29. Gao, Z.W.; Ding, S.X.; Cecati, C. Real-time fault diagnosis and fault-tolerant control. *IEEE Trans. Ind. Electron.* **2015**, *62*, 3752–3756. [CrossRef]
30. Gao, Z.W.; Sheng, S.W. Real-time monitoring, prognosis, and resilient control for wind turbine systems. *Renew. Energy* **2018**, *116*, 1–4. [CrossRef]
31. Tadisetty, S.; Matsui, K.; Shimada, H.; Gupta, R.N. Real time analysis and forecasting of strata caving behaviour during longwall operations. *Rock Mech. Rock Eng.* **2006**, *39*, 383–393. [CrossRef]

Article

Community-Based Link-Addition Strategies for Mitigating Cascading Failures in Modern Power Systems

Po Hu * and Lily Lee *

School of Electrical Engineering and Automation, Wuhan University, Wuhan 430072, China
* Correspondence: phu@whu.edu.cn (P.H.); 2014102070020@whu.edu.cn (L.L.)

Received: 9 December 2019; Accepted: 18 January 2020; Published: 21 January 2020

Abstract: The propagation of cascading failures of modern power systems is mainly constrained by the network topology and system parameter. In order to alleviate the cascading failure impacts, it is necessary to adjust the original network topology considering the geographical factors, construction costs and requirements of engineering practice. Based on the complex network theory, the power system is modeled as a directed graph. The graph is divided into communities based on the Fast–Newman algorithm, where each community contains at least one generator node. Combined with the islanding characteristics and the node vulnerability, three low-degree-node-based link-addition strategies are proposed to optimize the original topology. A new evaluation index combining with the attack difficulty and the island ratio is proposed to measure the impacts on the network under sequential attacks. From the analysis of the experimental results of three attack scenarios, this study adopts the proposed strategies to enhance the network connectivity and improve the robustness to some extent. It is therefore helpful to guide the power system cascading failure mitigation strategies and network optimization planning.

Keywords: power systems; complex network theory; Fast–Newman algorithm; link-addition strategy; cascading failures

---

1. Introduction

For smart grids, the advanced communication and information technology are employed to enhance the intelligence and automation of the power systems. Meanwhile, cyber threats are introduced to the physical systems triggering the self-organized criticality of the power system, leading to cascading failure propagation between networks even blackouts occurred [1–3]. As the scale of the smart grid expands, how to optimize the power system structure and effectively alleviate cascading failures has aroused public concern.

Modern power systems are dynamical systems featured by complexity and nonlinearity. For simplifying the model complexity, the complex network theory and the graph theory are introduced to demonstrate the network dynamics [4]. Besides, the characteristics of complex networks can be used to analyze the impacts on cascading propagation [5]. The larger the cluster coefficient (CC) of the network is, the wider the cascading failure propagation is. Moreover, the smaller the average path length (APL) of the network is, the deeper the cascading failure propagation is [6]. Statistics indicate that the power system is a typical sparse network owing to geographical location constraints and inadequate investment budgets [7]. As the power system expands, regional and long-distance power transmission lines are constructed to balance the regional generation capacity. With the increase in transmission lines, the APL increases slowly, while the regional CC is relatively large. Therefore, cascading failures can be easily propagated in large regions of the power system.

Previous studies have put forward the load-capacity model to analyze the cascading failure propagation. Cascading failure model of the power system based on the complex network theory combines with the characteristics of power flows [8]. System capacity and network connectivity affect the propagation of cascading failures [9]. An electrical path efficiency matrix is assisted with the assessment of power system influences and losses [10]. Based on the percolation theory [11], the remaining giant component indicates the robustness of the network. However, evaluation indexes of the existing studies are used to assess the connected component performance, which cannot be implemented for isolated islands. The power system can maintain islanding operations after attacks. Thus, the robustness index of the power system should contain all survival islands.

Additionally, relevant research focused on the mechanism of cascading failures. In the power system, cascading failure can be triggered by means of physical equipment malfunction or misoperation owing to weather or man-made, and intentional cyber-attacks. Power node or link failure caused by system hidden failures as well as large area blackouts caused by natural disasters exhibit random attacks (RA) to the power system. Adversaries can also attack specific targets. For example, high degree node attacks (HDNA) disconnect the highly connected substation to destroy the network connectivity. Moreover, cyber-attacks compromise communication data to control the power system operations, which can construct not only simultaneous attacks but also sequential attacks [12]. For example, a large area of new energy resources simultaneously disconnects from the backbone network, or some special targets are sequentially compromised by coordinated strategies. The current research indicates that vulnerability sequence attack (VSA) damages the network more seriously than simultaneous attacks [13], because VSA can collapse the whole network by attacking fewer nodes. The evolution of both logical and real values of system parameters can be analyzed by a hybrid attack graph under attack and recovery actions scenarios [14]. As simultaneous attacks and sequential attacks have diverse impacts on power systems, it is necessary to investigate the cascading failure propagation of multiple attack scenarios by using proper evaluation indexes.

However, vulnerability of topology is affected by the transmission efficiency, connectivity, and connected components [15], particularly the power flow distribution of power systems [16]. The topology of the power system is relatively inflexible and vulnerable to intentional attacks [17]. Diverse fault diagnosis technologies have applied to monitor, locate, and identify the faults, which need to handle a large amount of data and operate system resources [2,3]. The effective control chart technique could substantially decrease the loss caused by the diagnosis and correction [18]. Optimal nonlinear adaptive control reduced uncertainties and improved the robustness under different operation scenarios [19]. In order to decrease the network vulnerability, the network structure can be optimized by link-addition strategies to mitigate cascading failures [20]. Existing research proposes interlink addition strategy to increase connectivity density, in order to reduce cascade-safe region and improve the network connectivity [21]. For improving the network robustness, connectivity links and interlinks could be added simultaneously [22], while the construction costs are too high to realize [23]. Ji et al. [24] compared with various connectivity link addition strategies, for the purpose of verifying the feasibility of low-degree node link-addition strategy and improving the power network robustness. However, these link-addition strategies have focused on the pure topology evolution evaluating by using degree or betweenness indexes, without considering special characteristics of power systems.

Since the power system is managed in regions, isolated islands can maintain in operation. The Fast–Newman algorithm is introduced to divide the network topology into communities, thereby ensuring that the network can be effectively partitioned [25]. In power systems, the location of generators is the key factor for a valid community [8]. Besides, the load distribution has influences on the power generation dispatch and control strategy [26]. For providing sufficient power supply, the power system can be partitioned into communities following the power flow directions. Moreover, critical regions greatly affect the topology evolution, and the community partition of these regions seriously influences on the network vulnerability [27]. To achieve the reliability and preventive maintenance is another optimization goal [28]. Therefore, the community-based link-addition strategy is proposed to optimize the existing power network topology, in order to reduce investment budgets and alleviate the burden of load centers.

In summary, present researches have confirmed that the power system is affected by the community structure, but less attention is paid to the optimal community structure on mitigating cascading failure propagation. In order to address this issue, we propose an improved load-capacity model based on the islanding power flow distribution, in terms of the complex system and percolation theory. The island ratio is a measure of the robustness of power networks. For further demonstrating the difficulty of attacks, an evaluation indicator is introduced to assess the influence of the sequential attack. In order to optimize the original power system, three community-based link-addition strategies between low-degree nodes are therefore proposed to meet the requirements of engineering practice. This paper is of practical significance in how to optimize network topology and improve the network robustness of the power system.

The reminder of the paper is organized as follows. Section 2 presents the fundamental theoretical background on constructing a load-capacity model. Section 3 discusses the evaluation index. Section 4 describes the process of constructing link-addition strategy. Section 5 provides the simulation results and the corresponding analysis. Section 6 summarizes several concluding remarks and discusses the challenging issue. Lack of the period.

2. System Model

Based on the complex network theory, the power system is modeled as a directed graph $G_P = (V_P, E_P)$, with N nodes and without multiple edges or loops, where V_P and E_P are power nodes and lines, respectively. The power nodes are categorized as three types: generator nodes that generate electricity, load nodes that consume electricity, and substation nodes that transfer electricity. Particularly, one generator node carrying loads can be classified into the load node. The power lines are directed by the power flow changes over time. In order to decrease calculation complexity, this study ignores the differences in transmission lines, the transient voltage instability and phase angle mismatch. In this graph, the nodes and lines can be removed as a result of failures or attacks. It is assumed that the adversaries can manipulate the systematic information to construct malicious attacks of any target of the system.

In the power system, the real and reactive power injections are balanced at every node, as indicated in Equations (1) and (2). Moreover, the real and reactive power flows in transmission lines by following Kirchhoff's law, as expressed in Equations (3) and (4) [29].

Real and reactive power injection at node i:

$$P_i = V_i \sum_{j=1}^{N} V_j (G_{ij} \cos \theta_{ij} + B_{ij} \sin \theta_{ij}), \tag{1}$$

$$Q_i = V_i \sum_{j=1}^{N} V_j (G_{ij} \sin \theta_{ij} - B_{ij} \cos \theta_{ij}), \tag{2}$$

Real and reactive power flows from node i to node j are:

$$P_{ij} = V_i^2 G_{ij} - V_i V_j (G_{ij} \cos \theta_{ij} + B_{ij} \sin \theta_{ij}), \tag{3}$$

$$Q_{ij} = -V_i^2 B_{ij} - V_i V_j (G_{ij} \sin \theta_{ij} - B_{ij} \cos \theta_{ij}), \tag{4}$$

where P_i is the real power injection at the power node i, Q_i the reactive injection at the power nod i, P_{ij} the real power flow from node i to node j, Q_{ij} the reactive power flow from node i to node j, V the voltage magnitude, θ_{ij} the difference in the phase angle between power nodes i and node j, B_{ij} the admittance, G_{ij} the susceptance, and N the initial number of nodes, $i, j \in N$.

According to the power flow distribution, the power system capacity is assumed to be proportional to its initial states [30]. It is assumed that the power system is provided with moderately reactive power to compensate losses and avoid out-of-limit at the same voltage grade. The initial power flow capacity is the maximum power flow in transmission lines of Equation (5). The initial generation capacity is the maximum output of generators of Equation (6). The initial node capacity is the maximum sum of out flows $P_{outflow,ij}(i)$ and local loads $L_{load}(i)$ of node i of Equation (7).

$$\mathbb{C}_{PF} = \max(P_{ij}), \tag{5}$$

$$\mathbb{C}_{gen,i} = \max(P_{gen}(i)), \tag{6}$$

$$\mathbb{C}_{Node,i} = \max(\sum_{i,j \in N} P_{outflow,ij}(i) + L_{load}(i)), \tag{7}$$

So, the system capacity $\mathbb{C}_p$ is α times the initial states.

$$\mathbb{C}_p = \alpha(\mathbb{C}_{PF}, \mathbb{C}_{gen}, \mathbb{C}_{Node,i}), \tag{8}$$

where α is the tolerance parameter, $\alpha \geq 1$. In the model, the tolerance parameter α is a consistent one. It is assumed that the power system adopts the overcurrent protection mechanism. For simplicity, if the power flow exceeds the system capacity, the transmission lines trip off instantly without further automatic reclose.

3. Evaluation Index

(1) Cluster coefficient
CC indicates the network connectivity level between nodes and their neighboring nodes [31]. Assume that node i has a number of E_i links and k_i neighbors, while the maximum number links of these neighboring nodes is $n_i(n_i - 1)$. The CC of node i is shown as follows.

$$C(i) = \frac{2E_i}{n_i(n_i - 1)}, \tag{9}$$

Then, global CC of the network equals to the mean value of the local CC of all nodes

$$C = \sum_{i \in N} C(i)/N, \tag{10}$$

(2) Average path length
APL is a measure of network efficiency. Dijkstra algorithm [32] is used to find the shortest path from the source node i to the destination node j, then the average distance between two nodes is shown as follows.

$$L = \frac{1}{N(N-1)} \sum_{i \neq j \in N} d_{ij}, \tag{11}$$

In this study, d_{ij} is assumed to be the distance cost of one new connectivity link, which indicates the difficulty of adding one new link from one source node to the other destination node.

(3) Node vulnerability
Based on the percolation theory, nodes are functional only in a giant component, which is a maximal connected component of the graph. The number of nodes that belong to giant components owing to one node removal indicates the node vulnerability. In one network, although a number of nodes have the same vulnerability, node removal contributes various influences on the remained components. In literature [4], the node types and their locations are combined to further distinguish the most vulnerable node. If the nodes are in separate single loops, the node in the bigger single loop is more important than that of the smaller one. Since a line-shaped branch is generated after unlocking

the single loop, the longer the branch, the more the loss of nodes. If the nodes are in the same single loop or in different single loops of the same size, further investigation is required until the most critical node is located.

$$I_{r(i)} = \frac{N\prime}{N}, \forall length(r(i)) > length(r(\varphi)),\tag{12}$$

where $N\prime$ is the node number of the remaining giant component, φ the set of nodes with the same vulnerability, $r(i)$ the single loop where node i locates, and *length* stands for the length of the single loop, $i \in \varphi \in N$.

After part of nodes are removed from the network in a random or targeted manner, the remaining giant component ratio is used to estimate the network robustness [33]. However, the power system can maintain in islanding operations. Thus, the island ratio is the proportion of all survival isolated components of the power system.

$$I = \frac{\sum_x \Theta(x)}{N},\tag{13}$$

where Θ is the node number of one survival island, and x is the number of islands.

For assessing the influence of the network under sequential attacks, an evaluation indicator S is introduced to combine with the difficulty of attacks and the survivability of the network.

$$S = \tau \times I,\tag{14}$$

where τ is the number of sequential attacks, and S is a scalar without units.

4. Link-Addition Strategy

4.1. Fast–Newman Algorithm for Community Partition

According to the power system management, each community has at least one generator node to supply sufficient electricity, or it will fail to partition. The directed power system graph detects the valid community modularity by using the Fast–Newman algorithm [25].

$$Q = \frac{1}{2m}\sum_{ij} [A_{ij} - \frac{k_i k_j}{2m}]\delta(v_i, v_j),\tag{15}$$

where m is the link number, $2m$ the sum of degrees of the network, A the adjacent matrix, k the degree of a node, and $\delta(v_i, v_j)$ the function for judging the community of two nodes. If they are in the same community, it is 1, otherwise 0. The modularity Q ranges from $[-0.5, 1)$, the greater the modularity, the better the effect of community partition. Statistics show that when Q is between 0.3 and 0.7, communities will cluster effectively [34].

4.2. Low-Degree-Node-Based Link-Addition Strategy

One-degree node (leaf node) of the power system is easily removed, owing to its overloaded transmission line or neighboring node removal that suffers from disturbances or attacks. Through the addition of new links to the leaf nodes, the connectivity level of the network can be increased. This is because the removal of tree-shaped root nodes can cause a large area to be disconnected from the core component, and the leaf nodes of the most vulnerable nodes are critical for optimizing the power system topology. However, some leaf nodes are generator nodes, so it is unreasonable to connect two generators except one generator node carrying a heavy load. The newly added links cannot overlap the original links. Moreover, the new network has to ensure that each community has at least one generator node. In conclusion, three link-addition strategies are proposed to enhance the original network connectivity and decrease the vulnerability.

(1) Low-degree-node link-addition strategy (LDNLAS)

The strategy aims to optimize long-distance transmission line construction for solving the long-distance electricity transmission of the large scale power systems. Based on the community partition and node vulnerability of the original power system, the new links from one community to other communities satisfy the average shortest path. If the most vulnerable node has leaf nodes, new links are first added from them.

$$E_{LDNLAS} = \sum_{D_1} E_{st}\text{s.t. } \delta(v_s, v_t) = 0, s \in D_1, t \in N, D_1 \in N, s \neq t, \min L_{new} = \frac{1}{N(N-1)} \sum_{s \neq t} d_{st}, \quad (16)$$

where E_{st} is an additional link, s the low-degree nodes, t the leaf nodes, and D_1 the set of low-degree nodes that satisfy the average shortest path L_{new}.

(2) Nearest-neighboring-node link-addition strategy (NNNLAS)

The strategy aims to connect the nearest nodes to enhance the local network connectivity and density. Based on breadth-first search algorithm, the new links find the shortest distance between neighboring nodes. If new links have the same shortest distance, those who have the average shortest path will satisfy the requirement.

$$E_{NNNLAS} = \sum_{D_2} E_{st}, \text{ s.t. } neighbor(v_s, v_t), s \in D_2, t \in N, D_2 \in N, s \neq t, \min d_{st}, \quad (17)$$

where E_{st} is an additional link, t the leaf nodes, s the neighbor of leaf nodes that satisfy the shortest path d_{st}, and D_2 the set of neighboring nodes.

(3) Max-load-node link-addition strategy (MLNLAS)

The strategy aims to alleviate the heavy burden of load centers and balance electricity supply capacity. Combined with the community partition, the load centers get new electricity supply with other generator by new links. Moreover, the new links satisfy the average shortest path. If the leaf nodes are not generators, the new links will follow the LDNLAS.

$$E_{MLNLAS} = \sum_{D_3} E_{st}, \text{s.t. } s \in D_3, t \in N, s \neq t, \min L_{new} = \frac{1}{N(N-1)} \sum_{s \neq t} d_{st}, \quad (18)$$

where E_{st} is an additional link, t the leaf nodes, s the heavy load node, and D_3 the set of heavy load nodes in order.

5. Simulation Results and Data Analysis

In this section, the present study experiments with the data of IEEE 39-bus power system and establishes the simulation results in detail. The power flow calculation and the isolated island problems are solved using the MATPOWER 6.0 toolkit in MATLAB R2016a [35]. Based on the graph theory, the directed graph gets the average degree $\overline{D} = 2.359$, cluster coefficient $C = 0.0385$, and average path length $L = 4.749$, while the random network with the same $\overline{D}$, $C_{rand} \approx \overline{D}/N = 0.0605$ and $L_{rand} \approx \ln(N)/\ln(\overline{D}) = 4.2687$. The graph includes generator nodes ranging from 30 to 39, and it is partitioned into 5 communities according to the Fast–Newman algorithm. The modularity is $Q = 0.6125$, which indicates good community partition of this graph. Each community contains at least one generator node, which is shown as follows.

In Figure 1, communities are labelled by numbers and surrounded by an ellipse. Community 1 is the area of blue solid circles, community 2 the area of red squares, community 3 the area of magenta snowflakes, community 4 the area of green rhombuses, and community 5 the area of black stars.

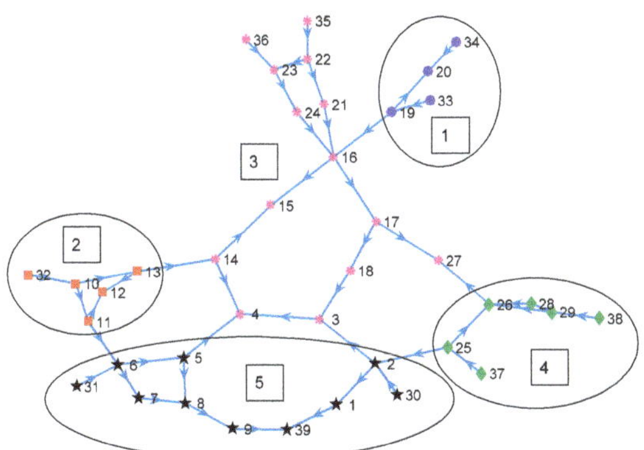

Figure 1. Communities of IEEE 39-bus system.

5.1. Generating Network

According to the principle of link-addition strategies, IEEE 39-bus system has 9 one-degree nodes ranging from node 30 to node 38. These leaf nodes without heavy loads are unnecessary to connect to each other, because they are all generator nodes. Therefore, the network has to add 9 additional links to get $\overline{D}_{new} = 2.8205$.

(1) LDNLAS Network

From Figure 1, the node importance of the original network is obtained to find the most vulnerable node 16 and 2 leaf nodes based on the Equation (9) in the same community. The low-degree nodes are randomly chosen to connect with these leaf nodes to find the average shortest path length. Following the rule, 9 links are added to the original network. In each step, the network can be partitioned into valid communities. The total cost of additional links is 53. See Table 1 for details.

Table 1. Connectivity link addition of LDNLAS.

New Link	Q	Community	C	L	d
35–7	0.6098	6	0.0385	4.5128	9
34–28	0.5920	5	0.0214	4.4143	7
32–9	0.5537	4	0.0214	4.3374	6
36–1	0.5176	4	0.0214	4.1916	8
38–15	0.4998	3	0.0214	4.0229	6
31–12	0.5229	4	0.0214	4.004	3
30–20	0.4664	3	0.0214	3.9096	7
33–21	0.4911	4	0.0214	3.8866	3
37–18	0.5127	4	0.0214	3.8475	4

The LDNLAS network detects 4 communities in Figure 2. Community 1 with 9 nodes is the area of blue solid circles, community 2 with 4 nodes is the area of red squares, community 3 with 15 nodes is the area of magenta snowflakes, and community 4 with 11 nodes is the area of green rhombuses.

The modularity of the LDNLAS network is $Q = 0.5127 \in [0.3, 0.7]$, which indicates the community partition is effective. $C = 0.0214$ is less than that of the original network, and $L = 3.8475$ is reduced to about 19%. Although the LDNLAS network reduces the aggregation degree than that of the original network, it improves the connectivity obviously.

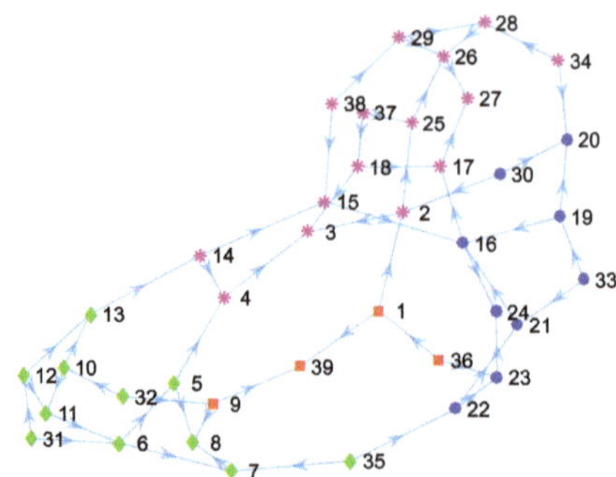

Figure 2. Communities of LDNLAS network.

(2) NNNLAS Network

Firstly, the neighbors of the leaf nodes are found. Owing to the symmetrical structure, several leaf nodes have the same shortest distance to their neighbors. The total cost of additional links is 22. See Table 2 for details.

Table 2. Connectivity link addition of NNNLAS.

New Link	Q	Community	C	L	d
35–21/36–24	0.6137	6	0.1239	4.6572	4
34–15	0.6122	5	0.1239	4.5466	4
30–1/31–7 33–20/38–28	0.6216	5	0.2692	4.529	8
32–12/37–27	0.6393	5	0.2692	4.4872	6

In Figure 3, the NNNLAS network detects 5 communities. Community 1 with 7 nodes is the area of blue solid circles, community 2 with 4 nodes is the area of red squares, community 3 with 12 nodes is the area of magenta snowflakes, community 4 with 7 nodes is the area of green rhombuses, and community 5 with 9 nodes is the area of black stars.

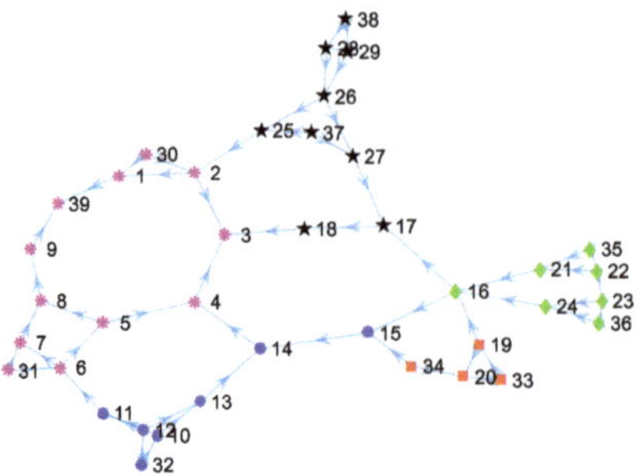

Figure 3. Communities of NNNLAS network.

The modularity of the NNNLAS network is $Q = 0.6393 \in [0.3, 0.7]$, which indicates the community partition is highly effective. $C = 0.2692$ is 7 times the original network, and $L = 4.4872$ is reduced to about 5%. Although the NNNLAS network enhances the aggregation degree enormously than that of the original network, it increases the connectivity level slightly.

(3) MLNLAS Network

First, the loads of the original network are ordered to select the first 9 load nodes. Then, new links are randomly added to the leaf nodes to satisfy the community partition principle and the average shortest path length. The total cost of additional links is 58. See Table 3 for details.

Table 3. Connectivity link addition of MLNLAS.

New Link	Q	Community	C	L	d
36–39	0.5776	4	0.0385	4.5304	9
34–8	0.5816	5	0.0385	4.363	9
38–20	0.5352	4	0.0385	4.1997	7
35–4	0.5272	4	0.0385	4.0513	6
32–16	0.5121	4	0.0385	3.8785	5
31–3	0.5274	4	0.0385	3.7787	4
37–15	0.4532	4	0.0385	3.6775	6
30–24	0.4458	3	0.0385	3.6086	6
33–29	0.4483	3	0.0342	3.5735	6

The MLNLAS network detects 3 communities in Figure 4. Community 1 with 13 nodes is the area of blue solid circles, community 2 with 12 nodes is the area of red squares, and community 3 with 14 nodes is the area of magenta snowflakes.

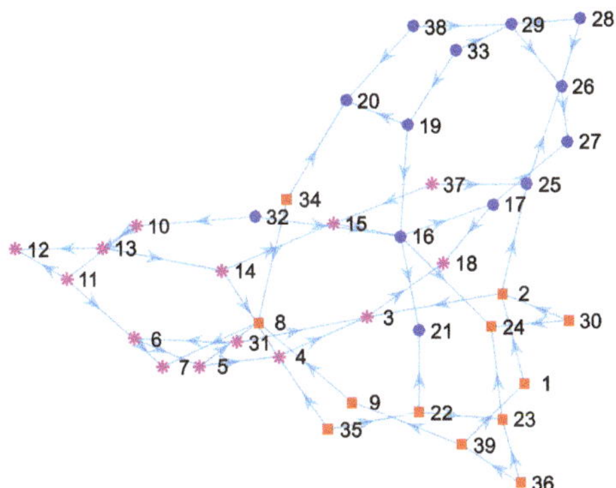

Figure 4. Communities of MLNLAS network.

The modularity of the MLNLAS network is $Q = 0.4483 \in [0.3, 0.7]$, which indicates the community partition is reasonable. $C = 0.0342$ is close to that of the original network, and $L = 3.5735$ is reduced to about 25%. Although the MLNLAS network decreases the aggregation degree than that of the original network, it increases effectively the connectivity level.

Three networks of the same additional links decrease the APL and increase the connectivity than that of the original network. NNNLAS network significantly improves the aggregation degree at the lowest cost; LDNLAS network effectively increases the connectivity with a higher cost than that of NNNLAS network; MLNLAS network dramatically improves the connectivity and alleviates the burdens of load centers, while the cost is the highest one of three strategies, and the community partition and aggregation degree are relatively weak.

5.2. Network Robustness Analysis

The robustness of networks is analyzed under three attack scenarios. Random node attacks and high-degree-node-based attacks are regarded as simultaneous attacks, while vulnerability-based attacks are sequential attacks. For reducing the influence of network capacity, this study assumes the universal system tolerance parameter $\alpha = 2$. Under the simultaneous attack scenarios, the component ratios are graphed with the distribution interval, median, 5%–95% position and mean at various attack ranges. Under the sequential attack scenarios, the component ratio curves are plotted by the number of attacks, and all remaining survival islands are demonstrated as directed graphs.

(1) RA Scenario

Random attack groups arc $C_{39}^4, C_{39}^8, C_{39}^{12}, C_{39}^{16}, C_{39}^{20}, C_{39}^{24}, C_{39}^{28}, C_{39}^{32}, C_{39}^{36}$, according to the attack ranges respectively. In one attack range, 1000 groups of data are selected to attack 4 networks, which is executed for 50 times to obtain the corresponding results.

From the distribution intervals of Figure 5, the maximum component ratios of the original network are all less than or equal to three new networks of any attack range. The less the range of distribution intervals, the more stable the cascading propagation; the greater the mean value, the better the network robustness. For further comparison, the mean and median values are shown in Figure 6.

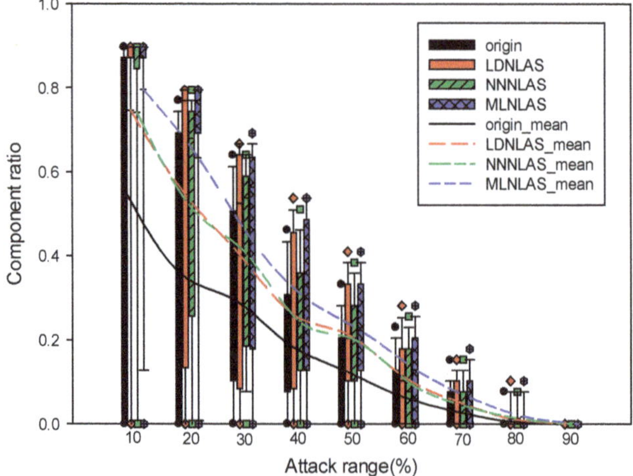

Figure 5. Component ratio under RAs.

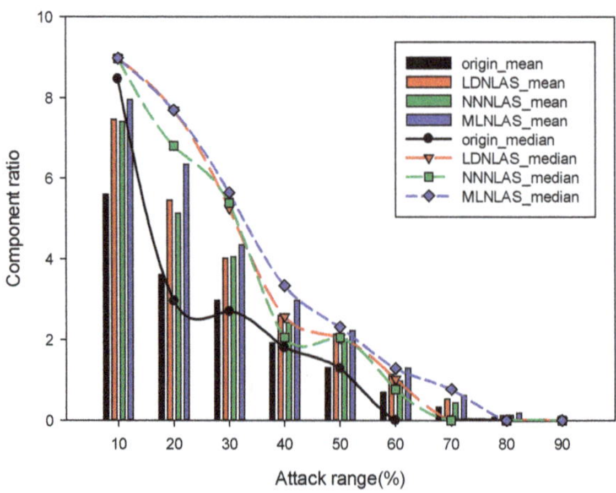

Figure 6. Mean and median values under RAs.

Observing the mean histogram and the median curve of Figure 6, the original network lefts fewer nodes when the attack range is up to 60%. The LDNLAS and NNNLAS networks survive up to 70% attack range, while the MLNLAS network can preserve in 80% attack range. Combined with the distribution intervals of Figure 5, the robustness of 4 networks orders is as follows: MLNLAS > LDNLAS > NNNLAS > original.

(2) HDNA

The nodes of networks are ordered in degrees. The attack range selects the nodes from the high degrees to the low ones. As the nodes with the same degree have a number of attack groups, the results can be obtained by traversing all attack groups of each attack range.

In Figure 7, when the attack range is up to 50%, the original network totally collapses, and the MLNLAS network lefts a few nodes. In contrast, the LDNLAS and NNNLAS networks remain a large number of nodes. Owing to the impacts of the highest degree nodes on the connectivity, the NNNLAS network losses the maximum nodes at 10% attack range of 4 networks. For further analysis, the mean and median values are shown in Figure 8.

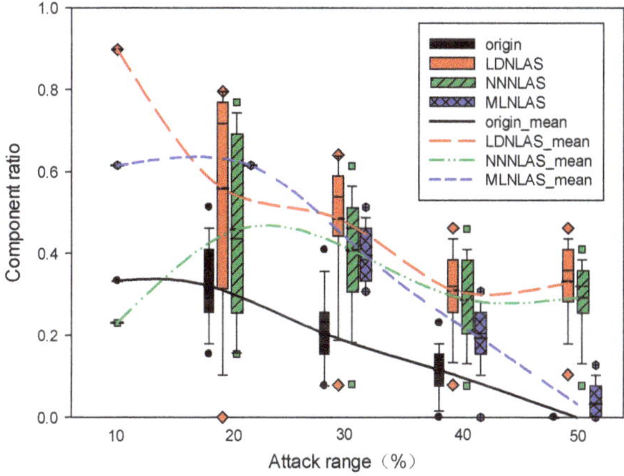

Figure 7. Component ratio under HDNAs.

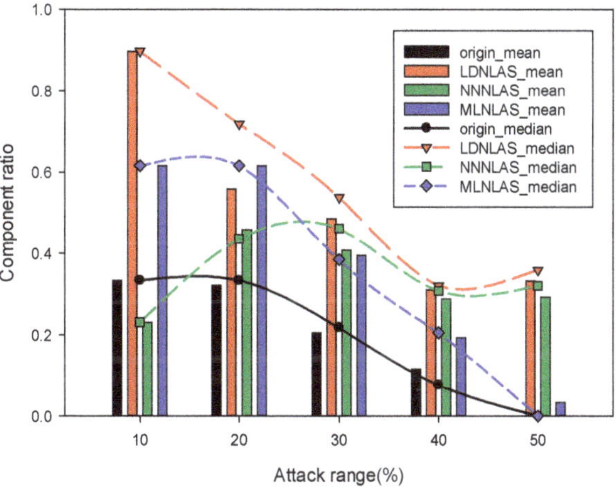

Figure 8. Mean and median values under HDNAs.

Combined with Figures 7 and 8, when the attack range reaches 20%, although the mean value of the LDNLAS network is smaller than that of the NNNLAS network, both the maximum value and the median value of the former are larger than the latter, which indicates that the mean value is smaller due to the influence of extreme value. Thus, the overall data should be larger than the latter. Attacking more than 20%, the robustness of the LDNLAS network is obviously superior to other 3 networks. Influenced by the community partition, when the attack range is more than 10%, the robustness of 4 networks orders as follows: LDNLAS > NNNLAS>MLNLAS > original.

(3) VSA

Based on the node vulnerability, one node is attacked each time. For comparing with the original, the attack originates from the most vulnerable node 16. The attack sequence of the original network is: 16–26–3–8–6; the attack sequence of the LDNLAS network is: 16–23–7–20–2–9–5–14; the attack sequence of the NNNLAS network is: 16–14–6–26; and the attack sequence of the MLNLAS network is: 16–13–6–8–26–3–22–2.

In Figure 9, the original network sequentially attacks 5 nodes (about 10%) splitting into 4 islands, and $S_{original} = 2.564$. The LDNLAS network sequentially attacks 8 nodes (about 20%) splitting into 3 islands, and $S_{LDNLAS} = 3.0768$. The NNNLAS network sequentially attacks 4 nodes (about 10%) splitting into 3 islands, and $S_{NNNLAS} = 1.9488$. The MLNLAS network sequentially attacks 8 nodes (about 20%) splitting into 4 islands, and $S_{MLNLAS} = 4.9232$.

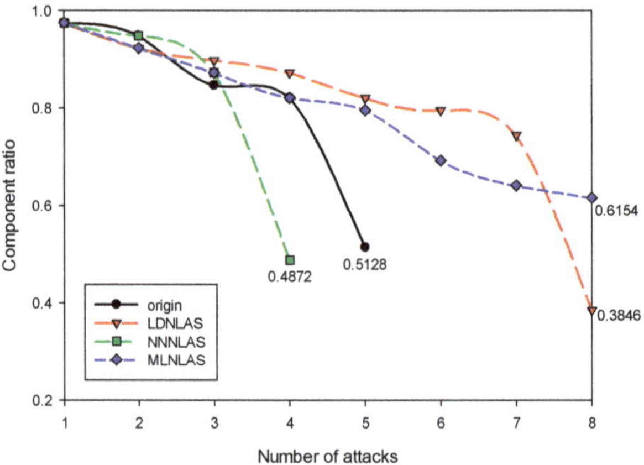

Figure 9. Component ratio under VSAs.

The remaining islands of sequential attacks are shown as follows.

From Figures 9 and 10, it is observed that the MLNLAS network is the most robust one of 4 networks. The LDNLAS network exhibits the difficulty of sequential attacks, while it is weak in islanding operations. The NNNLAS has the worst survivability under sequential attacks. In the sequential attack process, the more the attacks, the more difficult the implementation, and the more robust the network. Moreover, the network with few communities, a small CC and a short APL can resist the sequential attack more efficiently. Therefore, the robustness of 4 networks orders as follows: MLNLAS > LDNLAS > original > NNNLAS.

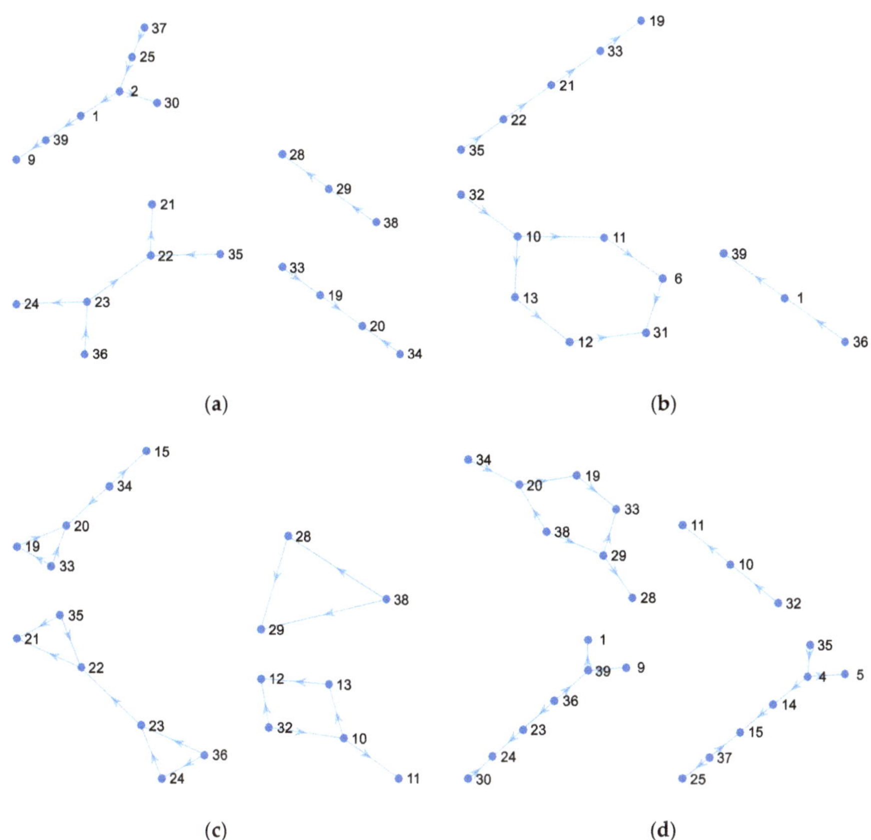

Figure 10. Remaining islands under VSAs (**a**) IEEE 39 system, (**b**) LDNLAS network, (**c**) NNNLAS network, and (**d**) MLNLAS network.

From the above analysis, LDNLAS gets the second largest link-addition cost of the three proposed strategies. The LDNLAS network obtains a shorter APL and smaller CC than the original network, which alleviates the depth of the cascading failure propagation. In fact, this network exhibits the best robustness against HDNAs, and the second best robustness against RAs and VSAs. Although this strategy requires slightly larger investments, it can resist both simultaneous attacks and sequential attacks, and enhance the connectivity of the long-distance transmission structure power system.

MLNLAS obtains the largest link-addition cost of the three proposed strategies. The MLNLAS network with the shortest APL enormously enhances the connectivity than that of the original network. Moreover, this network presents the best performance against RAs and VSAs. Although this strategy requires more investments, it optimizes the electricity supply to greatly alleviate the burdens of load centers. As Ref [36] says, it is difficult to gain the high robustness with the minimal cost simultaneously.

NNNLAS has the smallest link-addition cost of the three proposed strategies. The NNNLAS network with the largest CC improves the centralization of local area management and is robust to the simultaneous attacks. However, it cannot effectively decrease the network vulnerability against VSAs.

6. Conclusions

Cascading failure propagation can be alleviated by optimizing the network topology. Based on the community partition of the original network, three link-addition strategies are proposed to meet the requirements of engineering practices. It is thus useful to guide the power system planning to improve the network robustness.

From the analysis of simulation results, the three proposed strategies can improve the network connectivity by adding the same number of links. The MLNLAS network exhibits good robustness under RAs; the LDNLAS shows better performances than other networks under HDNAs; the MLNLAS network reveals highly survivability under sequential attacks.

In this study, the proposed strategies are beneficial for improving the robustness of the original network. The focus is on the influence on the power system. In the future work, the authors will continue to study optimal strategies to mitigate cascading failures and improve the robustness of smart grids.

Author Contributions: Conceptualization, L.L.; methodology, L.L.; software, L.L.; validation, L.L.; formal analysis, L.L.; investigation, L.L..; resources, L.L. and P.H.; data curation, L.L.; writing—original draft preparation, L.L.; writing—review and editing, L.L., P.H.; visualization, L.L.; supervision, P.H.; project administration, P.H.; funding acquisition, L.L. and P.H. All authors have read and agreed to the published version of the manuscript.

Funding: This research received no external funding.

Conflicts of Interest: The authors declare no conflict of interest.

Nomenclature

Indices

i, j, s, t	Index for node numbering.
x	Number of islands.
τ	Number of sequential attacks.
m	Number of links.

Constants

N	Numbers of system nodes.
α	System tolerance parameter.

Variables

P_i	Real power injection at the power node i.
Q_i	Reactive injection at the power node i.
P_{ij}	Real power flow from node i to node j.
Q_{ij}	Reactive power flow from node i to node j.
V	Voltage magnitude.
θ_{ij}	Difference in the phase angle between power nodes i and node j.
B_{ij}	Admittance matrix.
G_{ij}	Susceptance matrix.
$P_{outflow,ij}(i)$	Out flows of node i.
$L_{load}(i)$	Local loads of node i.
E_i	Links of node i.
n_i	Neighbors of node i.
$C(i)$	Cluster coefficient of node i.
d_{ij}	Shortest path from the source node i to the destination node j.
$N\prime$	Numbers of nodes of the remaining components.
$r(i)$	Single loop location of node i.
$I_{r(i)}$	Node importance of node i.
k_i	Degrees of node i.
A	Adjacent matrix.
v_i	Vertex i.
E_{st}	Additional link from the node s to the destination node t.

Sets and Functions

G_P	Directed graph of power system.
V_P	Power node set.
E_P	Power line set.
$\mathbb{C}_{PF}$	Power flow capacity function.
$\mathbb{C}_{gen,i}$	Generation capacity function.
$\mathbb{C}_{Node,i}$	Node capacity function.
$\mathbb{C}_p$	System capacity function.
C	Cluster coefficient function.
L	Average path length function.
length	Length function of a single loop.
φ	Nodes with the same vulnerability set.
I	Island ratio function.
Θ	Survival islands set.
S	Evaluation indicator.
Q	Community modularity function.
δ	Judging community function for two nodes.
E_{LDNLAS}	Low-degree-node link-addition strategy function.
D_1	Set of low-degree nodes that satisfy the average shortest path L_{new}.
E_{NNNLAS}	Nearest-neighboring-node link-addition strategy function.
D_2	Set of neighboring nodes
E_{MLDLAS}	Max-load-node link-addition strategy function.
D_3	Set of heavy load nodes in order.

References

1. Bretãs, A.S.; Bretas, N.G.; Carvalho, B.E. Further contributions to smart grids cyber-physical security as a malicious data attack: Proof and properties of the parameter error spreading out to the measurements and a relaxed correction model. *Int. J. Electr. Power Energy Syst.* **2019**, *104*, 43–51. [CrossRef]
2. Gao, Z.; Cecati, C.; Ding, S.X. A Survey of Fault Diagnosis and Fault-Tolerant Techniques—Part I: Fault Diagnosis with Model-Based and Signal-Based Approaches. *IEEE Trans. Ind. Electron.* **2015**, *62*, 3757–3767. [CrossRef]
3. Gao, Z.; Cecati, C.; Ding, S.X. A Survey of Fault Diagnosis and Fault-Tolerant Techniques Part II: Fault Diagnosis with Knowledge-Based and Hybrid/Active Approaches. *IEEE Trans. Ind. Electron.* **2015**, *62*, 1. [CrossRef]
4. Wang, Z.; Hill, D.J.; Chen, G.; Dong, Z.Y. Power system cascading risk assessment based on complex network theory. *Phys. A Stat. Mech. Its Appl.* **2017**, *482*, 532–543. [CrossRef]
5. Albert, R.; Barabasi, A.-L. Statistical mechanics of complex networks. *Rev. Mod. Phys.* **2001**, *26*, 1–7. [CrossRef]
6. Wang, S.; Liu, J. Robustness of single and interdependent scale-free interaction networks with various parameters. *Phys. A Stat. Mech. Its Appl.* **2016**, *460*, 139–151. [CrossRef]
7. Pagani, G.A.; Aiello, M. The Power Grid as a complex network: A survey. *Phys. A Stat. Mech. Its Appl.* **2013**, *392*, 2688–2700. [CrossRef]
8. Guo, W.; Wang, H.; Wu, Z. Robustness analysis of complex networks with power decentralization strategy via flow-sensitive centrality against cascading failures. *Phys. A Stat. Mech. its Appl.* **2018**, *494*, 186–199. [CrossRef]
9. Kinney, R.; Crucitti, P.; Albert, R.; Latora, V. Modeling cascading failures in the North American power grid. *Eur. Phys. J. B* **2005**, *46*, 101–107. [CrossRef]
10. Ren, H.-P.; Song, J.; Yang, R.; Baptista, M.S.; Grebogi, C. Cascade failure analysis of power grid using new load distribution law and node removal rule. *Phys. A Stat. Mech. Its Appl.* **2016**, *442*, 239–251. [CrossRef]
11. Motter, A.E.; Lai, Y.-C. Cascade-based attacks on complex networks. *Phys. Rev. E* **2002**, *66*. [CrossRef] [PubMed]
12. Liang, G.; Weller, S.R.; Zhao, J.; Luo, F.; Dong, Z.Y. The 2015 Ukraine Blackout: Implications for False Data Injection Attacks. *IEEE Trans. Power Syst.* **2017**, *32*, 3317–3318. [CrossRef]

13. Lee, L.; Hu, P. Vulnerability analysis of cascading dynamics in smart grids under load redistribution attacks. *Int. J. Electr. Power Energy Syst.* **2019**, *111*, 182–190. [CrossRef]

14. Ibrahim, M.; Alsheikh, A. Automatic Hybrid Attack Graph (AHAG) Generation for Complex Engineering Systems. *Processes* **2019**, *7*, 787. [CrossRef]

15. Wang, S.; Zhang, J.; Na, D. Multiple perspective vulnerability analysis of the power network. *Phys. A Stat. Mech. Its Appl.* **2017**, *492*. [CrossRef]

16. Wang, Z.; Chen, G.; Hill, D.J.; Dong, Z.Y. A power flow based model for the analysis of vulnerability in power networks. *Phys. A Stat. Mech. its Appl.* **2016**, *460*, 105–115. [CrossRef]

17. Zhao, S.; Maxim, A.; Liu, S.; De Keyser, R.; Ionescu, C.M. Distributed Model Predictive Control of Steam/Water Loop in Large Scale Ships. *Processes* **2019**, *7*, 442. [CrossRef]

18. Aslam, M.; Bantan, R.A.R.; Khan, N. Monitoring the Process Based on Belief Statistic for Neutrosophic Gamma Distributed Product. *Processes* **2019**, *7*, 209. [CrossRef]

19. Jiang, Y.; Jin, X.; Wang, H.; Fu, Y.; Ge, W.; Yang, B.; Yu, T. Optimal Nonlinear Adaptive Control for Voltage Source Converters via Memetic Salp Swarm Algorithm: Design and Hardware Implementation. *Processes* **2019**, *7*, 490. [CrossRef]

20. Peng, H.; Liu, C.; Zhao, D.; Han, J. Reliability analysis of CPS systems under different edge repairing strategies. *Phys. A Stat. Mech. Its Appl.* **2019**, *532*, 121865. [CrossRef]

21. Zio, E.; Sansavini, G. Modeling Interdependent Network Systems for Identifying Cascade-Safe Operating Margins. *IEEE Trans. Reliab.* **2011**, *60*, 94–101. [CrossRef]

22. Cui, P.; Zhu, P.; Wang, K.; Xun, P.; Xia, Z. Enhancing robustness of interdependent network by adding connectivity and dependence links. *Phys. A Stat. Mech. Its Appl.* **2018**, *497*, 185–197. [CrossRef]

23. Dong, G.; Gao, J.; Tian, L.; Du, R.; He, Y. Percolation of partially interdependent networks under targeted attack. *Phys. Rev. E* **2012**, *85*, 16112. [CrossRef] [PubMed]

24. Ji, X.; Wang, B.; Liu, D.; Chen, G.; Tang, F.; Wei, D.; Tu, L. Improving interdependent networks robustness by adding connectivity links. *Phys. A Stat. Mech. Its Appl.* **2016**, *444*, 9–19. [CrossRef]

25. Newman, M.E.J. Fast algorithm for detecting community structure in networks. *Phys. Rev. E* **2004**, *69*, 066133. [CrossRef]

26. Xue, S.; Che, Y.; He, W.; Zhao, Y.; Zhang, R. Control Strategy of Electric Heating Loads for Reducing Power Shortage in Power Grid. *Processes* **2019**, *7*, 273. [CrossRef]

27. Wang, S.; Zhang, J.; Zhao, M.; Min, X. Vulnerability analysis and critical areas identification of the power systems under terrorist attacks. *Phys. A Stat. Mech. Its Appl.* **2017**, *473*, 156–165. [CrossRef]

28. Bai, S.; Cheng, Z.; Guo, B. Maintenance Optimization Model with Sequential Inspection Based on Real-Time Reliability Evaluation for Long-Term Storage Systems. *Processes* **2019**, *7*, 481. [CrossRef]

29. Klempner, G.; Kerszenbaum, I. Operation and Control. In *Operation and Maintenance of Large Turbo-Generators*; John Wiley & Sons, Inc.: Hoboken, NJ, USA, 2005.

30. Motter, A.E. Cascade Control and Defense in Complex Networks. *Phys. Rev. Lett.* **2004**, *93*. [CrossRef]

31. Watts, D.J.; Strogatz, S.H. Collective dynamics of 'small-world' networks. *Nature* **1998**, *393*, 440–442. [CrossRef]

32. Knuth, D.E. A Generalization of Dijkstra's Algorithm. *Inf. Process. Lett.* **1977**, *6*, 1–5. [CrossRef]

33. Buldyrev, S.V.; Parshani, R.; Paul, G.; Stanley, H.E.; Havlin, S. Catastrophic cascade of failures in interdependent networks. *Nature* **2010**, *464*, 1025–1028. [CrossRef] [PubMed]

34. Newman, M.E.J. Modularity and community structure in networks. *Proc. Natl. Acad. Sci. USA* **2006**, *103*, 8577–8582. [CrossRef] [PubMed]

35. Zimmerman, R.D.; Murllo-Sanchez, C.E.; Thomas, R.J. MATPOWER: Steady-State Operations, Planning, and Analysis Tools for Power Systems Research and Education. *IEEE Trans. Power Syst.* **2011**, *26*, 12–19. [CrossRef]

36. Fan, W.-L.; Liu, Z.-G.; Hu, P. A High Robustness and Low Cost Cascading Failure Model Based On Node Importance In Complex Networks. *Mod. Phys. Lett. B* **2014**, *28*. [CrossRef]

Article

Automatic Hybrid Attack Graph (AHAG) Generation for Complex Engineering Systems [†]

Mariam Ibrahim [1,*] and Ahmad Alsheikh [1,2]

[1]   Department of Mechatronics Engineering, Faculty of Applied Technical Science, German Jordanian
    University, Amman 11180, Jordan; a.alsheikh@gju.edu.jo
[2]   Department of Mechanical Engineering and Mechatronics, Deggendorf Institute of Technology,
    94469 Deggendorf, Germany
[*]   Correspondence: mariam.wajdi@gju.edu.jo
[†]   This paper is an extended version of paper published in the international conference: 2018 10th International
    Conference on Electronics, Computers and Artificial Intelligence (ECAI), Iasi, Romania, 28–30 June 2018.

Received: 16 September 2019; Accepted: 21 October 2019; Published: 1 November 2019

Abstract: Complex Engineering Systems are subject to cyber-attacks due to inherited vulnerabilities in the underlying entities constituting them. System Resiliency is determined by its ability to return to a normal state under attacks. In order to analyze the resiliency under various attacks compromising the system, a new concept of Hybrid Attack Graph (HAG) is introduced. A HAG is a graph that captures the evolution of both logical and real values of system parameters under attack and recovery actions. The HAG is generated automatically and visualized using Java based tools. The results are illustrated through a communication network example.

Keywords: Hybrid Attack Graph; Level-of-Resilience; stability; topology

---

1. Introduction

As a result of the rapid advancement of complex engineering systems such as infrastructure, communications, energy systems, industrial automation, artificial intelligence, and cyber-physical systems, new research directions in modeling, monitoring, diagnosis, optimization, and control have emerged in recent years [1]. For instance, a control chart scheme for production processes was developed by [2] for monitoring the mean time between two events under the neutrosophic statistics using the belief estimator for the neutrosophic gamma distribution. The diagnosis and correction of many production problems which often cause huge loss to the production unit can substantially be improved with the utilization of the effective control chart technique.

In [3], the advantages of using Proportional-Integral (PI) controller for pH control in the raceway reactor during the whole day against traditional On/Off control were demonstrated. The paper also presented an event-based control architecture for Proportional–Integral–Derivative (PID) controllers. The objective is to tune a classical time-driven PI for pH control in the raceway reactor, and then to add event-based capabilities, but keeping the initial PI control design. The event-based systems allow a trade-off between control performance and control effort, which is perfect for the microalgae process in raceway reactors. The performed tests were oriented to establish a trade-off between control effort and control performance and present an alternative to traditional control.

The applicability of the Distributed Model Predictive Control (DiMPC) was investigated in [4] to deal with the constraints in the steam/water loop of a steam power plant. A comparison was conducted between the Decentralized Model Predictive Control (DeMPC), the Centralized Model Predictive Control (CMPC), and the DiMPC. The results showed the effectiveness of the DiMPC [5] designed an Optimal Nonlinear Adaptive Control (ONAC) strategy to achieve optimal parameter tuning of Nonlinear Adaptive Control (NAC) for Voltage Source Converter (VSC) operating in both rectifier mode and inverter mode where an optimal and robust control can be achieved under different operation scenarios.

A novel pole-zero cancelation method was proposed by [6] for Multi-Input Multi-Output (MIMO) temperature control in heating process systems. In the proposed method, the temperature differences and the transient response of each point can be controlled by considering the dead time and the coupling effect of the MIMO system. In [7], the Slow-Mode-Based Control (SMBC) method combined with decoupling and dead-time compensation was applied to the MIMO temperature control system. The temperature differences and the transient response of all points can be controlled and improved by making the output of the fast modes follow that of the slow mode. The results were then compared to the conventional PI control and gradient temperature control methods.

The Deep Deterministic Policy Gradient (DDPG) technique for the optimum boost control on a Variable Geometry Turbocharger (VGT)-equipped engine is implemented by [8]. The proposed DDPG algorithm is compared with a fine-tuned PID controller to validate its optimality. The results showed that the control performance based on the proposed DDPG algorithm can achieve a good transient control performance from scratch by autonomously learning the interaction with the environment, without relying on model supervision or complete environment models.

A framework was proposed by [9] to use the structural information in each Possible Conflict (PC) for fault diagnosis of complex industrial system to design a different kind of executable model. They proposed to build grey-box models based on a state space neural network architecture derived from that structural information in the PC, which links measurements with equations, and consequently with parameters related to faulty behavior. The state space Neural Networks (ssNN) were used to track the system behavior, but once a fault detection was confirmed, the structural information in the models and the consistency-based diagnosis paradigm were used to perform fault isolation.

Total decomposition of nonstationary variables for distributed monitoring of nonstationary industrial processes was handled by [10], in which different variable blocks were separated with both overlapping and nonoverlapping relationships considered, capturing different nonstationary characteristics. A two-level monitoring strategy was designed that can supervise both the local cointegration relationships and the interrelationship among different nonstationary blocks with enhanced interpretation of the nonstationary process.

A novel diagnosis framework was proposed by [11] for considering the deep feature learning and cross-domain feature distribution alignment simultaneously for industrial applications. Extending the Marginal Distribution Adaptation (MDA) to Joint Distribution Adaptation (JDA), the proposed framework can exploit the discrimination structures associated with the labeled data in source domain to adapt the conditional distribution of unlabeled target data, and thus, guarantee a more accurate distribution matching [12]. They reviewed over 220 technical research programs in total, with more attention on the recent developments of the fault diagnosis approaches and their applications during the last decade. Knowledge-based fault diagnosis, hybrid fault diagnosis, and active fault diagnosis were reviewed comprehensively. The distinctive advantages and various constrains of these diagnosis methods were commented on. A recent survey by [1] also summarized papers on monitoring and diagnosis for complex engineering systems.

The implementation of diagnostic and prognostic architectures can aid the implementation of advanced control algorithms in a resilient control system to recognize sensor degradation, as well as failures with industrial process equipment associated with the control algorithms [13]. A resilient control system is one that maintains state awareness and an accepted level of operational normalcy in response to disturbances, including threats of an unexpected and malicious nature [14]. As a result, it is hard to obtain true expectations about the consequences when a fault/attack occurs effecting or compromising the system. Hence, evaluating systems resilience in terms of stability, performance, and recovery time is crucial and valuable for cost management and design tradeoff. The traditional Attack graphs can generate various attack scenarios compromising the system in terms of violations of a security property. However, they are only concerned with tracking the logical changes in the system parameters under attacks, as captured by the pre and post-conditions. Therefore, the novelty of this work lies in introducing and generating automatically a Hybrid Attack Graph (HAG) using our new Automatic Hybrid Attack Graph (AHAG) Java based tool that combines logical and real values of system parameters. In fact, HAG can be used to provide additional information about the expected changes that could affect the system state by different attack scenarios constituting the graph. In such cases, attack scenarios could be compared for the worst attack scenario that would compromise a system. This can be done by determining the associated real values (i.e., the resilience levels as determined in [15]). The results are illustrated through two-communication networks example. The networks models and security properties (written using Architecture Analysis & Design Language (AADL) [16] and AADL Annex Assume Guarantee REasoning Environment (AGREE) plug-in [17] that relies on JKind model-checker tool [18]) are fed to the AHAG tool, which generates all possible attack scenarios of the systems model and visualizes the graphs using Unity software [19].

This paper extends the conference version [20] significantly by introducing the concept of the Hybrid Attack Graph (HAG), and a new developed tool for its automatic generation (AHAG). The tool is implemented on communication networks example. The remainder of this paper is organized as follows: Section 1.1 reviews the related work. Section 2 describes the model based attack graph implementation through illustrative communication networks example. Section 3 explains the Level-of-resilience assessment. Section 4 presents the Hybrid Attack Graph. Section 5 introduces our AHAG tool to automatically generate the Hybrid attack graph and shows the experimental results. Section 6 summarizes and presents certain future directions.

1.1. Related Work

Existing Attack graph generation tools can be summarized as follows [21]. One research implemented a tool that consisted of three main pieces: a model builder (it takes as input information about network topology, configuration, and a library of attack rules), an attack graph generator (SPIN), and a Graphical User Interface (GUI) for graphical presentation. Topological Vulnerability Analysis (TVA), Network Security Planning Architecture (NETSPA), and Multi-host Multi-stage Vulnerability Analysis (MULVAL) tools [22]. These tools can explore all potential methods an attacker can use to corrupt an enterprise network by determining the configuration information of the hosts and the network. The Cauldron tool implemented by [23] automatically mapped all paths of vulnerability by correlating, aggregating, normalizing, and fusing data from various assets. The Network Attack Graph GENeration (Naggen) tool is developed by [24] to generate attack graph. Other research [5] has illustrated how logical attack graph complexity can be simply elevated when a network becomes denser and larger [25]. They utilized the MULVAL tool, but they changed its engine so that the trace of evaluation was recorded and sent to a graph builder.

In [26], the New Symbolic Model Checker (NuSMV) model checker was implemented to develop Attack graph counterexamples. The value iteration method was also implemented to determine the reliability of Attack graph, which allowed designers to identify which nominal set of security defenses would assure systems safety. A model-checking-based Automated Attack Graph Generator and Visualizer (A2G2V) was proposed by [27]. The proposed A2G2V algorithm used existing model-checking tools, an architecture description tool, and C code to generate an attack graph that enumerates the set of all possible sequences in which atomic-level vulnerabilities can be exploited to compromise system security. The A2G2V tool required building three main functions: a counterexample parsing function, a cyclic testing function, and a Luster model editing function.

In [28], a heuristic algorithm was used to automatically determine the optimal security hardening through cost-benefit analysis [29]. They used multiple algorithms, such as Vulnerability Node Matching, Attack Graph Optimization, Maximum Loss Flow, Path Seeking, and Multi-Objective Augmented Road Sorting, to investigate the vulnerability, estimate the global path, and determine the optimal attack path. The surveys conducted by [30,31] illustrated the state-of-the-art technologies in Attack graph construction in computer networks, and their potential development challenges. Alerts were also analyzed from intrusion detection system, and how well the approaches scaled to larger networks.

A Hybrid Attack Graph (HAG) was modeled by [32], using linguistics and type extensions of traditional attack graph given the necessary inputs of asset, topology, and fact declarations in the typed grammar. The HAG was illustrated through an example of an attacker who is trying to compromise an automotive car by forcing it to drivetoward a wall. The tool automates the task of creating HAGs by compiling the inputs specified and making the appropriate connections. A Hybrid Attack Dependency Graph (HADG) was presented by [33], which allowed discretization into intervals of the reachable and related ranges of the system's state variables and their evolution over the execution of attacks with duration. The HAG generation software of [33] was used by [34] to model a Cyber physical System's attack for a smart grid in which an attacker has to obtain access to a Supervisory Control and Data Acquisition (SCADA) system to cause the transformer to overheat. Thus, transformer temperature was the continuous value in question and was discretized into intervals. The HAG was built by matching exploit patterns to a particular system state.

The concept of adding priorities to HAGs was introduced by [35]; if multiple exploit preconditions were met in a given state, only those with the highest priority value would be expanded into new states of the attack graph. Lower priority exploits would then only be considered if all exploits with higher priorities failed to meet their preconditions. By applying exploit priorities as a heuristic measure, attack graph states can be explored in a more strategic manner.

A novel Hybrid Attack Model (HAM) was introduced by [36] that combines Probabilistic Learning Attacker, Dynamic Defender (PLADD) game model and a Markov Chain model to simulate the planning and execution stages of a bad data injection attack in power grid. The hybrid model is shown to be capable of modeling long time-to-completion actions in the preparation stage and short time-to-completion actions in the execution stage. Table 1 summarizes the main characteristics and limitations of the existed studies in HAG generation as compared to our Automatic Hybrid Attack Graph (AHAG) tool.

Table 1. Main characteristics and limitations of the existed studies in Hybrid Attack Graph (HAG) generation.

	Main Characteristics	Limitations
HAG [32]	<ul><li>Stateful graph.</li><li>Captures blended attack vectors (comprising discrete and continuous exploit events).</li><li>Time is implemented using a single group of global exploits, each of which increments a class of assets' position in time depending upon its particular state.</li></ul>	<ul><li>Generation of attack graphs from nontrivial scenarios (with tens of hybrid assets) in acceptable time is daunting.</li></ul>

Table 1. *Cont.*

	Main Characteristics	**Limitations**
Hybrid Attack Dependency Graph (HADG) [33]	• Replaces the state transition graph with a dependency graph. • A hybrid attack takes a range of real values as preconditions and output a range of values as postconditions. • Provides the capability of modeling continuous state variables and their evolution over the execution of attacks with duration.	• The process for generating HADGs must be articulated and formalized and its performance characterized to better handle continuous variables. • Expansion of the exploit list and the asset parameters in the network file is required to fully explore the attack space.
HAG [36]	• Hybrid Attack Model (HAM) consists of both Probabilistic Learning Attacker, Dynamic Defender (PLADD) nodes and Markov state nodes. • The attacker must have control over all PLADD nodes, which represent the preparation for an attack, to be able to execute an attack by traversing through the Markov states. • Considers the time difference between the attacker's action to prepare for an attack in comparison to the attacker's action to execute for an attack.	• The number of time steps in the Markov Chain model per unit time in the PLADD model must be specified by a domain expert. • The PLADD and the Markov Chain models do not have the means to run a simulation where the time-to-completion of an action for attacker and defender can be significantly different in the preparation stage and the execution stage.
Automatic Hybrid Attack Graph (AHAG)	• Stateful graph. • Specifies attack patterns as sets of preconditions and postcondition. • The integration of level of resilience values captures the system response under sequential attack and recovery actions.	• Requires comprehensive overview of the system model and security/resiliency property. • Levels-of-Resilience values need to be determined under attacks.

2. Model-Based Attack Graph Implementation

2.1. Networked Systems Examples

The two images of Figure 1a,b show two communication networks with identical clients and services (Email, File Transfer Protocol (FTP), and Video), but with different topologies [15]. In both networks, the routers R_1, R_2, and R_3 are built with Routing Protocol (RIP) [37], where the traffic is rerouted under faults as a recovery action through a redundant path (if it exists) based on hop count [37]. In addition to that, R_1 is linked to three Local Area Networks (*LANs*); LAN_1, LAN_2, and LAN_3, where every *LAN* has 10 clients. R_2 is linked to LAN_4, which has 10 clients as well. Router R_3 has three more links, which connect it, respectively, to an Email server through the Internet, an FTP server, and a Video workstation. The second communication network, CN_2, has similar clients and services as CN_1, but has a different topology.

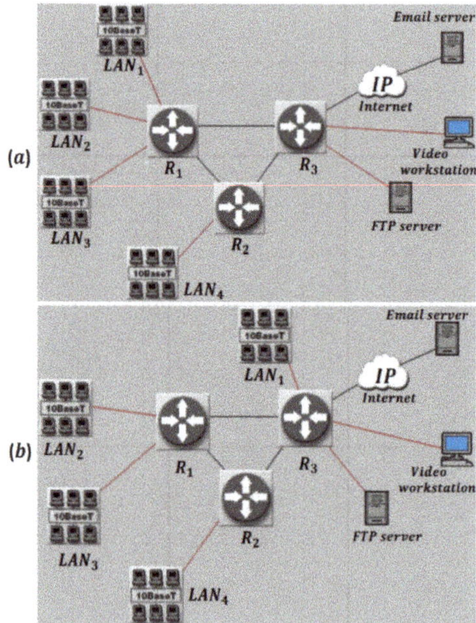

Figure 1. (a) Network CN_1 and (b) Network CN_2.

2.2. Formal System Description for Networked Examples

The generation of Attack Graph requires an overall formal description of the system model and security property being investigated, encoded in Architecture Analysis and Design Language (AADL) and checked using JKind. Here, the formal descriptions of CN_1 and CN_2, respectively, are given as follows [20].

1. Set of Routers R = 1, 2, 3, IP Cloud; Variable $I \in \{1, 2, 3, 4\}$ (static parameters).
2. Set of LANs N = 1, 2, 3, 4; Variable $k \in \{1, 2, 3, 4\}$ (static parameters).
3. Set of Service Providers S; Variable $s \in \{Ftp, Email, Video\}$ (static parameters).
4. Set of Connection Links $L \subseteq R \times R$, $R \times N$, $R \times S$; Labeled $l_{ij} \equiv$ Link is placed between component i and component j (static parameters).
5. System Connectivity C = L; Boolean $c_{ij} = 1$ if there is a connection between component i and component j (dynamic variables).
6. System Stability T; Boolean t = 1 if system is stable (dynamic variable).
7. System Performance $P \subseteq S$; Boolean $f_k = 1$ if ftp service is provided to LAN k, Boolean $e_k = 1$ if Email service is provided on LAN k and Boolean $v_k = 1$ if Video service is provided on LAN k (dynamic variables).
8. System recovery Action R; Variable $r \in \{p, a, d\}$, in case of normal operation $r = p$, in case of recovery action $r = a$, and in case of no action can be done, $r = d$ (dynamic variables).
9. Number of faulted Links that occur sequentially N; Variable $n \in \{0, 1, 2\}$, in case of no fault $n = 0$, in case of first fault $n = 1$, and in case of second fault $n = 2$ (dynamic variables).
10. Attack Instance $AI \subseteq A \times R \times R$, $A \times R \times N$, $A \times R \times S$, Labeled $a_{ij}{}^m \equiv$ Attack a on the Link between component i and component j, where $m \subseteq L$ is a sequence of the previous faulted link(s) if exists. (static parameters)
11. Pre-Attack conditions for CN_1:

- $\mathrm{Pre}(a_{13}) \equiv (c_{13} = 1) \wedge (t = 1) \wedge (r = p) \wedge (n = 0)$
- $\mathrm{Pre}(a_{12}) \equiv (c_{12} = 1) \wedge (t = 1) \wedge (r = p) \wedge (n = 0)$
- $\mathrm{Pre}(a_{23}) \equiv (c_{23} = 1) \wedge (t = 1) \wedge (r = p) \wedge (n = 0)$
- $\mathrm{Pre}(a_{23}^{13}) \equiv (c_{23} = 1) \wedge (r = a) \wedge (n = 1)$
- $\mathrm{Pre}(a_{23}^{12}) \equiv (c_{23} = 1) \wedge (r = p) \wedge (n = 1)$
- $\mathrm{Pre}(a_{13}^{23}) \equiv (c_{13} = 1) \wedge (r = a) \wedge (n = 1)$
- $\mathrm{Pre}(a_{13}^{12}) \equiv (c_{13} = 1) \wedge (r = p) \wedge (n = 1)$
- $\mathrm{Pre}(a_{12}^{13}) \equiv (c_{12} = 1) \wedge (r = a) \wedge (n = 1)$
- $\mathrm{Pre}(a_{12}^{23}) \equiv (c_{12} = 1) \wedge (r = a) \wedge (n = 1)$.

12. Post-Attack conditions for CN_1:

- $\mathrm{Post}(a_{13}) \equiv (c_{13} = 0) \wedge (r = a) \wedge (n = 1)$
- $\mathrm{Post}(a_{12}) \equiv (c_{12} = 0) \wedge (r = p) \wedge (n = 1)$
- $\mathrm{Post}(a_{23}) \equiv (c_{23} = 0) \wedge (r = a) \wedge (n = 1)$
- $\mathrm{Post}(a_{23}^{13}) \equiv (t = 0) \wedge (c_{23} = 0) \wedge (f_1 = f_2 = f_3 = f_4 = e_1 = e_3 = e_4 = v_2 = v_3 = 0) \wedge (r = d) \wedge (n = 2)$
- $\mathrm{Post}(a_{23}^{12}) \equiv (t = 0) \wedge (c_{23} = 0) \wedge (f_4 = e_4 = 0) \wedge (r = d) \wedge (n = 2)$
- $\mathrm{Post}(a_{13}^{23}) \equiv (t = 0) \wedge (c_{13} = 0) \wedge (f_1 = f_2 = f_3 = f_4 = e_1 = e_3 = e_4 = v_2 = v_3 = 0) \wedge (r = d) \wedge (n = 2)$
- $\mathrm{Post}(a_{13}^{12}) \equiv (t = 0) \wedge (c_{13} = 0) \wedge (f_1 = f_2 = f_3 = e_1 = e_3 = v_2 = v_3 = 0) \wedge (r = d) \wedge (n = 2)$
- $\mathrm{Post}(a_{12}^{13}) \equiv (t = 0) \wedge (c_{12} = 0) \wedge (f_1 = f_2 = f_3 = e_1 = e_3 = v_2 = v_3 = 0) \wedge (r = d) \wedge (n = 2)$
- $\mathrm{Post}(a_{12}^{23}) \equiv (t = 0) \wedge (c_{12} = 0) \wedge (f_4 = e_4 = 0) \wedge (r = d) \wedge (n = 2)$

13. Pre-Attack conditions for CN_2:

- $\mathrm{Pre}(a_{13}) \equiv (c_{13} = 1) \wedge (t = 1) \wedge (r = p) \wedge (n = 0)$
- $\mathrm{Pre}(a_{12}) \equiv (c_{12} = 1) \wedge (t = 1) \wedge (r = p) \wedge (n = 0)$
- $\mathrm{Pre}(a_{23}) \equiv (c_{23} = 1) \wedge (t = 1) \wedge (r = p) \wedge (n = 0)$
- $\mathrm{Pre}(a_{23}^{13}) \equiv (c_{23} = 1) \wedge (r = a) \wedge (n = 1)$
- $\mathrm{Pre}(a_{23}^{12}) \equiv (c_{23} = 1) \wedge (r = p) \wedge (n = 1)$
- $\mathrm{Pre}(a_{13}^{23}) \equiv (c_{13} = 1) \wedge (r = a) \wedge (n = 1)$
- $\mathrm{Pre}(a_{13}^{12}) \equiv (c_{13} = 1) \wedge (r = p) \wedge (n = 1)$
- $\mathrm{Pre}(a_{12}^{13}) \equiv (c_{12} = 1) \wedge (r = a) \wedge (n = 1)$
- $\mathrm{Pre}(a_{12}^{23}) \equiv (c_{12} = 1) \wedge (r = a) \wedge (n = 1)$.

14. Post-Attack conditions for CN_2:

- $\mathrm{Post}(a_{13}) \equiv (c_{13} = 0) \wedge (r = a) \wedge (n = 1)$
- $\mathrm{Post}(a_{12}) \equiv (c_{12} = 0) \wedge (r = p) \wedge (n = 1)$
- $\mathrm{Post}(a_{23}) \equiv (c_{23} = 0) \wedge (r = a) \wedge (n = 1)$
- $\mathrm{Post}(a_{23}^{13}) \equiv (t = 0) \wedge (c_{23} = 0) \wedge (f_2 = f_3 = f_4 = e_3 = e_4 = v_2 = v_3 = 0) \wedge (r = d) \wedge (n = 2)$
- $\mathrm{Post}(a_{23}^{12}) \equiv (t = 0) \wedge (c_{23} = 0) \wedge (f_4 = e_4 = 0) \wedge (r = d) \wedge (n = 2)$
- $\mathrm{Post}(a_{13}^{23}) \equiv (t = 0) \wedge (c_{13} = 0) \wedge (f_2 = f_3 = f_4 = e_3 = e_4 = v_2 = v_3 = 0) \wedge (r = d) \wedge (n = 2)$
- $\mathrm{Post}(a_{13}^{12}) \equiv (t = 0) \wedge (c_{13} = 0) \wedge (f_2 = f_3 = e_3 = v_2 = v_3 = 0) \wedge (r = d) \wedge (n = 2)$
- $\mathrm{Post}(a_{12}^{13}) \equiv (t = 0) \wedge (c_{12} = 0) \wedge (f_2 = f_3 = e_3 = v_2 = v_3 = 0) \wedge (r = d) \wedge (n = 2)$
- $\mathrm{Post}(a_{12}^{23}) \equiv (t = 0) \wedge (c_{12} = 0) \wedge (f_4 = e_4 = 0) \wedge (r = d) \wedge (n = 2)$

15. Initial state: $(t = 1) \wedge (c_{23} = c_{12} = c_{13} = 1) \wedge (f_1 = f_2 = f_3 = f_4 = e_1 = e_3 = e_4 = v_2 = v_3 = 1) \wedge (r = p) \wedge (n = 0)$. (Initially, the system is stable, normally operated, and no service outages).

16. The security/resiliency property φ is that both CN_1 and CN_2 are always stable under the given attacks/faults. This can then be written by a CTL formula: $\varphi \equiv AG(t = 1) \equiv AG(\neg (t = 0))$.

2.3. Attack Scenarios Implementation

In this section, we present the Attack Graphs resulted by running the JKind model checker for the encoded AADL CN_1 and CN_2 descriptions, respectively, against the security property φ [20]. JKind is an infinite state model checker for checking safety properties of synchronous systems [38], which are expressed in Lustre, a formally defined, declarative, and synchronous dataflow programming language for programming reactive systems [39]. The Verification is based on k-induction and property directed reachability using a back-end Satisfiability Modulo Theories (SMT) solver. A verified property is determined to be true for all runs of the system. A property violation is reported with an explicit Counter-Example (CE), which is given here as an attack scenario (a sequence of attack and recovery actions resulting in system disruption).

In our work, the CN_1 and CN_2 descriptive models included entities and their interfaces and connections, which were encoded using Architecture Analysis and Design Language (AADL), within the open-source integrated development environment (Osate2). The *AADL* models were embedded with the AGREE Annex plug-in [17] that is used to specify the component models and system-level security properties. AGREE also translates the AADL+Annex models and properties to Lustre language, which JKind can verify against a security property of concern and delivers the result as a CE, if it exists [40].

Figures 2 and 3 show the Attack graphs for CN_1 and CN_2, respectively [20], capturing the state evolution of the two networks dynamical variables given in the earlier formal description under attack instances. These graphs are visualized using *Unity* tool that supports two-dimensional (2D) and three-dimensional (3D) graphics, drag-and-drop functionality, and scripting using C# [41].

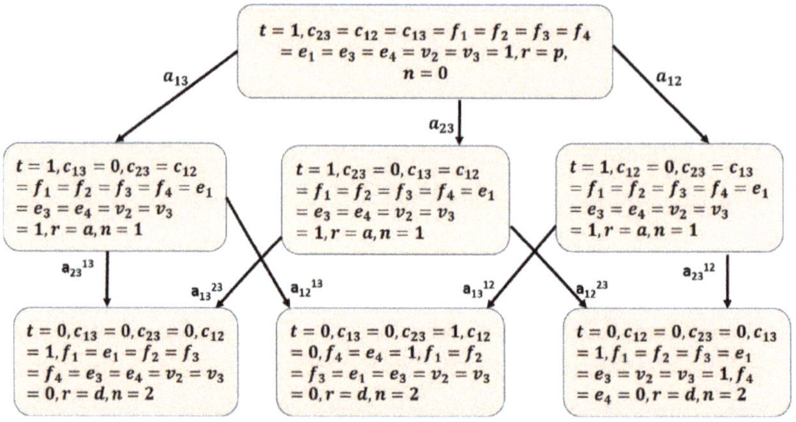

Figure 2. Network CN_1 Attack Graph.

From the obtained graphs it can be seen that both networks have six attack scenarios, resulting in networks loss of stability as determined by the unbounded traffic loss over time [42]. Each attack is a sequence of faults and recovery actions occurring sequentially as follow:

S_1: $a_{13} \rightarrow a_{23}^{13}$

S_2: $a_{13} \rightarrow a_{12}^{13}$

S_3: $a_{12} \rightarrow a_{23}^{12}$

S_4: $a_{23} \rightarrow a_{13}^{23}$

S_5: $a_{23} \rightarrow a_{12}^{23}$

S_6: $a_{12} \rightarrow a_{13}^{12}$

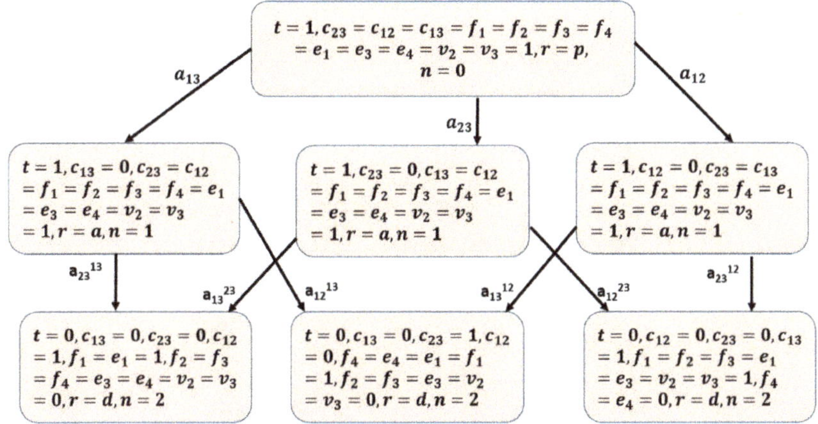

Figure 3. Network CN_2 Attack Graph.

3. Level-of-Resilience Assessment

Each path in the graph is a single attack scenario and has an associated Level-of-Resilience (LoR) [20]. The following definition can be utilized to identify the worst case Level-of-Resilience of a system with its Attack graph given. A system is the worst resilient to an attack scenario in the graph if it acquires the highest loss of stability, the highest loss of performance, or the highest recovery-time.

Definition 1 ([20]). *Given a system M and an attack graph A_G comprising a set of attack scenarios $S \equiv \cup S_i$, $i \in \{1, \dots, z\}$, where z is the number of attack scenarios, we say that LoR(M, S_i) is the worst if:*

$$[LoS_R(M, S_i) > LoS_R(M, S - S_i)]$$
$$\vee \; [[LoS_R(M, S_i) = LoS_R(M, S - S_i)$$
$$\wedge \; [LoP_R(M, S_i) > LoP_R(M, S - S_i)]]$$
$$\vee \; [[LoS_R(M, S_i) = LoS_R(M, S - S_i)]$$
$$\wedge \; [LoP_R(M, S_i) = LoP_R(M, S - S_i)]$$
$$\wedge \; [RT(M, S_i) > RT(M, S - S_i)]]$$

The next definition compares the LoR of many systems against an Attack scenario. A system is the most resilient to an attack scenario if this attack acquires a smallest loss of stability, a smallest loss of performance, or smallest recovery-time.

Definition 2 ([20]). *Given a set of systems $M \equiv \cup M_j$, $j \in \{1, \dots, y\}$, where y is the number of systems, and an attack scenario $S_i \in S$, we say that LoR(M_i, S_i) > LoR(M-M_i, S_i) if:*

$$[LoS_R(M_i, S_i) < LoS_R(M - M_i, S_i)]$$
$$\vee \; [[LoS_R(M_i, S_i) = LoS_R(M - M_i, S_i)]$$
$$\wedge \; [LoP_R(M_i, S_i) < LoP_R(M - M_i, S_i)]]$$
$$\vee \; [[LoS_R(M_i, S_i) = LoS_R(M - M_i, S_i)]$$
$$\wedge \; [LoP_R(M_i, S_i) = LoP_R(M - M_i, S_i)]$$
$$\wedge \; [RT(M_i, S_i) < RT(M - M_i, S_i)]].$$

4. Hybrid Attack Graph (HAG)

The states of Attack graph reflect the logical evolution of system parameters (e.g., system stability and services are either true or false) during attacks until reaching the final states where the security property φ is violated and hence are the worst states. As multiple attacks can reach the same final state, it is essential to identify the worst attack scenario. A Hybrid Attack Graph (HAG) associates real values to all attack scenarios terminating in these final states. These real values correspond to the Level-of-Resilience parameters determined from the system dynamical response. The detailed computations of LoS_R, eventual $\underline{LoP_R}$ and RT of networks CN_1 and CN_2 under A_G scenarios are given in [43], and [15]. We formally define HAG as follows.

Definition 3. *A Hybrid Attack Graph (HAG) of a system model M is a data structure representing a union of all attack paths comprising A_G annotated with the associated Levels-of-Resilience.*

$$HAG = A_G \cup LoR(M, A_G).$$

Algorithm 1 compares the Levels-of-Resilience, associated with the attacks comprising an Attack graph, and alerts the worst-case scenario.

Algorithm 1 Alerting Worst-Case Attack Scenario

INPUT: Attack graph (A_G) Comprising Attack Scenarios (S $\equiv$ $\cup S_i$, i $\in$ {1,..., z}) and associated LoR values
OUTPUT: Alert Worst Case $S_{[i]}$
Procedure:
for Case [LoS_R, LoP_R, RT]
    if $S_{[i]}${LoS_R} > S–$S_{[i]}$ {LoS_R}:
    Alert Worst Case $S_{[i]}$
    else if S[i] {LoP_R, Ftp} > S–$S_{[i]}$ {LoP_R, Ftp}
    Alert Worst Case $S_{[i]}$
    else if $S_{[i]}${LoP_R, Video} > S–$S_{[i]}$ {LoP_R, Video}
    Alert Worst Case $S_{[i]}$
    else if $S_{[i]}${LoP_R, Email} > S–$S_{[i]}${LoP_R, Email}
    Alert Worst Case S[i]
    else if $S_{[i]}${RT} > S–$S_{[i]}$ {RT}:
    Alert Worst Case $S_{[i]}$
    else
    no Alert

---

5. Automatic Hybrid Attack Graph (AHAG) Tool

In Section 2, the attack scenarios were generated by repeatedly running AGREE, and updating the security property in every run to exclude the previously generated attack scenarios. Here, the automatic generation of attack scenarios is presented.

The Automatic Hybrid Attack Graph (AHAG) tool was developed through NetBeans, an Integrated Development Environment (IDE) for Java [44]. *NetBeans* allows applications to be developed from a set of modular software components called modules. In addition to Java development, NetBeans has extensions for other languages like Personal Home Page (PHP), C, C++, and fifth version of Hyper Text Markup Language (HTML5). Maven, a build automation tool used primarily for Java projects, was selected as the main project for the AHAG tool in NetBeans [45].

When running an AGREE based JKind model-checker for Lustre models within Osate2, it can only produce one counterexample at a time. However, if executed more than once, it may repeat the same counterexample. The AHAG tool shown in Figure 4 confirms that a different new counterexample is produced (if exists) each time the JKind is called, and automatically, all possible attack scenarios are visualized using Unity software.

AHAG tool takes only the first Lustre model (a translation of the system model and the security property from AGREE). Then, it generates all possible combinations of potential attack scenarios (i.e., CEs) as (.lus) format files. Next, AHAG communicates with the model checker JKind through the Command Prompt Commands (CMD) to iteratively check the system model and the potential CE against the security property within the Lustre files. Doing so, AHAG generates all corresponding results as separate Excel files, illustrating if the potential CE is truly an attack scenario or not. Once all Excel files are generated, AHAG converts them to (.csv) files to be used later along with LoR values in Unity Visualizer.

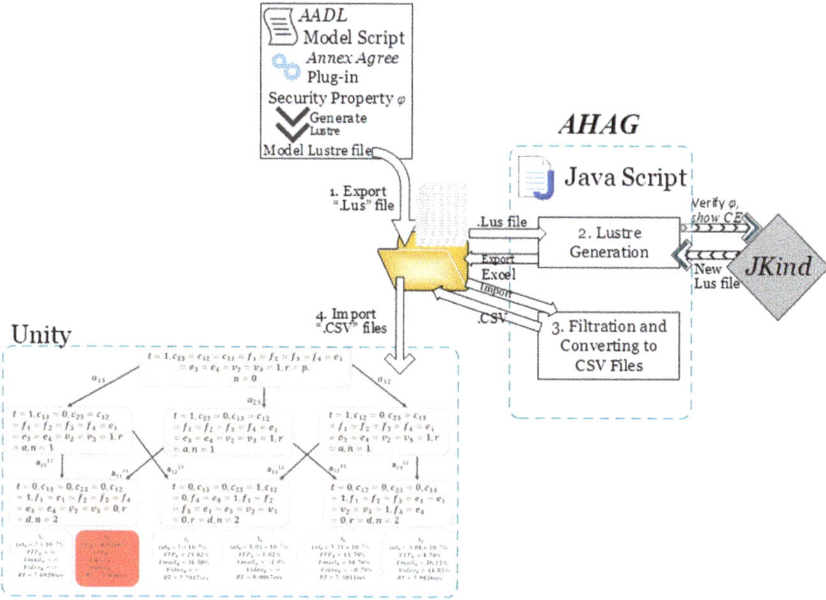

Figure 4. *AHAG* Workflow. AADL: Architecture Analysis & Design Language.

The AHAG Algorithm 2 is as follows. The user inserts the first Lustre Model as input and defines the attack instances/actions constituting the attack paths in a single-dimensional array. In addition to that, the user has to choose the maximum expected length n of an attack scenario. AHAG in turns, generates all potential combinations of attack scenarios A^n, where A is number of attack instances/actions. Each new potential attack scenario is stored as a variable CE_1 within (.lus) generated files. Next, JKind is called iteratively to check these files against the security property ϕ. If CE_1 violates the security property, then it is a true attack scenario, which belongs to the Attack graph, and the result is given as an Excel sheet (.xlsx). Otherwise, AHAG will reject the CE_1. Afterward, since it is easier for Unity visualizer to read (.csv) files, AHAG converts xlsx to csv files and feed them to Unity.

Algorithm 2 AHAG

INPUT: System Model (Luster 0.lus), Attack instances A[ ], maximum length (n)
OUTPUT: All attack scenario in (.csv)
Procedure:
    insert attack instances in the Single-Dimensional Array (A)
    set maximum length (n)
loop 1:
    CE_1 = possible combination from A[ ] of length n
    New Lustre = do new (.lus) copy (Lustre 0.lus)
    New Lustre = New Lustre + CE_1
    goto loop 1
loop 2:
    call JKIND through cmd
    New Lustre.xlsx= do result in (.xlsx) format
loop 3:
    New Lustre.csv = replace (New Lustre.xlsx) format to (.csv) format
    goto loop 3
    goto loop 2
loop 4:
    if New Lustre.csv contains CE_1 = False
    delete (New Lustre.csv)
    goto loop 4
generate violating attack scenarios

The generated attack scenarios were visualized using Unity visualizer tool and the Levels-of-Resilience values, including Level-of-Stability-Reduction (LoS_R), Level-of-Performance-Reduction (LoP_R) in the networks applications (given by FTP_R, $Email_R$, and $Video_R$), and the Recovery-Time (RT) are annotated to the attack scenarios terminating at the final states using Algorithm 1. This generates the HAG for CN_1 and CN_2, as shown in Figures 5 and 6, respectively. It can be seen that Attack scenario S_4 was chosen as the worst attack scenario (highlighted in red). The detailed computations of Levels-of-Resilience values for networks CN_1 and CN_2 are given in [43], and [15]. By comparing the generated HAGs to the traditional Attack graphs of Figures 2 and 3, respectively, it is clear how HAG can aid in differentiating attack scenarios terminating in the same state, yet with distinguishable resilience levels. Thus, network designers can have a better overview on the attack scenario that the network is most vulnerable to.

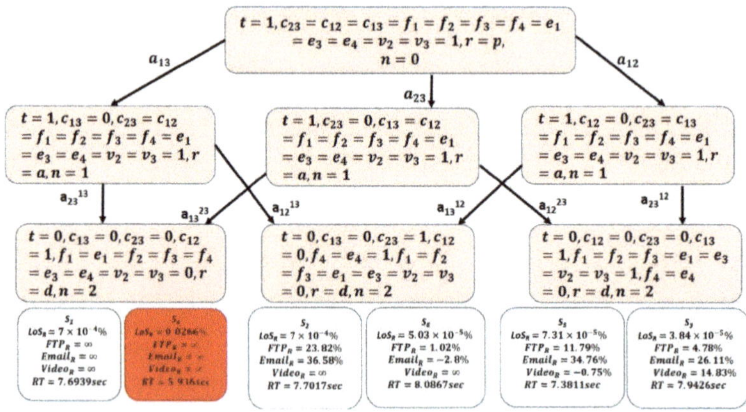

Figure 5. Network CN_1 Hybrid Attack Graph.

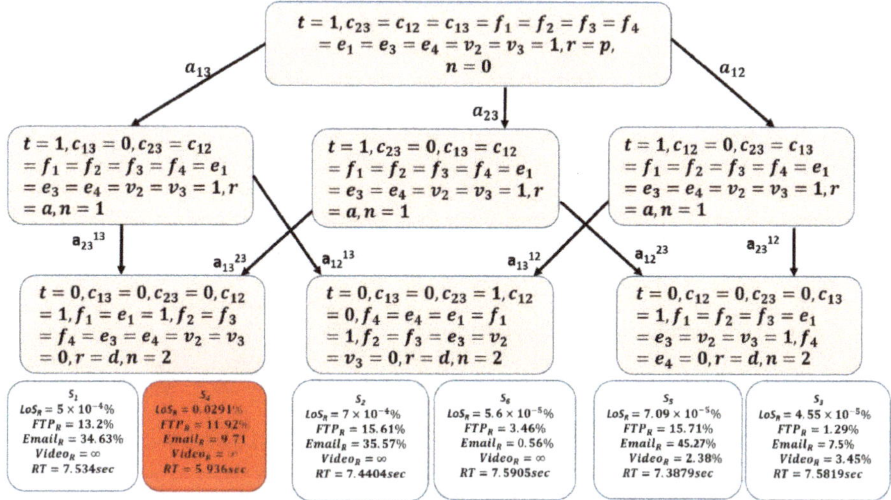

Figure 6. Network CN_2 Hybrid Attack Graph.

6. Conclusions

In this paper, we presented a new concept of Hybrid Attack Graph (HAG), which captures both the logical changes in the system parameters under attacks (as determined by the pre and post-conditions), and the real values of the levels of resilience parameters associated with the attacks constituting the graph. The nearest C-language based tool, Automated Attack Graph Generator and Visualizer (A2G2V), proposed by [27], interacts with the model-checker JKind for the generation of the attack paths one at a time, and with another tool Graph visualization (Graphviz) for visual display of the attack graph. Similar to our Automatic Hybrid Attack Graph (AHAG) tool, A2G2V tool requires one-time modeling effort to obtain the system description for components, connectivity, services, and their vulnerabilities. However, AHAG integrates the Levels-of-Resilience values obtained from system's dynamical response with the Attack graph. This automatically generates a HAG, which can aid system designers to investigate and select the worst LoR system design and its corresponding attack scenario from the graph. This ensures an appropriate defense and countermeasures placement in the system. The results were illustrated through a communication networks example.

Author Contributions: Conceptualization, M.I.; Methodology, M.I.; Software, M.I., A.A.; Validation, M.I., A.A.; Formal Analysis M.I.; Investigation, M.I., A.A.; Resources, M.I.; Data Curation, M.I., A.A.; Writing-Original Draft Preparation, M.I., A.A.; Writing-Review & Editing, M.I.; Visualization, M.I., A.A.; Supervision, M.I.; Project Administration, M.I.; Funding Acquisition, M.I.

Funding: This research was funded by Deanship of Graduation Studies and Scientific Research at the German Jordanian University for the seed fund SATS 02/2018.

Conflicts of Interest: The authors declare no conflict of interest.

References

1. Gao, Z.; Nguang, S.K.; Kong, D.X. Advances in Modelling, monitoring, and control for complex industrial systems. *Complexity* **2019**, *2019*, 2975083. [CrossRef]
2. Aslam, M.; Bantan, R.A.R.; Khan, N. Monitoring the Process Based on Belief Statistic for Neutrosophic Gamma Distributed Product. *Processes* **2019**, *7*, 209. [CrossRef]
3. Rodríguez-Miranda, E.; Beschi, M.; Guzmán, J.L.; Berenguel, M.; Visioli, A. Daytime/Nighttime Event-Based PI Control for the pH of a Microalgae Raceway Reactor. *Processes* **2019**, *7*, 247. [CrossRef]

4. Zhao, S.; Maxim, A.; Liu, S.; De Keyser, R.; Ionescu, C.M. Distributed Model Predictive Control of Steam/Water Loop in Large Scale Ships. *Processes* **2019**, *7*, 442. [CrossRef]

5. Jiang, Y.; Jin, X.; Wang, H.; Fu, Y.; Ge, W.; Yang, B.; Yu, T. Optimal Nonlinear Adaptive Control for Voltage Source Converters via Memetic Salp Swarm Algorithm: Design and Hardware Implementation. *Processes* **2019**, *7*, 490. [CrossRef]

6. Xu, S.; Hashimoto, S.; Jiang, W. Pole-Zero Cancellation Method for Multi Input Multi Output (MIMO) Temperature Control in Heating Process System. *Processes* **2019**, *7*, 497. [CrossRef]

7. Xu, S.; Hashimoto, S.; Jiang, W.; Jiang, Y.; Izaki, K.; Kihara, T.; Ikeda, R. Slow Mode-Based Control Method for Multi-Point Temperature Control System. *Processes* **2019**, *7*, 533. [CrossRef]

8. Hu, B.; Yang, J.; Li, J.; Li, S.; Bai, H. Intelligent Control Strategy for Transient Response of a Variable Geometry Turbocharger System Based on Deep Reinforcement Learning. *Processes* **2019**, *7*, 601. [CrossRef]

9. Pulido, B.; Zamarreño, J.M.; Merino, A.; Bregon, A. State space neural networks and model-decomposition methods for fault diagnosis of complex industrial systems. *Eng. Appl. Artif. Intell.* **2019**, *79*, 67–86. [CrossRef]

10. Zhao, C.; Sun, H.; Tian, F. Total Variable Decomposition Based on Sparse Cointegration Analysis for Distributed Monitoring of Nonstationary Industrial Processes. *IEEE Trans. Control Syst. Technol.* **2019**, 1–8. [CrossRef]

11. Han, T.; Liu, C.; Yang, W.; Jiang, D. Deep transfer network with joint distribution adaptation: A new intelligent fault diagnosis framework for industry application. *ISA Trans.* **2019**, in press. [CrossRef] [PubMed]

12. Gao, Z.; Cecati, C.; Ding, S.X. A Survey of Fault Diagnosis and Fault-Tolerant Techniques Part II: Fault Diagnosis with Knowledge-Based and Hybrid/Active Approaches. *IEEE Trans. Ind. Electron.* **2015**, *62*, 1. [CrossRef]

13. Ji, K.; Lu, Y.; Liao, L.; Song, Z.; Wei, D. Prognostics Enabled Resilient Control for Model-based Building Automation Systems. In Proceedings of the 12th Conference of International Building Performance Simulation Association, Sydney, Australia, 14–16 November 2011.

14. Rieger, C.G.; Gertman, D.I.; McQueen, M.A. Resilient Control Systems: Next Generation Design Research. In Proceedings of the 2nd IEEE Conference on Human System Interaction, Catania, Italy, 21–23 May 2009; Volume 9, pp. 632–636.

15. Ibrahim, M. A resiliency measure for communication networks. In Proceedings of the 8th International Conference on Information Technology (ICIT), Amman, Jordan, 17–18 May 2017; pp. 151–156.

16. SEI. *Architecture Analysis and Design Language*; SEI: Pittsburgh, PA, USA, 2004; Available online: http://standards.sae.org/as5506/ (accessed on 29 October 2019).

17. Rockwell-Collins; University of Minnesota. The Assume Guarantee Reasoning Environment. 2016. Available online: http://loonwerks.com/tools/agree.html (accessed on 29 October 2019).

18. Sheeran, M.; Singh, S.; Stålmarck, G. Checking Safety Properties Using Induction and a SAT-Solver. In *Proceedings of the Computer Vision—ECCV 2012; Austin, TX, USA, 1–3 November 2000*; Springer Science and Business Media LLC: Berlin, Germany, 2000; Volume 1954, pp. 127–144.

19. Download Unity, Unity3d. Available online: https://unity3d.com/get-unity/download (accessed on 29 October 2019).

20. Ibrahim, M.; Alsheikh, A. Assessing Level of Resilience Using Attack Graphs. In Proceedings of the 10th International Conference on Electronics, Computers and Artificial Intelligence (ECAI), Iasi, Romania, 28–30 June 2018; pp. 1–6.

21. Sheyner, O.; Wing, J. Tools for generating and analyzing attack graphs. In *Proceedings of the International Symposium on Formal Methods for Components and Objects*; Leiden, Germany, 4–7 November 2003, Springer: Berlin, Germany, 2003.

22. Ou, X.; Anoop, S. Attack graph techniques. In *Quantitative Security Risk Assessment of Enterprise Networks*; Springer: New York, NY, USA, 2012; pp. 5–8.

23. Jajodia, S.; Noel, S.; Kalapa, P.; Albanese, M.; Williams, J. Cauldron mission-centric cyber situational awareness with defense in depth. In Proceedings of the 2011—MILCOM 2011 Military Communications Conference, Baltimore, MD, USA, 7–10 November 2011.

24. Martın, B.; Lupu, E.C. Naggen: A Network Attack Graph Generation Tool. In Proceedings of the IEEE Conference on Communications and Network Security, Las Vegas, NV, USA, 9–11 October 2017.

25. Ou, X.; Boyer, W.F.; McQueen, M.A. A scalable approach to attack graph generation. In Proceedings of the 13th ACM Conference on Computer and Communications Security, Alexandria, VA, USA, 30 October–3 November 2006; p. 336.

26. Somesh, J.; Sheyner, O.; Wing, J. Two formal analyses of attack graphs. In Proceedings of the 15th IEEE Computer Security Foundations Workshop (CSFW-15), Cape Breton, NS, Canada, 24–26 June 2002.

27. Al Ghazo, A.T.; Ibrahim, M.; Ren, H.; Kumar, R. A2G2V: Automatic Attack Graph Generation and Visualization and Its Applications to Computer and SCADA Networks. *IEEE Trans. Syst. Man Cybern. Syst.* **2019**, 1–11. [CrossRef]

28. Wang, S.; Zhang, Z.; Kadobayashi, Y. Exploring attack graph for cost-benefit security hardening: A probabilistic approach. *Comput. Secur.* **2013**, *32*, 158–169. [CrossRef]

29. Huan, W. A Vulnerability Assessment Method in Industrial Internet of Things Based on Attack Graph and Maximum Flow. *IEEE Access* **2018**, *6*, 8599–8609.

30. Shandilya, V.; Simmons, C.B.; Shiva, S. Use of Attack Graphs in Security Systems. *J. Comput. Netw. Commun.* **2014**, *2014*, 1–13. [CrossRef]

31. Lippmann, R.P.; Ingols, K.W. *An Annotated Review of Past Papers on Attack Graphs*; Project Report IA-1; Massachusetts Institute of Technology, Lincoln Laboratory: Lexington, MA, USA, 2005.

32. Louthan, G.; Michael, H.; Phoebe, H.; Peter, H.; John, H. Hybrid extensions for stateful attack graphs. In Proceedings of the 9th Annual Cyber and Information Security Research Conference, Oak Ridge, TN, USA, 8–10 April 2014; p. 101.

33. Louthan, G.; Phoebe, H.; Peter, H.; John, H. Toward hybrid attack dependency graphs. In Proceedings of the 7th Annual Workshop on Cyber Security and Information Intelligence Research, Oak Ridge, TN, USA, 12–14 October 2011.

34. Hawrylak, P.J.; Haney, M.; Papa, M.; Hale, J. Using hybrid attack graphs to model cyber-physical attacks in the Smart Grid. In Proceedings of the 5th International Symposium on Resilient Control Systems (ISRCS), Salt Lake City, UT, USA, 14–16 August 2012; pp. 161–164.

35. Nichols, W.; Hawrylak, P.; Hale, J.; Papa, M. Introducing priority into hybrid attack graphs. In Proceedings of the 12th Annual Conference on Cyber and Information Security Research, Oak Ridge, TN, USA, 4–6 April 2017.

36. Chen, Y.-C.; Gieseking, T.; Campbell, D.; Mooney, V.; Grijalva, S. A Hybrid Attack Model for Cyber-Physical Security Assessment in Electricity Grid. In Proceedings of the 2019 IEEE Texas Power and Energy Conference (TPEC), College Station, TX, USA, 7–8 February 2019; pp. 1–6.

37. C. N. Academy. Routing Protocols and Concepts. Available online: https://www.netacad.com/web/aboutus/ccna-exploration (accessed on 29 October 2019).

38. An Infinite-State Model Checker for Safety Properties. Loonwerks. Available online: http://loonwerks.com/tools/jkind.html (accessed on 29 October 2019).

39. Halbwachs, N.; Paul, C.; Pascal, R.; Daniel, P. The synchronous data flow programming language LUSTRE. *Proc. IEEE* **1991**, *79*, 1305–1320. [CrossRef]

40. Carnegie-Mellon-University. Open Source AADL Tool Environment for the SAE Architecture. 2018. Available online: http://osate.github.io/index.html (accessed on 29 October 2019).

41. Craighead, J.; Burke, J. Using the unity game engine to develop SARGE: A case study. In Proceedings of the 2008 Simulation Workshop at the International Conference on Intelligent Robots and Systems (IROS 2008), Nice, France, 22–26 September 2008.

42. Àlvarez, C.; Blesa, M.J.; Serna, M. The robustness of stability under link and node failures. *Theor. Comput. Sci.* **2011**, *412*, 6855–6878. [CrossRef]

43. Riverbed Technology Inc. Opnet Modeler. Available online: http://mediacms.riverbed.com/documents/download.html (accessed on 29 October 2019).

44. HTML5 Web Development Support. NetBeans. Available online: https://netbeans.org/features/html5/index.html (accessed on 29 October 2019).

45. Böck, H. *The Definitive Guide to NetBeans™ Platform 7*; Apress: New York, NY, USA, 2011.

 *processes*

Article

Position Deviation Control of Drilling Machine Using a Nonlinear Adaptive Backstepping Controller Based on a Disturbance Observer

Huifu Ji [1,2,3] and Songyong Liu [1,2,*]

1 School of Mechatronic Engineering, China University of Mining and Technology, Xuzhou 221116, China; jihuifu@haut.edu.cn
2 Jiangsu Collaborative Innovation Center of Intelligent Mining Equipment, China University of Mining and Technology, Xuzhou 221116, China
3 School of Mechanical & Electrical Engineering, Henan University of Technology, Zhengzhou 450001, China
* Correspondence: liusongyong@cumt.edu.cn; Tel.: +86-0516-8359-0718

Abstract: Thin coal seam mining is a development direction to solve the problem of energy supply at this stage, which cannot be realized by small working space, low automation, and drilling deviation. In this paper, a nonlinear adaptive backstepping controller based on a disturbance observer is proposed and used on a coal auger for position tracking control to achieve directional drilling. Firstly, a nonlinear dynamic model for the deflection control mechanism is built with the consideration of parameter uncertainties and external disturbances. Then, the parameter uncertainty and external disturbance are regarded as a system compound disturbance. Furthermore, a disturbance observer is designed to estimate the system compound disturbance and a nonlinear adaptive backstepping controller was proposed to compensate the system compound disturbance. The upper bound of the compound disturbance, which can effectively reduce the chattering in the directional control process, cannot be obtained easily. A stability analysis of the DCM (deviation control mechanism) with the proposed controller is proved based on the Lyapunov theory. Finally, an electro-hydraulic servo displacement control experimental system with matlab xPC target rapid prototyping technology and a prototype experiment system is established to verify the effectiveness of the proposed control strategy. The experimental results indicate that the proposed controller can yield more satisfactory position tracking performance, such as parameter uncertainties and external disturbances, than the conventional proportion integral derivative (PID) controller and an adaptive backstepping controller. Using the control strategy, technical breakthrough on horizontal directional drilling can be realized for thin coal seam mining.

Keywords: deviation control; drilling machine; nonlinear adaptive backstepping controller disturbance observer; parameter uncertainties

Citation: Ji, H.; Liu, S. Position Deviation Control of Drilling Machine Using a Nonlinear Adaptive Backstepping Controller Based on a Disturbance Observer. *Processes* **2021**, *9*, 237. https://doi.org/10.3390/pr9020237

Received: 29 December 2020
Accepted: 24 January 2021
Published: 27 January 2021

1. Introduction

Thin coal seams, which are indispensable coal resources, are widely distributed around the world. The question of how to exploit these coal seams with high efficiency and mechanization has become the main development direction of coal excavation. The coal auger is a new type of thin seam mining equipment that has extensive prospects thanks to its characteristics of unmanned and non-supported coal face mining. Recently, a new type of coal auger working mechanism with five bits was used for further improving coal mining efficiency [1,2]. The drilling technology is shown in Figure 1. However, owing to the constraint reaction force of the coal wall, gravity, cutting resistance, and friction, the vibration of the working mechanism is relatively excessive, which may cause serious drilling deflection [3]. As a result, the efficiency of drilling and excavating is unbalanced. However, the excavation process has realized intelligent and rapid driving, based on supporting design optimization [4,5] and intelligent control [6].

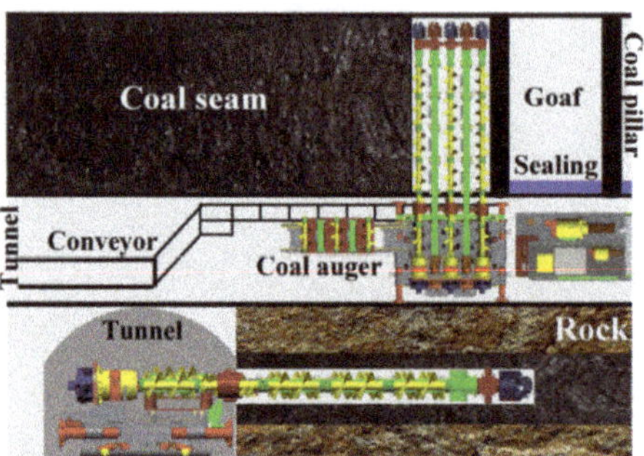

Figure 1. Mining technology of coal auger.

The dynamic characteristics and stability of the drilling mechanism are the main factors leading to deviation. There are many kinds of vibration forms—including vertical vibration [7], lateral vibration [8], torsional vibration [9], and coupling vibration [10]—in the actual drilling process that lead to deviation. For different structures of drilling mechanism, scholars have carried out extensive research on vibration characteristics to obtain the deviation behavior due to vibration [11–14]. However, the deflection in long-distance drilling cannot be solved by restraining the vibration behavior. Therefore, direct deflection detection and control methods have been extensively studied. Lueke et al. [15] analyzed the influence of borehole diameter, drill pipe pressure, bit structure, drilling depth, and geological conditions on deviation fluctuation in the process of directional drilling. It was pointed out that a deviation prevention device with a reasonable arrangement could realize deviation control in drilling. Manacorda G. [16] designed and put forward a kind of ground-penetrating radar equipment for a horizontal directional drilling (HDD) bit by using an angular displacement sensor and communication module. The equipment can provide the real-time position and pose information of the bit and provide a reference for controlling the drilling direction. Inyang I.J. [17] proposed a bilinear model attitude control method for directional drilling tools that describes the non-linear characteristics of directional drilling tools more accurately than the existing linear models are able to. Thus, by expanding the performance range, the attitude control of directional drilling tools was more effective, robust, and stable, which significantly reduced the effect of measurement delay and interference on the stability and performance. Kim J. [18] proposed a new hybrid rotary steering system aimed at the problem of uncontrollable borehole curvature in long-distance directional drilling. This system combines a traditional drill bit with a push bit and adopts a hybrid structure involving a hydraulic cylinder and spherical joint to achieve better maneuverability. Comprehensive analysis shows that deviation control in directional drilling was achieved by designing a special bit. Therefore, in view of the unique structure of the coal auger, a new hydraulic deviation correction mechanism was designed to achieve directional drilling. A structure diagram of the deviation control mechanism is shown in Figure 2.

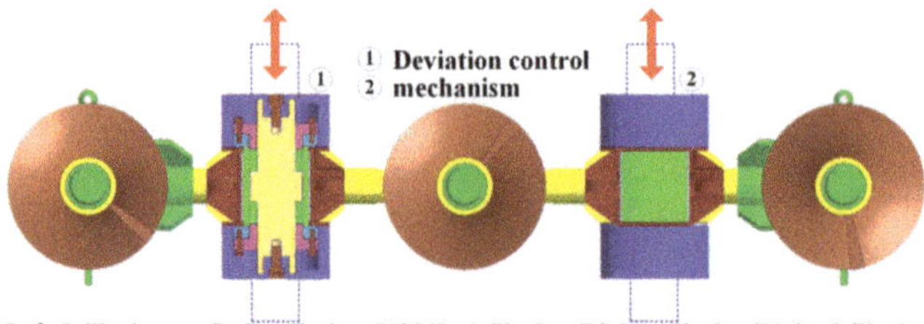

Figure 2. Structure diagram of the deviation control mechanism (DCM).

The DCM (deviation control mechanism) can be regarded as an electro-hydraulic servo control system. The position control of the DCM can be partially realized using a conventional proportional-integral (PI) controller [19] and backstepping controller [20]. However, the DCM is a complex nonlinear system with parameter uncertainties, such as servo-valve and hydraulic actuator dynamics [21], stiffness and damping differences [22], and external disturbances such as friction between the rod and bore of hydraulic cylinders [23]. All these factors make it difficult to obtain satisfactory position coordination performance with the conventional PI controller and backstepping controller, because these two controllers cannot adjust their control parameters in consideration of the parameter uncertainties and external disturbances of the DCM. In order to reduce the effects of the above-mentioned parameter uncertainties and external disturbances of the DCM, many control approaches, including adaptive sliding mode control [24], adaptive backstepping control [25], and a nonlinear disturbance observer (NDO) [26], were introduced. A nonlinear adaptive backstepping controller was employed to improve the position tracking performance, estimating uncertain parameters through adaptive laws derived by guaranteeing the stability of the DCM [27,27]. Choux M. et al. [28] developed an adaptive backstepping controller for a nonlinear hydraulic-mechanical system that handled uncertain parameters related to the internal leakage, friction, orifice equation, and oil characteristics. A new smooth and continuous sliding mode control law was proposed to solve the design conflicts between sliding mode control technology and backstepping adaptive control technology for position control by Ji X.H. [29]. An improved noise-alleviation method was proposed by Yang G.C. [30,31] to achieve the high-accuracy calculation of the standard sign function in direct adaptive control for high-precision position control. Zheng J.Z. [32] concerned a high-accuracy tracking control for hydraulic actuators with unmodeled flexible dynamics considering the structural flexibility of the mechanical components. However, external disturbances that usually exist in nonlinear systems are not taken into account during the controllers' design process. Guo K. [33,34] presented a nonlinear cascade controller based on an extended disturbance observer to track the desired position trajectory for electro-hydraulic single-rod actuators in the presence of both external disturbances and parameter uncertainties. Kasac J. [35] proposed a robust output tracking controller using external disturbances observer for reducing the measurement parameters of the control system. Hu X. [36] designed a dynamically positioned vessel exposed to unknown time-varying external disturbances, incorporating fuzzy logic systems (FLSs), projection operators, and the "robustifying" term into the vectorial backstepping technique. Teoh J.N. [37] took into consideration the rejection of narrow-band disturbances at two frequencies higher than the servo bandwidth in order to obtain a good positioning accuracy for a micro-actuator. A phase-stabilized feedback controller combined with a disturbance observer structure was applied to achieve the rejection of the two disturbances.

However, most of the research focuses on control accuracy using high-precision control elements. Our study tries to combine and improve these proposed controllers and use a control element with a slightly lower accuracy by achieving the required positioning control for easier industrial application. The main contribution of this work is to design a combined controller that consists of a nonlinear adaptive backstepping controller (NABC) and a disturbance observer (DO) for DCM and its implementation in a double-rope winding hoisting experimental system. The basic concept of ABSMC-DO is to regard the parameter uncertainty and external disturbance as a system compound disturbance. A disturbance observer was designed to estimate the system compound disturbance. A nonlinear adaptive backstepping controller was proposed to compensate for the system compound disturbance. The proposed control system does not need to accurately obtain the external disturbance, and has a good control effect for parameter uncertainty and uncertain nonlinearity. A stability analysis of the DCM with the proposed controller is performed based on the Lyapunov function. A series of experimental tests are carried out on the electro-hydraulic position control system with the characteristics of large load inertia and large external interference and a prototype to verify the availability of the proposed controller. Using the novel control method, the upper bound of the compound disturbance, which can effectively reduce the chattering in the directional control process, cannot be obtained easily.

2. Dynamic Model of the DCM

The structure of a servo valve-controlled hydraulic cylinder was adopted as the deviation control mechanism (DCM). The dynamic model of the DCM is shown in Figure 3. It is comprised of a double-rod cylinder, a 3–4-way servo valve, and a load force. The goal is to position track any specified motion trajectory as closely as possible. The nonlinear dynamical model of the DCM is given as follows [38].

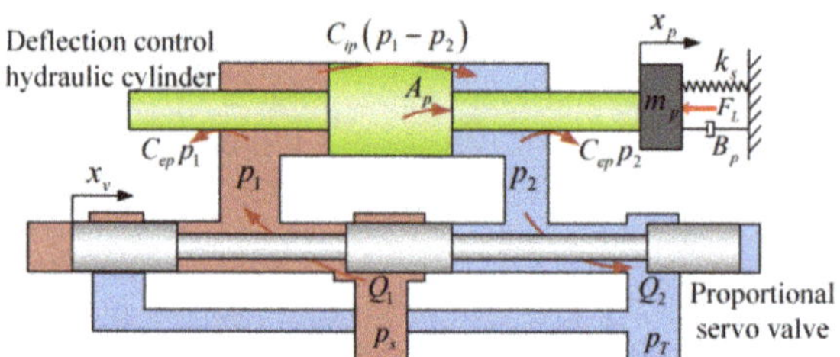

Figure 3. Dynamic model of the deviation control mechanism.

We defined the right shift of servo valve spool as positive. In this direction, the load flow Q_L of the servo-vale is defined as follows:

$$Q_L = \frac{1}{2}C_d\omega x_v\sqrt{\frac{p_S - \text{sgn}(x_v)p_L}{\rho}}.\tag{1}$$

where C_d is the flow coefficient of the servo valve, ω is the area gradient of the servo valve, x_v is the spool displacement of the servo valve, p_L is the differential pressure between the two chambers of the electro-hydraulic cylinder, $p_L = p_1 - p_2$, p_S is the supplied system oil pressure, ρ is the hydraulic oil density, and sgn is a symbolic function that is defined as follows:

$$\text{sgn}(\cdot) = \begin{cases} 1 & \text{if} \cdot > 0 \\ 0 & \text{if} \cdot = 0 \\ -1 & \text{if} \cdot < 0 \end{cases}.\tag{2}$$

As the dynamic response frequency of the servo valve is far higher than that of the correcting cylinder, without considering the dynamic model of the servo valve, the control accuracy of the system is less affected. The displacement of the servo valve spool and the control voltage can be regarded as a linear proportional relationship. Therefore, the approximation can be expressed as follows:

$$x_v = k_{xv}u,\tag{3}$$

where k_{xv} is a positive constant and u is the control voltage.

Considering the equivalent load elastic deformation during drilling, applying Newton's kinematics law, the force balance equation of the deflection control cylinder can be obtained:

$$m_p\ddot{x}_p = p_L A_p - B_p\dot{x}_p - k_s x_p - F_L,\tag{4}$$

where m_p is the total mass, B_p is the viscous damping coefficient, x_p is the spool displacement of the deflection control cylinder, k_s is the equivalent elastic stiffness coefficient of the drilling load, and F_L is the external load force acting on the control cylinder.

Considering the internal and external leakage of the deflection control cylinder, the load flow in two chambers of the deflection control cylinder is established as follows by applying the flow continuity equation:

$$Q_1 = A_p \dot{x}_p + \frac{V_{10} + A_p x_p}{\beta_e} \dot{p}_1 + C_{ip} p_L + C_{ep} p_1,\tag{5}$$

$$Q_2 = A_p \dot{x}_p - \frac{V_{20} - A_p x_p}{\beta_e} \dot{p}_2 + C_{ip} p_L - C_{ep} p_1,\tag{6}$$

where Q_1 is the flow rate (m^3/s) in intake chamber 1; Q_2 is the flow rate in the return chamber 2; C_{ip} is the internal leakage coefficient of the deflection control cylinder; C_{ep} is the external leakage coefficient of the deflection control cylinder; β_e is the effective bulk modulus; V_{10} and V_{20} are the initial volume of the oil-in cavity and oil-out cavity, respectively; and p_1 and p_2 are the oil pressures of the oil-in cavity and oil-out cavity, respectively.

Assuming that the cylinder initially stays in the middle position, $V_{10} = V_{20} = V_t/2$. V_t is the total volume of the oil-in cavity and oil-out cavity. Because $|A_p x_p| \ll V_t/2$, it can be neglected. The load flow is defined as the average flow of the oil-in cavity and oil-out cavity as follows:

$$Q_L = A_p \dot{x}_p + \frac{V_t}{4\beta_e} \dot{p}_L + C_{tp} p_L,\tag{7}$$

where C_{tp} is the total leakage coefficient and can be written as $C_{tp} = C_{ip} + C_{ep}/2$.

The displacement signal of the servo valve spool is used as the control input. Based on the derivation and discussion of the equations, the displacement, velocity, and acceleration of the deflection control hydraulic cylinder can be obtained. As the leakage coefficient and bulk modulus change with varying time and the external load of the drilling mechanism cannot be accurately obtained, the directional control system can be regarded as a strong nonlinear system with parameter uncertainty and disturbance uncertainty.

A state space equation was used for improving the control accuracy of the DCM based on the proposed dynamic model. State variables can be defined by $x = [x_1, x_2, x_3]^T = [x_p, \dot{x}_p, \ddot{x}_p]^T$. The directional control system can be presented in a state space form as follows:

$$\begin{cases} \dot{x}_1 = x_2 \\ \dot{x}_2 = x_3 \\ \dot{x}_3 = f(x) + g(x)u + d \end{cases},\tag{8}$$

where $f(x) = \theta_1 x_1 + \theta_2 x_2 + \theta_3 x_3$, $\theta_1 = -4\beta_e C_{tp} k_s/m_p V_t$, $\theta_2 = k_s/m_p + 4\beta_e(C_{tp} B_p + A_p^2)/m_p V_t$, $\theta_3 = B_p/m_p + 4\beta_e C_{tp}/V_t$, $g(x) = \sqrt{p_s - \mathrm{sgn}(u)p_L}\left(\frac{B_p}{m_p} + \frac{4\beta_e C_{tp} A_p}{m_p V_t}\right)$, $d = \frac{4\beta_e C_{tp}}{m_p V_{pt}} F_L - \frac{\dot{F}_L}{m_p}$.

The control system includes not only the nonlinearity of the hydraulic system, but also uncertain external disturbances, such as the friction force in the drilling process, the friction between the control cylinder and the piston rod, and the cutting load. The model uncertainty and external disturbances are regarded as a compound disturbance term, so the new system state space equation can be expressed as follows:

$$\begin{cases} \dot{x}_1 = x_2 \\ \dot{x}_2 = x_3 \\ \dot{x}_3 = f_0(x) + g_0(x)u + D \end{cases},\tag{9}$$

where $D = \Delta f(x) + \Delta g(x)u + d$ is the compound disturbance term of the deflection control system.

3. Controller Design

Position control systems based on the adaptive control method have been widely studied. However, it is difficult to obtain the optimal value of the compensation parameters when the external parameters change too much. A novel control method combining adaptive backstepping sliding mode control and an extended state observer was proposed to solve the above problems. The framework of the nonlinear adaptive backstepping control based on the disturbance observer (ABSMC-DO) for the DCM is shown in Figure 4. The basic concept of ABSMC-DO is to regard the parameter uncertainty and external disturbance as a system compound disturbance. A disturbance observer was designed to estimate the system compound disturbance. A nonlinear adaptive backstepping controller was proposed to compensate for the system compound disturbance. The stability of the overall system with the proposed control algorithm can be proved using Lyapunov analysis. The upper bound of directional vibration can be obtained without any disturbance. Using ABSMC-DO, the upper bound of the compound disturbance, which can effectively reduce the chattering in the directional control process, cannot be obtained easily.

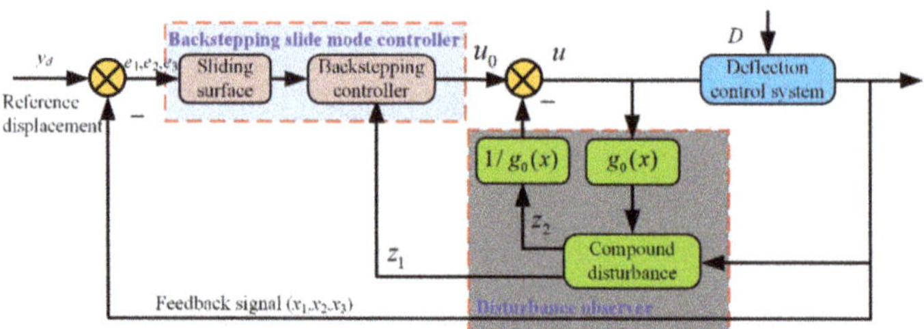

Figure 4. Control diagram of the adaptive backstepping control based on the disturbance observer (ABSMC-DO).

3.1. Adaptive Backstepping Slide Mode Controller

The target of this part is to obtain the value of control input Q_L to track the reference displacement with an adaptive backstepping controller. The controller design process can be divided into the following four steps.

Step 1: Define the displacement tracking error e_1 and the intermediate dummy variables of the system intermediate dummy variables as follows:

$$\begin{cases} e_1 = y - y_d = x_1 - y_d \\ e_2 = x_2 - a_1 \\ e_3 = x_3 - a_2 \end{cases}, \tag{10}$$

where y_d is the displacement reference value of the deviation control cylinder and a_1 and a_2 are intermediate dummy variables.

Step 2: Design the virtual control variable a_1 as follows:

$$a_1 = -k_1 e_1 + \dot{y}_d, \tag{11}$$

where $k_1 > 0$.

To ensure the displacement tracking error e_1 tends to zero, a semi-definite Lyapunov function can be defined as follows:

$$V_1 = \frac{1}{2}e_1^2. \tag{12}$$

The time derivative of V_1 can be given as follows considering Equation (11):

$$\dot{V}_1 = e_1\dot{e}_1 = e_1(\dot{x}_1 - \dot{y}_d) = e_1(e_2 + a_1 - \dot{y}_d) = -k_1 e_1^2 + e_1 e_2. \tag{13}$$

Step 3: To ensure the system is stable and convergent, $\dot{V}_1 \leq 0$. Therefore, a new Lyapunov function for e_2 is defined as follows:

$$V_2 = V_1 + \frac{1}{2}e_2^2. \tag{14}$$

By deriving Equation (14), we can obtained the following:

$$\dot{V}_2 = \dot{V}_1 + e_2\dot{e}_2 = -k_1e_1^2 + e_1e_2 + e_2(e_3 + a_2 - \dot{a}_1). \tag{15}$$

Then, the virtual control variable a_2 can be designed as follows:

$$a_2 = -k_2e_2 - e_1 + \dot{a}_1, \tag{16}$$

where $k_2 > 0$.

$\dot{V}_2$ is calculated as follows:

$$\dot{V}_2 = -k_1e_1^2 - k_2e_2^2 + e_2e_3. \tag{17}$$

Step 4: To ensure the Lyapunov function defined above has positive semi-definite properties, the synovial switching function is designed to make the system error converge to the synovial surface gradually as follows:

$$s = c_1e_1 + c_2e_2 + e_3, \tag{18}$$

where $c_1 > 0$, $c_2 > 0$, and satisfied Hurwitz polynomials $p(\xi) = \xi^2 + c_2\xi + c_1$.

The exponential reaching rate is employed in this work to make the control system approach a sliding surface:

$$\dot{s} = -\varepsilon\mathrm{sgn}(s) - k_3s, \tag{19}$$

where $\dot{s} = -ks$ is an exponential reaching term and parameters ε and k_3 are both positive.

The larger the parameters ε and k_3, the faster the convergence. However, too large parameters will cause system chattering and even lead to system instability.

Further, by deriving Equation (18), the following can be deduced:

$$\dot{s} = c_1\dot{e}_1 + c_2\dot{e}_2 + \dot{x}_3 - \dot{a}_2 = c_1\dot{e}_1 + c_2\dot{e}_2 + f_0(x) + g_0(x)u + D - \dot{a}_2. \tag{20}$$

The compound disturbance term in Equation (20) includes both parameter uncertainty and external disturbance. Therefore, the system controller is designed as follows:

$$u = \frac{1}{g_0(x)}(-\varepsilon s - k_3\mathrm{sgn}(s) - c_1\dot{e}_1 - c_2\dot{e}_2 - f_0(x) + \dot{a}_2 - D). \tag{21}$$

As $g_0(x)$ is a variable related to the parameters of the deflection-control hydraulic cylinder, the denominator of the designed controller (21) is not zero, which ensures its non-singularity. However, the unknown disturbance D in the controller directly affects the output performance and shaking characteristic. A disturbance observer was built to estimate and compensate the unknown disturbance D to avoid the control error and shaking.

3.2. Disturbance Observer

The compound disturbance is expanded based on the control principle of the extended state observer, so that the sliding surface in Equation (20) can be re-expressed as follows:

$$\begin{cases} \dot{s} = c_1\dot{e}_1 + c_2\dot{e}_2 + f_0(x) - \dot{a}_2 + g_0(x)u + s_1 \\ \dot{s}_1 = \dot{D} \end{cases}. \tag{22}$$

An extended state observer is designed for Equation (20) with a second-order form:

$$\begin{cases} e_1 = z_1 - s \\ \dot{z}_1 = z_2 + c_1\dot{e}_1 + c_2\dot{e}_2 + f_0(x) - \dot{a}_2 + g_0(x)u - \beta_1 e_1 \\ \dot{z}_2 = -\beta_1\eta(e,a,\delta) \end{cases}, \tag{23}$$

where z_1 and z_2 are the observer outputs, e_1 is the estimate error for the observer, and parameters β_1 and β_2 are greater than zero.

Function η is defined as follows:

$$\eta(e,a,\delta) = \begin{cases} |e_1|^a \text{sgn}(e_1), & |e_1| > \delta \\ \frac{e_1}{\delta^{1-a}}, & |e_1| \leq \delta \end{cases}, \tag{24}$$

where $0 < a < 1$ and $\delta > 0$.

Remark 1. *For the second-order extended state observer of Equation (23), there are appropriate design parameters β_1, β_2, a, and δ that make the observer steady.*

Furthermore, the observer estimation error equation is established as follows:

$$\begin{cases} \dot{e}_1 = e_2 - \beta_1 e_1 \\ \dot{e}_2 = -\dot{D} - \beta_2\eta \end{cases}. \tag{25}$$

The estimation error of the observer in steady state can be expressed as folows:

$$\begin{cases} \dot{e}_1 = e_2 - \beta_1 e_1 = 0 \\ \dot{e}_2 = -\dot{D} - \beta_2\eta = 0 \end{cases}. \tag{26}$$

According to Equations (24) and (26), the estimated error of the observer can be processed as follows:

$$\begin{cases} e_1 = -\eta^{-1}(\dot{D}/\beta_2) \\ e_2 = \beta_1 e_1 = -\beta_1\eta^{-1}(\dot{D}/\beta_2) \end{cases}. \tag{27}$$

Considering Equation (24), the observational errors can be divided into two cases:

$$\begin{cases} |e_1| = \left|\dot{D}\beta_2\right|^{1/a} & |e_1| > \delta \\ |e_2| = \beta_1\left|\dot{D}\beta_2\right|^{1/a} & |e_1| > \delta \\ |e_1| = \left|\dot{D}\delta^{1-a}\right|/\beta_2 & |e_1| < \delta \\ |e_2| = \beta_1\left|\dot{D}\delta^{1-a}\right|/\beta_2 & |e_1| < \delta \end{cases}. \tag{28}$$

Combined with Equations (27) and (28), it can be seen that reasonable values for parameters β_1, β_2, a, and δ determine the observer estimation error. Therefore, as long as the appropriate parameter values are selected, the state estimation error can be sufficiently small, so the observer output can converge to a certain critical region of s and D, which guarantees the observation accuracy of the compound disturbance.

3.3. Stability Analysis

The compound disturbance D can be replaced by the proposed extended state observer. Thus, the control law from Equation (21) can be rebuilt as follows:

$$u = \frac{1}{g_0(x)}(-\varepsilon s - k_3\text{sgn}(s) - c_1\dot{e}_1 + c_2\dot{e}_2 + f_0(x) - \dot{a}_2 + g_0(x)u - z_2). \tag{29}$$

Remark 2. *For the directional control system of Equation (8) with parameter uncertainties and external disturbances, the extended state observer of Equation (23) is used to estimate the compound disturbance D. If there are suitable parameters, β_1, β_2, a, and δ, the control system output will gradually tend to the sliding surface and keep a relative slip on the sliding surface, so that the position output of the system can accurately track the desired trajectory. The parameters must meet the following conditions:*

$$k_1 k_2 \varepsilon + k_1 \varepsilon c_2 - \frac{k_1 + \varepsilon c_1^2}{4} > 0. \tag{30}$$

Proof . Define the Lyapunov function:

$$V = V_2 + \frac{1}{2}s^2. \tag{31}$$

Combining Equations (13), (18) and (29), Equation (31) can be derived as follows:

$$\begin{aligned}
\dot{V} =\ & \dot{V}_2 + s\dot{s} \\
=\ & -k_1 e_1^2 - k_2 e_2^2 + e_2 e_3 + s(c_1 \dot{e}_1 + c_2 \dot{e}_2 + f_0(x) - \dot{a}_2 + g_0(x)u + D) \\
=\ & -k_1 e_1^2 - k_2 e_2^2 + e_2 e_3 + s(-k_3 s - \varepsilon \mathrm{sgn}(s) - z_2) \\
=\ & -k_1 e_1^2 - k_2 e_2^2 + e_2 e_3 - k_3 s^2 - \varepsilon |s| + s(D - z_2) \\
=\ & -\mathbf{e}^T \mathbf{Q} \mathbf{e} - \varepsilon |s| + s(D - z_2)
\end{aligned} \tag{32}$$

where $\mathbf{e} = [e_1, e_2, e_3]^T$ and

$$Q = \begin{bmatrix} k_1 + k_3 c_1 & c_1 c_2 k_3 & c_1 k_3 \\ c_1 c_2 k_3 & k_2 + c_2 k_3 & c_2 k_3 - \frac{1}{2} \\ c_1 k_3 & c_2 k_3 - \frac{1}{2} & k_3 \end{bmatrix}. \tag{33}$$

The designed parameters k_1, k_2, k_3, c_1, and c_2 are all greater than 0 and constants, which can ensure that the first- and second-order principal minor determinants of matrix Q are positive. The fact that the third-order principal minor determinant of matrix Q is positive can be guaranteed by Equation (30). Therefore, matrix Q satisfies the positive definite.

From Equation (33), the following can be concluded:

$$\begin{aligned}
\dot{V}\ & \leq -\lambda_{\min}(Q)\|\mathbf{e}\|^2 - k_3 |s| + s(D - z_2) \\
& \leq -\lambda_{\min}(Q)\|\mathbf{e}\|^2 - (k_3 - |e_2|)s
\end{aligned} \tag{34}$$

where $\lambda_{\min}(Q)$ is the minimum eigenvalue of matrix Q.

The tracking errors converge to a neighborhood of zero by assigning reasonable values for parameters β_1, β_2, a, and δ. Therefore, design parameter k_3 satisfies $k_3 > |e_2|$. Based on Equation (29), when $\dot{V} < 0$, the system will embody asymptotic stability.

Define $N = \mathbf{e}^T \mathbf{Q} \mathbf{e}$. When $k_3 > |e_2|$, $\dot{V} < -N$. The integral result is expressed as follows:

$$\int_0^t N dt \leq V(0) - V(t). \tag{35}$$

According to the boundedness of V and $\dot{V} < 0$, it can be deduced that $\lim\limits_{t \to \infty} \int_0^t N dt < \infty$. Therefore, $e_i \to 0$, $i = 1, 2, 3$ when $t \to \infty$ and $N \to 0$. Then, the sliding function s would be closer to zero, $s \to 0$. $\square$

Based on the Lyapunov stability criterion, the system will gradually approach the sliding surface and eventually converge to the sliding surface. Therefore, the output of the directional control system can effectively track the desired trajectory.

4. Experimental Research

4.1. Experimental Introduction

Figure 5 depicts an electro-hydraulic servo displacement control experimental system that is utilized to implement the proposed nonlinear adaptive backstepping control based on the disturbance observer (ABSMC-DO) designed in this paper and experiment on the deflection control testing system. The testing system is composed of a loading system and an actuating system. The loading hydraulic cylinder and actuating hydraulic cylinder are the actuators and are driven by the proportional servo valve (MA-DLHZO-TES-PS-040) manufactured by the YWL VTOZ company. The MTS magnetostrictive displacement sensor (RPS0500MD601V810050) with an accuracy of 0.1 mm is used to measure the hydraulic cylinder position. For measuring the pressure changes in real time, two pressure sensors (Mik-p300 and 0–20 Mpa, resolution ±0.5%) are installed on the oil inlet and outlet of the loading cylinder and actuating cylinder. All the executive elements and acquisition elements have relative control precision to realize industrial application. Different external loading signals, such as waves and sinusoidal steps, are imposed to the actuating cylinder. A tension pressure sensor is employed to measure the external loading with a repetitive accuracy of 0.05% FS (Full Scales).

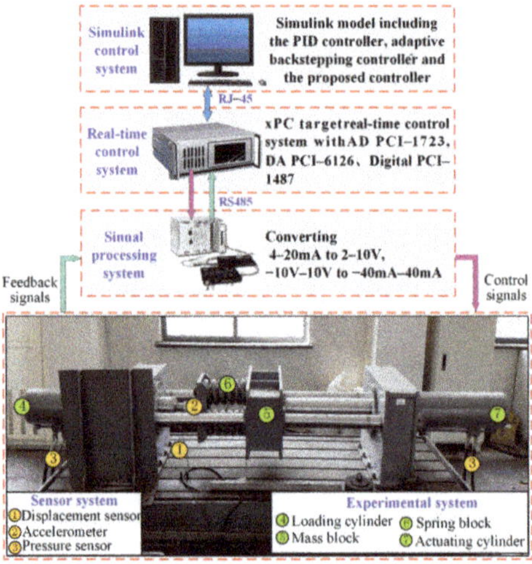

Figure 5. Electro-hydraulic servo displacement control experimental system. PID, proportion integral derivative.

The signals of each sensor are collected in the industrial computer through an analog to digital (A/D) board, PCI-1723. The 16-bit A/D board transforms the feedback analog signals measured by the sensors to digital signals and then sends the acquired digital signals to the controller after converting in signal modular. Using the RTX module of the industrial computer, the proposed control algorithm is programmed in Matlab/Simulink and then compiled and downloaded to the matlab xPC target real-time controller. The real-time analog control output signals that are produced by the 12-bit D/A board PCI-6126 and processed by the signal modular are sent to the two servo-valves to control the two electro-hydraulic cylinders.

To verify the effectiveness of the proposed control strategy for the DCM, experiments were carried out on the electro-hydraulic servo displacement control experimental system using the conventional proportion integral derivative (PID) controller, the adaptive backstepping controller (ABC), and the proposed ABSMC-DO for comparison. The main parameters of the electro-hydraulic servo displacement control experimental system are shown in Table 1.

Table 1. Main parameters of the electro-hydraulic servo displacement control experimental system.

Parameters	Symbol	Values	Units
Supply pressure	P_s	15×10^6	Pa
Effective bulk modulus	β_e	7.0×10^8	Pa
Volume of hydraulic cylinder	V	0.62×10^{-4}	m^3
Effective area of hydraulic cylinder	A_p	3.6×10^{-3}	m^2
Total mass	m_p	100	kg
Viscous damping coefficient	B_m	5652	N·m·s/rad
Flow coefficient of servo valve	C_d	0.602	——
Gain of servo valve	K_q	2.6×10^{-4}	m/A
Natural frequency of servo valve	ω_{sv}	150	rad/s
Area gradient of servo valve core	ω	1.63×10^{-3}	m

The control parameters of different control algorithms were obtained by many tests. The parameter tuning of the PID controller included the hash method, the attenuation curve method, the expanded critical proportional band method, and the response curve method. In the tests, PID parameter tuning was performed based on an attenuation curve to achieve the best control state. Meanwhile, the ideal parameters of ABC and the proposed ABSMC-DO could be obtained by the controlling variable referring to articles [21] and [23] as follows:

PID controller: $k_p = 0.0005$, $k_i = 0.04$, $k_d = 1.5$.

Adaptive backstepping controller: feedback gain $k_1 = 100$, $k_2 = 30$, $k_3 = 50$.

Nonlinear adaptive backstepping control based on disturbance observer: $k_1 = 200$, $k_2 = 320$, $k_3 = 40$, $\varepsilon = 0.5$, $c_1 = 1$, $c_2 = 2$, $\beta_1 = 4000$, $\beta_2 = 8000$, $a = 0.05$, $\delta = 0.1$.

4.2. Experimental Results

Deviation in the drilling process is caused by the fluctuation of the cutting load. The displacement of DCM should be adjusted in real time to ensure directional drilling at any time. Therefore, a random sinusoidal trajectory with an amplitude of 0.1 m and a frequency of 1.0 Hz was applied to simulate the tracking characteristics of the DCM in the drilling process. The experiments were carried out under the conditions of no disturbance, parameter uncertainty, external loading disturbance, and random signal, respectively.

4.2.1. Tracking Performance with No Disturbance

The tracking errors of different controllers with no disturbance are presented in Figure 6. The parameter estimation of the adaptive backstepping controller is shown in Figure 7.

It can be seen from Figure 6 that the displacement tracking error with the conventional PID is much bigger than that with the adaptive backstepping controller and the proposed ABSMC-DO. Owing to the adaptive law (Figure 7), the adaptive parameters of the ABC converge to a stable region in a short time. The tracking performances of the ABC and the proposed ABSMC-DO are both excellent with no disturbance.

4.2.2. Tracking Performance with Uncertain Parameters

The performance robustness with uncertain parameters was further measured by loading mass blocks of 20 kg on the hydraulic cylinder slider; the control parameters of the different controllers remained unchanged. The tracking performance was obtained as shown in Figure 8.

The tracking error of the PID controller increased significantly compared with the no disturbance conditions, resulting in a poor control effect. However, both the ABC and the proposed ABSMC-DO could achieve stability in a relatively short time because of their ability to self-adjust their parameters online. Even if the gain of the controller remained unchanged, a tracking response with a high accuracy could be achieved. The two controllers both had a high control accuracy and excellent tracking performance.

4.2.3. Tracking Performance with External Disturbance

It is a challenge to improve control accuracy owing to the real-time change in the external load in the cutting process. A random external force of 1000 N amplitude was applied at t = 1.5 s and removed at t = 4.5 s to imitate external disturbance. The tracking errors of different controllers are shown in Figure 9.

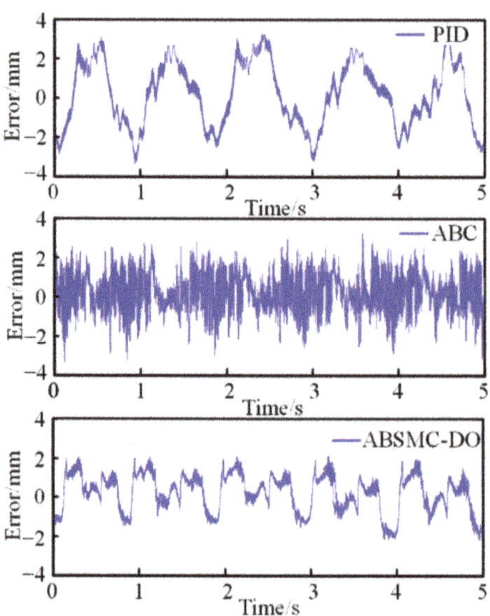

Figure 6. Tracking errors of different controllers with no disturbance. ABC, adaptive backstepping controller.

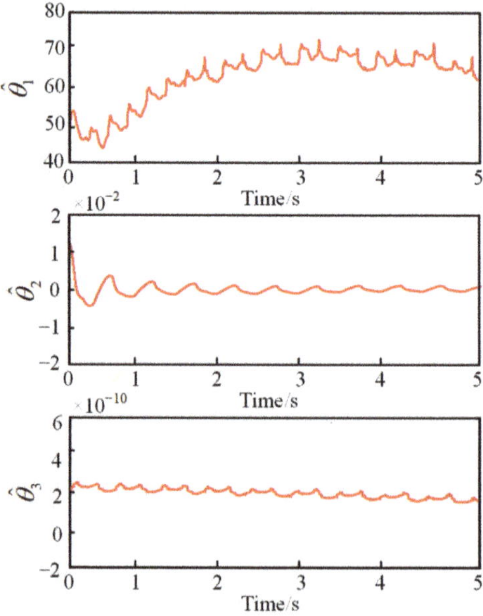

Figure 7. Parameter estimation of the adaptive backstepping controller.

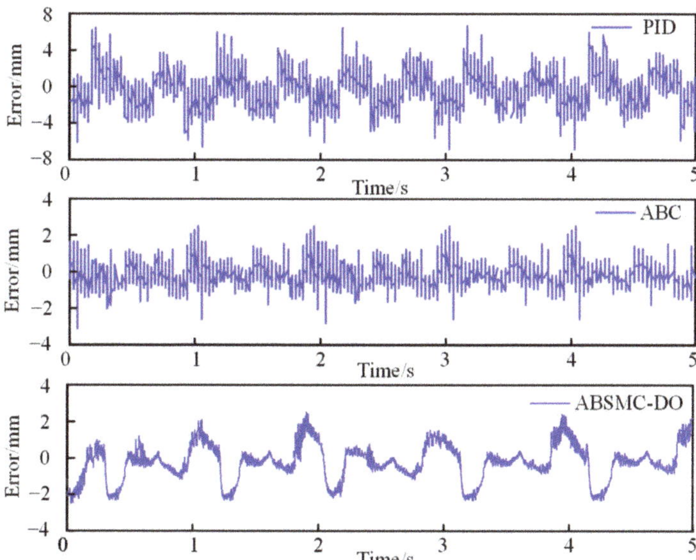

Figure 8. Tracking errors of different controllers with no disturbance.

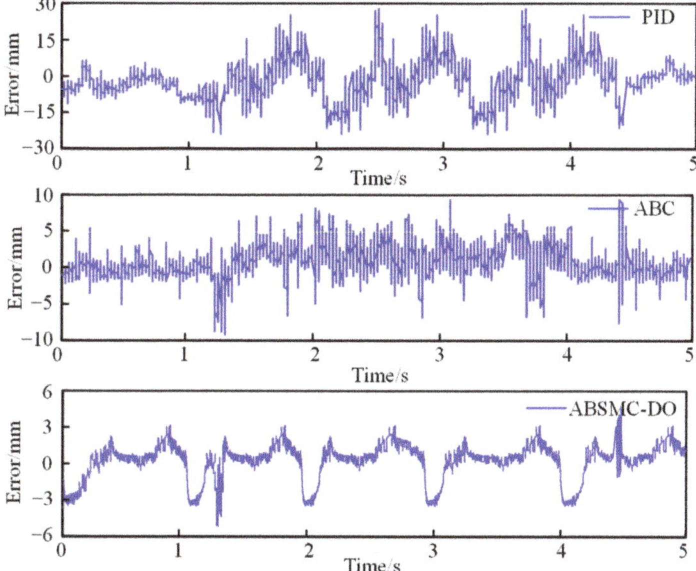

Figure 9. Tracking errors of different controllers with external disturbance.

The conventional PID controller was no longer stable and achieved a large error under external disturbance. The tracking error of ABC increased obviously when external disturbance existed. With the disappearance of the external disturbance, the error was controlled in a small range. However, the proposed ABSMC-DO controller was found to have a better tracking performance, except for the moment of imposing and removing external disturbance.

A quantitative evaluation of the control effect was carried out using the peak tracking errors and root mean square errors (RMSEs), as shown in Table 2.

Table 2. Position tracking performance index under different controllers. RMSE, root mean square error; PID, proportion integral derivative; ABC, adaptive backstepping controller; ABSMC-DO, adaptive backstepping control based on the disturbance observer.

Controller	No Disturbance		Uncertain Parameters		External Disturbance	
	Peak Error/mm	RMSE	Peak Error/mm	RMSE	Peak Error/mm	RMSE
PID	3.2	1.16	7.1	3.32	28.2	10.34
ABC	3.6	1.02	3.7	1.33	9.1	4.26
ABSMC-DO	2.3	0.84	2.9	1.14	4.7	1.62

It can be concluded that the peak errors and RMSEs of the PID controller are both the greatest under different external conditions. The ABC can reach a better control accuracy with no disturbance and uncertain parameters. The controller errors of the proposed ABSMC-DO controller still maintain the smallest under each experimental context, which proves that the designed controller has strong robustness.

4.2.4. Tracking Performance with Prototype Experiment

To realize the deviation control of the drill auger during the cutting process, a prototype experiment was carried out, as shown in Figure 10.

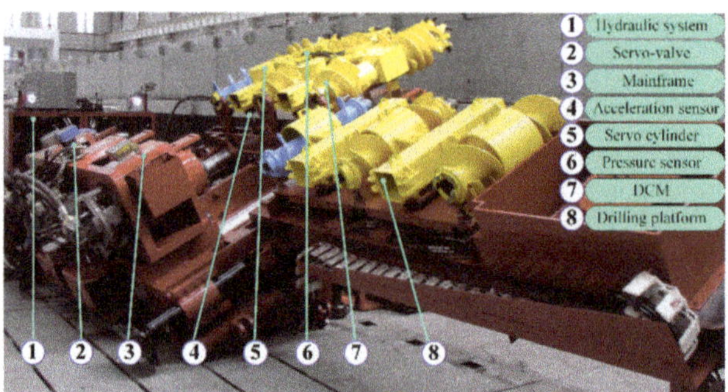

Figure 10. Experiment prototype for position deviation control.

In the actual drilling process, the control system cannot be added to the experimental prototype owing to the small space. Thus, the external disturbances and deviation position data cannot be collected in real time. To solve this problem, we carried out some related experiments [2] on the ground in the early stage and obtained relevant experimental data that can be applied to the experimental prototype to replace the downhole operation environment. This can ensure the reliability of the data. A low-pass filter was used to reduce the influence of noise. The position deviation control results are shown in Figure 11.

As we can see, the signal output waveform showed obvious distortion and phase lag with the conventional PID controller, whose amplitude deviation was the largest. Using the ABC and the proposed ABSMC-DO controller, the tracking effect improved significantly, which indicates that the nonlinear control effect was better in terms of its high response speed and high tracking accuracy.

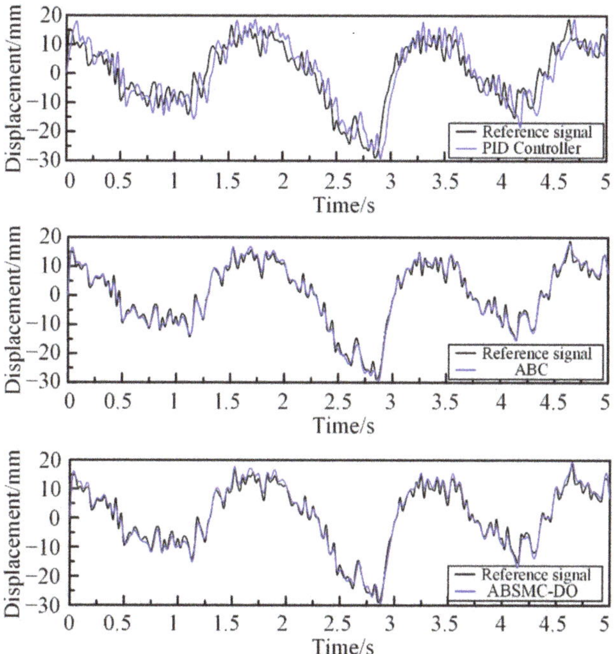

Figure 11. Position deviation control results of the prototype test.

Further, a normalization analysis was conducted to evaluate the deviation control effect quantitatively. The calculation formula of the normalized root mean square error is as follows:

$$NRMSE = \sqrt{\frac{1}{L}\sum_{i=1}^{L}[r_a(k) - y_a(k)]^2 / \max|r_a(k)|}, \tag{36}$$

where r_a is the value of the output signal, y_a is the value of the reference signal, and L is the length of the reference and output signals. Table 3 lists the RMSE (root mean squared error) of the position tracking with the different controllers. As can be seen, the NRMSE (normalized root mean squared error) based on the PID controller is the largest. However, the ABC and ABSMC-DO controller can achieve a better performance. The ABSMC-DO controller can realize position deviation control under different external conditions.

Table 3. Normalized performance index of position control under different controllers. NRMSE, normalized RMSE.

Controller	Max Signal	RMSE	NRMSE
PID	-0.0295	11.1009×10^{-3}	37.63%
ABC	-0.0295	5.7584×10^{-3}	19.52%
ABSMC-DO	-0.0295	4.2097	14.27%

5. Discussion

In this study, a combined controller consisting of a nonlinear adaptive backstepping controller and a disturbance observer for deviation control was established and verified. Some studies have focused on controller design using the backstepping control method [22,26] and the disturbance observer method [23,31,32], which indicates that these methods are significant. However, most of the available research focuses on control accuracy using high-precision control elements. Our studies tried to combine and improve these two controllers and use a control element with a slightly lower accuracy to achieve the required positioning control for easier industrial application. The results confirm that long-distance directional drilling can be realized under bad downhole working conditions, a finding that can be used in in oil drilling, deep sea exploitation, gas drainage, directional tunneling, and other fields. Besides this, our study shows that both the adaptive backstepping controller and disturbance observer controller cannot reach an excellent tracking performance individually in external disturbance, which is an important and assignable factor. The related research [28,30] has obtained consistent conclusions. Therefore, improving the control accuracy of a single controller is also one of the directions for further research.

However, prototype experiments in this paper were carried out using a ground test and the geological factors were neglected in this paper. Meanwhile, long-distance drilling over 100 m should be considered by fusing sins positioning technology for real-time control. Therefore, further investigation is necessary and will be discussed in future research.

6. Conclusions

In this work, a deflection control mechanism was designed to realize directional drilling. A nonlinear adaptive backstepping controller based on a disturbance observer consisting of an adaptive backstepping controller and a disturbance observer controller was firstly proposed for position tracking control, taking into account the parameter uncertainties and external disturbances. The stability of the overall system with the proposed controller was proven with the help of the Lyapunov theory. To verify the effectiveness of the proposed ABSMC-DO controller, an electro-hydraulic servo displacement control experimental system with matlab xPC target rapid prototyping technology was established. Then, a prototype experiment was conducted to prove the direction drilling performance. The experimental results indicate the following: (a) The displacement tracking error with no disturbance of the conventional PID was much larger than that with the adaptive backstepping controller and the proposed ABSMC-DO. The tracking performances of the ABC and the proposed ABSMC-DO were both excellent with no disturbance. (b) Besides this, the tracking error of the PID controller increased significantly with uncertain parameters, resulting in a poor control effect. However, both the ABC and the proposed ABSMC-DO could achieve stability in a relatively short time thanks to their ability to self-adjust their parameters online. (c) The conventional PID controller was no longer stable and achieved a large error under the conditions of external disturbance. The tracking error of the ABC increased obviously when there was external disturbance. With the disappearance of the external disturbance, the error was controlled in a small range. However, the proposed ABSMC-DO controller achieved a better tracking performance, except for the moment of imposing and removing external disturbance. (d) The NRMSE based on the PID controller was the largest with the prototype experiment. However, the ABC and ABSMC-DO controller could reach a better performance. The ABSMC-DO controller could realize position deviation control under different external conditions. All of the experimental results indicate that the proposed ABSMC-DO controller can yield a more satisfactory position tracking performance than the conventional PID controller and an adaptive backstepping controller.

Author Contributions: Writing—original draft, review, and editing, H.J.; data acquisition, S.L.; investigation, H.J.; methodology, S.L.; validation, H.J.; writing—review and editing, H.J. All authors have read and agreed to the published version of the manuscript.

Funding: This work is supported by the Xuzhou science and technology achievements transformation plan (KC20203), the Project Funded by the Priority Academic Program Development of Jiangsu Higher Education Institutions (PAPD).

Institutional Review Board Statement: Not applicable.

Informed Consent Statement: Not applicable.

Data Availability Statement: The data presented in this study are available on request from the corresponding author.

Conflicts of Interest: The authors declare no conflict of interest.

References

1. Yang, D.L.; Li, J.P.; Wang, Y.X.; Jiang, H.X. Research on vibration and deflection for drilling tools of coal auger. *J. Vibroeng.* **2017**, *19*, 4882–4897. [CrossRef]
2. Liu, S.; Ji, H.; Cui, X. Vibration and Deflection Behavior of a Coal Auger Working Mechanism. *Shock. Vib.* **2016**, *2016*, 6493859. [CrossRef]
3. Yuan, Y.; Chen, Z.; Xu, C.; Zhang, X.; Wei, H. Permeability enhancement performance and its control factors by auger mining of extremely thin coal seams. *J. Geophys. Eng.* **2018**, *15*, 2626–2641. [CrossRef]
4. Tang, B.; Cheng, H.; Tang, Y.Z.; Zheng, T.L.; Yao, Z.S.; Wang, C.B.; Rong, C.X. Supporting Design Optimization of Tunnel Boring Machines-Excavated Coal Mine Roadways: A Case Study in Zhangji, China. *Processes* **2020**, *8*, 46. [CrossRef]
5. Liu, Y.Y.; Sun, X.M.; Wang, J.W.; Li, J.B.; Sun, S.J.; Cui, X.B. Study on Three-Dimensional Stress Field of Gob-Side Entry Retaining by Roof Cutting without Pillar under Near-Group Coal Seam Mining. *Processes* **2019**, *7*, 552. [CrossRef]
6. Roman, M.; Josif, B.; Nikolay, D. Development of Position System of a Roadheader on a Base of Active IR-sensor. *Procedia Eng.* **2015**, *100*, 617–621. [CrossRef]
7. Cui, X.X.; Ji, H.F.; Lin, M.X.; Liu, S.Y. Vibration characteristic analysis of the multi-drilling mechanism. *J. Vibroeng.* **2014**, *16*, 2722–2734.
8. Cao, X.W.; Li, X.X.; Yue, F.T. Research and Application of Vibration Measurement While Drilling in the Undersea Coal Mine. *J. Coast. Res.* **2020**, *103*, 323–327. [CrossRef]
9. Li, D.M.; Zhang, Y.Q.; Li, J.; Xia, S. Investigation on sticking for drill rod based on finite element analysis and fuzzy proportion-integration-differentiation. *Adv. Mech. Eng.* **2019**, *11*. [CrossRef]
10. Taheran, F.; Monfared, V.; Daneshmand, S.; Abedi, E. Nonlinear vibration analysis of directional drill string considering effect of drilling mud and weight on bit. *J. Vibroeng.* **2016**, *18*, 1280–1287.
11. Bakhtiari-Nejad, F.; Hosseinzadeh, A. Nonlinear dynamic stability analysis of the coupled axial-torsional motion of the rotary drilling considering the effect of axial rigid-body dynamics. *Int. J. Non-Linear Mech.* **2017**, *88*, 85–96. [CrossRef]
12. Hosseinzadeh, A.; Bakhtiari-Nejad, F. A New Dynamic Model of Coupled Axial–Torsional Vibration of a Drill String for Investigation on the Length Increment Effect on Stick-Slip Instability. *J. Vib. Acoust.* **2017**, *139*, 061016. [CrossRef]
13. Tian, J.L.; Li, G.Y.; Dai, L.M.; Yang, L.; He, H.; Hu, S. Torsional Vibrations and Nonlinear Dynamic Characteristics of Drill Strings and Stick-Slip Reduction Mechanism. *J. Comput. Nonlinear Dyn.* **2019**, *14*, 081007. [CrossRef]
14. Liu, S.Y.; Cui, X.X.; Liu, X.H. Coupling vibration analysis of auger drilling system. *J. Vibroeng.* **2013**, *15*, 1442–1453.
15. Lueke, J.S.; Ariaratnam, S.T. Surface Heave Mechanisms in Horizontal Directional Drilling. *J. Constr. Eng. Manag.* **2005**, *131*, 540–547. [CrossRef]
16. Manacorda, G.; Miniati, D.; Dei, D.; Scott, H.F.; Pinchbeck, D. Development of a bore-head GPR for Horizontal Directional Drilling (HDD) equipment. In Proceedings of the XIII International Conference on Ground Penetrating Radar, Lecce, Italy, 21–25 June 2010.
17. Inyang, I.J.; Whidborne, J.F. Bilinear modelling, control and stability of directional drilling. *Control. Eng. Pract.* **2019**, *82*, 161–172. [CrossRef]
18. Kim, J.; Myung, H. Development of a Novel Hybrid-Type Rotary Steerable System for Directional Drilling. *IEEE Access* **2017**, *5*, 24678–24687. [CrossRef]
19. Gao, B.W.; Shao, J.P.; Yang, X.D. A compound control strategy combining velocity compensation with ADRC of electro-hydraulic position servo control system. *ISA Trans.* **2014**, *53*, 1910–1918. [CrossRef]
20. Lin, S.H.; An, G.C.; Huang, J.H.; Guo, Y. Robust Backstepping Control with Active Damping Strategy for Separating-Metering Electro-Hydraulic System. *Appl. Sci.* **2020**, *10*, 277. [CrossRef]
21. Feng, L.J.; Yan, H. Nonlinear Adaptive Robust Control of the Electro-Hydraulic Servo System. *Appl. Sci.* **2020**, *10*, 4494. [CrossRef]
22. Yang, M.X.; Zhang, Q.; Lu, X.L.; Wang, X. Adaptive Sliding Mode Control of a Nonlinear Electro-hydraulic Servo System for Position Tracking. *Mechanika* **2019**, *25*, 283–290. [CrossRef]
23. Wos, P.; Dindorf, R. Adaptive Control of the Electro-Hydraulic Servo-System with External Disturbances. *Asian J. Control.* **2013**, *15*, 1065–1080. [CrossRef]
24. He, Z.; Guan, W.-L.; Feng, J.; Gao, Q. Mathematical modelling and backstepping adaptive sliding mode control for multi-stage hydraulic cylinder. *Int. J. Model. Identif. Control.* **2018**, *30*, 322–342. [CrossRef]
25. Guo, Q.; Sun, P.; Yin, J.-M.; Yu, T.; Jiang, D. Parametric adaptive estimation and backstepping control of electro-hydraulic actuator with decayed memory filter. *ISA Trans.* **2016**, *62*, 202–214. [CrossRef]
26. Wang, W.P.; Wang, B. Disturbance observer–based nonlinear energy-saving control strategy for electro-hydraulic servo systems. *Adv. Mech. Eng.* **2017**, *9*, 11–18. [CrossRef]
27. Zhu, Z.-C.; Li, X.; Shen, G.; Wei, D. Wire rope tension control of hoisting systems using a robust nonlinear adaptive backstepping control scheme. *ISA Trans.* **2018**, *72*, 256–272. [CrossRef]
28. Choux, M.; Hovland, G. Adaptive Backstepping Control of Nonlinear Hydraulic-Mechanical System Including Valve Dynamics. *Model. Identif. Control.* **2010**, *31*, 35–44. [CrossRef]

29. Ji, X.; Wang, C.; Zhang, Z.; Chen, S.; Guo, X. Nonlinear adaptive position control of hydraulic servo system based on sliding mode back-stepping design method. *Proc. Inst. Mech. Eng. Part I J. Syst. Control. Eng.* **2020**. [CrossRef]

30. Yang, G.C.; Yao, J.Y. Nonlinear adaptive output feedback robust control of hydraulic actuators with largely unknown modeling uncertainties. *Appl. Math. Model.* **2020**, *79*, 824–842. [CrossRef]

31. Yang, G.Z.; Yao, J.Y. High-precision motion servo control of double-rod electro-hydraulic actuators with exact tracking performance. *ISA Trans.* **2020**, *103*, 266–279. [CrossRef]

32. Zheng, J.; Yao, J. Robust adaptive tracking control of hydraulic actuators with unmodeled dynamics. *Trans. Inst. Meas. Control.* **2019**, *41*, 3887–3898. [CrossRef]

33. Guo, K.; Wei, J.H.; Fang, J.H.; Feng, R.L.; Wei, J.H. Position tracking control of electro-hydraulic single-rod actuator based on an extended disturbance observer. *Mechatronics* **2015**, *27*, 47–56. [CrossRef]

34. Guo, K.; Wei, J.H.; Tian, Q.Y. Disturbance observer based position tracking of electro-hydraulic actuator. *J. Cent. South Univ.* **2015**, *22*, 2158–2165. [CrossRef]

35. Kasac, J.B.; Stevanovic, S.; Zilic, T.; Stepanic, J. Robust output tracking control of a quadrotor in the presence of external disturbances. *Trans. FAMENA* **2013**, *37*, 29–42. [CrossRef]

36. Hu, X.; Du, J.L.; Li, J.; Sun, Y. Asymptotic Regulation of Dynamically Positioned Vessels with Unknown Dynamics and External Disturbances. *J. Navig.* **2020**, *73*, 253–266. [CrossRef]

37. Teoh, J.N.; Du, C.; Guo, G.; Xie, L. Rejecting high frequency disturbances with disturbance observer and phase stabilized control. *Mechatronics* **2008**, *18*, 53–60. [CrossRef]

38. Van Nguyen, T.; Ha, C. Sensor Fault-Tolerant Control Design for Mini Motion Package Electro-Hydraulic Actuator. *Processes* **2019**, *7*, 89. [CrossRef]

Article

Intelligent Control Strategy for Transient Response of a Variable Geometry Turbocharger System Based on Deep Reinforcement Learning

Bo Hu [1,2,*], Jie Yang [1], Jiaxi Li [1], Shuang Li [1] and Haitao Bai [1]

[1]   Key Laboratory of Advanced Manufacturing Technology for Automobile Parts, Ministry of Education, Chongqing University of Technology, Chongqing 400054, China; j.young@2017.cqut.edu.cn (J.Y.); 11607990404@2016.cqut.edu.cn (J.L.); ls123456@2016.cqut.edu.cn (S.L.); Baiht@2017.cqut.edu.cn (H.B.)

[2]   State Key Laboratory of Engines, Tianjin University, Tianjin 300072, China

*   Correspondence: b.hu@cqut.edu.cn

Received: 15 August 2019; Accepted: 2 September 2019; Published: 6 September 2019

Abstract: Deep reinforcement learning (DRL) is an area of machine learning that combines a deep learning approach and reinforcement learning (RL). However, there seem to be few studies that analyze the latest DRL algorithms on real-world powertrain control problems. Meanwhile, the boost control of a variable geometry turbocharger (VGT)-equipped diesel engine is difficult mainly due to its strong coupling with an exhaust gas recirculation (EGR) system and large lag, resulting from time delay and hysteresis between the input and output dynamics of the engine's gas exchange system. In this context, one of the latest model-free DRL algorithms, the deep deterministic policy gradient (DDPG) algorithm, was built in this paper to develop and finally form a strategy to track the target boost pressure under transient driving cycles. Using a fine-tuned proportion integration differentiation (PID) controller as a benchmark, the results show that the control performance based on the proposed DDPG algorithm can achieve a good transient control performance from scratch by autonomously learning the interaction with the environment, without relying on model supervision or complete environment models. In addition, the proposed strategy is able to adapt to the changing environment and hardware aging over time by adaptively tuning the algorithm in a self-learning manner on-line, making it attractive to real plant control problems whose system consistency may not be strictly guaranteed and whose environment may change over time.

Keywords: self-learning; transient response; variable geometry turbocharger; deep reinforcement learning; deep deterministic policy gradient

---

1. Introduction

The concept of engine downsizing and down-speeding enables reductions in fuel consumption and CO_2 emissions from passenger cars in order to satisfy the greenhouse gas emission reduction targets set by the 2015 Paris Climate Change Conference [1,2]. These reductions are achieved by reducing pumping and friction losses at part-load operation. Conventionally, rated torque and power for downsized units are recovered by means of fixed-geometry turbocharging [3]. The transient response of such engines is, however, affected by the static and dynamic characteristics of the fixed-geometry turbo-machinery (especially when it is optimized for high-end torque) [4,5]. One feasible solution to this is the use of variable geometry turbocharger (VGT) technology, which is designed to enable the effective aspect ratio of the turbocharger to vary with different engine operating conditions (see Figure 1). This is because the best aspect ratio at high engine speeds is very different from the best aspect ratio at low engine speeds [6]. In engines equipped with VGT, and because part of the exhaust energy is used to accelerate the turbine shaft for boosting, engine transient response and fuel economy can be improved significantly [7].

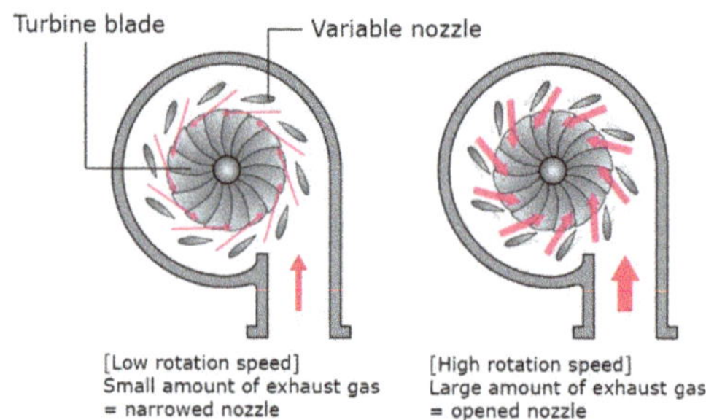

Figure 1. Variable geometry turbocharger (VGT) operating principle.

For diesel engines, VGT often interacts with an exhaust gas recirculation (EGR) system (which is for the engine's NOx emission reduction). This interaction increases the complexity of the VGT control problem. Furthermore, the time delay and hysteresis between the input and output dynamics of the diesel engine's gas exchange system make it difficult to accurately control the VGT [8]. Traditionally, the fixed-parameter structure proportion integration differentiation (PID) control is used in industry for VGT boost control, but the parameter setting processing is complicated and it is difficult to obtain satisfactory results, especially when the state of the control loop is altered [9–13]. There are other PID variants that include expert PID control [14], fuzzy PID control [15], and neural network-based PID control [16], etc. Although the variants are said to perform better if tuned well, those algorithms, respectively, need to acquire expert knowledge, construct fuzzy control decision tables, and tune complicated neural network parameters, and thus their widespread use for VGT boost control may be prohibited. Meanwhile, for complex industrial systems with high-order, large lag, strong coupling, nonlinear, and time-varying parameters (such as for VGT control systems [17–19]), the traditional control theory which relies on mathematical models is still immature, and some methods are too complicated and cannot be directly applied for industrial applications [20–22]. On the other hand, it may not be possible or feasible to develop a first-principle model for complex industrial processes. Furthermore, complex engineering systems are rather expensive, with a high requirement for system reliability and control performance. In this context, "model-free" intelligent algorithms (in the absence of a model with high fidelity) to achieve end-to-end learning and intelligent control while taking the industrial need of simplicity and robustness into consideration may provide an attractive alternative.

Reinforcement learning (RL), which is considered as one of three machine learning paradigms, focuses on how agents should act in the environment to maximize cumulative rewards (see Figure 2) [20]. Temporal-difference (TD) learning, which is a combination of dynamic programming (DP) ideas and the Monte Carlo idea, is considered to be the core and novelty of reinforcement learning [23]. In RL, there are two classes of the TD method—on-policy and off-policy. The most important on-policy algorithm includes Sarsa and Sarsa (λ), and one of the breakthroughs in off-policy reinforcement learning is known as Q-learning [24,25]. Deep reinforcement learning (DRL) is an area of machine learning that combines a deep learning approach and reinforcement learning (RL). This field of study was used to master a wide range of Atari 2600 games and its great success on AlphaGo, which was the first computer program to beat a human professional Go player, is a historic milestone in machine learning research [26]. Deep Q-network (DQN) based on value function and deep deterministic policy gradient (DDPG) based on policy gradient are the two latest DRL techniques. The DQN used on AlphaGo and AlphaGo Zero [27,28] uses only the original image of the game as input, and does not rely on the manual extraction of features. It innovatively combines deep convolutional neural networks with Q-learning to achieve human player control (it also achieved great success in Atari video games [29]). Although this algorithm achieves the generalization of continuous state space, it is theoretically only suitable for tasks in discrete action space. The DDPG strategy proposed by Lillicrap et al. [30] uses deep neural networks as approximators to effectively combine deep learning and deterministic strategy gradient algorithms [31]. It can cope with high-dimensional inputs, achieve end-to-end control, output continuous actions, and thus can be applied to more complex situations with large state spaces and continuous action spaces. To the authors' best knowledge, there is no literature that has applied DRL techniques to boost control problems for VGT-equipped engines. Furthermore, there seem to be few studies that analyze the DDPG algorithm on sequential decision control problems for industrial applications.

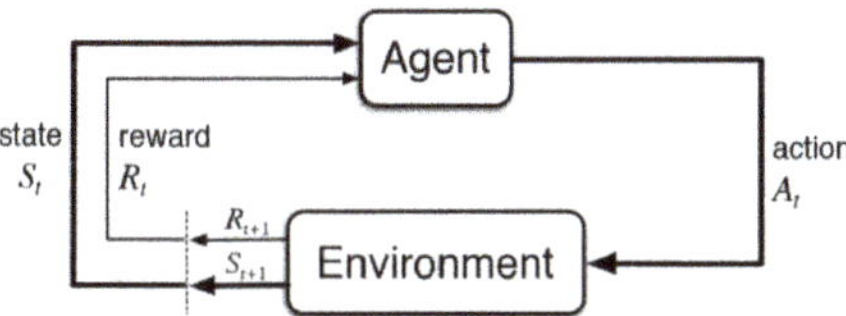

Figure 2. Basic idea and elements involved in a reinforcement learning formalization.

Based on the above discussion, it is appropriate to apply the DDPG techniques for the boost control on a VGT-equipped engine. In this paper, in order to achieve the optimum boost control performance, first the simulation model of a VGT-equipped diesel engine is introduced. Subsequently, a model-free DDPG algorithm is built to develop and finally form a strategy to track the target engine boost pressure under transient driving cycles by regulating the turbine vanes. Finally, the proposed DDPG algorithm is compared with a fine-tuned PID controller to validate its optimality. The rest of this article is structured as follows: Section 2 describes the mean value engine model (MVEM) of the VGT-equipped diesel engine. In Section 3, the DDPG-based framework is proposed to achieve the optimal boost control of the engine. In Section 4, the corresponding simulations are conducted to compare the proposed algorithm and a fine-tuned PID controller. Section 5 concludes the article.

2. Mean Value Engine Model Analysis

Mean value engine models (MVEMs) are useful for certain types of modeling where simulation speed is of primary importance, the details of wave dynamics are not critical, and bulk fluid flow is still important (for modeling turbocharger lag, etc.) [32]. A mean value engine model essentially contains a map-based cylinder model, which is computationally faster than a detailed cylinder. The simulation speed can be increased further by combining multiple detailed cylinders into a single mean value cylinder. In addition, many of the other flow components from the detailed model can be combined to create a simplified flow network of larger volumes.

The layout of the VGT-equipped diesel engine is illustrated in Figure 3. The detailed engine model was converted to a mean value model. As it is not the focus of this article, only a brief process summary is presented here. The mean value cylinder is defined by imposed values for indicated mean effective pressure (IMEP), volumetric efficiency, and exhaust gas temperature. These three quantities are predicted by neural networks (see Figure 4) that depend on seven input variables (intake manifold pressure and temperature, exhaust manifold pressure, EGR fraction, injection timing, fuel rate, and engine speed). Note here that each neural network (four in total) is trained using the data generated in the detailed simulation. The output of this training is an external file which contains the necessary neural network settings. Once the training has been completed, the neural network file can be called into the mean value model, which dramatically increases the computational efficiency. In addition, the friction mean effective pressure (FMEP) of the cranktrain is also calculated by a neural network dependent on the same seven variables.

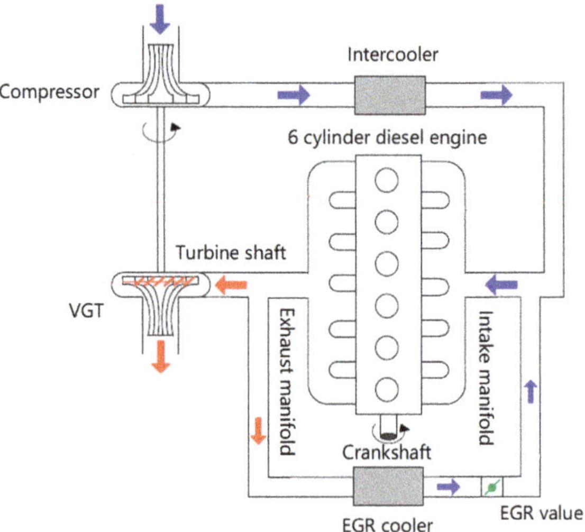

Figure 3. VGT-equipped diesel engine with an exhaust gas recirculation (EGR) system.

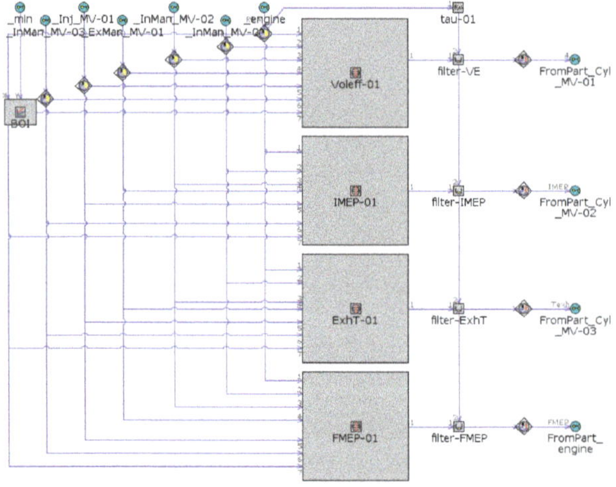

Figure 4. The proposed neural network that maps the mean value cylinder performance as a function of seven input variables in GT-SUITE.

The intake and exhaust systems are simplified into large "lumped" volumes so that system volume is conserved (with a loss of detailed wave dynamics). The large volumes allow large time steps to be taken by the solver. Pressure drops in the flow network are calibrated using restrictive orifice connections between the lumped volumes. Additionally, heat transfer rates are calibrated using the heat transfer multiplier in parts where heat transfer is significant (exhaust manifold). The intercooler and EGR cooler outlet temperatures are imposed, which allows the gas temperature to be imposed as it passes through the connection. This reduces the amount of volumes required and allows a reduction in potential for any instability in the solver caused by the high heat transfer rates in the heat exchangers at large time steps that are typical of mean value models. The mean value model results match the detailed results well (see Figure 5), and should provide sufficient accuracy for control system and vehicle transient studies. The mean value model runs approximately 150 times faster than the detailed model and runs faster than "real time", enabling it to be used for real-time hardware-in-the-loop (HIL) simulation.

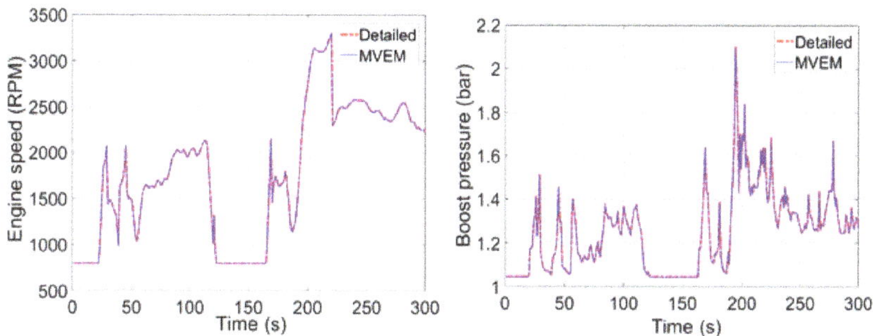

Figure 5. The first 300 s engine speed and boost pressure comparison for the US EPA FTP-72 (Federal Test Procedure) drive cycle.

The research engine was a 6 cylinder 3 L turbocharged direct injection (DI) diesel, with its GT-SUITE model seen in Figure 6. Advanced controllers should be used to dynamically control the position of the VGT rack in order to achieve the target boost pressure. It should be noted that the model was initially controlled by a fine-tuned PID controller, and the target boost pressure and the P and I gains were both mapped as a function of speed and requested load (implied by accelerator pedal position).

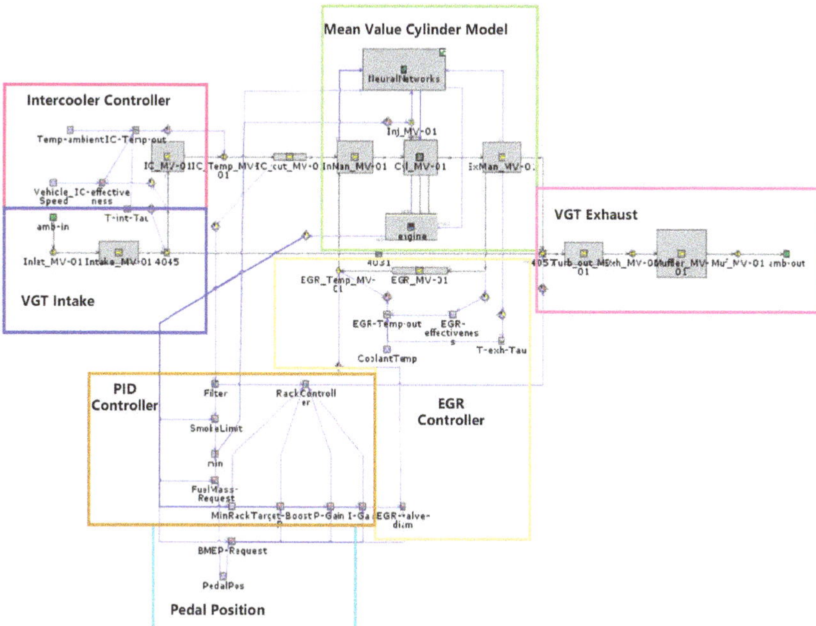

Figure 6. The GT-SUITE model layout of the 6 cylinder 3 L VGT-equipped diesel.

To analyze the transient behavior, the engine speed was imposed to match the prescribed vehicle speed profile from the FTP-72 driving cycle (see Figure 7). This transient engine speed profile was calculated using a simple kinematic mode simulation which can be seen in Figure 8. The same simulation provided the required brake mean effective pressure (BMEP) from the engine. Then, a separate detailed simulation was run with an injection controller to determine and store the transient pedal position required to achieve the requested BMEP.

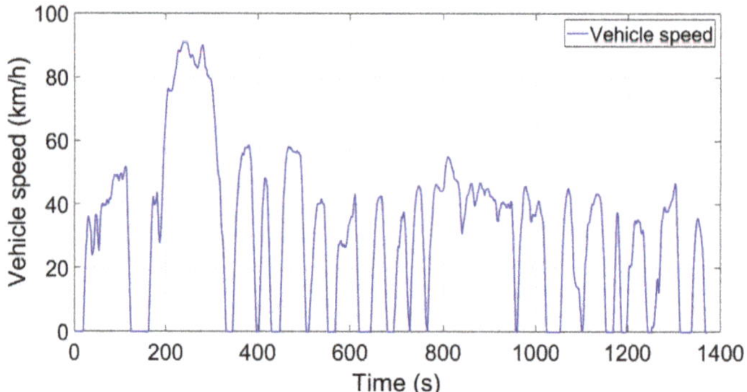

Figure 7. FTP-72 driving cycle.

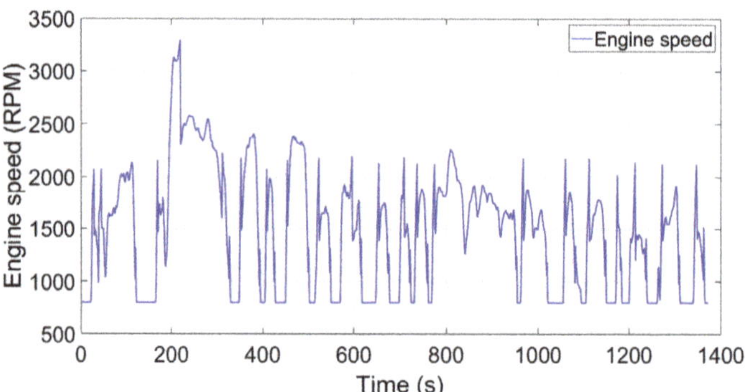

Figure 8. FTP-72 transient engine speed profile.

3. Deep Reinforcement Learning Algorithm

Model-free reinforcement learning is a technique for understanding and automating goal-directed learning and decision-making [33]. It differs from most other control algorithms in that it emphasizes on agents learning through direct interaction with the environment, without relying on model supervision or a complete environmental model [34]. As an interactive learning method, the main features of reinforcement learning are trial-and-error search and delayed return [35]. Figure 1 shows the interaction process between the agent and the environment. At any time step, the agent observes the environment in order to get the state S_t and then performs the action A_t. Afterward, the environment generates the next time S_{t+1} and R_t according to A_t. The probability that the process moves into its new state S_{t+1} is influenced by the chosen action and is given by the state transition function. Such a process can be described by Markov decision processes (MDPs) [36,37]. The goal of reinforcement learning is to formulate the problem as an MDP and find the optimal strategy [38]. The so-called strategy refers to the state-to-action mapping, which commonly uses the symbol policy π. It refers to a mapping on the action set for a given state s, that is, $\pi(a|s) = p[A_t = a|S_t = s]$. Reinforcement learning introduces a reward function to represent the return value at a certain time t, as follows:

$$G_t = r_{t+1} + \gamma r_{t+2} + \gamma^2 r_{t+3} + \cdots = \sum_{k=0}^{\infty} \gamma^k r_{t+k+1}, \tag{1}$$

where r represents an immediate reward and γ represents a discount factor which shows how important future returns are relative to current returns. The action–value function used in reinforcement learning algorithms describes the expected return after taking an action in state S_t and thereafter following policy π:

$$Q^\pi(s_t, a_t) = \sum_{r_{i \geq t}, s_{i>t} \sim E, a_{i>t} \sim \pi} [R_t | s_t, a_t] \tag{2}$$

Reinforcement learning makes use of the recursive relationship known as the Bellman equation:

$$Q^\pi(s_t, a_t) = \sum_{r_t, s_{t+1} \sim E} [r(s_t, a_t) + \gamma \sum_{a_{t+1} \sim \pi} [Q^\pi(s_{t+1}, a_{t+1})]] \tag{3}$$

The expectation depends only on the environment, which means that it is possible to learn Q^μ off-policy using transitions that are generated from a different stochastic behavior policy β. Q-learning, a commonly used off-policy algorithm, uses the greedy policy $\mu(s) = \text{argmax}_a Q(s,a)$. We consider function approximators parameterized by θ^Q, which we optimize by minimizing the loss, as follows:

$$L(\theta^Q) = \sum_{s_t \sim \rho^\beta, a_t \sim \beta, r_t \sim E} [(Q(s_t, a_t | \theta^Q) - y_t)^2] \tag{4}$$

where

$$y_t = r(s_t, a_t) + \gamma Q(s_{t+1}, \mu(s_{t+1}) | \theta^Q)$$

Recently, deep Q-network (DQN) adapted the Q-learning algorithm in order to make effective use of large neural networks as function approximators. Before DQN, it was generally deemed that it was difficult and unstable to use large nonlinear function approximators for learning value functions. Thanks to two innovations, DQN can use a function approximator to learn the value function in a stable and robust manner: (a) the network is trained off-policy with samples from a replay buffer to minimize correlations between samples and (b) the target Q-network is used to train the network to provide consistent goals during time difference (TD) backups.

The deterministic policy gradient (DPG) algorithm maintains a parameterized actor function, $\mu(s|\theta^\mu)$, which specifies the current policy by deterministically mapping states to a specific action. The critic $Q(s,a)$ is learned using the Bellman equation as in Q-learning. The actor is updated by applying the chain rule to the expected return from the start distribution J with respect to the actor parameters, as follows:

$$\begin{aligned}
\nabla_{\theta^\mu} J &\approx \sum_{s_t \sim \rho^\beta} [\nabla_{\theta^\mu} Q(s, a | \theta^Q)|_{s=s_t, a=\mu(s_t|\theta^\mu)}] \\
&= \sum_{s_t \sim \rho^\beta} [\nabla_a Q(s, a | \theta^Q)|_{s=s_t, a=\mu(s_t)} \nabla_{\theta_\mu} \mu(s|\theta^\mu)|_{s=s_t}]
\end{aligned} \tag{5}$$

Deep deterministic policy gradient (DDPG) combines the actor–critic (AC) approach based on deterministic policy gradient (DPG) [31] with insights from the recent success of deep Q-network (DQN). Although DQN has achieved great success in high-dimensional issues, like the Atari game, the action space for which the algorithm is implemented is still discrete. However, for many tasks of interest, especially physical industrial control tasks, the action space must be continuous. Note that if the action space is discretized too finely, the control problem eventually leads to excessive motion space, which is extremely difficult for the algorithm to learn. The DDPG strategy uses deep neural networks as approximators to effectively combine deep learning and deterministic strategy gradient algorithms. It can cope with high-dimensional inputs, achieve end-to-end control, output continuous actions, and thus can be applied to more complex situations with large state spaces and continuous action spaces. In detail, DDPG uses an actor network to tune the parameter θ^μ for the policy function, that is, decide the optimal action for a given state. A critic is used for evaluating the policy function estimated by the actor according to the temporal TD error (see Figure 9). One issue for DDPG is that it rarely explores actions. A solution is to add noise to the parameter space or the action space. It is claimed that adding to parameter space is better than to action space [39]. One commonly used noise is the Ornstein–Uhlenbeck random process. Algorithm 1 shows the pseudo-code of the proposed DDPG algorithm.

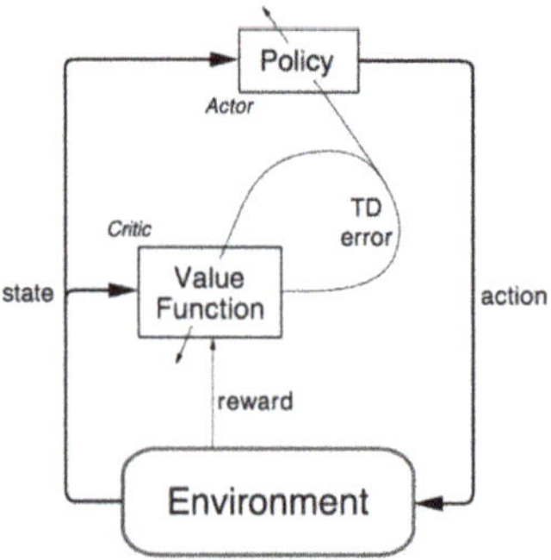

Figure 9. Actor–critic architecture.

Algorithm 1: Pseudo-code of the Deep deterministic policy gradient (DDPG) algorithm.

Randomly initialize critic network $Q(s, a|\theta^Q)$ and actor $\mu(s|\theta^\mu)$ with weights θ^Q and θ^μ.
Initialize target network Q' and μ' with weights $\theta^{Q'} \leftarrow \theta^Q$, $\theta^{\mu'} \leftarrow \theta^\mu$
Initialize replay buffer R
for episode = 1, M do
    Initialize a random process N for action exploration
    Receive initial observation state s_1
    for t = 1, T do
        Select action $a_t = \mu(s_t|\theta^\mu) + N_t$ according to the current policy and exploration noise
        Execute action a_t and observe reward r_t and observe new state s_{t+1}
        Store transition (s_t, a_t, r_t, s_{t+1}) in R
        Sample a random minibatch of N transitions (s_t, a_t, r_t, s_{t+1}) from R
        Set $y_i = r_i + \gamma Q'(s_{i+1}, \mu'(s_{i+1}|\theta^{\mu'})|\theta^{Q'})$
        Update critic by minimizing the loss: $L = \frac{1}{N}\sum_i (y_i - Q(s_i, a_i|\theta^Q))^2$
        Update the actor policy using the sampled gradient:

$$\nabla_{\theta^\mu} J \approx \frac{1}{N}\sum_i \nabla_a Q(s, a|\theta^Q)|_{a=s_i, a=\mu(s_i)} \cdot \nabla_{\theta^\mu} \mu(s|\theta^\mu)|_{s_i}$$

        Update the target networks:
$$\theta^{Q'} \leftarrow \tau\theta^Q + (1-\tau)\theta^{Q'}$$
$$\theta^{\mu'} \leftarrow \tau\theta^\mu + (1-\tau)\theta^{\mu'}$$

    end for
end for

TensorFlow is one of the widely used end-to-end open-source platforms for machine learning. In order to draw on the research findings of DRL in other research fields, especially to re-use the existing program code frameworks in machine learning, we used Python compatible with TensorFlow as the algorithm design language in this study. Meanwhile, in order to apply the DRL algorithm built in Python to the diesel engine environment, we proposed to use MATLAB/Simulink as the program interface, so that the two-way transmission among Python, MATLAB/Simulink, and GT-SUITE could be reached. The specific DDPG algorithm implementation and the corresponding co-simulation platform are shown in Figure 10. Key concepts of the DDPG-based boost control algorithm are formulated as follows.

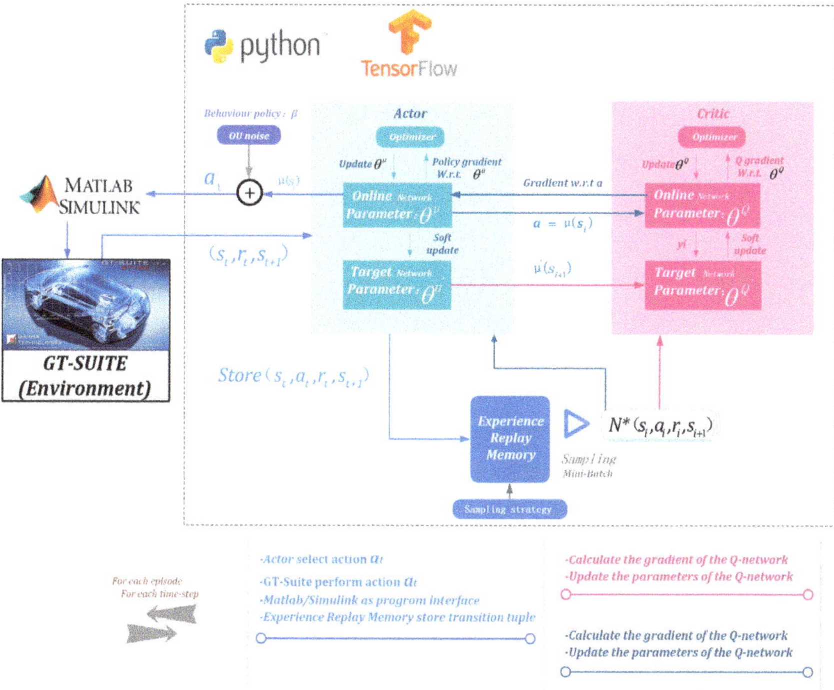

Figure 10. Deep deterministic policy gradient (DDPG) algorithm implementation and the corresponding co-simulation platform.

The engine speed, the actual boost pressure, the target boost pressure, and the current vane position were chosen to group a four-dimensional state space. It should be noted here that only a small number of states were chosen in this study in order to (a) facilitate the training process and (b) showcase the generalization ability of the DRL techniques. The vane position controlled by a membrane vacuum actuator was selected as the control action. Immediate reward is important in the RL algorithm because it directly affects the convergence curves and, in some cases, a fine adjustment of the immediate reward parameter can bring the final policy to the opposite poles. The agents always try to maximize the reward they can get by taking the optimal action at each time step because more cumulative rewards represent better overall control behavior. Therefore, the immediate reward should be defined based on optimization goals. The control objective of this work was to track the target boost pressure under transient driving cycles by regulating the vanes in a quick and stable manner. Keeping this objective in mind, the function of the error between the target and the current boost pressure and the rate of action change were defined as the immediate reward. The equation for the immediate reward is given as follows:

$$r_t = e^{-\frac{[0.95*|e(t)|+0.05*|I_t|]^2}{2}} - 1, \tag{6}$$

where r_t is the immediate reward generated when the state changes by taking an action at time t. $e(t)$ and I_t represent the error between the target and the current boost pressure and the rate of action change, respectively.

The corresponding DDPG parameters and the illustration of the actor–critic network are shown in Table 1 and Figure 11. In this study, the input layer of the actor network has four neurons, namely, the engine speed, the actual boost pressure, the target boost pressure, and the current vane position. There are three hidden layers each having 120 neurons. The output layer has one neuron representing the control action (i.e., the vane position). All these layers are fully connected. For the critic network, the input layer has an additional neuron, which is the control action implemented by the actor network, compared to that of the critic network. There is one hidden layer having 120 neurons. The output layer of the critic network has one neuron representing the value function of the selected action for the specific state. The network is trained for 50 episodes and each episode represents the first 80% time of the FTP-72 trips (1098s).

Table 1. DDPG parameters.

Parameters	Value
Learning rate for actor	0.0001
Learning rate for critic	0.0002
Reward discount factor γ	0.9
Soft replacement factor τ	0.01
Replay memory size	100,000
Mini-batch size	256
Action randomness decay	0.999995
Initial exploration	10

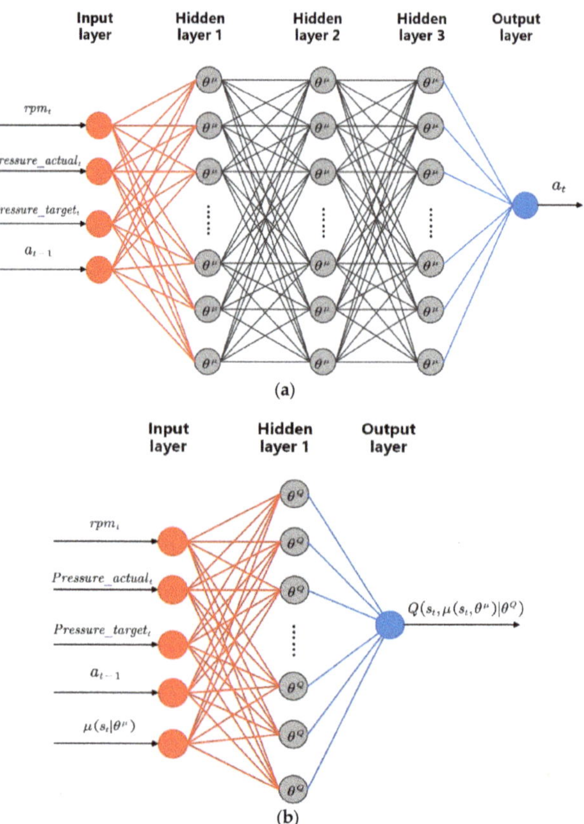

Figure 11. Illustration of the (a) actor network and (b) critic network.

4. Results and Discussion

In this article, the simulations were conducted based on an advanced co-simulation platform (see Figure 10). In order to validate its performance, the proposed DDPG algorithm was compared to a fine-tuned gain scheduled PID controller with both its P and I gains mapped as a function of speed and requested load. Without derivative action, a PI-controlled system may be less responsive, but it makes the system steadier at steady-state conditions (thus often adopted for industrial practice). It should be noted here that this PID controller adopted classic Ziegler–Nichols methods [40] to manually tune the control parameters, which took much effort and thus should be interpreted to represent a good control behavior benchmark. The US FTP-72 (Federal Test Procedure) driving cycle shown in Figure 7 was employed to verify the proposed strategy. The cycle simulated an urban route of 12.07 km with frequent stops and rapid accelerations. The maximum speed was 91.25 km/h and the average speed was 31.5 km/h. This transient driving cycle was selected because it mimics the real-world VGT environment system with large lag, strong coupling (especially with EGR) and nonlinear characteristics and thus, if a well-behaved control strategy in this environment is established, it should perform well in other driving cycles with more steady-state regions (such as the New European Driving Cycle (NEDC)). In this study, the first 80% time driving cycle was used to train the DDPG algorithm and the remaining data were destined for testing analysis. There are many different measures that can be used to compare the quality of controlled response. Integral absolute error (IAE), which integrates the absolute error over time, was used in this study to measure the control performance between the PID controller and the proposed DDPG algorithm.

Figure 12 shows the control performance using the fine-tuned PID controller. It can be seen that the actual boost pressure tracks the target boost pressure well at first glance. However, after zooming in on some operating conditions, a relatively large error can still be seen. This may be due to the turbo-lag, which cannot be improved from the control point of view (such as the time period from 10 s to 40 s, where although the VGT is already controlled to its minimum flow area for the fastest transient performance, it still exhibits lack of boost). Nevertheless, for most situations, taking the time period of 920 s to 945 s for example, there is still some room for the PID control strategy to improve. We note here that the results in Figure 12 are only a balance between control performance and tuning efforts, that is, a better control behavior can be achieved if the tuning process is made in a more finely manner, but more efforts and resources are required. In this research, the emphasis was not put on the final control performance comparison between PID and DRL theory, due to the fact that the structure behind each method is different and the control behavior, to a large extent, depends on how the control parameters are tuned. More focus was put on trying to solve the control problem in a self-learning manner and showcase good control adaptivity for the DRL approach.

The learning process of the DDPG algorithm can be seen in Figure 13. At the beginning of the learning, the cumulative rewards for the DDPG agent per episode were extremely low because (1) the agent (corresponding to the vane position actuator in the VGT boosting controller) only randomly selects actions in order to complete an extensive search process so as not to fall into local optimum and (2) the agent has no prior experience of what it should do for a specific state (thus the agent can only select the actions based on the initial DDPG parameters). After approximately 40 episodes, the algorithm has already been converged with the cumulative rewards, reaching a high value. This indicates that the agent has learned the experience to control the boost pressure. It should be noted that the learning process takes place only by direct interaction with the environment (in this case, the simulation software serves as the environment), without relying on model supervision or complete environment models, and a well-behaved control strategy is developed and finally formed autonomously from scratch. To answer the question of whether the learned controller was good enough, the control performance of the first 80% FTP-72 driving cycle using the final DDPG controller was compared with the performance based on the aforementioned fine-tuned PID controller. In Figures 12 and 14, it can be seen that both the PID and the proposed DDPG algorithm perform well at first glance, but after zooming in on some operating conditions, a large tracking disparity can still be seen. Although the PID controller seems to track the boost pressure with relatively small errors, the control performance based on the proposed DDPG algorithm outperforms that of the PID controller with almost excellent tracking behavior. The IAE value of the PID control performance is 41.72, whereas the value based on the proposed DDPG algorithm can be as low as 37.43.

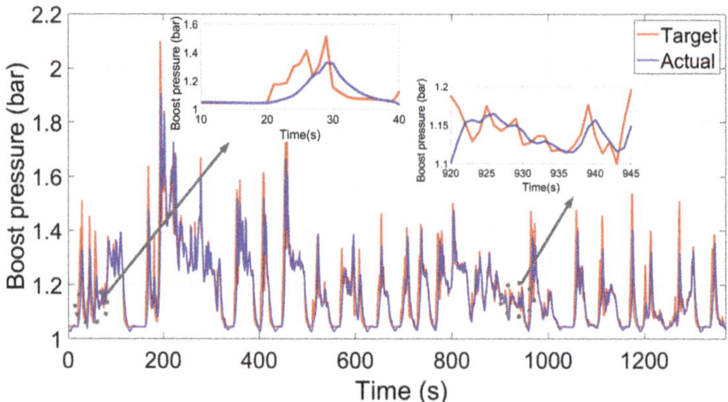

Figure 12. The fine-tuned gain-scheduled PID control performance.

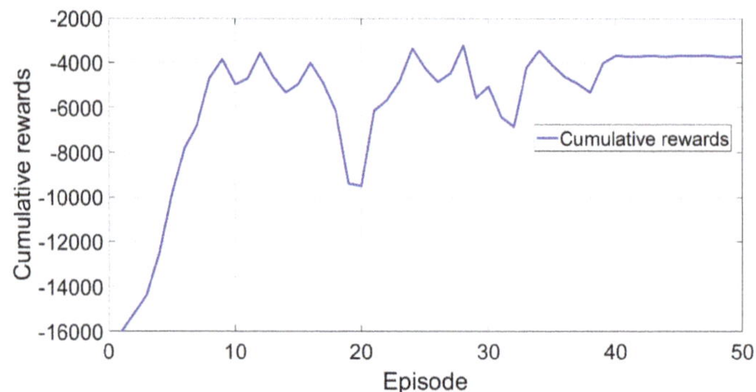

Figure 13. Track of all rewards per episode.

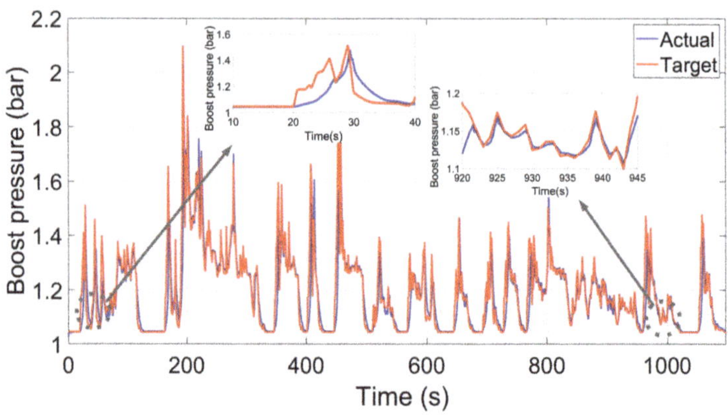

Figure 14. Control performance for the first 80% FTP-72 driving cycle using the final DDPG training parameters.

This difference is shown better in Figure 15, where the control performance comparison between the fine-tuned PID and the proposed DDPG from the time period of 920 s to 945 s is made. This time period was selected because it corresponds to the frequently used engine operating regions.

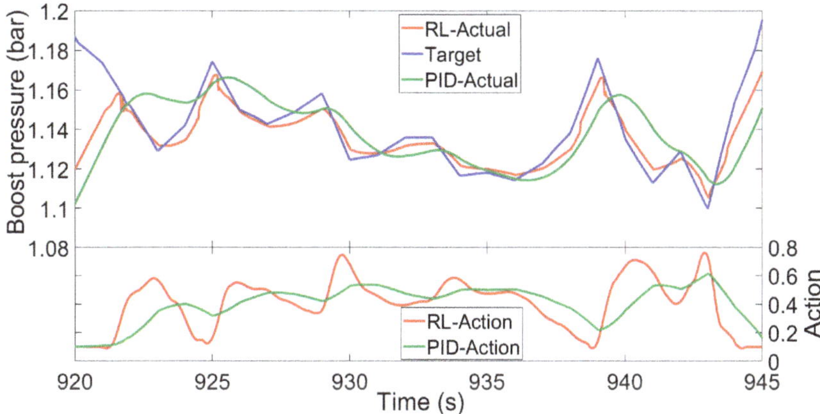

Figure 15. Control performance comparison between the fine-tuned PID and the proposed DDPG from the time period of 920 s to 945 s.

In order to showcase the generalization ability of the proposed DRL techniques, the control performance for the end 20% FTP-72 driving cycle based on the trained DDPG parameters was simulated. It can be seen in Figure 16 that although the control parameters were not trained based on this part of the driving cycle (i.e., some of the states may not have been visited in the previous training process), the performance still exhibits good control behavior. Compared to the same time period using the fine-tuned PID controller (already optimized for this time period), the control performance based on the proposed DDPG clearly performs better and the IAE of the PID and the proposed DDPG are 10.17 and 8.35, respectively.

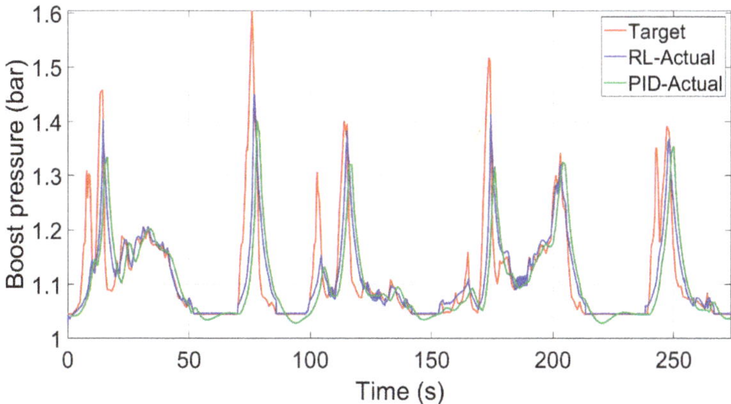

Figure 16. Control performance for the end 20% FTP-72 driving cycle based on the trained DDPG algorithm.

As the control strategy based on the proposed DDPG algorithm is able to achieve (or improve, depending on the tuning efforts of each algorithm) the control performance compared to a fine-tuned PID benchmark controller, it could replace the traditional PID controller for boosting control in the near future. Compared to the benchmark PID controller whose parameters traditionally require manual adjustment (thus the tuning efficiency is low), the control strategy based on DDPG is able to adaptively adjust the algorithm strategy in the learning process, which not only can save a lot of manpower resources, but also adapt more to the changing environment and hardware aging over time (thus being unbiased by modelling errors). To prove this, another simulation model with a different combustion and turbocharger model was used. This was a simplified replication of a real engine plant whose transient performance could be diverged from the simulation prediction mainly due to combustion and turbocharger modelling inaccuracy. Figure 17 shows the control performance using both the pre-trained algorithm (which indicates the off-line behavior) and the strategy after continuing on-line learning in the "real engine" simulation model. It can be seen that the off-line policy is able to achieve a relatively good control behavior and can be improved further by continuing its learning from the interaction with the new environment on-line. Thus, different from other studies whose control parameters optimized in the simulation platform, for most cases, are no longer valid in the experimental test, the control strategy based on the proposed DRL techniques can combine the simulation training and the experimental continuing training together in order to fully utilize the computational resources off-line and refine the algorithm in the experimental environment on-line.

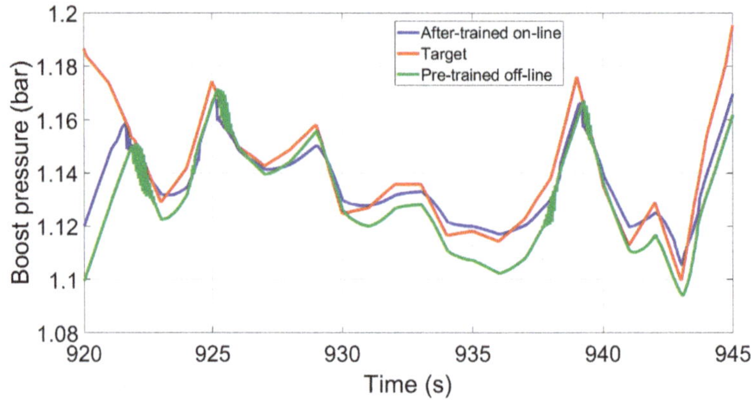

Figure 17. Control performance using the pre-trained off-line algorithm and the strategy after continuing on-line learning in the "real engine" simulation model.

Furthermore, because the learning process of the proposed DDPG algorithm distinguishes itself from other approaches by putting its emphasis on interacting with the transient environment, the final control performance is able to outperform that of the other approaches whose techniques are only based on the steady-state simulation or experimental control behavior. The most obvious example would be its capability to exceed the classic feedforward control which only responds to its control signal in a pre-defined way without responding to how the load reacts. It is known that most of the pre-defined map in a controller with feedforward function is calibrated in a steady-state environment in industry and is fixed for the entire product lifecycle. The proposed DDPG algorithm, however, because the control action adapts to the environment, is equivalent to the concept of the so-called automatic transient calibration.

5. Conclusions and Future Work

In this paper, a model-free self-learning DDPG algorithm was employed for the boost control of a VGT-equipped engine with the aim to compare the DDPG techniques with traditional control methods and to provide references for the further development of DDPG algorithms on sequential decision-control problems with other industrial applications. Using a fine-tuned PID controller as a benchmark, the results show that the control performance based on the proposed DDPG algorithm can achieve a good transient control performance from scratch by autonomously learning the interaction with the environment, without relying on model supervision or complete environment models. In addition, the proposed strategy is able to adapt to the changing environment and hardware aging over time by adaptively tuning the algorithm in a self-learning manner on-line, making it attractive to real plant control problems whose system consistency may not be strictly guaranteed and whose environment may change over time. This indicates that the control strategy based on the proposed DRL techniques can combine the simulation training and the experimental continuing training together in order to fully utilize the computational resources off-line and refine the algorithm in the experimental environment on-line. Future work may include applying DRL-based parallel computer architecture to boost the computational efficiency for the control problem with high-order, large lag, strong coupling, and nonlinear characteristics. Another interesting direction could be combining some look-ahead strategies with the proposed DRL techniques to accelerate the training process and improve the final control performance. The stability control between multiple reinforcement learning-based controllers will also be studied, which includes distributed RL and hierarchical RL.

Author Contributions: B.H. drafted the paper and provided the overall research ideas. J.Y. provided reinforcement learning strategy suggestions. S.L. and J.L. analyzed the data. H.B. revised the paper. These authors are contributed equally to this work.

Funding: This work was supported by the National Natural Science Foundation of China (Grant No. 51905061), the Chongqing Natural Science Foundation (Grant No. cstc2019jcyj-msxmX0097), the Science and Technology Research Program of Chongqing Municipal Education Commission (Grant No. KJQN201801124), the Venture & Innovation Support Program for Chongqing Overseas Returners (Grant No. cx2018135), and the Open Project Program of the State Key Laboratory of Engines (Tianjin University) (Grant No. k2019-02), to whom the authors wish to express their thanks.

Acknowledgments: The authors would like to thank the anonymous reviewers for their valuable comments and suggestions.

Conflicts of Interest: The authors declare no conflict of interest.

Data Availability: The data used to support the findings of this study are available from the corresponding author upon request.

Nomenclature

AC	Actor critic
BMEP	Brake mean effective pressure
DQN	Deep Q-network
DDPG	Deep deterministic policy gradient
DI	Direct injection
DP	Dynamic programming
DRL	Deep reinforcement learning
DPF	Diesel particulate filter
DOC	Diesel oxidation catalyst
EGR	Exhaust gas recirculation
FMEP	Friction mean effective pressure
FTP	Federal Test Procedure
FTP-72	Federal Test Procedure-72
HIL	Hardware-in-the-loop
IAE	Integral absolute error
FMEP	Friction mean effective pressure
IMEP	Indicated mean effective pressure
MDP	Markov decision process
MVEM	Mean value engine model
NEDC	New European Driving Cycle
PID	Proportion integration differentiation

References

1. Awad, O.I.; Ali, O.M.; Mamata, R.; Abdullaha, A.A.; Najafid, G.; Kamarulzamana, M.K.; Yusria, I.M.; Noora, M.M. Using fusel oil as a blend in gasoline to improve si engine efficiencies: A comprehensive review. *Renew. Sustain. Energy Rev.* **2017**, *69*, 1232–1242. [CrossRef]
2. Enang, W.; Bannister, C. Modelling and control of hybrid electric vehicles (A comprehensive review). *Renew. Sustain. Energy Rev.* **2017**, *74*, 1210–1239. [CrossRef]
3. Hu, B.; Akehurst, S.; Brace, C. Novel approaches to improve the gas exchange process of downsized turbocharged spark-ignition engines: A review. *Int. J. Engine Res.* **2015**, *17*, 1–24. [CrossRef]
4. Hu, B.; Turner, J.W.G.; Akehurst, S.; Brace, C.J.; Copeland, C. Observations on and potential trends for mechanically supercharging a downsized passenger car engine: A review. *Proc. Inst. Mech. Eng. Part D J. Automob. Eng.* **2016**, *231*, 435–456. [CrossRef]
5. Hu, B.; Akehurst, S.; Lewis, A.G.J.; Lu, P.; Millwood, D.; Copeland, C.; Chappell, E.; Freitas, A.D.; Shawe, J.; Burtt, D. Experimental analysis of the V-Charge variable drive supercharger system on a 1.0 L GTDI engine. *Proc. Inst. Mech. Eng. Part D J. Automob. Eng.* **2017**, *232*, 449–465. [CrossRef]
6. Tang, H.; Pennycott, A.; Akehurst, S.; Brace, C.J. A review of the application of variable geometry turbines to the downsized gasoline engine. *Int. J. Engine Res.* **2015**, *16*, 810–825. [CrossRef]
7. Zhao, D.; Winward, E.; Yang, Z.; Stobart, R.; Mason, B.; Steffen, T. An integrated framework on characterization, control, and testing of an electrical turbocharger assist. *IEEE Trans. Ind. Electron.* **2018**, *65*, 4897–4908. [CrossRef]
8. Oh, B.; Lee, M.; Park, Y.; Won, J.; Sunwoo, M. Mass air flow control of common-rail diesel engines using an artificial neural network. *Proc. Inst. Mech. Eng. Part D J. Automob. Eng.* **2013**, *227*, 299–310. [CrossRef]
9. Xu, S.; Hashimoto, S.; Jiang, W.; Jiang, Y.; Izaki, K.; Kihara, T.; Ikeda, R. Slow Mode-Based Control Method for Multi-Point Temperature Control System. *Processes* **2019**, *7*, 533. [CrossRef]
10. Xu, S.; Hashimoto, S.; Jiang, W. Pole-Zero Cancellation Method for Multi Input Multi Output (MIMO) Temperature Control in Heating Process System. *Processes* **2019**, *7*, 497. [CrossRef]
11. Sekour, M.; Hartani, K.; Draou, A.; Ahmed, A. Sensorless Fuzzy Direct Torque Control for High Performance Electric Vehicle with Four In-Wheel Motors. *J. Electr. Eng. Technol.* **2013**, *8*, 530–543. [CrossRef]
12. Zhang, X.; Sun, L.; Zhao, K.; Sun, L. Nonlinear Speed Control for PMSM System Using Sliding-Mode Control and Disturbance Compensation Techniques. *IEEE Trans. Power Electr.* **2012**, *28*, 1358–1365. [CrossRef]
13. Gao, R.; Gao, Z. Pitch control for wind turbine systems using optimization. estimation and compensation. *Renew. Energy* **2016**, *91*, 501–515. [CrossRef]
14. Chee, F.; Fernando, T.L.; Savkin, A.V.; Heeden, V.V. Expert PID Control System for Blood Glucose Control in Critically Ill Patients. *IEEE Trans. Inf. Technol. Biomed.* **2003**, *7*, 419–425. [CrossRef] [PubMed]
15. Savran, A. A multivariable predictive fuzzy PID control system. *Appl. Soft Comput.* **2013**, *13*, 2658–2667. [CrossRef]
16. Lopez-Franco, C.; Gomez-Avila, J.; Alanis, A.Y.; Arana-Daniel, N.; Villaseñor, C. Visual Servoing for an Autonomous Hexarotor Using a Neural Network Based PID Controller. *Sensors* **2017**, *17*, 1865. [CrossRef] [PubMed]
17. Omran, R.; Younes, R.; Champoussin, J. Neural networks for real-time nonlinear control of a variable geometry turbocharged diesel engine. *Int. J. Robust Nonlin.* **2008**, *18*, 1209–1229. [CrossRef]
18. Wahlstrom, J.; Eriksson, L.; Nielsen, L. EGR-VGT Control and Tuning for Pumping Work Minimization and Emission Control. *IEEE Trans. Contr. Syst. Technol.* **2010**, *18*, 993–1003. [CrossRef]
19. Zamboni, G.; Capobianco, M. Influence of high and low pressure EGR and VGT control on in-cylinder pressure diagrams and rate of heat release in an automotive turbocharged diesel engine. *Appl. Therm. Eng.* **2013**, *51*, 586–596. [CrossRef]
20. Sutton, R.S.; Barto, A.G. *Reinforcement Learning: An Introduction*, 2nd ed.; The MIT Press: Cambridge, MA, USA, 2018; pp. 97–113.
21. Gao, Z.; Saxen, H.; Gao, C. Special section on data-driven approaches for complex industrial systems. *IEEE Trans. Ind. Inform.* **2013**, *9*, 2210–2212. [CrossRef]
22. Gao, Z.; Nguang, S.K.; Kong, D.X. Advances in Modelling, Monitoring, and Control for Complex Industrial Systems. *Complexity* **2019**. [CrossRef]
23. Sutton, R.S. Learning to predict by the methods of temporal differences. *Mach. Learn.* **1988**, *3*, 9–44. [CrossRef]

24. Liu, T.; Zou, Y.; Liu, D.; Sun, F. Reinforcement Learning–Based Energy Management Strategy for a Hybrid Electric Tracked Vehicle. *Energies* **2015**, *8*, 7243–7260. [CrossRef]

25. Cao, H.; Yu, T.; Zhang, X.; Yang, B.; Wu, Y. Reactive Power Optimization of Large-Scale Power Systems: A Transfer Bees Optimizer Application. *Processes* **2019**, *7*, 321. [CrossRef]

26. Francois-Lavet, V.; Henderson, P.; Islam, R.; Bellemare, M.G.; Pineau, J. An introduction to deep reinforcement learning. *Found. Trends Mach. Learn.* **2018**, *11*, 245–246. [CrossRef]

27. Silver, D.; Aja, H.; Maddison, J.C.; Guez, A.; Sifre, L.; van den Driessche, G.; Schrittwieser, J.; Antonoglou, L.; Panneershelvam, V.; Lanctot, M.; et al. Mastering the game of go with deep neural networks and tree search. *Nature* **2016**, *529*, 484–489. [CrossRef]

28. Silver, D.; Schrittwieser, J.; Simonyan, K.; Antonoglou, I.; Huang, A.; Guez, A.; Hubert, T.; Baker, L.; Lai, M.; Bolton, A.; et al. Mastering the Game of Go without Human Knowledge. *Nature* **2017**, *550*, 354–359. [CrossRef]

29. Mnih, V.; Kavukcuoglu, K.; Silver, D.; Graves, A.; Antonoglou, I.; Wierstra, D.; Riedmiller, M. Playing Atari with Deep Reinforcement Learning. In Proceedings of the NIPS Deep Learning Workshop 2013, Lake Tahoe, NV, USA, 9 December 2013.

30. Lillicrap, T.P.; Hunt, J.J.; Pritzel, A.; Heess, N.; Erez, T.; Tassa, Y.; Silver, D.; Wierstra, D. Continuous control with deep reinforcement learning. In Proceedings of the International Conference on Learning Representations 2016, San Juan, PR, USA, 2–4 May 2016.

31. Silver, D.; Level, G.; Heess, N.; Degris, T.; Wierstra, D.; Riedmiller, M. Deterministic Policy Gradient Algorithms. In Proceedings of the 31st International Conference on International Conference on Machine Learning 2014, Beijing, China, 21–26 June 2014.

32. Nikzadfar, K.; Shamekhi, A.H. An extended mean value model (EMVM) for control-oriented modeling of diesel engines transient performance and emissions. *Fuel* **2015**, *154*, 275–292. [CrossRef]

33. Jiang, M.; Jin, Q. Multivariable System Identification Method Based on Continuous Action Reinforcement Learning Automata. *Processes* **2019**, *7*, 546. [CrossRef]

34. Zhang, D.; Gao, Z. Reinforcement learning–based fault-tolerant control with application to flux cored wire system. *Meas. Control* **2018**, *51*, 349–359. [CrossRef]

35. Kulkarni, T.D.; Narasimhan, K.R.; Saeedi, A.; Tenenbaum, J. Hierarchical Deep Reinforcement Learning: Integrating Temporal Abstraction and Intrinsic Motivation. In Proceedings of the 30th Conference on Neural Information Processing Systems 2016, Barcelona, Spain, 5–10 December 2016.

36. Howard, R.A. *Dynamic Programming and Markov Process*; The MIT Press: Cambridge, MA, USA, 1966.

37. Wang, Y.; Velswamy, K.; Huang, B. A Long-Short Term Memory Recurrent Neural Network Based Reinforcement Learning Controller for Office Heating Ventilation and Air Conditioning Systems. *Processes* **2017**, *5*, 46. [CrossRef]

38. Zhang, D.; Lin, Z.; Gao, Z. A Novel Fault Detection with Minimizing the Noise-Signal Ratio Using Reinforcement Learning. *Sensors* **2018**, *18*, 3087. [CrossRef] [PubMed]

39. Silver, D.; Lever, G. Better Exploration with Parameter Noise. Available online: https://openai.com/blog/better-exploration-with-parameter-noise/ (accessed on 28 May 2019).

40. ÅströmK, J.; Hägglund, T. *PID Controllers: Theory, Design, and Tuning*; Instrument Society of America: Research Triangle Park, NC, USA, 1995; Volume 2.

Article

Slow Mode-Based Control Method for Multi-Point Temperature Control System

Song Xu [1,2,†,‡], Seiji Hashimoto [1,2,*,‡], Wei Jiang [2,‡], Yuqi Jiang [1,‡], Katsutoshi Izaki [3], Takeshi Kihara [3] and Ryota Ikeda [3]

[1] Division of Electronics and Informatics, Gunma University, Kiryu 376-8515, Gunma, Japan
[2] Department of Electrical Engineering, Yangzhou University N.196 Huayang West Road, Yangzhou 225-000, China
[3] R & D Division, RKC Instrument Inc., Tokyo 146-8515, Japan
* Correspondence: hashimotos@gunma-u.ac.jp; Tel.: +81 277-30-1741
† Current address: 1-5-1 Tenjin-cho, Kiryu, Gunma, Japan.
‡ These authors contributed equally to this work.

Received: 10 July 2019; Accepted: 12 August 2019; Published: 14 August 2019

Abstract: In recent years, thermal processing systems with integrated temperature control have been increasingly needed to achieve high quality and high performance. In this paper, responding to the growing demands for proper transient response and to provide more accurate temperature controls, a novel slow-mode-based control (SMBC) method is proposed for multi-point temperature control systems. In the proposed method, the temperature differences and the transient response of all points can be controlled and improved by making the output of the fast modes follow that of the slow mode. Both simulations and experiments were carried out, and the results were compared to conventional control methods in order to verify the effectiveness of the proposed method.

Keywords: slow-mode-based control; multi-input multi-output (MIMO) temperature system; transient response; temperature differences

---

1. Introduction

In recent years, having benefited from the rapid progress in industrial process and its control system, thermal processing systems, especially those with temperature control, are attracting more attention for achieving high-quality and high-performance processing. In thermal processing, the temperature uniformity is one of the most basic and important requirements. For example, most of the applications of thermal processes are centered on the food industry, where requirements of temperature uniformity and faster transient response are strongly in demand.

Due to these requirements, various control methods have been proposed and introduced into temperature control systems. Among the various thermal processing techniques, the conventional proportional–integral–derivative (PID) controller has become the most commonly used control method benefiting from its simplicity and wide applicability [1–6]. However, due to the nonlinearity and the coupling influence of multi-point systems, it is difficult to realize precise temperature control based only on the conventional PID control method. Thus, the series the series connected fuzzy-proportional integral (SF-PI) control method is proposed to improve the control performance [7].

For the coupling difficulties as stated above, a novel decoupling compensation or quasi-decoupling method has been investigated in multi-input, multi-output (MIMO) temperature control systems by building an equivalent model of multiple SISO (single-input, single-output) systems [8–13]. However, if we cannot obtain a precise plant model, that decoupling compensation method would not work as efficiently as expected. Thus, the system identification methods have been applied to the temperature control system, and the most popular method is the step response method [14–16].

With the mathematical model of the system, the Model Predictive Control (MPC) method has been proposed on the heating system to provide precise control [17].

Moreover, although a precise plant model has been obtained and the decoupling compensation method has been applied, there remain some difficulties, such as: dead-time difference between the multiple points and the large time delay in the plant. These difficulties have a significant impact on the transient response and temperature difference. Thus, some advanced compensation methods have been proposed: feed forward compensation, data-driven tuning, and gradient temperature control [18–21].

Furthermore, for the excessively complex thermal process system, the mathematic model could not be obtained precisely. Thus, the data-driven approaches have been proposed as a potential method for controlling that system [22], and some optimization methods have also been proposed for these excessively complex processes [23].

Although after all the above methods have been introduced into the temperature control system in the thermal process, there are still some problems with the transient response and the closely controlled temperature of the multi-point temperature control system with dead-time difference and strong coupling effects. For example, in MIMO heating systems the temperature difference between each point should be under 5% of the reference temperature, no matter whether the transient response is fast or slow. Until now, the abovementioned method still cannot achieve the effect we expected in multi-point temperature control systems that have a large time constant and big dead-time difference. For example, if the system is composed of only a Smith predictive control method, the temperature difference may not be prevented. Also, if the system is composed of only a gradient control method, the transient response may not be drastically improved. Our previous work has proposed a pole-zero cancellation with reference model method for such a class of temperature control system [24]. This paper focuses on the multi-point temperature heating system, improving the transient response of each channel and reducing the temperature difference between each point, while also providing a systematic design method for the controller design, and a novel slow-mode-based control method is proposed for multi-point temperature control systems [25]. In the proposed method, the temperature differences and transient characteristics of all points can be controlled by making the output of the fast modes follow the output of the slow mode. In this paper, the SMBC method combined with decoupling and dead-time compensation is applied to the MIMO temperature control system, in addition to the theoretical analysis of the system being introduced, and simulations and experiments to verify the control efficiency of the proposed SMBC control method are described. The results are then compared to the conventional PI control and gradient temperature control methods.

2. Difficulties in Multi-Point Temperature Control

From the viewpoint of practical application of multi-point temperature control, well-tuned PID controllers with decoupling and dead-time compensations are used. The Ziegler–Nichols ultimate gain method and the constant heating rate method are representative heuristic methods based on experimental data [1,2]. These methods attempt to make the actual temperatures follow the reference values while minimizing the temperature differences between multiple points. However, in many cases, the following difficulties still exist: (1) The PID controller must be designed considering the effect of the dead-time difference of the controlled objects. (2) Although the temperature difference is somewhat decreased due to this consideration, a significant improvement in the transient response cannot be expected. (3) The transient response needs to be improved. (4) When the reference values of multiple points are different, it is difficult to control the individual setting times with the same temperature ratio.

3. Configuration of SMBC System

This section describes the proposed SMBC system. For convenience, the slow mode is defined as the plant system with the largest time constant, while the fast mode is defined as systems with smaller time constants. The block diagram of the proposed SMBC control system is shown in Figure 1. In the figure, r_s is the reference value; y_s is the output temperature of the slow mode; r_{f1} and r_{fn}, which are equal to y_s, are the reference values of the fast modes. In this proposal, the reference value of the fast modes is set to be the output value of the slow mode, and y_{f1} and y_{fn} are the outputs of the fast modes. C_s is defined as the controller of the slow mode, C_{f1} and C_{fn} are indicate as the controllers of the fast modes.

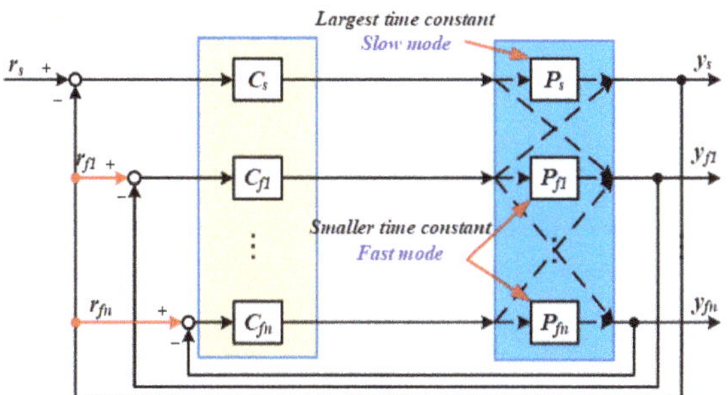

Figure 1. Block diagram of slow-mode-based control (SMBC) control system.

For simplicity, the proposed SMBC control method was applied to a simplified multi-point temperature control system with two inputs and two outputs. The block diagram is shown in Figure 2, where r and y indicate the reference and the output temperatures respectively. The subscripts 1 (Ch1: channel 1) and 2 (Ch2: channel 2) refer to the slow mode and the fast mode, respectively. The steady-state gain, time constant, and dead time of the plant are expressed by K, T, and L respectively. The proposed SMBC structure can be divided into seven blocks.

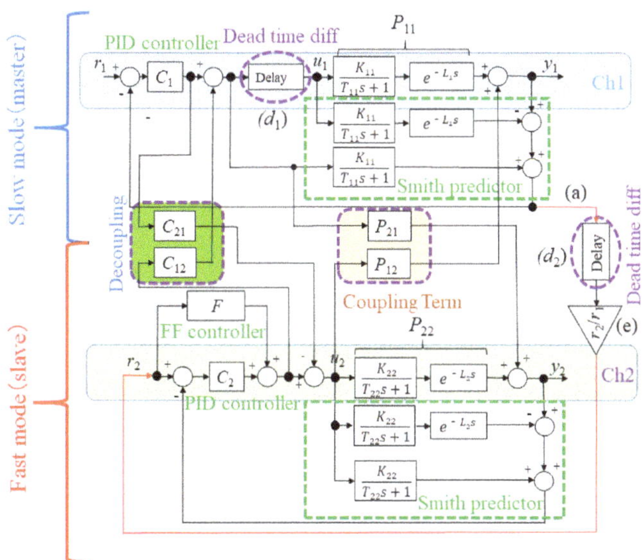

Figure 2. Overview of SMBC structure.

3.1. Structure in Which the Controlled Objects Are under Strong Coupling Effectiveness

Since the controlled plant has been simplified to a two inputs, two outputs system, the two channels are strongly under coupled effectiveness. The diagram of the two channels is shown in

Figure 3, where P_{11} and P_{22} indicate the controlled objects of Ch1 and Ch2, respectively, and P_{12} and P_{21} indicate the respective coupling terms between these two channels.

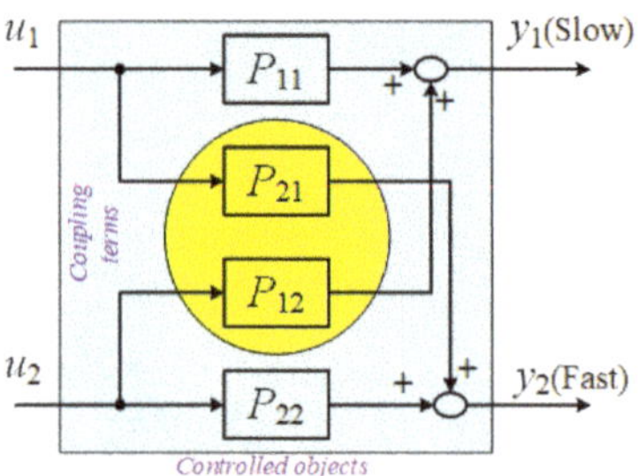

Figure 3. Block diagram of controlled objects.

The transfer function of both the controlled objects and the coupling terms are expressed as a first-order plus time delay system as (1), where the plant parameters K, T, and L have already been defined.

$$P(s) = \frac{K}{Ts+1}e^{-Ls} \tag{1}$$

3.2. Structure in Which Fast-Mode Outputs Follow the Slow-Mode Output

To make the temperature difference between two points close to zero, the output of the slow response mode is used as the reference value for the fast response modes, so that the fast mode outputs follow the slow mode, as is shown in Figure 4.

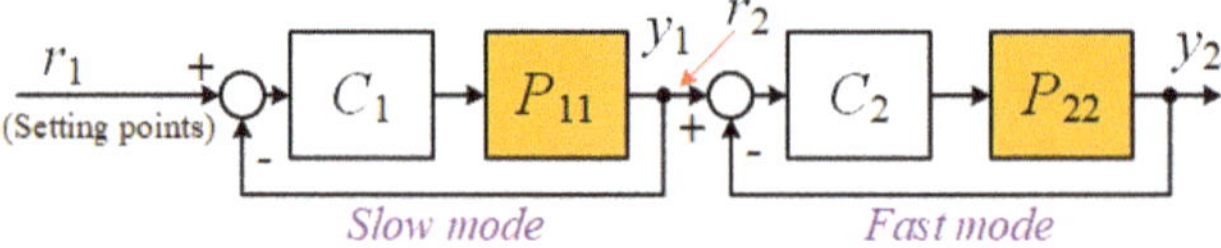

Figure 4. Fast mode output following the slow mode.

3.3. Structure for Obtaining a Fast-Mode Reference Value

In the proposed control structure, the Smith predictor method is introduced for not only dead time compensation but for reference generation, as shown in Figure 5. The model output without dead time, which is indicated by y_{1n}, is used as the reference value r_2 for the fast mode. This structure allows avoidance of a further delay of the fast mode response due to the delay of the slow-mode output. In this proposal, the model based method has been introduced for both Smith predictor and feed-forward compensation, and the model is approximated by coupled SISO systems with time delay.

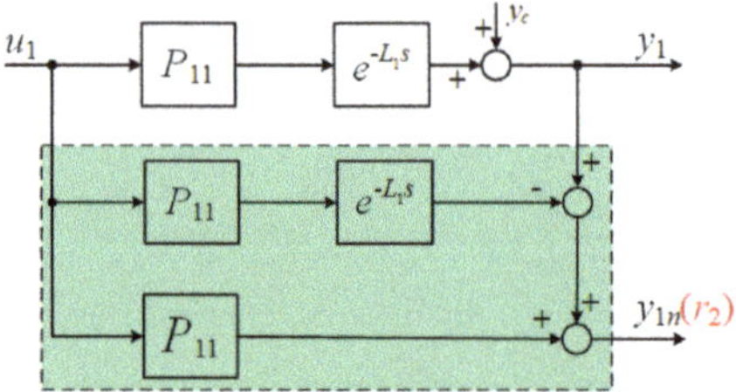

Figure 5. Block diagram of Smith predictor compensation.

3.4. Structure in Which Decoupling Compensation Is Introduced to Compensate the Coupling Effect

The coupling influence affects the temperature of two points; thus, decoupling compensation was introduced into the system to reduce the influence of the coupling term between the two points. The structure of the decoupling compensation is shown in Figure 6, where the C_{21} and C_{12} are the decoupling compensators. The decoupling compensators were designed as (2) and (3), respectively. In this proposal, the decoupling compensators consider the transient response between the two points using only the high-frequency gains.

$$C_{12} = P_{11} P_{12}^{-1} \tag{2}$$

$$C_{21} = P_{21} P_{22}^{-1} \tag{3}$$

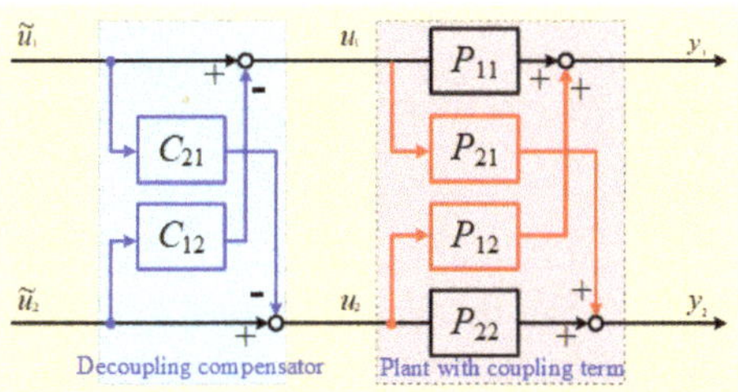

Figure 6. Block diagram of decoupling compensation.

3.5. Structure of Feed-Forward Compensation to Follow Fast-Mode Reference Value

A feed-forward path was added to compensate for the dynamic delay of the fast-mode system, so that the fast-mode output can follow the slow-mode output without delay. This compensation makes the temperature differences between slow mode and fast modes extremely small. Its structure is shown in Figure 7.

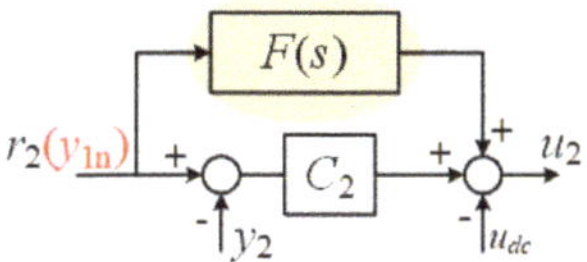

Figure 7. Structure of feed-forward compensation.

3.6. Structure of Compensation for Dead-Time Difference between Fast and Slow Modes

After the compensations described above are made, a temperature difference still remains in the outputs of both modes due to the dead-time difference between the fast and slow modes. To avoid this problem, when the dead time of the slow mode (L_1) is larger than that of the fast mode (L_2), the reference value of the fast mode is delayed by the difference ($L_1 - L_2$), as indicated by (d_1) in Figure 2. This compensation causes both outputs to have the same dead time, that is, L_1. Conversely, when L_1 is smaller than L_2, the delay of the slow mode is included as ($L_2 - L_1$), as indicated by (d_2) in Figure 2, so that both outputs have the same dead time, that is, L_2. This effectively reduces the temperature difference between the two modes. The dead-time difference compensation of both the fast and slow modes is shown in Figure 8.

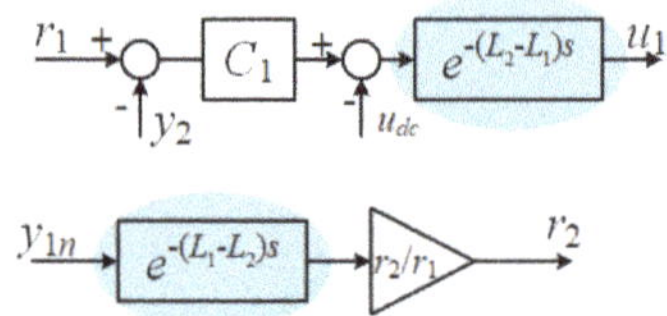

Figure 8. Dead-time difference compensation.

3.7. Structure Enabling Control of the Output Ratio of Two Modes

In the proposed SMBC method, the output ratio can be controlled even when the reference temperatures at multiple points are different. The gain block (r_2/ r_1), as indicated by (e) in Figure 2, which is located after the reference (r_2), corresponds to this part. The fast-mode output follows different references from the slow-mode reference while maintaining a constant ratio. As a result, the setting time and ratio of each output's trajectory can be precisely matched in each mode.

4. Controlled Objects Identification

In this proposal, to precisely verify the control efficiency of the SMBC method, the controlled object was based on a real plant system; thus, a step response experiment needed to be performed to obtain the transfer function of the controlled objects. Figure 9 shows the experimental setup for the multi-point temperature control system with strong coupling effect, which use a DSP as the temperature controller. The system has four coupling channels, and each channel has two independent heaters and one temperature sensor. The temperature can be controlled through controlling the duty ratio of the PWM signal.

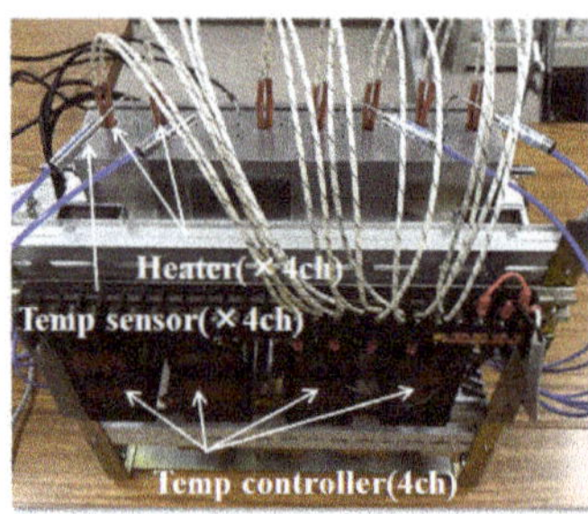

Front view

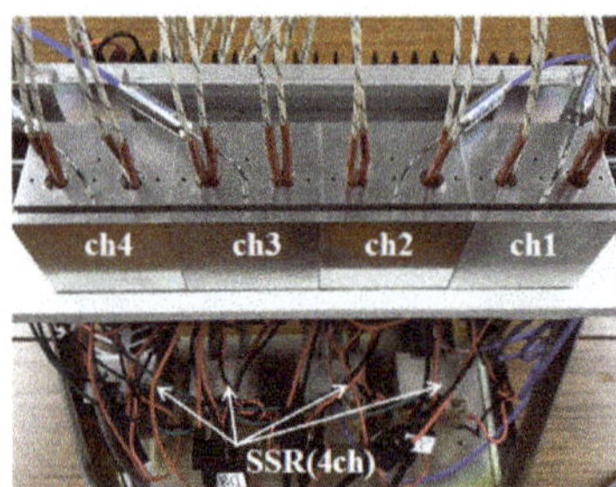

Rear view

Figure 9. Experimental setup.

In this study, the two channels Ch1 and Ch2 were used as the controlled objects to apply the SMBC control method. A step response method was introduced to these two channels for controlled object identification, and the identification results for the paired coupling terms are shown in Figure 10.

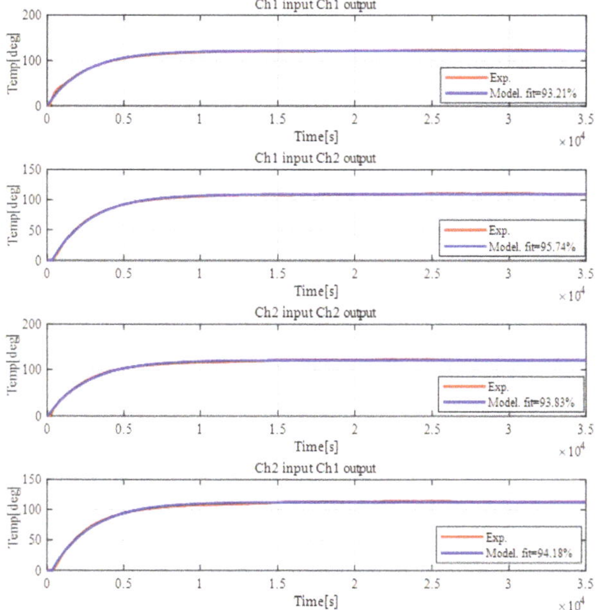

Figure 10. System identification results.

Thus, the identification results can be finalized as (4). The system controller parameters design, system simulation, and experiments are all based on the identified plant transfer functions.

$$P(s) = \begin{bmatrix} P_{11}(s) & P_{12}(s) \\ P_{21}(s) & P_{22}(s) \end{bmatrix} = \begin{bmatrix} \frac{4.34}{1639s+1}e^{-30s} & \frac{1.91}{5848s+1}e^{-125s} \\ \frac{1.67}{3984s+1}e^{-150s} & \frac{4.52}{1211s+1}e^{-25s} \end{bmatrix} \tag{4}$$

5. System Simulation Results

To verify the control efficiency of the proposed control SMBC method, a simulation was carried out in the MATLAB/SIMULINK environment with the controlled objects described above. The results of the simulation focused on the transient response and the temperature difference between the two channels. Control efficiency was verified by comparison with the conventional PI and gradient control methods. In simulation, the controllers C_1 and C_2 were calculated as (5) and (6), respectively. Hence, one of the most important factors is the stability of the controller, and there have already existed several methods for controller stability analysis [26,27], however, as the PID controller in this proposed system is designed based on the Ziegler–Nichols method (step response tuning method), the stability has been ensured [28]. The parameters K_p, T_i, and T_d are decided by the plant parameters K, T, and L which have already been defined.

$$C_1 = \frac{1639s + 1}{178s} \tag{5}$$

$$C_2 = \frac{1211s + 1}{178s} \tag{6}$$

Also, as already mentioned before, a decoupling compensation was added by considering only the high-frequency gain between the two channels, where the decoupling controllers C_{21} and C_{12} are as (7) and (8), respectively.

$$C_{21} = 0.1580 \tag{7}$$

$$C_{12} = 0.0875 \tag{8}$$

The feed-forward compensation of the fast mode reference delay time was designed as (9) and the dead time difference compensation between the slow mode and the fast mode were designed as (10). Hence, in this simulation the reference temperature of the two channels stays the same, so that the temperature difference ratio is set as 1, as shown in (11).

$$C_{FF} = \frac{1211s + 1}{4.52s + 4.52} \tag{9}$$

$$L_2 - L_1 = 5s \tag{10}$$

$$r_2/r_1 = 1 \tag{11}$$

To realize the SMBC control structure, the simulation was divided into two phase. Phase 1 was the SMBC control system with Smith predictor compensation. Phase 2 was the SMBC control system with Smith, feed-forward, and decoupling compensation. A step signal was applied to the slow-mode reference (Ch1). Simulation results for the two phases are shown in Figures 11 and 12, respectively. To evaluate the proposed SMBC control method, a simulation of the conventional PI control method and gradient control method with decoupling compensation were also carried out. Simulation results for the two methods are shown in Figures 13 and 14 respectively.

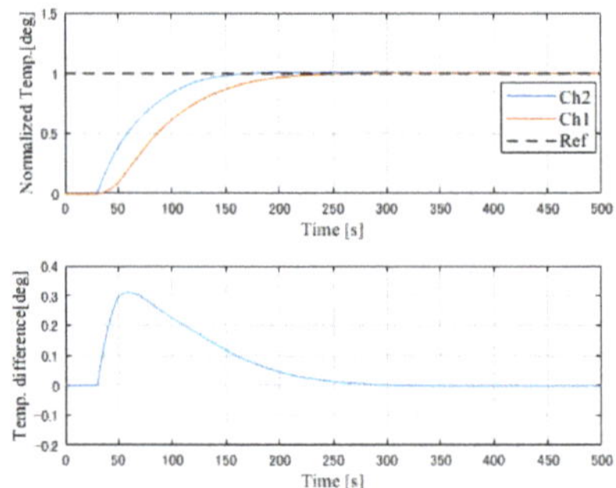

Figure 11. SMBC with Smith predictor compensation.

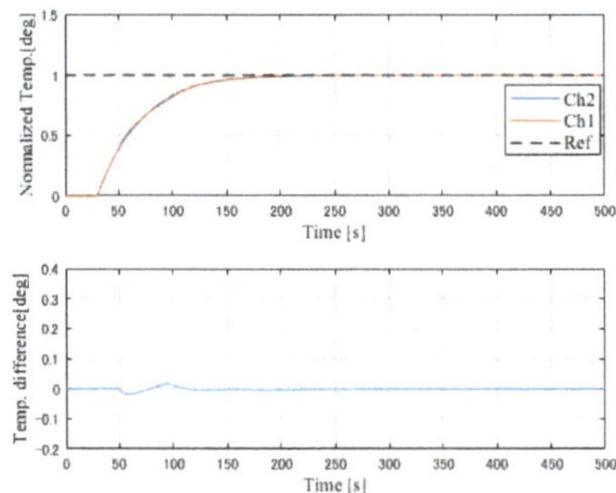

Figure 12. SMBC with Smith, feed-forward, and decoupling compensation.

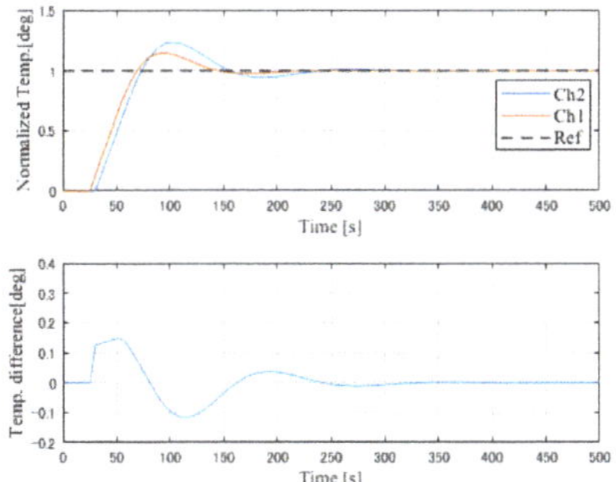

Figure 13. Conventional proportional integral (PI) control system.

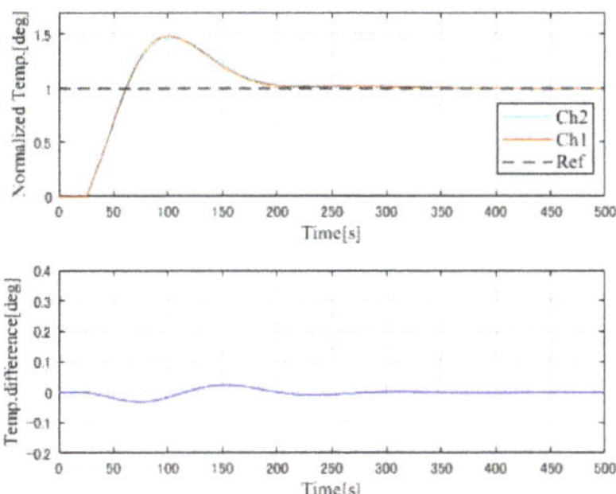

Figure 14. Gradient temperature control system.

The analysis of the simulation results can be devided into two phases. Phase 1: The transient response, the response time of both PI control method and gradient control system is about 60 *s*, similar to the proposed SMBC method, however, the conventional PI control system has an overshoot as 0.3 *deg C* (30% of the reference value) and the gradient control system has an overshoot as 0.5 *deg C* (50% of the reference value). For the SMBC with Smith preditive control system, there is no overshoot but with an asynchronous response between two channels, and it can be compensated by introducing feed-forward and decoupling compensation. Thus, the transient response has been improved. Phase 2: The temperature difference between the controlled two channels, as shown in the simulation results, yielded the maximum temperature difference of the conventional PI control system as 0.18 *deg C* (18% of the reference value) and that of gradient control system is about 0.05 *deg C* (5% of the reference value). However, the temperature difference of SMBC control system is only 0.01 *deg C* (1% of the reference value) and quickly dropped to 0 *deg C*, where we can observe that the temperature difference has been reduced. Thus, the control efficience of the proposed SMBC method has been evaluated.

6. Experimental Results

Experiments with the proposed SMBC control method were carried out using parameter values identical to those used in the simulation, and the experimental setup was the same as that shown in Figure 9. According to the identified plant transfer function, Ch1 was the slow response mode due to its larger time constant and Ch2 was the fast response mode with a smaller time constant. The experiments were carried out by using Ch1 output as the fast mode (Ch2) reference and making the Ch2 output follow the output of Ch1. The experiments were divided into the same two phases used in the simulation. Also, the step signal was applied to the slow-mode reference (Ch1). The experimental results for both phases are shown in Figures 15 and 16, respectively. To evaluate the proposed SMBC control method, experiments with a conventional PI control system were also carried out. The results are shown in Figure 17.

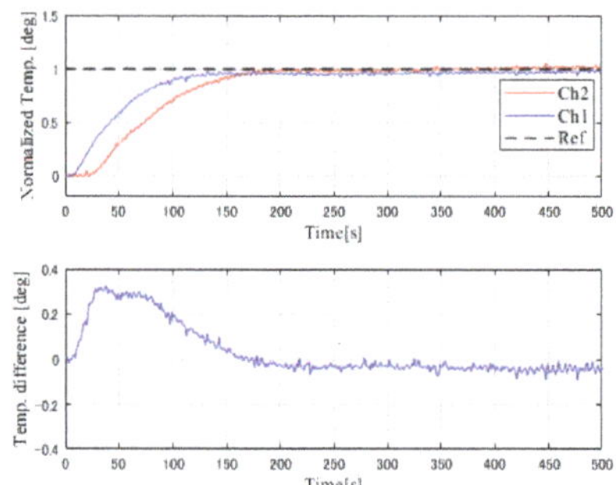

Figure 15. SMBC with Smith predictor compensation.

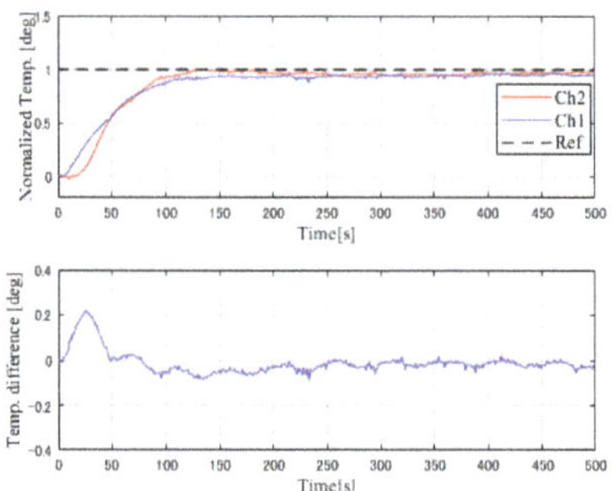

Figure 16. SMBC with Smith, feed-forward, and decoupling compensations.

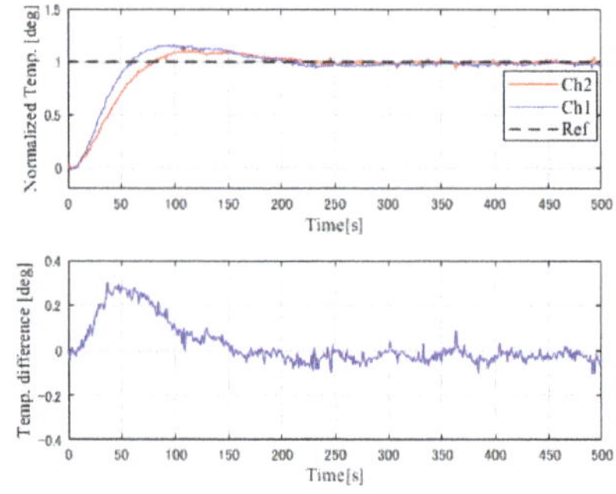

Figure 17. Conventional PI and decoupling control.

The analysis of the experimental results also can be divided into two phases. Phase 1: Transient response, from the results, the transient response of all control systems is simular as about 60 s rising time of each channel. However, the conventional PI control system has an overshoot of 0.3 deg C (30% of the reference value), while the proposed SMBC control system had no overshoot. Although the SMBC with Smith predictor compensation has an asynchronous response between the two channels, it can be improved by introducing feed-forward and decoupling compensation by which we could state that the transient response of both channel has been improved. Phase 2: The temperature difference. The conventional PI control system has a maximum temperature difference of 0.3 *deg C* (30% of the reference value), while that of proposed SMBC with Smith, feed-forward and decoupling compensation system is about 0.21 *deg C* (21% of the reference value) and quickly drops to 0 *deg C*, where the temperature difference has been reduced. As a result, the simulation and experimental results are shown to be similar, indicating a positive evaluation for the proposed SMBC control method.

7. Conclusions

In this paper, a novel SMBC method was proposed for the multi-point temperature control system. The controlled object was defined as a two inputs, two outputs temperature control system. A step-by-step introduction of the proposed control system was investigated. The system identification was carried out to obtain the transfer functions of the controlled objects, and the Ziegler–Nichols (step response) method was applied to the controller design. Simulation of the proposed SMBC control system was carried out in the MATLAB/SIMULINK environment, and the experiments were carried out with a DSP control platform. The effectiveness of the proposed SMBC method was verified by comparing the results to those for a conventional PI control system and gradient temperature system.

8. Patents

This section is not mandatory, but may be added if there are patents resulting from the work reported in this manuscript.

Author Contributions: Conceptualization, S.H. and S.X.; methodology, S.H.; software, S.X. and Y.J.; validation, S.X., S.H., Y.J. and W.J.; formal analysis, S.H. and S.X.; investigation, S.H.; resources, W.J.; data curation, S.X. and Y.J.; writing—original draft preparation, S.X.; writing—review and editing, S.H.; visualization, W.J.; supervision, W.J.; project administration, S.H. K.I., T.K. and R.I.; funding acquisition, S.H., K.I., T.K., R.I. and W.J.

Funding: This research received no external funding.

Acknowledgments: In this section you can acknowledge any support given which is not covered by the author contribution or funding sections. This may include administrative and technical support, or donations in kind (e.g., materials used for experiments).

Conflicts of Interest: The authors declare no conflict of interest.

References

1. Suda, N. *PID Control*; Asakura Publishing: Tokyo, Japan, 1992.
2. Oshima, M. *Process Control Systems*; Corona Publishing: New York, NY, USA, 2003.
3. Goodwin, G.C.; Fraebe, S.F.; Salgado, M.E. *Control System Design*; Prentice Hall: Upper Saddle River, NJ, USA, 2001.
4. Morari, M.; Zafiriou, E. *Robust Process Control*; Prentice Hall: Upper Saddle River, NJ, USA, 1992.
5. Maciejowski, J.M. *Multivariable Feedback Design*; Addison Wesley: Boston, MA, USA, 1989 .
6. Park, C.J. Dynamic Temperature Control with Variable Heat Flux for High Strength Steel. *Int. J. Control Autom. Syst.* **2012**, *10*, 659–665. [CrossRef]
7. Ko, J.S.; Huh, J.H.; Kim, J.C. Improvement of Temperature Control Performance of Thermoelectric Dehumidifier Used Industry 4.0 by the SF-PI Controller. *Processes* **2019**, *7*, 98. [CrossRef]
8. Liu, L.; Tian, S.; Xue, D.Y.; Zhang, T.; Chen, Y.Q.; Zhang, S. A Review of Industrial MIMO Decoupling Control. *Int. J. Control Autom. Syst.* **2019**,*17*, 1246–1254. [CrossRef]
9. Song, B.Q.; Mills, J.K.; Liu, Y.H.; Fan, C. Z. Nonlinear Dynamic Modeling and Control of a Small-Scale Helicopter. *Int. J. Control Autom. Syst.* **2010**, *8*, 534–543. [CrossRef]
10. Deng, M.; Bi, S. Operator-based robust nonlinear control system design for MIMO nonlinear plants with unknown coupling effects. *Int. J. Control* **2010**, *83*, 1939–1946. [CrossRef]
11. Li, G.M.; Tsang, K.M. Concurrent Relay-PID Control for Motor Position Servo Systems. *Int. J. Control Autom. Syst.* **2007**, *5*, 234–242.
12. Wang, Q.G.; Zhang, Y.; Chiu, M.S. Decoupling internal model control for multi-variable systems with multiple time delays. *Chem. Eng. Sci.* **2002**, *57*, 115–124. [CrossRef]

13. Pop, C.I.; Ionescu, C.M.; Keyser, R.D. Time delay compensation for the secondary processes in a multi-variable carbon isotope separation unit. *Chem. Eng. Sci.* **2012**, *80*, 205–218. [CrossRef]

14. Fujimori, A.; Ohara, S. Order Reduction of Plant and Controller in Closed Loop Identification based on Joint Input-Output Approach. *Int. J. Control Autom. Syst.* **2017**, *15*, 1217–1226. [CrossRef]

15. Yao, Y.; Yang, K.; Huang, M.; Wang, L. A state-space model for dynamic response of indoor air temperature and humidity. *Build. Environ.* **2013**, *64*, 26–37. [CrossRef]

16. Seong, Y.B.; Cho, Y.H. Development and evaluation of applicable optimal terminal box control algorithms for energy management control systems. *Sustainability* **2016**, *8*, 1151. [CrossRef]

17. Ganesh, H.S.; Edgar, T.F.; Baldea, M. Model Predictive Control of the Exit Part Temperature for an Austenitization Furnace'. *Processes* **2016**, *4*, 53. [CrossRef]

18. Jeng, J.C.; Chang, Y.J.; Lee, M.W. Novel design of dynamic matrix control with enhanced decoupling control performance. *Comput. Aided Chem. Eng.* **2014**, *44*, 541–546.

19. Nanno, I.; Tanaka, M.; Matsunaga, N.; Kawaji, S. On Performance of the Gradient Temperature Control Method for Uniform Heating. *IEEJ Trans. EIS* **2004**, *124*, 1606–1612. [CrossRef]

20. Matsunaga, N.; Kawaji, S.; Tanaka, M.; Nanno, I. A Novel Approach of Thermal Process Control for Uniform Temperature. In Proceedings of the 16th IFAC World Congress, Prague, Czech Republic, 3–8 July 2005; Volume 38, pp. 111–116.

21. Xu, D.Z.; Song, X.Q.; Jiang, B.; Yang, W.L.; Yan, W.X. Data-driven Sliding Mode Control for MIMO systems and Its Application on Linear Induction Motors. *Int. J. Control Autom. Syst.* **2019**, *17*, 1717–1725. [CrossRef]

22. Data-driven approaches for complex industrial systems. *IEEE Trans. Ind. Inform.* **2013**, *9*, 2210–2212. [CrossRef]

23. Gao, Z.; Nguang, S.K.; Kong, D.X. Advances in Modelling, Monitoring, and Control for Complex Industrial Systems. *Complexity* **2019**. [CrossRef]

24. Xu, S.; Hashimoto, S.; Jiang, W. Pole-Zero Cancellation Method for Multi Input Multi Output (MIMO) Temperature Control in Heating Process System. *Processes* **2019**, *7*, 497. [CrossRef]

25. Hashimoto, S.; Liu, Y.; Yoshida, K.; Izaki, K.; Kihara, T.; Ikeda, R.; Jiang, W. A Novel Multi-point Temperature Control Method Based on Slow Response Mode. In Proceedings of the IECON 2017—43rd Annual Conference of the IEEE Industrial Electronics Society, Beijing, China, 29 October–1 November 2017; pp. 5623–5627.

26. Gao, R.; Gao, Z. Pitch control for wind turbine systems using optimization, estimation and compensation. *Renew. Energy* **2016**, *91*, 501–515. [CrossRef]

27. Wang, J.; Tse, N.; Gao, Z. Synthesis on PI-based pitch controller of large wind turbines generator. *Energy Convers. Manag.* **2011**, *52*, 1288–1294. [CrossRef]

28. Ziegler, J.G.; Nichols, N.B. Optimum settings for automatic controllers. *Trans. ASME* **1942**, *64*, 759–768. [CrossRef]

Article

Pole-Zero Cancellation Method for Multi Input Multi Output (MIMO) Temperature Control in Heating Process System

Song Xu [1,2,†], Seiji Hashimoto [1,2,*,†] and Wei Jiang [2]

[1]   Division of Electronics and Informatics, Gunma University, Kiryu 3768515, Japan
[2]   Department of Electrical Engineering, Yangzhou University, N.196 Huayang West Road, Yangzhou 225000, China
*   Correspondence: hashimotos@gunma-u.ac.jp; Tel.: +81-0277-30-1741
†   These authors contributed equally to this work.

Received: 3 July 2019; Accepted: 23 July 2019; Published: 1 August 2019

Abstract: With the rapid development of industrial technology, the multi-point (multi-input multi-output) heating processing systems with integrated temperature control have been increasingly needed to achieve high-quality and high-performance processing. In this paper, in response to the demand for proper transient response and to provide more accurate temperature controls, a novel pole-zero cancelation method is proposed for multi-input multi-output (MIMO) temperature control in heating process systems. In the proposed method, the temperature differences and transient characteristics of all points can be improved by compensating dead time difference and coupling effect together by matrix compensation and pole-zero cancelation with the feedforward reference model. Both simulations and experiments were carried out. The results were compared to the well-tuned conventional PI control system and PI plus decoupling compensation system to evaluate the control efficiency of the proposed method.

Keywords: MIMO temperature control in heating process system; pole-zero cancelation; temperature difference; transient response; dead time

---

1. Introduction

Recently, industrial processes, such as thermal processes, manufacturing processes, producing processes and so on, are becoming more and more important, and have higher requirement for the operation performance. A thermal process system, as one of the most complex processes, has a wide range of applications in the industrial field, especially in the food processes. In the thermal processes, especially the multi-point (multi-input multi-output) thermal processing systems, are playing a more and more important role in the industrial application fields. Thus, the demands for the multi-point system to achieve high-quality and high-performance processing has been severely raised. As a result, a lot of control methods have been proposed and introduced into the multi-point temperature control systems. Among the various thermal processing techniques, the conventional proportional–integral–derivative (PID) control method has become the most commonly used control method because of its simplicity, efficiency, and wide applicability. However, due to the nonlinearity and large time delay of the control objects, the performance of the only PID control system may not satisfy the expected requirements [1–5]. So the series connected fuzzy-proportional integral (SF-PI) control method has been proposed to improve the control performance [6]. In addition, to achieve more precise control, the mathematical model of the control objects needs to be obtained. Presently, a lot of system identification methods have been introduced to build the mathematical model of the multi-point objects. One of the most popular methods is the step response method [7–10]. With the mathematical model of the system, the Model Predictive Control (MPC) method has been proposed to the heating system to provide precise control [11].

Although with the mathematical model, the controller can be well designed and the system can be precisely controlled, the coupling influence inside the multi-point system still has a significant impact on the transient response of each point. As a result, a novel decoupling compensation or quasi-decoupling method has been investigated into the multi-point system through building an equivalent model of each point [12–16].

Moreover, even after successfully getting the precise system model and introducing the well-designed controllers with delay time compensation and decoupling compensation, there still are some difficulties that will be caused by the worsening conditions such as dead time difference, disturbance, model perturbation and so on. These difficulties will have a great impact on the transient response and the steady-state stability of the controlled system. Thus more and more advanced compensation methods have been proposed to compensate the influence of each point and uniform the output of multi-point systems such as feedforward compensation, the data-driving method and the gradient control method [17–22].

Furthermore, for the excessively complex thermal process system, the mathematic model cannot be obtained precisely, thus, the data-driven approaches have been proposed as a potential method for controlling that system [23], and some optimization methods have also been proposed for these excessively complex processes [24].

Although after all the above methods been introduced into the temperature control system in the thermal process, there still are some problems with the transient response and the closely controlled temperature of the multi-point temperature control system with dead-time difference and strong coupling effects. For example, in MIMO heating systems, the temperature difference between each point should be under 5% of the reference temperature no matter whether transient response is fast or slow. Until now the above method mentioned still cannot achieve the effect we expected in multi-point temperature control systems that have a large time constant and big dead-time difference.

This proposal, focusing on the multi-point temperature control system which has the large time constant and big dead-time difference, to improve the transient response of each channel and to reduce the temperature difference between each point, also to provide an auto design method for the controller, a novel pole-zero cancelation method has been proposed for MIMO temperature control system. In the proposed method, the temperature differences and the transient response of each point can be controlled by considering the dead time and the coupling effect of the MIMO system. The rest of the paper is organized as follows: Section 2 describes the difficulties of in the MIMO temperature control. In Section 3, the step by step design of the system configuration will be introduced. Section 4 shows the simulation results of the proposed control method and the experimental results will be present in Section 5, both in simulation and experiments, the results are compared to the well-tuned conventional PI control system and PI plus decoupling control system. Finally, a simple conclusion is made in Section 7.

2. Difficulties in MIMO Temperature Control

From the viewpoint of the practical application of multi-point temperatures control, fine-tuned PID controllers with decoupling and dead time compensations are used. The Ziegler-Nichols ultimate gain method and the constant heating rate method are representative heuristic methods based on experimental data (1), (2). These methods attempt to make the actual temperatures follow the reference values while minimizing the temperature differences between multiple points. However, in many cases, the following difficulties still exist: (1) The PID controller must be designed considering the effect of the dead-time of the controlled objects. (2) Although the temperature difference is somewhat decreased due to this consideration, a significant improvement in the transient response cannot be expected. (3) Improving the transient response causes the disturbance response for deteriorating.

(4) Finally, when the reference values of multiple points are different, it is difficult to control the individual settling times with the same temperature ratio.

3. Configuration of MIMO Control System

This section describes the configuration of the proposed multi-input multi-output (MIMO) control system, with a pole-zero cancelation method. Figure 1 shows the block diagram of the proposed MIMO control system. For simplicity, the controlled object is defined as two-inputs two-outputs temperature system. In the figure, r_1 and r_2 indicate the reference of the MIMO system while y_1 and y_2 are the output temperature, respectively. The control system configuration can be divided into five parts: Part 1 describes the simplified controlled objects; Part 2 will somewhat compensate the coupling effect and dead time difference of the two channels; Part 3 will introduce the pole-zero cancellation with reference model to convert the complex system to almost equal to the reference model; Part 4 is the main controller of the system; Part 5 indicates the anti-wind-up compensation that will not only compensate for the saturation of the control input but also ensure the uniformity of the temperature by setting difference maximum saturation.

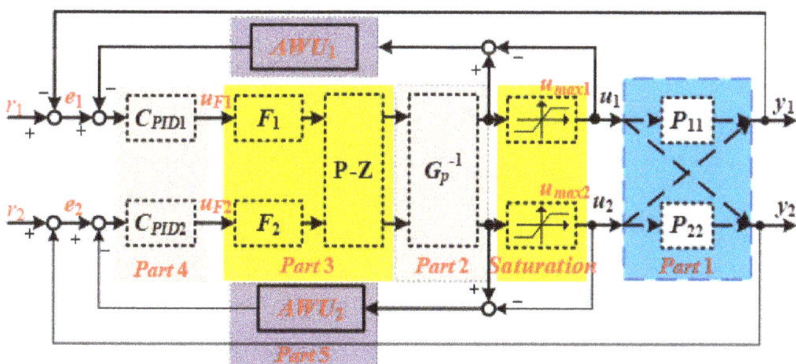

Figure 1. Block diagram of pole-zero cancellation for two-input two-output system.

3.1. Part 1: MIMO System Controlled Object With Strong Coupling Effectiveness

The control system is designed based on multi-input multi-output(MIMO) temperature system with strong coupling influence, the schematic block diagram of the coupled system is shown in Figure 2, where u_1 and u_2 are defined as the inputs of Ch1 and Ch2, respectively. In addition, y_1 and y_2 indicate the output of Ch1 and Ch2, respectively. The coupling terms between the two channels are obtained as P_{21} and P_{12}, respectively.

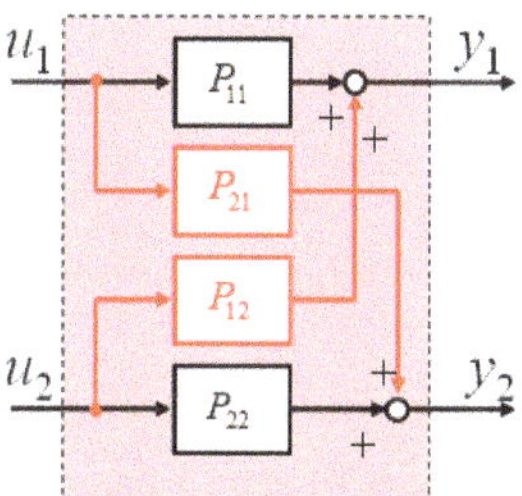

Figure 2. Block diagram of coupled system.

3.2. Part 2: Compensation for Dead Time Difference and Decoupling

In this paper, the controlled objects can be defined as a first-order plus time delay(FOPTD) system, shown as (1), and can be approximated to (2) based on Pade approximation method, which leads easiness for dead-time compensation. Considering the characteristic of the transient response of the temperature system, there is a difference in the delay time d between the multipoint temperature outputs. As a result, a temperature difference remains in the outputs of different channels.

$$P(s) = \frac{K}{Ts+1}e^{-ds} \tag{1}$$

$$P(s) \approx \frac{K}{Ts+1} \bullet \frac{1}{ds+1} \tag{2}$$

Same as the delay time difference, the coupling influence also affects the temperature of both channels, compensation of the coupling term needs to be introduced, the block diagram of decoupling control and the compensated system are shown in Figure 3.

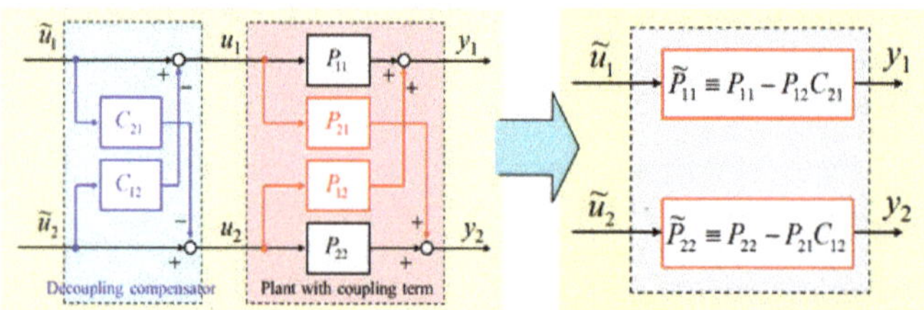

Figure 3. Block diagram of decoupling compensation and compensated system.

In this proposal, a matrix gain compensation method has been introduced to compensate delay time difference and coupling effects together. Taking G_p as the system characteristic matrix, including the coupling and delay time gain of the system. The compensation of the G_p based on the inverse of the matrix G_p^{-1}, the matrix can be obtained by giving the step signal to the two channels one by one after natural cooling. The measurement method of the characteristic G_p for the MIMO-controlled object is shown in Figure 4.

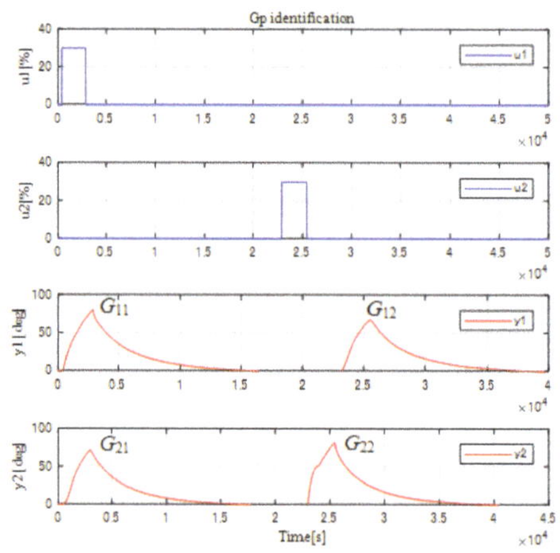

Figure 4. System matrix gain G_p identification.

The matrix gain G_p can be calculated as (3).

$$G_p = \begin{bmatrix} G_{11} & G_{12} \\ G_{21} & G_{22} \end{bmatrix} \tag{3}$$

The delay time difference and coupling compensator G_c can be obtained as (3) by the inverse of the G_p.

$$G_c = G_p^{-1} \tag{4}$$

For this compensation gain G_c, it can somewhat compensate delay time difference and coupling influence together, therefore, the temperature difference can be reduced.

3.3. Part 3: Plor-Zero Cancelation With Feedforward Reference Model

As introduced above, after the matrix compensation, the MIMO-controlled object can be treated as non-interference system, therefore the feedforward reference model with pole-zero cancellation is designed for this system. The block diagram and the simplified system are shown in Figure 5, where the F_1 and F_2 are reference models which can provide an expected response reference for the controlled systems, by using this, after the pole-zero cancellation, the system can be equaled as the reference model. In addition, u_{F1} and u_{F2} are the inputs of two channels, respectively, while y_1 and y_2 indicate the outputs of Ch1 and Ch2 respectively. The F_1 and F_2 can be designed as (5) and (6), respectively, and the order of the controllers is decided by the inverse of the plant transfer function (in this paper 1st order). In this proposal, to make it easy for the design of the controllers, F_1 and F_2 are designed as $F_1 = F_2$, thus, $w_1 = w_2$.

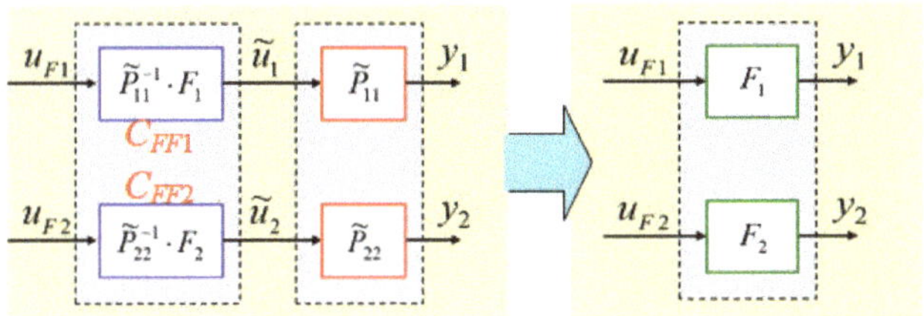

Figure 5. Block diagram of feedforward reference model with pole-zero cancellation and simplified system.

$$F_1 = \frac{\omega_1}{s + \omega_1} \tag{5}$$

$$F_2 = \frac{\omega_2}{s + \omega_2} \tag{6}$$

3.4. Part 4: PID Controller Design

In part 3, defining the equivalent plant as $F_1 = F_2$, the same PID controllers can be designed ($C_{PID1} = C_{PID2}$). Hence, one of the most important factors is the stability of the controller, and there have already existed several methods for controller stability analysis [25,26], however, the PID controller in this proposed system is designed based on the Ziegler-Nichols method (step response tuning method), the stability has been ensured [27]. The PID control system diagram is shown in Figure 6.

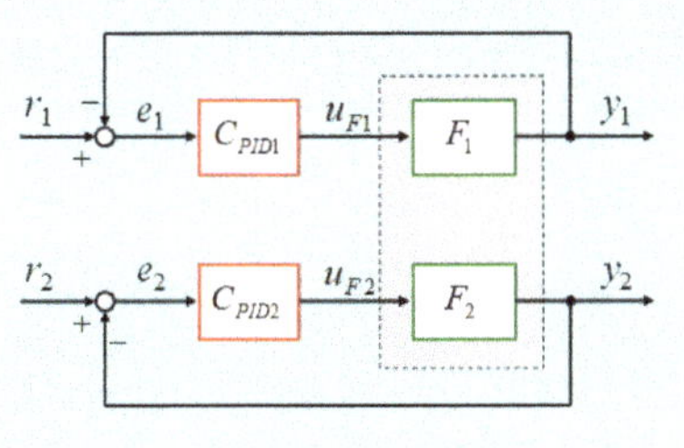

Figure 6. Block diagram of PID control system.

3.5. Part 5: Anti-Wind-Up Compensation for Control Input Saturation

In system design, the control inputs u_1 and u_2 are under a saturation from 0 to u_{max}, thus, to improve the effectiveness of the controller, the anti-wind-up compensation has been added to the control loop. The structure of the anti-wind-up compensation is shown in Figure 7. In addition, due to the difference transient response characteristics of the two channels, the relationship between these two channels is as (7), where u_{max1} and u_{max2} indicate the maximum saturation of Ch1 and Ch2, respectively, K_{11} and K_{22} indicates the steady-state gain of Ch1 and Ch2, respectively. By setting this, it can somewhat make the two channels' response be precisely the same even the controllers are working under the saturation.

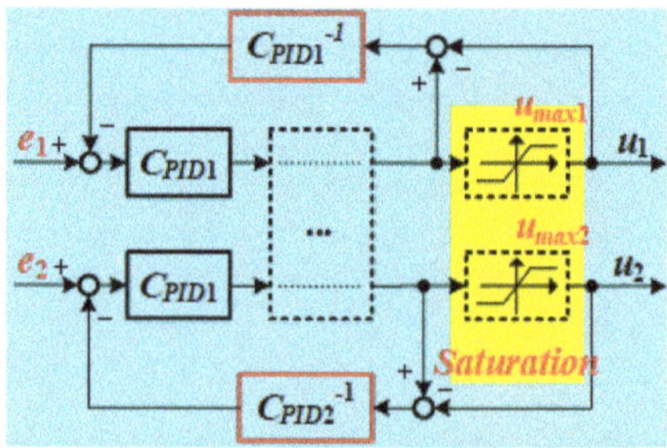

Figure 7. Structure of anti-wind-up compensation.

$$u_{max1} = \frac{K_{11}}{K_{22}} * u_{max2} \tag{7}$$

4. System Simulation

In the simulation, the control object $P(s)$ of the MIMO temperature system is defined as a two-input and two-output vectors as (8), which is identified by the step response method.

$$P(s) = \begin{bmatrix} \dfrac{4.0992}{2627s+1}e^{-200s} & \dfrac{3.7036}{2536s+1}e^{-326s} \\ \dfrac{2.2805}{2580s+1}e^{-324s} & \dfrac{4.1268}{2461s+1}e^{-150s} \end{bmatrix} \tag{8}$$

Each factor is represented by FOPTD, in which the parameters are calculated by the step response system identification method, the factor(1,1) can be defined as Ch1 while the factor(2,2) can be defined as Ch2. Figure 8 shows the setpoint tracking results for the control inputs and temperature outputs for the system obtained using the conventional PI control method.

Figure 9 shows the temperature difference, $(y_1 - y_2)$, and the mean temperature, $(y_1 + y_2)/2$. As is shown in Figure 9, the overshoot of the mean temperature reaches 45%, and the maximum temperature difference becomes 23 °C.

On the other hand, the set point tracking results of the proposed pole-zero cancellation method are shown in Figure 10, and Figure 11 shows the mean temperature and temperature difference, where there is no overshoot of the mean temperature, the maximum temperature difference is also controlled to 8 °C.

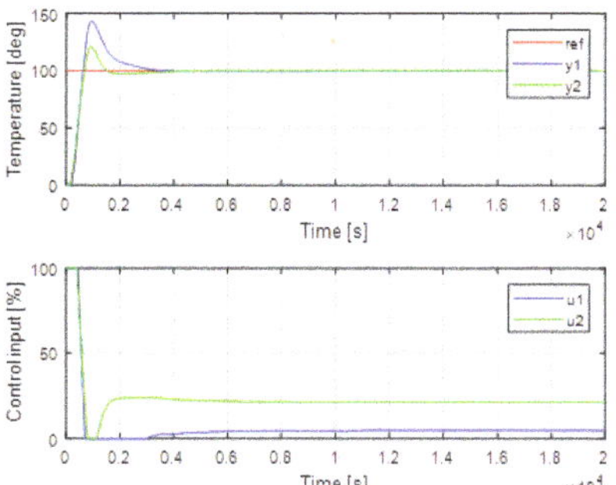

Figure 8. Tracking results for conventional PI control.

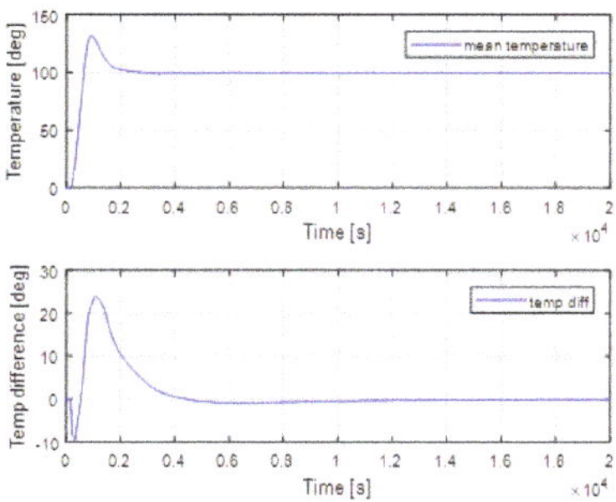

Figure 9. Mean temperature and temperature difference results for conventional PI control.

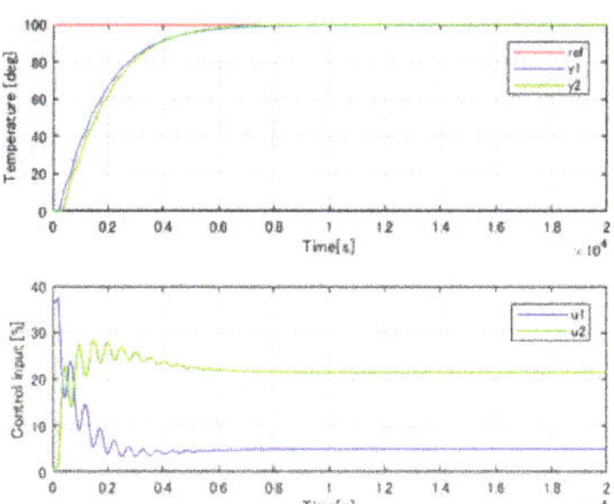

Figure 10. Tracking results for proposed pole-zero cancellation method.

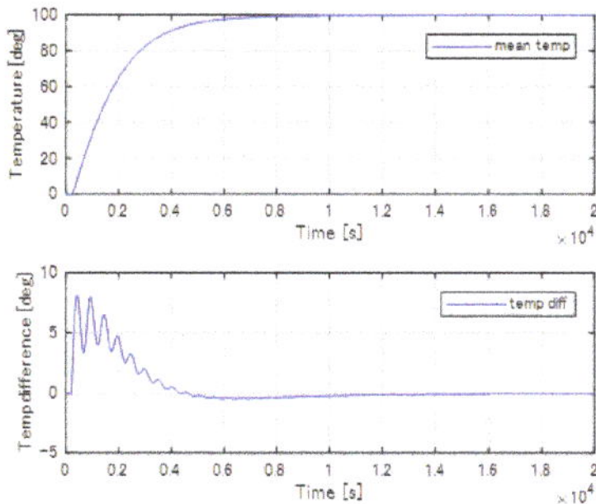

Figure 11. Mean temperature and temperature difference results for proposed pole-zero cancellation method.

5. Experimental Results

Experiments with the proposed pole-zero cancelation control method were carried out using parameter values identical to those used in the simulation. Figure 12 shows the experimental setup for the MIMO temperature control system with strong coupling effect.

The experimental setup equips DSP as the temperature controller. The system has four coupling channels, each channel has two independent heaters and one temperature sensor. The temperature sensor can transfer temperature 0–400 °C to 0–10 VDC output voltages. In addition, the heaters are driven by PWM signals. The temperature can be controlled by controlling the duty ratio of the PWM signals.

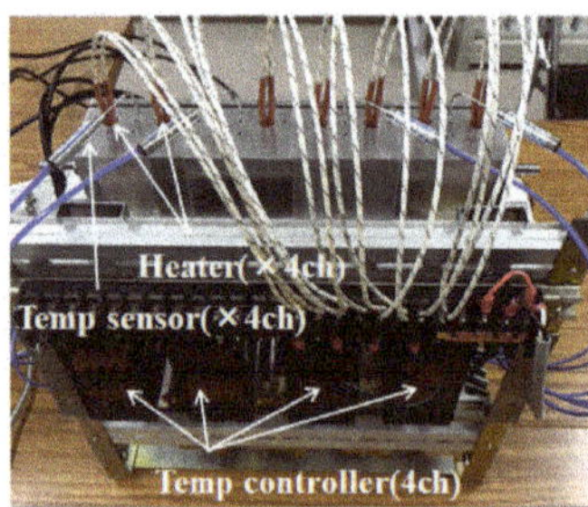

Front view

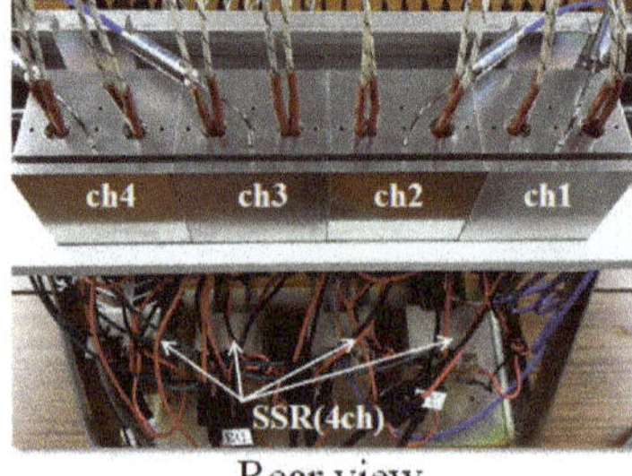

Rear view

Figure 12. Experimental setup.

In this proposal, the two channels Ch1 and Ch2 are used as the control objects to apply the proposed pole-zero cancelation control method. Just as in the simulation, the control objects can be identified as (9).

$$P(s) = \begin{bmatrix} \dfrac{4.0992}{2627s+1}e^{-200s} & \dfrac{3.7036}{2536s+1}e^{-326s} \\ \dfrac{2.2805}{2580s+1}e^{-324s} & \dfrac{4.1268}{2461s+1}e^{-150s} \end{bmatrix} \tag{9}$$

According to the identified plant model, the two channels are under strong coupling effects, and with 100 s delay time difference between them.

The matrix compensator G_c is designed by only considering the steady state gain of each part as (10).

$$G_c = \begin{bmatrix} 0.0979 & -0.0901 \\ -0.0901 & 0.0970 \end{bmatrix} \tag{10}$$

After the matrix compensation, the feedforward reference models F_1 and F_2 can be designed as (11). As introduced previously, the feedforward reference models F_1 and F_2 are designed to provide an expected reference response. In this proposal, to simplify the controlled system, the reference models are designed as $F_1 = F_2$.

$$F_1 = F_2 = \frac{0.2}{s+0.2} \tag{11}$$

Thus, the controllers of these two channels are designed by Ziegler-Nichols method (step response method), $C_{PID1} = C_{PID2}$ as (12)

$$C_{PID1} = C_{PID2} = \frac{0.04s+1}{25s} \tag{12}$$

In the experiments, to verify the control efficiency of the proposed method, the results were compared to well-tuned conventional PI control system and conventional PI plus decoupling compensation system. The experiments were carried out by controlling the temperature of the two channels from room temperature (24 °C) to 100 °C. The results of the conventional PI control are shown in Figure 13, while the results of the conventional PI plus decoupling compensation are shown in Figure 14, the experiment results of the proposed pole-zero cancelation method are shown in Figure 15. In addition, the results of the compared mean temperature and the temperature difference between the two channels are shown in Figure 16.

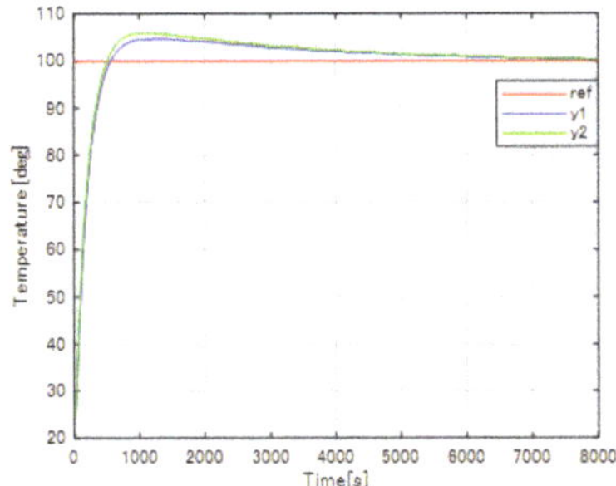

Figure 13. Experimental results of conventional PI control.

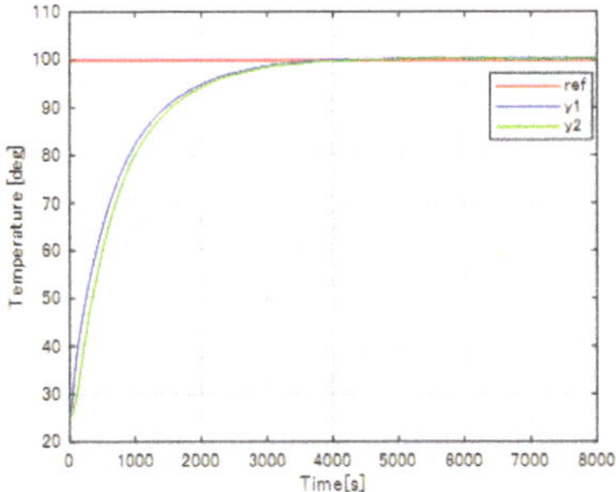

Figure 14. Experimental results of conventional PI plus decoupling compensation.

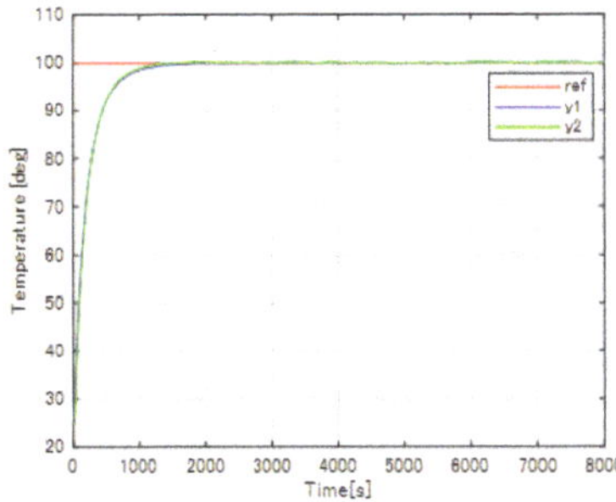

Figure 15. Experimental results of proposed pole-zero cancelation control.

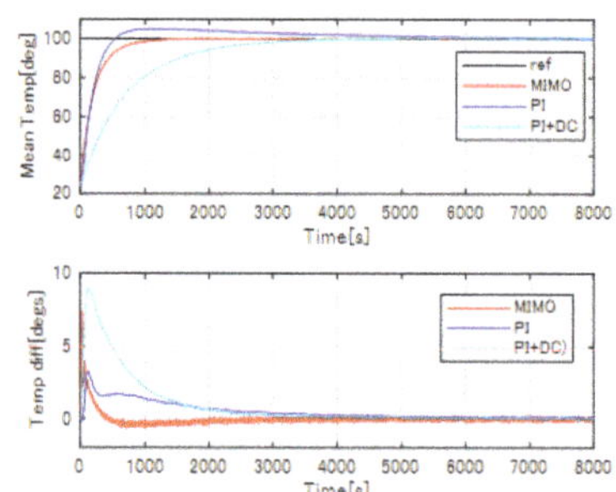

Figure 16. Compared results of mean temperature and temperature difference.

From the experimental results, the well-tuned conventional PI control system has the fastest transient response, while the conventional PI plus decoupling compensation system has the slowest response. However, the conventional PI control has an overshoot of 8 °C (8% of the reference value), and the conventional PI plus decoupling compensation has the biggest maximum temperature difference as 9 °C (9% of the reference temperature). By introducing the proposed pole-zero cancelation method, the overshoot and the maximum temperature difference can be improved. As shown in the figures above, the transient response of the proposed pole-zero cancelation control system is as fast as the well-tuned conventional PI control system about 250 s rising time while the PI plus decoupling compensation has almost 1000 s rising time. In addition, the proposed method has no overshoot, the maximum temperature difference has been reduced to 4 °C (4% of the reference value), and quickly drops to 0 °C, the temperature uniformity has been realized.

6. Discussion

In this proposal, compared to the conventional PI and PI plus decoupling compensation, the advantages of the proposed pole-zero cancellation method with reference model can be divided into two phases. Phase 1: The transient response has been improved to about 250 s rising time as fast as the well-tuned PI control, almost 750 s shortened compared to the PI plus decoupling compensation, and no overshoot compared to the conventional PI control (8% overshoot); Phase 2: The maximum temperature difference between these two channels has been minified almost half that of PI plus decoupling compensation, and the temperature difference quickly goes down to zero compared to the conventional PI control, temperature uniformity has been realized. However, if the system has plant perturbation, the proposed method may not work as efficiently as expected.

7. Conclusions

In this paper, a novel pole-zero cancelation method was proposed for the MIMO temperature control in heating process system. The main purposes of the proposed method are to ensure proper transient responses and to provide more closely controlled temperatures. In the proposed method, the temperature differences and transient characteristics of all points were controlled by considering the delay time difference and the coupling term together with matrix gain compensation G_c and investigating the pole-zero cancellation with feedforward LPFs to the control loop. After pole-zero cancelation compensation, the controllers were designed by the Ziegler-Nichols method. The simulations of the proposed control system were carried out in the MATLAB/SIMULINK environment and the experiments were done based on the DSP controlled system platform. The effectiveness of the proposed control method was evaluated by comparing the results to those for a well-tuned conventional PI control system and PI plus decoupling compensation system.

8. Patents

This section is not mandatory, but may be added if there are patents resulting from the work reported in this manuscript.

Author Contributions: Conceptualization, S.H. and S.X.; methodology, S.H.; software, S.X.; validation, S.X., S.H. and W.J.; formal analysis, S.H. and S.X.; investigation, S.H.; resources, W.J.; data curation, S.X. and W.J.; writing—original draft preparation, S.X.; writing—review and editing, S.H.; visualization, W.J.; supervision, W.J.; project administration, S.H.; funding acquisition, S.H. and W.J.

Funding: This research received no external funding.

Conflicts of Interest: The authors declare no conflict of interest.

References

1. Suda, N. *PID Control*; Asakura Publishing: Tokyo, Japan, 1992.
2. Oshima, M. *Process Control Systems*; Corona Publishing: Tokyo, Japan, 2003.
3. Goodwin, G.C.; Fraebe, S.F.; Salgado, M.E. *Control System Design*; Prentice Hall: Upper Saddle River, NJ, USA, 2001.
4. Morari, M.; Zafiriou, E. *Robust Process Control*; Prentice Hall: Upper Saddle River, NJ, USA, 1992.
5. Maciejowski, J.M. *Multivariable Feedback Design*; Addison Wesley: Boston, MA, USA, 1989.
6. Ko, J.-S.; Huh, J.-H.; Kim, J.-C. Improvement of Temperature Control Performance of Thermoelectric Dehumidifier Used Industry 4.0 by the SF-PI Controller. *Processes* **2019**, *7*, 98. [CrossRef]
7. Ganesh, H.S.; Edgar, T.F.; Baldea, M. Model Predictive Control of the Exit Part Temperature for an Austenitization Furnace. *Processes* **2016**, *4*, 53. [CrossRef]
8. Garrido, J.; Lara, M.; Ruz, M.L.; Vazquez, F.; Alfaya, J.A.; Morilla, F. Decentralized PID control with inverted decoupling and superheating reference generation for efficient operation: Application to the Benchmark PID 2018. *IFAC-PapersOnLine* **2018**, *51*, 710–715 . [CrossRef]
9. Gilbert, A.F.; Yousef, A.; Natarajan, K.; Deighton, S. Tuning of PI controllers withone-way decoupling in MIMO systems based on finite frequency response data. *J. Process Control* **2003**, *13*, 553–567. [CrossRef]
10. Waller, M.; Waller, J.B.; Waller, K.V. Decoupling revised. *Ind. Eng. Chem. Res.* **2003**, *42*, 4575–4577. [CrossRef]
11. Deng, M.; Bi, S. Operator-based robust nonlinear control system design for MIMO nonlinear plants with unknown coupling effects. *Int. J. Control* **2010**, *83*, 1939–1946. [CrossRef]
12. Wang, Q.G.; Zou, B.; Zhang, Y. Decoupling Smith delay compensator design formultivariable systems with multiple time delays. *Chem. Eng. Res. Des.* **2000**, *78 Pt. A*, 565–572. [CrossRef]
13. Wang, Q.G.; Zhang, Y.; Chiu, M.S. Decoupling internal model control for multi-variable systems with multiple time delays. *Chem. Eng. Sci.* **2002**, *57*, 115–124. [CrossRef]
14. Pop, C.I.; Ionescu, C.M.; Keyser, R.D. Time delay compensation for the secondaryprocesses in a multivariable carbon isotope separation unit. *Chem. Eng. Sci.* **2012**, *80*, 205–218. [CrossRef]
15. Coakley, D.; Raftery, P.; Keane, M. A review of methods to match building energy simulation models to measured data. *Renew. Sustain. Energy Rev.* **2014**, *37*, 123–141. [CrossRef]
16. Yao, Y.; Yang, K.; Huang, M.; Wang, L. A state-space model for dynamic response of indoor air temperature and humidity. *Build. Environ.* **2013**, *64*, 26–37. [CrossRef]
17. Seong, Y.B.; Cho, Y.H. Development and evaluation of applicable optimal terminal box control algorithms for energy management control systems. *Sustainability* **2016**, *8*, 1151. [CrossRef]
18. Jeng, J.C.; Chang, Y.J.; Lee, M.W. Novel design of dynamic matrix control with enhanced decoupling control performance. *Comput. Aided Chem. Eng.* **2014**, *44*, 541–546.
19. Nanno, I.; Tanaka, M.; Matsunaga, N.; Kawaji, S. On Performance of the Gradient Temperature Control Method for Uniform Heating. *IEEJ Trans. EIS* **2004**, *124*, 1606–1612. [CrossRef]
20. Matsunaga, N.; Kawaji, S.; Tanaka, M.; Nanno, I. A Novel Approach of Thermal Process Control for Uniform Temperature. *IFAC Proc. Vol.* **2005**, *38*, 111–116. [CrossRef]
21. Tomaru, T.; Mori, Y. Design Method of Minimum-Phase State Decoupling Control with Feedforward Compensation. In Proceedings of the SICE Annual Conference 2007, Takamatsu, Japan, 17–20 September 2007; IEEE: Takamatsu, Japan, 2007; 1C07-1.

22. Ishii, Y.; Masuda, S. Data-driven Update of The Free Parameter of The Youla-Kucera Parametrization in Disturbance Attenuation FRIT based on Variance Evaluation. In Proceedings of the 2016 International Conference on Advanced Mechatronic Systems (ICAMechS), Melbourne, VIC, Australia, 30 November–3 December 2016; IEEE: Melbourne, Australia, 2016; pp. 11–16.
23. Data-driven approaches for complex industrial systems. *IEEE Trans. Ind. Inform.* **2013**, *9*, 2210–2212. [CrossRef]
24. Gao, Z.; Nguang, S.K.; Kong, D.X. Advances in Modelling, Monitoring, and Control for Complex Industrial Systems. *Complexity* **2019**. [CrossRef]
25. Gao, R.; Gao, Z. Pitch control for wind turbine systems using optimization, estimation and compensation. *Renew. Energy* **2016**, *91*, 501–515. [CrossRef]
26. Wang, J.; Tse, N.; Gao, Z. Synthesis on PI-based pitch controller of large wind turbines generator. *Energy Convers. Manag.* **2011**, *52*, 1288–1294. [CrossRef]
27. Ziegler, J.G.; Nichols, N.B. Optimumsettingsforautomaticcontrollers. *Trans. ASME* **1942**, *64*, 759–768.

Article

Optimal Nonlinear Adaptive Control for Voltage Source Converters via Memetic Salp Swarm Algorithm: Design and Hardware Implementation

Yueping Jiang [1], Xue Jin [1], Hui Wang [1], Yihao Fu [1], Weiliang Ge [1], Bo Yang [2,*] and Tao Yu [3]

[1] NARI Technology Co. Ltd., Nanjing 211106, China
[2] Faculty of Electric Power Engineering, Kunming University of Science and Technology, Kunming 650500, China
[3] College of Electric Power, South China University of Technology, Guangzhou 510640, China
* Correspondence: yangbo_ac@outlook.com; Tel.: +86-183-1459-6103

Received: 9 July 2019; Accepted: 25 July 2019; Published: 1 August 2019

Abstract: Voltage source converter (VSC) has been extensively applied in renewable energy systems which can rapidly regulate the active and reactive power. This paper aims at developing a novel optimal nonlinear adaptive control (ONAC) scheme to control VSC in both rectifier mode and inverter mode. Firstly, the nonlinearities, parameter uncertainties, time-varying external disturbances, and unmodelled dynamics can be aggregated into a perturbation, which is then estimated by an extended state observer (ESO) called high-gain perturbation observer (HGPO) online. Moreover, the estimated perturbation will be fully compensated through state feedback. Besides, the observer gains and controller gains are optimally tuned by a recent emerging biology-based memetic salp swarm algorithm (MSSA), the utilization of such method can ensure a desirably satisfactory control performance. The advantage of ONAC is that even though the operation conditions are constantly changing, the control performance can still be maintained to be globally consistent. In addition, it is noteworthy that in rectifier mode only the reactive power and DC voltage are required to be measured, while in inverter mode merely the reactive power and active power have to be measured. At last, in order to verify the feasibility of ONAC in practical application, a hardware experiment is implemented.

Keywords: optimal nonlinear adaptive control; memetic salp swarm algorithm; voltage source converter; perturbation observer; hardware experiment

1. Introduction

The rapidly growing energy demand has driven the fast deployment and development of renewable energy which involves advanced electronic devices to generate power. Generally speaking, such electrical power usually needs to be converted, e.g., DC (direct current)-DC(direct current) boost or buck, AC (alternating current)-DC (rectifier mode), DC-AC (inverter mode), etc., to provide a proper power supply to satisfy various power demanders. In the past decade, thanks to the advancement of modern electronics technology, studies on voltage source converters (VSC) utilizing insulated gate bipolar transistors (IGBT) have gained considerable attentions and interests around the globe [1]. VSC has many advantages over the conventional current source converter (CSC), e.g., decoupled control of the active power and reactive power and the communication process can successfully be achieved without external voltage source supply. Several typical VSC applications can be traced to VSC-based high voltage direct current (VSC-HVDC) systems [2], doubly fed induction generation (DFIG) [3], permanent magnetic synchronous generator (PMSG) [4], and photovoltaic (PV) inverter [5]. So far, VSC plays a major role in distributed generation (DG) systems [6–8] to ensure a reliable power supply.

In practice, it often requires VSC to operate rapidly to ensure the reactive power and DC voltage can be regulated in time when it operates at rectifier mode, while achieving the regulation of the reactive power and active power when it operates at inverter mode. Hence, the proper design of control systems plays an important role in VSC operation. Conventional proportional–integral–derivative (PID) control has acquired wide applications, since the one-point linearization is always used when the control parameters are tuned, once the operation points change the overall control performance of PID controller may decline significantly [9]. Such issue become particularly severe for DG systems as they often operate under stochastic atmospheric conditions, such as random wind speed variation, solar irradiation, and temperature change.

To resolve this challenging issue, many advanced control strategies have been developed, such as feedback linearization control (FLC) [10] applied in VSC-HVDC systems, which fully removes the nonlinearities of the system to guarantee a global control consistency under different operation conditions. Nevertheless, FLC requires a system model which can precisely describe the characteristics of the system and fully state measurements, thus it lacks robustness for the parameters of ideal model change and is difficult to be implemented in practice. To enhance the robustness, sliding-mode control (SMC) [11] was applied to VSC-HVDC system. However, chattering may emerge from SMC due to the use of discontinuous functions in the controller loop, together with an increased scale of controller structure complexity. Meanwhile, in reference [12], a passivity-based sliding-mode control (PBSMC) scheme was employed for rotor-side VSC of PMSG to efficiently undertake maximum power point tracking (MPPT) under different wind profiles. However, it requires many state measurements and an accurate system model which is hard to achieve in reality, thus making its implementation questionable. Moreover, a robust sliding-mode control (RSMC) strategy was developed on the rotor-side VSC of DFIG to enhance the robustness under modelling uncertainties [13], which largely increased the complexity and resulted in an overconservativeness due to the upper bound of perturbation used, which does not often occur in practice. In addition, model predictive control (MPC) has been applied on PV inverters to achieve a rapid power regulation in the presence of modelling uncertainties [14], which, however, needs to employ extra control loops, thus, the system complexity becomes significant. Besides, literature [15] adopted an effective control strategy, in which a disturbance estimator-based predictive current controller is utilized for PV inverters to estimate the unknown PV inverter parameters and modelling uncertainties. However, it introduces a large number of estimator gains for which tuning is difficult.

Basically, the aforementioned nonlinear approaches have a relatively complex structure, thus their implementation is usually difficult. Based on extended state observer (ESO) [16–18], a nonlinear adaptive control (NAC) strategy for VSC operating in both rectifier mode and inverter mode is proposed in this paper. The combined influence of system nonlinearities, uncertain parameters, unmodelled dynamics, and time-varying external disturbances can be synthesized into a perturbation, which can be estimated through a high-gain perturbation observer (HGPO) [19–21]. NAC only requires the measurement of DC voltage and reactive power for rectifier mode, and active power and reactive power for inverter mode. Thus, it provides the merit of inherently easy implementation in real systems.

However, NAC involves many parameters that need to be properly tuned, e.g., (a) observer gains which determine the perturbation estimation performance and (b) controller gains which influence the control performance. These gains are interactively coupled, thus, the optimal tuning of them is worth studying. To handle this issue, this paper aimed to design an optimal NAC (ONAC) strategy to achieve optimal parameter tuning of NAC for VSC, such that an optimal and robust control can be achieved under different operation scenarios. In particular, a newly emerging biology based metaheuristic algorithm was utilized, called memetic salp swarm algorithm (MSSA) [22], to tune the parameters of NAC more optimally and efficiently, such that the control performance of NAC can be optimized. More specifically, the salp chains adopted by MSSA are multiple independent ones, such a strategy can better balance exploration and exploitation, thus, a high-quality optimum can be efficiently found.

Besides this, a novel regroup operation using a virtual population is employed to globally coordinate different salp chains, such that the convergence stability can be considerably improved.

Lastly, the proposed ONAC is compiled and downloaded into the dSPACE processor as the real-time controller for VSC, in which the I/O interface of dSPACE simulator enables the real-time sampling of inputs from measurements and output control signals with IGBT converters. Several case studies were done to verify the hardware practicality of ONAC based inverter controller and rectifier controller.

The originality and novelty of this work can be summarized as follows:

(a) The large parameter tuning burden of conventional NAC can be efficiently resolved via MSSA, which can efficiently and effectively seek the optimal controller and observer parameters with a high stability;
(b) Various modelling uncertainties can be effectively estimated by HGPO in the real-time, which is then fully compensated by the controller online, such that great robustness can be realized;
(c) A dSpace based hardware experiment was undertaken which validates the implementation feasibility of ONAC.

The structure of this paper is organized as follows: Section 2 presents the VSC modelling; Section 3 provides the ONAC design; in Section 4, ONAC design for VSC is given; Section 5 presents the experimental results of ONAC; lastly, Section 6 concludes the whole paper.

2. VSC Modelling

VSC can be operated in either rectifier mode (which can achieve the regulation of the reactive power and DC voltage) or inverter mode (which can achieve the regulation of the reactive power and active power). Figure 1 presents a typical structure of VSC. Here, the balanced condition has been taken into consideration, i.e., the parameters of three phases on the AC side are totally the same and the amplitude of voltages and currents are identical, while each phase shifts 120° between themselves. When the angular frequency is ω, the rectifier dynamics in d-q rotational frame can be expressed as [9]:

$$\begin{cases} \frac{di_{d1}}{dt} = -\frac{R_1}{L_1}i_{d1} + \omega i_{q1} + u_{d1} \\ \frac{di_{q1}}{dt} = -\frac{R_1}{L_1}i_{q1} - \omega i_{d1} + u_{q1} \\ \frac{dV_{dc1}}{dt} = \frac{3u_{sq1}i_{q1}}{2C_1 V_{dc1}} - \frac{i_L}{C_1} \end{cases} \tag{1}$$

where the rectifier is linked to the AC grid by the equivalent resistance R_1 and inductance L_1, respectively, C_1 represents the capacitance of DC bus, $u_{d1} = \frac{u_{sd1} - u_{rd}}{L_1}$ and $u_{q1} = \frac{u_{sq1} - u_{rq}}{L_1}$. i_L denotes the DC current that flows out of the rectifier.

The inverter dynamics in d-q rotational frame are written as:

$$\begin{cases} \frac{di_{d2}}{dt} = -\frac{R_2}{L_2}i_{d2} + \omega i_{q2} + u_{d2} \\ \frac{di_{q2}}{dt} = -\frac{R_2}{L_2}i_{q2} - \omega i_{d2} + u_{q2} \\ \frac{dV_{dc2}}{dt} = \frac{3u_{sq2}i_{q2}}{2C_2 V_{dc2}} + \frac{i_L}{C_2} \end{cases} \tag{2}$$

where the inverter is linked to the AC grid by the equivalent resistance R_2 and inductance L_2, respectively, C_2 represents the capacitance of DC bus, $u_{d2} = \frac{u_{sd2} - u_{rd}}{L_2}$ and $u_{q2} = \frac{u_{sq2} - u_{rq}}{L_2}$. i_L denotes the DC current that flows into the inverter.

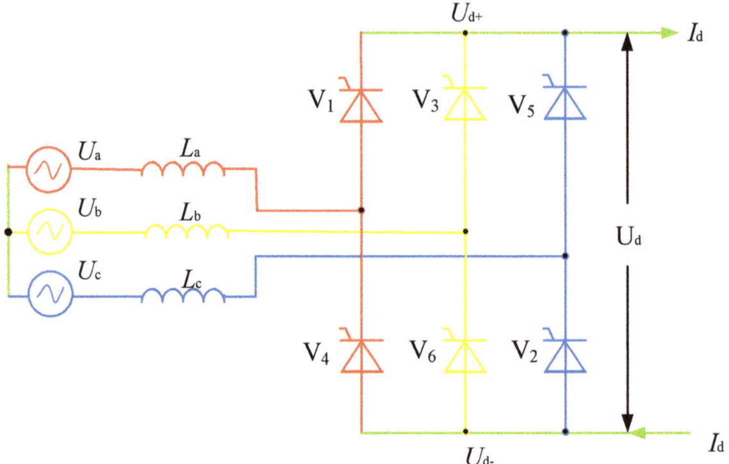

Figure 1. The structure of voltage source converters (VSC).

On the rectifier side, the q-axis is defined to share the same phase with the AC grid voltage u_{s1}. Similarly, the q-axis and the AC grid voltage u_{s2} are defined to share the same phase on the inverter side. Under this circumstance, the values of u_{sd1} and u_{sd2} can be regarded as 0, while the values of u_{sq1} and u_{sq2} are deemed as identical with the magnitude of u_{s1} and u_{s2}. Hence, the reactive and active power can be described by

$$\begin{cases} P_1 = \frac{3}{2}\left(u_{sq1}i_{q1} + u_{sd1}i_{d1}\right) = \frac{3}{2}u_{sq1}i_{q1} \\ Q_1 = \frac{3}{2}\left(u_{sq1}i_{d1} + u_{sd1}i_{q1}\right) = \frac{3}{2}u_{sq1}i_{d1} \\ P_2 = \frac{3}{2}\left(u_{sq2}i_{q2} + u_{sd2}i_{d2}\right) = \frac{3}{2}u_{sq2}i_{q2} \\ Q_2 = \frac{3}{2}\left(u_{sq2}i_{d2} + u_{sd2}i_{q2}\right) = \frac{3}{2}u_{sq2}i_{d2} \end{cases} . \tag{3}$$

3. Optimal Nonlinear Adaptive Control

3.1. Nonlinear Adaptive Control

The standard form of an uncertain nonlinear system can be expressed as follows:

$$\begin{cases} \dot{x} = Ax + B(a(x) + b(x)u + d(t)) \\ y = x_1 \end{cases} \tag{4}$$

where $x = [x_1, x_2, \cdots, x_n]^{\mathrm{T}} \in R^n$ represents the state variable vector; $y \in R$ and $u \in R$ denote the system output and control input, respectively; $a(x)$: $R^n \mapsto R$ and $b(x)$: $R^n \mapsto R$ represent the unknown smooth functions which describe the aggregation influence of nonlinearities, parameter uncertainties, and unmodelled dynamics; and $d(t)$: $R^+ \mapsto R$ means the time-varying external disturbances.

The perturbation of system (4) can be obtained by [19–21]

$$\Psi(x, u, t) = a(x) + (b(x) - b_0)u + d(t) \tag{5}$$

where b_0 represents a constant control gain defined by users, by which the uncertainties of the control gain $b(x)$ can be synthesized into the perturbation.

From system (4), the final state, e.g., x_n, can be further expressed as

$$\dot{x}_n = a(x) + (b(x) - b_0)u + d(t) + b_0 u = \Psi(x, u, t) + b_0 u. \tag{6}$$

Besides, the perturbation term can be expressed by defining an extended state, e.g., $x_{n+1} = \Psi(x, u, t)$. Further, system (4) can be described by

$$
\begin{cases}
y = x_1 \\
\dot{x}_1 = x_2 \\
\quad\vdots \\
\dot{x}_n = x_{n+1} + b_0 u \\
\dot{x}_{n+1} = \dot{\Psi}(\cdot)
\end{cases}
\tag{7}
$$

Under the two assumptions given in [19–21], the extended state vector can be represented by $x_e = [x_1, x_2, \cdots, x_n, x_{n+1}]^T$.

A.1 The selection of b_0 needs to satisfy certain constraint: $\left|b(x)/b_0 - 1\right| \le \theta < 1$, where θ is a positive constant.

A.2 The function $\Psi(x, u, t) : R^n \times R \times R^+ \mapsto R$ and $\dot{\Psi}(x, u, t) : R^n \times R \times R^+ \mapsto R$ are locally Lipschitz in their arguments with a bounded domain of interest, with $\Psi\,(0,0,0) = 0$ and $\dot{\Psi}\,(0,0,0) = 0$.

In this paper, $\widetilde{x} = x - \hat{x}$ denotes the estimation error of x and $\hat{x}$ means the estimation of x, while x^* represents the reference of variable x. We designed an $(n + 1)$th-order HGSPO (high-gain state and perturbation observer) for the expanded system (8) to evaluate the states and perturbation at the same time, as follows [19–21]

$$
\dot{\hat{x}}_e = A_0 \hat{x}_e + B_1 u + H(x_1 - \hat{x}_1)
\tag{8}
$$

where $H = \left[\alpha_1/\varepsilon, \alpha_2/\varepsilon^2, \ldots, \alpha_n/\varepsilon^n, \alpha_{n+1}/\varepsilon^{n+1}\right]^T$, thickness layer boundary of observer $0 \le \varepsilon \ll 1$, thus a high-gain can be successfully obtained in this way; and $\alpha_i, i = 1, 2, \cdots, n + 1$ represents the gains of Luenberger observer—proper choice of its value can help to ensure the poles of polynomial $s^{n+1} + \alpha_1 s^n + \alpha_2 s^{n-1} + \cdots + \alpha_{n+1} = (s + \lambda_\alpha)^{n+1} = 0$ are located at the open left-half complex plane at $-\lambda_\alpha$, as follows

$$
\alpha_i = C_{n+1}^i \lambda_\alpha^i, i = 1, 2, \cdots, n + 1
\tag{9}
$$

where λ_α represents the observer root utilized for ensuring the convergence of observer.

The NAC for system (4) under the estimation of states and perturbation can be written as

$$
u = \frac{1}{b_0}\left[x_1^{*(n+1)} - \hat{\Psi}(\cdot) - K(\hat{x} - x^*)\right]
\tag{10}
$$

where $K = [k_1, k_2, \cdots, k_n]$ represents the feedback control gain which makes the matrix $A_1 = A\text{-}BK$ Hurwitzian.

Note that the closed-loop system stability of NAC has been provided in previous studies, interested readers can refer to literatures [19–21] for more details.

3.2. Memetic Salp Swarm Algorithm

MSSA is an emerging and promising metaheuristic algorithm which mimics the food hunting behavior of salp chains in deep oceans. Note that only the main idea of MSSA is presented in this section while more details are introduced in literature [21] for interested readers.

3.2.1. Optimization Framework

The memetic computing framework is utilized in MSSA for the purpose of improving the searching efficiency of SSA, which consists of two steps shown in Figure 2, as follows [22]:

- Local search in each chain: Several various salp chains are connected in parallel to constitute MSSA, in which a single salp chain consists of a host of salps. At each iteration, on the basis of foraging behavior of SSA, all the salp chains will undertake a local search while this process is independently carried out in each salp chain.

- Global coordination in virtual population: The whole species group of MSSA can merely be considered as a large number of memes, in which each meme is deemed as a unit of cultural evolution [23]. In particular, positive strategy is adopted in the selection of memes to enhance the communication ability among the salps in MSSA (memetic salp swarm algorithm). Besides, each individual can keep its own physical characteristics stable during the process of global coordination. Therefore, the salps gather together can be considered as a virtual population, where all individuals need to rearrange into many new salp chains, thus various salp chains can achieve a global coordination.

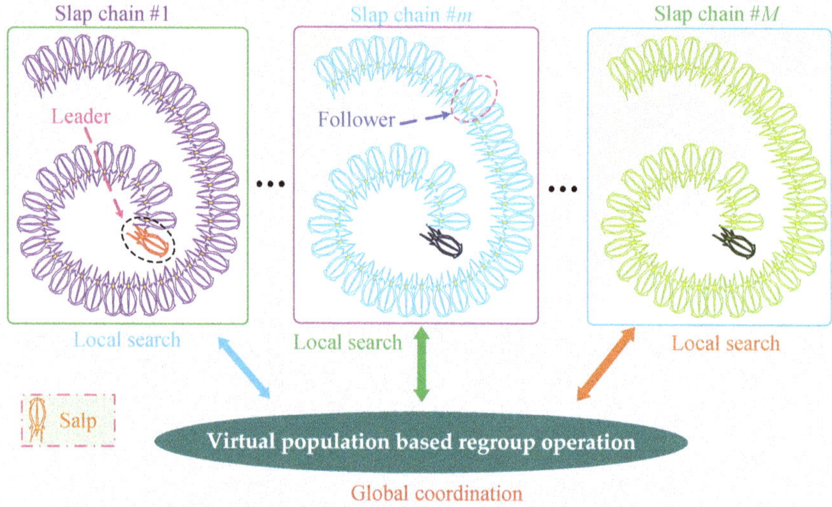

Figure 2. Optimization framework of memetic salp swarm algorithm (MSSA).

3.2.2. Local Search in Each Chain

The salp chain is composed of two groups, i.e., leader and follower. As illustrated by Figure 2, the leader always stays at the first position of the chain to decide the moving direction of the swarm, while all the other salps act as the followers to execute the command. In SSA (salp swarm algorithm), the position of each salp can be mathematically represented. Basically, the leader's position in the mth salp chain during the foraging process is expressed as [24]:

$$x_{m1}^j = \begin{cases} F_m^j + c_1\left(c_2\left(ub^j - lb^j\right) + lb^j\right), & \text{if } c_3 \geq 0 \\ F_m^j - c_1\left(c_2\left(ub^j - lb^j\right) + lb^j\right), & \text{if } c_3 < 0 \end{cases} \tag{11}$$

where the superscript j denotes the jth dimension of the searching space; x_{m1}^j represents the leader's position in the mth salp chain; F_m^j means the position of the food source; lb^j and ub^j represent the lower and the upper bounds of the jth dimension, respectively; c_1, c_2, and c_3 represent the random numbers, while c_2 and c_3 are uniformly distributed in the range of 0–1, respectively.

Since c_1 acts as the most vital parameter in Equation (11) as it directly influences the trade-off between global exploration and local exploitation, for the purpose of realizing a more appropriate balance, it has been designed based on the iteration number, which can be written as [24]:

$$c_1 = 2e^{-\left(\frac{4k}{k_{\max}}\right)^2}.$$ (12)

where k represents the number of current iteration and $k_{\max}$ denotes the maximum iteration number.

Moreover, the position of the followers is updated as follows [24]:

$$x_{mi}^j = \frac{1}{2}\left(x_{mi}^j + x_{m,i-1}^j\right), \quad i = 2, 3, \ldots, n; \; m = 1, 2, \ldots, M$$ (13)

where x_{mi}^j represents the position of the ith salp in the mth salp chain; n denotes the population size of each salp chain; and M means the salp chains number, respectively.

3.2.3. Global Coordination in Virtual Population

For the purpose of enhancing the stability of convergence, the virtual population need to be rearranged and generate various new salp chains based on the fitness values of the salps. In particular, the regroup operation of the original population in MSSA is achieved according to the classical memetic algorithm called shuffled frog leaping algorithm (SFLA) [25], which is a widely utilized population partition principle. Therefore, through arranging the fitness value in descending order the maximum optimization of regroup operation can be realized. Here, the best solution is allocated to the salp chain #1, then the salp chain #2 will capture the second-best solution, and so on. Hence, the mth salp chain can update the swarm according to

$$Y^m = \left\{(x_{mi}, f_{mi})\,|\,x_{mi} = X(m + M(i-1), :), \quad f_{mi} = F(m + M(i-1)), \atop i = 1, 2, \cdots, n\}, \quad m = 1, 2, \cdots, M\right.$$ (14)

where x_{mi} represents the position vector of the ith salp in the mth chain; f_{mi} denotes the fitness value of the ith salp in the mth chain; F means the fitness value set of all the salps in descending order; and X represents the relevant position vector set of all the salps, i.e., a position matrix, by

$$X = \begin{bmatrix} x_1^1 & x_1^2 & \cdots & x_1^d \\ x_2^1 & x_2^2 & \cdots & x_2^d \\ \vdots & \vdots & \ddots & \vdots \\ x_{n \times M}^1 & x_{n \times M}^2 & \cdots & x_{n \times M}^d \end{bmatrix}$$ (15)

where d represents the number of dimensions; and each row vector denotes the position vector of each salp.

3.2.4. Optimization Procedure of MSSA

In conclusion, Table 1 demonstrates the overall process of MSSA. When solving a practical optimization issue, the design of fitness function takes on the most important part in the whole procedure, and it needs to take the related objective function and constraints into consideration.

Table 1. The optimization procedure of MSSA (memetic salp swarm algorithm).

1.	Initialize the optimization parameters;
2.	Determine the fitness function according to the specific optimization problem;
3.	Set $k: = 1$;
4.	**WHILE** (the termination condition is not met)
5.	**FOR1** $m: = 1$ to M
6.	**FOR2** $i: = 1$ to n
7.	Calculate the fitness value of the ith salp in the mth salp chain;
8.	**END FOR2**
9.	**END FOR1**
10.	Implement the regroup operation based global coordination in virtual population by (14) and (18);
11.	**FOR3** $m: = 1$ to M
12.	Determine the food source of the mth salp chain;
13.	Update the position of the leader in the mth salp chain by (13) and (12);
14.	**FOR4** $i: = 2$ to n
15.	Update the position of the ith salp (i.e., the follower) in the mth salp chain by (13);
16.	**END FOR4**
17.	**END FOR3**
18.	Set $k: = k + 1$;
19.	**END WHILE**

4. ONAC Design for VSC Based Rectifier and Inverter

4.1. Rectifier Controller Design

Select system output $y_r = [y_{r1}, y_{r2}]^T = [Q_1, V_{dc1}]^T$, Q_1^* and V_{dc1}^* act as the given references of the DC voltage and reactive power, respectively. Throughout this paper, denote $\hat{x}_i$ and $\widetilde{x}_i = x_i - \hat{x}_i$ as the estimate and estimate error of system state x_i, respectively. Define the tracking error as $e_r = [e_{r1}, e_{r2}]^T = [Q_1 - Q_1^*, V_{dc1} - V_{dc1}^*]^T$, differentiate e_r for rectifier (1) and repeat such operation until the control input can appear clearly, as follows

$$\begin{bmatrix} \dot{e}_{r1} \\ \ddot{e}_{r2} \end{bmatrix} = \begin{bmatrix} f_{r1} - \dot{Q}_1^* \\ f_{r2} - \ddot{V}_{dc1}^* \end{bmatrix} + Br \begin{bmatrix} u_{d1} \\ u_{q1} \end{bmatrix} \tag{16}$$

where

$$\begin{cases} f_{r1} = \frac{3u_{sq1}}{2}\left(-\frac{R1}{L1}i_{d1} + \omega i_{q1}\right) \\ f_{r2} = \frac{3u_{sq1}}{2C_1 V dc1}\left[-\omega i_{d1} - \frac{R_1}{L_1}i_{q1} - \frac{i_{q1}}{V dc1}\left(\frac{3u_{sq1}i_{q1}}{2C_1 V dc1} - \frac{i_L}{C_1}\right)\right] - \\ \frac{1}{2R_0 C_1}\left(\frac{3u_{sq1}i_{q1}}{2C_1 V_{dc1}} - \frac{i_L}{C_1} - \frac{3u_{sq2}i_{q2}}{2C_2 V_{dc2}} - \frac{i_L}{C_2}\right) \end{cases} \tag{17}$$

and

$$Br = \begin{bmatrix} \frac{3u_{sq1}}{2L_1} & 0 \\ 0 & \frac{3u_{sq1}}{2C_1 L_1 V_{dc1}} \end{bmatrix}. \tag{18}$$

The determinant of matrix B_r is obtained as $|B_r| = 9u_{sq1}^2 / \left(4C_1 L_1^2 V_{dc1}\right)$, which is nonzero in the entire operation range of the rectifier. Note that the only condition for zero value of $|B_r|$ is that $u_{sq1} = 0$, such condition corresponds to the case that the connected AC power grid voltage becomes zero. In practice, when the power gird voltage is below 0.2 p.u., the protection devices of VSC are activated and disconnect/stop the VSC working [8,9]. It is the same to $|B_i|$ under inverter mode. As a result, VSC will not encounter this extreme case in practice. Hence, the system (16) can be regarded as linearizable, let $v_r = [v_{r1}, v_{r2}]^T$ be the control input of the linear system

$$\begin{bmatrix} v_{r1} \\ v_{r2} \end{bmatrix} = -\begin{bmatrix} k_{r1}e_{r1} \\ k'_{r1}e_{r2} + k'_{r2}\dot{e}_{r2} \end{bmatrix} \tag{19}$$

where k_{r1}, k'_{r1}, and k'_{r2} represent the positive feedback gains.

Under the assumption that the nonlinearities are all unknown, the perturbations $\Psi_{r1}(\cdot)$ and $\Psi_{r2}(\cdot)$ can be written as

$$\begin{bmatrix} \Psi r1(\cdot) \\ \Psi r2(\cdot) \end{bmatrix} = \begin{bmatrix} fr1 \\ fr2 \end{bmatrix} + (Br - B_{r0}) \begin{bmatrix} u_{d1} \\ u_{q1} \end{bmatrix} \tag{20}$$

where the constant control gain B_{r0} can be expressed as

$$Br0 = \begin{bmatrix} b_{r10} & 0 \\ 0 & b_{r20} \end{bmatrix}. \tag{21}$$

Then, system (16) can be represented by

$$\begin{bmatrix} \dot{e}_{r1} \\ \ddot{e}_{r2} \end{bmatrix} = \begin{bmatrix} \Psi r1(\cdot) \\ \Psi r2(\cdot) \end{bmatrix} + Br0 \begin{bmatrix} u_{d1} \\ u_{q1} \end{bmatrix} - \begin{bmatrix} \dot{Q}^*_1 \\ \ddot{V}^*_{dc1} \end{bmatrix}. \tag{22}$$

Define $z_{11} = Q_1$, a second-order HGPO is given by

$$\begin{cases} \dot{\hat{z}}_{11} = \hat{\Psi}r1(\cdot) + \frac{\alpha r1}{\varepsilon}(z_{11} - \hat{z}_{11}) + br10u_{d1} \\ \dot{\hat{\Psi}}_{r1}(\cdot) = \frac{\alpha r2}{\varepsilon^2}(z_{11} - \hat{z}_{11}) \end{cases}. \tag{23}$$

Define $z'_{11} = V_{dc1}$ and $z'_{12} = \dot{z}'_{11}$, a third-order high-gain state and perturbation observer (HGSPO) is obtained by

$$\begin{cases} \dot{\hat{z}}'_{11} = \hat{z}'_{12} + \frac{\alpha'_{r1}}{\varepsilon}(z'_{11} - \hat{z}'_{11}) \\ \dot{\hat{z}}'_{12} = \hat{\Psi}_{r2}(\cdot) + \frac{\alpha'_{r2}}{\varepsilon^2}(z'_{11} - \hat{z}'_{11}) + br20uq1 \\ \dot{\hat{\Psi}}_{r2}(\cdot) = \frac{\alpha'_{r3}}{\varepsilon^3}(z'_{11} - \hat{z}'_{11}) \end{cases} \tag{24}$$

where $\alpha r1$, $\alpha r2$, α'_{r1}, α'_{r2}, and α'_{r3} represent positive constants with $1 \gg \varepsilon > 0$.

The NAC-based rectifier controller based on the estimations of perturbation can be given by

$$\begin{bmatrix} u_{d1} \\ u_{q1} \end{bmatrix} = B_{r0}^{-1}\left(\begin{bmatrix} -\hat{\Psi}r1(\cdot) \\ -\hat{\Psi}r2(\cdot) \end{bmatrix} + \begin{bmatrix} \dot{Q}^*_1 \\ \ddot{V}^*_{dc1} \end{bmatrix} + v_r \right). \tag{25}$$

4.2. Inverter Controller Design

Select system output $y_i = [y_{i1}, y_{i2}]^T = [Q_2, P_2]^T$, Q^*_2 and P^*_2 act as the provided references of the active power and reactive power, respectively. The tracking error can be defined as $e_i = [e_{i1}, e_{i2}]^T = [Q_2 - Q^*_2, P_2 - P^*_2]^T$, differentiate e_i for inverter (2) and repeat such operation until the control input appears, as follows

$$\begin{bmatrix} \dot{e}_{i1} \\ \dot{e}_{i2} \end{bmatrix} = \begin{bmatrix} fi1 - \dot{Q}^*_2 \\ fi2 - \dot{P}_2 \end{bmatrix} + Bi \begin{bmatrix} u_{d2} \\ u_{q2} \end{bmatrix} \tag{26}$$

where

$$\begin{cases} fi1 = \frac{3u_{sq2}}{2}\left(-\frac{R2}{L2}i_{d2} + \omega i_{q2}\right) \\ fi2 = \frac{3u_{sq2}}{2}\left(-\frac{R2}{L2}i_{q2} + \omega i_{d2}\right) \end{cases} \tag{27}$$

and

$$Bi = \begin{bmatrix} \frac{3u_{sq2}}{2L2} & 0 \\ 0 & \frac{3u_{sq2}}{2L2} \end{bmatrix}. \tag{28}$$

The determinant of matrix B_i is obtained as $|B_i| = 9u_{sq2}^2/(4L_2^2)$, which is nonzero in the whole operation range of the inverter. Such that system (26) can be regarded as linearizable. Let $v_i = [v_{i1}, v_{i2}]^T$ be the control input of the linear system

$$\begin{bmatrix} v_{i1} \\ v_{i2} \end{bmatrix} = - \begin{bmatrix} k_{i1}e_{i1} \\ k'_i e_{i2} \end{bmatrix} \tag{29}$$

where k_{i1} and k'_{i1} represent positive feedback gains.

Under the assumption that the nonlinearities are all unknown, the perturbations $\Psi_{r1}(\cdot)$ and $\Psi_{r2}(\cdot)$ can be written as

$$\begin{bmatrix} \Psi_{i1}(\cdot) \\ \Psi_{i2}(\cdot) \end{bmatrix} = \begin{bmatrix} f_{i1} \\ f_{i2} \end{bmatrix} + (B_i - B_{i0}) \begin{bmatrix} u_{d2} \\ u_{q2} \end{bmatrix} \tag{30}$$

where B_{i0} represents the constant control gain, as follows

$$B_{i0} = \begin{bmatrix} b_{i10} & 0 \\ 0 & b_{i20} \end{bmatrix}. \tag{31}$$

Then system (26) can be rewritten as

$$\begin{bmatrix} \dot{e}_{i1} \\ \dot{e}_{i2} \end{bmatrix} = \begin{bmatrix} \Psi_{i1}(\cdot) \\ \Psi_{i2}(\cdot) \end{bmatrix} + B_{i0} \begin{bmatrix} u_{d2} \\ u_{q2} \end{bmatrix} - \begin{bmatrix} \dot{Q}_2^* \\ \dot{P}_2^* \end{bmatrix}. \tag{32}$$

Similarly, define $z_{21} = Q_2$ and $z'_{21} = P_2$, two second-order HGPOs are designed as

$$\begin{cases} \dot{\hat{z}}_{21} = \hat{\Psi}_{i1}(\cdot) + \frac{\alpha_{i1}}{\varepsilon}(z_{21} - \hat{z}_{21}) + b_{i10}u_{d2} \\ \dot{\hat{\Psi}}_{i1}(\cdot) = \frac{\alpha_{i2}}{\varepsilon^2}(z_{21} - \hat{z}_{21}) \end{cases} \tag{33}$$

$$\begin{cases} \dot{\hat{z}}'_{21} = \hat{\Psi}_{i2}(\cdot) + \frac{\alpha'_{i1}}{\varepsilon}(z'_{21} - \hat{z}'_{21}) + b_{i20}u_{q2} \\ \dot{\hat{\Psi}}_{i2}(\cdot) = \frac{\alpha'_{i2}}{\varepsilon^2}(z'_{21} - \hat{z}'_{21}) \end{cases} \tag{34}$$

where α_{i1}, α_{i2}, α'_{i1}, and α'_{i2} denote positive constants.

The NAC-based inverter controller on the basis of the estimations of perturbation can be given by

$$\begin{bmatrix} u_{d2} \\ u_{q2} \end{bmatrix} = B_{i0}^{-1} \left(\begin{bmatrix} -\hat{\Psi}_{i1}(\cdot) \\ -\hat{\Psi}_{i2}(\cdot) \end{bmatrix} + \begin{bmatrix} \dot{Q}_2^* \\ \dot{P}_2^* \end{bmatrix} + v_i \right). \tag{35}$$

4.3. Optimal Control Parameter Tuning

The controller gains in Equations (19), (25), (29), and (35) and observer gains in Equations (23)–(24) and (33)–(34) are optimally tuned through MSSA under reactive power and active power regulation.

The ultimate purpose of the optimization is to maximumly reduce the tracking error of DC link voltage, reactive power, active power, perturbation estimation error, as well as the total control costs, the model for which can be described by

$$\text{Minimize } J(x) = \int_0^T [|P_i - P_i^*| + |Q_i - Q_i^*| + |V_{dc} - V_{dc}^*| + \omega_{i1}|\tilde{\Psi}_{ri}| + \omega_{i2}|\tilde{\Psi}_{ii}| + \omega_{i3}|u_{di}| + \omega_{i4}|u_{qi}|] \, dt$$

$$\text{subject to } \begin{cases} k_{i1}^{min} \leq k_{i1} \leq k_{i1}^{max} \\ k_{i1}'^{min} \leq k_{i1}' \leq k_{i1}'^{max} \\ k_{r1}^{min} \leq k_{r1} \leq k_{r1}^{max} \\ k_{ri}'^{min} \leq k_{ri}' \leq k_{ri}'^{max} \\ \alpha_{ii}^{min} \leq \alpha_{ii} \leq \alpha_{ii}^{max} \\ \alpha_{ri}^{min} \leq \alpha_{ri} \leq \alpha_{ri}^{max} \\ \alpha_{ii}'^{min} \leq \alpha_{ii}' \leq \alpha_{ii}'^{max} \\ \alpha_{rj}'^{min} \leq \alpha_{rj}' \leq \alpha_{rj}'^{max} \\ u_{di}^{min} \leq u_{di} \leq u_{di}^{max} \\ u_{qi}^{min} \leq u_{qi} \leq u_{qi}^{max} \end{cases} , \, i = 1, 2 \text{ and } j = 1, 2, 3. \tag{36}$$

where ω_{i1} and ω_{i2} are two weights used to measure the magnitude of perturbation estimation error which are set to be 0.01; while ω_{i3} and ω_{i4} are the weights utilized to measure the magnitude of overall control costs which usually are set to be 0.25; note that the selection of weighting factor value is depended on the magnitude of controller output, perturbation estimation error, and control error. Normally, one should examine their magnitudes and make them have the same order of magnitude. In this paper, through trial and error, it can be found that perturbation estimation is 120 times higher (peak value occurs due to peaking phenomenon of HGPO) than that of control error while controller output is around six times higher than that of control error. Thus, their values are chosen as aforementioned to scale them to have the same order of magnitude. Lastly, $T = 5$ s represents the experiment time, while the control costs are restricted by the limits.

Lastly, the overall control structure of ONAC for VSC is demonstrated in Figures 3 and 4, respectively, in which the space vector pulse width modulation (SVPWM) is utilized for modulating the control inputs [26].

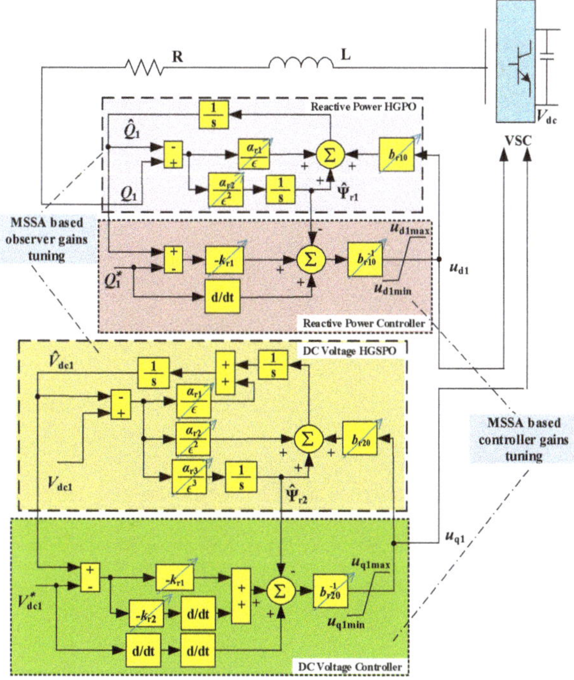

Figure 3. The overall optimal nonlinear adaptive control (ONAC)-based control framework of rectifier.

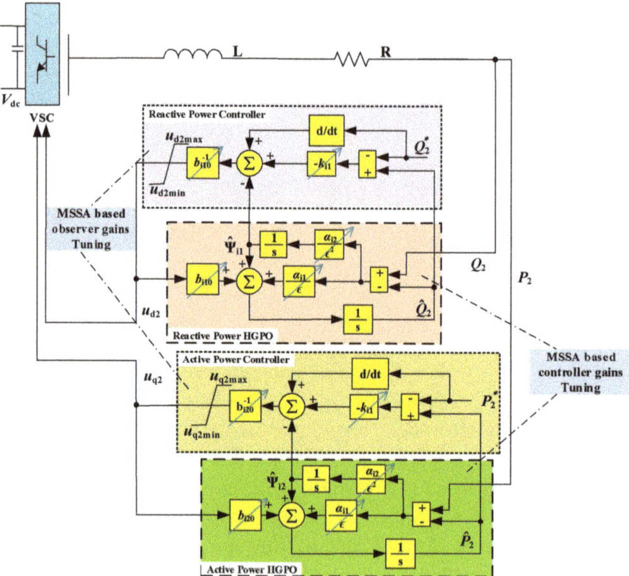

Figure 4. The overall ONAC-based control framework of inverter.

5. Experiment Results

5.1. Experimental Platform

For the purpose of verifying the practicability of ONAC, a hardware experiment was carried out while the experimental platform is illustrated by Figure 5. A strong AC/DC power supply was used to provide stable three phase AC and DC voltages. The power supply was connected with the converter by an IGBT converter [27] with an inductive filter. IGBT converters can be driven by pulse width modulation (PWM) signals. Compared with the conventional PWM or sinusoidal pulse width modulation (SPWM) techniques, the space vector modulation (SVM) can increase 15% more of the maximum output voltage and reduce the switching times. The voltage/current isolators measured the voltages and currents, which were then sent to the dSPACE simulator through analogue/digital (A/D) blocks [28].

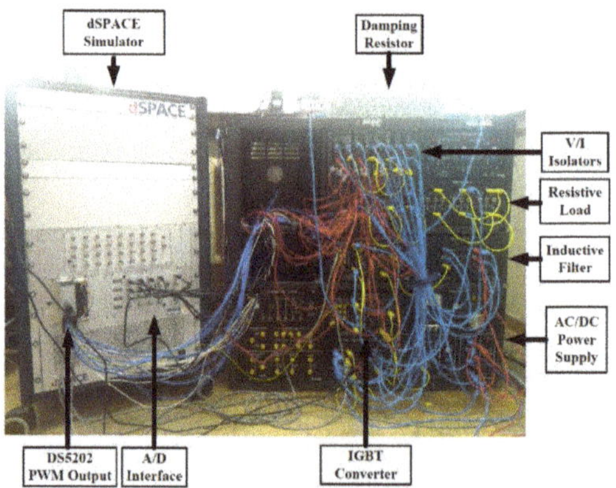

Figure 5. The experimental platform based on the dSPACE platform.

The dSPACE simulator had quad-core AMD (advanced micro devices) processors (DS1104) and operated with DS5202 ACMC board, which was capable of generating high-frequency PWM [29] signals and providing high-speed A/D interfaces. The IGBT converter, inductive filter, voltage/current isolators, and AC/DC power supply were all from the Lab-Volt company, which provides various equipment with the user-friendly interfaces, accurate system parameters, and reliable hardware protections. The DC bus of the IGBT converter was protected by adopting a 100 Ω damping resistor with 1 kW rated power connected to the damping circuit built in to the IGBT converter to avoid potential damages caused by overvoltage or overcurrent in the DC bus. Finally, Table 2 demonstrates the guaranteed measurement error of DS1104 board.

Table 2. The error specification of quad-core AMD (advanced micro devices) processors (DS 1104) board [30].

Component Parameters	A/D Converter	D/A Converter
Offset error	±5 mV	±1 mV
Gain error	Multiplexed channels: ±0.25%	±0.1%
	Parallel channels: ±0.5%	
Offset drift	40 μV/K	130 μV/K
Gain drift	25 ppm/K	25 ppm/K
Signal-to-noise ratio	Multiplexed channels: >80 dB	>80 dB
	Parallel channels: >65 dB	

The controller was embedded in the dSPACE platform (dSPACE Inc., Paderborn, Germany) shown in Figure 5, which measured the DC voltage, reactive power, and active power as inputs. Then, the PWM signals were generated with various duty cycles as controller outputs to the IGBT converter. Furthermore, the sampling frequency f_s and the PWM frequency f_{PWM} with SVPWM are given in Table 3. Here, sampling frequency was determined by the performance of industrial microprocessors of VSC, which usually ranges from 1 kHz to 3 kHz. Note that a lower sampling frequency might degrade the control precision while a higher sampling frequency might increase the computation burden. To make a trade-off between the control precision and computation burden, this paper adopted the median sampling frequency (2 kHz) to validate the controller implementation feasibility. MSSA was run 30 times and the optimal outcomes (the controller gains and observer gains leading to the lowest fitness function) were undertaken in ONAC. The parameters of the controller are provided in Table 4, while the control inputs were limited as $|u_{di}| \leq 50$ V and $|u_{qi}| \leq 50$ V, $i = 1,2$, respectively. Based on SVPWM and ignoring the resistance in the steady state, the minimum DC voltage must satisfy the following inequality to ensure the converter is controllable and can work properly as [16]

$$V_{dci} \geq \sqrt{3}\sqrt{(u_{sdi} + \omega Lii_{qi})^2 + (u_{sqi} - \omega Lii_{di})^2} \tag{37}$$

At last, the IGBT converter was used as the DC/AC converter, the antiparallel diodes were combined with the IGBT converters such that it could be operated as either a rectifier or an inverter.

Table 3. System parameters used in the experiment.

Rated Active Power	P_0	300 W
Rated rms voltage	V_0	30 V
Rated rms current	I_0	10 A
Rated frequency	f_0	50 Hz
Filter inductance	$L_1{}', L_2{}'$	60 mH
PWM frequency	f_{PWM}	2 kHz
Sampling frequency	f_s	10 kHz
DC bus capacitance	$C_1{}', C_2{}'$	1320 μF
Load resistance	R_L	1200 Ω

Table 4. Optimized controller gains and observer gains tuned by MSSA.

	Rectifier Controller		
Controller gains	$k_{r1} = 40$	$k_{r1}' = 400$	$k_{r2}' = 40$
	$b_{r10} = 105$	$b_{r20} = -3000$	
Observer gains	$\alpha_{r1} = 80$	$\alpha_{r2} = 1600$	$\alpha_{r1}' = 120 = 120$
	$\alpha_{r2}' = 4800$	$\alpha_{r3}' = 6.4 \times 10^4$	$\varepsilon = 0.1$
	Inverter controller		
Controller gains	$k_{i1} = 20$	$k_{i1}' = 20$	$b_{i10} = 100$
	$b_{i20} = -100$		
Observer gains	$\alpha_{i1} = 60$	$\alpha_{i2} = 80$	$\alpha_{i1}' = 60$
	$\alpha_{i2}' = 900$	$\varepsilon = 0.1$	

5.2. Rectifier Controller Experiment

The configuration of the rectifier controller experiment is illustrated by Figure 6, in which the DC bus was connected to a resistive load to achieve a specific evaluation of control performance when the active power flowed from the AC grid to the DC cable.

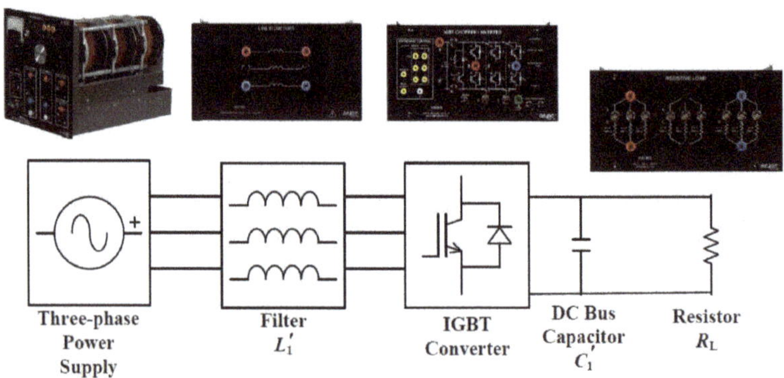

Figure 6. The configuration of the rectifier controller experiment.

The effectiveness of the ONAC-based rectifier controller was tested at first, in which the initial DC voltage V_{dc1} and reactive power Q_1 were adjusted to be 100 V and 0 Var, respectively. The DC voltage remained at 100 V for the whole period of the experiment. The reference of reactive power was set to be 15 Var, then dropped to 5 Var and was restored to 15 Var. After reactive power was stabilized at 15 Var, a resistive load R_L was connected to the DC bus given in Figure 6 to represent the DC cable. Then a 50% voltage drop of the AC grid occured to evaluate the system transient performance.

The experiment results are provided in Figure 7. It demonstrates that both the reactive power and DC voltage were rapidly regulated, while the DC voltage was able to be restored after the AC grid voltage drop occurred, which validates the effectiveness of the ONAC-based rectifier controller.

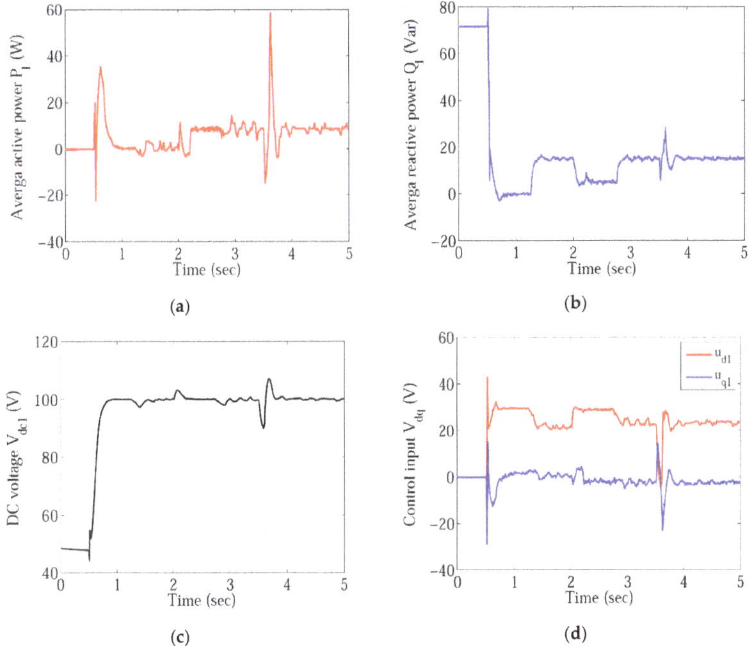

Figure 7. System responses obtained in the rectifier controller experiment: (**a**) Average active power P_1; (**b**) average active power Q_1; (**c**) DC voltage V_{dc1}; (**d**) control input V_{dq}.

5.3. Inverter Controller Experiment

The configuration of the inverter controller experiment is shown in Figure 8, where a DC power supply was utilized to output a stable DC voltage (representing a constant DC voltage regulated by the rectifier controller) such that the IGBT converter could work properly.

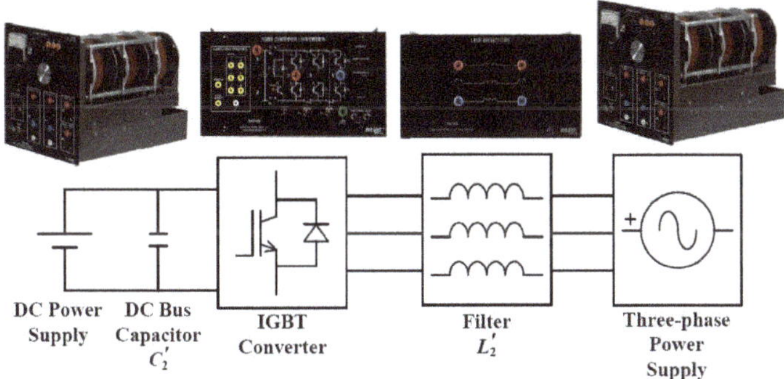

Figure 8. The configuration of the inverter controller experiment.

The effectiveness of the ONAC-based inverter controller was then tested, the initial value of reactive power and active power were zero. The DC voltage V_{dc2} was maintained at 60 V for the whole period of the experiment. The controller was activated to increase the reactive power Q_2 to be 20 Var at first; after reactive power Q_2 was stabilized the active power P_2 was increased to be 10 W, then the reactive power Q_2 was decreased to be 10 Var. At last a 33.3% voltage drop of the AC grid occurred to evaluate the system transient performance.

The experiment outcomes are given in Figure 9. Obviously, the reactive power and active power were able to be regulated independently, and the system could also be restored after the voltage drop, which verified the effectiveness of the ONAC-based inverter controller.

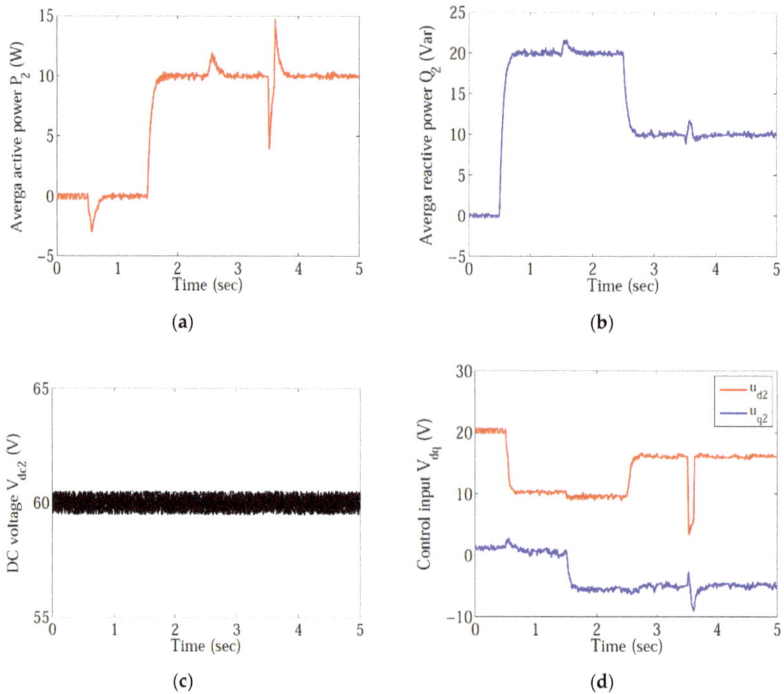

Figure 9. System responses obtained in the inverter controller experiment. (**a**) Average active power P_2; (**b**) average active power Q_2; (**c**) DC voltage V_{dc2}; (**d**) control input V_{dq}.

5.4. Set point Tracking

In order to evaluate the set point tracking performance of ONAC, both the simulation (set point curve) and experiment are compared in Figure 10. One can clearly observe that the experiment can closely track the set point, while their tiny difference resulted from the transmission delay of control signals or measurement noises during the experiment.

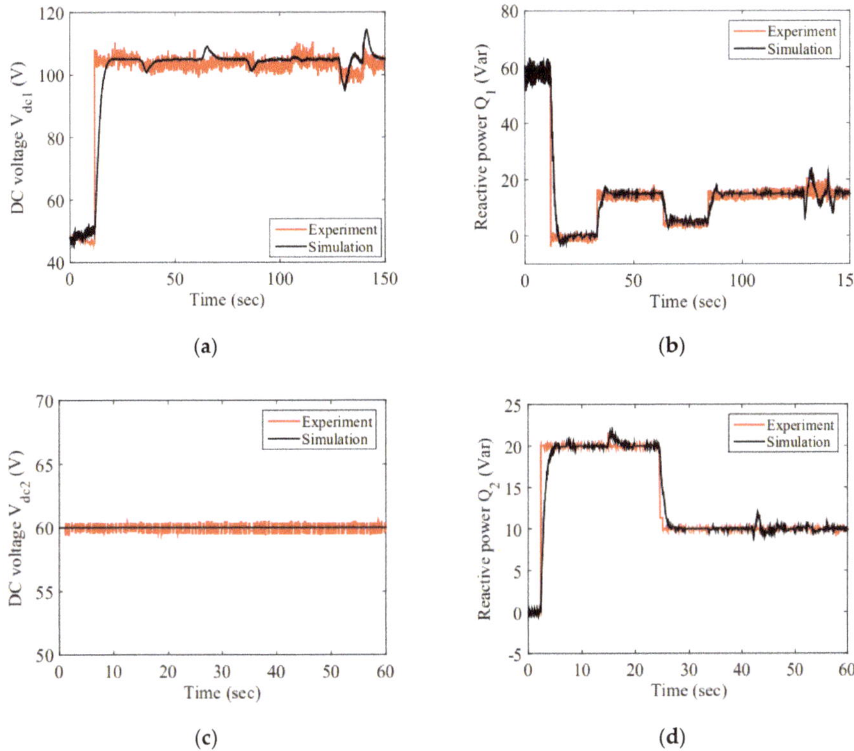

Figure 10. System responses obtained under set point tracking. (**a**) DC voltage V_{dc1} of rectifier; (**b**) average reactive power Q_1 of rectifier; (**c**) DC voltage V_{dc2} of inverter; (**d**) average reactive power Q_2 of inverter.

5.5. Disturbance Rejection

Disturbances are very common in many industrial processes which often have a malignant impact on the predesigned control performance of the studied system. As a result, disturbance rejection performance is a very crucial property for advanced controller design in industries [31–33]. This test aimed to evaluate the disturbance rejection performance of ONAC, while the perturbation estimation performance of HGPO under 20% voltage drop lasting for 15 ms (t = 0.1 s–0.115 s) of VSC under rectifier mode was recorded and illustrated in Figure 11. As shown in Figure 11, both perturbations Ψ_{r1} and Ψ_{r2} could be rapidly estimated by HGPO s under the voltage drop of VSC as the perturbation estimation error could be efficiently converged to 0. As a result, the disturbance was able to be effectively rejected as its real-time estimate could be rapidly obtained and then fully compensated by ONAC. Here, only the rectifier results are provided as it involved both the second-order HGPO (Figure 11a) and third-order HGPO (Figure 11b), while similar results could be found in inverter mode. Note that the transient relatively large variation of perturbation when the disturbance occurred was due to the high gains used in HGPO.

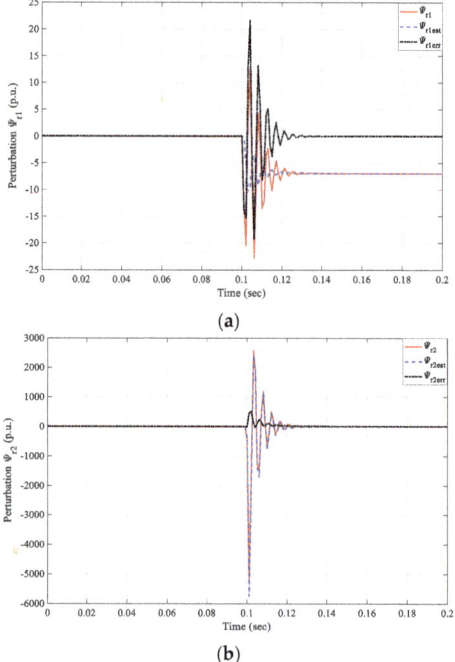

(a)

(b)

Figure 11. Disturbance rejection performance obtained under 20% voltage drop of rectifier. (**a**) Perturbation Ψ_{r1} and (**b**) perturbation Ψ_{r2}.

5.6. Robustness to Parameter Uncertainties

Lastly, the robustness to parameter uncertainties was investigated in this section. Here, a series of plant model mismatches of equivalent resistance R_1 and inductance L_1 within ±20% variation around their nominal value were undertaken. Then, a 50% voltage drop lasting 100 ms at VSC was applied, in which the peak value of output power $|P|$ was recorded. Figure 12 shows that the variation of peak value of active power $|P|$ obtained by FLC [10] and ONAC were 73.4% and 26.5%, which clearly demonstrated that ONAC provided the highest robustness against VSC parameter uncertainties thanks to the real-time compensation of perturbation, while FLC was vulnerable to parameter uncertainties as it required an accurate VSC system model.

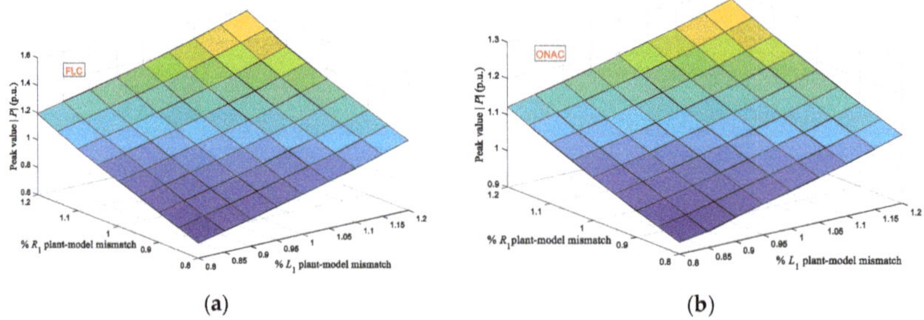

(a)   **(b)**

Figure 12. Peak value of active power $|P|$ obtained under a 50% voltage drop lasting 100 ms at VSC with 20% variation of the equivalent resistance R_1 and inductance L_1 of FLC and ONAC. (**a**) FLC and (**b**) ONAC.

6. Conclusions

This paper designed an ONAC-based rectifier controller and ONAC based inverter controller. It can provide significant robustness as various modelling uncertainties were aggregated into a perturbation, which was rapidly evaluated by HGPO and then compensated for by the proposed controller. Besides, ONAC merely needs the active power and reactive power, as well as DC voltage need to be measured, while a precise VSC system model is not required, thus the proposed controller can be readily implemented in practice. Moreover, the controller gains and observer gains are optimally tuned by an emerging and promising biology based metaheuristic algorithm called MSSA, which can effectively achieve an appropriate trade-off between global exploration and local exploitation, thus a satisfactory control performance and perturbation estimation performance could be guaranteed. Finally, for the sake of verifying the practicability of the proposed controller, a hardware experiment utilizing dSPACE platform was undertaken. The experiment results showed that ONAC could achieve satisfactory control performance.

Future studies will apply ONAC-based VSC on renewable energy systems, e.g., PV inverter, wind energy conversion systems, as well as electric energy storage systems.

Author Contributions: Preparation of the manuscript has been performed by Y.J., X.J., H.W., Y.F., W.G., B.Y., and T.Y.

Funding: The authors gratefully acknowledge the support of National Natural Science Foundation of China (51777078), and Yunnan Provincial Basic Research Project-Youth Researcher Program (2018FD036).

Conflicts of Interest: The authors declare no conflict of interest.

Nomenclature

Variables		*Abbreviations*	
R_1, R_2	equivalent resistance, Ω	VSC	voltage source converter
L_1, L_2	equivalent inductance, mH	CSC	current source converter
i_{d1}, i_{d2}	d-axes components of the line currents, A	VSC-HVDC	VSC based high voltage direct current
i_{q1}, i_{q2}	q-axes components of the line currents, A	DFIG	doubly fed induction generation
u_{d1}, u_{d2}	d-axes components of the line voltages, V	PMSG	permanent magnetic synchronous generator
u_{q1}, u_{q2}	q-axes components of the line voltages, V	PV	photovoltaic
V_{dc1}, V_{dc2}	DC voltages, V	DG	distributed generation
u_{sd1}, u_{sd2}	d-axes components of the respective AC grid voltages, V	PID	proportional-integral-derivative
u_{sq1}, u_{sq2}	q-axes components of the respective AC grid voltages, V	FLC	feedback linearization control
C_1, C_2	capacitance of DC bus, μF	SMC	sliding-mode control
u_{rd}, u_{id}	d-axes components of the converter input voltages, V	PBSMC	passivity-based sliding-mode control
u_{rq}, u_{iq}	q-axes components of the converter input voltages, V	MPPT	maximum power point tracking
U_a, U_b, U_c	abc three-phase voltages at AC side, V	RSMC	robust sliding-mode control
L_a, L_b, L_c	abc three-phase inductances at AC side, mH	MPC	model predictive control
$V_1 \sim V_6$	#1-#6 valves of IGBT	ESO	extended state observer
U_d	DC-link voltage, V	NAC	nonlinear adaptive control
I_d	DC-link current, A	HGPO	high-gain perturbation observer
P_1, P_2	active powers transmitted from the AC grid to the VSC, W	MSSA	memetic salp swarm algorithm
Q_1, Q_2	reactive powers transmitted from the AC grid to the VSC, W	PWM	pulse width modulation
System parameters		SPWM	sinusoidal pulse width modulation
P_0	rated active power, W	SVM	space vector modulation
V_0	rated rms voltage, V	SSA	salp swarm algorithm
I_0	rated rms current, A	DC	direct current
f_0	rated frequency, Hz	AC	alternating current
L_1', L_2'	filter inductance, mH	*MSSA parameters*	
f_{pwm}	PWM frequency, kHz	c_1, c_2, c_3	random numbers
f_s	Sampling frequency, kHz	n	population size of each salp chain
C_1', C_2'	DC bus capacitance, μF	M	number of salp chains
R_L	load resistance, Ω	k_{max}	maximum iteration number

References

1. He, Y.Q.; Chen, Y.H.; Yang, Z.Q.; He, H.B.; Li, L. A review on the influence of intelligent power consumption technologies on the utilization rate of distribution network equipment. *Prot. Control Mod. Power Syst.* **2018**, *3*, 183–193. [CrossRef]

2. Ruan, S.Y.; Li, G.J.; Jiao, X.H.; Sun, Y.Z.; Lie, T. Adaptive control design for VSC-HVDC systems based on backstepping method. *Electr. Power Syst. Res.* **2007**, *77*, 559–565. [CrossRef]

3. Nguyen, T.T. A rotor-sync signal-based control system of a doubly-fed induction generator in the shaft generation of a ship. *Processes* **2019**, *7*, 188. [CrossRef]

4. Li, J.H.; Wang, S.; Ye, L.; Fang, J.K. A coordinated dispatch method with pumped-storage and battery-storage for compensating the variation of wind power. *Prot. Control Mod. Power Syst.* **2018**, *3*, 21–34. [CrossRef]

5. Han, P.P.; Fan, G.J.; Sun, W.Z.; Shi, B.L.; Zhang, X.A. Research on identification of LVRT characteristics of photovoltaic inverters based on data testing and PSO algorithm. *Processes* **2019**, *7*, 250. [CrossRef]

6. Yang, B.; Zhang, X.S.; Yu, T.; Shu, H.C.; Fang, Z.H. Grouped grey wolf optimizer for maximum power point tracking of doubly-fed induction generator based wind turbine. *Energy Convers. Manag.* **2017**, *133*, 427–443. [CrossRef]

7. Yang, B.; Jiang, L.; Wang, L.; Yao, W.; Wu, Q.H. Nonlinear maximum power point tracking control and modal analysis of DFIG based wind turbine. *Int. J. Electr. Power Energy Syst.* **2016**, *74*, 429–436. [CrossRef]

8. Dash, P.K.; Patnaik, R.K.; Mishra, S.P. Adaptive fractional integral terminal sliding mode power control of UPFC in DFIG wind farm penetrated multimachine power system. *Prot. Control Mod. Power Syst.* **2018**, *3*, 79–92. [CrossRef]

9. Ruan, S.Y.; Li, G.J.; Peng, L.; Sun, Y.Z.; Lie, T.T. A nonlinear control for enhancing HVDC light transmission system stability. *Int. J. Electr. Power Energy Syst.* **2007**, *29*, 565–570. [CrossRef]

10. Moharana, A.; Dash, P.K. Input-output linearization and robust sliding-mode controller for the VSC-HVDC transmission link. *IEEE Trans. Power Deliv.* **2010**, *25*, 1952–1961. [CrossRef]

11. Li, S.; Haskew, T.A.; Xu, L. Control of HVDC light system using conventional and direct current vector control approaches. *IEEE Trans. Power Electron.* **2010**, *25*, 3106–3118.

12. Yang, B.; Yu, T.; Shu, H.C.; Zhang, Y.M.; Chen, J.; Sang, Y.Y.; Jiang, L. Passivity-based sliding-mode control design for optimal power extraction of a PMSG based variable speed wind turbine. *Renew. Energy* **2018**, *119*, 577–589. [CrossRef]

13. Yang, B.; Yu, T.; Shu, H.C.; Dong, J.; Jiang, L. Robust sliding-mode control of wind energy conversion systems for optimal power extraction via nonlinear perturbation observers. *Appl. Energy* **2018**, *210*, 711–723. [CrossRef]

14. Tang, R.L.; Wu, Z.; Fang, Y.J. On figuration of marine photovoltaic system and its MPPT using model predictive control. *Sol. Energy* **2017**, *158*, 995–1005. [CrossRef]

15. Mohomad, H.; Saleh, S.A.M.; Chang, L. Disturbance estimator-based predictive current controller for single-phase interconnected PV systems. *IEEE Trans. Ind. Appl.* **2017**, *53*, 4201–4209. [CrossRef]

16. Yao, J.; Jiao, Z.; Ma, D. Extended-state-observer-based output feedback nonlinear robust control of hydraulic systems with backstepping. *IEEE Trans. Power Electron.* **2014**, *61*, 6285–6293. [CrossRef]

17. Xue, W.; Bai, W.; Yang, S.; Song, K.; Huang, Y.; Xie, H. Adrc with adaptive extended state observer and its application to air–fuel ratio control in gasoline engines. *IEEE Trans. Ind. Electron.* **2015**, *62*, 5847–5857. [CrossRef]

18. Cui, R.X.; Chen, L.P.; Yang, C.G.; Chen, M. Correction to extended state observer-based integral sliding mode control for an underwater robot with unknown disturbances and uncertain nonlinearities. *IEEE Trans. Ind. Electron.* **2017**, *64*, 6785–6795. [CrossRef]

19. Wu, Q.H.; Jiang, L.; Wen, J.Y. Decentralized adaptive control of interconnected non-linear systems using high gain observer. *Int. J. Control* **2004**, *77*, 703–712. [CrossRef]

20. Jiang, L.; Wu, Q.H.; Wen, J.Y. Decentralized nonlinear adaptive control for multi-machine power systems via high-gain perturbation observer. *IEEE Trans. Circuits Syst. I Regul. Pap.* **2004**, *51*, 2052–2059. [CrossRef]

21. Chen, J.; Jiang, L.; Wei, Y.; Wu, Q.H. Perturbation estimation based nonlinear adaptive control of a full-rated converter wind turbine for fault ride-through capability enhancement. *IEEE Trans. Power Syst.* **2014**, *29*, 2733–2743. [CrossRef]

22. Yang, B.; Zhong, L.E.; Yu, T.; Li, H.F.; Zhang, X.S.; Shu, H.C.; Sang, Y.Y.; Jiang, L. Novel bio-inspired memetic salp mswarm algorithm and application to MPPT for PV systems considering partial shading condition. *J. Clean. Prod.* **2019**, *215*, 1203–1222. [CrossRef]

23. Eusuff, M.; Lansey, K.; Pasha, F. Shuffled frog-leaping algorithm: A memetic meta-heuristic for discrete optimization. *Eng. Optim.* **2006**, *38*, 129–154. [CrossRef]

24. Mirjalili, S.; Gandomi, A.H.; Mirjalili, S.Z.; Saremi, S.; Faris, H.; Mirjalili, S.M. Salp swarm algorithm: A bio-inspired optimizer for engineering design problems. *Adv. Eng. Softw.* **2017**, *114*, 163–191. [CrossRef]

25. Eusuff, M.M.; Lansey, K.E. Optimization of water distribution network design using the shuffled frog leaping algorithm. *J. Water Resour. Plan. Manag.* **2015**, *129*, 210–225. [CrossRef]

26. Bharatiraja, C.; Jeevananthan, S.; Latha, R. FPGA based practical implementation of NPC-MLI with SVPWM for an autonomous operation PV system with capacitor balancing. *Int. J. Electr. Power Energy Syst.* **2014**, *61*, 489–509. [CrossRef]

27. Xia, Y.; Gou, B.; Xu, Y. A new ensemble-based classifier for IGBT open-circuit fault diagnosis in three-phase PWM converter. *Prot. Control Mod. Power Syst.* **2018**, *3*, 364–372. [CrossRef]

28. Smith, M. DSpace: An Open Source Dynamic Digital Repository. 2003. Available online: https://dspace.mit.edu/handle/1721.1/29465 (accessed on 9 July 2019).

29. Holmes, D.G.; Lipo, T.A. *Pulse Width Modulation for Power Converters: Principles and Practice*; John Wiley & Sons: Hoboken, NJ, USA, 2003.

30. Single-Board Hardware /DS1104 R&D Controller Board. Available online: https://www.dspace.com/shared/data/pdf/2018/dSPACE_DS1104_Catalog2018.pdf (accessed on 25 May 2019).

31. Gao, Z.; Saxen, H.; Gao, C. Special Section on Data-Driven Approaches for Complex Industrial Systems. *IEEE Trans. Ind. Inform.* **2013**, *9*, 2210–2212. [CrossRef]

32. Gao, Z.; Nguang, S.K.; Kong, D.X. Advances in Modelling, monitoring, and control for complex industrial systems. *Complexity* **2019**, *2019*. [CrossRef]

33. Cao, H.Z.; Yu, T.; Zhang, X.S.; Yang, B.; Wu, Y.X. Reactive power optimization of large-scale power systems: A transfer bees optimizer application. *Processes* **2019**, *7*, 321. [CrossRef]

Article

Distributed Model Predictive Control of Steam/Water Loop in Large Scale Ships

Shiquan Zhao [1,2,3,*], **Anca Maxim** [1,3,4], **Sheng Liu** [2], **Robin De Keyser** [1,3] and **Clara M. Ionescu** [1,3,5]

[1] Research Group on Dynamical Systems and Control, Department of Electrical Energy, Metals, Mechanical Constructions and Systems, Ghent University, 9052 Ghent, Belgium
[2] College of Automation, Harbin Engineering University, Harbin 150001, China
[3] Core Lab EEDT—Energy Efficient Drive Trains, Flanders Make, 9052 Ghent, Belgium
[4] Department of Automatic Control and Applied Informatics, "Gheorghe Asachi" Technical University of Iasi, Blvd. Mangeron 27, 700050 Iasi, Romania
[5] Department of Automation, Technical University of Cluj-Napoca, Memorandumului Street No.28, 400114 Cluj-Napoca, Romania
* Correspondence: Shiquan.Zhao@ugent.be; Tel.: +32-048-645-2783

Received: 12 June 2019; Accepted: 5 July 2019; Published: 11 July 2019

Abstract: In modern steam power plants, the ever-increasing complexity requires great reliability and flexibility of the control system. Hence, in this paper, the feasibility of a distributed model predictive control (DiMPC) strategy with an extended prediction self-adaptive control (EPSAC) framework is studied, in which the multiple controllers allow each sub-loop to have its own requirement flexibility. Meanwhile, the model predictive control can guarantee a good performance for the system with constraints. The performance is compared against a decentralized model predictive control (DeMPC) and a centralized model predictive control (CMPC). In order to improve the computing speed, a multiple objective model predictive control (MOMPC) is proposed. For the stability of the control system, the convergence of the DiMPC is discussed. Simulation tests are performed on the five different sub-loops of steam/water loop. The results indicate that the DiMPC may achieve similar performance as CMPC while outperforming the DeMPC method.

Keywords: distributed model predictive control; steam power plant; steam/water loop; multi-input and multi-output system; loop design

---

1. Introduction

The steam/water loop is an important part of a steam power plant, which plays a role in feed water supply and recycling processes. It is a highly complex and constrained system with multiple variables and interactions [1]. Meanwhile, due to the harsh and challenging operating environment (sea winds, sea waves and sea currents) and various operating modes (automatic start-up, reverse, stop, setting speed, emergency stop and reduction of revolutions) [2], there are difficulties to design a controller that delivers satisfactory performance for the steam/water loop. In order to design an effective approach overcoming the difficulties mentioned above and improving their liability and flexibility of the system, the feasibility of a distributed model predictive control is studied in this paper.

Nowadays, the major concern of the steam power plant is not only the tracking performance, but also other performances such as the consumed energy or safety in terrible conditions. Apart from realizing the load tracking performance, the controllers should also fulfill the flexible requirements for each sub-loop. A general way to improve the flexibility is to apply distributed controllers in the system [3]. Also, the multiple controllers improve the reliability of the system [4]. Concurrently, in order to deal with the constraints in the steam/water loop, model predictive control is selected to maintain good performance for each sub-loop. In fact, such methods are readily matured for various applications in the Industry 4.0 paradigm [5]. For the particular class of complex processes with varying time delays, the simple loop control has been proposed and successfully tested in real life processes [6]. However, the system under investigation in this paper is a multivariable process. Hence, the single loop control does not suffice to guarantee good performance of all sub-system loops.

In recent years, many advanced control strategies have been developed to improve the performance of steam power plants, such as: Advanced PID control [7–10]; backstepping control [11–13]; sliding mode control [14–18]; robust control [19,20]; adaptive disturbance rejection control [21–23]. As it has the advantage in dealing with the inputs and outputs constraints, a considerable amount of research has been conducted in the application of model predictive control (MPC) for complex systems in general and for steam power plants, in particular. To overcome the control problems arising from load disturbances and intrinsic nonlinearity, an extended state observer based fuzzy MPC was proposed for the ultra-supercritical boiler-turbine unit [24]. By incorporating both the plant-wide economic process optimization and regulatory process control into a hierarchical control structure, a hierarchical MPC was proposed [25]. A stable fuzzy model predictive controller was designed to solve the superheater steam temperature control problem in a power plant [26], in which the effect of modeling mismatched and unknown plant behavior variations were overcome by the use of a disturbance term and steady-state target calculator. Liu proposed an economic model predictive controller by directly utilizing the economic index of the boiler-turbine system as the cost function. The method realized the economic optimization as well as the dynamic tracking [27].

However, the methods mentioned above are mainly concerned with the loop control of drum steam pressure, electric power and steam–water density in the boiler-turbine system. Most of the concerns concern the tracking performance of the system installed on land. The steam power plant in this paper is installed on ships, and the steam/water loop is depicted in Figure 1 with five sub-loops: Drum water level control loop; deaerator water level control loop; deaerator pressure control loop; exhaust manifold pressure control loop; and condenser water level loop. Also, the steam/water loop in large scale ships has special characteristics compared with those installed on land: (i) Receiving more disturbances from the ocean waves; (ii) smaller capacity; (iii) multiple operation points [28]. Hence, effective control strategies are required to ensure the performance of the steam/water loop.

A linear extended predictive self-adaptive control centralized MPC framework was studied in our previous work. The effects of the predictive horizon and the control horizon on the control system were summarized [28,29]. It was concluded that the predictive horizon be selected according to the system dynamics and the control horizon be selected to ensure a good trade-off between the tracking performance and the computational complexity.

Today's industry processes, for instance, those in heavy industries (e.g., papermaking, steelmaking, petrochemical and power generation), are becoming more and more complex [30], and are generally composed of a couple of interconnected systems, and possess high nonlinear and stochastic dynamics [31]. Due to the interplay between high performance local control and interactions between subsystem to subsystem, the centralized model predictive control (CMPC) is largely viewed as impractical, inflexible and unsuitable for control of such complex processes [32]. For example, as the authors have five sub-loops in the complete steam/water process, it is necessary to ensure the reliability of the system. This may be achieved by localizing control functions near each sub-loop with remote monitoring and supervision (i.e., distributed control) instead of one central controller (i.e., centralized control). Hence, the applicability of the distributed model predictive control (DiMPC) is investigated in this paper. A comparison is conducted between the decentralized model predictive control (DeMPC), the CMPC and the DiMPC, and the results show the effectiveness of the DiMPC.

The convergence issue is discussed to ensure the stability of the DiMPC. It is proved that the DiMPC, in our study, satisfies certain conditions, under which the properties of the CMPC problem are enjoyed by the DiMPC. Then, by the means provided in [33], the stability of the CMPC is proved, meanwhile, the stability of the DiMPC is guaranteed.

The rest of the paper is organized as follows: Section 2 describes the steam/water loop, and the modeling of the system is shown. The details about the DiMPC, CMPC and DeMPC are introduced in Section 3. Section 4 shows the results and analysis. Finally, some conclusions are drawn in Section 5.

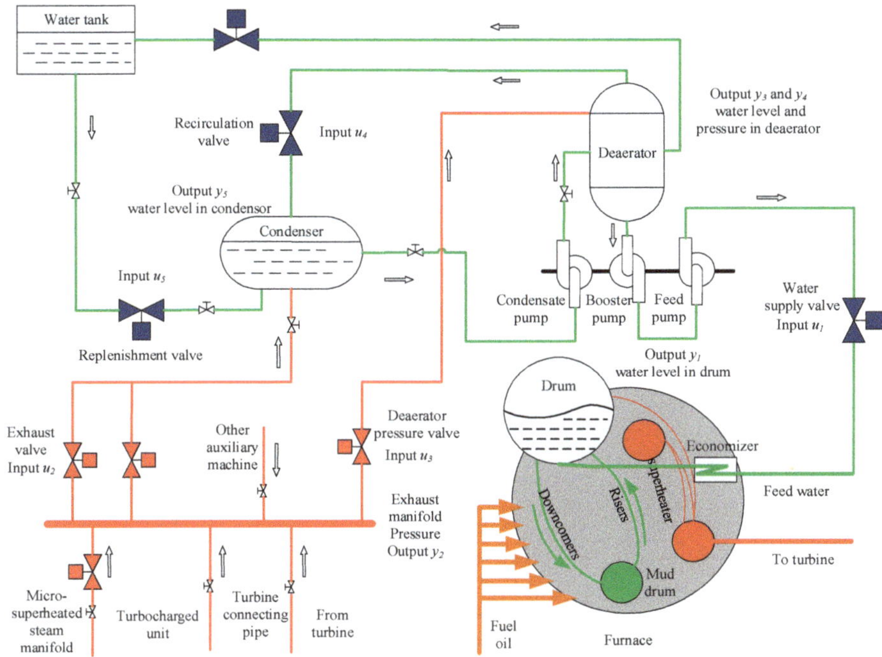

Figure 1. Scheme of a complete steam/water loop [28] (reproduced with permission from Zhao, S.; Maxim, A.; Liu, S.; De Keyser, R.; and Ionescu, C, Processes; published by MDPI, 2018).

2. Description of the Steam/Water Loop

As shown in Figure 1, the steam/water loop is composed of two main loops, in which one is the water loop indicated by green line and the other is steam loop indicated by red line. The system works as follows. Firstly, the water from the water tank goes to the condenser. Then it is deoxygenated in the deaerator. After being pumped to boiler, the feed water goes into the mud drum due to its high density. The feed water is turned into a mixture of steam and water in the risers. Following this, the steam is separated from the mixture and heated in the superheater. Finally, the steam with a certain pressure and temperature services in the steam turbine. The used steam is sent back to the exhaust manifold and most of the steam is condensed in the condenser, while the remaining part services in the deaerator for deoxygenation [28]. The sub-loops have strong interactions between each other, such as the water level between the deaerator and the condenser, the pressure between the deaerator and the exhaust manifold system. Hence, there are challenges to obtain a desired controller for the steam/water loop.

In order to explore the characteristics of the steam/water loop, staircase experiments are conducted around the operating point on the system. The normalized outputs and corresponding static gains are shown in Figure 2; Figure 3, respectively. In the experiment, every 10% step changes are imposed in one input variable, while keeping the other inputs constant. The results show that the static gains change considerably along with the input changes, which indicates the nonlinearity of the system.

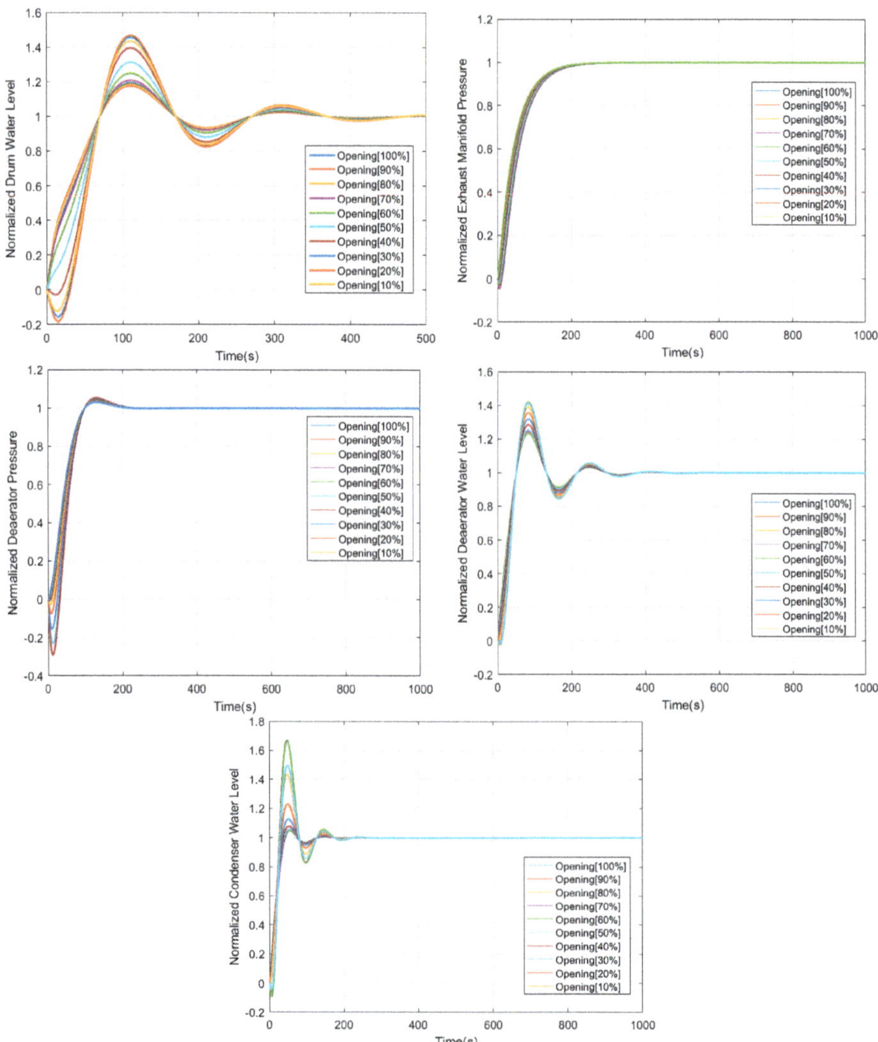

Figure 2. The normalized outputs of the steam/water loop with 10% step change in the input variables.

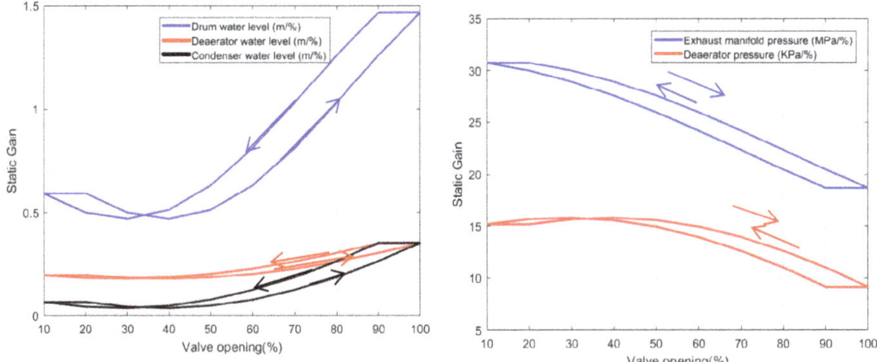

Figure 3. Static gain under different input changes. (The figures on left hand indicate the static gain for drum water level loop, deaerator water level loop, and condenser water level loop; and on the right hand indicate the static gain for deaerator pressure loop, and exhaust manifold pressure loop).

By linearization around the operating point, the model of the system is obtained as shown in (1), with five inputs and five outputs. The input vector $\mathbf{u} = [u_1, u_2, u_3, u_4, u_5]$ contains the positions of the valves that control the flow rates of feedwater to the drum (u_1), exhaust steam from the exhaust manifold (u_2), exhaust steam to the deaerator (u_3), water from the deaerator (u_4) and water to the condenser (u_5), respectively. The output vector $\mathbf{y} = [y_1, y_2, y_3, y_4, y_5]$ contains the values of the water level in drum (y_1), pressure in exhaust manifold (y_2), water level (y_3) and pressure (y_4) in deaerator, and water level of condenser (y_5), respectively. The ranges and operating points of the output variables are listed in Table 1, and the operating points are obtained according to a real large-scale ship.

$$
\begin{bmatrix} y_1 \\ y_2 \\ \vdots \\ y_5 \end{bmatrix} = \begin{bmatrix} G_{11} & G_{12} & \cdots & G_{15} \\ G_{21} & G_{22} & \cdots & G_{25} \\ \vdots & \vdots & \ddots & \vdots \\ G_{51} & G_{52} & \cdots & G_{55} \end{bmatrix} \begin{bmatrix} u_1 \\ u_2 \\ \vdots \\ u_5 \end{bmatrix} \tag{1}
$$

where $G_{11} = \dfrac{0.0000987}{(s+0.1131)(s+0.0085+0.032j)(s+0.0085-0.032j)}$, $G_{22} = \dfrac{0.7254}{(s+1.2497)(s+0.0223)}$,

$G_{23} = \dfrac{-0.5}{(s+1.9747)(s+0.0253)}$, $G_{33} = \dfrac{0.0132}{(s+0.0265+0.0244j)(s+0.0265-0.0244j)}$, $G_{34} = \dfrac{-0.009}{(s+0.0997)(s+0.0411)}$,

$G_{41} = \dfrac{-0.0008}{(s+0.012+0.126j)(s+0.012-0.126j)}$, $G_{44} = \dfrac{0.0005152}{(s+0.012+0.038j)(s+0.012-0.038j)}$,

$G_{54} = \dfrac{-0.00015}{(s+0.0175+0.0179j)(s+0.0175-0.0179j)}$, $G_{55} = \dfrac{0.00147}{(s+0.025+0.0654j)(s+0.025-0.0654j)}$, and other transfer functions $G_{12} = G_{13} = \ldots = G_{53} = 0$.

Table 1. The parameters used in the steam/water loop.

Output Variables	Operating Points	Range	Units
Drum water level	1.79	[1.39–2.19]	m
Exhaust manifold pressure	100.03	[87.03–133.8]	MPa
Deaerator pressure	30	[24.9–43.86]	KPa
Deaerator water level	0.7	[0.489–0.882]	m
Condenser water level	0.5	[0.32–0.63]	m

The rates and amplitudes of the five inputs are constrained to:

$$\begin{cases} -0.007 \leq \frac{du_1}{dt} \leq 0.007 & 0 \leq u_1 \leq 1 \\ -0.01 \leq \frac{du_2}{dt} \leq 0.01 & 0 \leq u_2 \leq 1 \\ -0.01 \leq \frac{du_3}{dt} \leq 0.01 & 0 \leq u_3 \leq 1 \\ -0.007 \leq \frac{du_4}{dt} \leq 0.007 & 0 \leq u_4 \leq 1 \\ -0.007 \leq \frac{du_5}{dt} \leq 0.007 & 0 \leq u_5 \leq 1 \end{cases} \tag{2}$$

The inputs units are normalized percentage values of the valve opening (i.e., 0 represents a fully closed valve, and 1 is completely opened). Additionally, the input rates are measured in percentage per second.

3. Control Strategies

3.1. Introduction for Centralized MPC (CMPC)

The following is a short summary of the extended prediction self-adaptive control (EPSAC) and more details are found in [34]. Consider a discrete system described below:

$$y(k) = x(k) + w(k) \tag{3}$$

where k is the discrete-time index; $y(k)$ indicates the measured output sequence of system; $x(k)$ is the output sequence of model; and $w(k)$ is the model/process disturbance sequence. The output of the model $x(k)$ depends on the past outputs and inputs, and can be expressed generically as:

$$x(k) = f[x(k-1), x(k-2), \ldots, u(k-1), u(k-2), \ldots] \tag{4}$$

In EPSAC, the future input consists of two parts:

$$u(k+l|k) = u_{base}(k+l|k) + \delta u(k+l|k) \tag{5}$$

where $u_{base}(k+l|k)$ indicates lth predicted basic future control scenario and $\delta u(k+l|k)$ indicates the optimizing future control actions, and they are all based on the states and inputs of time k. Then, following the results of lth predicted output is obtained by applying (5) as the control effort.

$$y(k+l|k) = y_{base}(k+l|k) + y_{opt}(k+l|k) \tag{6}$$

where $y_{base}(k+l|k)$ is the effect of base future control and $y_{opt}(k+l|k)$ is the effect of optimizing future control actions $\delta u(k|k), \ldots, \delta u(k+N_c-1|k)$. The part of $y_{opt}(k+l|k)$ can be expressed as a discrete time convolution as follows:

$$y_{opt}(k+l|k) = h_l\delta u(k|k) + h_{l-1}\delta u(k+1|k) + \ldots + g_{l-N_c+1}\delta u(k+N_c-1|k) \tag{7}$$

where $h_1, \ldots h_{N_p}$ are impulse response coefficients; $g_1, \ldots g_{N_p}$ are the step response coefficients; N_c, N_p are the control horizon and prediction horizon, respectively. Thus, the following formulation is obtained:

$$\mathbf{Y} = \overline{\mathbf{Y}} + \mathbf{GU} \tag{8}$$

with, $\mathbf{Y} = [y(k+N_1|k)\dots y(k+N_p|k)]^T$, $\mathbf{U} = [\delta u(k|k)\dots\delta u(k+N_c-1|k)]^T$, $\overline{\mathbf{Y}} = [y_{base}(k+N_1|k)\dots y_{base}(k+N_P|k)]^T$ and

$$\mathbf{G} = \begin{bmatrix} h_{N_1} & h_{N_1-1} & \cdots & g_{N_1-N_c+1} \\ h_{N_1+1} & h_{N_1} & \cdots & \cdots \\ \cdots & \cdots & \cdots & \cdots \\ h_{N_P} & h_{N_P-1} & \cdots & g_{N_P-N_c+1} \end{bmatrix} \tag{9}$$

where N_1 indicates the time-delay in the system.

The disturbance term $w(k)$ is defined as a filtered white noise signal [30]. When there is no information concerning the noise, the disturbance model used in (3) is chosen as an integrator, to ensure zero steady-state error in the reference tracking experiment:

$$w(k) = w(k-1) + e(k) \tag{10}$$

where $e(k)$ denotes the white noise sequence.

In order to apply EPSAC for a multiple-input and multiple-output (MIMO) system, the individual error of each output is minimized separately. The cost function for the steam/water system with five sub-loops is as follows:

$$J_i = \sum_{l=N_1}^{N_P} [r_i(k+l|k) - y_i(k+l|k)]^2 \ (i = 1,2,\dots,5) \tag{11}$$

where r_i ($i = 1, 2, \dots, 5$) are the setpoints for the five loops.

By defining $\mathbf{G}_{ij}$ as the influence from *jth* input to *ith* output, (11) is rewritten as:

$$(\mathbf{R}_i-\mathbf{Y}_i)^T(\mathbf{R}_i-\mathbf{Y}_i)=(\mathbf{R}_i-\overline{\mathbf{Y}}_i-\sum_{j=1}^{5}\mathbf{G}_{ij}\mathbf{U}_j)^T(\mathbf{R}_i-\overline{\mathbf{Y}}_i-\sum_{j=1}^{5}\mathbf{G}_{ij}\mathbf{U}_j) \tag{12}$$

with $\mathbf{R}_i$ denotes the reference for loop i, and $\mathbf{Y}_i$ denotes the predicted output for loop i.

Taking constraints from inputs and outputs into account, the process to find the minimum cost function becomes an optimization problem which is called quadratic programming.

$$\min_{\mathbf{U}_i} J_i(\mathbf{U}_i)=\mathbf{U}_i^T\mathbf{H}_i\mathbf{U}_i+2\mathbf{f}_i^T\mathbf{U}_i+c_i \text{ subject to } \mathbf{AU} \leq \mathbf{b} \tag{13}$$

$$\text{with} \begin{cases} \mathbf{H}_i=\mathbf{G}_{ii}^T\mathbf{G}_{ii}\mathbf{f}_i=-\mathbf{G}_{ii}^T(\mathbf{R}_i-\overline{\mathbf{Y}}_i-\sum_{j=1,j\neq i}^{5}\mathbf{G}_{ij}\mathbf{U}_j) \\ c_i = (\mathbf{R}_i-\overline{\mathbf{Y}}_i-\sum_{j=1,j\neq i}^{5}\mathbf{G}_{ij}\mathbf{U}_j)^T(\mathbf{R}_i-\overline{\mathbf{Y}}_i-\sum_{j=1,j\neq i}^{5}\mathbf{G}_{ij}\mathbf{U}_j) \end{cases}$$

where $\mathbf{A}$ is a matrix; $\mathbf{b}$ is a vector according to the constraints and $\mathbf{U}_i$ is the input for sub-loop i.

Figure 4 shows the conceptual representation of the centralized MPC [35]. To get the optimal solutions for sub-loop i, the interaction $\{u_{j\in N_i}\}$, $N_i = \{j \in N : \widetilde{G}_{ij} \neq 0\}$ from other sub-loops is taken into account as shown in (13).

Hence, the optimal centralized solution $\mathbf{U} = [\mathbf{U_1}\ \mathbf{U_2}\ \mathbf{U_3}\ \mathbf{U_4}\ \mathbf{U_5}]$ is obtained by solving the following global cost function:

$$J = \sum_{i=1}^{5} p_i J_i \tag{14}$$

where J_i are defined in (11), and p_i are weighting factors. In our case, the p_i are chosen as the values which can normalize the cost function J_i.

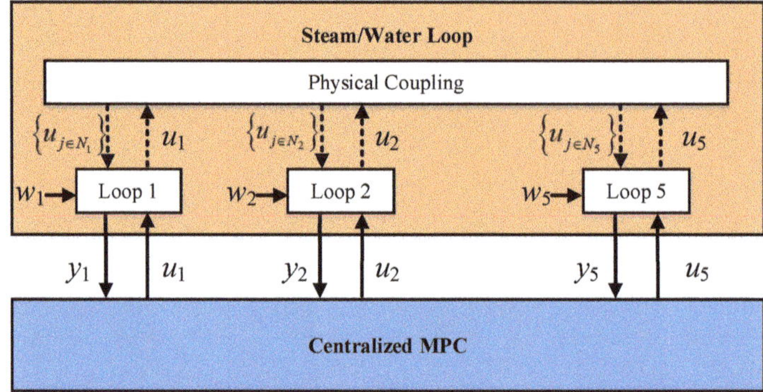

Figure 4. A conceptual representation of the centralized model predictive control (MPC) architecture [33] (Reproduced with permission from Maxim, A.; Copot, D.; De Keyser, R.; and Ionescu, C.M., Journal of Process Control; published by Elsevier, 2018).

3.2. Proposed Distributed MPC (DiMPC)

It is noteworthy to mention that the centralized approach implies that all the information regarding to all the sub-systems (or sub-loops) is gathered in a single controller, as showed in Figure 4. The advantage is straightforward since the cost function (14) has an optimal solution. However, if one sub-loop malfunctions, then the entire steam/water loop collapses, with serious consequences for safety of the large scale ship.

One solution is provided by the distributed MPC (DiMPC) method, that regards all the sub-systems as independent modules which are controlled by an individual controller. Through the communication network, the inherent interactions are considered.

Thus, the same local cost function (13) is locally minimized by each controller, in which the coupling term $\sum_{j=1,j\neq i}^{5} \mathbf{G}_{ij}\mathbf{U}_j$ is computed with the input trajectory $\mathbf{U}_j$ received from the neighbors, and several iterations are performed until the local optimal solution is reached. For the sake of clarity, conceptual representation of the distributed MPC architecture is shown in Figure 5, and a pseudo-code is provided:

Algorithm 1 The Iterative DiMPC

Step 1: Sub-loop I receives an optimal local control action $\delta\mathbf{U}_i$ at the iterative time as *iter* = 0 according to the EPSAC, and the local control action $\delta\mathbf{U}_i$ can be rewritten as $\delta\mathbf{U}_i^{iter}$, where $\delta\mathbf{U}_i$ indicates the vector of the optimizing future control actions with length of N_{ci};

Step 2: The $\delta\mathbf{U}_j^{iter}$ ($j \in N_i$) is communicated to the loop i, and the $\delta\mathbf{U}_i^{iter+1}$ is calculated again with the $\delta\mathbf{U}_j^{iter}$ from other loops;

Step 3: If the termination conditions $\|\delta\mathbf{U}_i^{iter+1} - \delta\mathbf{U}_i^{iter}\| \leq \varepsilon_i \vee iter + 1 > \overline{iter}$ are reached, the $\mathbf{U}_i^{iter+1}$ is adopted, where ε_i is the positive value and $\overline{iter}$ indicates the upper bound of the number of iteration times.

Otherwise, the iter is set as *iter* = *iter* + 1, and return to Step 2;

Step 4: Calculate the optimal control effort as $\mathbf{U}_t = \mathbf{U}_{base} + \delta\mathbf{U}^{iter}$, and the control effort is applied to the system;

Step 5: Set $t = t + 1$, return to Step 1.

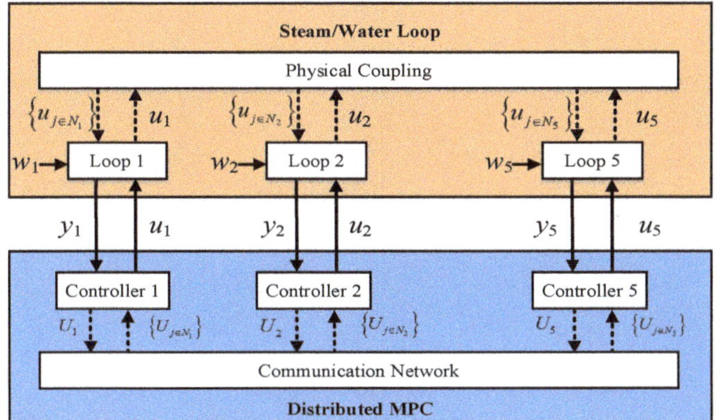

Figure 5. A conceptual representation of the distributed MPC architecture [33] (Reproduced with permission from Maxim, A.; Copot, D.; De Keyser, R.; and Ionescu, C.M., Journal of Process Control; published by Elsevier, 2018).

3.3. Classical Decentralized MPC (DeMPC)

In Figure 6, the conceptual representation of the DeMPC is presented. When comparing with the distributed strategy from Figure 5, it can be seen that the main difference is given by the fact that the controllers do not exchange information, although the physical coupling remains.

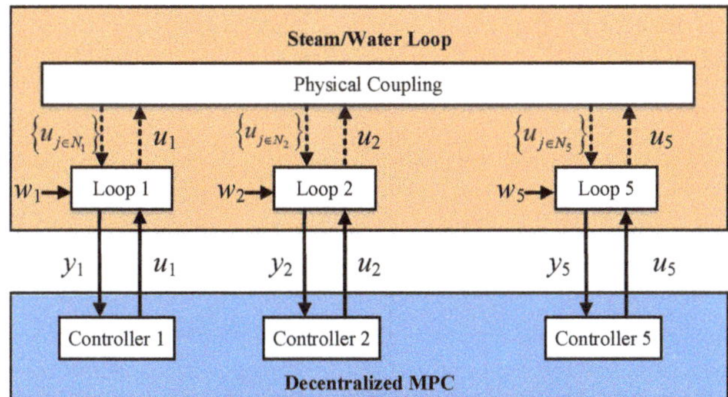

Figure 6. A conceptual representation of the decentralized MPC architecture [33] (Reproduced with permission from Maxim, A.; Copot, D.; De Keyser, R.; and Ionescu, C.M., Journal of Process Control; published by Elsevier, 2018).

Hence, the local cost function to be minimized by each controller is:

$$\min_{\mathbf{U}_i} \mathbf{J}_i(\mathbf{U}_i) = \mathbf{U}_i{}^T \mathbf{H}_i \mathbf{U}_i + 2\mathbf{f}_i{}^T \mathbf{U}_i + c_i \text{ subject to } \mathbf{A}\mathbf{U} \leq \mathbf{b} \tag{15}$$

$$\text{with} \begin{cases} \mathbf{H}_i = \mathbf{G}_{ii}^T \mathbf{G}_{ii} \mathbf{f}_i = -\mathbf{G}_{ii}^T (\mathbf{R}_i - \overline{\mathbf{Y}}_i) \\ c_i = (\mathbf{R}_i - \overline{\mathbf{Y}}_i)^T (\mathbf{R}_i - \overline{\mathbf{Y}}_i) \end{cases}$$

which is derived from (13), by removing the coupling influence between sub-loops.

3.4. Multiple Objective Distributed Model Predictive Control (MODiMPC)

Nowadays, setpoint tracking is not the only target for the control system. For some fast dynamic systems, the computing speed of the control strategies has an important influence. In order to improve the computing speed, a small loss in tracking performance is made to realize the fast computing speed. The scheme of the MODiMPC is shown in Figure 7. There are three layers of structure, in which the priority is shown as safety > tracking performance > energy.

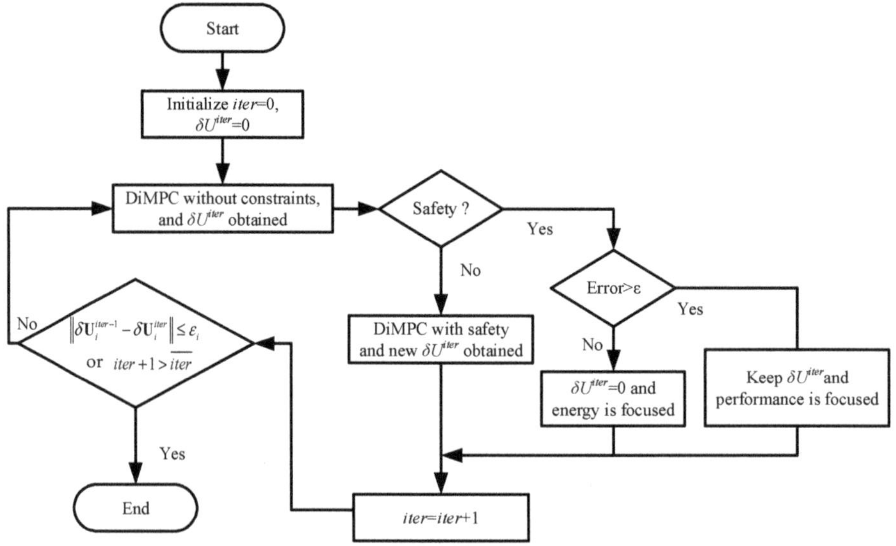

Figure 7. Flow chart for the MODiMPC.

The algorithm starts from initial zero conditions and computes the optimal control effort as an unconstrained solution of the optimization problem that aims to minimize the tracking error.

If the predicted inputs or outputs are not safe (generally out of the hard constraints), the constraints is included in the optimization to ensure safety. If the variables are in the safety interval, then according to the tracking error condition, two options are available, namely: (i) Error > ε, the focus is on performance, and the control effort is kept $\delta \mathbf{U}_i^{iter}$, (i.e., the control effort is kept to the one which minimizes the cost function with (11)); or (ii) Error < ε, the focus is on energy, and the control effort is kept $\delta \mathbf{U}_i^{iter} = 0$ (i.e., the actuator does not need to do any change), where ε is a tolerance error, and in this paper, it is chosen as 1% of the upper bound of the corresponding output. According to the end conditions ($\|\delta \mathbf{U}_i^{iter+1} - \delta \mathbf{U}_i^{iter}\| \leq \varepsilon_i \vee iter + 1 > \overline{iter}$) of the DiMPC, the procedure stops or continues to obtain a new result.

Due to the hydraulic cylinder being linked with the valve in the steam/water loop, the frequent changes in valves mean frequent changes in the hydraulic cylinder, which results in a large energy costs. In this sense, the energy is saved if the valves do not need to do any change under the condition Error < ε.

In the traditional MPC, the influence from constraints has always been considered to obtain the optimal inputs for the system. In our study, the quadratic programming is applied. However, the setpoint does not always change during the operation of the system, and most of the time, the system operates at a stable operating point. During this kind of period, the only thing to be considered is energy, in which the control effort is always kept the same as the last sampling time. Hence, no optimization process exists anymore, and there is a huge reduction in computing time.

3.5. Convergence Issue

In order to analyze the stability of the optimal solution of the distributed control system, first the convergence issue is discussed. A standard MPC formulation is written in a form of a series of static optimization problems shown as follows:

$$SP_k : \min_{S} J(S) \tag{16}$$

$$\text{s.t. } M(S) = 0$$
$$C(S) \leq 0$$

where S is the vector of the decision variables, including the state variables X and the control variables U. $M(S)$ is the prediction model and $C(S)$ denotes the constraints. Although our process model has an input-output formulation, it can be easily translated into a state-space definition.

In the DiMPC, Equation (16) is decompositioned into subproblems. For the sub-loop i:

$$SP_{ki} : \min_{S_i} J_i(S_i, S_i^{nei}) \tag{17}$$

$$\text{s.t. } M_i(S_i, S_i^{nei}) = 0$$
$$C_i(S_i, S_i^{nei}) \leq 0$$

where S_i is the vector of the decision variables for sub-loop i, and S_i^{nei} is the vector of the variables for the sub-loops that have interaction with sub-loop i. M_i and C_i meet the conditions: $\cup_i S_i = S, \cup_i M_i = M, \cup_i C_i = C$.

According to [35], if the distributed control methodologies satisfy some conditions, the properties that can be proved for the equivalent CMPC problem are enjoyed by the solution obtained using the DiMPC implementation. Also, the convergence issue of the DiMPC is equal to the CMPC. The optimal results from the DiMPC converge to the global optimal point. The conditions are listed as follows:

(1) The sub-loops can completely cover the full large system;
(2) J_i and C_i are convex;
(3) The sub-loops work sequentially;
(4) The starting point is in the interior of the feasible region;
(5) Each sub-loop cooperates with its neighbors in that it broadcasts its latest iteration to these neighbors;
(6) Each sub-loop uses the same optimal method to generate its iterations.

However, the conditions 2 and 3 are over strict as many systems are nonconvex and have nonlinearity in reality. Further, [36,37] show that these two conditions can be relaxed to nonconvex optimal problems with nonlinearity.

Moreover, the convergence of the DiMPC is further analyzed using the study given in [33]. Hence, starting from the unconstrained optimal solution of the distributed algorithm, the idea is to rewrite it with a recursive matrix formulation. After some matrix manipulation, a compact description is obtained:

$$U^*(k) = F(k) - \hat{H} U^*(k-1) \tag{18}$$

where $U^*(k)$ consists of the optimal sequences of all the sub-loops, computed at sample time k, while $U^*(k-1)$ are the shifted optimal trajectories computed at the previous sampling time $k-1$, with the last term doubled, to ensure the dimensions consistencies. The term $F(k)$ is variable and computed at each sampling instant using the prediction error, whereas $\hat{H}$ is a constant term that is computed off-line in the initialization stage of the algorithm (see [33] for further details).

Note that using Equation (18), the convergence of the local optimal solutions can be checked by verifying that all the eigenvalues from $\hat{H}$ are inside the unit circle. Additionally, Equation (18) can be reformulated in the classical system approach as:

$$U^*(k) = (I - q^{-1}\hat{H})F(k) \tag{19}$$

where q^{-1} is an operator that shifts the data backward one sampling period, $F(k)$ is regarded as the system's input, while $U^*(k)$ denotes the system's output. It is noteworthy to mention that Equation (19) can be used to analyze the stability of the optimal solution $U^*(k)$ at sampling time k, in the classical linear time-invariant framework by verifying that all the eigenvalues of the system equivalent matrix $(I - q^{-1}\hat{H})$ are inside the unit circle. Furthermore, if this condition is satisfied on the equality case, it results in the optimal solution of the distributed algorithm being marginally stable.

Hence, using this simple approach, the evolution of the system is computed using $F(k)$ as the system's input, which is calculated using the prediction error from each sampling period. Although this is an analytical approach to recursively place the system's progression in time, it can be straightforwardly computed in an automatic manner, using the simulation tools available for a control engineer. Moreover, all the computations are computed in a distributed manner, since using Equation (18), each sub-loop computes the optimal trajectories of the coupling neighbors, and knowing this information, it computes its own optimal trajectory at each sampling instant.

4. Simulation Results and Analysis

According to our previous work, the parameter configuration for the EPSAC method is shown in Table 2.

Table 2. The parameters applied in various MPC controllers.

Controllers	N_c	T_s	N_p	N_1	N_s
DeMPC CMPC DiMPC	$N_{c1} = 4, N_{c2} = 1, N_{c3} = 1,$ $N_{c4} = 4, N_{c5} = 6$ samples	5 s	$N_{p1} = 20; N_{p2} = 15; N_{p3} = 15;$ $N_{p4} = 20; N_{p5} = 20$ samples	1	300

Where the T_s is the sampling time; $N_{c1}, N_{c2}, \ldots, N_{c5}$ are control horizons; (the control horizons were selected by finding a good trade-off between tracking performance and computation time for each loop), $N_{p1}, N_{p2}, \ldots, N_{p5}$ are prediction horizons of the five loops, respectively (the prediction horizons were selected taking into account the specific transient dynamics for each loop); N_s is the number of the samples. The step setpoints are provided in Table 3. In the experiments, the initial condition was set at the operating point of the steam/water loop.

Table 3. Step setpoints changes in the experiments.

Time (s)	2–300	300–600	600–900	900–1200	1200–1500
Drum Water Level (m)	2	2	2	2	2
Exhaust Manifold Pressure(MPa)	100.03	116	116	116	116
Deaerator Pressure (KPa)	30	30	35	35	35
Deaerator Water Level(m)	0.7	0.7	0.7	0.8	0.8
Condenser Water Level(m)	0.5	0.5	0.5	0.5	0.6

The simulation results are shown in Figure 8, including the system outputs and the corresponding control efforts. In order to test which case provides the best result, performance indexes in an average value for the five sub-loops were compared including integrated absolute relative error (*IARE*), integral secondary control output (*ISU*), ratio of integrated absolute relative error (*RIARE*), ratio of integral secondary control output (*RISU*) and combined index (*J*). These indexes are calculated with the following expressions:

$$IARE = \frac{1}{5} \sum_{i=1}^{5} \sum_{k=0}^{N_s-1} |r_i(k) - y_i(k)| / r_i(k) \tag{20}$$

$$ISU = \frac{1}{5} \sum_{i=1}^{5} \sum_{k=0}^{N_s-1} (u_i(k) - u_{ssi}(k))^2 \tag{21}$$

$$RIARE(C_2, C_1) = \frac{IARE(C_2)}{IARE(C_1)} \tag{22}$$

$$RISU(C_2, C_1) = \frac{ISU(C_2)}{ISU(C_1)} \tag{23}$$

$$J(C_2, C_1) = \frac{w_1 RIARE(C_2, C_1) + w_2 RISU(C_2, C_1)}{w_1 + w_2} \tag{24}$$

where u_{ssi} is the steady state value of the ith input; C_1, C_2 are the compared controllers and the weighting factors w_1 and w_2 in (24) are chosen as $w_1 = w_2 = 0.5$.

As depicted in Figure 8, the CMPC had similar performance as the DiMPC, and both outperformed the DeMPC. This conclusion is not only valid for this process, but also for other processes, since the DeMPC strategy does not take the interactions into account, which lead to some severe fluctuations when the setpoint changes in other variables. Although the control efforts in the DiMPC are obtained separately, the performance is still good due to the iteratively communication between the controllers for each sub-loop. In real life operations, where the addition of the effects of noise and stochastic disturbances are needed, perhaps with adding periodic disturbances from sea dynamics, the DeMPC may even lead to instability in the overall system.

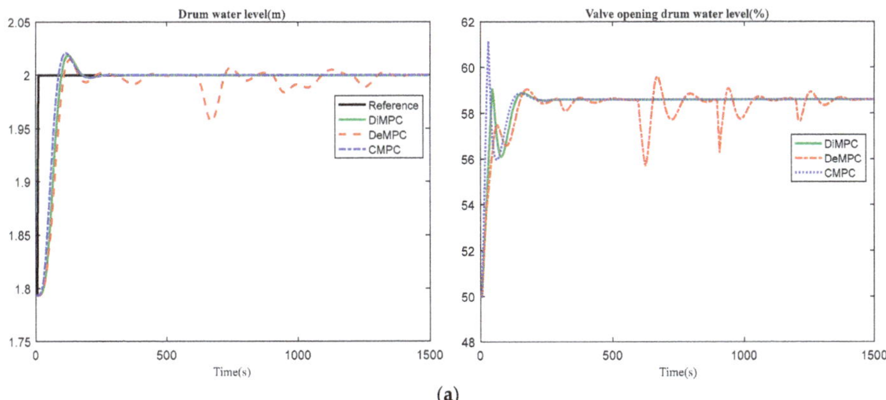

(**a**)

Figure 8. *Cont.*

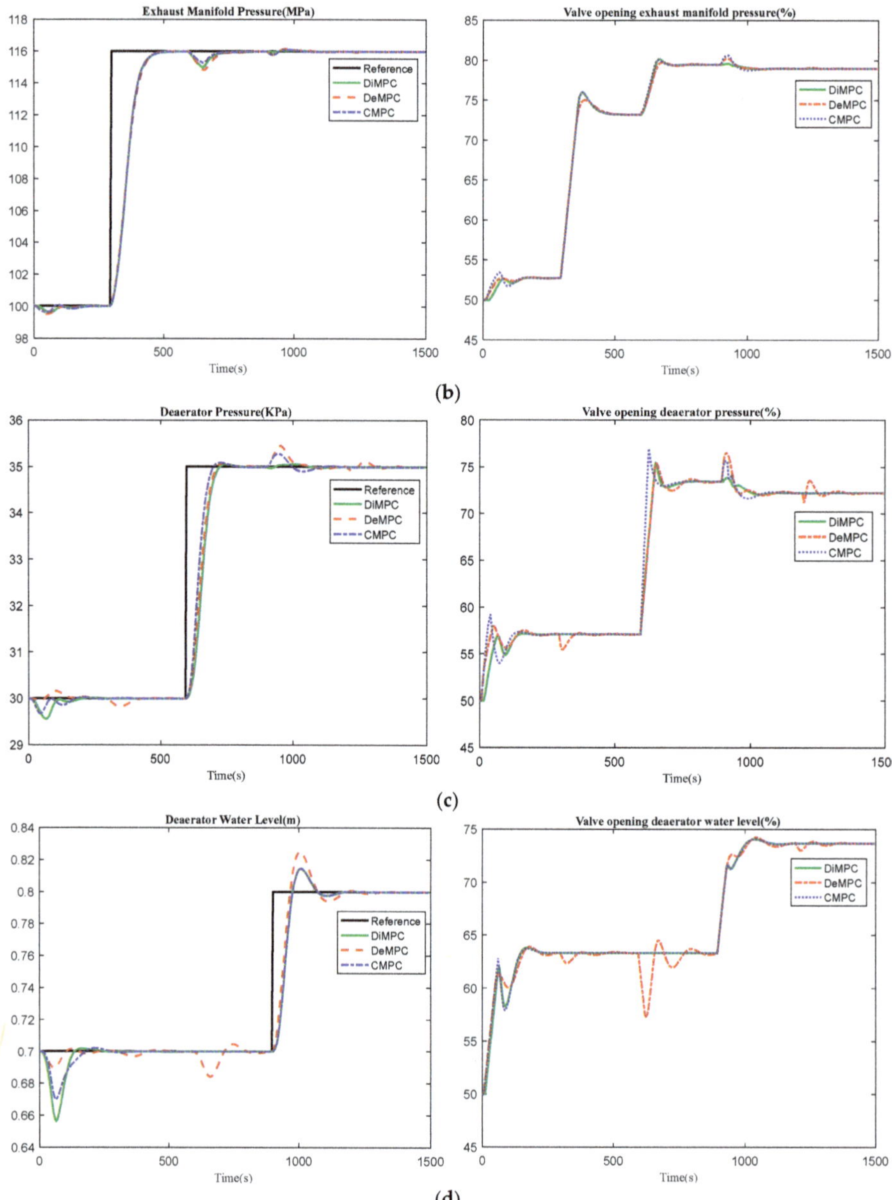

(b)

(c)

(d)

Figure 8. *Cont.*

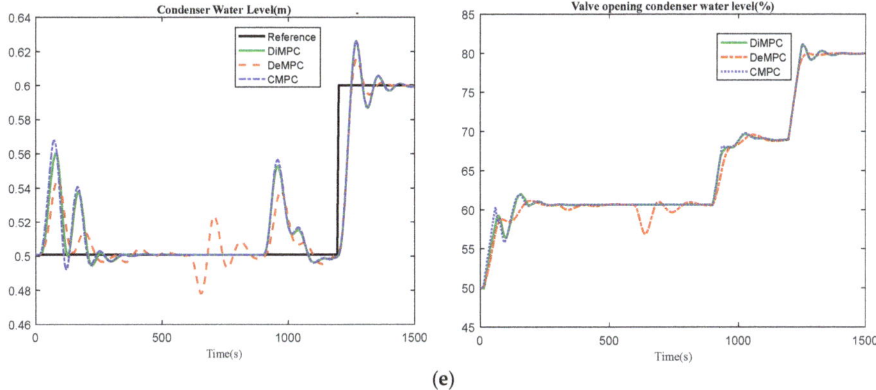

(e)

Figure 8. The responses of the steam/water loop under the DeMPC, CMPC and DiMPC for (**a**) drum water level control loop, (**b**) deaerator water level control loop, (**c**) deaerator pressure control loop, (**d**) condenser water level control loop and (**e**) exhaust manifold pressure control loop (The figures on left hand indicate the outputs, and on the right hand indicate the inputs).

The iterations are shown in Figure 9 during the optimization of DiMPC. In the algorithm, the two conditions to end the iteration are designed as: (i) The difference between consecutive optimal inputs fulfill the condition $\|\delta\mathbf{U}_i^{iter+1} - \delta\mathbf{U}_i^{iter}\| \leq 0.002$; (ii) the maximum iteration time is five, i.e., $iter > 5$.

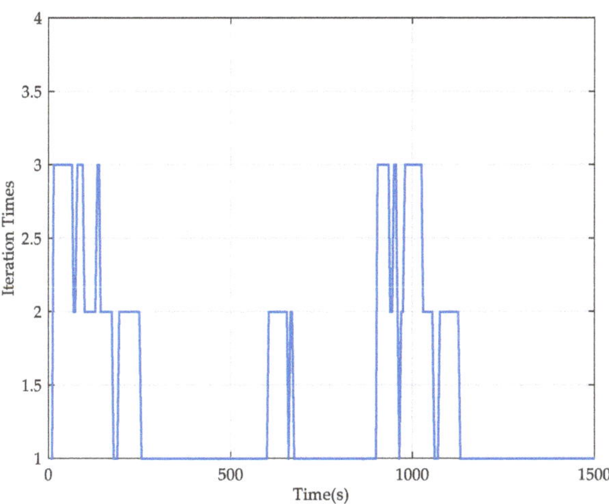

Figure 9. The iteration times when the DiMPC is applied to the steam/ water loop.

The same conclusion is also obtained according to the numerical values shown in Tables 4 and 5. As the index *J* implies, the DiMPC and CMPC have similar results. However, there is only one controller in the CMPC, which means the system may be out of service if there is any problem with the controller. On the contrary, the DiMPC has more ability in fault-tolerance and flexibility without much performance loss compared with the CMPC. As the DiMPC only needs a part of the entire model, it is much easier to find a feasible solution, while CMPC needs the entire model to obtain all the solutions at one time. In this context, the DiMPC has better robust performance than the CMPC. Hence, the system model required for the DiMPC can be less accurate than the CMPC. In an industry context, a staggering 60–70% of the project time is spent on model development, while the rest is claimed for the controller design and validation [38]. Hence, any reduction in identification time requirement greatly diminishes overall loop maintenance control related costs. The analysis given in this paper provides a trade-off solution, yet with acceptable performance but with great yields for cost reduction in control design and validation time.

Table 4. Performance indexes for *IARE* and *ISU*.

Index	DiMPC	DeMPC	CMPC
IARE	2.5117	2.7507	2.4795
ISU	0.4672	0.4686	0.4937

Table 5. Performance indexes for *RIARE* and *RISU*.

Index	DiMPC VS DeMPC	DeMPC VS CMPC	CMPC VS DiMPC
RIARE	0.8959	1.1417	1.0011
RISU	0.8863	1.1629	1.0290
J	0.8911	1.1523	1.0151

The results for the DiMPC and the multiple objective distributed model predictive control (MODiMPC) are shown in Figure 10. The performance indexes are shown in Table 6. The computing time for MODiMPC is 2.81 s and for the DiMPC 29.36 s, respectively.

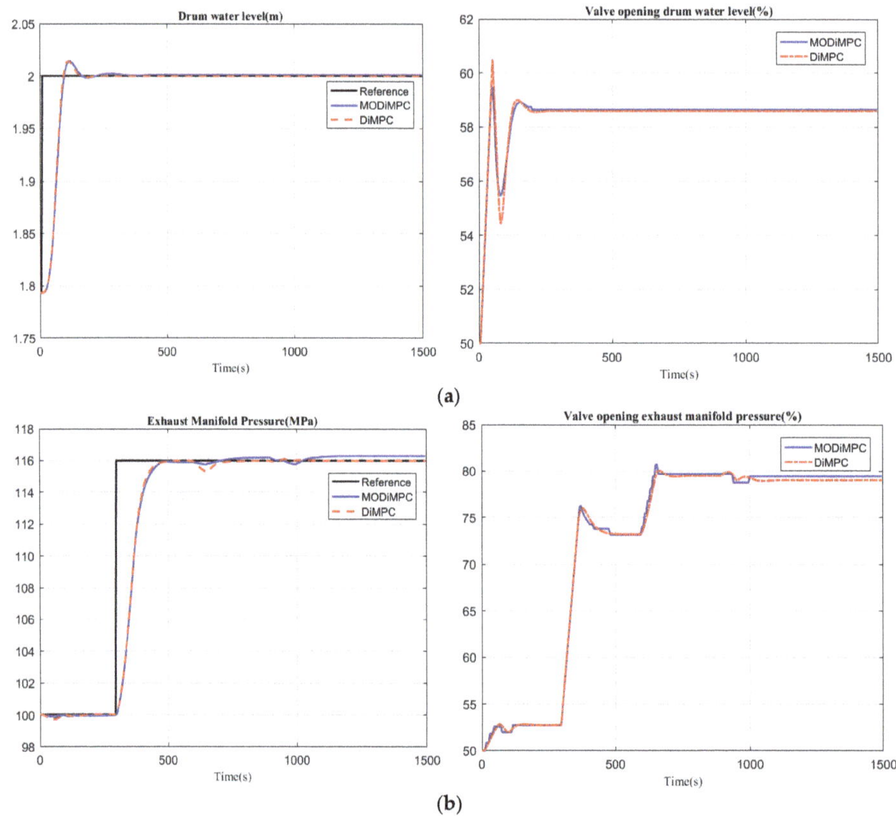

Figure 10. *Cont.*

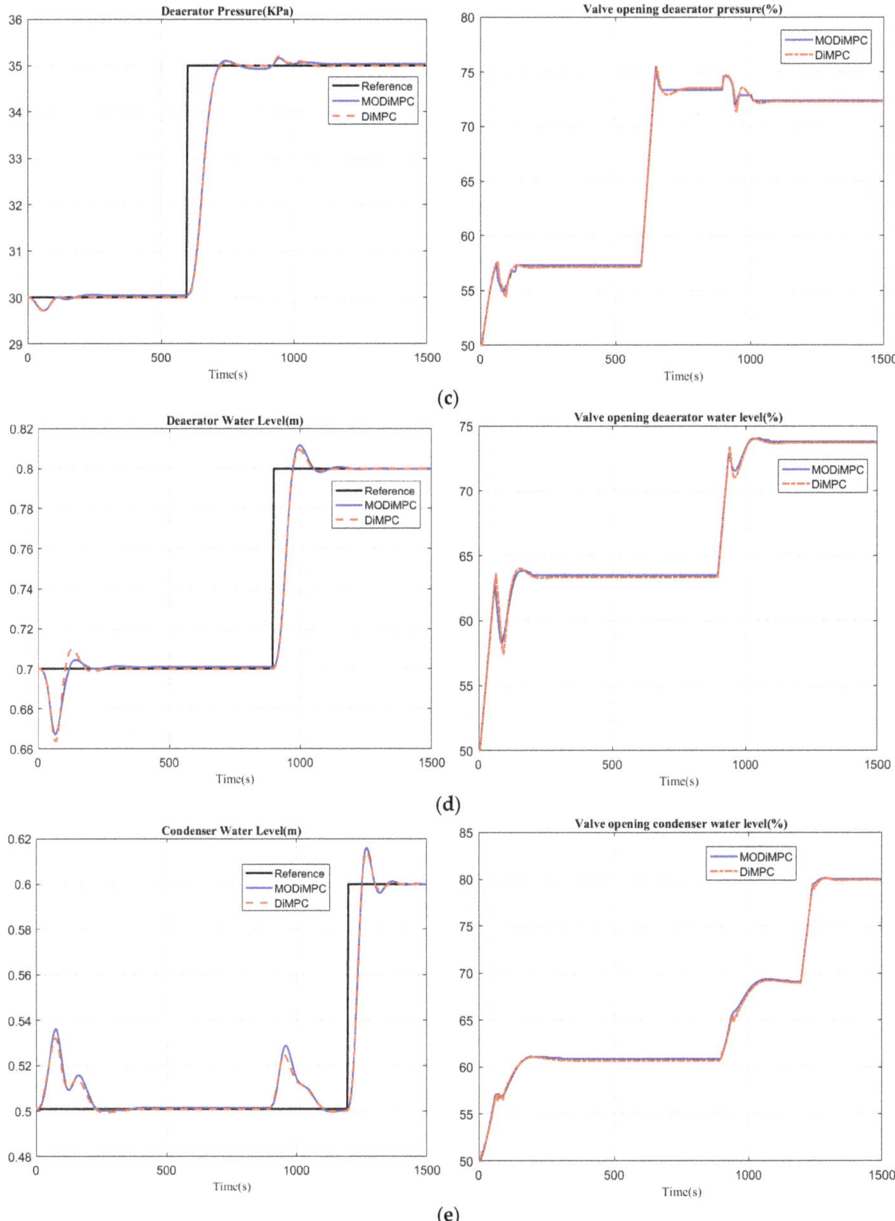

Figure 10. The responses of the steam/water loop under the MODiMPC and DiMPC for (**a**) drum water level control loop, (**b**) deaerator water level control loop, (**c**) deaerator pressure control loop, (**d**) condenser water level control loop and (**e**) exhaust manifold pressure control loop (The figures on left hand indicate the outputs, and on the right hand indicate the inputs).

Table 6. Performance indexes for *IARE* and *ISU*.

	MODiMPC	DiMPC
IARE	2.6289	2.5117
ISU	0.4434	0.4672

It can be seen from the results that there is a large improvement in the computing time after the MODiMPC was applied and without too much loss in tracking performance.

As previously mentioned, in order to guarantee the stability of the optimal solution in a DiMPC framework, the convergence of the optimal solution was firstly discussed. Due to the fact that there are only constraints in input variables in our study, the feasibility of the steam/water loop belongs to the trivial case (according to [33], solving the optimization problem the existence of a feasible solution is ensured at each sampling period). The H_i matrices for the five loops are calculated as follows (for more details about convergence issue, please refer to the reference [33]):

$$H_1 = \begin{bmatrix} 0.9613 & -0.0352 & -0.0308 & -0.0639 \\ -0.0352 & 0.9673 & -0.0293 & -0.0716 \\ -0.0308 & -0.0293 & -0.9730 & -0.0781 \\ -0.0639 & -0.0716 & -0.0781 & 0.2062 \end{bmatrix}, H_2 = 0.0003, H_3 = 0.0025,$$

$$H_4 = \begin{bmatrix} 0.9866 & -0.0125 & -0.0113 & -0.0224 \\ -0.0125 & 0.9876 & 0.0114 & -0.0298 \\ -0.0113 & -0.0114 & -0.9888 & -0.0369 \\ -0.0224 & -0.0298 & -0.0369 & 0.4255 \end{bmatrix},$$

$$H_5 = \begin{bmatrix} 0.9788 & -0.0189 & -0.0159 & -0.0119 & -0.0077 & 0.0145 \\ -0.0189 & 0.9789 & 0.0189 & -0.0160 & -0.0121 & 0.0106 \\ -0.0159 & -0.0189 & -0.9788 & -0.0190 & -0.0160 & 0.0041 \\ -0.0119 & -0.0160 & -0.0190 & 0.9788 & -0.0188 & -0.0049 \\ -0.0077 & -0.0121 & -0.0160 & -0.0188 & 0.9792 & -0.0158 \\ 0.0145 & 0.0106 & 0.0041 & -0.0049 & -0.0158 & 0.4403 \end{bmatrix}$$

By the eig function in MATLAB (R2016b, Mathworks, Natick, MA, USA, 2016), the eigenvalues are calculated in a centralized manner for the steam/water loop, and the maximum value is $\rho_{\max} < 1$, which indicates that the DiMPC is convergent. In the steam/water loop, the five sub-loops cover the full system, and in the DiMPC the information is exchanged iteratively. Hence, the conditions 1 and 3–6 are satisfied. In order to cover the worst circumstances, the sufficient conditions in Section 3.5 tend to be conservative, and in some cases the convexity are not necessary [37]. Hence, it is concluded that the DiMPC has the same convergence as the CMPC, and the convergence of the DiMPC is guaranteed.

5. Conclusions

Regarding the multiple sub-loops in the steam/water loop, this paper introduced a distributed model predictive control based on the EPSAC framework. Different types of the MPC were applied to the steam/water loop system, including the DeMPC, CMPC and DiMPC. According to the simulation results, the DiMPC had similar performance with the CMPC, and outperformed the DeMPC. Due to the multiple controllers in the DiMPC strategy, the DiMPC had better performance of fault-tolerance and flexibility than the CMPC which improved the reliability of the steam/water loop. By proving equivalence in stability between the DiMPC and the CMPC, and the stability of the CPMC, the stability of the DiMPC is guaranteed. Meanwhile, a multiple objective MPC was proposed, and the computing speed was improved without too much loss in tracking performance.

Author Contributions: Methodology, R.D.K., A.M. and S.Z.; software, S.Z., A.M. and S.L.; formal analysis, S.Z.; writing—original draft preparation, S.Z. and A.M.; writing—review and editing, S.Z., A.M., R.D.K., S.L., and C.M.I.; supervision, C.M.I., S.L. and R.D.K.; funding acquisition, S.Z., S.L. and C.M.I.

Funding: Shiquan Zhao acknowledges the financial support from Chinese Scholarship Council (CSC) under grant 201706680021 and the Co-funding for Chinese PhD candidates from Ghent University under grant 01SC1918. Part of this project is funded by a special research fund of Ghent University, MIMOPREC STG020-18 (Ionescu). This work is partly supported by the Fundamental Research Funds for the Central Universities under grant 3072019CF0408, 3072019CFT0403.

Conflicts of Interest: The authors declare no conflicts of interest.

Nomenclature

EPSAC	Extended Prediction Self Adaptive Control
MPC	Model Predictive Control
CMPC	Centralized Model Predictive Control
DiMPC	Distributed Model Predictive Control
DeMPC	Decentralized Model Predictive Control
MOMPC	Multiple Objective Model Predictive Control
MODiMPC	Multiple Objective Distributed Model Predictive Control
ISU	Integral Secondary control output
RISU	Ratio of Integral Secondary control output
IARE	Integrated Absolute Relative Error
RIARE	Ratio of Integrated Absolute Relative Error

References

1. Mazur, K.; Wydra, M.; Ksiezopolski, B. Secure and time-aware communication of wireless sensors monitoring overhead transmission lines. *Sensors* **2017**, *17*, 1610. [CrossRef] [PubMed]
2. Smierzchalski, R. Simulation system for marine engine control room. In Proceedings of the 2008 11th International Biennial Baltic Electronics Conference, Tallinn, Estonia, 6–8 October 2008; pp. 281–284.
3. Vandermeulen, A.; van der Heijde, B.; Helsen, L. Controlling district heating and cooling networks to unlock flexibility: A review. *Energy* **2018**, *151*, 103–115. [CrossRef]
4. Morstyn, T.; Hredzak, B.; Demetriades, G.D.; Agelidis, V.G. Unified distributed control for DC microgrid operating modes. *IEEE Trans. Power Syst.* **2016**, *31*, 802–812. [CrossRef]
5. Maxim, A.; Copot, D.; Copot, C.; Ionescu, C.M. The 5W's for Control as Part of Industry 4.0: Why, What, Where, Who, and When—A PID and MPC Control Perspective. *Inventions* **2019**, *4*, 10. [CrossRef]
6. Copot, D.; Ghita, M.; Ionescu, C.M. Simple Alternatives to PID-Type Control for Processes with Variable Time-Delay. *Processes* **2019**, *7*, 146. [CrossRef]
7. Haji, V.H.; Monje, C.A. Fractional order fuzzy-PID control of a combined cycle power plant using Particle Swarm Optimization algorithm with an improved dynamic parameters selection. *Appl. Soft Comput.* **2017**, *58*, 256–264. [CrossRef]
8. Puchalski, B.; Duzinkiewicz, K.; Rutkowski, T. Multi-region fuzzy logic controller with local PID controllers for U-tube steam generator in nuclear power plant. *Arch. Control Sci.* **2015**, *25*, 429–444. [CrossRef]
9. Magdy, G.; Mohamed, E.A.; Shabib, G.; Elbaset, A.A.; Mitani, Y. SMES based a new PID controller for frequency stability of a real hybrid power system considering high wind power penetration. *IET Renew. Power Gener.* **2018**, *12*, 1304–1313. [CrossRef]
10. Salehi, A.; Safarzadeh, O.; Kazemi, M.H. Fractional order PID control of steam generator water level for nuclear steam supply systems. *Nucl. Eng. Des.* **2019**, *342*, 45–59. [CrossRef]
11. Xi, Y.; Yu, X.; Wang, Y.; Li, Y.; Huang, J. Robust Nonlinear Adaptive Backstepping Coordinated Control for Boiler-Turbine Units. In Proceedings of the 2018 IEEE 27th International Symposium on Industrial Electronics (ISIE), Cairns, Australia, 13–15 June 2018; pp. 1167–1172.
12. Cai, J.; Sun, L. Direct Fuzzy Backstepping Control for Turbine Main Steam Valve of Multi-machine Power System. In Proceedings of the 2015 2nd International Conference on Information Science and Control Engineering, Shanghai, China, 24–26 April 2015; pp. 702–706.

13. Roy, T.K.; Mahmud, M.A.; Shen, W.; Oo, A.M. Non-linear adaptive coordinated controller design for multimachine power systems to improve transient stability. *IET Gener. Trans. Distrib.* **2016**, *10*, 3353–3363. [CrossRef]

14. Rinaldi, G.; Cucuzzella, M.; Ferrara, A. Sliding mode observers for a network of thermal and hydroelectric power plants. *Automatica* **2018**, *98*, 51–57. [CrossRef]

15. Kenne, G.; Fombu, A.M.; de Dieu Nguimfack-Ndongmo, J. Coordinated excitation and steam valve control for multimachine power system using high order sliding mode technique. *Electr. Power Syst. Res.* **2016**, *131*, 87–95. [CrossRef]

16. Ansarifar, G.R.; Akhavan, H.R. Sliding mode control design for a PWR nuclear reactor using sliding mode observer during load following operation. *Ann. Nucl. Energy.* **2015**, *75*, 611–619. [CrossRef]

17. Moradi, H.; Saffar-Avval, M.; and Bakhtiari-Nejad, F. Sliding mode control of drum water level in an industrial boiler unit with time varying parameters: A comparison with H_∞-robust control approach. *J. Process. Contr.* **2012**, *22*, 1844–1855. [CrossRef]

18. Ansarifar, G.R. Control of the nuclear steam generators using adaptive dynamic sliding mode method based on the nonlinear model. *Ann. Nucl. Energy* **2016**, *88*, 280–300. [CrossRef]

19. Wu, J.; Nguang, S.K.; Shen, J.; Liu, G.; Li, Y.G. Robust H_∞ tracking control of boiler-turbine systems. *ISA Trans.* **2010**, *49*, 369–375. [CrossRef] [PubMed]

20. Ghabraei, S.; Moradi, H.; Vossoughi, G. Design & application of adaptive variable structure &H_∞ robust optimal schemes in nonlinear control of boiler-turbine unit in the presence of various uncertainties. *Energy* **2018**, *142*, 1040–1056.

21. Sun, L.; Li, D.; Hu, K.; Lee, K.Y.; Pan, F. On tuning and practical implementation of active disturbance rejection controller: A case study from a regenerative heater in a 1000 MW power plant. *Ind. Eng. Chem. Res.* **2016**, *55*, 6686–6695. [CrossRef]

22. Sun, L.; Hua, Q.; Shen, J.; Xue, Y.; Li, D.; Lee, K.Y. Multi-objective optimization for advanced superheater steam temperature control in a 300 MW power plant. *Appl. Energy* **2017**, *208*, 592–606. [CrossRef]

23. Sun, L.; Hua, Q.; Li, D.; Pan, L.; Xue, Y.; Lee, K.Y. Direct energy balance based active disturbance rejection control for coal-fired power plant. *ISA Trans.* **2017**, *70*, 486–493. [CrossRef]

24. Zhang, F.; Wu, X.; Shen, J. Extended state observer based fuzzy model predictive control for ultra-supercritical boiler-turbine unit. *Appl. Therm. Eng.* **2017**, *118*, 90–100. [CrossRef]

25. Kong, X.; Liu, X.; Lee, K.Y. Nonlinear multivariable hierarchical model predictive control for boiler-turbine system. *Energy* **2015**, *93*, 309–322. [CrossRef]

26. Wu, X.; Shen, J.; Li, Y.; Lee, K.Y. Fuzzy modeling and predictive control of superheater steam temperature for power plant. *ISA Trans.* **2015**, *56*, 241–251. [CrossRef] [PubMed]

27. Liu, X.; Cui, J. Economic model predictive control of boiler-turbine system. *J. Process Control* **2018**, *66*, 59–67. [CrossRef]

28. Zhao, S.; Maxim, A.; Liu, S.; De Keyser, R.; Ionescu, C. Effect of Control Horizon in Model Predictive Control for Steam/Water Loop in Large-Scale Ships. *Processes* **2018**, *6*, 265. [CrossRef]

29. Zhao, S.; Cajo, R.; De Keyser, R.; Liu, S.; Ionescu, C.M. Nonlinear predictive control applied to steam/water loop in large scale ships. In Proceedings of the 12th IFAC Symposium on Dynamics and Control of Process Systems, Including Biosystems, Florianopolis, Brazil, 24–26 April 2019.

30. Gao, Z.; Saxen, H.; Gao, C. Special section on data-driven approaches for complex industrial systems. *IEEE Trans. Ind. Inform.* **2013**, *9*, 2210–2212. [CrossRef]

31. Gao, Z.; Nguang, S.K.; Kong, D.X. Advances in Modelling, Monitoring, and Control for Complex Industrial Systems. *Complexity* **2019**. [CrossRef]

32. Venkat, A.N.; Rawlings, J.B.; Wright, S.J. Stability and optimality of distributed model predictive control. In Proceedings of the 44th IEEE Conference on Decision and Control, Seville, Spain, 15 December 2005; pp. 6680–6685.

33. Maxim, A.; Copot, D.; De Keyser, R.; Ionescu, C.M. An industrially relevant formulation of a distributed model predictive control algorithm based on minimal process information. *J. Process Control* **2018**, *68*, 240–253. [CrossRef]

34. De Keyser, R. Model based predictive control for linear systems, in: UNESCO Encyclopaedia of Life Support Systems. *Control Syst. Robot. Autom.* **2009**, *11*, 24–58.

35. Camponogara, E.; Jia, D.; Krogh, B.H.; Talukdar, S. Distributed model predictive control. *IEEE Control Syst. Mag.* **2002**, *22*, 44–52.
36. Camponogara, E. Controlling Networks with Collaborative Nets. Ph.D. Thesis, Carnegie Mellon University, Pittsburgh, PA, USA, 2000.
37. Talukdar, S.; Camponogara, E. Collaborative nets. In Proceedings of the 33rd Annual Hawaii International Conference on System Sciences, Maui, HI, USA, 7 January 2000; p. 9.
38. Starr, K.D. *Single Loop Control Methods*; ABB Inc.: Westerville, OH, USA, 2015.

Article

Daytime/Nighttime Event-Based PI Control for the pH of a Microalgae Raceway Reactor

Enrique Rodríguez-Miranda [1], Manuel Beschi [2], José Luis Guzmán [3,*], Manuel Berenguel [3] and Antonio Visioli [1]

[1] Department of Mechanical and Industrial Engineering, University of Brescia, 25123 Brescia, Italy; e.rodriguezmiran@unibs.it (E.R.-M.); antonio.visioli@unibs.it (A.V.)
[2] Institute of Intelligent Industrial Technologies and Systems for Advanced Manufacturing, National Research Council of Italy, 20133 Milan, Italy; Manuel.Beschi@stiima.cnr.it
[3] Department of Computer Science, University of Almería, 04120 Almería, Spain; beren@ual.es
* Correspondence: joseluis.guzman@ual.es

Received: 12 March 2019; Accepted: 19 April 2019; Published: 28 April 2019

Abstract: In this paper, a new solution to improve the traditional control operation of raceway microalgae reactors is presented. The control strategy is based on an event-based method that can be easily coupled to a classical time-driven proportional-integral controller, simplifying the design process approach. The results of a standard Proportional-Integral (PI) controller, as well as of two event-based architectures, are presented in simulation and compared with each other and with traditional On/Off control. It is demonstrated that the event-based PI controller—operating during the whole day instead of only during daytime—achieves a better performance by reducing the actuator effort and saving costs related to gas consumption.

Keywords: microalgae; raceway; control problem; PID; event-based

---

1. Introduction

The importance of environmental sustainability and the rising of renewable energies promote the development of new energy sources, such as microalgae reactors. These biological processes have become very popular nowadays due to their great potential to produce biofuels and high-value products. The biomass obtained in these processes can be useful for applications such as cosmetics at the pharmaceutical industry, or even products for agriculture and aquaculture coupled to wastewater treatment [1]. Among many advantages, these biological systems stand out for the capability of microalgae to produce valuable compounds from photosynthesis while consuming CO_2 and nutrients, even coming from flue gases and wastewater.

There are mainly two types of reactors: tubular photobioreactors, for high-value microalgae strains, and raceway or open reactors. This second type of reactor is the most common one on an industrial scale due to its operation simplicity and its low maintenance costs [2,3].

Biological systems have complex dynamics, difficult to control, and can be affected by several variables. Microalgae growth depends on temperature, solar radiation, pH and dissolved oxygen [4]. The incidence of solar radiation and temperature conditions are determined by the reactor design. Therefore, the controlled variables are pH and dissolved oxygen, both of which are disturbed by solar radiation [5]. Concretely, pH is more critical in the growth process because of its direct influence on the photosynthesis process, and thus it is the controlled variable analysed in this work.

Traditionally, microalgae reactors are operated by On/Off controllers that work only during the daytime period. In this way, the pH of the microalgae varies significantly during the whole day, especially during nighttime, being harmful to the algae. Furthermore, this type of control does not achieve enough efficiency in productivity to compete against other biofuel sources in the energy market. Therefore, it is necessary to improve the control algorithms that allow the critical variables of the system to be maintained at optimum values while reducing the cost of biomass production. The Proportional-Integral-Derivative (PID) controllers are widely used in industry with satisfactory control results (performance) and can be used for these type of processes. Different examples can be found in the literature. An example of a linear Proportional-Integral (PI) controller with feedforward for pH control in tubular bioreactors can be found in [6]. A robust PID controller for pH control in raceway reactors based on Quantitative Feedback Theory (QFT) has been used in [7]. On the other hand, event-based control is gaining great interest for these kinds of processes, where the sensors are at a considerable distance from the control device. Moreover, these sensors can be wireless, so the use of event-based controllers would increase their life span. In [8], a controller with a sensor deadband produces a considerable reduction of CO_2 losses in a microalgae tubular reactor. Another example of the application of event-based control can be seen in [9], where an event-based Generalized Predictive Controller (GPC) with a disturbance compensation approach is used for the effective use of CO_2 in a raceway reactor. Subsequently, this GPC scheme was improved and combined with a selective control for dissolved oxygen [5]. Thus, through the event-based control paradigm, the desired trade-off between control performance and control effort was achieved. In addition, the actuator provides only the amount of gas needed, keeping the pH at an optimum value while reducing gas consumption.

In this work, the advantages of using PI control during the whole day (daytime and nighttime periods) against traditional On/Off control (performed only during daytime) are demonstrated. The greatest advantage that can be found in the use of PI control is the reduction in CO_2 consumption, despite operating the reactor during the whole day. In this way, the necessary amount of CO_2 must be injected to avoid losses to the atmosphere, while maintaining the pH at certain levels allowing the growth of the algae. Another advantage is the reduction in the Integrated Absolute Error (IAE) for pH, which helps to maintain the pH close to the set-point throughout daytime and nighttime. However, On/Off control leaves the pH to evolve free during nighttime, increasing the IAE value.

Moreover, this paper presents an event-based control architecture for PID controllers. The objective is to tune a classical time-driven PI for pH control in the raceway reactor, and then to add event-based capabilities but keeping the initial PI control design. The event-based systems allow a trade-off between control performance and control effort, being perfect for the microalgae process in raceway reactors. Thus, this type of event-based control improves the PI controller behaviour so that the CO_2 consumed can be further reduced at the expense of a slight degradation of the pH, also reducing the associated valve control effort.

One of the best-known event-based sampling methods is the so-called Send-On-Delta (SOD) sampling [10,11], a deadband or level-crossing sampling method where a node samples the signal only when it changes of a fixed value with respect to the previously sampled value. A modified version of the SOD technique, called Symmetric-Send-on-Delta (SSOD) sampling [12–14], applied to a Proportional-Integral (PI) controller, has been used in this work. In particular, two control schemes from this sampling method, called SSOD-PI and PI-SSOD have been proposed. The difference between both architectures depends on were the triggering function is applied in the control loop. This method improves the SOD sampling with a symmetric deadband mapping. The stability properties and limit cycles conditions for first-order-plus-dead-time (FOPDT) processes are highlighted in [12].

Therefore, this work focuses on a simulation study of the event-based SSOD approach combined with a PI controller in a dynamic model for microalgae production [15]. Two control architectures of the SSOD event-based method (SSOD-PI and PI-SSOD) are compared to an On/Off control method, which is traditionally used to operate these kinds of reactors for pH control. The performed tests were oriented to establish a trade-off between control effort and control performance and present an alternative to traditional control. In addition, the advantages and disadvantages over classical time-driven PI control techniques are presented.

The paper is organized as follows. The control scheme is presented in Section 2, where the control problem and the SSOD architecture are also described. The controller design, results, and discussion obtained from the simulation study are presented in Section 3. Finally, conclusions are drawn in Section 4.

2. Material and Methods

This section describes the control problem, the Symmetric-Send-on-Delta approach and two configurations applied to a PI controller.

2.1. Microalgae Raceway Reactor

The reactor simulator model represents a raceway reactor located at "Estación Experimental Las Palmerillas" owned by the *Fundación Cajamar* Cajamar, Almería, Spain. This reactor has a total surface area of 100 m^2, formed of two channels of 50 m that are connected by 1 m width U-shaped bends (Figure 1). The raceway reactor is operated at 0.2 m constant depth (20 m^3 total volume), as recommended in [16] to achieve the best overall hydraulic performance in terms of power consumption. Mixing is made with a 1.2 m diameter paddlewheel with eight blades of marine plywood, operated by an electric motor. Otherwise, carbonation is carried out in a sump (1 m deep and 0.65 m wide) at 1.8 m from paddlewheel. Three plate membrane diffusers at the bottom inject flue gas in the sump.

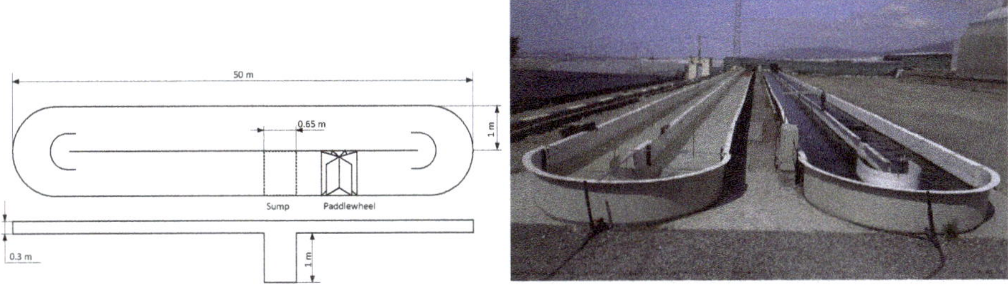

(a) Reactor scheme (top and side view).   (b) Reactor view.

Figure 1. Raceway reactor used in this work for simulation.

2.2. Reactor Dynamic Model

For the simulation results presented, a dynamic model of microalgae production in raceway reactors has been used [15]. The model is based on fundamental principles instead of empirical equations and takes into account mass balances, thermodynamic relationships and biological phenomena. This model can be used to predict the evolution of the main reactor variables, such as biomass concentration, pH and dissolved oxygen. It has been calibrated and validated using experimental data from a 100 m^2 pilot-scale raceway reactor, as can be seen in [15].

The culture growth is modelled as a function of the photosynthesis rate. The main parameter that determined the photosynthesis rate is the available light, which is based on different parameters such as external irradiance, culture characteristic, and reactor design [17]. The following equation represents the available light:

$$I_{av}(t,x) = \frac{I_0(t)}{K_a C_b(t,x)h}(1 - exp(-K_a C_b(t,x)h)), \tag{1}$$

where t is the time, x the space, I_0 is the solar irradiance on an obstacle-free horizontal surface, K_a is the extinction coefficient, C_b is the biomass concentration, and h is the liquid height on the channels.

The photosynthesis rate is modelled with the available light by the following equation [4]:

$$P_{O_2}(t,x) = (1 - \alpha_s) \frac{P_{O_{2,max}} I_{av}(t,x)^n}{K_i exp(I_{av}(t,x)m) + I_{av}(t,x)^n} \left(1 - \left(\frac{[O_2](t,x)}{K_{O_2}}\right)^z\right)$$

$$\left(B_1 exp\left(\frac{-C_1}{pH(t,x)}\right) - B_2 exp\left(\frac{-C_2}{pH(t,x)}\right)\right) - \alpha_s R_{O_2},$$

(2)

where P_{O_2} is the photosynthesis rate, α_s is a solar distributed factor that represents the shadow projection on the perpendicular axis of the reactor walls, $P_{O_{2,max}}$ is the maximum photosynthesis rate for microalgae under culture conditions, n is the form exponent, K_i and m are form factors for the exponential function of average irradiance. Furthermore, B_1 and B_2 are pre-exponential factors for the pH influence on the photosynthesis rate, and C_1 and C_2 are the activation energies of the Arrhenius model. R_{O_2} is a respiration constant that represents the respiration phenomenon.

The influence of pH and dissolved oxygen in the photosynthesis rate equation is of special attention, being the main variables of the system that have to be controlled.

These equations describe the biological part of the system, while the mixing and the gas–liquid mass transfer are expressed by several balances formulated in terms of Partial Differential Equations (PDE).

In [18], the pH is related to other variables such dissolved oxygen, carbonate or bicarbonate by several equilibrium equations. From these balances' equations, a prediction of pH along time and space can be obtained. In the following equation, the inorganic carbon concentration ($[C_T]$) is modelled taking into account the microalgae photosynthesis process and transport phenomena due to recirculation along the raceway reactor. We assumed constant velocity (v), and constant cross-sectional area due to the multiplication between the liquid height (h) and the channel width (w):

$$wh\frac{\partial [C_T](t,x)}{\partial t} = -whv\frac{\partial [C_T](t,x)}{\partial x} + wh\frac{P_{CO_2}(t,x)Cb(t,x)}{M_{CO_2}}$$
$$+ whK_{laCO_{2_c}}([CO_2^*](t,x) - [CO_2](t,x))$$

(3)

where P_{CO_2} is the carbon consumption rate, M_{CO_2} is the molecular weight of CO_2, and $K_{laCO_{2_c}}$ is the mass transfer coefficient for CO_2. The term $[CO_2]$ represents the carbon dioxide in the liquid phase and $[CO_2^*]$ represents the equilibrium concentration in the gas phase, which can be calculated by Henry's law taking into account the CO_2 properties in the air.

As for the inorganic carbon concentration, a balance equation for the dissolved oxygen is presented:

$$wh\frac{\partial [O_2](t,x)}{\partial t} = -whv\frac{\partial [O_2](t,x)}{\partial x} + wh\frac{P_{O_2}(t,x)Cb(t,x)}{M_{O_2}}$$
$$+ whK_{laO_{2_c}}([O_2^*](t,x) - [O_2](t,x)),$$

(4)

where P_{O_2} is the photosynthesis rate previously described, M_{O_2} is the molecular weight of the oxygen, and $K_{laO_{2_c}}$ is the volumetric gas–liquid mass transfer coefficient for oxygen. The term $([O_2^*](t,x) - [O_2])$ represents the driving force, where the equilibrium concentration in the gas phase $[O_2^*]$ can be calculated by Henry's law, as a function of the oxygen concentration in the gas phase.

Other mass balances can be applied to the paddlewheel and sump of the raceway reactor and represented by Ordinary Differential Equations (ODE) expressions. In the sump, CO_2 is injected in a gaseous form to control pH, and the air is injected to remove dissolved oxygen accumulated in the reactor.

The oxygen balance is established from the relationship of the gases to the nitrogen molar ratio due its solubility is approximately zero and can be considered constant:

$$\frac{dYO_{2,out}(t)}{dt} = -\frac{Q_{gas}}{V_s(1-\varepsilon_s(t))}(YO_{2,out}(t) - YO_{2,in}(t))$$
$$- K_{laO_{2_s}}\frac{V_{mol}}{y_{N_2}}\frac{(1-\varepsilon_s(t))}{\varepsilon_s(t)}([O_2^*](t) - [O_2](t))_{lm},$$

(5)

where YO_2 is the oxygen to nitrogen molar ratio in the gas phase, defined in the inlet and outlet of the sump, V_{mol} is the molar volume under reactor conditions, and y_{N_2} is the nitrogen molar fraction.

An analogous mass balance can be described for the carbon dioxide, where YCO_2 is the carbon dioxide to nitrogen molar ratio in the gas phase:

$$\frac{dYCO_{2,out}(t)}{dt} = -\frac{Q_{gas}}{V_s(1-\varepsilon_s(t))}(YCO_{2,out}(t) - YCO_{2,in}(t))$$
$$-K_{laCO_{2s}}\frac{V_{mol}}{y_{N_2}}\frac{(1-\varepsilon_s(t))}{\varepsilon_s(t)}([CO_2^*](t) - [CO_2](t))_{lm}. \tag{6}$$

2.3. Microalgae Strain

The microalgae strain used corresponds to *Scenedesmus almeriensis* (CCAP 276/24). This microalga is characterized by a high growth rate, withstanding temperature up to 45 °C and pH values from 7 up to 10, although its optimum conditions are 30 °C and pH around 8. In particular, in this work, a set-point of 7.8 will be considered.

2.4. Control Problem

To maintain the microalgae culture in good conditions inside the raceway reactor, it is necessary to maintain two main variables at their optimal values, the pH and dissolved oxygen. These variables are controlled by injecting CO_2 and air, respectively, into the sump. However, the pH has a more critical role than the dissolved oxygen and it will be the controlled variable in this work.

The control problem consists of maintaining the pH of the culture at a given level. The injection of CO_2 coming from flue gases reduces the pH level due to the formation of carbonic acid, while the photosynthesis process increases the pH due to consuming CO_2 and producing O_2.

When more CO_2 is injected in excess and it cannot be completely dissolved in the water, it is released into the atmosphere, being harmful to the environment. Thus, an adequate control algorithm is required to look for a trade-off between the pH control and the CO_2 injections. Furthermore, reducing the number of CO_2 injections in the reactor implies a reduction of the costs and an increment of the life-span of the electronic valves. Summarizing, the control scheme is presented in the following way: the process output is the culture pH, the aperture of the flue gas injection valve is the manipulated variable, and the solar radiation acts as the main disturbance.

The On/Off controllers are widely used for pH control in raceway reactors as in other industrial processes, due to its simplicity. Its behaviour is a relay with hysteresis and represents the most simple feedback controller that can be used to control a process. This type of control is suitable for processes that have two states (open and close) because the controller switches the control variable between two states (on or off), depending on the set-point error with respect to the controlled variable. However, this type of control is characterized by low accuracy and pH oscillations due to the changes in the control signal, causing a negative influence on microalgae. The On/Off valve is opened and flue gases are injected until the pH measure decreases below the set-point; then, the control valve is closed until the pH reaches a value above the set-point, and so on.

On the other hand, PI control has proven to be advantageous over traditional On/Off control, as shown in [6,7]. This type of control allows the pH to be maintained close to the set-point, reducing the characteristic oscillations of On/Off control and, in addition, improving the control signal of the valve.

Recently, the use of event-based PI controllers is increasing thanks to the advantages they have over the use of classical PI controllers. Using event-based architectures applied to PI controllers allows the reduction the CO_2 consumption in the injections, granting a smooth degradation in pH. Moreover, it is also possible to reduce the control effort in the valve and, in addition, to increase the battery life span of wireless sensors.

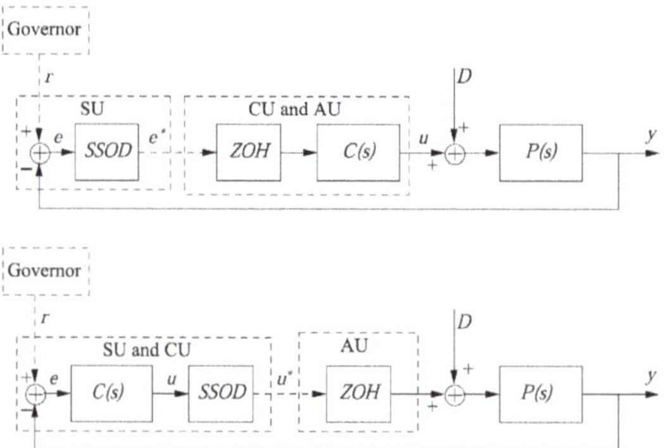

Figure 2. Symmetric-Send-On-Delta (SSOD) control architecture schemes: Symmetric-Send-On-Delta-Proportional-Integral (SSOD-PI) controller (top) and Proportional-Integral-Symmetric-Send-On-Delta (PI-SSOD) controller (bottom). Dashed blocks: Governor, Sensor Unit (SU), Control Unit (CU) and Actuator Unit (AU).

2.5. Symmetric-Send-on-Delta Method

The event-based method, applied to a PI controller, used in this work is derived from the Send-On-Delta method [10,11]. This derivation is called Symmetric-Send-On-Delta [12], and, generally, it can be seen as a relay behaviour with hysteresis.

The behaviour can be described with the following equation:

$$v^*(t) = \begin{cases} (i+1)\Delta\beta & \text{if } \frac{v(t)}{\Delta} \geq (i+1) \text{ and } v^*(t^-) = i\Delta\beta \\ i\Delta\beta & \text{if } \frac{v(t)}{\Delta} \in [(i-1),(i+1)] \text{ and } v^*(t^-) = i\Delta\beta \\ (i-1)\Delta\beta & \text{if } \frac{v(t)}{\Delta} \leq (i-1) \text{ and } v^*(t^-) = i\Delta\beta \end{cases} . \tag{7}$$

In [12], it has been demonstrated that the parameter Δ influences the system tolerance without affecting the system stability. For this reason, it has to be properly tuned to establish a trade-off between the increments of the steady-state error and the decrements of the number of events.

2.6. Event-Based Control Architectures

Two different architectures are implemented with the SSOD technique based on the event-triggered data exchange position in the control loop and are shown in Figure 2. In this figure, the dashed blocks are as follows:

- Governor: establishes the set-point for the process.
- Sensor Unit (SU): composed by the hardware associated with the sensors.
- Control Unit (CU): which corresponds to the controller hardware and software that computes the control signal.
- Actuator Unit (AU): which receives the control signal and applies it to the actuator.

These blocks can be coupled depending on the control configuration, being possible to use wireless devices when are far from the main control.

2.6.1. SSOD-PI Scheme

The first architecture is called SSOD-PI controller because the SSOD block is placed in the SU, before the PI controller in the control loop. In this scheme (Figure 2 top), the CU and the AU are located in the same machine while the SU is located separately. Thus, the communication from the sensor could be wireless. The control action is computed by the controller at a regular sampling rate, considering the last received sampled error. One of the benefits of this configuration is the reduction on the communication between the sensor and the control unit, improving the life span of the batteries of wireless sensors.

2.6.2. PI-SSOD Scheme

In the other scheme, the SSOD block is located after the PI controller in the control loop and it is called PI-SSOD. In this case, the SU and the CU are placed in the same machine, while the AU is located separately and can be powered independently. The control action is sent to the actuator, which holds the last received control signal until the next data exchanging triggering event. Analogous as the SSOD-PI scheme, this configuration presents a reduction of the number of changes in the control action, thus the actuator wear can be reduced.

3. Results and Discussion

This section discusses the results obtained in simulation with the application of two SSOD event-based architectures for the pH control problem in a raceway simulator [15]. Several experiments were performed with different solar radiation profiles during five days to observe how the event-based controller reacts to changes in the photosynthesis rate and in the pH variable. An evaluation of the pH referred to the Integrated Absolute Error, to the control effort and to the CO_2 consumption associated with the injection time have been carried out.

The aim is to establish a comparison between the traditional On/Off controller architecture and the SSOD event-based method. An initial comparison was made between the traditional On/Off control operating during daytime and a PI control architecture that operates both during daytime and nighttime periods. Afterwards, further comparisons have been made with the SSOD event-based control architectures (SSOD-PI and PI-SSOD) from the initially designed PI controllers and applied to the combined daytime-nighttime solution. Results about the stability of SSOD-PI and PI-SSOD for FOPDT processes can be found in [12].

Usually, the pH control is performed only during the diurnal period because of its influence on the photosynthesis process. Thus, the system is working in open-loop during the night period to save CO_2 injections and thus saving costs and CO_2 losses. However, in this work, the control scheme will be evaluated for the whole day in order to analyse how the (event-based) control approach can contribute to control the system also during the night and without increasing the costs too much. Notice that the pH presents different dynamics at the diurnal and nocturnal periods. Thus, two different PI controllers will be designed for diurnal and nocturnal situations.

In a previous work [19], two different linear models were identified for both daytime and nighttime periods:

$$G(s)_{daytime} = \frac{-4.100}{3390 \cdot s + 1} \cdot e^{-100 \cdot s}, \tag{8}$$

$$G(s)_{nighttime} = \frac{-1.619}{2171 \cdot s + 1} \cdot e^{-100 \cdot s}. \tag{9}$$

It is worth noting that the normalized dead time of the system, that is, the ratio between the dead time and the time constant, in both cases, is very small, i.e., the process is lag dominant. For this reason, among the wide variety of PID tuning rules [20], a Simple-Internal-Model-Control (SIMC) tuning rule [21,22] has been selected, which states that the PI parameters have to be selected as:

$$K_p = \frac{1}{k} \frac{T}{\lambda + \theta},$$

$$T_i = min\{T, 4(\lambda + \theta)\},$$

where θ is the dead time of the process, T is its time constant and λ is the desired closed-loop time constant. The value of λ should be greater than or equal to the system dead time. The closed-loop time constant, in the two cases, has actually been fixed as 0.2 times the open-loop time constants, which results in $\lambda = 678$ (daytime) and $\lambda = 434.2$ (nighttime) seconds. These values result, indeed, to be much higher than the dead times, ensuring a high robustness to the system, which is desirable by considering the linealization procedure of the system model (2)–(6). Then, the following PI controllers were obtained:

$$C(s)_{daytime} = -1.063 \cdot \left(1 + \frac{1}{3112 \cdot s}\right),$$

$$C(s)_{nighttime} = -2.511 \cdot \left(1 + \frac{1}{2137 \cdot s}\right).$$

Obviously, when tuning a PI controller, it is important to ensure the asymptotic stability of the system [23–25]. For this reason, the analysis provided in [12] to determine the parameter stability region for which there are no limit cycles for the two systems (8) and (9) have been performed. Results are shown in Figures 3 and 4 and they confirm that, for both controllers, the parameters have been selected so that the avoidance of limit cycles and instability is ensured. Moreover, the tuning is also robust as the parameters are far from the border of the region.

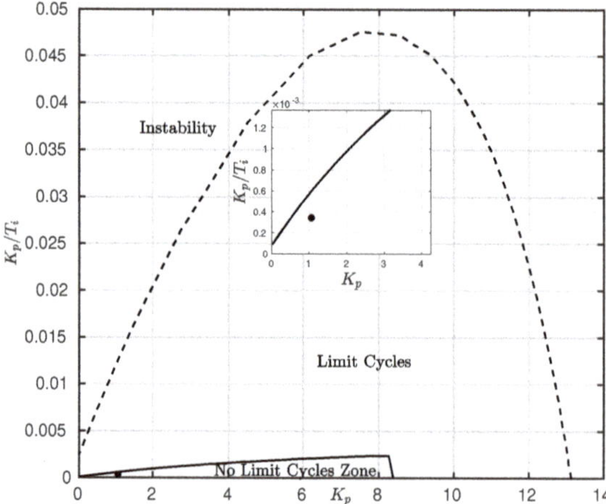

Figure 3. Stability region for daytime controller.

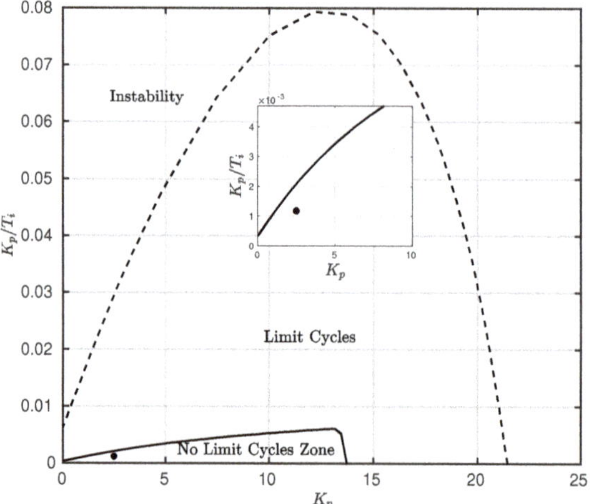

Figure 4. Stability region for nighttime controller.

Tables 1–3 show the performance indexes for the control architectures tested during the five days. In these tables, IAE is the Integrated Absolute Error (also during daytime and nighttime), TV_u is the total variation in the control signal, $E_y(day)$ and $E_y(night)$ are the number of events during daytime and nighttime periods, IT is the CO_2 injection time to the reactor (also during daytime and nighttime) and Gas is the CO_2 consumption.

3.1. PI Controller Results

The first scenario shows the simulation results considering two control architectures, a traditional On/Off controller (red) operated during daytime (the system is in open loop during the nighttime) and a combined daytime-nighttime PI control (black). Figure 5 shows a two days simulation where it can be seen how the pH oscillates around set-point (7.8) during daytime for the On/Off controller. These oscillations range from 7.73 to 7.9 while, for the PI controller, the pH remains close to the set-point. Another remarkable fact is the variation of the pH during nighttime, which, in the case of On/Off control, can reach values of 8.7. On the opposite, the nighttime PI controller keeps the pH around set-point in that period.

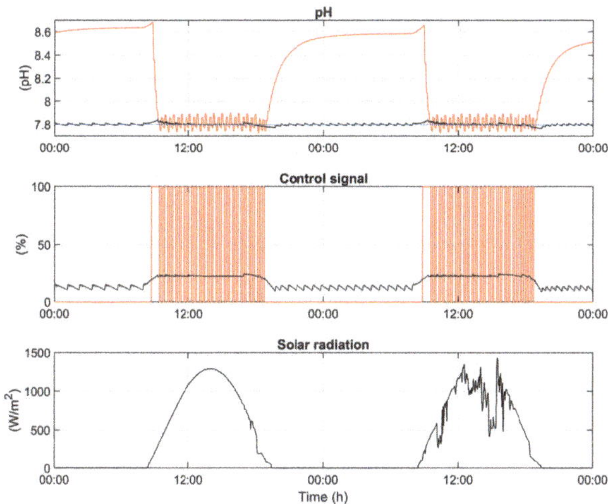

Figure 5. Two days comparison between traditional On/Off daytime control (red) and PI control during daytime and nighttime (black).

Referring to the control effort, the control signal graphic of Figure 5 represents the opening of the injection valve for both control schemes. In this graphic, it can be verified that, in the case of the On/Off controller, the valve opens completely in short periods of time, which causes a great control effort. In the case of the PI controller, the maximum opening of the valve is around 23% and 12.5% for the daytime and nighttime periods, respectively.

In Table 1, the indexes for these two control schemes during the five days can be seen. Starting with the IAE, it is very remarkable how the PI controller reduces this error by up to 98.40% with respect to the On/Off control, since it also operates during the nighttime. Observing this error independently between the daytime and nighttime periods, reductions of the order of 88.21% and 99.08% respectively for the IAE can be achieved. From the point of view of the total variation in the control signal, the PI control reduces this variation by 94.16%, due to a more accurate control during the daytime and even the nighttime. This improvement in the control signal is reflected in a reduction of the control effort. The most remarkable comparison that can be made is about injection time and gas consumption. The PI controller reduces these values by 13.78% compared to the On/Off control, despite operating during the entire day and not only during the daytime period.

Table 1. Performance On/Off and Proportional-Integral (PI) control indexes.

Indexes	On/Off Controller	PI Controller
IAE	185,513.10	2972.96
$IAE_{daytime}$	11,590.10	1366.30
$IAE_{nighttime}$	173,923.00	1606.66
TV_u	274.00	15.99
IT (min)	1856.47	1600.70
Gas (m^3)	1.86	1.60
$IT_{daytime}$ (min)	1856.47	851.05
$IT_{nighttime}$ (min)	-	749.65

3.2. SSOD-PI Controller Results

The second scenario presents the results of the SSOD-PI control architecture for five days. Figure 6 represents the SSOD-PI approach for the pH control problem with two different Δ values ($\Delta = 0.01$ and $\Delta = 0.05$ for daytime and $\Delta = 0.05$ for nighttime). By analysing the graphics of Figure 6, a comparison can be established between both Δ values, which represent the lowest and higher values studied for this SSOD configuration. Regarding pH (first plot), during the night, both signals show similar behaviour, due to a Δ value of 0.05. On the contrary, during the daytime period, the change in the tolerance produced by the Δ parameter is appreciated. With a Δ value of 0.01 (black), it can be seen that the pH it kept around set-point with a peak at the beginning of the daytime period caused by the dilution process (microalgae raceway reactor works in continuous mode, where biomass is extracted every morning at the same hour and medium is injected to keep the same volume in the reactor). On the other hand, with a Δ value of 0.05 (red), this dilution peak is increased due to a lower tolerance caused by a higher Δ value.

Moreover, at the end of the daytime, a peak is observed, which is produced by a decrease in solar radiation and greater tolerance in the event-band. For the control signal (second plot), during the nighttime period, the response of the valve is the same in both cases. Regarding the daytime, the control signal corresponding to a Δ value of 0.05 shows a slower rise at the beginning of the daytime period and a higher average value with respect to the Δ value of 0.01. In relation to the events (third and fourth plots), a decrease in the number of events can be appreciated when the value of the Δ parameter increases, as it can be seen in Table 2.

Table 2. Performance Symmetric-Send-On-Delta-Proportional-Integral (SSOD-PI) indexes.

Indexes	SSOD-PI Δ				
	0.01	0.02	0.03	0.04	0.05
IAE	9418.45	9743.28	10,196.29	10,805.44	11,484.70
$IAE_{daytime}$	2089.36	2416.45	2899.45	3309.99	3974.87
$IAE_{nighttime}$	7329.09	7326.83	7296.84	7495.45	7509.81
TV_u	11.38	11.39	11.86	11.82	12.41
E_y (daytime)	132	55	43	32	26
E_y (nighttime)	91	93	95	93	96
IT (min)	1537.70	1541.90	1253.50	1545.70	1542.97
Gas (m^3)	1.54	1.54	1.54	1.55	1.54
$IT_{daytime}$ (min)	852.42	857.53	864.33	867.03	865.70
$IT_{nighttime}$ (min)	685.28	684.37	679.97	678.67	677.27

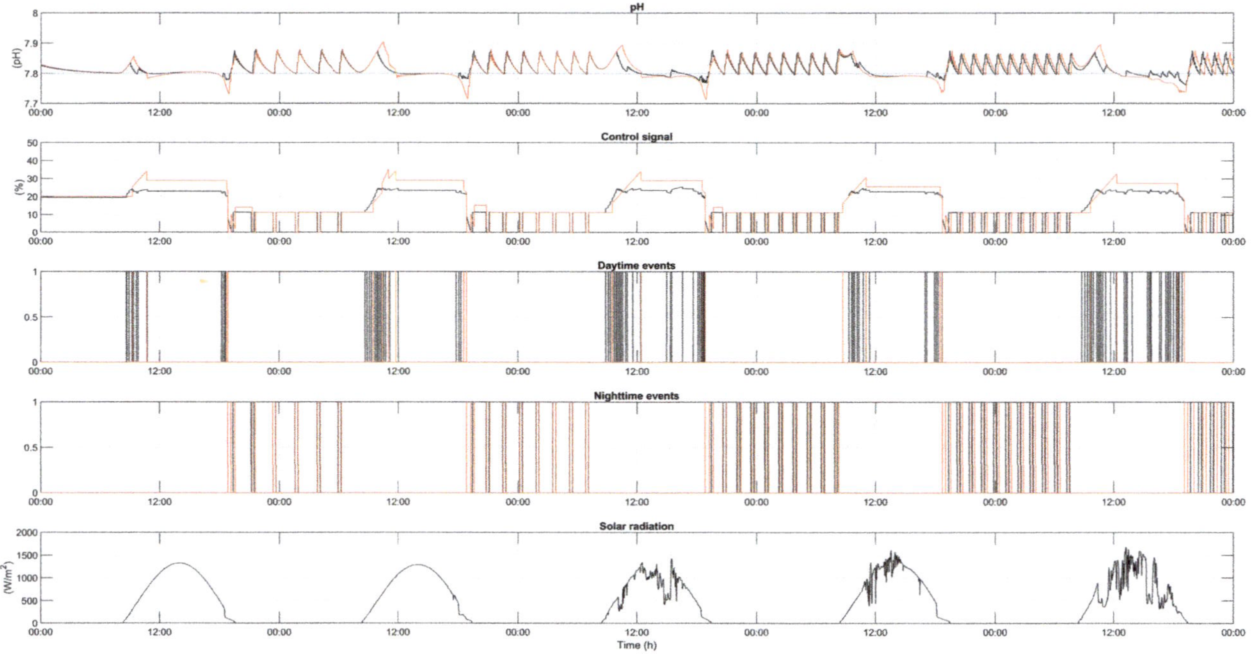

Figure 6. Results with the SSOD-PI controller. $\Delta = 0.01$ (black) and $\Delta = 0.05$ (red). First plot: pH output. Second plot: control signal. Third plot: diurnal events. Fourth plot: nocturnal events. Fifth plot: solar radiation.

The SSOD-PI performance indexes are presented in the Table 2, where the test carried out with different Δ values (Δ = 0.01, 0.02, 0.03, 0.04 and 0.05) are collected for five days each. The IAE error increases during the daytime period as the Δ value increases, while for the nighttime similar values result for all cases. This is due to the fact that a fixed Δ value was used for the nighttime period while a value from 0.01 to 0.05 was used for the daytime, thus increasing the error tolerance. Therefore, due to this increase in the error tolerance, the number of events—or communications between the sensor and the control unit—will be reduced with the increase of the Δ value, as can be seen in the table, especially during the daytime period. The total variation increases slightly in every case due to the tolerance imposed by the Δ value, which makes the control signal during the daytime period more constant but higher in the case of Δ = 0.05. Regarding the injection time and gas consumption, there is no significant variation between cases, ranging from 1537.70 min (Δ = 0.01) to 1545.70 maximum (Δ = 0.04), with the highest consumption during the daytime period.

3.3. PI-SSOD Controller Results

The last scenario presents the results of the PI-SSOD control architecture for five days. Figure 7 shows the simulation performed using different Δ values (Δ = 0.001 and 0.01) to compare both results. In the top plot, the pH behaviour during the whole experiment is presented, where the pH reaches a higher value during the dilution process on daytime for the Δ value of 0.01. This dilution process acts as a disturbance at the same hour for every simulated day and the controller reacts later than for a Δ value of 0.001 because the event-based control method has a higher tolerance regarding the step in the control signal. On the other hand, for the Δ value of 0.001, this peak is smaller and the pH remains around the set-point. In the nighttime period, a Δ value of 0.006 for the PI controller was used for all the experiments, to maintain the pH oscillating close to the set-point.

In the second plot, the combined control signal applied to the actuator is shown. This signal is similar during the nighttime period for both simulations, and during the start of the daytime period, it can be seen that the signal corresponding to Δ value of 0.01 begins to act later than for the Δ value of 0.001.

The daytime and nighttime events are presented in the third and fourth plots, respectively. It can be seen that, during the daytime, there are fewer events than during nighttime, as it can be checked in Table 3, because of the small deadband in the nighttime period.

Table 3 shows the performance indexes for all Δ cases studied (Δ = 0.001, 0.003, 0.005, 0.007 and 0.01) during a period of five days. Comparing the IAE, it can be appreciated that it increases as the Δ value increases because the increments in the control signal are greater. The error during nighttime is similar in all cases, due to the same Δ value used in that period. The total variation does not show a regular increase or decrease, and it varies depending on the Δ value because the control signal is similar in all cases. The only observed difference is a later actuation during the daytime period. The daytime events are reduced as the Δ value increases and they remain similar during nighttime as the IAE and the nighttime events. The injection time and gas consumption do not show much variation in the analyzed cases. In the case of Δ = 0.01, the reduction in the injection time due to the dilution peak above the set-point is compensated with an increase due to the opposite effect, the period in which the pH remains below the set-point. Thus, gas consumption is similar to the case of Δ = 0.001, where the pH remains close to the set-point.

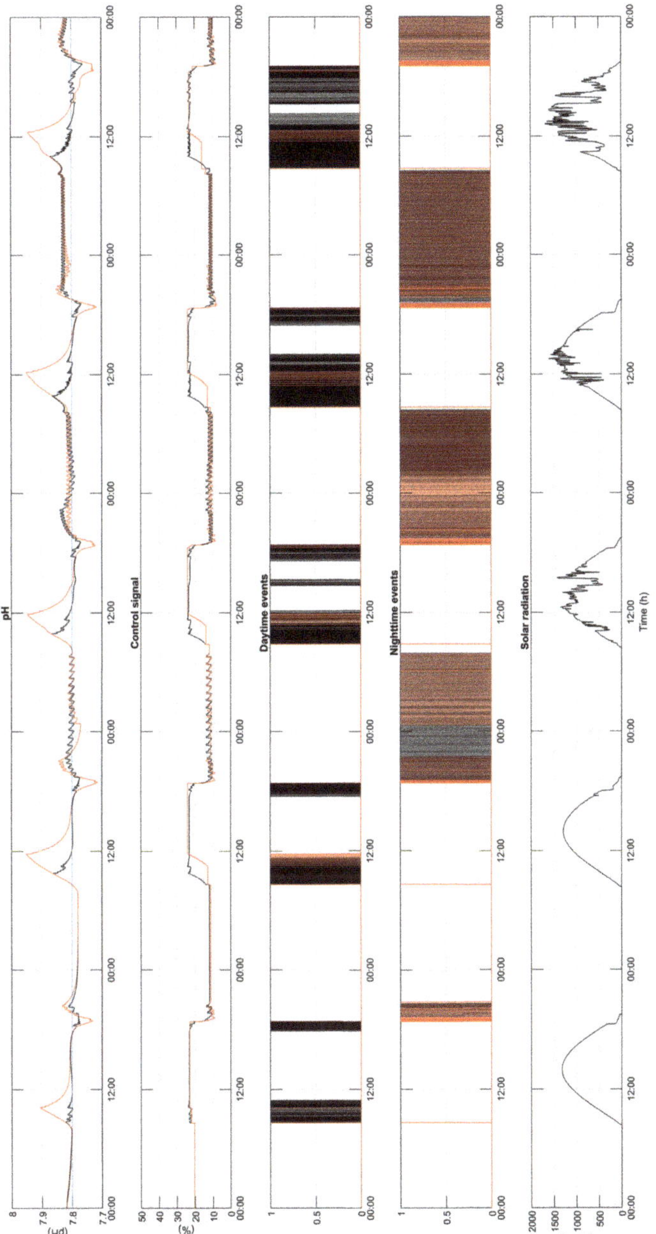

Figure 7. Results with the PI-SSOD controller. $\Delta = 0.001$ (black) and $\Delta = 0.01$ (red). First plot: pH output. Second plot: control signal. Third plot: diurnal events. Fourth plot: nocturnal events. Fifth plot: solar radiation.

Table 3. Performance Proportional-Integral-Symmetric-Send-On-Delta (PI-SSOD) indexes.

Indexes	PI-SSOD Δ				
	0.001	0.003	0.005	0.007	0.01
IAE	6643.80	7339.80	8109.32	9260.24	12,799.96
$IAE_{daytime}$	2093.44	2782.69	3555.31	4462.89	7847.62
$IAE_{nighttime}$	4550.36	4557.11	4554.01	4797.35	4952.34
TV_u	7.73	9.43	6.31	5.85	6.09
E_y (daytime)	1226	332	161	88	68
E_y (nighttime)	700	1040	720	688	724
IT (min)	1577.60	1585.30	1579.00	1570.30	1540.03
Gas (m^3)	1.58	1.59	1.58	1.57	1.54
$IT_{daytime}$ (min)	849.03	856.63	851.63	845.23	818.90
$IT_{nighttime}$ (min)	728.57	728.67	727.37	725.07	721.13

3.4. Discussion

As verified from the first scenario comparing the On/Off control with the PI control architecture, this last scheme brings a series of benefits to the pH control problem with respect to the traditional one. It provides an improvement in the CO_2 usage even operating the reactor during both daytime and nighttime, while the On/Off control was used only during the diurnal period.

On the other hand, the event-based control architectures based on the SSOD method and coupled with a classical time-driven PI (SSOD-PI and PI-SSOD) controller also count with the benefits that improve the reactor control compared to the traditional On/Off control. Comparing these event-based control architectures with respect to the PI control, certain advantages are obtained accompanied by a series of disadvantages related to the trade-off between tolerance and control accuracy. Looking at Tables 1–3, the first advantage for the event-based schemes is the reduction of the CO_2 consumption, at the expense of degrading the pH control performance slightly by increasing the IAE values. This reduction is obviously more relevant when the plant operates for long periods of time (i.e., months). The second advantage is related to the control effort, which is reduced due to control tolerances established in both daytime and nighttime periods, as can be seen, by the total variation and injection time.

Finally, the two event-based control architectures (SSOD-PI and PI-SSOD) show some differences between them regarding the event-based architectures and the Δ values suitable for its operation. One of the main differences that can be seen between the two architectures is that the Δ parameter has a greater influence on the PI-SSOD scheme since small variations of this parameter showed greater changes in IAE, as from Tables 2 and 3. The PI-SSOD scheme shows a greater reduction in the total variation of the control signal with respect to the SSOD-PI. On the contrary, the SSOD-PI scheme shows less number of events, being more suitable for use with wireless sensors. Regarding the injection time, it can be appreciated that, for the PI-SSOD architecture, it is slightly higher than for the SSOD-PI one. These differences are due to the event-based triggering method in both cases, where the Δ parameter does not work in the same way, and therefore, it must be selected regarding the needs of the process.

4. Conclusions

In this paper, two event-based control architectures derived from the SSOD method have been presented. Two event-based PI controllers (for daytime and nighttime periods) with different Δ values were evaluated to establish a comparison with classical time-based PI controllers and traditional On/Off control. The main objective is to reduce CO_2 injection and control effort in the reactor maintaining a satisfactory pH control within established thresholds.

Simulation tests with both event-based architectures have been carried out obtaining promising results in the response of pH and gas consumption. These proposed event-based controllers provide improvements in the Integrated Absolute Error, gas consumption and control effort with respect to the traditional On/Off control for this process. The use of classical PI controllers improves traditional control, and then the use of event-based controllers coupled with PI ones offers even more advantages. One of the most remarkable results is the need to operate the reactor throughout the whole day (daytime and nighttime periods) versus the daytime operation of the traditional control to improve the pH control and save CO_2. The tolerance in the output signal for the daytime and nighttime periods was better controlled, reducing CO_2 consumption and improving the control effort at the same time, which results in benefits in terms of cost savings and maintenance.

The results have shown that the output tolerance can be varied by adjusting the Δ parameter, becoming another tuning variable in the control architecture. The higher this parameter is, the greater the process tolerance, reducing the number of events at the expense of degrading control performance.

The use of the proposed event-based controllers can improve the biomass production and would reduce the costs of the operations. In general, these controllers allow the use of wireless sensors/actuators and its implementation does not require a special hardware for the controller, as the control code is fairly simple. It is also worth stressing that event-based control somehow mimics the manual control of an operator, who usually changes the value of the manipulated variable when an event occurs. For this reason, it is likely that this strategy is well accepted by operators. Future work will consist of the implementation of the proposed control strategies in a real raceway reactor.

Author Contributions: E.R.-M. participates in the implementation, validation and analysis of the presented data. He also contributed in the writing—original draft preparation. M.B. (Manuel Beschi) participated in the writing—original draft preparation and the design of the proposed event-based control methodology. A.V., M.B. (Manuel Berenguel) and J.L.G. contributed in writing—review and editing, formal analysis and funding acquisition. Moreover, A.V. participated in the evaluation of the event-based control approaches, and M.B. (Manuel Berenguel) and J.L.G. in the implementation and analysis of the simulator for the microalgae raceway reactor.

Funding: This work has been partially funded by the following projects: DPI2014-55932-C2-1-R and DPI2017-84259-C2- 1-R (financed by the Spanish Ministry of Science and Innovation and EU-ERDF funds), and the European Union's Horizon 2020 Research and Innovation Program under Grant Agreement No. 727874 SABANA.

Abbreviations

The following abbreviations are used in this manuscript:

FOPDT	First-Order-Plus-Dead-Time
QFT	Quantitative Feedback Theory
IAE	Integrated Absolute Error
IT	Injection Time
ODE	Ordinary Differential Equation
PDE	Partial Differential Equation
PID	Proportional-Integral-Derivative
PI-SSOD	Proportional-Integral-Symmetric-Send-On-Delta
SIMC	Simple-Internal-Model-Control
SOD	Send-On-Delta
SSOD	Symmetric-Send-On-Delta
SSOD-PI	Symmetric-Send-On-Delta-Proportional-Integral
TV	Total Variation

References

1. Bahadar, A.; Bilal Khan, M. Progress in energy from microalgae: A review. *Renew. Sustain. Energy Rev.* **2013**, *27*, 128–148. [CrossRef]
2. Oswald, W.J.; Golueke C.G. Biological transformation of solar energy. *Appl. Microbiol.* **1960**, *2*, 223–262.
3. Weissman, J.; Goebel R. *Design And Analysis of Pond System for the Purpose of Producing Fuels*; Solar Energy Research Institute, U.S. Department of Energy, Microbial Products, Inc.: Fairfield, CA, USA, 1987.

4. Costache, T.A.; Acién, F.G.; Morales, M.M.; Fernández-Sevilla, J.M.; Stamatin, I.; Molina E. Comprehensive model of microalgae photosynthesis rate as a function of culture conditions in photobioreactors. *Appl. Microbiol. Biotechnol.* **2013**, *97*, 7627–7637. [CrossRef]

5. Pawlowski, A.; Mendoza, J.L.; Guzmán, J.L.; Berenguel, M.; Acién, F.G.; Dormido S. Selective pH and dissolved oxygen control strategy for a raceway rector within an event-based approach. *Control Eng. Pract.* **2013**, *44*, 209–218. [CrossRef]

6. Fernández, I.; Peña, J.; Guzmán, J.L.; Berenguel, M.; Acién F.G. Modelling and control issues of pH in tubular photobioreactors. *IFAC Proc.* **2010**, *43*, 186–191. [CrossRef]

7. Hoyo, A.; Guzmán, J.L.; Moreno, J.C.; Berenguel M. Control Robusto con QFT del pH en un Fotobiorreactor Raceway. In Proceedings of the Jornadas de Automática, Gijón, Spain, 6–8 September 2017.

8. Pawlowski, A.; Fernández, I.; Guzmán, J.L.; Berenguel, M.; Acién, F.G.; Normey-Rico J.E. Event-based predictive control of pH in tubular photobioreactors. *Comput. Chem. Eng.* **2014**, *65*, 28–39. [CrossRef]

9. Pawlowsli, A.; Mendoza, J.L.; Guzmán, J.L.; Berenguel, M.; Acién, F.G.; Dormido S. Effective utilization of flue gases in raceway reactor with event-based pH control for microalgae culture. *Bioresour. Technol.* **2014**, *170*, 1–9. [CrossRef] [PubMed]

10. Miskowicz M. Send-on-delta: An event-based data reporting strategy. *Sensors* **2006**, *6*, 49–63. [CrossRef]

11. Vasyutynskyy, V.; Kabitzsh K. Implementation of PID controller with send-on-delta sampling. In Proceedings of the International Control Conference, Glasgow, UK, 8–11 May 2006.

12. Beschi, M.; Dormido, S.; Sánchez, J.; Visioli A. Characterization of symmetric send-on-delta PI controllers. *J. Process Control* **2012**, *22*, 1930–1945. [CrossRef]

13. Beschi, M.; Dormido, S.; Sánchez, J.; Visioli A. Tuning rules for event-based SSOD-PI controllers. In Proceedings of the 20th Mediterranean Conference on Control & Automation, Barcelona, Spain, 3–6 July 2012; pp. 1073–1078.

14. Pawlowski, A.; Beschi, M.; Guzmán, J.L.; Visioli, A.; Berenguel, M.; Dormido S. Application of SSOD-PI and PI-SSOD event-based controllers to greenhouse climatic control. *ISA Trans.* **2016**, *65*, 525–536. [CrossRef] [PubMed]

15. Fernández, I.; Acién, F.G.; Guzmán, J.L.; Berenguel, M.; Mendoza J.L. Dynamic model of an industrial raceway reactor for microalgae production. *Algal Res.* **2016**, *17*, 67–78. [CrossRef]

16. Mendoza, J.L.; Granados, M.R.; de Godos, I.; Acién, F.G.; Molina, E.; Banks, C.J.; Heaven S. Fluid-dynamic characterization of real-scale raceway reactors for microalgae production. *Biomass Bioenergy* **2013**, *54*, 267–275. [CrossRef]

17. Acién, F.G.; Fernández, J.M.; Molina-Grima E. Photobioreactors for the production of microalgae. *Rev. Environ. Sci. Biotechnol.* **2013**, *12*, 2–21.

18. Camacho, F.; Acién, F.G.; Sánchez, J.; García, F.; Molina E. Prediction of dissolved oxygen and carbon dioxide concentration profiles in tubular photobioreactors for microalgal culture. *Biotechnol. Bioeng.* **1999**, *62*, 71–86.

19. Rodríguez Miranda, E.; Beschi, M.; Guzmán, J.L.; Berenguel, M.; Visioli A. Application of a Symmetric-Send-On-Delta Event-Based Controller for a Microalgal Raceway Reactor. Available online: https://ecc19.eu (accessed on 12 March 2019)

20. O'Dwyer, A. *Handbook of PI and PID Controller Tuning Rules*; Imperial College Press: London, UK, 2009.

21. Skogestad, S. Simple analytic rules for model reduction and PID controller tuning. In Proceedings of the 2nd IFAC Conference on Advances in PID Control, Brescia, Italy, 28–30 March 2003; pp. 291–309.

22. Grimholt, C.; Skogestad S. Optimal PI-Control and verification of the SIMC tuning rule. In Proceedings of the 2nd IFAC Conference on Advances in PID Control, Brescia, Italy, 28–30 March 2012; pp. 11–22.

23. Tan, N.; Kaya, I.; Yeroglu, C.; Atherton, D. Computation of stabilizing PI and PID controllers using the stability boundary locus. *Energy Convers. Manag.* **2006**, *47*, 3045–3058. [CrossRef]

24. Gao, R.; Gao, Z. Pitch control for wind turbine systems using optimization, estimation and compensation. *Renew. Energy* **2016**, *91*, 501–515. [CrossRef]

25. Turksoy, O.; Ayasun, S.; Hames, Y.; Sonmez, S. Gain-phase margins-based delay-dependent stability analysis of pitch control system of large wind turbines. *Trans. Inst. Meas. Control* **2019**. [CrossRef]

Article

Controller Design and Control Structure Analysis for a Novel Oil–Water Multi-Pipe Separator

Sveinung Johan Ohrem [1,*,†], Håvard Slettahjell Skjefstad [2], Milan Stanko [2] and Christian Holden [1]

[1]   Department of Mechanical and Industrial Engineering, Faculty of Engineering,
      NTNU Norwegian University of Science and Technology, 7491 Trondheim, Norway;
      christian.holden@ntnu.no
[2]   Department of Geoscience and Petroleum, Faculty of Engineering,
      NTNU Norwegian University of Science and Technology, 7491 Trondheim, Norway;
      havard.s.skjefstad@ntnu.no (H.S.S.); milan.stanko@ntnu.no (M.S.)
*   Correspondence: sveinung.j.ohrem@ntnu.no
†   Current address: Institutt for Maskinteknikk og Produksjon, NTNU, 7491 Trondheim, Norway.

Received: 8 March 2019; Accepted: 28 March 2019; Published: 2 April 2019

Abstract: To enable more efficient production of hydrocarbons on the seabed in waters where traditional separator equipment is infeasible, the offshore oil and gas industry is leaning towards more compact separation equipment. A novel multi-pipe separator concept, designed to meet the challenges of subsea separation, has been developed at the Department of Geoscience and Petroleum at the Norwegian University of Science and Technology. In this initial study, a control structure analysis for the novel separator concept, based on step-response experiments, is presented. Proportional-integral controllers and model reference adaptive controllers are designed for the different control loops. The proportional-integral controllers are tuned based on the well-established simple internal model control tuning rules. Both control methods are implemented and tested on a prototype of the separator concept. Different measurements are controlled, and results show that the performance of the separator under varying inlet conditions can be improved with proper selection of control inputs and measurements.

Keywords: process control; separation; oil and gas

---

1. Introduction

In mature oil fields on the Norwegian Continental Shelf, the amount of water extracted in 2016 accounted for more than twice the amount of produced oil [1]. This produced water is transported topside for separation and cleaning. Eventually, the water treatment capacity of the topside facilities will be reached which causes a bottleneck in the production and leaves a substantial part of the hydrocarbon processing capacity left unused. Furthermore, a high amount of water in the well stream will cause a loss of pressure in the transportation pipelines, leading to a lower production. Removing the water on the seabed frees up capacity at the topside facility, which can be utilized for new tie-ins to existing fields.

In offshore oil and gas production, the processing of oil and gas on the seabed is also considered an enabler for more efficient liquid boosting, longer range gas compression from subsea to onshore, cost efficient hydrate management, more efficient riser slug depression, and access to challenging field developments [2].

Large vessels, referred to as gravity separators, are commonly used offshore for separation of oil, water and gas. These separators are robust and have a high performance, but they are not suited for use at ultra deep waters ($\geq$3000 m) due to the required size, which makes the installation and maintenance economically challenging [2]. Detailed descriptions on modelling and control of gravity separators can be found in [3,4].

A novel separator concept not relying on vessels, but rather on separation in multiple pipes (Figure 1), was recently developed [5]. This separator has been dubbed the MPPS, the Multiple Parallel Pipe Separator. The reduced diameter of the pipes compared to that of vessels makes the pipe separator better suited for installation at deeper waters. A prototype of the separator has been built at the Department of Geoscience and Petroleum at the Norwegian University of Science and Technology and the steady-state performance has been evaluated [6].

Currently, the separator laboratory does not have a control system and all valves are opened and closed manually via a LabView human–machine interface. Automatic control of key variables in the separator is important, as it helps counteract the effects of external disturbances, enables tracking of setpoints, enables optimal operation, and ensures that safety requirements are met.

The separator is equipped with several sensors providing measurements that may serve as controlled variables (CVs). Two valves are used as inputs, or manipulated variables (MVs). It is not straightforward to select a CV, as some variables may be more difficult to control and more sensitive to disturbances. This issue is addressed in Section 2.5 of this paper. Furthermore, some variables may be difficult or impossible to control directly, and, hence, finding a secondary variable that is easier to control and has an effect on the primary control variable can be very helpful.

In [7], a control design study was performed for a complete subsea separation system including a pipe separator. The liquid level in the pipe separator was chosen as the CV, and Proportional-Integral-Derivative (PID) controllers were used in all control loops. The authors state that controlling the system is challenging due to strong interactions between process components, constraints in valve openings and opening/closing speed of the valves. The study does not go into detail about the tuning of the controllers, nor is a control structure analysis presented.

The same is true for the work presented in [8]. Here, PID controllers, tuned by trial and error, was used to achieve the desired performance. The level of the oil/water interface in the pipe separator was controlled. The authors also stated that a control system should be able to adapt to varying operating conditions.

Other previous control-related work on pipe separators [9–11] do not go into detail on the selected control structure or control algorithm used. In this paper, the pairing of the MVs and CVs is analyzed. A detailed presentation of the Proportional-Integral (PI) controller tuning, and a comparison of the separator efficiency when using different candidate CVs is presented. Furthermore, model reference adaptive control is applied to a pipe separator. Adaptive control schemes have seen applications in process control [12] and offshore oil and gas production [13–15], but the authors have not been able to locate any prior published work on adaptive control of pipe separators.

Although the pipe separator used in this study is different from those used in [7–11], it is the authors' belief that the results are transferable and that the results presented here can serve as a basis for future control design of pipe separators.

This work contains an initial control structure analysis and an initial controller design for the MPPS. The purpose is to investigate, analyze and test several control structures, hence both a conventional PI controller and an adaptive controller is developed and tested in the laboratory. The PI controller tuning is based on the well-established simple internal model control (SIMC) tuning rules [16]. The performance of the different control structures and controllers are qualitatively and quantitatively compared and a basis for future work is established.

2. Materials and Methods

2.1. Separator Concept

The separator being tested in this paper is the Multiple Parallel Pipe Separator (MPPS), a multi-pipe arrangement for oil–water bulk separation. The concept was previously presented in [5,6], where the reader can find detailed information on design considerations and performance evaluation. Experiments are carried out on a two-pipe 150.6 mm ID prototype, which is depicted in Figure 1.

Figure 1. The Multiple Parallel Pipe Separator (MPPS) prototype.

An oil–water mixture enters at the separator inlet ($\dot{Q}_{in}$), where the flow is divided into two parallel and identical separator branches. The fluids pass through the horizontal pipe segments, where they separate and are then extracted through their respective outlets. Water is extracted through the water extraction line ($\dot{Q}_{e_w}$), while oil is extracted through the oil extraction line ($\dot{Q}_{e_o}$). As seen in Figure 1, an inclined extraction section is utilized in the design. This is to increase the water holdup in the horizontal pipe sections prior to extraction and to slow down and build up water close to the water extraction point.

The inlet has a tangential configuration and is fitted with novel phase-rearranging internals. Detailed information on the inlet configuration can be found in [6]. The total horizontal length of the separator prototype is 6.1 m.

2.2. Test Facility

The test fluids used in the separator are distilled water with added wt% 3.2 NaCl, and Exxsol D60 model oil. To prevent bacterial growth, 750 ppm of the biocide IKM CC-80 was added to the water. Furthermore, 0.015 g/L of the colorant Oil Red O has been added to the Exxsol D60 for phase distinction. The test fluid properties are given in Table 1.

Table 1. Test fluid properties at 20 °C.

Fluid	Density (kg/m^3)	Viscosity (cP)
Water	1020.0	1.0
Exxsol D60	792.2	1.4

A piping and instrumentation diagram (P&ID) of the test facility is given in Figure 2. The storage tank is a gravity separator with a diameter of 1.2 m and a length of 5.5 m. It has a capacity of 6 m^3 and serves as a baseline separator. The gravity separator provides two clean phase outlets (water and Exxsol D60), which are connected to a pump manifold.

The pump manifold consists of four centrifugal pumps, two of which are used at any given time. The pumps used for the presented experiments each have a flow capacity of 100–700 L/min and a maximum head specification of 55 m. The pumps are controlled by 0–50 Hz frequency converters, where 50 Hz constitutes a maximum rpm of 2900. Two flow lines are connected to the pump manifold, one for each phase.

Both flow lines are fitted with a Coriolis flow meter measuring flow rate (FT.1/2) and density (DT.1/2). The flow meters allow accurate adjustment of the desired inlet flow and water cut (WC), as well as monitoring of phase purities.

Figure 2. Lab facility piping and instrumentation diagram (P&ID).

Downstream from the Coriolis flow meters, the flow lines merge in a Y-junction to a multiphase flow line. The multiphase flow line is a 67.8 mm ID transparent polyvinyl chloride (PVC) pipe, which consists of a 5.5 m long straight section, a 720 mm radius 180° turn, and a secondary 5 m long straight section down to the inlet of the MPPS prototype.

Static gauge pressure (PT.1) and temperature (TT.1) are measured at the MPPS inlet. A ball-type inlet choke valve (VT.1) is fitted two meters upstream from the MPPS inlet. The differential pressure (dPT.1) is measured across the valve. For all presented experiments, VT.1 has been 100% open with zero pressure loss over the inlet choke valve.

Two return lines, one for water and one for Exxsol D60, are connected to the MPPS prototype. The return lines are fitted to their respective separator outlets $\dot{Q}_{e_w}$ and $\dot{Q}_{e_o}$. Static pressure transducers (PT.2/3) are fitted to each return line, and a third Coriolis flow meter (FT.3/DT.3) is mounted on the water return line. This allows tracking of the amount of water extracted from the separator, as well as calculation of the purity of the water extracted. Detailed information on logged and calculated parameters will follow in the next section.

A second dP transducer (dPT.2) is installed at the water extraction point of the MPPS prototype. This measures the dP over the inclined extraction pipe, serving as a proxy level indicator for water in the section. An illustration of the sensor mounting is given in Figure 3. The connector lines are filled with water, and the left side connection is the positive side, hence the dP measurement will be zero when the entire inclined section is filled with water and increase with the amount of oil present. The dP transducer is, unfortunately, working in the extreme end of its range and thus the measurements are quite noisy. Furthermore, low-frequency waves form in the pipeline leading up to the incline, causing a continuous disturbance on the dP measurement.

Lastly, both return lines are fitted with control valves for pressure and extraction rate control. The water return line is fitted with an electrically controlled ball valve (VT.2), while the oil return line is fitted with a pneumatic membrane valve (VT.3).

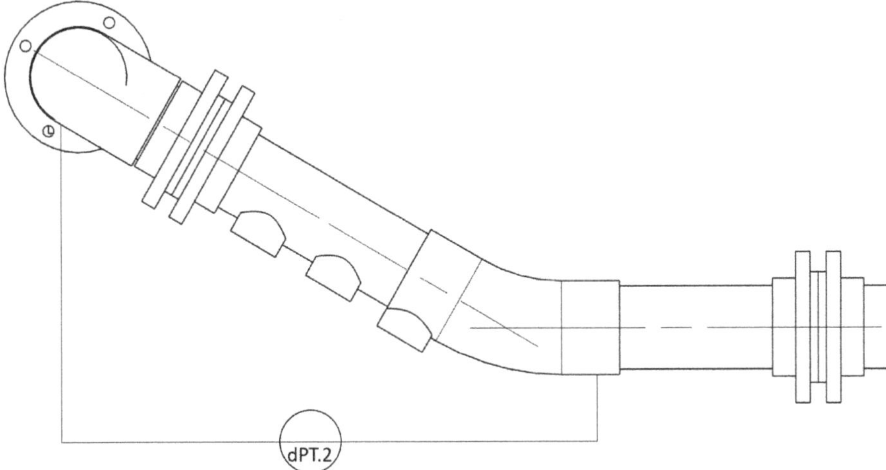

Figure 3. dPT.2 installation.

2.3. Test Parameters

All recorded test parameters are listed in Table 2. The table includes tag names, parameter names, parameter units and specified sensor range. The values for VT.2 and VT.3 are specified values sent to the valves by the controllers, not the actual measured position of the valves.

Table 2. Recorded parameters.

Tag	Parameter	Unit	Range
FT.1	$\dot{Q}_1$	L/min	0–1000
FT.2	$\dot{Q}_2$	L/min	0–1000
FT.3	$\dot{Q}_3$	L/min	0–1000
DT.1	ρ_1	kg/m^3	750–1050
DT.2	ρ_2	kg/m^3	750–1050
DT.3	ρ_3	kg/m^3	750–1050
PT.1	P_1	barg	0–6
PT.2	P_2	barg	0–6
PT.3	P_3	barg	0–6
dPT.2	dP_2	mbar	0–50
TT.1	T_1	°C	−30–122
VT.2	Valve 2	% Closed	0–100
VT.3	Valve 3	% Closed	0–100

As mentioned in Section 2.2, three Coriolis flow meters are used to adjust inlet flow rate and water cut, monitor phase purities, determine the amount of water extracted from the MPPS prototype, and the purity of the extracted water. The WC in the respective flow lines is determined by

$$\mathrm{WC}_i = \frac{\rho_i - \rho_o}{\rho_w - \rho_o} \ . \tag{1}$$

Here, ρ_i is the measured density at DT.1/2/3, while ρ_w and ρ_o are the pre-determined temperature-corrected densities of the water and Exxsol D60, respectively. For pure-phase feed streams, WC$_1$ should be equal to 100% while WC$_2$ should be equal to 0%. From calculated WC and measured flow rates, the actual WC in the multiphase transport line (WC$_{in}$) is calculated as

$$\text{WC}_{in} = \frac{\text{WC}_1 \dot{Q}_1 + \text{WC}_2 \dot{Q}_2}{\dot{Q}_1 + \dot{Q}_2}, \tag{2}$$

where $\dot{Q}_1$ and $\dot{Q}_2$ are the water and oil flow, respectively, before mixing. When running experiments, $\dot{Q}_1$ and $\dot{Q}_2$ are adjusted such that the desired total flow and WC_{in} are reached.

The amount of water extracted from the MPPS prototype is determined by the extraction rate (ER). The ER is the flow rate through the water extraction line divided by the flow rate in the water feed line:

$$\text{ER} = \frac{\dot{Q}_3}{\dot{Q}_1}, \tag{3}$$

where $\dot{Q}_3$ is the flow at the water outlet of the MPPS.

As the test loop is a closed system, the water and Exxsol D60 phases will be contaminated over time. Microscopic droplets of water will be dispersed in the oil and vice versa. In order to give a performance measurement that is independent of occurring contamination, the WC ratio is calculated. The WC ratio is equal to the WC at the water extraction line (WC_3) divided by the WC at the water feed line (WC_1):

$$\text{WC}_r = \frac{\text{WC}_3}{\text{WC}_1}. \tag{4}$$

A WC_r equal to 100% means that the extracted water from the MPPS prototype is of equal quality to the water, leaving the baseline separator prior to being mixed with the oil. A WC_r of 100% is thus the upper limit on the purity that can realistically be achieved by the MPPS prototype.

2.4. System Identification

A dynamic model of the system is very helpful when designing controllers. A classical way to identify the dynamic relations between a manipulated variable and a control variable is to perform a step response experiment and calculate the transfer function. In this work, the procedure from [16] was followed, and it was assumed that the dynamic model between each input and output could be described by a first-order plus time delay transfer function on the form

$$\frac{y}{u} = G(s) = \frac{ke^{-\theta s}}{\tau s + 1}, \tag{5}$$

where y is the output, u is the input, and s is the Laplace variable. The transfer function variables describing the dynamic response, k, τ and θ are of special interest. These variables represent the plant gain, the time constant and the time delay, respectively. The plant gain provides the steady-state output of the plant, for a specific input, and is given by

$$k = \frac{\Delta y}{\Delta u}. \tag{6}$$

The time constant, τ, is the time it takes the output to reach 63% of the steady-state value, and the time delay, θ, is the amount of time it takes the input to cause a reaction on the output.

The step response experiments were performed with one valve at a time, with the other valve in a fixed position, and at a fixed inlet flow rate and inlet WC. The valve openings, inlet flow rate and inlet WCs are listed in Table 3.

Some of the measurements contained significant measurement noise, hence the measured values where filtered using a 1st order Butterworth filter before the parameter analysis was performed. The transfer function between each input and output was then calculated and validated by comparing it to the original response. If any deviations were present, the transfer function variables were tuned manually to improve the fit.

Table 3. Inlet conditions and changes in valve openings used in step response experiments.

Output	Δ VT.2 (% Closed)	VT.3 (% Closed)	Flow Rate (L/min)	WC_{in} (%)
WC_r	30	80	450	50
PT.1	30	70	450	50
dPT.2	30	70	450	50

Output	Δ VT.3 (% Closed)	VT.2 (% Closed)	Flow Rate (L/min)	WC_{in} (%)
WC_r	−30	20	450	50
PT.1	20	50	350	50
dPT.2	20	50	350	50

The step response experiments were performed on the two inputs VT.2 and VT.3, and three measurements were chosen as candidate CVs. The pressure PT.1 is a necessary CV for safety reasons. A measurement that gives a direct indication of the separator efficiency is the the water cut ratio WC_r, and hence this is also a candidate CV. The laboratory is not equipped with a level measurement sensor, instead the pressure drop dPT.2 over the incline is used for this purpose. The level is often used as an CV in previous work, as mentioned in the Introduction, and will also be a candidate CV in this work.

From the step response experiments, the following transfer functions were identified:

$$\frac{WC_r}{VT.2} = G_1(s) = \frac{0.2423e^{-10s}}{20.9s + 1}, \tag{7}$$

$$\frac{PT.1}{VT.2} = G_2(s) = \frac{0.0026e^{-10s}}{14.1s + 1}, \tag{8}$$

$$\frac{dPT.2}{VT.2} = G_3(s) = \frac{-0.0739e^{-10s}}{23.1s + 1}, \tag{9}$$

$$\frac{WC_r}{VT.3} = G_4(s) = \frac{-0.5962e^{-4s}}{2s + 1}, \tag{10}$$

$$\frac{PT.1}{VT.3} = G_5(s) = \frac{0.0156e^{-4s}}{8.3s + 1}, \tag{11}$$

$$\frac{dPT.2}{VT.3} = G_6(s) = \frac{0.2722e^{-4s}}{13.2s + 1}. \tag{12}$$

A comparison between the measured response, the filtered response, and the transfer function response is shown in Figure 4. Here, we see that some of the responses could be better described by a second-order transfer function. In particular, the transfer functions between VT.2 and the different outputs (note the second order dynamics in WC_r in Figure 4a not captured by $G_1(s)$ and the overshoot in dPT.2 in Figure 4e not captured by $G_3(s)$). However, for the control study in this work, it is assumed that a first order model is sufficient. The fluctuations present are purely caused by measurement and process noise.

2.5. Control Structure Analysis

The separator is a multiple input multiple output (MIMO) system with two inputs and several possible outputs. It is not straightforward to pair an input with an output, and hence a relative gain array (RGA) analysis ([17], Section 3.4) was performed. The RGA provides a measure of interactions between the inputs and outputs and [17,18] recommends pairing inputs and outputs such that the rearranged system has an RGA matrix close to identity. Furthermore, negative steady-state RGA elements should be avoided. The RGA for a square system on the form

$$y = G(s)u \tag{13}$$

is found by calculating the element-wise matrix product

$$\text{RGA}(G) = G_0 G_0^{-T}, \tag{14}$$

where $G_0 = G(0)$ is the steady-state transfer function matrix of $G(s)$ in Equation (13).

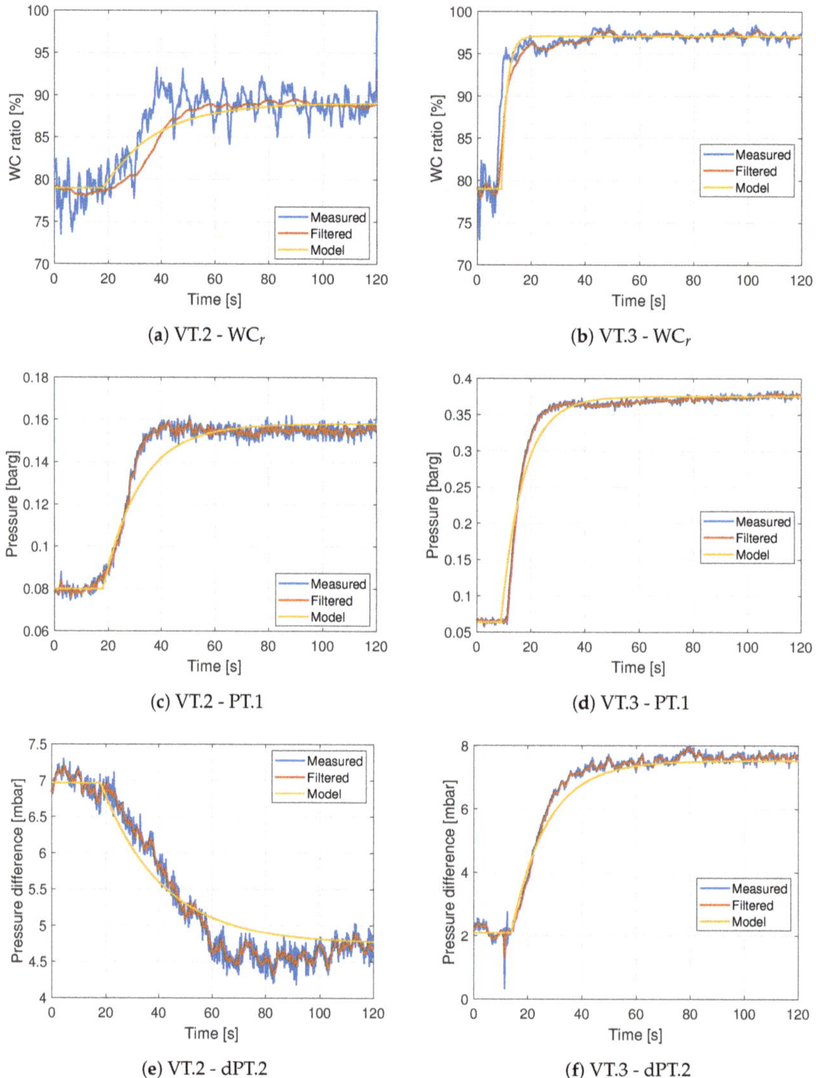

Figure 4. Step response comparison between measured signal, filtered signal and transfer functions.

Another parameter to consider when investigating the pairing of CVs and MVs is the Niederlinski index (NI) ([18], Section 2.2.1)

$$\text{NI} = \frac{\det\left[G(0)\right]}{\prod_{i=1}^{n} g_{ii}(0)}, \tag{15}$$

where g_{ii} are the diagonal elements of $G(0)$. If the open-loop system is stable (which is the case here), one should select pairings corresponding to positive NI values [19]; otherwise, the closed-loop system will be unstable ([18], Th. 1). The RGA matrices and the NI for the separator is shown in Table 4. From the RGA analysis, it is clear that VT.2 should be paired with WC_r or dPT.2, and VT.3 with PT.1, as this corresponds to the pairing closest to 1. The NI is positive for both possible pairings, hence no pairing will lead to an unstable system.

Table 4. Relative gain array (RGA) and Niederlinski index (NI) for the separator.

Input	Output	RGA		NI
VT.2	WC_r	0.7082	0.2918	1.4121, 3.4269
VT.3	PT.1	0.2918	0.7082	
VT.2	dPT.2	0.6185	0.3815	1.6169, 2.6209
VT.3	PT.1	0.3815	0.6185	

2.6. Controller Design

A multivariable system, such as the one investigated here, could benefit from a multivariable control scheme. However, since this is an initial control study, the developed controllers are decoupled, single-loop controllers.

2.6.1. PI Control

A PI controller has the form

$$u(s) = k_c \left(1 + \frac{1}{\tau_I s} \right) (r - y), \tag{16}$$

where k_c is the proportional gain, τ_I is the integral time in seconds, r is the reference and y is the measured output to be controlled. PI controllers are developed for the separator by applying the SIMC tuning rules [16]. The SIMC tuning rules states that the proportional gain and integral time of the PI controller should be chosen as

$$k_c = \frac{1}{k} \frac{\tau}{\tau_c + \theta}, \tag{17}$$

$$\tau_I = \min \left(\tau, (\tau_c + \theta) \right), \tag{18}$$

where k_c is the proportional gain, τ_I is the integral time, and τ_c is the desired closed-loop time constant, which is the only tuning parameter. For tight and robust control, Ref. [16] recommends choosing $\tau_c = \theta$. Although the RGA analysis recommended a specific pairing of inputs and outputs, PI controllers are developed and tested for all configurations. The parameters used in the PI controllers are found in Section 3.2.1.

A derivative part could have been added to the controllers, but this would have required a measurement of the derivative of the CV. This is not available but could have been calculated numerically. However, as the control objective is to keep the CVs at steady-state, the derivative of the CV would be close to zero when operating at steady-state and the contribution would only be from the measurement noise. Derivative action is uncommon in process control applications where the plants are stable with overdamped responses and first-order dominant dynamics (which is the case here), since the performance improvement is usually too small compared to the added complexity [17,20].

2.6.2. Adaptive Control

As an alternative to conventional PI control, a model reference adaptive controller (MRAC) ([21], Section 6.2.2) was implemented and tested for the two control configurations recommended by the RGA analysis, i.e., VT.3 controls PT.1 and VT.2 controls either WC_r or dPT.2.

When using MRAC, the controller parameters are automatically updated such that the error between the measured variable and the output of a reference model is reduced. This could be very beneficial if the process parameters change over time or at different operating points, which may lead to poor control when using a controller with fixed gains. The MRAC structure is different from a PI controller structure, i.e., there is no proportional and integral gain in the MRAC, but rather two parameters that try to approximate a value that causes the closed-loop system dynamics to be equal to the reference model dynamics. Hence, using the proportional and integral gain of a PI controller as initial values in an MRAC is not necessarily helpful. A schematic of the MRAC is shown in Figure 5.

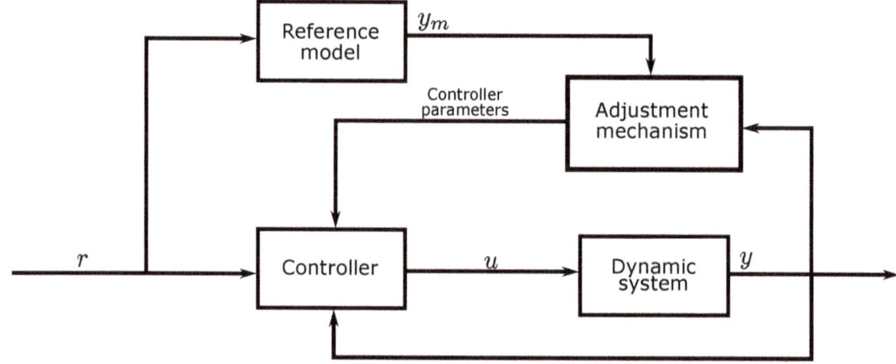

Figure 5. Model reference adaptive control (MRAC) schematic.

The MRAC has the form

$$u = -k_a(t)y + l_a(t)r, \tag{19}$$

where $k_a(t)$ and $l_a(t)$ are time-varying gains, updated by the adaptive laws

$$\dot{k}_a = \gamma_k ey \operatorname{sign}(k), \tag{20}$$

$$\dot{l}_a = -\gamma_l er \operatorname{sign}(k), \tag{21}$$

where γ_k, γ_l are adaptation gains, y is the measured value to be controlled, r is the reference and $\operatorname{sign}(k)$ is either 1 or -1. The error signal $e = y - y_m$, where y_m is the output of the reference model

$$\dot{x}_m = a_m x_m + b_m r, \tag{22}$$

$$y_m = x_m, \tag{23}$$

where $a_m < 0$ and b_m are chosen by the operator and specify the desired closed-loop dynamics of the system.

The only system knowledge required by the MRAC is the sign of k. It can be shown ([21], Section 6.2.2) that, if $y = x$ is the state in a linear system, the controller given by Equations (19)–(23) causes $y \to y_m$ asymptotically for $\gamma_k, \gamma_l > 0$. The parameters used in the MRAC are found in Section 3.2.2.

3. Results

In total, six experiments were carried out in the laboratory: four experiments with PI controllers and two experiments with MRAC. When using PI control, all control pairings were tested, but with MRAC only the pairings recommended by the RGA analysis was carried out.

3.1. LabView Implementation

The laboratory is controlled through a computer running LabView 2015 (National Instruments, Austin, TX, USA) on Windows 7 (Microsoft, Redmond, WA, USA) with an Intel i7 4770S (Santa Clara, CA, USA), 3.1 GHz processor and 16 GB RAM. The PI controllers could be readily implemented in the LabView 2015 block diagram through existing PI controller blocks.

The MRAC, however, had to be implemented using an add-on for LabView called MathScript Module. This module allows the user to write code, and execute it at each iteration of the LabView program. To calculate the MRAC input in Equation (19), the differential equations in Equations (20)–(22) must be solved. This was done using first-order Euler integration with a step length of 0.0001 s. A dead band was introduced to the adaptation algorithms, i.e., the adaptation was stopped if the error was less than 5% of the setpoint for dPT.2 and PT.1 and less than 0.5% for WC_r.

3.2. Controller Tuning

3.2.1. PI Controllers

A transfer function model is only an approximation of the real system dynamics. It was assumed that the transfer functions were of first order. The controller parameters may require re-tuning if the transfer function models differ significantly from the real dynamics (they may change with operating point and inlet conditions) or if nonlinearities in the valve openings are not considered. Furthermore, there are interactions between the control loops; hence, a multivariable controller would probably be a better choice. This, however, has not been studied in this work.

It was found during initial testing that the choice of $\tau_c = \theta$ was too aggressive for VT.2 and, hence, $\tau_c = 30$ s was chosen for this valve. For VT.3, $\tau_c = \theta$ could only be used when controlling WC_r. Otherwise, $\tau_c = 10$ s was used. The controller parameters for the different PI controllers are listed in Table 5.

Table 5. PI controller parameters.

MV	CV	τ_c	k_c	τ_I
VT.2	WC_r	30 (s)	2.15 (-)	20.8 (s)
VT.2	PT.1	30 (s)	135.2 (1/barg)	14.1 (s)
VT.2	dPT.2	30 (s)	−7.82 (1/mbar)	23.1 (s)
VT.3	WC_r	4 (s)	−0.424 (-)	2.04 (s)
VT.3	PT.1	10 (s)	38.1 (1/barg)	8.28 (s)
VT.3	dPT.2	10 (s)	3.464 (1/mbar)	13.2 (s)

3.2.2. Model Reference Adaptive Controller

It was found during testing in the laboratory that the initial values $k_a(0)$ and $l_a(0)$ as well as the adaptation gains γ_k, γ_l had to be chosen with care. This is due to the fact that time delays and measurement noise was ignored when the controller was derived. The adaptive controller parameters used are listed in Table 6.

Table 6. Model reference adaptive controller (MRAC) parameter values.

Experiment	MV	CV	a_m (1/s)	b_m (1/s)	γ_k (1/s)	γ_l (1/s)	$k_a(0)$	$l_a(0)$
5	VT.2	WC_r	−1/30	1/30	0.00002	0.00002	0 (-)	0 (-)
	VT.3	PT.1	−1/10	1/10	100	100	50 (1/barg)	50 (1/barg)
6	VT.2	dPT.2	−1/30	1/30	0.01	0.01	0 (1/mbar)	0 (1/mbar)
	VT.3	PT.1	−1/10	1/10	20	20	50 (1/barg)	50 (1/barg)

3.3. Test Sequence and Control Objectives

All experiments presented here were performed on the same day. Prior to each experiment, the lab was operated at nominal inlet conditions for 5 min, i.e., a flow rate of 350 L/min and 60% WC. The valves were manually set to VT.2 = 30% closed and VT.3 = 70% closed. This led to an initial pressure PT.1 $\sim$ 0.1 barg, an initial WC_r $\sim$ 99% and an initial dPT.2 $\sim$ 1.9 mbar when the controllers were activated. The static pressure is initialized at 0.1 barg in order to see how the controllers, and especially the adaptive controllers, perform during an initial transient. A setpoint of 0.4 barg for PT.1 was necessary as the valve controlling the pressure could then operate in the middle of its range, and not saturate, when the inlet flow rate was high.

The inlet conditions were varied in order to emulate situations that may occur in a subsea oil/water separator. The inlet variables available for manipulation are the total liquid flow and the inlet WC. Table 7 shows the different inlet conditions and the scenarios they emulate.

The main control objectives in all experiments are to maintain the desired pressure PT.1 and to keep WC_r as high as possible. The latter is important in order to ensure that the water quality is high enough for the downstream water cleaning equipment. A setpoint of 99% is chosen for the WC_r. Looking only at WC_r may, however, be misleading, since it says nothing about how much water is extracted. For this, the ER is used. It is important to maintain a high ER while maintaining a high WC_r. If the ER is very low, almost no liquid is leaving the separator through the water outlet. In this case, the WC_r may be high, but a lot of water is leaving through the oil outlet.

In Experiments 1 and 2, the WC_r is controlled directly using either VT.2 or VT.3. In Experiments 3, 4 and 6, the dP between the bottom of the outlet incline and the top of the outlet incline is controlled. This serves as a proxy level measurement of the water level in the incline. When controlling the dP, and the level, by proxy, a buffer of water is built up in the incline, making the WC_r more robust to inlet variations. It was found through image analysis that a dP of 2 mbar gave stable oil and water layers (see the Appendix A) and the buffer volume was assumed sufficient. Hence, this setpoint is used in the controllers. Figure 6 shows a sketch of this.

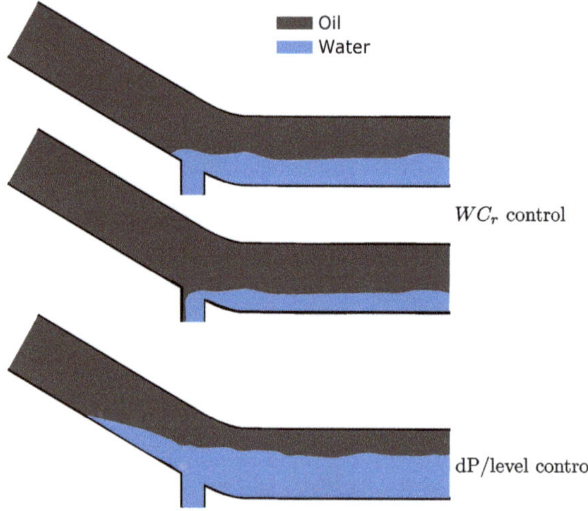

Figure 6. When controlling WC_r directly, no buffer volume of water is present in the inclined section. Hence, oil breakthrough into the water outlet is more frequent when comparing to the dP/level control.

Table 7. Inlet conditions used in all experiments.

Total Flow (L/min)	WC_{in} (%)	Time Stamp (s)	Situation
350	60	0–480	Nominal conditions
350	80	480–840	Water breakthrough
400	74	840–900	New well introduced, step 1
450	67	900–960	New well introduced, step 2
500	60	960–1020	New well introduced, step 3
450	40	1020–1380	Shut down of old well

3.4. Experiment 1. PI Control VT.2–PT1, VT.3–WC_r

Figure 7 shows the results of Experiment 1 where the recommended pairings from the RGA analysis were not used. Since both outputs depend on both inputs, both controllers need to work continuously to counteract the effects caused by the other controller. Furthermore, when WC_r is higher than the setpoint, the controller will close VT.3 to decrease WC_r. This causes the ER to increase above 100% and oil starts flowing through the water outlet. Due to time delays, this effect suddenly causes a drop in WC_r which then has to be counteracted. These drops in WC_r are clearly seen in Figure 7b.

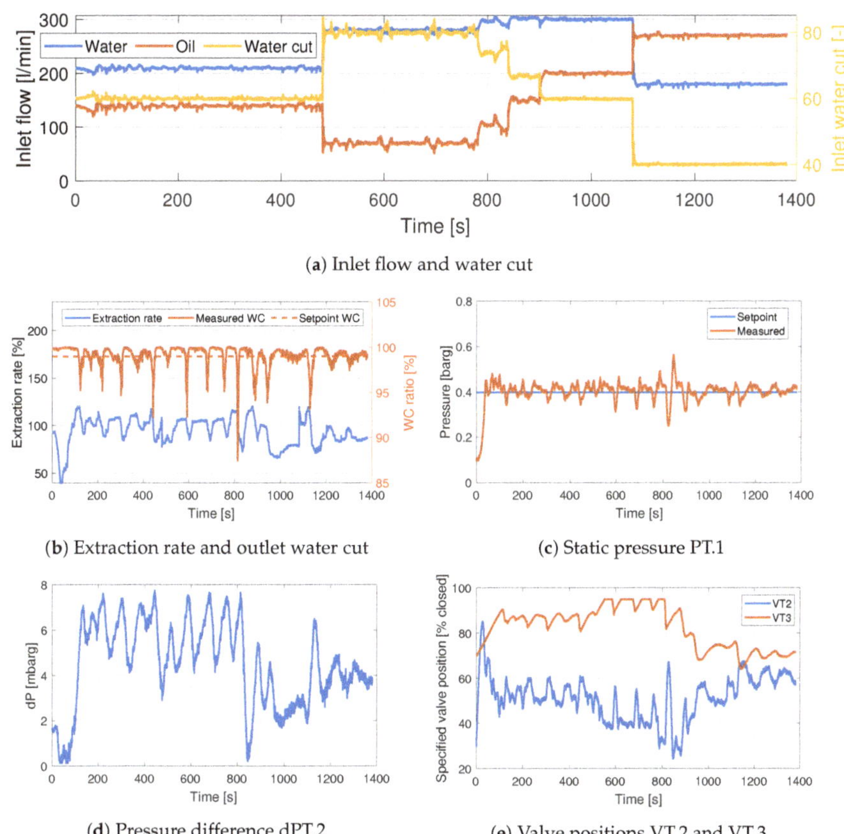

(**a**) Inlet flow and water cut

(**b**) Extraction rate and outlet water cut

(**c**) Static pressure PT.1

(**d**) Pressure difference dPT.2

(**e**) Valve positions VT.2 and VT.3

Figure 7. Experiment 1. PI controllers on VT.2 and VT.3 where VT.2 controls PT.1 and VT.3 controls WC_r.

The pressure controller is able to keep the pressure around the setpoint. The controller handles the steps in inlet flow and inlet WC quite well, but the effect of the other control loop is quite clear.

Notice how the differential pressure dPT.2 oscillates (Figure 7d) due to the lack of control. The dP is quite high, indicating a large amount of oil in the incline. Numerical values for the performance can be found in Table 8.

3.5. Experiment 2. PI Control VT.2–WC$_r$, VT.3–PT.1

Figure 8 shows the results of Experiment 2. Here, the recommended pairings from the RGA analysis were used, and the results show that PT.1 is controlled much better than in Experiment 1. The variations in WC$_r$ are less frequent, but this is caused by the slow controller operating VT.2. As can be seen in Figure 8e, from $t \sim 500$ s to $t \sim 700$ s, the VT.2 valve opens very slowly. This causes the extraction rate to increase slowly (it increased much faster in Experiment 1) towards the point where a drop in WC$_r$ happens. This occurs at $t \sim 700$ s, causing the valve to close again. Since the variations in WC$_r$ happen less frequently, the effect on PT.1 from VT.2 opening and closing is also less than in Experiment 1, which may explain why PT.1 is better controlled in Experiment 2. It should be noted, however, that VT.3 is much faster than VT.2, hence the controller could possibly be able to counteract the influence of VT.2 even if the oscillations had been more frequent. The differential pressure dPT.2 oscillates here as well (Figure 8d), due to the lack of control and the dP is quite high, indicating a large amount of oil in the incline. Numerical values for the performance can be found in Table 8.

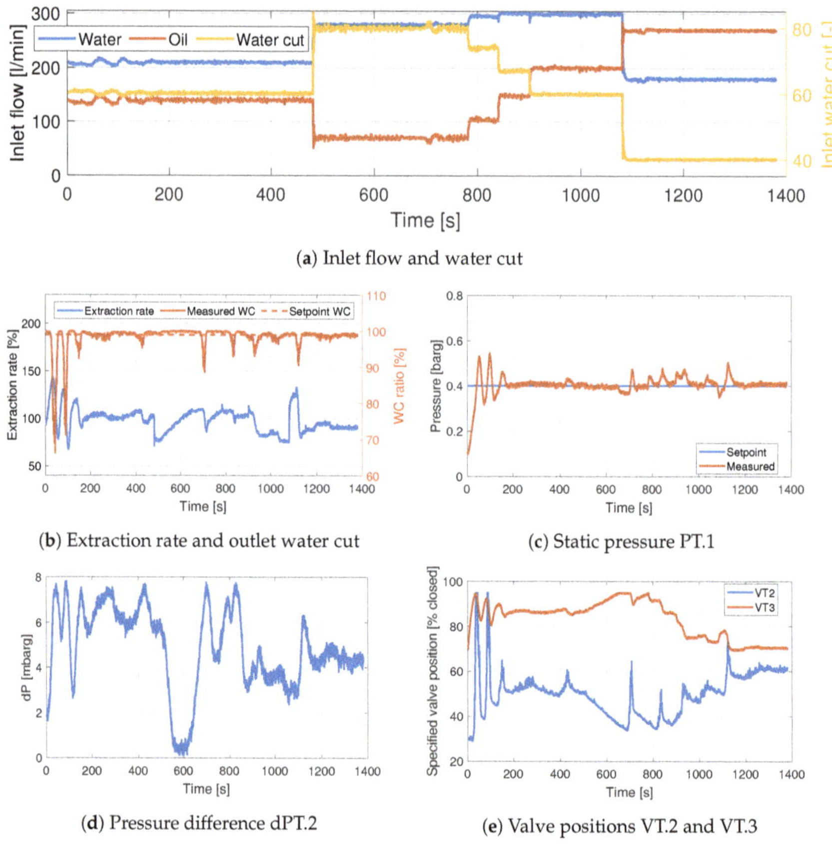

(**a**) Inlet flow and water cut

(**b**) Extraction rate and outlet water cut

(**c**) Static pressure PT.1

(**d**) Pressure difference dPT.2

(**e**) Valve positions VT.2 and VT.3

Figure 8. Experiment 2. PI controllers on VT.2 and VT.3, with the pairing recommended by the RGA analysis.

The initial transient is quite oscillatory. This is caused by the large initial error in the PT.1 and the fact that no reference filter is used. After PT.1 stabilizes, so does WC_r. During the steps in inlet conditions, both controllers are able to keep the controlled variable close to the setpoint. The variations in PT.1 are smaller than in Experiment 1.

3.6. Experiment 3. PI Control VT.2–PT.1, VT.3–dPT.1

In this experiment, WC_r is not used in the controller. Instead, the dP is controlled to a fixed setpoint. Figure 9 shows the results. Here, we see that the behaviour of the WC_r and the ER is not as in Experiments 1 and 2. Since the dP is controlled, a buffer volume is established in the incline. This buffer volume functions as a filter for the disturbances occurring at the inlet. The WC_r has very few drops below 99% in this experiment (Figure 9b). Numerical values for the performance can be found in Table 8.

The effects of measurement noise in the dP transducer are clearly seen in Figure 9d.

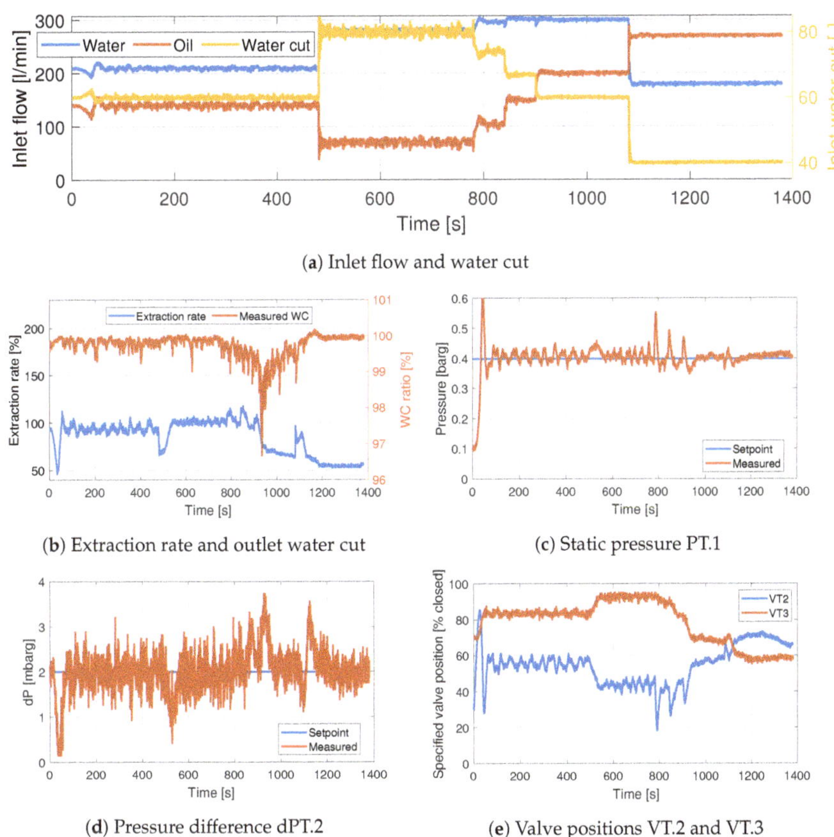

(**a**) Inlet flow and water cut

(**b**) Extraction rate and outlet water cut

(**c**) Static pressure PT.1

(**d**) Pressure difference dPT.2

(**e**) Valve positions VT.2 and VT.3

Figure 9. Experiment 3. PI controllers on VT.2 and VT.3 with VT.2 controlling PT.1 and VT.3 controlling dPT.2.

3.7. Experiment 4. PI Control VT.2–dPT.2, VT.3–PT.1

This experiment uses the control configuration recommended by the RGA analysis. The results of the experiment are shown in Figure 10. The results are quite similar to those found in Experiment 3, but the oscillations in PT.1 during the disturbances are smaller. It is also clear that VT.2 is changing significantly more than VT.3 in Experiment 3. The variations and initial overshoot in dPT.2 are larger in this experiment, but this is caused by VT.2 being much slower than VT.3. The behaviour of the WC_r and the ER is quite similar to Experiment 3, but the undershoot at $t \sim 950$ s is smaller in this experiment. Numerical values for the performance can be found in Table 8.

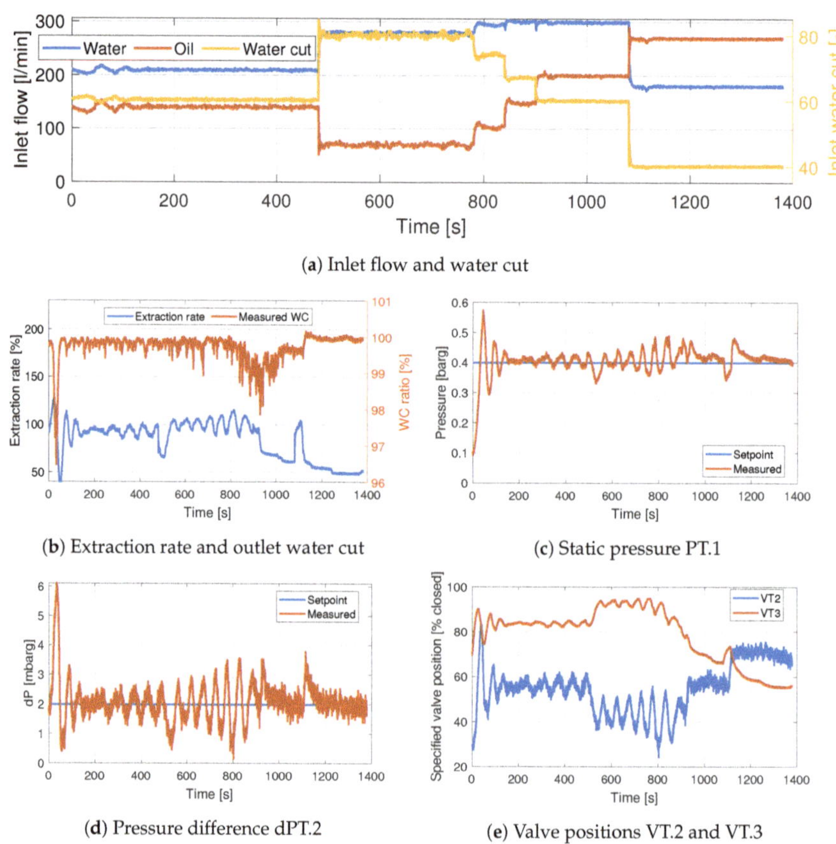

(**a**) Inlet flow and water cut

(**b**) Extraction rate and outlet water cut

(**c**) Static pressure PT.1

(**d**) Pressure difference dPT.2

(**e**) Valve positions VT.2 and VT.3

Figure 10. Experiment 4. PI controllers on VT.2 and VT.3 with VT.2 controlling dPT.2 and VT.3 controlling PT.1, as recommended by the RGA analysis.

3.8. Experiment 5. Adaptive Control VT.2–WC_r, VT.3–PT.1

The model reference adaptive controller was first tested on the recommended control configuration with VT.2 controlling WC_r and VT.3 controlling PT.1. The results are shown in Figure 11. The pressure is controlled very well when using the MRAC. Note that the reference signal in Figure 11c is the output of the reference model given in Equations (22) and (23), hence the signal is filtered and the initial response has much less overshoot compared to Experiment 2. The WC_r controller is quite slow, hence the extraction rate increases slowly to the level where the drops in WC_r occur. When the drops do happen, they are approximately equal to the drops experienced in Experiment 2.

The adaptive gains are shown in Figure 11f,g. The gains for VT.2 are initialized at 0, but the gains for VT.3 are initialized at 50. This was found through trial and error to be a good initial value for VT.3. Numerical values for the performance can be found in Table 8.

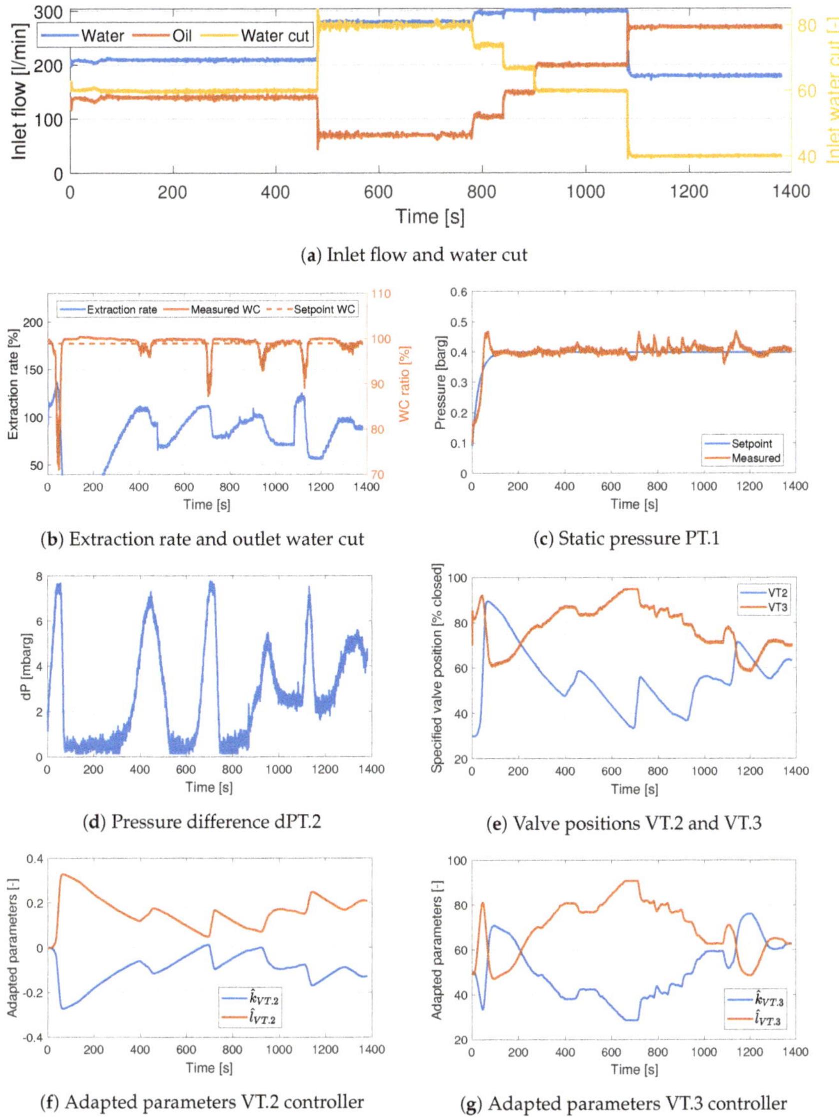

(**a**) Inlet flow and water cut

(**b**) Extraction rate and outlet water cut

(**c**) Static pressure PT.1

(**d**) Pressure difference dPT.2

(**e**) Valve positions VT.2 and VT.3

(**f**) Adapted parameters VT.2 controller

(**g**) Adapted parameters VT.3 controller

Figure 11. Experiment 5. MRAC on VT.2 and VT.3 with VT.2 controlling WC_r and VT.3 controlling PT.1, as recommended by the RGA analysis.

3.9. Experiment 6. Adaptive Control VT.2–dPT.2, VT.3–PT.1

The final experiment used MRAC for VT.2 and VT.3, with VT.2 controlling dPT.2 and VT.3 controlling PT.1. From Figure 12b, it is clear that the WC_r and ER have similar behaviour to that shown in Experiments 3 and 4. The pressure PT.1 is controlled quite well, though with some increases in oscillations compared to that observed in Experiment 5, caused by the need for a lower adaptation gain in this experiment.

The pressure difference controller has trouble bringing dPT.2 to the reference after the first change in inlet conditions happens at $t = 480$ s. This could be caused by the large initial overshoot caused by the zero initialization of the adaptive parameters. As can be seen from Figure 12f,g, the adaptive parameters start changing direction at $t = 480$ s, but, since γ_k and γ_l had to be chosen to be quite small because of the slow valve, the adaptation takes a long time. Furthermore, the changes in PT.1 is causing the pressure difference dPT.2 to change. Since the VT.2 controller (and valve) is so slow, it is unable to bring dPT.2 to the reference as can be seen in Figure 12d.

Numerical values for the performance can be found in Table 8.

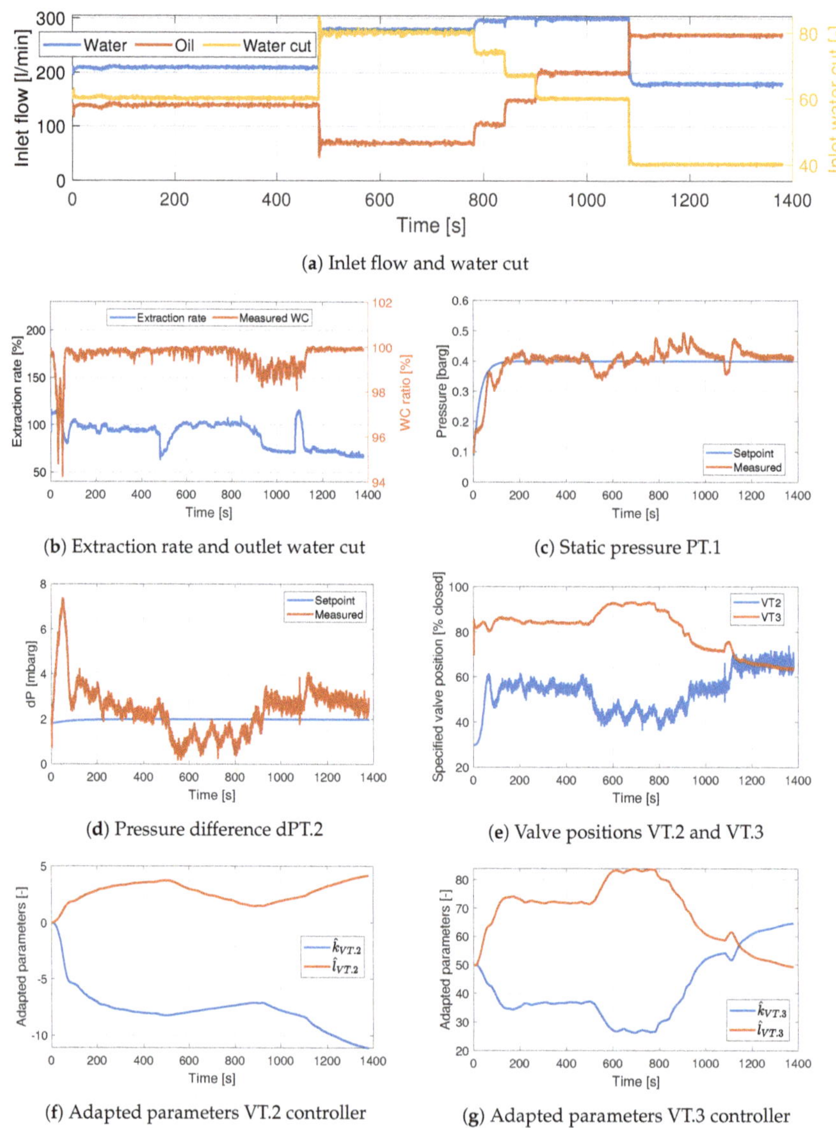

(**a**) Inlet flow and water cut

(**b**) Extraction rate and outlet water cut

(**c**) Static pressure PT.1

(**d**) Pressure difference dPT.2

(**e**) Valve positions VT.2 and VT.3

(**f**) Adapted parameters VT.2 controller

(**g**) Adapted parameters VT.3 controller

Figure 12. Experiment 6. MRAC on VT.2 and VT.3 with VT.2 controlling dPT.2 and VT.3 controlling PT.1, as recommended by the RGA analysis.

3.10. Numerical Comparison

Table 8 shows a numerical comparison of the values of interest from Experiments 1–6. The table includes the mean, root-mean-square (RMS), standard deviation (STD) and median, as well as the integrated absolute error (IAE) for the variables being controlled. The initial transient has an effect on all these numbers, hence the values are also calculated from $t = 200$ s rather than from $t = 0$ to exclude this effect. These values are showed in parentheses.

Table 8. Numerical comparison of Experiments 1–6. Values in parentheses are calculated after the initial transient is over, i.e., from $t = 200$ s.

Experiment	Variable	Mean	RMS	STD	Median	IAE
1	WC_r	99.09 (99.03)	99.09 (99.03)	1.17 (1.21)	99.43 (99.37)	1085.6 (929.2)
	PT.1	0.40 (0.41)	0.41 (0.41)	0.05 (0.03)	0.41 (0.41)	41.72 (29.03)
	ER	93.96 (94.64)	95.10 (95.5)	14.64 (12.73)	95.99 (95.86)	-
2	WC_r	98.38 (98.92)	98.44 (98.93)	3.46 (1.25)	99.25 (99.24)	1803.9 (901.44)
	PT.1	0.40 (0.40)	0.41 (0.41)	0.045 (0.02)	0.41 (0.41)	31.79 (17.09)
	ER	98.17 (96.7)	98.9 (97.3)	12.36 (10.5)	99.44 (98.32)	-
3	WC_r	99.73 (99.72)	99.73 (99.72)	0.33 (0.35)	99.85 (99.84)	-
	PT.1	0.40 (0.41)	0.40 (0.41)	0.048 (0.02)	0.41 (0.41)	31.27 (17.87)
	dPT.2	2.02 (2.07)	2.07 (2.10)	0.44 (0.40)	2.02 (2.04)	417.24 (336.46)
	ER	84.58 (83.80)	86.41 (85.78)	17.69 (18.34)	91.50 (90.97)	-
4	WC_r	99.72 (99.74)	99.72 (99.74)	0.37 (0.29)	99.83 (99.83)	-
	PT.1	0.41 (0.41)	0.41 (0.41)	0.047 (0.026)	0.41 (0.41)	39.34 (26.13)
	dPT.2	2.08 (2.03)	2.19 (2.10)	0.68 (0.52)	2.04 (2.04)	619.55 (459.9)
	ER	83.98 (82.63)	86.34 (85.01)	20.05 (19.97)	91.97 (91.47)	-
5	WC_r	98.83 (99.04)	98.88 (99.06)	2.89 (1.68)	99.74 (99.69)	1861.8 (1254.5)
	PT.1	0.40 (0.40)	0.40 (0.41)	0.04 (0.014)	0.40 (0.41)	23.78 (12.76)
	ER	78.71(84.67)	84.10 (86.92)	29.60 (19.65)	84.60 (87.71)	-
6	WC_r	99.59 (99.67)	99.60 (99.67)	0.55 (0.36)	99.80 (99.83)	-
	PT.1	0.40 (0.41)	0.40 (0.42)	0.053 (0.025)	0.41 (0.41)	43.25 (26.31)
	dPT.2	2.42 (2.16)	2.65 (2.30)	1.10 (0.79)	2.49 (2.36)	1222.3 (831.00)
	ER	89.61 (87.75)	90.60 (88.73)	13.38 (13.15)	94.50 (93.59)	-

4. Discussion

From the RGA analysis, it was found that VT.3 should control PT.1 in all cases. The numerical data from Experiments 1 and 2 (Table 8) shows that the control structure proposed by the RGA analysis improves the ER. The values for WC_r is slightly worse in Experiment 2, but if the initial transient is ignored the difference is reduced. The slow valve VT.2 is controlling WC_r in Experiment 2, which may explain why the values are worse, as it takes this valve more time to reduce the error compared to VT.3.

The numerical data from Experiments 3 and 4 (Table 8) show that controlling dPT.2 rather than WC_r improves the separator performance. The mean, median and RMS of the WC_r are higher and the STD is much lower in both Experiments 3 and 4 compared to Experiments 1 and 2. This comes at the cost of a lower extraction rate. This could probably be improved by finding a better setpoint for dPT.2. The differences between Experiment 3 (not RGA) and Experiment 4 (RGA) are very small when looking at WC_r. Experiment 4 has slightly lower values in mean, median and RMS and slightly higher in STD, but, if the initial transient is ignored, the values are slightly better than in Experiment 3 (except for median). Overall, Experiment 3 has slightly better values than Experiment 4. This is the opposite of what one might expect based on the results of the RGA analysis. The RGA analysis, however, is only based on steady state behaviour and does not consider time-delays or transients. The results may indicate that the transfer functions used in the RGA analysis are significantly different from the real dynamics, i.e., the model identification in Section 2.4 may be insufficiently accurate.

The performance of the adaptive controllers are approximately equal to the performance of the PI controllers. Experiments 5 and 6 must be compared with Experiments 2 and 4, respectively.

The WC_r is slightly higher in Experiment 5 compared to Experiment 2, but the ER is much lower in Experiment 5. This is caused by the very low adaptation gains chosen in Experiment 5, which causes VT.2 to close very slowly and, hence, less variations are present in WC_r. The pressure control, however, is slightly improved when using the adaptive controller.

The values from Experiment 6 are very similar to those from Experiment 4. The WC_r is slightly worse, but the ER is higher. The adaptive gains for the dPT.2 controller again had to be chosen very low, which causes sluggish control of dPT.2. This again affects the static pressure due to the interactions between the two CVs, and hence the performance is reduced for both dPT.2 and PT.1. Comparing Experiments 5 and 6, it is again clear that controlling dPT.2 rather than WC_r improves the efficiency.

According to the information shown in Table 8, the performance of the PI controller and the adaptive controller is approximately equal. However, aspects such as implementation and ease of operation should also be considered. The PI controllers could be easily implemented in the LabView block diagram, but the tuning required step response experiments and some trial and error. The MRAC, however, did not require a step response model, but the implementation required a custom script and the tuning was largely based on trial and error and the experience of the operators. The adaptation gain for the controller operating the slow valve VT.2 had to be very low, which may have negatively affected the end result. Improving the performance significantly with tuning, however, would be difficult due to the constraints imposed by the slow valve. Finding suitable adaptation gains and initial values for the MRAC was not trivial.

The SIMC tuning rules was chosen for the PI controllers, due to its simplicity and proven efficiency for first-order plus time delay systems [16], but other tuning methods specifically designed for tuning decentralized PI controllers with two inputs and two inputs exist [22,23]. These methods may reduce the interactions between the control loops and lead to tighter control during transients, at the cost of a more complex tuning procedure.

A multivariable controller (adaptive or not) would probably outperform both controllers as it would better compensate for the interactions between the control loops. Implementing this is suggested as future work.

A model predictive controller (MPC) would also be a natural next step. The MPC can calculate the optimal setpoints and inputs while also handling the input and variable constraints. Implementing an MPC is also suggested as future work.

5. Conclusions

This paper presented a control structure analysis and controller design for a novel multi-pipe separator concept developed at the Department of Geoscience and Petroleum at the Norwegian University of Science and Technology. The control structure analysis gives an indication of which outputs to pair with which inputs, and the controller design for the conventional PI controllers is based on the well established SIMC tuning rules. Step response experiments were performed to gather data for the dynamic models of the different input/output relations in the separator. The dynamic models were assumed to be of first order with a time delay, but second order models and models accounting for the interactions between the states would probably yield better results, considering the measured system responses. Model reference adaptive controllers were also developed for the separator. The performances of the PI and adaptive controllers were quite similar, but the adaptive controller does not require a step response model in the tuning procedure. Due to a lack of tuning rules, however, the adaptive controller was quite difficult to tune. Furthermore, the adaptation gains in the MRAC had to be chosen very small due to the slow control valve VT.2. A faster valve would probably improve the results.

It was found that controlling the dP over the incline in the separator, and the water/oil interface level, by proxy, yielded a more stable water cut ratio on the water outlet, which was the primary control objective. This is due to dP control establishing a buffer volume of water in the incline, unlike when controlling WC_r directly.

The separator is a multiple input, multiple output (MIMO) system and would probably benefit from a multivariable controller rather than two decoupled controllers. Model predictive control could potentially improve the results even more, as the separator is subject to several constraints and control objectives. Finding the optimal setpoints and outputs within the constraints is key for efficient operation. This is suggested for future work.

Author Contributions: Conceptualization, S.J.O., H.S.S., M.S. and C.H.; methodology, S.J.O. and H.S.S.; software, S.J.O. and H.S.S.; validation, S.J.O. and H.S.S.; formal analysis, S.J.O. and H.S.S.; investigation, S.J.O. and H.S.S.; resources, H.S.S. and M.S.; data curation, S.J.O. and H.S.S.; writing—original draft preparation, S.J.O. and H.S.S.; writing—review and editing, S.J.O., H.S.S., M.S. and C.H.; supervision, M.S. and C.H.

Funding: This research was carried out as a part of SUBPRO, a Research-based Innovation Center within Subsea Production and Processing. The authors gratefully acknowledge the financial support from SUBPRO, which is financed by the Research Council of Norway, major industry partners, and NTNU.

Conflicts of Interest: The authors declare no conflict of interest.

Abbreviations

The following abbreviations are used in this manuscript:

MPPS	Multiple parallel pipe separator
CV	Controlled variable
MV	Manipulated variable
SIMC	Simple internal model control
ppm	Parts per million
P&ID	Piping and instrumentation diagram
PVC	Polyvinyl chloride
MRAC	Model reference adaptive control
ER	Extraction rate
PID	Proportional, Integral, Derivative
WC	Water cut
PT	Pressure transmitter
dPT	Differential pressure transmitter
RMS	Root mean square
STD	Standard deviation
IAE	Integrated absolute error

Appendix A

The incline was photographed under varying inlet conditions and with varying pressure difference over the incline. The photos show that a pressure difference setpoint around 2 mbar will give stable oil and water layers and a decent water buffer in the incline. The photos are shown in Figures A1–A4.

Q_{tot} [l/min]	WC_{in} [%]	$dPT.2$ [mbar]	WC_r [%]	ER [%]	
299.83	30.09	1.22	99.93	45.96	
300.04	30.11	1.67	99.91	60.11	
300.09	30.11	2.24	99.83	75.78	

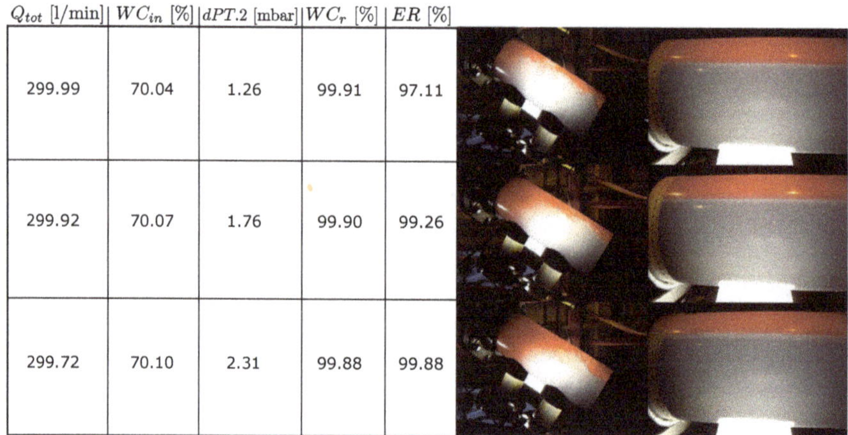

Figure A1. 300 L/min inlet flow and 30% water cut.

Q_{tot} [l/min]	WC_{in} [%]	$dPT.2$ [mbar]	WC_r [%]	ER [%]	
299.99	70.04	1.26	99.91	97.11	
299.92	70.07	1.76	99.90	99.26	
299.72	70.10	2.31	99.88	99.88	

Figure A2. 300 L/min inlet flow and 70% water cut.

Q_{tot} [l/min]	WC_{in} [%]	$dPT.2$ [mbar]	WC_r [%]	ER [%]	
499.94	30.25	1.76	100.02	12.18	
499.98	30.11	2.21	100.0	24.55	
499.96	30.08	3.25	98.12	58.36	

Figure A3. 500 L/min inlet flow and 30% water cut.

Q_{tot} [l/min]	WC_{in} [%]	$dPT.2$ [mbar]	WC_r [%]	ER [%]
499.99	69.83	1.16	99.47	75.44
499.91	69.89	1.70	99.17	80.89
500.15	69.81	2.16	98.80	86.72

Figure A4. 500 L/min inlet flow and 70% water cut.

References

1. Norwegian Oil and Gas Association. *Environmental Report—Environmental Work by the Oil and Gas Industry, Facts and Development Trends*; Technical Report; Norwegian Oil and Gas Association: Stavanger, Norway, 2017.
2. Hannisdal, A.; Westra, R.; Akdim, M.R.; Bymaster, A.; Grave, E.; Teng, D.T. Compact separation technologies and their applicability for subsea field development in deep water. In Proceedings of the Offshore Technology Conference, Houston, TX, USA, 30 April–3 May 2012; Offshore Technology Conference: Houston, TX, USA, 2012.
3. Backi, C.J.; Skogestad, S. A simple dynamic gravity separator model for separation efficiency evaluation incorporating level and pressure control. In Proceedings of the 2017 American Control Conference (ACC), Seattle, WA, USA, 24–26 May 2017; pp. 2823–2828.
4. Backi, C.J.; Krishnamoorthy, D.; Skogestad, S. Slug handling with a virtual harp based on nonlinear predictive control for a gravity separator. *IFAC-PapersOnLine* **2018**, *51*, 120–125. [CrossRef]
5. Skjefstad, H.S.; Stanko, M. An Experimental Study of a Novel Parallel Pipe Separator Design for Subsea oil–water Bulk Separation. In Proceedings of the SPE Asia Pacific Oil and Gas Conference and Exhibition, Brisbane, Australia, 23–25 October 2018; Society of Petroleum Engineers: Richardson, TX, USA, 2018.
6. Skjefstad, H.S.; Stanko, M. Experimental performance evaluation and design optimization of a horizontal multi-pipe separator for subsea oil–water bulk separation. *J. Pet. Sci. Eng.* **2019**, *176*, 203–219. [CrossRef]
7. Pereira, R.M.; Campos, M.C.M.M.D.; de Oliveira, D.A.; de Souza, R.D.S.A.; Orlowski, R.; Duarte, D.G.; Raposo, G.M.; Lillebrekke, C.; Ljungquist, D.; Carvalho, A.; et al. SS: Marlim 3 Phase Subsea Separation System: Controls Design Incorporating Dynamic Simulation Work. In Proceedings of the Offshore Technology Conference, Houston, TX, USA, 30 April–3 May 2012; Offshore Technology Conference: Houston, TX, USA, 2012.
8. Li, Z.; Olson, M.; Rayachoti, V.; Gupte, P.; Pierre, F.; Gul, K. Subsea Compact Separation: Control System Design. In Proceedings of the Offshore Technology Conference, Houston, TX, USA, 5–8 May 2014; Offshore Technology Conference: Houston, TX, USA, 2014.
9. Olson, M.; Grave, E.; Juarez, J.; Gul, K. Performance Testing of an Integrated, Subsea Compact Separation System with Electrocoalescence for Deepwater Applications. In Proceedings of the Offshore Technology Conference, Houston, TX, USA, 4–7 May 2015; Offshore Technology Conference: Houston, TX, USA, 2015.
10. Grave, E.; Olson, M. Design and Performance Testing of a Subsea Compact Separation System for Deepwater Applications. *Oil Gas Facil.* **2014**, *3*, 16–23. [CrossRef]
11. Sagatun, S.I.; Gramme, P.; Horgen, O.J.; Ruud, T.; Storvik, M. The Pipe Separator-Simulations and Experimental Results. In Proceedings of the Offshore Technology Conference, Houston, TX, USA, 5–8 May 2008; Offshore Technology Conference: Houston, TX, USA, 2008.

12. Minh, V.T. Modeling and control of distillation column in a petroleum process. In Proceedings of the 2010 5th IEEE Conference on Industrial Electronics and Applications, Taichung, Taiwan, 15–17 June 2010; pp. 259–263.

13. Campos, M.; Takahashi, T.; Ashikawa, F.; Simões, S.; Stender, A.; Meien, O. Advanced anti-slug control for offshore production plants. *IFAC-PapersOnLine* **2015**, *48*, 83–88. [CrossRef]

14. Ohrem, S.J.; Holden, C.; Jahanshahi, E.; Skogestad, S. $\mathcal{L}_1$ adaptive anti-slug control. In Proceedings of the 2017 American Control Conference (ACC), Seattle, WA, USA, 24–26 May 2017; pp. 444–449.

15. Ohrem, S.J.; Kristoffersen, T.T.; Holden, C. Adaptive feedback linearizing control of a gas liquid cylindrical cyclone. In Proceedings of the 2017 IEEE Conference on Control Technology and Applications (CCTA), Mauna Lani, HI, USA, 27–30 August 2017; pp. 1981–1987.

16. Skogestad, S.; Grimholt, C. The SIMC method for smooth PID controller tuning. In *PID Control in the Third Millennium*; Springer: London, UK, 2012; pp. 147–175.

17. Skogestad, S.; Postlethwaite, I. *Multivariable Feedback Control: Analysis and Design*; John Wiley & Sons Ltd.: Hoboken, NJ, USA, 2007; Volume 2.

18. Wang, Q.G.; Ye, Z.; Cai, W.J.; Hang, C.C. *PID Control for Multivariable Processes*; Springer: Berlin, Germany, 2008.

19. Hovd, M.; Skogestad, S. Pairing criteria for decentralized control of unstable plants. *Ind. Eng. Chem. Res.* **1994**, *33*, 2134–2139. [CrossRef]

20. Visioli, A. *Practical PID Control*; Springer Science & Business Media: Berlin, Germany, 2006.

21. Ioannou, P.A.; Sun, J. *Robust Adaptive Control*; PTR Prentice-Hall: Upper Saddle River, NJ, USA, 1996; Volume 1.

22. Vázquez, F.; Morilla, F.; Dormido, S. An iterative method for tuning decentralized PID controllers. In Proceedings of the 14th IFAC World Congress, Beijing, China, 5–9 July 1999; pp. 491–496.

23. Huang, H.P.; Jeng, J.C.; Chiang, C.H.; Pan, W. A direct method for multi-loop PI/PID controller design. *J. Process Control* **2003**, *13*, 769–786. [CrossRef]

Article

Improvement of Refrigeration Efficiency by Combining Reinforcement Learning with a Coarse Model

Dapeng Zhang [1] and Zhiwei Gao [2,*]

[1] School of Electrical and Information Engineering, Tianjin University, Tianjin 300072, China; zdp@tju.edu.cn
[2] Faculty of Engineering and Environment, University of Northumbria, Tyne and Wear NE1 8ST, UK
* Correspondence: zhiwei.gao@northumbria.ac.uk

Received: 2 November 2019; Accepted: 15 December 2019; Published: 17 December 2019

Abstract: It is paramount to improve operational conversion efficiency in air-conditioning refrigeration. It is noticed that control efficiency for model-based methods highly relies on the accuracy of the mechanism model, and data-driven methods would face challenges using the limited collected data to identify the information beyond. In this study, a hybrid novel approach is presented, which is to integrate a data-driven method with a coarse model. Specifically, reinforcement learning is used to exploit/explore the conversion efficiency of the refrigeration, and a coarse model is utilized to evaluate the reward, by which the requirement of the model accuracy is reduced and the model information is better used. The proposed approach is implemented based on a hierarchical control strategy which is divided into a process level and a loop level. The simulation of a test bed shows the proposed approach can achieve better conversion efficiency of refrigeration than the conventional methods.

Keywords: data-driven methods; reinforcement learning; coarse model; refrigeration

1. Background and Motivation

Along with the comfortable environment, the air-conditioning brings a problem of energy consumption. According to the U.S. Energy Information Administration, the energy consumption of buildings for residential and commercial users accounts for 20.1% of the global energy consumption worldwide. Of this amount, the buildings' heating, ventilation, and air conditioning (HVAC) systems can account for up to 50% of total building energy demand [1,2]. To minimize overall system energy input or operating cost while still maintaining the satisfying indoor thermal comfort and healthy environment all kinds of control algorithms are employed on HVAC. In [3], model-based monitoring of the occupants' thermal state by predictive controlling was proposed. In [4], the supervisory and optimal controls were summarized and classified into the model-based methods which used physical models, gray-box models, and black-box models, and the model-free category which used expert systems and pure learning approaches. In [5] the existed methods were grouped into the hard control (HC), such as basic controls involving PID control, optimal control, nonlinear control, robust or H∞ control, and adaptive control, and the soft control (SC) involving neural networks (NNs), fuzzy logic (FL), genetic algorithms (GAs), and other evolutionary methods. Moreover, hybrid control resulting from the fusion of SC and HC may achieve better performance.

Most of today's air-conditioning systems work based on the principle of compression refrigeration in which refrigeration is the core of the air-conditioning system [6]. The theoretical optimal conversion efficiency is used in the design of air-conditioning based on the specified loads, according to refrigeration theory. However, it is inevitable in practice for performance degradation due to the inconsistencies between design requirements and operational conditions. For example, a public place will provide a little refrigerating output to respond to small load at night, meanwhile it will need a large number of refrigerating output for the varying flow of people. The conventional control algorithm will result in the low operational conversion efficiency in this time-varying and uncertain condition because as we know the efficiency of compression refrigeration is confined to the working point and affected by the loads.

With the development of modern electronic and measurement technologies, such as supervisory control and data acquisition (SCADA), and smart sensors, the refrigeration system is prone to obtain abundant data which provide the information about consistency in the current refrigeration states and the operational environment. It is promising for the data-driven methods to improve the conversion efficiency by directly using these data. The reinforcement learning (RL) is a famous data-driven method which is regarded as a model-free supervisory control method in [5] and a soft control (SC) in [6], is about learning from interaction and how to behave in order to achieve a goal. The basic idea of reinforcement learning is simply to capture the most important aspects of an agent, which includes the sensation, the action, and the goal [7]. A reward rule is then employed to encourage the agent to seek the goal by adjusting its action with exploration and exploitation. Finally, the agent owns the optimal action to adapt to the surroundings by trials and errors. In this way, the RL can achieve optimal action based on the current states and environment. Along with incredible success in games [8], the RL has attracted great interest in various industries, such as robot [9,10], fault detection [11], and fault-tolerant control [12,13].

The RL has been introduced to HVA and some excellent results have been published recently. Baeksuk Chu et al. proposed an actor-critic architecture and natural gradient method with an improvement of the efficiency by the recursive least-squares (RLS) method aiming at two objectives: Maintaining an adequate level of pollutants and minimizing power consumption [14]. Pedro Fazenda et al. followed the idea of a multi-objective optimal supervisory control and studied the application of a discrete and continuous reinforcement-learning-based supervisory control approach, which actively learned how to appropriately schedule thermostat temperature set-points [15]. In [16], a multi-grid method of Q-learning was addressed to handle the problem that online reinforcement learning often suffered from slow convergence and faults in early stages. In [17], a suitable learning architecture was proposed for an intelligent thermostat that comprised a number of different learning methods, each of which contributed to creating a complete autonomous thermostat capable of controlling an HVAC system and proposed a novel state action space formalism to enable RL successfully. In [18], a ventilated facade with PCM was controlled using a reinforcement learning algorithm. In [19], a reinforcement learning control (RLC) approach was presented for optimal control of low exergy buildings. An improved reinforcement learning controller was designed in [20] to obtain an optimal control strategy of blinds and lights, which provided a personalized service via introducing subject perceptions of surroundings gathered by a novel interface as the feedback signal. In [21], a model-free actor-critic reinforcement learning (RL) controller was addressed using a variant of artificial recurrent neural networks called long-short-term memory (LSTM) networks. A model-free Q-learning was addressed in [22] that made optimal control decisions for HVAC and window systems to minimize both energy consumption and thermal discomfort. Recently, the latest methods on RL, for example, deep reinforcement learning control [23,24], regularized fitted Q-iteration approach [25], proximal actor critic [26], and auto-encoder [27,28] were appeared in air-conditioning to minimize both energy consumption and thermal discomfort. However, all the studies do not involve the time-varying and uncertain loads.

The RL is classified as online learning and offline learning, according to the resource of training data [29]. The online learning will make sense in the refrigeration system for the reason of overcoming the loads time-varying and uncertainty. There are two challenges: (1) The new states should be obtained only after the actions have acted on the system. It is a risk for the system because no one knows the results of these actions. (2) The conventional RL uses an errors and trials method to train, which is a forced way to the unknown environment but with low efficiency due to little information between states and actions. In this study, a coarse model is proposed to help solve both problems. We consider a coarse model based on a fact that the human has grasped the principles of refrigeration conversion. It is possible to build a coarse model to discover the relations of state variables according to the principles based on the figures and charts through a large number of experiences. On the one hand, it will bring more information between states and actions to speed up the learning process. On the other hand, it will forecast the effects of actions as a simulation. We give up the accurate model for the reasons of the system complexity and surrounding disturbances. In addition, the RL only needs the reward as a basis of action adjustment in the process of seeking for the goal. The reward can be obtained from the tendency of performance indictor which reduces to the requirement of model accuracy. As a result, it is enough for a coarse model to be used for RL. Based on a coarse model, the computer simulation can be approximated to estimate the next states instead of the real system which reduces the risk of unknown states after a trial action.

Motivated by the above discussions, a novel approach combines the reinforcement learning with a coarse model in which we take the conversation efficiency as a goal of reinforcement learning, the measurement data of refrigeration as the sensation of agent, and the control valve of refrigeration as the action of agent, and learn by Q-algorithm, is proposed to improve the conversion efficiency of refrigeration and the advantages are summarized as follows:

1. This is an online control strategy with data-driven control and learning simultaneously which reduces the requirement of model accuracy.
2. It has the ability to get the optimal action by exploration and exploitation to achieve a better conversion efficiency which fits the current operating conditions.
3. The risk on the system in the learning process, which is complex and unknown in advance is prevented mostly by a hierarchical control of the process level, which makes use of the existing knowledge of refrigeration and the loop level, which implements the tracking control.

The remaining parts of this paper are organized as follows: The aim of the study and the preliminaries on a compression refrigerating system are introduced in Section 2. The fundamentals of reinforcement learning algorithms are reviewed in Section 3. Based on Sections 2 and 3, a reinforcement learning control algorithm for compression refrigerating systems is proposed in Section 4, and the case studies are addressed in Section 5. The paper is ended by conclusions in Section 6.

2. Preliminaries

2.1. Aim of the Research

For a compression refrigerating system, the compressor drives refrigerants to a circulatory flow to transfer the heating of cryogenic medium to the high-temperature medium via a phase change of refrigerant. In this process, the temperature is the key indicator of reflecting efficiency which is a comprehensive result of the interaction of mass, pressure, process, and loads, so the coefficient of performance (COP) at time k is defined as

$$COP(k) = \frac{T_c(k)}{T_c(k) - T_e(k)} \tag{1}$$

where T_c and T_e are the evaporation temperature and the condensation temperature at sampling k, respectively. Our aim is to maximize the coefficient of performance of the refrigeration cycle during a period of time *l*

$$COP = \max\left\{ \sum_{k=1}^{l} \frac{T_c(k)}{T_c(k) - T_e(k)} \right\} \tag{2}$$

2.2. Principle of Compression Refrigerating System

A compression refrigerating system is always made up of four elements: The compressor, the evaporator, the condenser, and the throttle mechanic. The compressor provides power for the circulation of the refrigerant. The evaporator is employed to implement the heating change process between the refrigeration and the chilled water. The condenser is used to complete the heating change process between the refrigeration and the cooling water or air cooling system. The throttle mechanic (mainly electronic expansion valve) is applied to adjust the mass flow rate of the refrigerant in the circulation. Here, the principle of each element is reviewed for the integrity according to [30–33] and more details can be found in [30–33] and the references therein.

2.2.1. Principle of the Evaporator

The scheme of the evaporator process is dictated in Figure 1.

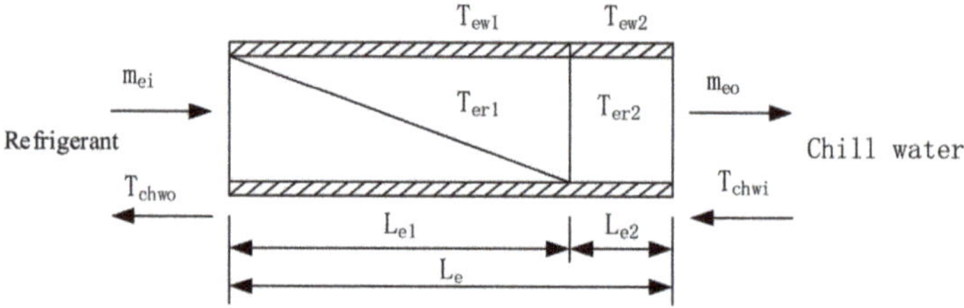

Figure 1. The scheme of the two-phase zone.

The inner region of the evaporator is divided into a two-phase zone where the refrigerant lives in the combination of liquid and gas and a superheated zone where the liquid refrigerant evaporates into a gas with absorbing the heating and finally becomes the superheated steam. The refrigerant is moving from the two-phase zone to the superheated zone. An average void fraction correlation is used to express the proportion of gas in the two-phase zone whose definition is an integrating of void-age along the pipeline direction

$$\overline{\gamma} = \frac{1}{x_2 - x_1} \int_{x_1}^{x_2} \gamma(x) d(x) \tag{3}$$

where γ is a void fraction correlation with a form of formula (4)

$$\gamma = \frac{1}{1 + \left(\frac{1-x}{x}\right)\left(\frac{\rho_g}{\rho_l}\right)S} \tag{4}$$

in which x is the vapor quality, ρ_g and ρ_l are the density of saturated steam and the density of saturated liquid, respectively, kg/m^3, S is the slip ratio.

According to the law of mass conservation, the law of momentum conservation and the law of energy conservation, one will obtain the state equation of evaporator as formula (5)

$$E(x_e)\dot{x}_e = f(x_e, u_e) \tag{5}$$

$$\begin{bmatrix} E_{11} & E_{12} & 0 & 0 & 0 \\ E_{21} & E_{22} & E_{23} & 0 & 0 \\ E_{31} & E_{32} & E_{33} & 0 & 0 \\ 0 & 0 & 0 & E_{44} & 0 \\ E_{51} & 0 & 0 & 0 & E_{55} \end{bmatrix} \begin{bmatrix} \dot{L}_{e1} \\ \dot{P}_e \\ \dot{h}_{eo} \\ \dot{T}_{ew1} \\ \dot{T}_{ew2} \end{bmatrix} = \begin{bmatrix} \dot{m}_{ei}(h_{ei} - h_g) + \alpha_{ei1} A_{ei}\left(\frac{L_{ei1}}{L_{etotal}}\right)(T_{ew1} - T_{er1}) \\ \dot{m}_{eo}(h_g - h_{eo}) + \alpha_{ei2} A_{ei}\left(\frac{L_{ei2}}{L_{etotal}}\right)(T_{ew2} - T_{er2}) \\ \dot{m}_{ei} - \dot{m}_{eo} \\ \alpha_{ei1} A_{ei}(T_{er1} - T_{ew1}) + \alpha_{eo} A_{eo}(T_{chw} - T_{ew1}) \\ \alpha_{ei2} A_{ei}(T_{er2} - T_{ew2}) + \alpha_{eo} A_{eo}(T_{chw} - T_{ew2}) \end{bmatrix}$$

$$E_{11} = \rho_1(h_1 - h_g)(1 - \overline{\gamma})A_{ecs}$$

where

$$E_{12} = \left[\left(\frac{d\rho_1 h_1}{dP_e} - \frac{d\rho_1}{dP_e}h_g\right)(1 - \overline{\gamma}) + \left(\frac{d\rho_g h_g}{dP_e} - \frac{d\rho_g}{dP_e}h_g\right)(\overline{\gamma} - 1)\right]A_{ecs}L_{e1}$$

$$E_{21} = \rho_{e2}\left(h_g - h_{e2}\right)A_{ecs}$$

$$E_{22} = \left[\left(\left(\frac{\partial \rho_{e2}}{\partial P_e}\bigg|_{h_{e2}}\right) - \frac{1}{2}\left(\frac{\partial \rho_{e2}}{\partial h_{e2}}\bigg|_{P_e}\right)\left(\frac{dh_g}{dP_e}\right)\right)\left(h_{e2} - h_g\right) + \frac{1}{2}\left(\frac{dh_g}{dP_e}\right)\rho_{e2} - 1\right]A_{ecs}L_{e2}$$

$$E_{23} = \left[\frac{1}{2}\left(\frac{\partial \rho_{e2}}{\partial h_{e2}}\bigg|_{P_e}\right)\left(h_{e2} - h_g\right) + \frac{\rho_{e2}}{2}\right]A_{ecs}L_{e2}$$

$$E_{31} = \left[\left(\rho_g - \rho_{e2}\right) + \left(\rho_1 - \rho_g\right)(1 - \overline{\gamma})\right]A_{ecs}$$

$$E_{32} = \left[\left(\left(\frac{\partial \rho_{e2}}{\partial P_e}\bigg|_{h_{e2}}\right) + \frac{1}{2}\left(\frac{\partial \rho_{e2}}{\partial h_{e2}}\bigg|_{P_e}\right)\left(\frac{dh_g}{dP_e}\right)\right)L_{e2} + \left(\frac{d\rho_1}{dP_e}(1 - \overline{\gamma}) + \left(\frac{d\rho_g}{dP_e}\right)\overline{\gamma}\right)L_{e1}\right]A_{ecs}$$

$$E_{33} = \frac{1}{2}\left(\frac{\partial \rho_{e2}}{\partial h_{e2}}\bigg|_{P_e}\right)A_{ecs}L_{e2}, \; E_{34} = \left(C_p\rho V\right)_{ew},$$

$$E_{51} = -\left(C_p\rho V\right)_{ew}\frac{T_{ew2} - T_{ew1}}{L_{e2}}, \; E_{55} = \left(C_p\rho V\right)_{ew}$$

The meanings of states and parameters are seen in Table 1.

Table 1. Description of model symbols.

Letter	Name	Subscript	Name	Subscript	Name
T	Temperature, $^\circ C$	k	Compressor	$ei2$	Inner of superheated zone of evaporator
M	Mass, Kg	c	Condenser	$er2$	Refrigerant in superheated zone of evaporator
ρ	Density, Kg/m^3	e	evaporator	$ew2$	Tube wall of superheated zone of evaporator
N	Power, W	val	Expansion valve	$ei1$	Inner of two-phase zone of evaporator
η	Efficiency	r	refrigerant	$er1$	Refrigerant in two-phase zone of evaporator
K	Heat transfer coefficient, $W/m^2 \cdot {}^\circ C$	w	Tube wall, water	$ew1$	Tube wall of two-phase zone of evaporator
α	Heat transfer coefficient, $W/m^2 \cdot {}^\circ C$	i	Input/inner	$ci2$	Inner of superheated zone of condenser
μ	Dynamic viscosity, $Pa \cdot s$	o	Output/outer	$cr2$	Refrigerant in superheated zone of condenser
A	Area, m^2	sc	supercool	$cw2$	Tube wall of superheated zone of condenser
m	Mass flow, $Kg \cdot s$	sh	superhot	$ci1$	Inner of two-phase zone of condenser
C	Specific heat capacity, $KJ/Kg \cdot {}^\circ C$	th	Theoretical value	$cr1$	Refrigerant in two-phase zone of condenser
v	Specific volume, m^3/Kg	g	Gas	$cw1$	Tube wall of two-phase zone of condenser
h	Specific enthalphy, KJ/Kg	l	Liquid	$ci3$	Tube wall of supercool zone of condenser

2.2.2. Principle of the Condenser

The condenser inner region is divided into a two-phase zone, a superheated zone, and a supercool zone. For each zone, the partial differential equations have been built according to the law of conservation of mass, the law of conservation of momentum and the law of conservation of energy based on the properties of refrigerants and the experience equations. Therefore, the state equation of the condenser, obtained by integrating along the pipeline direction, is given as follows:

$$C(x_c)\dot{x}_c = f(x_c, u_c) \tag{6}$$

$$
\begin{bmatrix}
C_{11} & 0 & C_{13} & 0 & 0 & 0 & 0 \\
C_{21} & C_{22} & C_{23} & C_{24} & 0 & 0 & 0 \\
C_{31} & C_{32} & C_{33} & C_{34} & 0 & 0 & 0 \\
C_{41} & C_{42} & C_{43} & C_{44} & 0 & 0 & 0 \\
C_{51} & 0 & 0 & 0 & C_{55} & 0 & 0 \\
0 & 0 & 0 & 0 & 0 & C_{66} & 0 \\
C_{71} & 0 & 0 & 0 & 0 & 0 & C_{77}
\end{bmatrix}
\begin{bmatrix}
\dot{L}_{c1} \\ \dot{L}_{c2} \\ \dot{P}_c \\ \dot{h}_{co} \\ \dot{T}_{cw1} \\ \dot{T}_{cw2} \\ \dot{T}_{cw3}
\end{bmatrix}
=
\begin{bmatrix}
\dot{m}_{ci}(h_{ci} - h_g) + \alpha_{ci1}A_{ci}\left(\frac{L_{c1}}{L_{ctotal}}\right)(T_{cw1} - T_{cr1}) \\
\dot{m}_{ci}h_g - \dot{m}_{co}h_l + \alpha_{ci2}A_{ci}\left(\frac{L_{c2}}{L_{ctotal}}\right)(T_{cw2} - T_{cr2}) \\
\dot{m}_{co}(h_l - h_{co}) + \alpha_{ci3}A_{ci}\left(\frac{L_{c3}}{L_{ctotal}}\right)(T_{cw3} - T_{cr3}) \\
\dot{m}_{ci} - \dot{m}_{co} \\
\alpha_{ci1}A_{ci}(T_{cr1} - T_{cw1}) + \alpha_{co}A_{co}(T_a - T_{cw1}) \\
\alpha_{ci2}A_{ci}(T_{cr2} - T_{cw2}) + \alpha_{co}A_{co}(T_a - T_{cw2}) \\
\alpha_{ci3}A_{ci}(T_{cr3} - T_{cw3}) + \alpha_{co}A_{co}(T_a - T_{cw3})
\end{bmatrix}
$$

where

$$C_{11} = \rho_{c1}\left(h_{c1} - h_g\right)A_{ecs}$$

$$C_{13} = \left[\left(\left.\frac{\partial \rho_{e1}}{\partial P_c}\right|_{h_{c1}}\right)h_{c1} + \frac{1}{2}\frac{dh_g}{dP_c}\left(\left(\left.\frac{\partial \rho_{c1}}{\partial P_{c1}}\right|_{P_c}\right)(h_{c1} - h_g) + \rho_{c1}\right) - 1\right]A_{ccs}L_{c1}$$

$$C_{21} = \left(\rho_{c1}h_g - \rho_{c3}h_1\right)A_{ecs}, \quad C_{22} = \left[\left(\rho_g h_g - \rho_1 h_1\right)\bar{\gamma} + (\rho_1 - \rho_{c3})h_1\right]A_{ccs}$$

$$
\begin{aligned}
C_{23} = &\left[\left(\left.\frac{\partial \rho_{c1}}{\partial P_c}\right|_{h_{c1}}\right) + \frac{1}{2}\left(\left.\frac{\partial \rho_{c1}}{\partial P_{c1}}\right|_{P_c}\right)\frac{dh_g}{dP_c}\right]h_g A_{ccs}L_{c1} \\
&+ \left(\frac{d\rho_1 h_1}{dP_c}(1 - \bar{\gamma}) + \frac{d\rho_g h_g}{dP_e}(\bar{\gamma} - 1)\right)A_{ccs}L_{c2} \\
&+ \left[\left(\left.\frac{\partial \rho_{c3}}{\partial P_c}\right|_{h_{c3}}\right) + \frac{1}{2}\left(\left.\frac{\partial \rho_{c3}}{\partial P_{c3}}\right|_{P_c}\right)\frac{dh_1}{dP_c}\right]h_1 A_{ccs}L_{c3}
\end{aligned}
$$

$$C_{24} = \frac{1}{2}\left(\left.\frac{\partial \rho_{c3}}{\partial h_{c3}}\right|_{P_c}\right)h_1 A_{ccs}L_{c3}, \quad C_{31} = \rho_{c3}(h_1 - h_{c3})A_{ccs}, \quad C_{32} = \rho_{c3}(h_1 - h_{c3})A_{ccs}$$

$$C_{33} = \left[\left(\left(\left.\frac{\partial \rho_{c3}}{\partial P_c}\right|_{h_{c3}}\right) + \frac{1}{2}\frac{dh_1}{dP_c}\left(\left.\frac{\partial \rho_{c3}}{\partial P_{c3}}\right|_{P_c}\right)\right)(h_{c3} - h_1) + \frac{1}{2}\frac{dh_1}{dP_c}\rho_{c3} - 1\right]A_{ccs}L_{c3}$$

$$C_{34} = \left[\frac{1}{2}\left(\left.\frac{\partial \rho_{c3}}{\partial h_{c3}}\right|_{P_c}\right)(h_{c3} - h_1) + \frac{1}{2}\rho_{c3}\right]A_{ccs}L_{c3}$$

$$C_{41} = (\rho_{c1} - \rho_{c3})A_{ccs}, \quad C_{42} = \left[\left(\rho_g - \rho_1\right)\bar{\gamma} + (\rho_1 - \rho_{c3})\right]A_{ccs}$$

$$
\begin{aligned}
C_{43} = &\left[\left(\left.\frac{\partial \rho_{c1}}{\partial P_c}\right|_{h_{c1}}\right) + \frac{1}{2}\left(\left.\frac{\partial \rho_{c1}}{\partial h_{c1}}\right|_{P_c}\right)\frac{dh_g}{dP_c}\right]A_{ccs}L_{c1} + \left(\frac{d\rho_1}{dP_c}(1 - \bar{\gamma}) + \frac{d\rho_g}{dP_e}\bar{\gamma}\right)A_{ccs}L_{c2} \\
&+ \left[\left(\left.\frac{\partial \rho_{c3}}{\partial P_c}\right|_{h_{c3}}\right) + \frac{1}{2}\left(\left.\frac{\partial \rho_{c3}}{\partial h_{c3}}\right|_{P_c}\right)\frac{dh_1}{dP_c}\right]A_{ccs}L_{c3}
\end{aligned}
$$

$$C_{44} = \frac{1}{2}\left(\left.\frac{\partial \rho_{c3}}{\partial h_{c3}}\right|_{P_c}\right)A_{ccs}L_{c3}, \quad C_{51} = \left(C_p\rho V\right)_{cw}\frac{T_{cw2} - T_{cw1}}{L_{c1}}, \quad C_{55} = \left(C_p\rho V\right)_{cw}$$

$$C_{66} = \left(C_p\rho V\right)_{cw}, \quad C_{71} = \left(C_p\rho V\right)_{cw}\frac{T_{cw2} - T_{cw3}}{L_{c3}}, \quad C_{72} = \left(C_p\rho V\right)_{cw}\frac{T_{cw2} - T_{cw3}}{L_{c3}}, \quad C_{77} = \left(C_p\rho V\right)_{cw}$$

The meanings of the states and parameters are provided in Table 1.

2.2.3. Principle of the Compressor

The refrigerant mass flow rate of the compressor outlet is

$$\dot{m}_{com} = f\eta_{vol}V_{com}\rho_g \tag{7}$$

where $\dot{m}_{com}$ is the refrigerant mass flow rate of the compressor outlet, kg/s, f is the working frequency, Hz, V_{com} is the theoretical gas delivery determined by the producer, m^3/s, ρ_g is the density of refrigerant at the entrance of compressor, kg/m^3, η_{vol} is the volume efficiency of the compressor which follows the formula (8)

$$\eta_{vol} = 0.98 - 0.085\left[\left(\frac{P_c}{P_e}\right)^{\frac{1}{k}} - 1\right]$$

(8)

where P_c and P_e are the pressure of condensation and evaporation, respectively, k is the polytrophic exponent.

2.2.4. Principle of the Electronic Expansion Valve

The mass flow of the electronic expansion valve is

$$\dot{m}_e = C_v k \sqrt{\rho_1(P_c - P_e)}$$

(9)

where C_v and k is the flow coefficient and opening of the expansion valve, ρ_1 is the density of refrigerant at the entrance of expansion valve, kg/m^3.

3. Reinforcement Learning

3.1. Reinforcement Learning

Reinforcement learning has been paid much attention during the past decades [34,35], whose basics are reviewed as follows. Consider the Markov decision process MDP $(X, \mathcal{A}, \mathcal{P}, \mathcal{R})$, where X is a set of states and $\mathcal{A}$ is a set of actions or controls. The transition probabilities $\mathcal{P}: X \times X \times \mathcal{A} \times X \rightarrow [0,1]$ represent each state $x \in X$ and action $a \in \mathcal{A}$, the conditional probability $P(x(k+1), x(k), a(k)) = \Pr\{x(k+1)|x(k+1), a(k)\}$ of transitioning to state $x(k+1) \in X$ where the MDP is in state $x(k)$ and takes action $a(k)$. The cost function $\mathcal{R}: X \times \mathcal{A} \times X \rightarrow \mathcal{R}$ is the expected immediate cost $R_k(x(k+1), x(k), a(k))$ paid after transition to state $x(k+1) \in X$, given that the MDP starts in state $x(k) \in X$ and takes action $a(k) \in \mathcal{U}$. The value of a policy $V_k^\pi(x(k))$ is defined as the conditional expected value of the future cost $E_\pi\{\sum_{i=k}^{k+T} \gamma^{i-k} R_i\}$, $\mathcal{R}_i \in R$ when starting in state $x(k)$ at time k and following policy $\pi(x, a)$, The target is to find the optimal actions that maximize the value of the future cost $V_k^\pi(x)$

$$V_k^\pi(x) = E_\pi\left\{\sum_{i=k}^{k+T} \gamma^{i-k} R_i\right\}$$

$$= \sum_a \pi(x, a) \sum_{x(k+1)} P(x(k+1), x(k), a(k))[R_k(x(k+1), x(k), a(k))$$

$$+ \gamma E_\pi\left\{\sum_{i=k+1}^{k+T} \gamma^{i-(k+1)} R_i\right\}]$$

$$= \sum_a \pi(x, a) \sum_{x(k+1)} P(x(k+1), x(k), a(k))\left[R_k(x(k+1), x(k), a(k)) + \gamma V_{k+1}^\pi(x(k+1))\right]$$

(10)

with $T = \infty$, the value function $V_k^\pi(x)$ for the policy $\pi(x, a)$ satisfies the Bellman equation:

$$V_k^\pi(x) = \sum_a \pi(x, a) \sum_{x(k+1)} P(x(k+1), x(k), a(k))[R_k(x(k+1), x(k), a(k))$$

$$+ \gamma V_{k+1}^\pi(x(k+1))]$$

(11)

To a deterministic system $\sum_a \pi_k(x, a) \sum_{x(k+1)} P(x(k+1), x(k), a(k)) = 1$, so the optimal actions can be gained by alternating the policy evaluation (12) and policy improvement (13) according to the following two equations:

$$V_k(x) = R_k(x(k+1), x(k), a(k)) + \gamma V_k(x(k+1))$$

(12)

$$\pi_k(x,a) = \underset{\pi}{\text{argmax}}R_k(x(k+1),x(k),a(k)) + \gamma V_k(x(k+1)) \tag{13}$$

where γ is a discount factor, $0 \le \gamma < 1$ in order to be convergent.

3.2. Q-Algorithm

In fact, formulas (12) and (13) cannot be solved directly because the need the information of $V_k(x(k+1))$ that no one knows. A Q-algorithm proposed by Watkins [34] provides an effective solution by substituting function Q. The Q-algorithm defines the evaluation function $Q(x(k),a(k))$ as the maximum discounted cumulative reward that can be achieved from state $x(k)$ and $a(k)$ as the first action:

$$Q(x(k),a(k)) \overset{\text{def}}{=} R_k(x(k),a(k)) + V^*(x(k+1)) \tag{14}$$

where the state $x(k+1)$ comes from $x(k)$ and $a(k)$, $V^*(x(k+1))$ is the optimization of $V_k(x(k+1))$ beginning with $x(k+1)$ and following the optimal actions.

Denote the optimum of Q as Q^*, therefore one has

$$Q^*(x(k),a(k))$$
$$= \underset{a(k)}{\max}[R_k(x(k),a(k)) + V^*(x(k+1))] = R^*(x(k),a(k)) + V^*(x(k+1)) \tag{15}$$
$$= V^*(x(k),a(k))$$

where the superscript $*$ expresses the optimal values.

It is seen from formula (15) that $Q^*((x(k),a(k))$ is equivalent to $V^*(x(k),a(k))$ with the same action. Therefore, the optimal actions $a^*(k)$ can be obtained by the value iteration following the formula:

$$Q(x(k),a(k)) \leftarrow Q(x(k),a(k))$$
$$+\alpha[R(x(k),a(k)) + \gamma \text{max}Q(x(k+1),a(k)) - Q(x(k),a(k))] \tag{16}$$

where α is a learning rate, $0 \le \alpha < 1$, and the states $x(k+1)$ will be got at the next sampling time $k+1$. By using this iteration formula, it will finally converge to steady state by continuously adjusting the action $a(k+1)$ and one obtains the responding control $a^*(k)$.

$$a^*(k) = \underset{a}{\text{argmax}}Q(x(k),a(k)) \tag{17}$$

A proof on strict convergence of Q-valued function (16) was given by Watskin [35], whose theorem is rewritten here as Lemma 1.

Define $n^i(x, a)$ as the index of the ith time that action a is tried in state x.

Lemma 1 [35]: *Given bounded rewards $|R_n| \le C$, learning rates $0 \le \alpha_n < 1$, and $\sum_{i=1}^{\infty}\alpha_{n^i(x,a)} = \infty, \sum_{i=1}^{\infty}\left[\alpha_{n^i(x,a)}\right]^2 < \infty, \forall x,a$, then $Q_n(x,a) \to Q^*(x,a)$ as $n \to \infty$, $\forall x,a$ with probability 1.*

4. The Proposed Approach

The scheme of the proposed approach is indicted as Figure 2. It is a hierarchical control of an integer of the process level and the loop level. The process level targets the conversion efficiency of the refrigerating system by adjusting action variables whose optimization a^* is the reference of the loop level. The loop level as a basic control circuit implements the reference control by adjusting the control variables u whose elements are the compressor frequency and the electronic expansion valve opening, respectively.

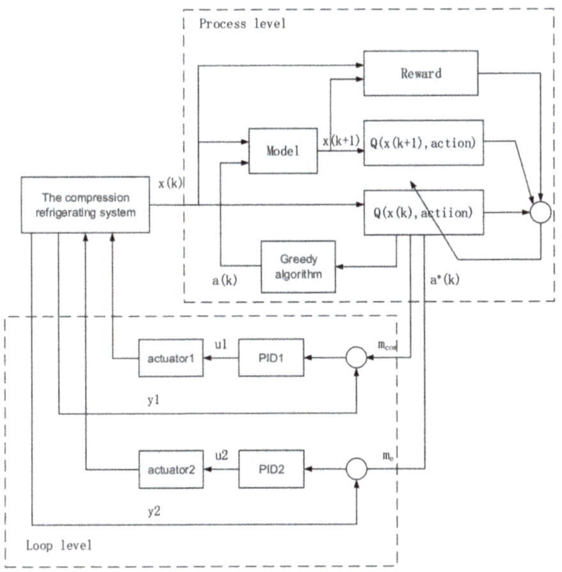

Figure 2. The scheme of the proposed approach.

4.1. Process Level

4.1.1. States Vector, Action Vector, and Cost Function

As discussed in Section 2.2, there are seven states in evaporation and five states in condensation with a complex description and high order nonlinearity. Some hypotheses are considered for the sake of simplicity. (1) The evaporator and the condenser are regarded as a two-phase zone with the extent of superheat in evaporator being estimated by the local energy conservation because the most area in heating change belongs to the two-phase zone, according to experiences. (2) Only the temperature of heating exchange is changing in the thermodynamic equation and only the length of the two-phase zone is changing in the mass conversation equation. (3) The void fraction is considered as a constant based on the fact of a small change in the working condition. (4) The heat transfer coefficient and the refrigerant physical parameter, which can be obtained by the identification technique are supposed to be constant though they are different depending on conditions. Based on the above hypotheses the states vector is reduced as

$$x = \begin{bmatrix} x_1 & x_2 & x_3 & x_4 & x_5 \end{bmatrix}^T = \begin{bmatrix} T_e & l_e & T_c & l_c & T_{wa} \end{bmatrix}^T \tag{18}$$

where l_e is the length of the two-phase zone in evaporator, m, T_{wa} is the tube wall temperature of the condenser. It is important to point out that the temperature states can be measured directly by sensors and the length states can be obtained by soft-sensor with the degree of superheat.

The action vector of the process is

$$a = [m_e, m_{com}]^T \tag{19}$$

where m_e and m_{com} are the mass flow of the expansion valve refrigerant and of the compressor refrigerant which are controlled by the opening of the expansion valve K and the frequency of the compressor f, according to formula (7) and formula (9), respectively.

The cost function is selected as the goal under an action $a(k) \in \mathcal{U}$

$$R_k(x(k), a(k)) = \left. \frac{T_c(k)}{T_c(k) - T_e(k)} \right|_{a(k)} \tag{20}$$

As a result, the optimal actions $a^*(k)$ can be obtained by Q-algorithm following the formulas (16) and (17).

4.1.2. Exploitation and Exploration

The RL gets the optimal action according to the reward from the interaction with the environment. There are two ways called the exploitation, which seeks the optimal action according to the best rewards of Q-value and the exploration, which seeks the optimal action by dispatching randomly at a certain probability in order to escape the local optimization of exploitation or find a new unknown optimization. The ε-greedy method is proposed to carry out the exploitation and the exploration with the form of formula (21)

$$a = \begin{cases} \text{argmax} Q(s, a), & p < \varepsilon \\ \text{rand}(U), & p \geq \varepsilon \end{cases} \tag{21}$$

where ε is a pre-set probability (usually as a large probability event), p is the probability of action selection, and U is the action space.

4.1.3. Procedure of the Proposed Algorithm

Based on the selection of action vector, state vector, and the cost function, the procedure of the proposed algorithm that begins with state vector $x(k)$ is summarized as procedure 1.
Procedure 1.

Step 1: Select an action $u(k)$ randomly.
Step 2: Receive immediate reward $R(x(k), a(k))$ according to formula (20).
Step 3: Get the new state $x(k+1)$ according to the coarse model and compute value function according to formula (10).
Step 4: Step 4: Renew the reward value Q based on current state $x(k)$

$$Q(x(k), a(k)) = R(x(k), a(k)) + \gamma \min_{a(k+1)} Q(x(k+1), a(k+1)) \tag{22}$$

Step 5: Adjust a new control vector $a(k+1)$ with ε-greedy method according to (21).
Step 6: Repeat step 1 to step 5 until it is convergent.
Step 7: Get the optimal $a^*(k)$ according to the best reward

$$a^*(k) = \text{argmax}_{a(k)} Q(x(k), a(k)) \tag{23}$$

Step 8: Apply $a^*(k)$ to the refrigeration system as the reference of the loop level and then get $x(k+1)$ under the control u.
Step 9: Replace $x(k)$ by $x(k+1)$ and go to step 1.

4.2. Loop Level

The technology of loop control has been very mature and a simple PID is proposed to achieve this function for each loop whose framework is in Figures 3 and 4. Here m_{com} and m_e are the reference variables which come from the process level by RL. The u1 and u2 are the control variables which exert the effect on the system by the associated actuators that are parts of the compression refrigeration system. The y1 and y2 are the system outputs which are obtained by the sensors.

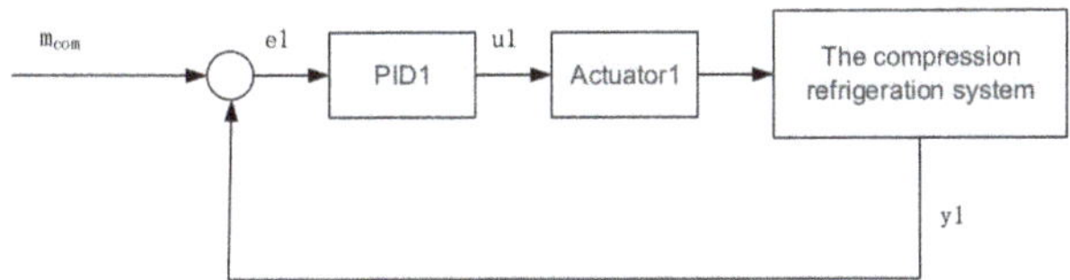

Figure 3. The framework of m_{com} loop.

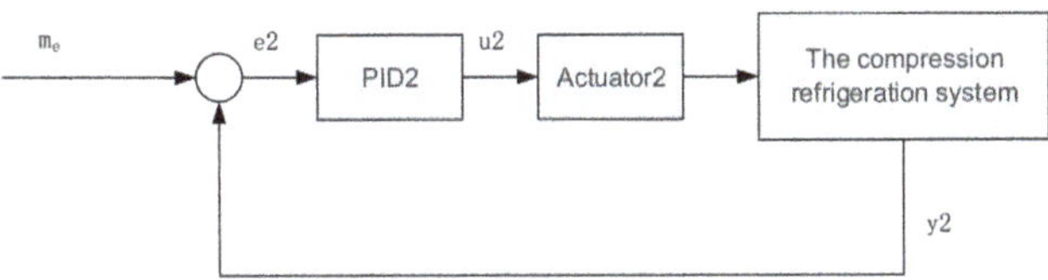

Figure 4. The framework of m_e loop.

5. Case Studies

Our test-bed of a compression refrigeration system is seen in Figure 5, and its structure is dedicated in Figure 6 [36,37]. The liquid refrigerant stored in the tank goes into the heating interexchange through the desiccator, where they gasify by absorbing the heat. The gas comes back to the condenser driven by a compressor and becomes liquid in the stored tank. The exchanged cool air goes into the compound air conditioning plant to adjust the environment. The main devices and their specifications are listed in Table 2.

Table 2. The devices and specifications.

No.	Name of Devices	Specifications
1	Compressor	Semi-closed piston compressor, rated speed 1450 rpm, 7.5 Kw/10 HP, rated discharge 38.25 m^3/h, rated refrigeration capacity 30 KW
2	Frequency Converter for Compressor	Capacity 11 KVA, rated current 17 A, rated voltage 380 V, frequency 0–50 Hz, V/F signal 0–10 V
3	Condenser	Borehole heat exchangers, heat exchange 35 Kw, heat transfer area 3.4 m^2, heat transfer coefficient K = 970 W/m^2·°C
4	Evaporator	Borehole heat exchangers, heat exchange 30 kW, heat transfer area 3.02 m^2, heat transfer coefficient K = 1455 W/m^2·°C
5	Electronic Expansion Valve	Flow coefficient Kvs = 0.63 m^3/h, refrigerating capacity 74 kW, caliber DN15, AC24 V, control signal 0–10 V
6	Cooling Water Pump	Multistage centrifugal pump: Rated power 2.2 kW, rated speed 2840 rpm, lift 32 m, flow 8.4 m^3/h
7	Frequency Converter for Cooling Water Pump	Rated voltage 380 V, frequency 0–50 Hz, V/F signal 0–10 V
8	Chilled Water Pump	Single-stage centrifugal pump, rated power 1.5 kW, rated speed 2840 rpm, lift 27.4 m, flow 7.8 m^3/h
9	Frequency Converter for Refrigerated Water Pump	Rated voltage 380 V, frequency 0-50 Hz, V/F signal 0-10 V
10	The Cooling Tank	Volume 1 m^3
11	The Chilled Tank	Volume 1 m^3
12	Reservoir	Volume 20 L

The temperature was obtained by the temperature transmitter of Pt100 with the range of 0–30 °C, 0–60 °C, and 0–100 °C under the degree of accuracy ±0.1 °C (class A). The temperature of the evaporator and condenser was limited within the range of −5–15 °C and 30–50 °C, respectively. The pressure was obtained by the pressure transmitter with the range of 0–2.5 MPa, 0–1.6 MPa under the degree of accuracy 0.25%. The electromagnetic flowmeters with the degree of accuracy 0.5% were added to measure the m$_{com}$ and m$_e$, which were the action vectors of the process level. The traditional coefficient of performance (COP) was obtained by the stable state of the compression refrigeration system. However, the loads are often uncertain which makes the compression refrigeration system stay in the transient process. As a result, a 30 kW tunable stainless steel electric heater was applied to simulate a thermal load with the range from 22.5 to 30 kW and the sampling time was selected as 1 min. In the experiment, we adjusted the electric heater by changing the resistance value through rotating an adjustment button. In this way, we simulated the load climb by running the adjustment button in one direction and the load uncertainty by random changing the direction.

Figure 5. The test-bed of a compression refrigeration system.

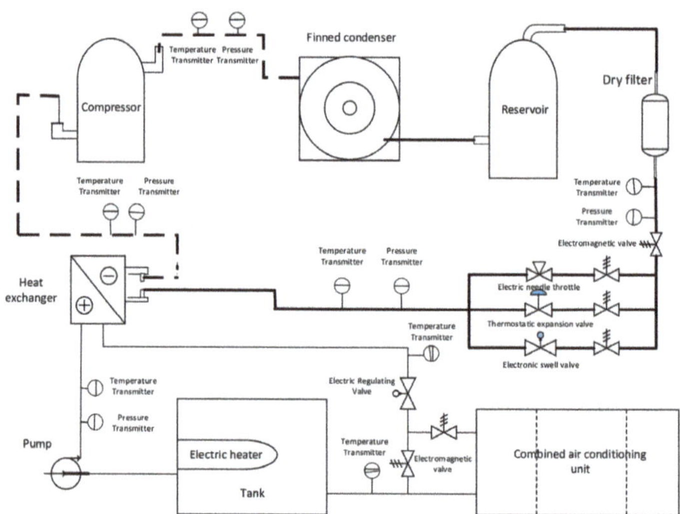

Figure 6. The structure of the compression refrigeration air-conditioning system.

5.1. Effect of Model Accuracy on Performance

A coarse model was built based on the principle of compression refrigeration in the process level with the following form:

$$
\begin{cases}
\dot{T}_e = c_0 m_{com} + c_1 l_e (T_w - T_e) \\
\dot{l}_e = c_2 (m_e - m_{com}) \\
\dot{T}_c = c_3 m_{com} + c_4 l_c (T_{wa} - T_c) \\
\dot{l}_c = c_5 (m_{com} - m_e) \\
\dot{T}_{wa} = c_6 (T_c - T_{wa}) + c_7 (T_{wa} - T_a)
\end{cases}
\tag{24}
$$

where the variables of T_e, T_w, T_c, l_e, l_c, T_{wa}, T_a, m_e and m_{com} are the same meaning in Table 1. The coefficients c_0-c_7 are obtained based on collected data by the technology of the system identification [33]. The values are seen in Table 3.

Table 3. The value of parameters in the coarse model [33].

C_0	C_1	C_2	C_3	C_4	C_5	C_6	C_7
−1672	1	5	1	456	0.3	−1	1

Taking the values from Table 3 as the criterion (called the first coarse model) we built the second coarse model by changing the value of coefficient c0 from −1672 to −1500, the values with the rate 4.58%, and the value of coefficient c4 from 456 to 450. The control algorithm follows the procedure 1. The evolution of states, the instantaneous efficiency and the action variables are seen in Figures 7–9. The refrigerant circulated between gas and liquid with endothermic and exothermic reactions. The evaporation temperature was the external temperature of the refrigerant during the evaporation of the evaporator which usually is constantly pre-set according to the environmental requirement. It is seen from Figure 7 that the evaporation temperature Te (either of the first coarse model (blue curve) or of the second coarse model) fluctuated around 7 °C. The condensing temperature is prone to be affected by the working pressure. More energy changing will produce higher pressure which will increase the condensing temperature. It is seen from Figure 7 that the condensing temperature Tc (either of the first coarse model (blue curve) or of the second coarse model) was rising slowly to meet the requirement of the loads from 22.5 to 30 kW. Different from the conventional coefficient of performance (COP) which is defined in the steady state, our COP is extended to the whole process which includes both the transient and steady. The transient is often prevailing due to the resistance change. Therefore, the mean of the COP during a certain period of time is meaningful according to our definition. The curve of instantaneous efficiency in Figure 8 shows the differences between coarse models on the system. There was an instantaneous error between the first coarse model (blue curve) and the second coarse model (green curve). However, the average efficiency of both coarse models during 200 min was 1.2681 and 1.2685, respectively, with the error of 0.03%. We calculated the residual to evaluate the robustness of both coarse models. First, we made a polynomial fitting by the least square method with the same style of cubic curves (no significant error change over this order) according to the standard procedure [36] and obtained the estimation $\hat{COP}$.

$$\hat{COP} = p_1 \cdot COP^3 + p_2 \cdot COP^2 + p_3 \cdot COP + p_4 \tag{25}$$

where the coefficients are shown in Table 4.

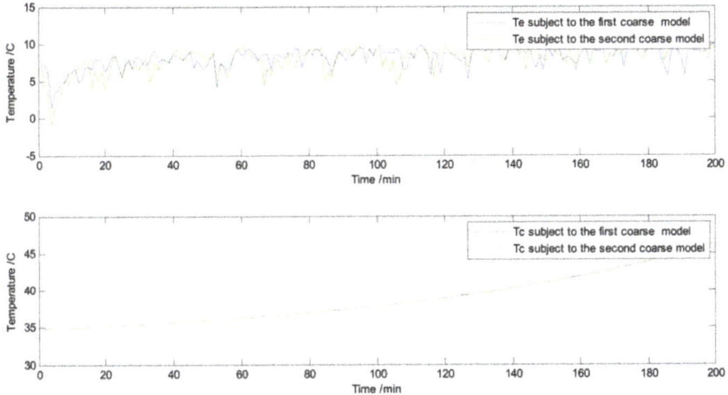

Figure 7. The evolution of states Te and Tc.

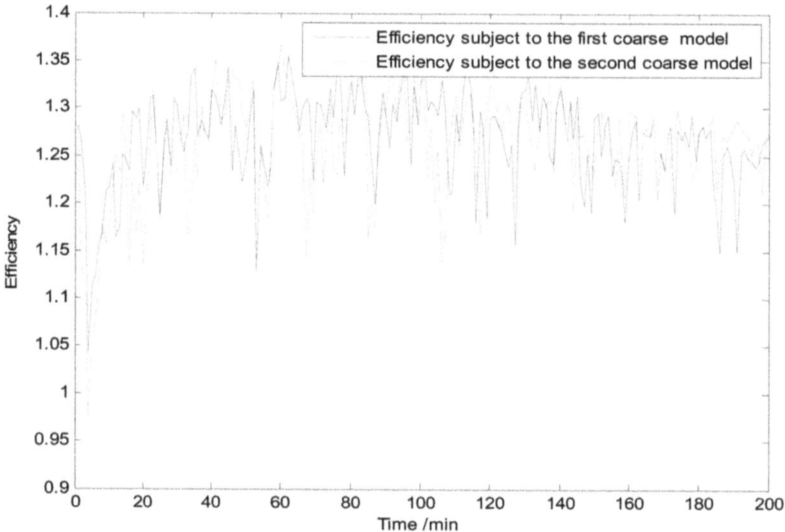

Figure 8. The instantaneous efficiency.

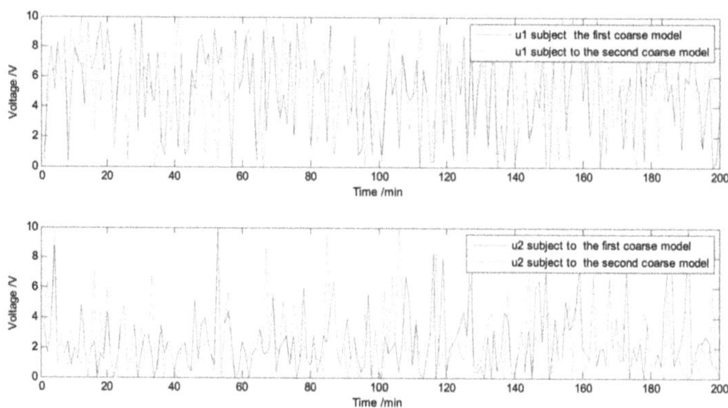

Figure 9. The action variables.

Table 4. Coefficients of the polynomial fitting.

	p_1	p_2	p_3	p_4
The first coarse model	7.9673×10^{-8}	3.1662×10^{-5}	0.0034666	1.1841
The second coarse model	1.0584×10^{-7}	3.9588×10^{-5}	0.0043562	0.69583

Therefore, the norm of residual (σ) is obtained according to the formula (26):

$$\sigma = \sqrt{\frac{\sum_{k=1}^{l}\left(COP(k) - C\hat{O}P(k)\right)^2}{l-1}} \tag{26}$$

where l is the number of examples. The responding norms of residual are $\sigma_1 = 0.60575$ and $\sigma_2 = 0.67583$, respectively, which means the coarse model has a good robustness due to the approximation norm of residual. As a result, we can evaluate the effect of model accuracy on the performance of efficiency by comparing the result of the COP with changing the value of parameters. The mean change of 0.03%

compared with the parameter's change of 4.58% indicates only a small amount of the impact among the coarse models.

Figure 9 gives the action variables of m_e and m_{com}. It can be seen that there were noticeable fluctuations in m_e and m_{com}. It is due to the RL implementation and load variations. We adopted the classical table mode to map the relation between the states and the actions. However, limited by our computer's capability (Intel(R) Core(TM)2 Duo CPU E7300 @2.66GHz @2.67GHz RAM 4.00GB) the actions were divided into 1K pieces which causes some errors between the acquired actions and the ideal actions. The errors were somewhat enlarged in the process of evaluating the action value because we took each episode to last 200 steps instead of $T = \infty$. Another important factor is the influence of the load variations. The final actions of the RL algorithm were determined by the initial states and the current loads. The load variation was simulated by sequentially rotating the adjustment button which lead to the continuous transition.

5.2. Comparison with the Conventional Approach

There is no method to deal with the coefficient of performance (COP) under the temperature changing by load variation. The predominant approach in HVAC is to keep the outlet temperature matching the loads by automatically adjusting the flow with an electronic expansion valve and maintaining the constant speed of the compressor. We compared the proposed method with this conventional approach. Figures 10–12 show the evolution of states Te and Tc, the instantaneous efficiency, and the control variables. The blue curves and the red curves represent the results of the proposed approach and the conventional approach, respectively. It is seen from Figure 11 that the Tc and Te with the conventional approach and the proposed approach are a similar tendency to meet the requirement of loads during the test period. Te with the proposed approach shows large fluctuations due to the coarse model leading to a rough estimation of states. While the Te with the conventional approach shows a good smooth curve because it is under classical control. The Tc has small fluctuations because the accuracy of Tc is better.

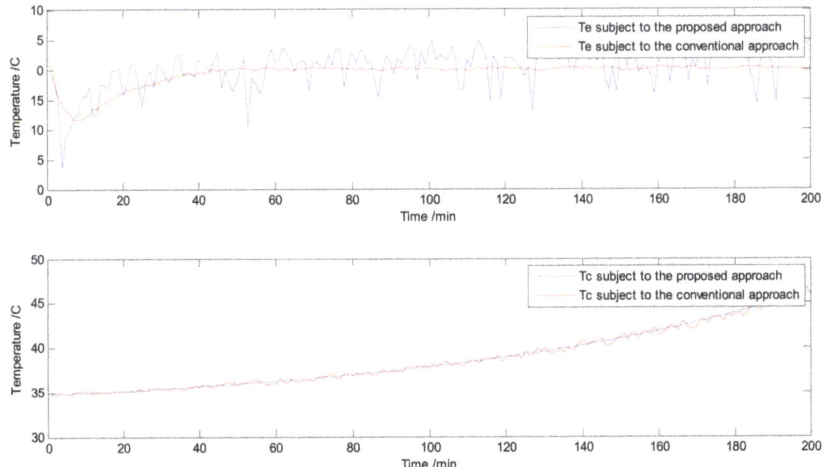

Figure 10. The evolution of states T_e and T_c.

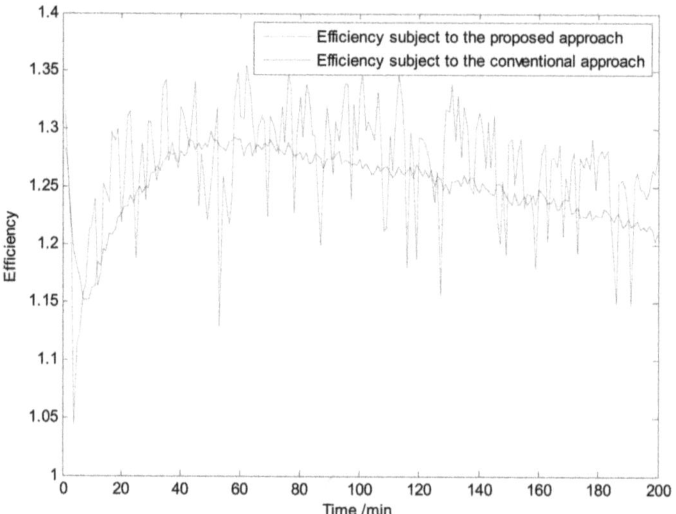

Figure 11. The instantaneous efficiency.

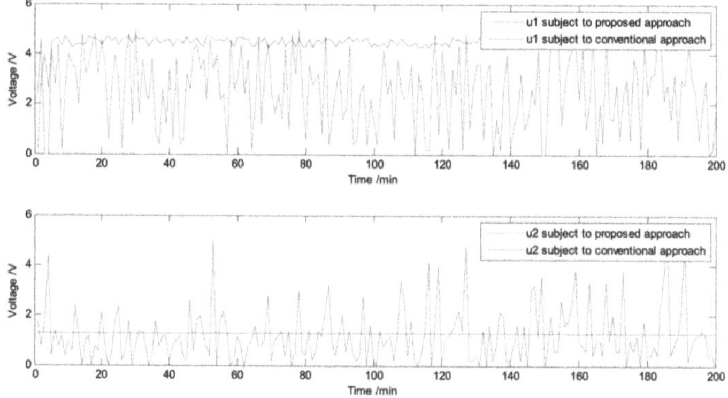

Figure 12. The control variables.

There is a small fluctuation of instantaneous efficiency for the conventional approach because it tries to keep the temperature bias of inlet and outlet by taking control. Our proposed method releases the control of temperature within the range of limitation, so the instantaneous efficiency formulated from the T_e and Tc shows bigger fluctuations, even sometimes low efficiency and sometimes high efficiency. We tried to reduce this fluctuation by adjusting the parameters. First, we changed the pre-set probability ε from 0.7 to 0.9, which is the general value for errors and trials method. However, there was no improvement of fluctuation and efficiency. Considering the requirement of real time, we did not decrease under 0.7, which means that 30% more searches were used to explore, but in fact, the new states can be computed according to the coarse model, which means it does not need a high probability to explore. Instead, we tried to increase the value of ε and get a good COP (1.34% better than the conventional method) even by setting ε as 0.99, though there was no improvement in the fluctuation. We also tried to adjust the learning rate α that is a compromise between the speed and the accuracy of training. If it is too small, the training will take more time. If it is too big. the training will reduce accuracy. There is no theoretical basis so far in reinforcement learning but with an empirical value of 0.95–0.98. It was noticed that the simulation shows the fluctuations were still present. A representative is shown in Figure 11. It is vague for instantaneous efficiency in Figure 11. However, as with an energy conversion process, we are more concerned with refrigeration efficiency over a period of time. The average efficiency was proposed as the metrics. The average efficiency with the conventional approach (red curve) and with the proposed approach (blue curve) was 1.2513 and 1.2681, respectively. The proposed approach will raise the efficiency by 1.34% more than the conventional approach on average during the period of 200 min.

The control variables are shown in Figure 12. As we have pointed out in 4.2, the controller parameters of PID1 and PID2 should be predesigned. The noticeable fluctuations in our proposed method cause difficulties in regulating the parameters of PID. It is fortunate that the refrigeration is a slow second-order system and the constant parameters of PID have a wide feasible scope around the working point. The values of parameters are prepared group by group in the table based on the previous experiments by considering the system stability. Therefore, the controller will directly call the group of parameter values from this preparatory table to track the references of actions provided by the process level which are fluctuating due to the partition error of actions and the load variation.

6. Conclusions

In this paper, we have discussed an online data-driven approach to improve the conversion efficiency of refrigeration system under the condition of load variation. A reinforcement learning approach that has an ability of seeking the unknown environment is proposed to find the optimal actions based on the online data in the process level and its risk on the system because the training process is complex and unknown in advance due to the warning from the computer simulation. A coarse model is developed to evaluate the action value which reduces the dependence of traditional control methods on model accuracy. Finally, the actions are achieved as the pre-set variables by implementing the single loop control. The simulation shows the proposed algorithm is better than the conventional methods in the conversion efficiency of the refrigeration system from the viewpoint of the average, although larger fluctuations are noticeable. How to reduce this fluctuation is our further study.

Author Contributions: Conceptualization and methodology, D.Z. and Z.G. Writing—original draft preparation, D.Z. Writing—review and editing, Z.G.

Funding: This research was funded by the National Science Foundation of China under grant 61673074.

Acknowledgments: The authors would like to acknowledge the research support from the School of Electrical Engineering and Automation at Tianjin University, the National Science Foundation of China under grant 61673074; and the Faculty of Engineering and Environment at the University of Northumbria, Newcastle.

Conflicts of Interest: The authors declare no conflict of interest.

References

1. Consumption & Efficiency. Available online: https://www.eia.gov/consumption (accessed on 1 October 2019).
2. Abdul, A.; Farrokh, J.S. Theory and applications of HVAC control systems-a review of model predictive control (MPC). *Build. Environ.* **2014**, *72*, 343–355.
3. Youssef, A.; Caballero, N.; Aerts, J.M. Model-Based Monitoring of Occupant's Thermal State for Adaptive HVAC Predictive Controlling. *Processes* **2019**, *7*, 720. [CrossRef]
4. Wang, S.; Ma, Z. Supervisory and optimal control of building HVAC systems: A review. *HVAC R Res.* **2008**, *14*, 3–32. [CrossRef]
5. Naidu, D.S.; Rieger, C.G. Advanced control strategies for heating ventilation air conditioning and refrigeration systems an overview part I hard control. *HVAC R Res.* **2011**, *17*, 2–21. [CrossRef]
6. Shang, Y. Critical Stability Analysis, Optimization and Control of a Compression Refrigeration System. Ph.D. Thesis, Tianjin University, Tianjin, China, 2016.
7. Sutton, R.; Barto, A. *Reinforcement Learning: An Introduction*; The MIT Press Cambridge: Cambridge, UK, 2005.
8. Silver, D.; Huang, A.; Maddison, C.J.; Guez, A.; Sifre, L.; Van Den Driessche, G.; Schrittwieser, J.; Antonoglou, I.; Panneershelvam, V.; Lanctot, M.; et al. Mastering the game of Go with deep neural networks and tree search. *Nature* **2016**, *529*, 484. [CrossRef]

9. Vamvoudakis, K.G.; Lewis, F.L. Online actor-critic algorithm to solve the continuous-time infinite horizon optimal control problem. *Automatica* **2010**, *46*, 878–888. [CrossRef]

10. Jens, K.; Andrew, B.J.; Jan, P. Reinforcement learning in robotics: A survey. *Int. J. Robot. Res.* **2013**, *32*, 1238–1274.

11. Zhang, D.; Gao, Z. Reinforcement learning-based fault-tolerant control with application to flux cored wire system. *Meas. Control.* **2018**, *51*, 349–359. [CrossRef]

12. Zhang, D.; Lin, Z.; Gao, Z. Reinforcement-learning based fault-tolerant control. In Proceedings of the 15th International Conference on Industrial Informatics (INDIN), Emden, Germany, 24–26 November 2017.

13. Zhang, D.; Lin, Z.; Gao, Z. A novel fault detection with minimizing the noise-signal ratio using reinforcement learning. *Sensors* **2018**, *18*, 3087. [CrossRef]

14. Baeksuk, C.; Jooyoung, P.; Daehie, H. Tunnel ventilation controller design using an RLS-based natural actor-critic algorithm. *Int. J. Precis. Eng. Manuf.* **2010**, *11*, 829–838.

15. Fazenda, P.; Veeramachaneni, K.; Lima, P.; O'Reilly, U.M. Using reinforcement learning to optimize occupant comfort and energy usage in HVAC systems. *J. Ambient Intell. Smart Environ.* **2014**, *6*, 675–690. [CrossRef]

16. Li, B.; Xia, L. A multi-grid reinforcement learning method for energy conservation and comfort of HVAC in buildings. In Proceedings of the IEEE International Conference on Automation Science and Engineering, Gothenburg, Sweden, 24–28 August 2015.

17. Enda, B.; Stephen, L. Autonomous HVAC control, a reinforcement learning approach. In Proceedings of the European Conference on Machine Learning and Principles and Practice of Knowledge Discovery in Databases (ECML PKDD), Porto, Portugal, 7–11 September 2015; Volume 9286.

18. Gracia Cuesta, A.D.; Fernàndez Camon, C.; Castell, A.; Mateu Piñol, C.; Cabeza, L.F. Control of a PCM ventilated facade using reinforcement learning techniques. *Energy Build. (SI)* **2015**, *106*, 234–242. [CrossRef]

19. Yang, L.; Nagy, Z.; Goffin, P.; Schlueter, A. Reinforcement learning for optimal control of low exergy buildings. *Appl. Energy* **2015**, *156*, 577–586. [CrossRef]

20. Cheng, Z.; Zhao, Q.; Wang, F.; Jiang, Y.; Xia, L.; Ding, J. Satisfaction based Q-learning for integrated lighting and blind control. *Energy Build.* **2016**, *127*, 43–55. [CrossRef]

21. Wang, Y.; Velswamy, K.; Huang, B. A long-short term memory recurrent neural network based reinforcement learning controller for office heating ventilation and air conditioning systems. *Processes* **2017**, *5*, 46. [CrossRef]

22. Chen, Y.; Norford, L.K.; Samuelson, H.W.; Malkawi, A. Optimal control of HVAC and window systems for natural ventilation through reinforcement learning. *Energy Build.* **2018**, *169*, 195–205. [CrossRef]

23. Wei, T.; Wang, Y.; Zhu, Q. Deep reinforcement learning for building HVAC control. In Proceedings of the 54th ACM/EDAC/IEEE Design Automation Conference (DAC), Austin, TX, USA, 18–22 June 2017.

24. Zhang, Z.; Lam, K.P. Practical implementation and evaluation of deep reinforcement learning control for a radiant heating system. In Proceedings of the 5th Conference on Systems for Built Environments, Shenzhen, China, 7–8 November 2018.

25. Farahmand, A.M.; Nabi, S.; Grover, P.; Nikovski, D.N. Learning to control partial differential equations: Regularized fitted Q-iteration approach. In Proceedings of the 55th IEEE Conference on Decision and Control (CDC), Las Vegas, NV, USA, 12–14 December 2016.

26. Wang, Y.; Velswamy, K.; Huang, B. A novel approach to feedback control with deep reinforcement learning. In Proceedings of the 10th IFAC Symposium on Advanced Control of Chemical Processes (ADCHEM), Shenyang, China, 25–27 July 2018.

27. Ruelens, F.; Iacovella, S.; Claessens, B.; Belmans, R. Learning agent for a heat-pump thermostat with a set-back strategy using model-free reinforcement learning. *Energies* **2015**, *8*, 8300–8318. [CrossRef]

28. Valladares, W.; Galindo, M.; Gutierrez, J.; Wu, W.C.; Liao, K.K.; Liao, J.C.; Lu, K.C.; Wang, C.C. Energy optimization associated with thermal comfort and indoor air control via a deep reinforcement learning algorithm. *Build. Environ.* **2019**, *155*, 105–117. [CrossRef]

29. Kaelbling, L.K.; Littman, M.L.; Moore, A.W. Reinforcement learning: A survey. *J. Artif. Intell. Res.* **1996**, *4*, 237–285. [CrossRef]

30. Ma, Z.; Yao, Y. *Air Conditioning Design of Civil Buildings*, 3rd ed.; Chemical Industry Press: Beijing, China, 2015.

31. Tahat, M.A.; Ibrahim, G.A.; Probert, S.D. Performance instability of a refrigerator with its evaporator controlled by a thermostatic expansion-valve. *Appl. Energy* **2001**, *70*, 233–249. [CrossRef]

32. Aprea, C.; Mastrullo, R.; Renno, C. Performance of thermostatic and electronic expansion valves controlling the compressor. *Int. J. Energy Res.* **2006**, *30*, 1313–1322. [CrossRef]

33. Xue, H. Active Disturbance Rejection Control of Compression Refrigeration System. Master's Thesis, Tianjin University, Tianjin, China, 2016.
34. Watldns, C.J.C.H. Learning from Delayed Rewards. Ph.D. Thesis, University of Cambridge, Cambridge, UK, 1989.
35. Watkins, C.J.C.H.; Dayan, P. Q-learning. *Mach. Learn.* **1992**, *8*, 279–292. [CrossRef]
36. Zhang, H. Research on Internal Model Control Strategy of Compression Refrigerating System. Master's Thesis, Tianjin University, Tianjin, China, 2013.
37. Xiao, D. *Theory of System Identification with Application*; Tsinghua University Press: Beijing, China, 2014.

Article

Supporting Design Optimization of Tunnel Boring Machines-Excavated Coal Mine Roadways: A Case Study in Zhangji, China

Bin Tang [1,2,3,*], Hua Cheng [2,*], Yongzhi Tang [4], Tenglong Zheng [2], Zhishu Yao [2], Chuanbing Wang [3,4] and Chuanxin Rong [2]

[1] State Key Laboratory of Mining Response and Disaster Prevention and Control in Deep Coal Mines, Anhui University of Science and Technology, 168 Taifeng St, Huainan 232001, China
[2] School of Civil Engineering and Architecture, Anhui University of Science and Technology, 168 Taifeng St, Huainan 232001, China; shlzheng@aust.edu.cn (T.Z.); zsyao@aust.edu.cn (Z.Y.); chxrong@aust.edu.cn (C.R.)
[3] Ping'an Coal Mining Institute of Engineering Technology Co., Ltd., No. 6 Building, Zhihui Valley, Huainan 232001, China
[4] Huainan Mining Industry (Group) Co., Ltd., No. 1 Dongshanzhong Rd, Huainan 232001, China; tang_yz1962@163.com (Y.T.); cbwang_hmi@163.com (C.W.)
* Correspondence: btang@aust.edu.cn (B.T.); hcheng@aust.edu.cn (H.C.); Tel.: +86-554-6668713 (B.T.)

Received: 17 October 2019; Accepted: 20 December 2019; Published: 1 January 2020

Abstract: Tunnel Boring Machines (TBMs) are a cutting-edge excavating equipment, but are barely applied in underground coal mines. For TBM excavation projects involving the Zhangji coal mine, the surrounding rock properties, stress field, cross section geometry, as well as the excavation-induced stress path of TBM-excavated coal mine roadways are different from those of traditional tunnels or roadways. Consequently, traditional roadway supporting technologies and experiences cannot be relied on for this project. In order to research an appropriate supporting pattern for a TBM-excavated coal mine roadway, first of all, the constitutive model of roadway surrounding rocks was derived, and a rock failure criterion was proposed based on rock mechanical tests. Secondly, a three-dimension finite element model was established and computer simulations under three different supporting patterns were conducted. Stress redistribution, roadway convergence, and excavation damage zone ranges of surrounding rocks under three different support patterns were analyzed and an optimal support design of the TBM-excavated roadway was made based on simulation results. During roadway excavation, convergence gauge and rock bolt dynamometers were installed for monitoring roadway convergence and the axial forces of rock bolts. The in-situ monitoring results verified the validity of roadway supporting designs.

Keywords: tunnel boring machine; roadway supporting; constitutive model; failure criterion; numerical simulation; in-situ monitoring

---

1. Introduction

In China, coal is the major energy source and it accounts for about 68% of the total primary energy consumption. As the coal resources within shallow grounds are depleting, mining operation have been moving to increasing deep grounds in recent years [1]. In China, the average depth of coal mining operations has reached up to 556 m and it is increasing at a rate of 8 to 12 m per year. For metal mines, the depths of mining operations are even higher. For example, the Mponeng gold mine in South Africa extends 4250 m below the surface [2]. As the depth of mining works increased significantly, potential risks such as the collapse of roadways surrounding rocks, high coalbed methane emissions and outbursts, as well as groundwater inrushing increased dramatically [3–6]. Consequently, an increasing amount of coal mines have started excavating permanent or semi-permanent roadways within rock layers rather than coal seams to obtain increased safety during mining and roadway excavation operations [6]. Compared with coal, rocks have higher strength and it is beneficial to increase the stability of roadways and reduce risks. On the other hand, the higher strength of rocks also increases the difficulties of roadway excavation and slows down the excavation speed. At present, excavation speeds in rock layers are significantly slower than these in coal seams (40 to 100 m/mth in rock layers vs. 120 to 300 m/mth in coal seams) [7]. Typically in mining operations, roadways in rock layers include main roadways and other auxiliary roadways for transportation, ventilation, coalbed methane drainage, and groundwater drainage roadways purposes. They should be finished before the coal seam roadways start being used for excavation. Obviously, a whole mining operation would be delayed by a low penetration rate in rock layers.

Increasing the safety and speed of coal mine roadway excavation is the key factor for safe and effective mining operations. For over one hundred years, roadways or tunnels were excavated by using drilling and blasting technology. While intensive application of explosives increases potential risks in roadway excavation, and toxic smoke also results in health hazards for miners. Since the 1960s, mechanical excavation equipment (roadheader, continue miner, etc.) were introduced in roadway excavation. The use of these machines dramatically increased safety and excavation speeds. Nevertheless, these machines are not able to break hard rocks [8]. TBMs (Tunnel Boring Machines) have a higher rock breakage capacity. Moreover, rock breakage, rock chips loading, and roadway support can be conducted simultaneously by using TBMs [9]. TBMs have been widely applied in tunneling, though application cases of TBMs in coal mines have barely been reported [10].

After excavation, roadway rock deformation or damage is encountered. Therefore, these roadways need reinforcement and support to ensure the safety of excavation works. Much research has proposed the optimization of coal mine roadway support design based on the results of numerical simulation or in-situ monitoring [11]. Based on the simulation results of FLAC3D, Wang proposed full cross-section anchor-grouting reinforcement technology and applied it in underground coal mine roadways with loose and fractured surrounding rocks based on simulation results of FLAC3D, while Cao resolved the problems of lateral wall collapse and severe floor heaving in roadway support involving fractured rock layers [12,13]. Huang proposed a concrete-filled steel tubular support structure to eliminate the problems of large-scale deformation of deep roadways [14]. Stone presented a designing methodology of roadway support based on statistical evaluation of in-situ monitoring results [15]. Yang evaluated the rock mass properties of roadways surrounding rocks by using the geological strength index (GSI) and established the numerical model for roadways using UDEC (Universal Distinct Element Code). Deformation and stress behavior of roadways under different support conditions was obtained and a "bolt-cable-mesh-shotcrete + shell" combined support mode was subsequently proposed [16]. When compared with traditional roadways or tunnels, TBM-excavated roadways have some different characteristics on surrounding rock properties (TBMs are typically used in hard rock strata), for cross section geometry (rectangular or straight wall arch cross section of traditional method vs. circular cross section of TBM method), as well as for excavation-induced stress paths (dynamic blasting loading vs. static loading of machine cutting). These differences mean te support pattern and parameters of TBM-excavated coal mine roadways still need to be settled.

This paper proposed a support design methodology for TBM-excavated roadways in deep coal mines. First of all, the very specific in-situ stress parameters were obtained by in-situ measurement. Secondly, an innovative rock constitutive model which takes anisotropic damage and failure into consideration was established, and a non-linear failure criterion obtained from previous laboratory rock tests rather than conventional universal linear failure criteria was also used to further enhance calculation precision. Thirdly, a proposed constitutive model and failure criterion were introduced into computer simulation software by using new DLL (dynamic link library) files of simulation software and roadway support design was made based on simulation results. Finally, the in-situ monitoring of TBM-excavated roadway was implemented in Zhanji coal mine, Huainan, China and the proposed support design was verified by monitoring data [17–19]. The study results indicated that the roadway surrounding rocks remained stable after excavation and roadway supporting, and the supporting design is able to fulfill requirements of ensuring roadway stability and enhancing the efficiency of supporting roadways.

2. Engineering Background

Zhangji coal mine is a large scale underground colliery located at Huainan, Anhui province, China (shown in Figure 1a) with an annual output capacity of 13 million tons of raw coal. The working horizontal extends from −450 m to −1000 m. Zhangji coal mine covers seven mining areas which are the East-I, East-II, East-III, West-I, West-II, West-III, and North-I mining areas. The primary minable coal seams within Zhangji coal mine include No. 13-1, No. 11-2, No. 8, No. 6, as well as No. 1 and the total thickness of coal seams is 21.08 m. The measured resources and minable reserves of Zhangji coal mine are 17.78 billion and 8.77 billion tons of raw coal, respectively [20].

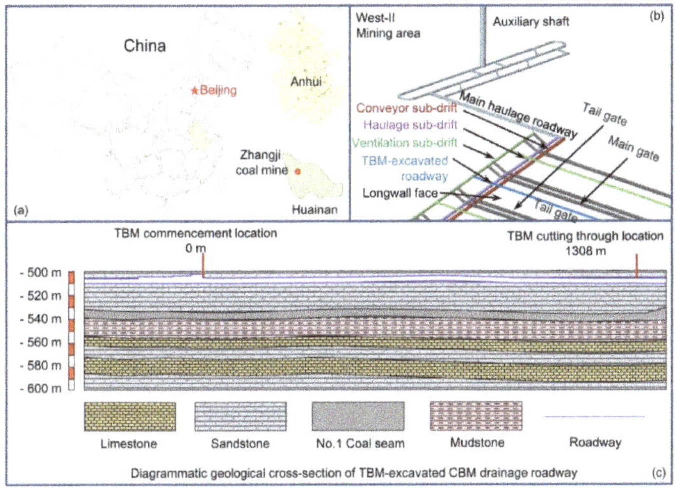

Figure 1. Simplified map of Zhangji coal mine and layout of a Tunnel Boring Machine (TBM)-excavated coalbed methane (CBM) drainage roadway. (**a**): location of Zhangji coal mine (**b**): position of TBM-excavated roadway (**c**): geological sketch of TBM-excavated roadway.

The West-II mining area is situated in west wing of Zhangji coal mine with the No.1 coal seam as its primary minable coal seam, and the mining operations within the West-II mining area are conducted from −480 m to −625 m. The No.1 coal seam has an average thickness of 6.5 m, a dip angle of 3° to 5°, and a calorific value of 6000 kcal/kg. The coal is classified as high volatile bituminous coal and coking coal and the thickness of overburden is about 500 m. The coal seam gases (mainly include CH_4, CO, H_2S, CO_2, etc.) content of the No.1 coal seam is 8.05 to 11.22 m^3/t and the gas pressure of coalbed methane (CBM) ranges from 1.4 to 4.35 MPa. Generally, gas content greater over 8 m^3/t or a gas pressure greater than 0.74 MPa is considered sufficient to initiate an outburst if other conditions are favorable. Therefore, the No.1 coal seam is considered to be an outburst-prone coal seam. In order to eliminate coal and gas outburst risk, CBM drainage roadways were excavated prior to coal mining operations.

2.1. Description of Working Site

The TBM-excavated roadway is the overlying CBM drainage roadway of 1413A longwall panel in Zhangji coal mine, China (Figure 1b). The roadway has a length of 1598 m, a diameter of 4.5 m, and a buried depth of 505 m (shown in Figure 1c). In total, 1308 m section of the roadway was excavated by a gripper TBM, which was located in the West-II panel north mining area. The objective of the roadway is conducting CBM drainage works, thereby eliminating the potential risk of explosion and gases outburst when the longwall is retreating [21].

The overlying CBM drainage roadway was excavated 25 m above the No.1 coal seam, and the horizontal distance from the overlying drainage roadway to main gate is 30 m. Before the excavation of the main gate and tail gate, the overlying drainage roadway was constructed, and downcast boreholes were drilled from overlying drainage roadway to the coal seam in order to reduce pre-drainage and eliminate outburst risks in roadway excavation works. After coal extraction works started, the longwall face advanced forward, the coal roof collapsed, and fractures generated between the coal seam and the No.1 overlying drainage roadway, meaning the CBM could be extracted from the No.1 overlying drainage roadway. The overlying CBM drainage roadway lies on the roof of the coal seam with a length of 1500 m, and the geometry of the overlying drainage roadway is a circle with a diameter of 4.5 m. Taking the reduction of construction cost and the improvement of efficiency of roadway support works into account, the TBM-excavated roadway was supported by rock bolts instead of the segments that were typically used in TBM-excavated tunnels [22].

2.2. Rock Property

Geological settings have been studied through a series of measurements and surveying. Coal measure strata lie in the Shanxi formation of Permian and Taiyuan formation of Carboniferous, which consists of coal, medium sandstone, fine sandstone, siltstone, and argillaceous sandstone (shown in Figure 1).

Along the roadway alignment, the strata consist of medium sandstone, fine sandstone and siltstone, and the UCS (uniaxial compressive strength) which varies between 41.6 and 105.7 MPa. The roadway was excavated through solid sandstone strata.

The roadway is located in a 27-m-thick sandstone strata and No.1 coal seam is 25 to 30 m beneath the roadway. The overburden of roadways consists of clay layers, sand layers, mudstone, and sandstone strata. The total overburden thickness is 480 to 500 m (shown in Figure 1).

2.3. In-Situ Stress Field

In-situ stress is a principal element which causes rock deformation and failure, and its magnitude and orientation significantly affect the stability of roadways [23,24]. In order to investigate the in-situ stress of Zhangji coal mine, borehole stress relief measurements were conducted in the mine's roadways by Anhui University of Science and Technology. The measurement results suggest that the in-situ stress field is controlled by tectonic stress. The stress filed parameters had been obtained through analysis: Vertical stress is approximately 14.3 MPa and is related to a overburden of approximately 500 m. The maximum horizontal stress is 21.6 MPa and oriented 130.72°, and the minimum horizontal stress is 13.2 MPa and oriented 223.55°. The stress field surveying results are shown in Table 1.

Table 1. Parameters of in-situ stress field.

Stress Components	σ_1 [1]	σ_3 [2]	σ_v [3]
Magnitude (MPa)	21.6	13.2	14.3
Orientation (°)	130.72	223.55	

[1] maximum horizontal stress; [2] minimum horizontal stress; [3] vertical stress.

3. A Constitutive Model and Failure Criterion of Surrounding Rocks under TBM Excavation

A constitutive model and failure criterion play a indispensable role in studying stress, strain, and failure distribution behaviors of roadways surrounding rocks. The constitutive model defines the stress-strain relationship of rock materials and a failure criterion is developed to predict the failure of rock materials. The constitutive model and failure criterion should be able to reproduce material behavior, stress condition, as well as loading and unloading paths [25,26]. Therefore, universal constitutive models and failure criteria cannot fit all mechanical behaviors of rocks or rock-like materials. In previous research, universal constitutive models and failure criteria such as Mohr-Coulomb model and failure criterion were widely used. The mechanical performance of rock materials is influenced by both rock lithological characters and stress paths. Even for the same rock material, the difference in loading or unloading patterns results in varying mechanical properties for rocks. Universal constitutive models ignore these influences, meaning applications of these universal constitutive models and failure criteria results in large modeling errors, especially under complex loading conditions or stress paths because most universal constitutive models and failure criteria are established based on simple uniaxial loading tests [27]. In this paper, damage factors along the maximum and minimum principle stresses were added into consideration and a constitutive model considering anisotropic damage was proposed. The failure criterion of the surrounding rocks was obtained from rock mechanical tests. In subsequent research works, the proposed constitutive model and failure criterion will be introduced into computer simulation software.

3.1. Constitutive Model

Previous constitutive models barely take the microscopic damage of rock materials into consideration. Few constitutive models use a single damage factor or damage coefficient to define the influences of microscopic damage on the mechanical behaviors of rocks. In loading and unloading processes, the microscopic damage development behaviors are different along the maximum and minimum principle stress directions. In this research, an anisotropic tensor and damage tensor were introduced into the constitutive model for representing anisotropy and damage factors along the maximum and minimum principle stress direction. Based on lab test results and working experiences from working sites, hard rocks barely show plastic behaviors and macrocosmic damage happens suddenly. The stress-strain relationship is mainly affected by the generation and development of microcosmic damage and failure. The model was established based on an elastic model. After adding damage factors along maximum and minimum principle stresses directions, as well as some anisotropic parameters, the constitutive model was able to represent influences of damage on the stress-strain relationship. In the proposed constitutive model, microscopic damage along the maximum and minimum principle stress directions was indicated by two damage factors. Therefore, the anisotropic properties (different damage development behaviors along the maximum and minimum principle stress directions) of brittle rocks are able to be represented by the proposed constitutive model. The proposed constitutive model was subsequently applied in computer simulations by rebuilding the constitutive model and failure criterion files using the simulation software.

First of all, if rock is regarded as an isotropic material, then the stress-strain relationship can be expressed as:

$$\sigma_{ij} = C_{ijrs}\varepsilon_{rs} \tag{1}$$

where σ_{ij} is the component of the Cauchy stress tensor, ε_{rs} means the component of the stain tensor, and C_{ijrs} indicates the component of the elastic stiffness tensor.

$$C_{ijkl} = \lambda^e \delta_{ij}\delta_{kl} + \mu^e(\delta_{ik}\delta_{jl} + \delta_{il}\delta_{jk}) \tag{2}$$

where λ^e and μ^e are Lamé's first parameter and Lamé's second parameter, and δ represents the Kronecker delta.

Effective stress $\widetilde{\sigma}$ can be expressed as:

$$\widetilde{\sigma} = (I - D)^{-1}\sigma = M(D)\sigma \tag{3}$$

$$M(D) = \begin{bmatrix} \frac{1}{1-D_1} & 0 & 0 \\ 0 & \frac{1}{1-D_2} & 0 \\ 0 & 0 & \frac{1}{1-D_3} \end{bmatrix} \tag{4}$$

where, D is damage tensor, D_1, D_2, and D_3 are components of the damage tensor on coordinate directions, and I is the unit tensor.

An effective compliance tensor can be expressed as an inversed effective stiffness tensor:

$$\widetilde{S} = \widetilde{C}^{-1} = C^{-1}M(D) \tag{5}$$

$$\varepsilon^e = C^{-1}\widetilde{\sigma} = C^{-1}M(D)\sigma = C^{-1}(I-D)^{-1}\sigma \tag{6}$$

where $\widetilde{S}$ is an effective compliance tensor, ε^e is the elastic strain, C indicates the stiffness tensor of undamaged rocks, and $\widetilde{C}$ means the stiffness tensor of rocks.

For coordinate directions:

$$\left\{ \begin{array}{c} \varepsilon^e_1 \\ \varepsilon^e_2 \\ \varepsilon^e_3 \end{array} \right\} = \frac{1}{E} \left[\begin{array}{ccc} \frac{1}{1-D_1} & \frac{-\nu}{1-D_2} & \frac{-\nu}{1-D_3} \\ \frac{-\nu}{1-D_1} & \frac{1}{1-D_2} & \frac{-\nu}{1-D_3} \\ \frac{-\nu}{1-D_1} & \frac{-\nu}{1-D_2} & \frac{1}{1-D_3} \end{array} \right] \left\{ \begin{array}{c} \sigma_1 \\ \sigma_2 \\ \sigma_3 \end{array} \right\} \tag{7}$$

where, $\varepsilon^e_1, \varepsilon^e_2$, and ε^e_3 are strain components on coordinate directions, σ_1, σ_2, and σ_3 are stress components on coordinate directions, ν is the Poisson's ratio, and E is an elastic module.

Based on the hypothesis of elastic energy equivalence, we get:

$$\widetilde{C}^{-1}(D) = (I-D)^T C^{-1}(I-D) \tag{8}$$

where D is fourth order asymmetric damage tensor.

$$D = \left[\begin{array}{cccccc} D_1 & 0 & 0 & 0 & 0 & 0 \\ \frac{\nu}{1-\nu}D_1 & 0 & 0 & 0 & 0 & 0 \\ \frac{\nu}{1-\nu}D_1 & 0 & 0 & 0 & 0 & 0 \\ 0 & 0 & 0 & 0 & 0 & 0 \\ 0 & 0 & 0 & 0 & D_3 & 0 \\ 0 & 0 & 0 & 0 & 0 & D_3 \end{array} \right] \tag{9}$$

Then the relationship of effective stress components and stress components can be expressed as:

$$\left\{ \begin{array}{c} \widetilde{\sigma}_{11} \\ \widetilde{\sigma}_{22} \\ \widetilde{\sigma}_{33} \\ \widetilde{\sigma}_{23} \\ \widetilde{\sigma}_{13} \\ \widetilde{\sigma}_{12} \end{array} \right\} = \left[\begin{array}{cccccc} \frac{D_1}{1-D_1} & 0 & 0 & 0 & 0 & 0 \\ \frac{\nu}{1-\nu}\frac{D_1}{1-D_1} & 1 & 0 & 0 & 0 & 0 \\ \frac{\nu}{1-\nu}\frac{D_1}{1-D_1} & 0 & 1 & 0 & 0 & 0 \\ 0 & 0 & 0 & 1 & 0 & 0 \\ 0 & 0 & 0 & 0 & \frac{1}{1-D_3} & 0 \\ 0 & 0 & 0 & 0 & 0 & \frac{1}{1-D_3} \end{array} \right] \left\{ \begin{array}{c} \sigma_{11} \\ \sigma_{22} \\ \sigma_{33} \\ \sigma_{23} \\ \sigma_{13} \\ \sigma_{12} \end{array} \right\} \tag{10}$$

Γ can then be set as anisotropic tensor:

$$\Gamma = \left[\begin{array}{cccccc} 1 & 0 & 0 & 0 & 0 & 0 \\ \frac{\nu}{1-\nu} & 0 & 0 & 0 & 0 & 0 \\ \frac{\nu}{1-\nu} & 0 & 0 & 0 & 0 & 0 \\ 0 & 0 & 0 & 0 & 0 & 0 \\ 0 & 0 & 0 & 0 & \xi & 0 \\ 0 & 0 & 0 & 0 & 0 & \xi \end{array} \right] \tag{11}$$

The degree of anisotropy can be expressed as:

$$Q = (1-\gamma)\Gamma + \gamma I \tag{12}$$

where γ is the anisotropic coefficient: when $\gamma = 0$, the damage of rock is fully anisotropic, while if $\gamma = 1$, the damage of rock will be fully isotropic. Q is the fourth order tensor used to represent the degree of anisotropy, Γ is the anisotropic tensor, and ξ is the material coefficient.

Under asymmetrical loading, the damage tensor is:

$$D = Q \cdot D_1 + RRQR^T R^T D_3 \tag{13}$$

where D_1 is the damage factor along the maximum principle stress direction, D_3 is the damage factor along the minimum principle stress direction, and R is the rotation tensor.

The constitutive model of rocks considering anisotropic damage can be expressed by Equation (14).

$$\sigma = \widetilde{C}\varepsilon = (I - D)C\varepsilon \tag{14}$$

where σ indicates stress and ε represents strain.

3.2. Non-Linear Failure Criterion

The Mohr-Coulomb criterion is a universal criterion which can be used for all types of rocks and it usually is treated as a failure criterion for brittle rocks. Although the Mohr-Coulomb criterion had been widely applied since the early 1900s and continues to be used at present, it still has some limitations [28]. One of the most significant limitations is the failure envelope of the Mohr-Coulomb criterion. The conventional failure envelope of the Mohr-Coulomb criterion is typically considered as a straight line, which means the cohesion c and friction angle ϕ of the rock are treated as constant values. Some scholars found that the cohesion and friction angle of rocks are not constant. With the formation and development of micro-cracks, the initial cohesion is lost and reaches its final value. Moreover, generation of a micro-shear plane also results in changing of frictional strength [29,30]. Consequently, the linear strength envelope usually is not able to present the rock failure behaviors from the excavation site. Applying improper failure criterion in calculations will result in large modeling errors.

Tang conducted triaxial cyclic loading-unloading tests of rock specimens which are sampled from excavation working sites [31]. The rock specimens are loaded and unloaded cyclically until rock specimens damage occurs when reproducing and simulating the stress path of TBM excavation. The tests were conducted under four different confining pressure levels (10, 15, 25, and 30 MPa) due to the ground pressure ranges for 15 to 30 MPa in deep-buried coal mine roadways. The initial axial force in four tests was set to be higher than uniaxial compressive strength and lower than triaxial compressive strength under four different confining pressures (250, 320, 380, and 390 kN). Because the cutter head of TBM applies thrust forces on rock faces circularly during excavation processes, the processes of rock breakage and excavation essentially are cyclic loading-unloading processes of rocks. Therefore, the initial confine pressures were reduced from initial levels (10, 15, 25, and 30 MPa) to zero with unloading speed of 0.05 MPa/s. If no damage occurred on rock specimens after unloading, the confine pressure is added to the initial level and the axial force is set to be 20 kN higher. The loading-unloading processes were repeated until rock specimens were damaged.

The strength envelopes are supposed to be common tangents of adjacent circles according to Mohr's Theory. Various slope angles of different common tangents between adjacent circles indicated that the single linear strength envelope are not capable of representing strength behavior under all confined pressure conditions. The strength envelope must be a curve and the function of this curved strength envelope is the failure criterion of rock material, namely the criterion of macrocosmic damage or failure of rocks. Various curves were used to fit the Mohr circles and finally the power function curve was chosen as the strength envelope due to its highest value of determination coefficients (R^2). A strength envelope was fitted based on test results and Mohr circles (shown in Figure 2), and the function of the non-linear strength envelope is shown in Equation 15:

$$\tau = 6.946\sigma^{0.77} + 23.72 \tag{15}$$

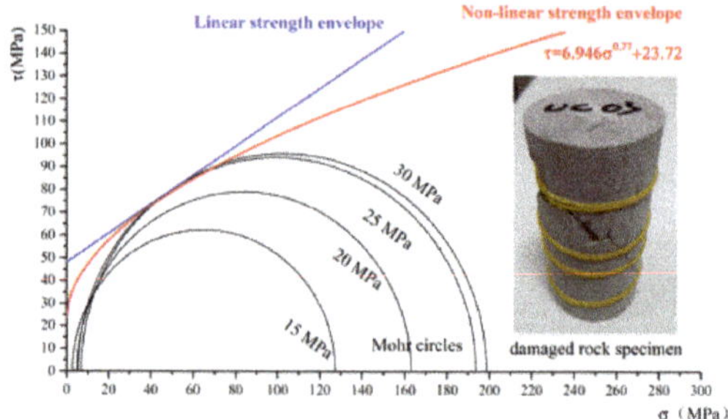

Figure 2. Strength envelopes of Mohr circles.

The determination coefficients (R^2) of proposed non-linear failure criterion and conventional linear failure criterion are 0.989 and 0.911, respectively. It can be seen from Figure 2 that the strength envelope of proposed non-linear failure criterion has a much better performance on fitting Mohr circles than that of conventional linear failure criterion.

4. Numerical Simulation

The model used in this study was established by Fast Lagrangian Analysis of Continua (FLAC). A huge amount of research on coal mine roadway support and surrounding rock stability has indicated that the mechanisms of surrounding rock failure after excavations can be extremely complex [24,32]. The successful application of computer simulations therefore requires a set of detailed parameters of rock properties and in-situ stress fields. Moreover, the constitutive model and failure criterion of surrounding rocks under very specific stress conditions and stress paths should be studied to ensure the precision of simulation results. In-situ stress and rock property parameters have been obtained by in-situ measurement and laboratory testing. Proposed constitutive models and failure criterion have been introduced in FLAC3D simulation software by using an alternative DLL (dynamic link library) files. The Numerical implementation processes are shown in Figure 3.

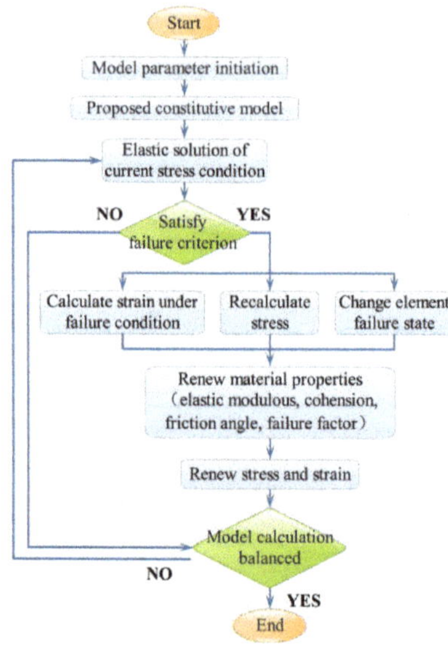

Figure 3. Numerical implementation processes of proposed constitutive model and failure criterion.

4.1. Model Calibration

The rock specimens for laboratory tests were obtained from a TBM excavation working site. Triaxial compression tests were conducted in order to obtain rock property parameters and the stress-strain relationship. The triaixal tests were also simulated using software and the test results of computational simulation were compared with these of lab tests for verifying the feasibility of the proposed constitutive model and failure criterion.

The results of lab tests and computational simulation are illustrated in Table 2, where σ_1 indicates the axial stress when there is rock specimen damage, σ_3 represents confinement pressure, ε_1 denotes the axial strain, and ε_3 means the lateral strain. The simulation results suggested that the deviation between lab tests and simulations is negligible. The constitutive model, failure criterion, and material parameters are able to represent the mechanical behaviors of rock specimens in lab tests.

Table 2. Model calibration results.

σ_1 (MPa)	σ_3 (MPa)	ε_1		ε_3	
		Simulation	Lab Tests	Simulation	Lab Tests
127.39	15	0.0169	0.0151	0.0048	0.0042
163.06	20	0.0171	0.0154	0.0052	0.0062
193.63	25	0.0175	0.0155	0.0045	0.0037
198.72	30	0.0193	0.0189	0.0043	0.000028

4.2. D Finite Difference Model

The sizes of numerical models were decided based on the Saint Venant principle. According to the Saint Venant principle and construction experience, the stress redistribution, surrounding rock deformation, and other effects of excavation out of a domain with 3 to 5 times the roadway radius are negligibly small [33]. The diameter of the roadway is 4.53 m, therefore the model size is taken as 60 m × 120 m × 60 m (X, Y, Z axis direction). Stress and displacement distribution of the roadway surrounding rocks out of the model range can be considered as being barely influenced by roadway excavation. The model consists of 40,000 elements and 41,041 grid points, while the excavation part was modeled using cylinder elements with six grid points and the surrounding rock was modeled using radcylinder elements and eight grid points. The benefit of modeling surrounding the rock using radcylinder elements is that elements and grid points are intensive in surrounding areas of the roadway and are extensive in the areas far from the roadway. This significantly reduces the computation, thereby ensuring precision of simulation in near-roadway areas. The model geometry is shown in Figure 4. Prior to numerical simulation, the obtained power function failure criterion was introduced with the software by rewriting a build-in constitutive model for the simulation software.

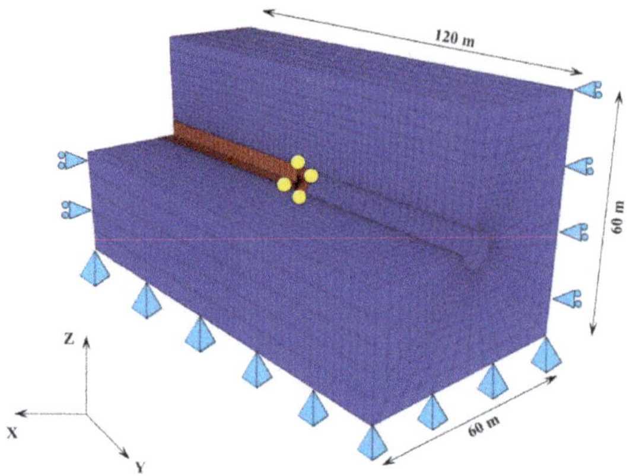

Figure 4. Computational model: Roadway, excavation section (red region), and monitoring points (golden spots).

In this model, rock parameters were gained from the previous rock mechanical tests. Model element lengths were set as 1.0 m along the longitudinal direction of the roadway. As the rock mechanical properties within the model range are similar, the model was considered as being homogeneous for getting higher computational efficiency with acceptable error rates. The details of rock properties are provided in Table 3.

Table 3. Rock properties.

Density (kg/m^3)	Elastic Modulus (10^4 MPa)	Bulk Modulus (10^4 MPa)	Shear Modulus (10^4 MPa)	Passion Ratio	Cohesion (MPa)	Friction Angle (°)
2500	4.5	2.5	1.875	0.20	9.0	45

Rock bolts were modeled using cable elements. The rock bolt parameters input are shown in Table 4. Three different support patterns (unsupported, roof support, and roof-sides support) are applied in the model. The stress field redistribution and displacement of surrounding rock under three support conditions were simulated and analyzed and the most feasible support layout can be decided based on simulation results. Support layouts are shown in Figure 5. In engineering practices, the TBM is equipped with two rock bolters that have moving and rotating functions. The rock bolters drill boreholes on surrounding rocks and rock bolts are fixed in the boreholes by resin cartridges. The tunnel excavation and rock bolt installation works are conducted simultaneously.

Table 4. Rockbolt mechanical properties.

Diameter (mm)	Length (mm)	Elastic Modulus (10^4 MPa)	Tensile Strength (MPa)	Passion Ratio	Rigidity of Resin (kN/m)	Friction Angle of Resin (°)
20	2000	20.6	345	0.30	1000	54

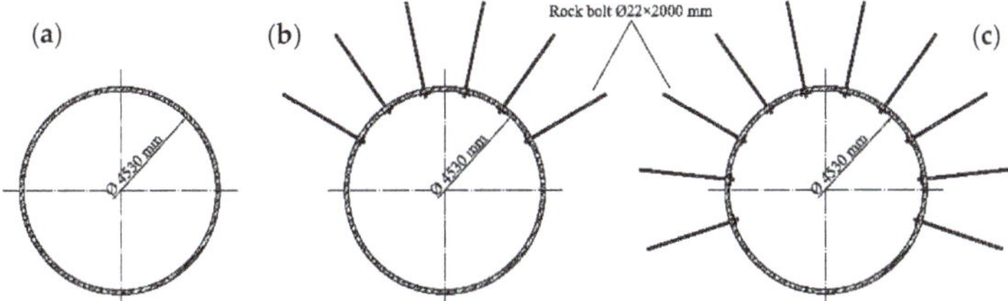

Figure 5. Three support patterns: (**a**) Pattern 1, unsupported; (**b**) Pattern 2, Roof support; (**c**) Pattern 3, roof-sides support.

4.3. Simulation Procedure

The Y-axis in Figure 6 is the excavation direction, with excavation starting from +Y and moving towards the −Y direction. The simulation procedure is:

1. Setup the initial boundary conditions and the initial stress conditions. The in-situ stress filed parameters had been obtained through instrument works, while the maximum and minimum stress components are not perpendicular to the roadway alignment. While in FLAC3D software, stress boundary conditions are only able to be set by applying stresses that are normal to model boundaries. Therefore, for easier boundary conditions setting, both the maximum and minimum stresses need to be decomposed into two components: being parallel and perpendicular to the roadway alignment direction. As shown in Table 5, σ_1 and σ_3 indicates measured in-situ principle stresses, σ_x and σ_y means stress applied on model boundaries. Four lateral sides and the bottom of the model were fixed by roller supports and pinned supports, respectively. Then stresses were applied on six boundaries of the model based on stress decomposition results. The horizontal in-situ stress had been decomposed along the alignment of the roadway in the model. The orientation of the roadway is 32°, and the stress decomposition results are shown in Figure 6 and Table 5. The magnitude of stresses which are applied on the top and bottom, front and back, and lateral boundaries of the model are 14.3 MPa, 12.71 MPa and 23.91 MPa, respectively. Figure 6 illustrates the relationship between the roadway orientation, in-situ tectonic stresses, stress applied on model boundaries, and coordinate axes of the numerical model. Black and green lines indicate roadway alignment and coordinate axes of the numerical model. Blue and red arrows represent in-situ tectonic principle stresses, and stresses applied on model boundaries. Radial lengths of arrows indicate the magnitudes of stresses. Moreover, 0, 90, 180, and 270 degree orientations indicate geodetic north, east, south, and west directions, respectively.

2. The roadway was set in the middle of the model was 30 m away from the model boundary to eliminate the boundary effect. Excavate a 1.0 m section along the excavation direction of the roadway by setting the excavation region as a null model (A null model can be assigned to zones to represent material that is removed or excavated).

3. Set a cable element in a new excavation region representing rock bolts after excavation (this step was not conducted in support pattern 1).

4. Record the stress and displacement on all monitor points for analysis. Monitoring points are set on the surface of roadway and in surrounding rock.

5. Repeat step 2 to step 4 until the excavation work is finished.

Table 5. Stress decomposition results.

In-Situ Stress			Stress Applied on the Model		
Stress Components	Magnitude (MPa)	Orientation of Stresses (°)	Stress Components	Magnitude (MPa)	Orientation (°)
σ_1	21.6	130.72	σ_x	23.91	122
σ_3	13.2	223.55	σ_y	12.71	32
σ_v	14.3	-	σ_z	14.3	-

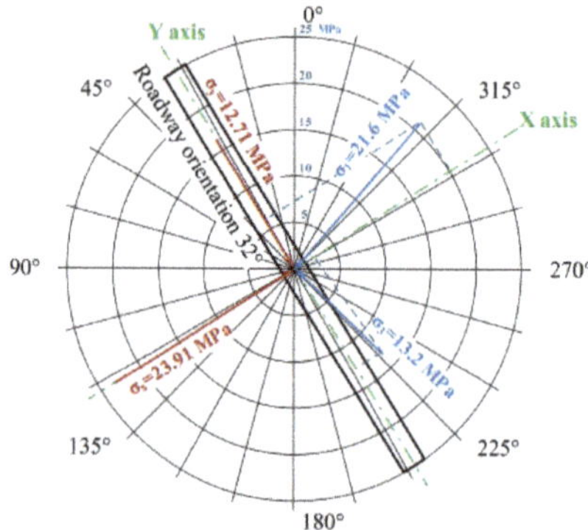

Figure 6. Decomposing tectonic stress along the roadway alignment.

All the simulation procedures had been conducted under three support conditions (unsupported, roof support, and roof-sides support), and the stress and displacement on monitor points was recorded and analyzed. The initial support design of the TBM-excavated roadway was made based on the simulation results.

4.4. Simulation Results Analysis

The model was launched under three conditions: no support, roof support and roof-and-ribs support. Stress redistribution, surrounding rock displacement, roadway convergence, and stress concentration under three conditions were recorded and analyzed.

4.4.1. Stress of Surrounding Rocks

As can be seen from Figure 7, after excavation, the stress filed of surround rock near the roadway is redistributed. The horizontal stress on the roadway roof under three support patterns (no support, roof support, and roof-sides support) are 41.33 MPa, 41.35 MPa, and 40.55 MPa, respectively, the horizontal stress on the roadway floor under the three support patterns are 41.67 MPa, 44.37 MPa, and 41.47 MPa, respectively, while vertical stress on roadway sides under three support patterns are 20.9 MPa, 22.3 MPa, and 24.4 MPa, respectively. The calculated stresses indicated that the horizontal stress is concentrated on the roof and floor of the roadway, and the vertical stress is concentrated on lateral sides.

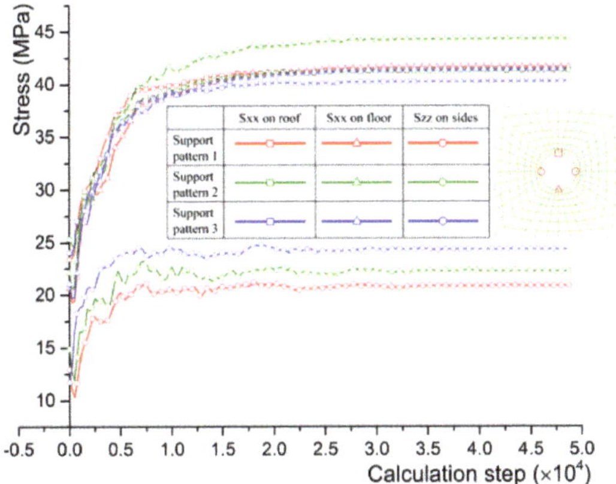

Figure 7. Stress distribution of the roadway surrounding rocks: Sxx indicates horizontal stress and Szz represents vertical stress.

4.4.2. Displacements of Surrounding Rocks

Figure 8 illustrates that, under the tectonic stress, the displacements on the roadway roof under three support patterns are: 4 mm, 2.2 mm, and 2.1 mm. The displacements on the roadway floor under three support patterns are: 2.29 mm, 2.3 mm, and 2.5 mm. The horizontal displacement on roadway lateral sides under three support patterns is: 7.63 mm, 7.2 mm, and 2.01 mm. The simulation results suggest that after rock bolt supporting, the surrounding rock of roadway deformation was controlled significantly, from 4 mm to about 2 mm on the roof and from 7.63 mm to around 2 mm on lateral sides. When the roof and lateral sides were supported and displacements on those areas were controlled, the displacement on the roadway bottom increased slightly.

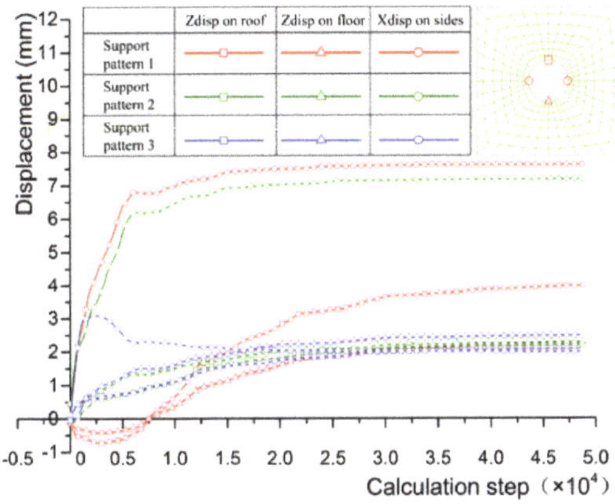

Figure 8. Displacement distribution of the roadway surrounding rocks: Xdisp indicates horizontal displacement and Zdisp represents vertical displacement.

4.4.3. Excavation Damage Zone Distribution

The EDZ under three support patterns are shown in Figure 9. The EDZ under support pattern 1 is approximately circular with the largest depth of 2.2 m. In pattern 2, the range of EDZ shrinks to 0.9 m for the maximum due to the rock bolts support system. Meanwhile in support pattern 3, the EDZ is distributed mainly on the rib sides and bottom of the roadway and the maximum depth is

0.77 m. The EDZ range shrunk significantly after obtaining support, while there was no significant difference on the EDZ range under the roof support and roof-sides support pattern.

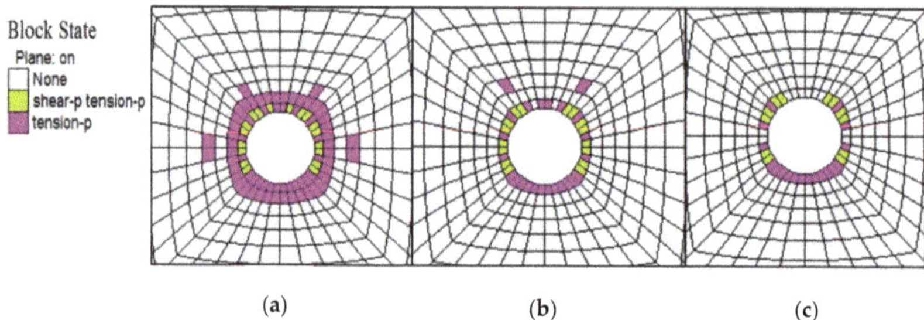

Figure 9. Excavation damage zone (EDZ) distribution under three support patterns: (**a**) EDZ distribution under support pattern 1; (**b**) EDZ distribution under support pattern 2; (**c**) EDZ distribution under support pattern 3.

4.5. Decision Making of Roadway Supporting

The initial support pattern was decided based upon the simulation results. After support, the stress redistribution changed slightly, while the roadway convergence and the range of EDZ decreased significantly. Between support patterns 1 and 2, there was no significant difference on roadway convergence and EDZ range. Compared with support pattern 2, support pattern 3 leads to a larger support range and higher safety. Support pattern 2 meets the requirement of roadway support according to simulation results and in-situ practices. Therefore, support pattern 2 which is the roof support pattern, was been chosen for the roadway supporting.

5. In-SITU Monitoring

In order to verify the precision of modeling simulations and study the surrounding rock deformation behavior and bolt stress, monitoring stations were set in the roadway [34]. Each station contains four rock bolt dynamometers and a set of convergence monitors. The rock bolt dynamometers were installed on the bottom of the rock bolt for monitoring the axial force of rock bolts. Roadway convergence behavior can be studied through measuring distance changing between two convergence monitor spots by laser geodimeter. The monitoring station layout is shown in Figure 10.

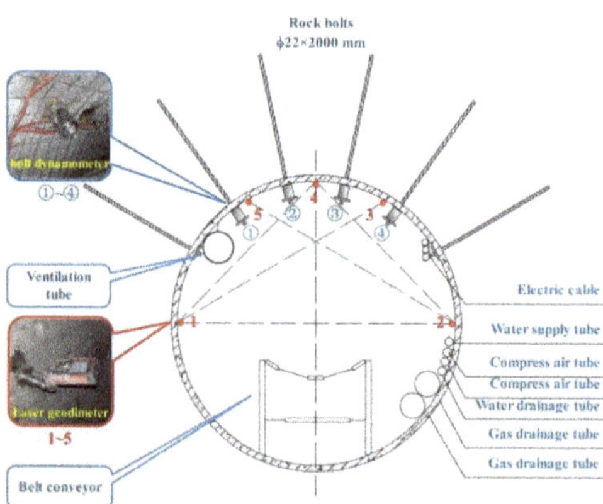

Figure 10. Monitoring station layout: 1–5 are convergence monitor spots and laser geodimeter; ①–④ indicate No.1 to No.4 rock bolt dynamometers.

5.1. Rock Bolt Axial Force Monitoring

Figure 11 shows the monitoring results of bolt axial force. The largest bolt axial force (62 kN) appeared on the No.2 rock bolt 20 days after excavation. Monitoring data illustrated that after rock bolts were installed, the bolt axial force increased rapidly on the rock bolts installed near the roof of roadway in 10 days, then the increasing speed slowed down and the bolt axial force remained stable after 20 days. In contrast, the axial force of rock bolts installed near the shoulders and lateral sides of the roadway decreased rapidly in 10 days and then leveled off at around 45 kN. Monitoring data suggested that the surrounding rock reached a stable state and the support design is reasonable.

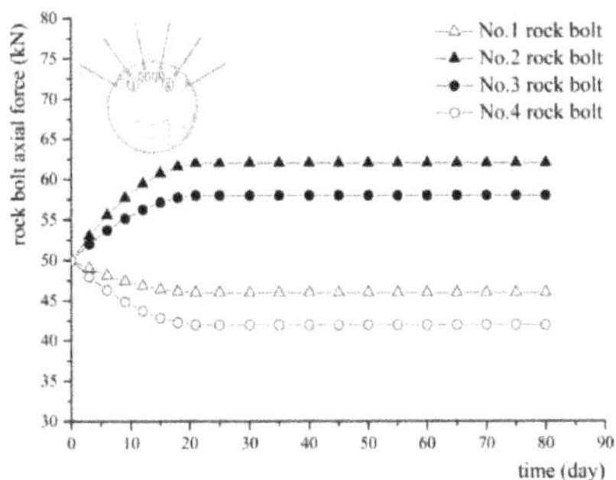

Figure 11. Monitoring results for the bolt axial force. ①–④ indicate No.1 to No.4 rock bolt dynamometers.

5.2. Roadway Convergence Monitoring

Convergence monitors were installed on the upper hemisphere of the roadway because the conveyor belt was installed on the floor. The roadway convergence is shown in Figure 12, the lateral convergence shot up to 8 mm within 20 days after roadway excavation, and then the increasing of convergence slowed down gradually and stabilized at 12 mm after 30 days. The monitoring data indicates that the surrounding rock deformation had been controlled significantly under support pattern 2 (roof support). The convergence is extremely small when considering that the roadway width is 4.53 m.

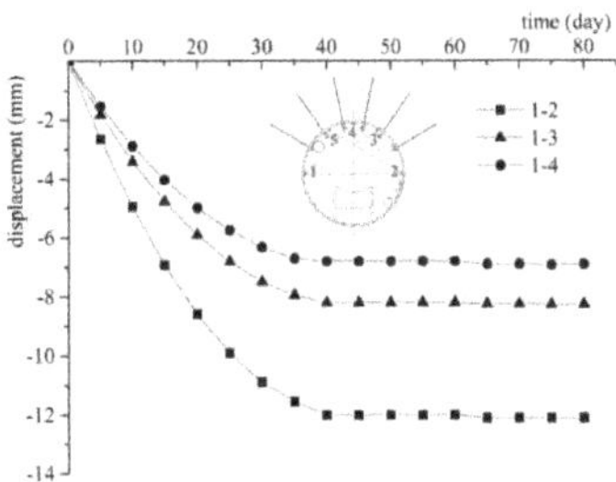

Figure 12. Monitoring results of roadway convergence. Red numbers 1–5 represent convergence monitor spots.

6. Conclusions

This paper presents a case study on optimization of the support design of a TBM-excavated roadway. The CBM drainage roadway of a 1413A longwall panel in Zhangji coal mine was excavated by a TBM and the deformation and failure behaviors of the roadway surrounding the rocks was different from these of conventional roadways due to different roadway geometries and excavation methods.

A modified constitutive model and failure criterion of the roadway surrounding rocks was introduced within a numerical simulation to increase calculation precision. The optimized support design was proposed based on results of numerical simulation and it was verified by in-situ monitoring results. When the roof of the roadway was supported by rock bolts, the roadway convergence stabilized at 12 mm in 30 days after excavation and the axial force of rock bolts finally leveled off at a safety range after the installation of rock bolts. The calculation and monitoring results of roadway convergence were very close (14.4 and 12 mm). Considering the diameter of the roadway (4.5 m), the calculation accuracy was accepted. Field applications suggested that the surrounding rock reached a stable state and the support design is appropriate. Supporting the roof of TBM-excavated coal mine roadways enables the safety of roadway excavation and speed of roadway supporting to be balanced.

This paper only focuses on the constitutive model and failure criterion of Zhangji sandstone under a cyclic loading-unloading stress path. The constitutive models and failure criterions of other rocks or under other stress paths can be obtained by using the research methodology proposed in this paper. The customized constitutive model and failure criterion can increase the simulation accuracy and efficiency of excavation safety controlling.

Author Contributions: Conceptualization, B.T. and H.C.; Data curation, C.R.; Formal analysis, B.T.; Funding acquisition, B.T.; Investigation, H.C.; Methodology, B.T. and T.Z.; Project administration, Y.T.; Resources, Y.T. and C.W.; Software, B.T.; Supervision, B.T. and H.C.; Validation, Z.Y. and C.W.; Visualization, B.T.; Writing–original draft, B.T.; Writing–review & editing, H.C. All authors have read and agreed to the published version of the manuscript.

Funding: This research was funded by the National Natural Science Foundation of China (Grant No. 51804006), China Postdoctoral Science Foundation (Grant No. 2018M642503), Youth Fund of Anhui University of Science and Technology (Grant No. QN2017222), and Talent Introduction Fund of Anhui University of Science and Technology (Grant No. 11667). The authors gratefully acknowledge financial support of the above-mentioned agencies.

Conflicts of Interest: The authors declare no conflict of interest.

References

1. Huang, X.; Liu, Q.; Liu, B.; Liu, X.; Pan, Y.; Liu, J. Experimental Study on the Dilatancy and Fracturing Behavior of Soft Rock Under Unloading Conditions. *Int. J. Civ. Eng.* **2017**, *15*, 921–948. [CrossRef]
2. Li, Z.; Wang, E.; Ou, J.; Liu, Z. Hazard evaluation of coal and gas outbursts in a coal-mine roadway based on logistic regression model. *Int. J. Rock Mech. Min. Sci.* **2015**, *80*, 185–195. [CrossRef]
3. Erdogan, H.H.; Duzgun, H.S.; Selcuk-Kestel, A.S. Quantitative hazard assessment for Zonguldak Coal Basin underground mines. *Int. J. Min. Sci. Technol.* **2019**, *29*, 453–467. [CrossRef]
4. Krause, E.; Karbownik, M. Tests of methane desorption and emission from samples of hard coal in the context of mine closures through flooding. *J. Sustain. Min.* **2019**, *18*, 127–133. [CrossRef]
5. Kordos, J. Tests of new method of monitoring endogenous fire hazard in hard coal mines. *J. Sustain. Min.* **2019**, *18*, 134–141. [CrossRef]
6. Malkowski, P. The impact of the physical model selection and rock mass stratification on the results of numerical calculations of the state of rock mass deformation around the roadways. *Tunn. Undergr. Space Technol.* **2015**, *50*, 365–375. [CrossRef]
7. Zheng, Y.L.; Zhang, Q.B.; Zhao, J. Challenges and opportunities of using tunnel boring machines in mining. *Tunn. Undergr. Space Technol.* **2016**, *57*, 287–299. [CrossRef]
8. Mandal, P.K.; Das, A.J.; Kumar, N.; Bhattacharjee, R.; Tewari, S.; Kushwaha, A. Assessment of roof convergence during driving roadways in underground coal mines by continuous miner. *Int. J. Rock Mech. Min. Sci.* **2018**, *108*, 169–178. [CrossRef]
9. Huang, X.; Liu, Q.; Shi, K.; Pan, Y.; Liu, J. Application and prospect of hard rock TBM for deep roadway construction in coal mines. *Tunn. Undergr. Space Technol.* **2018**, *73*, 105–126. [CrossRef]
10. Gong, Q.; Liu, Q.; Zhang, Q. Tunnel boring machines (TBMs) in difficult grounds. *Tunn. Undergr. Space Technol. Inc. Trenchless Technol. Res.* **2016**, *57*. [CrossRef]
11. Li, Q.; Yang, R.; Li, J.; Wang, H.; Wen, Z. Strength and Cost Analysis of New Steel Sets as Roadway Support Project in Coal Mines. *Adv. Mater. Sci. Eng.* **2018**, *2018*, 3927843. [CrossRef]

12. Wang, F.; Zhang, C.; Wei, S.; Zhang, X.; Guo, S. Whole section anchor–grouting reinforcement technology and its application in underground roadways with loose and fractured surrounding rock. *Tunn. Undergr. Space Technol. Inc. Trenchless Technol. Res.* **2016**, *51*, 133–143.

13. Cao, R.; Cao, P.; Lin, H. Support technology of deep roadway under high stress and its application. *Int. J. Min. Sci. Technol.* **2016**, *26*, 787–793. [CrossRef]

14. Huang, W.P.; Yuan, Q.; Tan, Y.L.; Wang, J.; Liu, G.L.; Qu, G.L.; Li, C. An innovative support technology employing a concrete-filled steel tubular structure for a 1000-m-deep roadway in a high in situ stress field. *Tunn. Undergr. Space Technol.* **2018**, *73*, 26–36. [CrossRef]

15. Stone, R. Design of primary ground support during roadway development using empirical databases. *Int. J. Min. Sci. Technol.* **2016**, *26*, 131–137. [CrossRef]

16. Yang, S.Q.; Chen, M.; Jing, H.W.; Chen, K.F.; Meng, B. A case study on large deformation failure mechanism of deep soft rock roadway in Xin'An coal mine, China. *Eng. Geol.* **2016**, *217*, 89–101. [CrossRef]

17. Gao, Z.W.; Kong, D.X.; Gao, C.H. Modeling and control of complex dynamic systems: Applied mathematical aspects. *J. Appl. Math.* **2012**, *2012*, 869792. [CrossRef]

18. Gao, Z.W.; Nguang, S.K.; Kong, D.X. Advances in Modelling, Monitoring, and Control for Complex Industrial Systems. *Complexity* **2019**, *2019*, 2975083. [CrossRef]

19. Gao, Z.W.; Saxen, H.; Gao, C.H. Guest Editorial: Special section on data-driven approaches for complex industrial systems. *IEEE Trans. Ind. Inform.* **2013**, *9*, 2210–2212. [CrossRef]

20. Tang, B.; Cheng, H.; Tang, Y.; Yao, Z.; Rong, C.; Wang, X.; Xue, W.; Guo, S. Experiences of gripper TBM application in shaft coal mine: A case study in Zhangji coal mine, China. *Tunn. Undergr. Space Technol.* **2018**, *81*, 660–668. [CrossRef]

21. Tang, B.; Cheng, H. Application of Distributed Optical Fiber Sensing Technology in Surrounding Rock Deformation Control of TBM-Excavated Coal Mine Roadway. *J. Sens.* **2018**, *2018*, 8010746. [CrossRef]

22. Huang, X.; Liu, Q.; Chen, L.; Pan, Y.; Liu, B.; Kang, Y.; Liu, X. Cutting force measurement and analyses of shell cutters on a mixshield tunnelling machine. *Tunn. Undergr. Space Technol.* **2018**, *82*, 325–345. [CrossRef]

23. Kang, Y.; Liu, Q.; Gong, G.; Wang, H. Application of a combined support system to the weak floor reinforcement in deep underground coal mine. *Int. J. Rock Mech. Min. Sci.* **2014**, *71*, 143–150. [CrossRef]

24. Kang, Y.; Liu, Q.; Xi, H.; Gong, G. Improved compound support system for coal mine tunnels in densely faulted zones: A case study of China's Huainan coal field. *Eng. Geol.* **2018**, *240*, 10–20. [CrossRef]

25. Sun, X.; Liu, Y.; Wang, J.; Li, J.; Sun, S.; Cui, X. Study on Three-Dimensional Stress Field of Gob-Side Entry Retaining by Roof Cutting without Pillar under Near-Group Coal Seam Mining. *Processes* **2019**, *7*, 552. [CrossRef]

26. Xie, B.; Yan, Z.; Du, Y.; Zhao, Z.; Zhang, X. Determination of Holmquist–Johnson–Cook Constitutive Parameters of Coal: Laboratory Study and Numerical Simulation. *Processes* **2019**, *7*, 386. [CrossRef]

27. Li, G.; Ma, F.; Liu, G.; Zhao, H.; Guo, J. A Strain-Softening Constitutive Model of Heterogeneous Rock Mass Considering Statistical Damage and its Application in Numerical Modeling of Deep Roadways. *Sustainability* **2019**, *11*, 2399. [CrossRef]

28. Liu, X.; Liu, Q.; Kang, Y.; Yucong, P. Improved Nonlinear Strength Criterion for Jointed Rock Masses Subject to Complex Stress States. *Int. J. Geomech.* **2018**, *18*, 04017164. [CrossRef]

29. Hajiabdolmajid, V.; Kaiser, P.K.; Martin, C.D. Modelling Brittle Failure of Rock. *Int. J. Rock Mech. Min. Sci.* **2002**, *39*, 731–741. [CrossRef]

30. Hajiabdolmajid, V.; Kaiser, P.K. Brittleness of rock and stability assessment in hard rock tunneling. *Tunn. Undergr. Space Technol.* **2003**, *18*, 35–48. [CrossRef]

31. Tang, B.; Hua, C.; Tang, Y.; Yao, Z.; Li, M.; Zheng, T.; Jian, L. Excavation damaged zone depths prediction for TBM-excavated roadways in deep collieries. *Environ. Earth Sci.* **2018**, *77*, 165. [CrossRef]

32. Yan, G.; Ma, Z.; Hu, Y. Study On Numerical Simulation of the Arrangement about Mining Roadway with Short Distance. *Procedia Eng.* **2011**, *26*, 301–310.

33. Tsinidis, G.; Pitilakis, K.; Madabhushi, G. On the dynamic response of square tunnels in sand. *Eng. Struct.* **2016**, *125*, 419–437. [CrossRef]
34. Tang, B.; Cheng, H.; Tang, Y.; Yao, Z.; Rong, C.; Xue, W.; Lin, J. Application of a FBG-Based Instrumented Rock Bolt in a TBM-Excavated Coal Mine Roadway. *J. Sens.* **2018**, *2018*, 8191837. [CrossRef]

Article

Optimal Siting and Sizing of Distributed Generation Based on Improved Nondominated Sorting Genetic Algorithm II

Wei Liu [1,*], Fengming Luo [2], Yuanhong Liu [1,3,*] and Wei Ding [1]

[1] School of Electrical Information and Engineering, Northeast Petroleum University, Daqing 163318, China; vivi_ting@stu.nepu.edu.cn

[2] Lorentech (Beijing) Co., Ltd., Beijing 100000, China; LFM1220@126.com

[3] Faculty of Engineering and Environment, Northumbria University, Newcastle NE1 8ST, UK

* Correspondence: liuwei@nepu.edu.cn (W.L.); liu.yuanhong@northumbria.ac.uk (Y.L.)

Received: 19 October 2019; Accepted: 2 December 2019; Published: 13 December 2019

Abstract: With the development of distributed generation technology, the problem of distributed generation (DG) planning become one of the important subjects. This paper proposes an Improved non-dominated sorting genetic algorithm-II (INSGA-II) for solving the optimal siting and sizing of DG units. Firstly, the multi-objective optimization model is established by considering the energy-saving benefit, line loss, and voltage deviation values. In addition, relay protection constraints are introduced on the basis of node voltage, branch current, and capacity constraints. Secondly, the violation constrained index and improved mutation operator are proposed to increase the population diversity of non-dominated sorting genetic algorithm-II (NSGA-II), and the uniformity of the solution set of the potential crowding distance improvement algorithm is introduced. In order to verify the performance of the proposed INSGA-II algorithm, NSGA-II and multiple objective particle swarm optimization algorithms are used to perform various examples in IEEE 33-, 69-, and 118-bus systems. The convergence metric and spacing metric are used as the performance evaluation criteria. Finally, static and dynamic distribution network planning with the integrated DG are performed separately. The results of the various experiments show the proposed algorithm is effective for the siting and sizing of DG units in a distribution network. Most importantly, it also can achieve desirable economic efficiency and safer voltage level.

Keywords: distributed generation (DG); INSGA-II; multi-objective optimization; potential crowding distance; static and dynamic planning

---

1. Introduction

In recent years, distributed generation systems using renewable energy technologies such as wind power generation and photovoltaic power generation have become one of the hotspots of research at home and abroad. After the distributed generation (DG) units are connected to the distribution network (DN), the structure, operation mode, and control strategy of the DN will undergo tremendous changes [1,2]. The research shows that DGs provide more flexibility and expansibility for distribution network. When DGs are connected to the distribution network (DN), the performances of DGs are most important for the power quality, reliability, and security of DN. In addition, DGs with renewable energy technology can effectively reduce system line loss and transmission congestion [3–5]. Despite the above advantages, if the placement and sizing of the DGs are improperly selected, it may cause a series of power system safety hazards such as power flow reverse, insufficient voltage stability, and malfunction of the protection device [6,7]. Therefore, the placement and sizing of the DGs has become one of the important subject.

The optimal planning of distributed generation sizing and siting is critical to ensure the operational performance of a distribution network in terms of power quality, voltage stability, reliability and profitability. DG planning problems are often defined as multi-objective and multi-constrained optimization problems. A number of review papers [8–10] have surveyed the optimization techniques for optimal DG planning in power distribution networks. The aforementioned review papers mainly focused on the discussion of various computational methods and metaheuristic algorithms.

Nowadays, industrial systems, such as papermaking, steelmaking, petrochemical, and power generation, are becoming more and more complex. In such cases, data-driven models based on novel nonlinear signal processing and data analysis techniques may provide an attractive alternative [11]. Therefore, it is paramount but challenging to develop effective techniques in modelling, monitoring, and control for complex industrial systems [12,13].

Among many optimization methods, multi-objective heuristic algorithms are the main means to solve multi-objective optimization problems today because they can effectively balance multiple objectives for optimal search [14]. At present, the main objectives of DG planning include the lowest investment cost of DG, the lowest environmental pollution, the optimal voltage quality of the power grid, and the minimum power loss. Partha Kayal et al. aimed to minimize the reactive power loss of the DN, to maximize the system voltage stability, and to establish a multi-objective optimization model for power systems [15]. Although it can effectively reduce the network loss and improve the voltage distribution, planning with only two objectives does not reflect the real condition. R. Li et al. established a multi-objective optimization model considering economy, voltage quality, and network loss and used the distance entropy multi-objective particle swarm optimization algorithm to solve the optimization problem [16]. However, there is a lack of environmental energy-saving factors in economic costs; M. Esmaili and T. Wang et al. also introduced voltage safety margins [17] and pollutant emissions [18] as optimization models, respectively. However, the objective function of optimization varies greatly in the timescale. For example, the voltage and reactive power change rapidly, and the pollutant emission often has a long timescale, which makes it difficult to use a unified model to optimize. Therefore, this paper establishes a multi-objective optimization model for economy, environmental protection, safety, and reliability.

Constraint processing is a key part of engineering optimization problems, and the treatment of constraints is the key to solving constrained multi-objective optimization problems. Commonly used constraint handling methods include rejection of infeasible solutions, penalty functions, and various correction algorithms. In fact, in the iterative process, infeasible solution is always difficult to be rejected, especially in the case of small feasible region, and repeated trial and error will affect the speed of solution; other modified algorithms are always designed for specific problems. The penalty function method [19] is the most classical method to deal with constraints, but the penalty factor has the same weight problem and is not easy to grasp, resulting in large errors in planning results.

In this paper, a new operator is proposed to address the degree of default. This method does not need to set parameters in advance but deals with feasible and infeasible solutions by considering the number and degree of violation of constraints. This method can effectively combine with multi-objective optimization algorithm to improve the diversity of populations and to avoid the algorithm falling into local optima.

At present, the planning methods of multi-objective optimization problems can be divided into weighted single-objective optimization algorithm and multi-objective optimization algorithm. In order to obtain the unique solution, scholars transform the multi-objective into single-objective optimization model by the weight method and solve it by combining with the single-objective optimization algorithm. This method makes the calculation convenient by reducing the dimension of the problem. For example, H. Su introduced improved simulated annealing particle swarm optimization [20] and established a multi-objective model to maximize DG utilization and to minimize system losses and environmental pollution. Finally, the DG injection model is optimized by this algorithm. However, the selection of weights depends on experience. Multi-objective functions are both interrelated and independent. The subjectivity of weight selection makes the calculation result unsatisfactory. Aiming at the above problems, multi-objective optimization algorithm is introduced to solve the DG planning problem [21]. J. Neale solved the reconfiguration problem of the DN with the integrated DG by using non-dominated sorting genetic algorithm-II (NSGA-II) [22]. This method can effectively improve the operation performance of DN. A. Ameli used the multi-objective particle swarm optimization (MOPSO) [23] with adaptive grid method to optimize the distribution of energy supply system electric vehicle (EV) in DG technology. The results show the effectiveness of MOPSO programming; R. S. Maciel proposed an evolutionary particle swarm optimization algorithm (MEPSO) [24] as a multi-objective optimization algorithm for DG, of which the planning scheme can ultimately improve the objectives. However, in the multi-objective optimization algorithm, there is a common problem; that is, the evaluation information of solution set is insufficient due to crowding distance, which leads to the elimination of potential high-quality solutions in the truncation process. The distribution of the final algorithm planning results is uneven.

In addition, the mutation operator of NSGA-II will gradually lose the ability of local search as the dimension of solution variable increases, so the population updating strategy in the fireworks algorithm [25] is introduced to simultaneously improve the mutation operator and the diversity of solution set. Based on the effective Pareto-optimal set of multi-objective optimization problems, decision makers cannot get a unique solution. Therefore, this paper introduces an unbiased compromise strategy [26] to choose a best compromise from the Pareto-optimal set.

According to the above analysis, a lack of the operation state research planning of DN with the integrated DG considers the relay protection. Actually, when the DG is connected to the DN, the DN that is originally powered by a single system power becomes a network with multiple power structures, causing a change in the magnitude and direction of the short-circuit current during the failure. The action affects the safe and stable operation of the DN. Therefore, this paper increases the short-circuit current constraint condition and compares the result with the optimization result without considering the short-circuit constraint, so as to prove the necessity of considering the relay protection condition. At present, with the development of technology, distributed energy resources are bound to penetrate into the fields of industry, commerce, and urban and rural residents. The problem of DG planning with timing characteristics needs to be solved urgently. Therefore, the paper studies the planning of various examples and plans the static state and dynamic running state of the IEEE 33-bus system separately to provide a reasonable DG planning for decision makers.

The main contributions of this work are as follows:

(1) An improved non-dominated sorting genetic algorithm (INSGA-II) for placement and sizing of DGs is proposed. The violation constrained index and improved mutation operator are proposed to increase the population diversity of NSGA-II, and the uniformity of the solution set of the potential crowding distance improvement algorithm under the 3D objective function is introduced.

(2) The energy-saving benefit of DGs is considered, and a multi-objective optimization model is established.

(3) The restraint condition of relay protection is added; that is, the size of short-circuit current is limited, so that the protection device will not malfunction.

(4) Convergence metric and spacing metric are employed to evaluate the performance of INSGA-II, NSGA-II, and MOPSO.

(5) The effectiveness of the algorithm in different cases is verified.

The proposed algorithm is applied to solve the static and dynamic planning of the DN with the integrated DG units. The rest of this article is organized as follows. Section 2 presents the mathematical programming model. Section 3 briefly introduces the algorithms needed in the planning scheme and proposes INSGA-II. Section 4 provides the comparison of the simulation results of the proposed algorithm in various cases and the performance of the algorithm. Section 5 concludes our work.

2. Problem Formulation

2.1. Objective Functions

Three objectives should be taken into account when establishing multi-criteria optimization model of DN with the integrated DG units: (1) Improving the energy-saving benefit of the system; (2) reducing system line losses; and (3) reducing the node voltage deviation.

1. Maximizing annual energy-saving benefit

Maximizing annual energy-saving benefit of DN with the integrated DG units can be expressed as follows:

$$\max f_1 = Z_{\text{cost}}^{\text{NODG}} - Z_{\text{cost}}^{\text{DG}} \tag{1}$$

where $Z_{\text{cost}}^{\text{NODG}}$ is the total annual cost of DN without DGs and $Z_{\text{cost}}^{\text{NODG}}$ is the total annual cost of DN with DGs.

Total annual costs of DN excluding DGs can be expressed as follows:

$$Z_{\text{cost}}^{\text{NODG}} = C_{\text{loss}} + C_{\text{b}} \tag{2}$$

where C_{loss} is the annual loss of DN without DGs and C_{b} is the annual purchase cost of DN without DGs.

$$C_{\text{loss}} = \sum_{j=1}^{k}\left(C_{\text{p}} \cdot \tau_{\max} \cdot \Delta P_j\right) \tag{3}$$

where k is the number of branches in the DN; C_{p} is the unit price of electricity consumed per unit ($/kWh); $\tau_{\max}$ is the annual maximum load loss hour (h) of the DN; and ΔP_j is the active power loss (kW) of the jth branch.

$$C_{\text{b}} = C_{\text{p}} \cdot P_{\text{load}} \cdot T_{\max} \tag{4}$$

where P_{load} is the DN total active load and $T_{\max}$ is the DN annual maximum load utilization hours.

The total annual cost of DN with DGs can be expressed as follows:

$$Z_{\text{cost}}^{\text{DG}} = C_{\text{dgm}} + C_{\text{ploss}} + C_{\text{B}} - C_{\text{sub}} \tag{5}$$

where C_{dgm} is the annual investment and maintenance cost of DGs; C_{ploss} is the annual cost of DN with DGs (calculated in the same way as C_{loss}); C_{B} is the annual purchase cost of DN with DGs; and C_{sub} is the financial subsidy of new energy generation.

$$C_{\text{dgm}} = \sum_{i=1}^{n}\left(\frac{r(1+r)^t}{(1+r)^t - 1} \cdot C_{dgi} + C_{mi}\right) \cdot P_{dgi} \tag{6}$$

where r is the annual interest rate; t is the planning period; C_{dgi} is the ith distributed generation equipment investment cost ($/kWh); C_{mi} is the annual operation and maintenance cost ($/kWh) for the ith distributed generation; and P_{dgi} is the installed capacity (kW) of the ith distributed generation.

$$C_{\text{B}} = C_{\text{P}} \cdot \left(P_{\text{load}} - \sum_{i=1}^{n} \lambda_i \cdot S_{dgi}\right) \cdot T_{\max} \tag{7}$$

In order to reflect the benefits of DGs in environmental protection, the government has put forward policy support to DGs, that is, financial subsidies for distributed generation. It can be expressed as follows:

$$C_{\text{sub}} = C_{\text{sp}} \cdot \sum_{i=1}^{n} P_{dgi} \cdot T_{\max} \tag{8}$$

where C_{sp} is the subsidy amount (\$) of distributed generation unit.

2.   Minimize line losses

The existence of line losses will lead to line heating, which will accelerate the aging of insulated lines, will reduce the insulation of lines, and will ultimately lead to the risk of leakage. Reducing line losses can improve energy efficiency of electrical equipment or processes.

Lowering the line losses can improve the energy utilization efficiency of the energy-using equipment or process, and it is also one of the important measures of energy saving. The objective function can be expressed as follows [27]:

$$\min f_2 = \sum_{i,j \in n} g_{ij}\left(U_i^2 + U_j^2 - 2U_i U_j \cos \theta_{ij}\right) \tag{9}$$

where n is the total number of DN nodes; g_{ij} is the admittance of the branch (i, j); U_i and U_j are the voltage amplitudes of branches i and j respectively; and θ_{ij} is the voltage phase angle.

3.   Minimize the total voltage deviation

With the increase of load, the system voltage stability will deteriorate gradually, even voltage collapse, resulting in a system in danger. Therefore, the voltage deviation is one of the important indexes to evaluate the operation safety and power quality of the system. In addition, the increase of network node voltage can effectively reduce the reactive power loss of the system. The objective function can be expressed as follows:

$$\min f_3 = \sum_{i=0}^{N} \frac{\left|U_i - U_i^{specified}\right|}{U_i^{max} - U_i^{min}} \tag{10}$$

where U_i is the node ith voltage real value of the DN when DG is incorporated and $U_i^{specified}$ is the voltage rated value. This paper assumes that the rated voltage is 1 (p.u).

2.2. Influence of DG on Relay Protection of DN

DN is generally equipped with three-stage current protection. After the DG is connected to the DN, the system changes from single power supply to multiple power supply. When the transmission line is short-circuited, the magnitude and direction of the short-circuit current will change.

Figure 1 shows a typical simple DN. L_1 and L_2 represent feeders; T_1 represents a transformer; PD_1, PD_2, PD_3, and PD_4 are protective devices of the corresponding feeders; and DG is connected to the DN from node C.

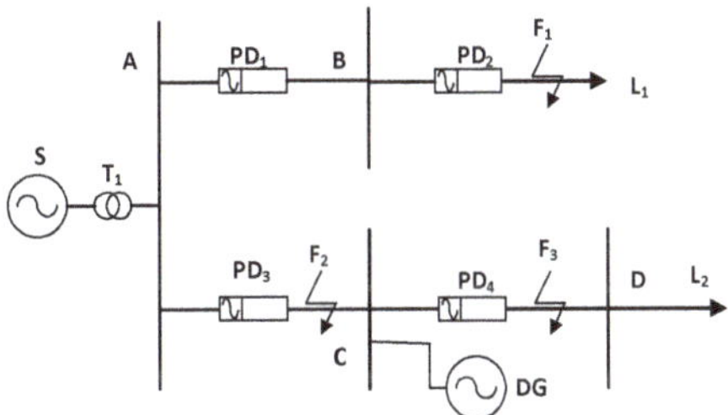

Figure 1. Distribution network system diagram of distributed generation (DG).

Assuming that the DG is connected, the short-circuit fault can be divided into the following three cases:

1.   The adjacent feeder connected to the DG is short-circuited

When the short-circuit fault F_1 occurs at the end of the adjacent feeder, the power source S and DG provide the overlapping short-circuit current to PD_1 and PD_2. Compared without DG connection, the short-circuit current increases and its value may be larger than the original I segment setting value of PD_1, which makes PD_1 malfunction, resulting in the loss of coordination between PD_1 and PD_2. Simultaneously, PD_3 will also flow through the reverse current provided by DG. When the capacity of DG is larger, the value of reverse current may be larger than the original I segment setting value of PD_3, resulting in PD_3 malfunction.

2. The upstream of DG access is short-circuited

When the DG upstream feeder AC is short-circuit fault F_2, PD_4 can operate normally. However, when PD_3 is activated, the island downstream of the DG will form an island operation. At the same time, the DG will always provide short-circuit current to the fault point, which will affect the reliability of the system.

3. The downstream of DG access is short-circuited

When DG downstream feeder CD short-circuit fault F_3 occurs, protections PD_1 and PD_2 can operate normally. Although the short-circuit current through PD_3 is only supplied by the system power S, the fault current will be smaller than before due to DG connection, which may cause the backup protection of PD_3 to refuse to operate. The short-circuit current through PD_4 is provided by the system power S and DG together. The increase of short-circuit current may increase the protection distance of PD_4, thus losing the selectivity.

When there are multiple DGs in the system, the analysis method is similar. From the above analysis, it can be seen that the access of DG will have adverse effects on the reliability of the relay protection and system of DN, so it is necessary to consider the restraint of relay protection when optimizing the configuration of DG. In the traditional three-stage current protection, the I-stage instantaneous current quick-break protection is the most important, so this paper focuses on the analysis of the impact of DG on the I-stage of the original current protection of the PD in the DN.

2.3. Constraint Condition

DG connected to a distribution network has great influence on power flow, voltage distribution, branch current, and power quality of the system. In order to design a reasonable programming model, the following equality constraints and inequality constraints are included in this paper.

1. Equality constraint

Below are the power flow equations constraints.

$$\begin{cases} P_{DGi} - P_{Li} - U_i \sum_{k \in i}^{N_1} U_k (G_{ik} \cos \theta_{ik} + B_{ik} \sin \theta_{ik}) = 0 \\ Q_{DGi} - Q_{Li} - U_i \sum_{k \in i}^{N_1} U_k (G_{ik} \sin \theta_{ik} + B_{ik} \cos \theta_{ik}) = 0 \end{cases} \tag{11}$$

where P_{DGi} and Q_{DGi} are the ith node active and reactive power of the DGs injected; P_{Li} and Q_{Li} are the ith node active and reactive powers of load output; U_k is the voltage amplitude of all kth nodes connected to ith nodes; G_{ik} is branch conductance; B_{ik} is the branch susceptance; θ_{ik} is the difference between the voltage angles of ith node and kth node; and N_1 is the number of branches associated with the ith node.

2. Inequality constraints

In order to make the planning scheme meet the power system operation standards, it is usually required that the node voltage, current, power, and permeability satisfy the constraints after DGs are connected to the grid.

a. Node voltage constraints

$$U_i^{\min} \le U_i \le U_i^{\max} \tag{12}$$

where U_i represents the voltage amplitude of the ith node, and $U_i^{\min}$ and $U_i^{\max}$ represent the upper and lower limits of the ith nodes voltage amplitude, respectively.

b. Branch current constraints

$$I_{ij} \le I_{ij}^{\max} \tag{13}$$

where I_{ij} is the current of branch (i, j) and $I_{ij}^{\max}$ is the maximum current allowed of branch (i, j).

c. Single DG access capacity constraints

$$\begin{cases} P_{DGi}^{\min} \le P_{DGi} \le P_{DGi}^{\max} \\ Q_{DGi}^{\min} \le Q_{DGi} \le Q_{DGi}^{\max} \end{cases} \tag{14}$$

where P_{DGi} and Q_{DGi} are the active power and reactive power of the DG connected to the ith node, respectively.

d. Distributed generation access total capacity constraints

$$\sum_i^N S_{DGi} \le 0.3 S_L \tag{15}$$

where $\sum S_{DGi}$ is the total capacity of distributed generation and S_L is the maximum load value of DN.

e. Short-circuit current constraints

$$\begin{cases} K_{sen,i}^{\mathrm{III}} = \dfrac{I_{k,i+1.\min}^{(2)}}{I_{set,i}^{\mathrm{III}}} \ge 1.2 \\ I_{set,i}^{\mathrm{I}} > I_{k,i.\max}^{(3)} \\ I_{set,i}^{\mathrm{I}} > I_{rk,i-1.\max}^{(3)} \\ I_{set,i}^{\mathrm{II}} > I_{rk,i-1.\max}^{(3)} \\ I_{set,i}^{\mathrm{III}} > I_{rk,i-1.\max}^{(3)} \\ I_{set,j}^{\mathrm{I}} > I_{k,j.\max}^{(3)} \end{cases} \tag{16}$$

where $K_{sen,i}^{\mathrm{III}}$ is the sensitivity coefficient of current III segment protection of branch i as backup protection; $I_{set,i}^{\mathrm{I}}$, $I_{set,i}^{\mathrm{II}}$, and $I_{set,i}^{\mathrm{III}}$ are the setting values of the current I, II, and III protections of branch i, respectively; $I_{set,j}^{\mathrm{I}}$ is the setting value of the current I segment protection of the branch j of the feeder line of the DG; $I_{k,i-1.\max}^{(3)}$ is the maximum reverse current through branch i when three-phase short circuit occurs at the end of the upstream of branch i; $I_{k,i.\max}^{(3)}$ and $I_{k,j.\max}^{(3)}$ are the maximum short-circuit currents through branch i of the terminal three-phase short circuit of branch i and branch j, respectively; and $I_{k,i+1.\max}^{(2)}$ is the minimum short-circuit current through branch i of the two-phase short circuit at the end of the downstream of branch i.

3. Overview formulation

From the above analysis, the multi-objective function and constraints can be described as follows:

$$\min\left[f_1(x_s, x_c), f_2(x_s, x_c), \cdots, f_{N_{obj}}(x_s, x_c) \right] \tag{17}$$

$$s.t. \quad h(x, x) = 0, i = 1, \cdots, p \tag{18}$$

$$g_i(x_s, x_c) \leq 0, \ i = 1, 2, \cdots, q \tag{19}$$

where f_i is the ith objective function, N_{obj} is the number of objective functions, x_s and x_c denote the state vector and the control vector respectively, h_i is the equation constraint, p is the number of equality constraints, g_i is the inequality constraint, and q is the number of inequality constraints.

x_c is composed of n variables, and n represents the number of nodes of the optimized network, where each variable represents two components: the DG installation placement and capacity. For example, a variable is assigned a value indicating the placement of the variable to installation DG. The value of the variable indicates the installation capacity. If DG is not installed, the corresponding variable value is 0. x_c can be illustrated as follows:

$$x_c = [(L_{DG1}, P_{DG1}), (L_{DG2}, P_{DG2}), \cdots, (L_{DGN}, P_{DGN})] \tag{20}$$

where L_{DGi} indicates that the ith node is the DG installation location and P_{DGi} represents the DG installation capacity of the ith node.

x_s is the network parameter that x_c is calculated from the power flow. Variables such as line losses, node voltage, and branch current are used to calculate multi-objective functions and to determine the satisfaction of constraints. x_s can be illustrated as follows:

$$x_s = [P', Q', U', I', \theta'] \tag{21}$$

where P', Q', U', θ', and I' represent the network feeder active power vector, reactive power vector, node voltage vector, voltage argument vector, and branch current vector respectively after DG is connected.

2.4. Establish PQ Mode of DG

The DG injects its power output into the grid through power electronics. Typically, for a PQ model, which shows the active power (P) versus reactive power (Q) called a PQ model. DG is modeled as a constant power factor and negative load model.

In this case, the DG is modeled as a constant power source. P_{DG} is the actual power output specified for the DG model. The load at ith node with the DG device is modified as Equations (22)–(23).

$$P'_{loadi} = P_{loadi} - P_{DGi} \tag{22}$$

$$Q_{DGi} = P_{DGi} \cdot \tan\left(\cos^{-1}(\varphi)\right) \tag{23}$$

where P_{loadi} is the original network load power of ith node, P_{DGi} and Q_{DGi} are the active and reactive power of DG at the ith node, P'_{loadi} is the active power after installing DG, and $\cos \varphi$ is the power factor.

In general, constrained problems can be solved using either deterministic or stochastic algorithms. However, deterministic approaches such as feasible direction and generalized gradient descent require strong mathematical properties of the objective function such as continuity and differentiability. In cases where these properties are absent, evolutionary computation, such as NSGA-II, offers reliable alternative methods [13].

3. Improved NSGA-II Algorithm

3.1. NSGA-II Algorithm Overview

NSGA-II was proposed by Deb, K. in 2002 [28], which is one of the most popular multi-objective optimization algorithms. It uses simulated binary crossover and polynomial variation to introduce non-dominated sorting and crowding distance operator instead of NSGA. The sharing radius of the algorithm ensures the diversity of the population and the uniformity of the Pareto-optimal set and introduces the elite retention strategy and the elimination strategy, which improves the operation speed and stability of the algorithm.

3.2. Dominant, Non-Dominated, Pareto-Optimal Set and Crowding Distance

NSGA-II introduces Pareto-optimal set and crowding distance to replace the fitness of traditional intelligent optimization algorithm. The solution set is divided into dominated solution set and non-dominated solution set according to the dominant relationship among the solutions. The rank of the solution is from 1 to n: the better the quality, the smaller the value. Solutions of the same rank are divided into pros and cons by comparing the size of the crowding distance.

Multi-objective optimization problems can be described as follow:

$$\min f_i(x), \quad i = 1, 2, \ldots, N_{obj}, \quad x \in I \tag{24}$$

where I is a feasible solution space.

Definition 1. *If solution x_1 dominates x_2, then the following definitions must be met.*

$$\forall i, j \in \left[1, 2, \cdots, N_{obj}\right], \\ \underset{i \neq j}{\exists} f_i(x_1) < f_i(x_2) \vee f_j(x_1) = f_j(x_2) \tag{25}$$

Definition 2. *If solution x_1 does not dominate x_2, then the following definitions must be met.*

$$\forall i, j \in \left[1, 2, \cdots, N_{obj}\right], \\ \underset{i \neq j}{\exists} f_i(x_1) \leq f_i(x_2) \wedge f_j(x_1) > f_j(x_2) \tag{26}$$

Equations (25) and (26) in Definitions 1 and 2 are derived from Reference [29].

Definition 3. *Pareto-optimal set is composed of mutually non-dominated solution sets, and ranks are composed of mutually dominated solution sets.*

Definition 4. *Assume that $P = \{x_1, x_2, \cdots, x_n\}$ is a Pareto-optimal set and that the crowding distance represents the distribution density of other solution sets around the solution. The larger the solution distance at the same layer, the better the solution distribution, that is, the better the solution set diversity. Its calculation formula is as follows:*

$$cd_i^k = \begin{cases} \dfrac{f_{i+1}^k - f_{i-1}^k}{f_{max}^k - f_{min}^k}, & if \quad index\left(x_i^k\right) \in [2, n-1] \\ \infty, & otherwise \end{cases} \tag{27}$$

$$cd_i = \sum_{k=1}^{m} cd_i^k \tag{28}$$

where cd_i^k is the crowding distance on the kth objective function of x_i, m is the number of objective functions, index (x_i^k) is the sort index of x_i on the kth objective function, f_{i+1}^k is the value of objective function corresponding to the last solution on the axis of the kth objective function, and f_{max}^k and f_{min}^k represent the maximum and minimum values of the kth objective function, respectively.

3.3. INSGA-II: Improved Mutation Operator

Because mutations in genetics often lead to worse results, the probability of mutation operations in genetic algorithm (GA) is usually set to be very small. In GA, crossover and mutation operations have two functions: global search and local search. When dealing with low-dimensional convex problems, only the crossover operator can solve these problems well. However, when dealing with high-dimensional nonconvex problems, it often falls into the local optimal solution. For this reason, the mutation operator is improved by referring to the population-updating method in the fireworks algorithm [25]; the new mutation individual is created by Equations (29)–(31).

$$d_i = vd \cdot rand \tag{29}$$

$$S = randperm(1, d_i) \tag{30}$$

$$u'_j = \begin{cases} u_j^{mu}, if \ j \in S \\ u_j, otherwise \end{cases} \tag{31}$$

where d_i denotes the dimension of the ith solution for mutation, vd denotes the variable dimension, *rand* denotes the random number between [0, 1], S denotes the index for performing the mutation operation, u_j^{mu} denotes the jth gene performing the mutation operation, and u'_j represents the value of the jth gene after the mutation operation.

Crossover and mutation operations often produce solutions beyond the feasible space. The general method is to assign solutions less than the lower limit of feasible region to the lower limit, and the solution that exceeds the upper limit of the feasible region is given the upper limit value. This operation is simple to perform but reduces the diversity of solution sets. According to Equation (32), solutions beyond the boundary can be effectively mapped back to the feasible space.

$$\hat{u}_i = u_i^{min} + |u_i|\% \left(u_i^{max} - u_i^{min} \right) \tag{32}$$

where u_i^{max} and u_i^{min} represent the upper limit and lower limit of the gene, respectively, and % is the remainder operation.

3.4. INSGA-II: Violation Constrained Index

In view of the subjectivity of penalty factor selection, the violation constrained index (VCI) is proposed when dealing with infeasible solutions.

The method divides the solution set into feasible solution and infeasible solution by calculating the error degree (ED) of the constraint condition of the solution.

The VCI calculation formula is as follows:

$$ED(x_s, x_c) = \left| h(x_s, x_c) \right| + \max(g(x_s, x_c), 0) \tag{33}$$

$$VCI_i = \sum_j^C \frac{ED_j^i - ED_j^{min}}{ED_j^{max} - ED_j^{min}} \tag{34}$$

where ED_j^{max} and ED_j^{min} represent the maximum and minimum error degree of infeasibility under the jth constraint, respectively, and C indicates the number of constraints. The classification process of the solution set is performed according to Equation (35).

$$S\_t = \begin{cases} 0, if \ all \ VCI_i = 0 \\ 1, if \ all \ VCI_i \ != \ 0 \\ \frac{1}{2}, otherwise \end{cases} \tag{35}$$

where S_t denotes the type of the solution set. The default index can be effectively used in combination with the non-dominated sorting strategy. For example, when the S_t is 0 or 1/2, the feasible solution part is given the true rank and the crowding distance, and the rank of the infeasible solution is sorted in descending order according to VCI_i and continues to be sorted from the feasible solution.

3.5. INSGA-II: Improved Crowding Distance Operator

Figure 2 shows that, after solutions a–h pass the truncation strategy, a, d, e, g and h are preserved but the eliminated solution c has a better crowding distance than the solution d, which is due to the traditional crowding distance information being missing, and the truncation strategy eliminates the solution set with small crowding distances at one time, resulting in uneven distribution of the final set. The calculation of the traditional crowding distance only considers the spatial distribution of two adjacent solutions on the axis of the objective function without considering that the effect of deleting the neighborhood solution leads to the elimination of potential high-quality solutions. Finally, a potential crowding distance is proposed. The potential crowding distance in the two-dimensional space is simple to calculate. As shown in Figure 2a, when the solution map is re-target functions 1 and 2, the neighborhoods are the same in the solution set and only the order changes. The truncation process is shown in Figure 2b,c:

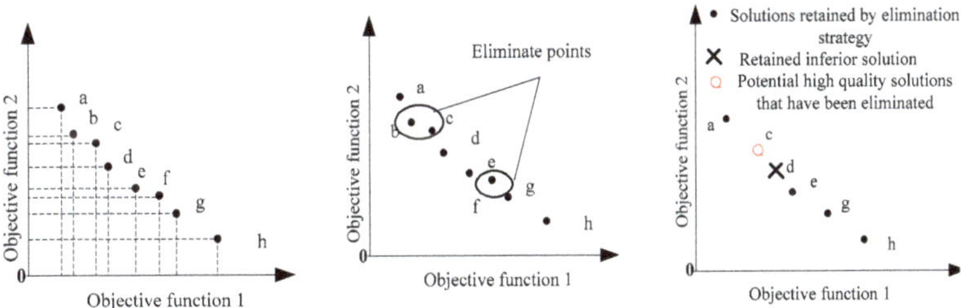

(**a**) Individual distribution chart   (**b**) Eliminating points chart   (**c**) Truncation process chart

Figure 2. Pareto-optimal set of truncation strategy.

Therefore, the formula for calculating the potential crowding distance in two-dimensional space is as follows:

$$\begin{cases} \Delta d_i^1 = \dfrac{f_{i-1}^1 - f_{i-2}^1}{f_{max}^1 - f_{min}^1} + \dfrac{f_{i+2}^2 - f_{i+1}^2}{f_{max}^2 - f_{min}^2}, \\[2mm] \Delta d_i^2 = \dfrac{f_{i+2}^1 - f_{i+1}^1}{f_{max}^1 - f_{min}^1} + \dfrac{f_{i-1}^2 - f_{i-2}^2}{f_{max}^2 - f_{min}^2}, \\[2mm] pd_i = cd_i + max\left(\Delta d_i^1, \Delta d_i^2\right) \end{cases} \tag{36}$$

where Δd_i^1 and Δd_i^2 denote the increments of the crowding distance after the adjacent solution of the kth solution is deleted, f_{i-1}^1 and f_{i-1}^2 denote the neighborhood solutions mapped to the kth solution on the objective function 1, and pd_i denotes the potential crowding distance of the kth solution.

The objective function above three-dimensions does not have the symmetric relationship of the two-dimensional objective function. Each solution no longer has only the same two adjacent solution sets on each objective function axis but has (2–n) different solutions. Therefore, the potential crowding distance of the solution under the 3D objective function is calculated as follows:

$$pd_i = cd_i + max(k\_\Delta d_i), k \in \Phi \tag{37}$$

$$k\Delta d_i = \sum_{j=1}^{m} 1^j \left[1^l \left(\frac{f_{i-1}^j - f_{i-2}^j}{f_{max}^j - f_{min}^j} \right) + 1^r \left(\frac{f_{i+2}^j - f_{i+1}^j}{f_{max}^j - f_{min}^j} \right) \right] \tag{38}$$

where Φ denotes the set of adjacent fields mapped to the ith solution on each objective function axis; m denotes the number of objective functions; $k\Delta d_i$ denotes the increment generated by the crowding distance of the ith solution after the kth solution in the neighborhood set is eliminated; and 1^j, 1^l, and 1^r respectively indicate that the kth solution belongs to the neighborhood of the ith solution or belongs to the left neighborhood solution or the right neighborhood solution on the jth objective function.

3.6. INSGA-II: Improved Crowding Distance

NSGA-II uses the tournament selection operator for population reproduction, and the selection criteria considers the rank and crowding distance. The specific selection process is described in Reference [27]. Since the improved NSGA-II algorithm increases the potential crowding distance, it needs to involve the crowding distance for the selection operation. After some modifications, the selection rules are as follows:

$$newx_k = \begin{cases} x_i, if \ cd_i > cd_j \ and \ pd_i > pd_j \\ x_j, if \ cd_j > cd_i \ and \ pd_j > pd_i \\ x_i \ or \ x_j, otherwise \end{cases} \tag{39}$$

where $newx_k$ denotes the selected kth descendant solution, and x_i and x_j respectively denote two different solutions in the parent. Improved selection strategy comprehensively considers the current crowding distance and potential crowding distance. Therefore, it can select a better solution and can avoid the defects of insufficient diversity of the solution set when the truncation strategy is executed.

3.7. Unbiased Compromise Strategy

In order to solve the cumbersome scheme of the Pareto-optimal set, this paper uses an unbiased compromise strategy based on fuzzy set, which uses unbiased member parameter ω to evaluate the quality of each individual in the Pareto-optimal set. The calculation formula is as follows:

$$\omega^* = \max_{i=1,\cdots,n} \left(\frac{\sum\limits_{j=1}^{m} \omega_j^i}{\sum\limits_{i=1}^{n} \sum\limits_{j=1}^{m} \omega_j^i} \right) \tag{40}$$

where ω_j^i denotes the unbiased parameter of the ith solution in the Pareto-optimal set on the jth objective function; f_j^{min} and f_j^{max} denote the minimum and maximum values of the jth objective function corresponding to the Pareto-optimal set, respectively; f_j^i denotes the value of the jth objective function corresponding to the ith solution in the Pareto-optimal set; n denotes the number of objective functions; m denotes the number of solutions in the Pareto-optimal set; and ω^* denotes the unbiased member parameters of the comprehensive optimal solution.

3.8. Optimized Configuration Process Based on Improved NSGA-II

The proposed algorithm optimizes the placement and sizing process configuration of DG, as follows:

Algorithm 1. The algorithm of the proposed INSGA-II

Input: Set IEEE-33 distribution network-related parameters, the number of target functions, the number of variables, the size of the population, the maximum number of iterations, etc.
Outputs: Comprehensive optimal solution x^*.
1: Initialization population $P(x_c)_t$, set the number of iterations t = 0;
2: $P(x_s)_t$ is calculated by forward and backward substitution method, and then, the objective function values f_1, f_2, f_3, and S_t are calculated;
3: Algorithm iteration start, while $t < t_{max}$ do;
4: The solution set is classified by S_t, feasible solution to perform non-dominated sorting strategy, and the infeasible solution to calculate DCI and sorted;
5: Generation of children $Q(x_c)_t$ from the parent $P(x_c)_t$ by improving selection operations and improving genetic manipulation;
6: $Q(x_c)_t$ performs power flow calculation, mixes with $P(x_c)_t$ to form a mixed population $R(x_c)_t$ and $R(x_c)_t$ for non-dominated sorting strategy, and calculates the crowding distance according to Equations (37) and (38);
7: Selecting a new parent population $R(x_c)_t$ of size from $P(x_c)_{t+1}$ by a truncation strategy;
8: $t = t + 1$;
9: End while;
10: Using unbiased compromise strategy to select comprehensive optimal solution x^*;
11: Return x^*.

---

4. Experiment and Results

4.1. Experiment Setup and Description

In order to better verify the effectiveness of the proposed algorithm, IEEE 33-, 69-, and 118-bus systems are introduced as simulation examples to compare performance with current mainstream multi-objective optimization algorithms, considering the dynamic and static characteristics of the DGs and DN. Hardware parameters used in the experiment are as follows: Intel (R) Core (TM) i5-3337 CPU 1.80 GHz, memory: 4.00 GB, and simulation software: Matlab-2014a (The MathWorks, Inc.3 Apple Hill Drive, Natick, MA, USA, 2014).

4.2. IEEE 33-Bus System Case Simulation Experiment

An IEEE 33-bus system topology diagram is shown in Figure 3, where the system includes 33 nodes and 32 branches [30]. The 0 node is assumed to be a balanced node with a voltage reference of 12.66 kV, total power load of 3.17 MW, and reactive power 2.30 MVAr.

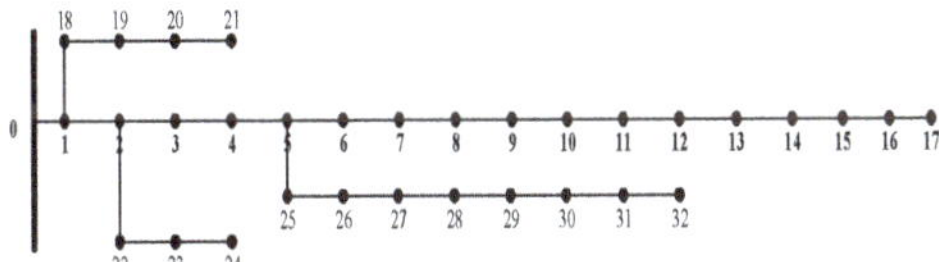

Figure 3. IEEE 33-bus system topology diagram.

Multi-objective function parameter setting: The maximum installation capacity of each node of DG is 1 MW, the annual maximum load utilization time and the annual maximum line loss hour of DGs are 4500 h, the investment cost of micro-turbine (MT) unit equipment is 0.07 million $/kW, annual operation and maintenance cost is 0.009 $/kWh, wind turbine generator (WTG) group unit equipment investment cost is 0.114 million $/kW, annual operation and maintenance costs are 0.004 $/kWh, MT and WTG unified as a PQ model for processing, power factor is 0.9, the unit price of the loss is 0.071 $/kWh, the financial subsidy per unit of electricity is 0.019 $/kWh, and the conversion factor of the annual equipment investment cost of the distributed power source is obtained by 3% of the annual interest rate. We use the algorithm of the proposed INSGA-II for the following series of analysis.

4.2.1. Analysis of Optimization Results with Different Parameters

As shown in Table 1, in order to improve the operation speed of the algorithm and the accuracy of the solution, experimental comparisons of the experimental hyperparameters are made. Compared with experiment 1 and experiment 2, the optimization results are improved with the increase of iteration times.

Table 1. Simulation results of INSGA-II with different parameters.

Experiment	Population Size (NP)	Iterations (tmax)	Time (s)	Energy-Saving Benefit (Million $)	Voltage Deviation (p.u)	Line Loss (MW)
1	50	100	26.71	0.193	5.289	0.081
2	50	200	52.46	0.196	5.273	0.084
3	100	100	83.67	0.197	5.254	0.082
4	200	100	107.91	0.199	5.267	0.081

Comparing experiments 1, 3, and 4, the optimization effects are improved with the increase of population size but the time increases. The population selection of 100 and the number of iterations of 100 are taken as the hyperparameters of the algorithm.

Through large numbers of experiments, subsequent experiments decided to set improved NSGA-II parameters: population size is 100, number of iterations is 100, and crossover probability is 0.7.

4.2.2. Analysis of DG Installation Capacity Optimization Results

The DG optimization configuration results of the improved NSGA-II are shown in Table 2.

Table 2. INSGA-II optimization results.

Case	DG Location (Bus Number)	DG Sizes (MW)	Energy-Saving Benefit (Million $)	Voltage Deviation (p.u)	Line Loss (MW)
1	-	-	0	19.6024	0.2017
2	5/6/17/32	0.2379/0.2391/0.6965/0.4451	0.1976	5.2453	0.0797
3	6/17/24/32	0.2148/0.5013/0.2062/0.4126	0.1781	4.9348	0.0613

As shown in Table 2, Case-1 has no DG, Case-2 is installation without short-circuit current constraints (other constraints are considered), and Case-3 is installation with short-circuit current constraints (other constraints are considered). Compared with Case-1, the planning of Case-2 and -3 obtained by INSGA-II can effectively improve the objective function. Case-2 can produce energy-saving benefits of about $0.197 million, the voltage deviation is improved by 73.24%, and the line loss is reduced by 69.60%. Case-3 can produce energy-saving benefits of about $0.193 million, the voltage deviation is improved by 74.83%, and the line loss is reduced by 60.49%. Comparison of Case-2 and -3 show that, under the influence of short-circuit current protection constraints, the capacity of DGs decrease and voltage deviation and line losses are improved. In addition, when the DGs are close, the short-circuit current of some branches will be very high, so the Case-3 optimization results do not show that the DG access points are close. The above results confirm the rationality of considering short-circuit current constraints.

As shown in Figure 4a,b, Case-2 and -3 can improve the voltage amplitude of each node and line losses by installing DG appropriately. Because most of the installation nodes of Case-3 are terminal nodes of the system, the supporting effect of the node voltage is stronger [31]. Active network loss is usually determined by node voltage and branch resistance. Distribution network voltage level is lower and R/X value is larger, so it will lead to larger network losses. When DG is reasonably connected, the voltage fading will be improved [31]. In the case of the conditions, Case-3 has a greater effect on node voltage support than Case-2, so the line losses are better.

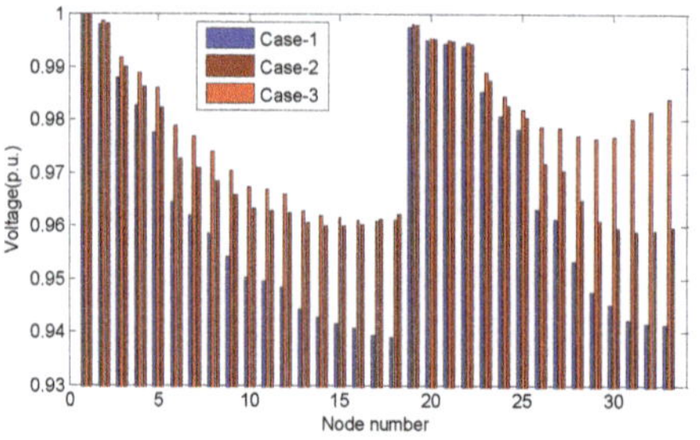

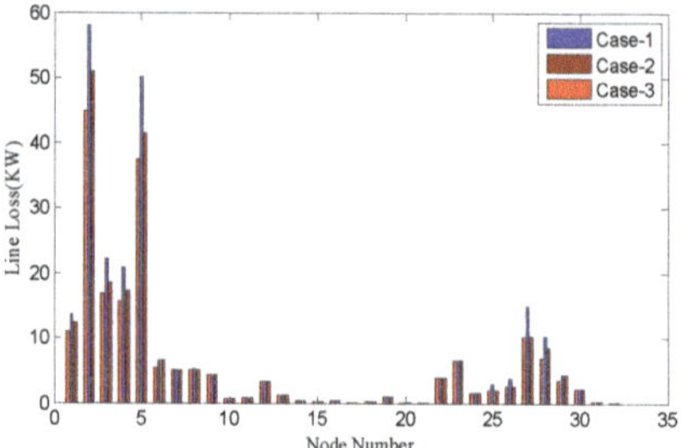

(**a**) The voltage amplitude of each node　　　　(**b**) The line loss of each node

Figure 4. Results of nodal voltage optimization.

4.3. Performance Analysis of INSGA-II Algorithm

In order to verify the optimization performance of the proposed algorithm, INSGA-II, NSGA-II, and MOPSO are used to analyze and compare the planning results of the IEEE 33-bus system independently. The parameters of NSGA-II are set as INSGA-II, the inertia factor is 0.8, the local speed factor is 0.1, and the global speed factor is 0.1; other parameters are the same as INSGA-II.

As shown in Figure 5, NSGA-II and MOPSO have uneven distribution of Pareto solutions due to the defects of crowding distance operators. MPSO optimizes by choosing leaders and by archiving mechanism. Similar to NSGA-II, its selection process also depends on a dominant relationship, so crowding distance is also needed to participate in it. When solving the high-dimensional nonconvex problem, the solution set is easy to fall into the local optimal solution and the population diversity is insufficient. Obviously, the uniformity of Pareto-optimal set obtained by INSGA-II is the best of the three algorithms, which verifies the validity of the potential crowding distance and improves the mutation operator to improve the optimization ability of the algorithm.

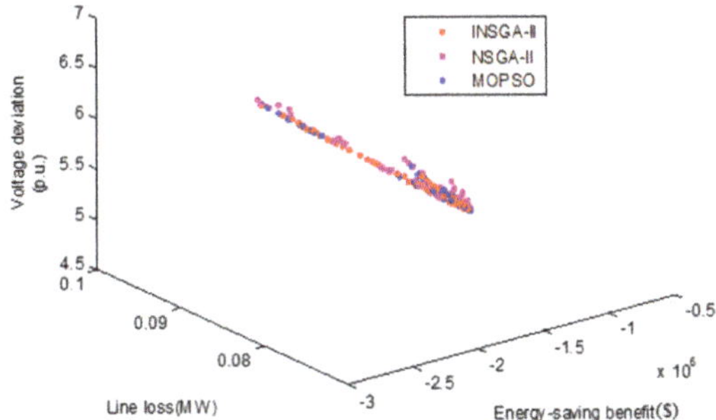

Figure 5. Pareto-optimal set of three algorithms.

In order to verify the performance of the improved algorithm more intuitively, the convergence metric [29] and the spacing metric [32] are used to compare the three optimization algorithms.

(1)   C_mertic

Convergence metric can be expressed as follows:

$$I_C(X', X'') = \frac{|\{a'' \in X''; \exists a' \in X' : a' \prec a''\}|}{|X''|} \tag{41}$$

where X' and X'' are respectively two solution sets, a' and a'' are solutions belonging to two solution sets, p is the dominant symbol, and $|X''|$ is the number of solutions in the solution set X''. If $I_C(X', X'')$ is 1, all solutions in the solution set X' dominate the solution in the solution set X'', and if $I_C(X', X'')$ is equal to 0, all solutions in the solution set X'' are not dominated by the solution set X'.

As shown in Table 3, about 32.39% of NSGA-II and 33.04% of MOPSO are dominated by INSGA-II. In contrast, only about 6.41% and 7.92% of INSGA-II are dominated by NSGA-II and MOPSO, respectively. In addition, the computational complexity O (mn³) is similar, so the running time of the three algorithms is almost the same, which proves that the improved crowding distance operator can effectively improve the quality of the solution set and can ensure operation efficiency.

Table 3. Convergence metric of each algorithm.

Algorithm	NSGA-II	INSGA-II	MOPSO	Time(s)
NSGA-II	–	6.41%	11.21%	52.87
INSGA-II	32.39%	–	33.04%	53.04
MOPSO	15.73%	7.92%	–	53.12

(2)   S_mertic

Spacing metric [27] can be expressed as follows:

$$I_S = \sqrt{\frac{1}{N-1}\sum_{i=1}^{N}\left(\bar{d}-d_i\right)^2} \tag{42}$$

where N represents the number of solutions, d_i represents the shortest distance from the ith individual to the rest of the solution, and $\bar{d}$ represents the mean of all individuals d_i. A smaller spacing metric means that the solution distribution in the Pareto solution is more uniform. The zero value of the interval metric means that all solutions in the Pareto-optimal set are equally spaced.

As shown in Figure 6, in order to visually describe the uniformity of the distribution of the solution sets of different algorithms, the diversity indices obtained by each algorithm running 30 times independently are represented by box diagrams, in which each box represents the distribution of the measurement values of the distance between the solution sets of different algorithms. The upper quartile line at the top of each box diagram and the lower quartile line at the bottom represent the boundary values except outliers. If there is an abnormal value, use "+" to identify it. The median value is the red line in the rectangle box. Compared with the other two algorithms, INSGA-II has smaller median values (see red line in the middle) and minimum values (see quartile line below). Therefore, it is further explained that the distribution of INSGA-II solutions is more uniform and has better diversity.

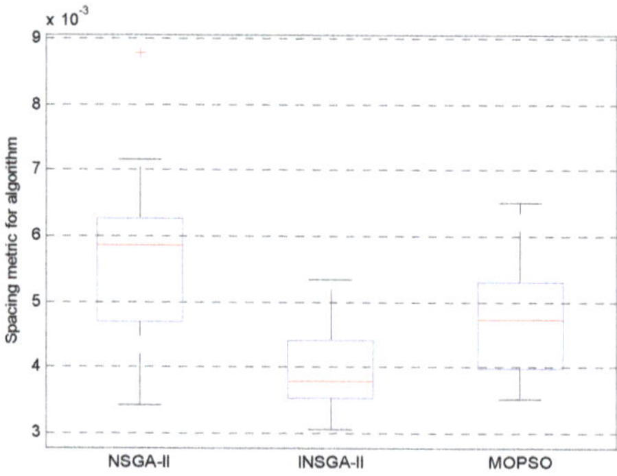

Figure 6. Three algorithms' spacing metric.

(3)　Analysis of the relationship between objective functions

As shown in Figure 7a, the energy-saving benefit has a nonlinear relationship with line loss, and the degree of line loss improvement will have an extreme point, which will not decrease with the increase of DG capacity but will be damaged. In view of Figure 7b, the voltage increases with the increase of energy-saving benefits. The results of DG optimization show that the capacity of DG-mounted nodes is too large to reduce the total voltage deviation correction effect. Therefore, the energy-saving benefits have a linear relationship with the voltage deviation when the permeability of DG is satisfied. Line loss and voltage deviation also show a nonlinear relationship in Figure 7c, and the two restrict each other.

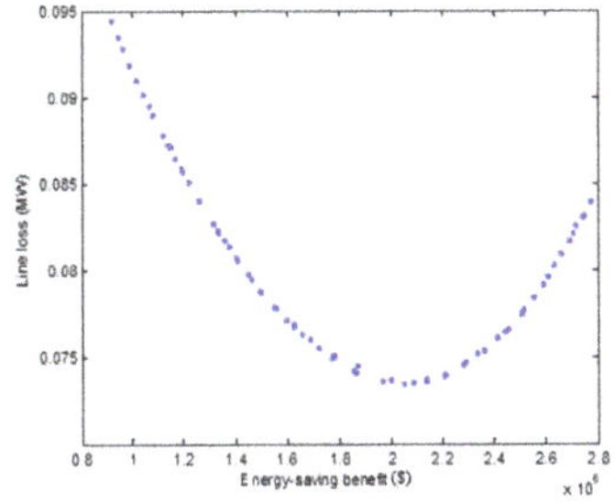

(**a**) The energy-saving benefit relationship with line loss

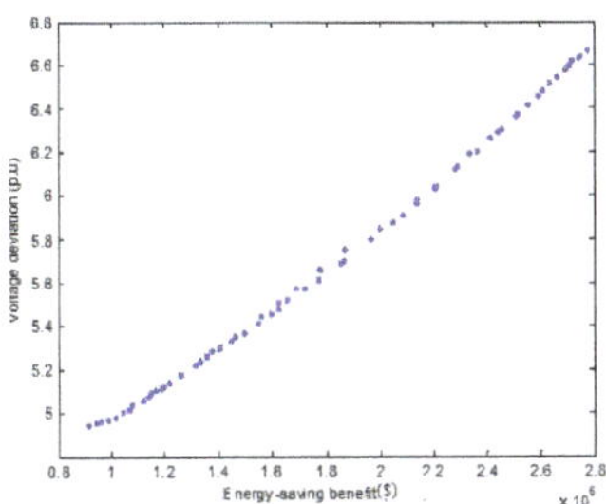

(**b**) The energy-saving benefit relationship with voltage deviation

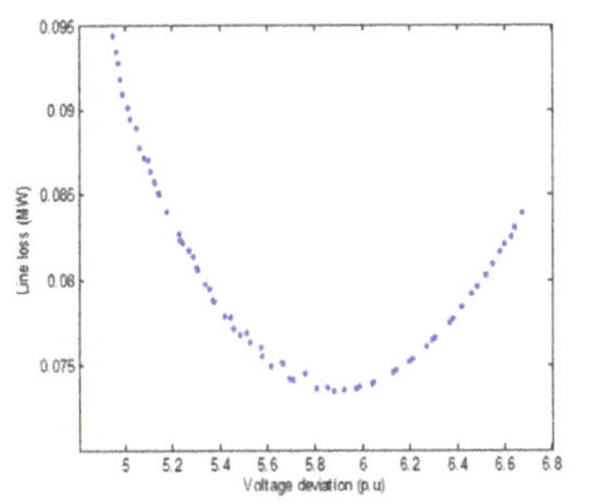

(**c**) The line loss relationship with voltage deviation

Figure 7. Relationship between different objective functions.

4.4. Case Analyses of IEEE 33-, 69-, and 118-Bus Systems

In order to prove the improvement degree of the NSGA-II algorithm on network optimization results of different node systems, the IEEE 33-, 69- and 118-bus systems are introduced. The parameters of the IEEE 69- and 118-bus systems are described in References [33,34].

Table 4 shows the optimization results of the INSGA-II algorithm on three examples. For the objective function, with the increase of network complexity, the improvement degree of the objective function of the three DNs increases. For example, IEEE 33-bus, 69-bus, and 118-bus are improved by 74.38%, 77.39%, and 77.32%, respectively, in terms of voltage deviation. On the hand, it also proves the importance of reasonable installation of DG. On the other hand, most of the installation nodes are the end of the system, which also proves that DGs can improve the voltage deviation better than other nodes.

Table 4. Results of different network optimization.

Test System	DG Location (Bus Number)	DG Sizes (MW)	Optimal Objectives			
			Optimization	Energy-Saving Benefit (Million $)	Voltage Deviation (p.u)	Line Loss (MW)
33-bus	6/17/24/32	0.2148/0.5013/0.2062/0.4126	Before/	0	19.6024	0.2017
			After	0.1781	4.9348	0.0797
69-bus	27/35/39/56/52	0.4266/0.2618/0.3446/0.5426/0.5698	Before/	0	31.6747	0.2206
			After	0.4876	7.1614	0.0609
118-bus	3/17/27/18/80/116	0.7787/0.4544/0.7240/0.6483/0.1027/0.2698	Before/	0	54.3018	0.6427
			After	0.8161	12.3112	0.10272

4.5. Experiments on Load and DG Output Considering Annual Time Series Changes

With the increase of installations and capacity of DG in the future, considering only static load and DG operation status, it may not be applicable to other load levels of permeability constraints and may even lead to reverse power flow, voltage limit, original system protection measure failures, and other phenomena, so that the system is threatened.

To this end, this section optimizes the configuration scheme to consider the DG annual output change and load change trend and the revised IEEE 33-bus system-related data reference [34]. The data was recorded every hour, so it is assumed that the DG output is equal every hour and that the load demand is constant. The four-season output level and load demand of wind turbines are shown in Figure 8a,b.

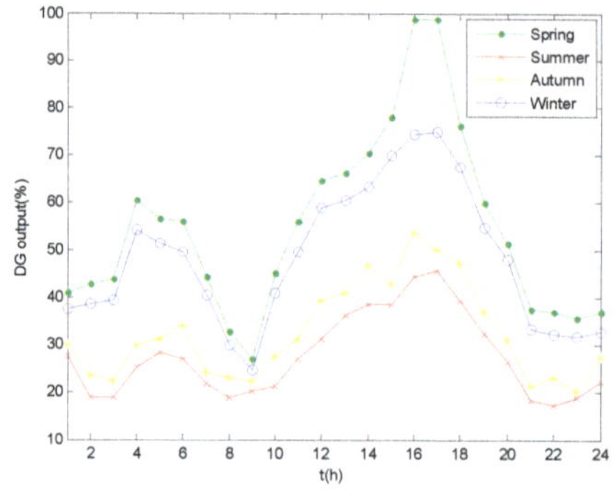

(**a**) The four-season output level

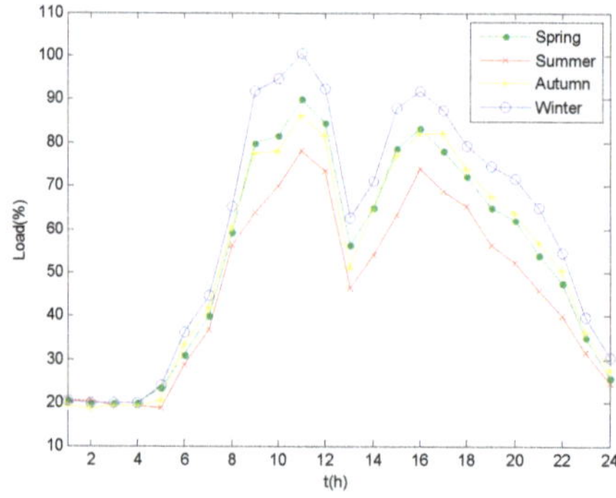

(**b**) The four-season load demand

Figure 8. Time series data of wind turbine output and load.

As seen in Figure 8a,b, the wind speed is higher in spring and winter and the output of wind turbine is also improved. On the contrary, the output of wind turbine in summer and autumn is small and the seasonal data of the load has changed but the overall change is not large. On the one hand, if the optimization is based on the summer and autumn season time data, it may lead to waste of DG capacity installation. On the other hand, optimizing the configuration in spring and winter will destroy the network penetration rate and cause economic and security crisis. Here, we choose to optimize the four seasons separately and consider them comprehensively. The four seasonal DG optimization results using INSGA-II are shown in Table 5.

Table 5. Four quarterly independent optimization results.

Seasonal Condition	DG Location (Bus)	DG Sizes (MW)	Optimal Objectives			
			Optimization	Energy-Saving Benefit (Million $)	Voltage Deviation (p.u)	Line Loss (MW)
Spring	-	-	Before/	0	471.6464	1.5656
	6/17/24/32	0.2148/0.5013/0.2062/0.4126	After	0.0056	121.3572	1.5326
Summer	-	-	Berfore/	0	472.3140	0.3733
	7/17/24/32	0.1962/0.6129/0.1135/0.4544	After	0.0065	67.9313	0.3599
Autumn	-	-	Before/	0	472.2005	0.5149
	6/17/24/32	0.1308/0.5136/0.1012/0.4227	After	0.0067	83.6967	0.5019
Winter	-	-	Before/	0	471.8158	1.1628
	6/17/24/32	0.2029/0.5364/0.1910/0.4313	After	0.0583	119.9812	1.1352

As shown in Table 5, due to the relatively stable output of the wind turbines in summer and autumn, it is possible to seek a relatively high-quality configuration result. The spring and winter seasons are relatively unsatisfactory because of the large fluctuations in output. The combined installation capacity of the dynamic optimization scheme should meet the DG permeability constraint of the remaining time period. In each season, it is quite difficult for a DG configuration scheme to improve the optimization function while satisfying the different constraints of 24 h. Therefore, the installation capacity of each season is almost close to the maximum installation capacity, which is a compromise that is comprehensively considered to optimize the time-series data of each objective function. In order to meet the annual operating constraints, the combination of the smallest installation capacity on each quarter node is selected as the planning scheme.

5. Conclusions

Considering the influence of DG access on the economy, security, and reliability of DN, this paper establishes a multi-objective optimization model to minimize the line losses and voltage deviation and to maximize the annual energy benefit. For DN with integrated DG units, equality constraints and inequality constraints, which include power-flow equality, nodal voltage, branch current, DG capacity, and short-circuit current, are considered in the optimization model. The principle of installation of DG is summarized through experiments and power system theory, and the violation constraint indicators are proposed to solve the infeasible solution.

The mutation operator, crowding distance operator, and selection operator of traditional NSGA II are improved to increase the population diversity and consistency in the optimal allocation. The introduction of an unbiased, compromised strategy can quickly give decision makers a comprehensive plan to weigh each goal.

The different examples of the IEEE 33-, 69-, and 118-bus systems are analyzed and optimized, and different super parameters are compared experimentally. The performance of the proposed INSGA-II is verified by comparing NSGA-II and MOPSO algorithms. The IEEE 69- and 118-bus systems are selected as the verification case. The proposed algorithm is applied to solve the static and dynamic characteristics of the DN with the integrated DGs. Finally, the IEEE 33-bus system is modified and optimized by using the four season data of wind turbine and load. The results indicate that the proposed method can achieve better precision and diversity and can provide a good configuration plan for decision makers under the premise of meeting the annual penetration rate.

In practice, the choice of the best site may not always be feasible due to many reality constraints. However, the optimization and analysis here suggest that considering multi-objectives helps to decide siting and sizing of DG units for the decision maker. The optimization planning of a multi-distributed generation access distribution system combined with geographic information system (GIS) technology will be further investigated.

Author Contributions: Conceptualization, W.L. and F.L.; methodology, W.L.; software, F.L.; validation, Y.L., F.L., and W.D.; formal analysis, W.L.; investigation, F.L.; resources, F.L.; data curation, W.D.; writing—original draft preparation, F.L.; writing—review and editing, W.L. and Y.L.; visualization, W.D.; supervision, W.L.; project administration, W.L.; funding acquisition, W.L.

Funding: This research was funded by Natural Science Foundation of Heilongjiang Province, China, grant number E201332.

Acknowledgments: The authors would like to thank the editors and the reviewers for their constructive comments and suggestions.

Conflicts of Interest: The authors declare no conflict of interest. The funders had no role in the design of the study; in the collection, analyses, or interpretation of data; in the writing of the manuscript; or in the decision to publish the results.

Abbreviations

Z_{cost}^{NODG}	total annual cost of DN without DGs
Z_{cost}^{NODG}	total annual cost of DN with DGs
C_{loss}	annual loss of DN without DGs
C_b	annual purchase cost of DN without DGs.
C_p	unit price of electricity consumed per unit
τ_{max}	annual maximum load loss hour (h) of the DN
ΔP_j	active power loss (kW) of the jth branch
P_{load}	DN total active load
P'_{loadi}	the active power after installing DG
T_{max}	DN annual maximum load utilization hours
C_{dgm}	annual investment and maintenance cost of DGs
C_{ploss}	annual cost of DN with DGs
C_B	annual purchase cost of DN with DGs
C_{sub}	financial subsidy of new energy generation
C_{dgi}	distributed generation equipment investment cost ($/kWh)
C_{mi}	annual operation and maintenance cost ($/kWh) for the ith distributed generation
P_{dgi}	installed capacity (kW) of the ith distributed generation
C_{sp}	subsidy amount ($) of distributed generation unit
g_{ij}	admittance of the branch (i, j)
$U_i^{specified}$	voltage-rated value
$P_{DGi}Q_{DGi}$	ith node active and reactive power of the DGs injected
$P_{Li}Q_{Li}$	ith node active and reactive powers of load output
B_{ik}	the branch susceptance
ΣS_{DGi}	the total capacity of distributed generation
S_L	maximum load value of DN
$K_{sen,i}^{III}$	the sensitivity coefficient of current III segment protection of branch i as backup protection
$I_{set,i}^{I}$	the setting values of the current I protection of the branch i, respectively
$I_{k,i-1.max}^{(3)}$	the maximum reverse current through branch i
L_{DGi}	the ith node is the DG installation location
cd_i^k	the crowding distance on the kth objective function of x_i
f_{i+1}^k	the value of objective function corresponding to the last solution on the axis of the kth objective function

ED_j^{max} the maximum error degree of infeasibility under the jth constraint

S_t the type of the solution set

$newx_k$ the selected kth descendant solution

ω_j^i the unbiased parameter of the ith solution in the Pareto-optimal set on the jth objective function

References

1. Costa, P.M.; Matos, M.A. Loss allocation in distribution networks with embedded generation. *IEEE Trans. Power Syst.* **2004**, *19*, 384–389. [CrossRef]
2. Alvehag, K.; Awodele, K. Impact of Reward and Penalty Scheme on the Incentives for Distribution System Reliability. *IEEE Trans. Power Syst.* **2014**, *29*, 386–394. [CrossRef]
3. Delfanti, M.; Falabrett, D.; Merlo, M. Dispersed generation impact on distribution network losses. *Electric Power Syst. Res.* **2013**, *97*, 10–18. [CrossRef]
4. Shao, C.; Wang, X.; Shahidehpour, M.; Wang, X. Partial Decomposition for Distributed Electric Vehicle Charging Control Considering Electric Power Grid Congestion. *IEEE Trans. Smart Grid* **2017**, *8*, 75–83. [CrossRef]
5. Hakimi, S.M.; Moghaddas-Tafreshi, S.M. Optimal Planning of a Smart Microgrid Including Demand Response and Intermittent Renewable Energy Resources. *IEEE Trans. Smart Grid* **2014**, *5*, 2889–2900. [CrossRef]
6. Aghaei, J.; Muttaqi, K.M.; Azizivahed, A.; Gitizadeh, M. Distribution expansion planning considering reliability and security of energy using modified PSO (Particle Swarm Optimization) algorithm. *Energy* **2014**, *65*, 398–411. [CrossRef]
7. Esau, Z.; Jayaweera, D. Reliability assessment in active distribution networks with detailed effects of PV systems. *J. Mod. Power Syst. Clean Energy* **2014**, *2*, 59–68. [CrossRef]
8. Ehsan, A.; Yang, Q. Optimal integration and planning of renewable distributed generation in the power distribution networks: A review of analytical techniques. *Appl. Energy* **2018**, *210*, 44–59. [CrossRef]
9. Duong, M.Q.; Pham, T.D.; Nguyen, T.T.; Doan, A.T.; Tran, H.V. Determination of Optimal Location and Sizing of Solar Photovoltaic Distribution. *Energies* **2019**, *12*, 174. [CrossRef]
10. Gkaidatzis, P.A.; Bouhouras, A.S.; Sgouras, K.I.; Doukas, D.I.; Christoforidis, G.C.; Labridis, D.P. Generation Units in Radial Distribution Systems Efficient RES Penetration under Optimal Distributed Generation Placement Approach. *Energies* **2019**, *12*, 1250. [CrossRef]
11. Gao, Z.; Saxén, H.; Gao, C. Data-driven approaches for complex industrial systems. *IEEE Trans. Ind. Inf.* **2013**, *9*, 2210–2212. [CrossRef]
12. Gao, Z.; Nguang, S.K.; Kong, D. Advances in modelling, monitoring, and control for complex industrial systems. *Complexity* **2019**. [CrossRef]
13. Gao, Z.; Kong, D.; Gao, C. Modeling and control of complex dynamic systems: Applied mathematical aspects. *J. Appl. Math.* **2012**, *2012*. [CrossRef]
14. Niknam, T.; Doagou-Mojarrad, H. Multiobjective economic/emission dispatch by multiobjective -particle swarm optimization. *IET Gener. Transm. Distrib.* **2012**, *6*, 363–377. [CrossRef]
15. Kaya, P.; Chanda, C.K. Placement of wind and solar based DGs in distribution system for power loss minimization and voltage stability improvement. *Electr. Power Energy Syst.* **2013**, *53*, 795–809.
16. Li, R.; Ma, H.; Wang, F.; Wang, Y.; Liu, Y.; Li, Z. Game optimization theory and application in distribution system ex pansion planning, including distributed generation. *Energies* **2013**, *7*, 1101–1124. [CrossRef]
17. Esmaili, M. Placement of minimum distributed generation units observing power losses and voltage stability with network constraints. *IET Gener. Transm. Distrib.* **2013**, *7*, 813–821. [CrossRef]
18. Wang, T.; He, X.; Huang, T.; Li, C.; Zhang, W. Collective neurodynamic optimization for economic emission dispatch problem considering valve point effect in microgrid. *Neural Netw.* **2017**, *93*, 126–136. [CrossRef]
19. Ding, T.; Bo, R.; Li, F.; Gu, Y.; Guo, Q.; Sun, H. Exact penalty function based constraint relaxation method for optimal power flow considering wind generation uncertainty. *IEEE Trans. Power Syst.* **2015**, *30*, 1546–1547. [CrossRef]
20. Su, H. Siting and sizing of distributed generators based on improved simulated annealing particle swarm optimization. *Environ. Sci. Pollut. Res.* **2019**, *26*, 17927–17938. [CrossRef]
21. Pandi, R.; Zeineldin, H.; Xiao, W. Determining optimal location and size of distributed generation resources considering harmonic and protection coordination limits. *IEEE Trans. Power Syst.* **2013**, *28*, 1245–1254. [CrossRef]

22. Neale, J.; Nettleton, S.; Pickering, L.; Fischer, J. Pareto optimal reconfiguration of power distribution systems using a genetic algorithm based on NSGA-II. *Energies* **2013**, *6*, 1439–1455.

23. Ameli, A.; Bahrami, S.; Khazaeli, F.; Haghifam, M.R. A multiobjective particle swarm optimization for sizing and placement of DGs from DG owner's and distribution company's viewpoints. *IEEE Trans. Power Deliv.* **2014**, *29*, 1831–1840. [CrossRef]

24. Maciel, R.S.; Rosa, M.; Miranda, V.; Padilha-Feltrin, A. Multi-objective evolutionary particle swarm optimization in the assessment of the impact of distributed generation. *Electr. Power Syst. Res.* **2012**, *89*, 100–108. [CrossRef]

25. Imran, A.M.; Kowsalya, M. A new power system reconfiguration scheme for power loss minimization and voltage profile enhancement using fireworks algorithm. *Int. J. Electr. Power Energy Syst.* **2014**, *62*, 312–322. [CrossRef]

26. Everett, L.J.; Bierl, C.; Master, S.R. Unbiased statistical analysis for multi-stage proteomic search strategies. *J. Proteome Res.* **2010**, *9*, 700–707. [CrossRef]

27. Sheng, W.X.; Liu, K.Y.; Meng, X.L. Optimal placement and sizing of distributed generation via an improved nondominated sorting genetic algorithm II. *IEEE Trans. Power Deliv.* **2015**, *30*, 569–578. [CrossRef]

28. Deb, K.; Pratap, A.; Sameer, A.; Meyarivan, T. A fast and elitist multiobjective genetic algorithm: NSGA-II. *IEEE Trans. Evol. Comput.* **2002**, *6*, 182–197. [CrossRef]

29. Yammani, C.; Maheswarapu, S.; Matam, S.K. A Multi-objective Shuffled Bat algorithm for optimal placement and sizing of multi distributed generations with different load models. *Int. J. Electr. Power Energy Syst.* **2016**, *79*, 120–131. [CrossRef]

30. Kashem, M.A.; Ganapathy, V.; Jasmon, G.B.; Buhari, M.I. A novel method for loss minimization in distribution networks. In Proceedings of the International Conference on Electric Utility Deregulation and Restructuring and Power Technologies, London, UK, 4–7 April 2000; pp. 251–255.

31. Zhao, X.Y.; Wang, X.X.; Feng, Y.F. Siting and Sizing of Distributed Generation in Micro-Grid. *Smart Grid* **2015**, *5*, 137–147. [CrossRef]

32. Helbig, M.; Engelbrecht, A. Issues with performance measures for dynamic multi-objective optimization. *Inf. Sci.* **2013**, *250*, 17–24. [CrossRef]

33. Mohamed, A.; Salehi, V.; Ma, T.; Mohammed, O.A. Real-time energy management algorithm for plug-in hybrid electric vehicle charging parks involving sustainable energy. *IEEE Trans. Sustain. Energy* **2014**, *5*, 577–586. [CrossRef]

34. Pena, I.; Martinez-Anido, C.B.; Hodge, B.S. An Extended IEEE 118-bus Test System with High Renewable Penetration. *IEEE Trans. Power Syst.* **2017**, *33*, 280–289. [CrossRef]

Article

Flexible Flow Shop Scheduling Method with Public Buffer

Zhonghua Han [1,2,3,4], **Chao Han** [1,*], **Shuo Lin** [1], **Xiaoting Dong** [1,5] **and Haibo Shi** [2,3,4]

[1] Faculty of Information and Control Engineering, Shenyang Jianzhu University, Shenyang 110168, China
[2] Department of Digital Factory, Shenyang Institute of Automation, the Chinese Academy of Sciences(CAS), Shenyang 110016, China
[3] Key Laboratory of Network Control System, Chinese Academy of Sciences, Shenyang 110016, China
[4] Institutes for Robotics and Intelligent Manufacturing, Chinese Academy of Sciences, Shenyang 110016, China
[5] Department of Equipment Engineering, Sichuan College of Architectural Technology, Deyang 618000, China
* Correspondence: sfcc1717@163.com; Tel.: +86-24-2469-0045

Received: 5 August 2019; Accepted: 29 September 2019; Published: 1 October 2019

Abstract: Actual manufacturing enterprises usually solve the production blockage problem by increasing the public buffer. However, the increase of the public buffer makes the flexible flow shop scheduling rather challenging. In order to solve the flexible flow shop scheduling problem with public buffer (FFSP–PB), this study proposes a novel method combining the simulated annealing algorithm-based Hopfield neural network algorithm (SAA–HNN) and local scheduling rules. The SAA–HNN algorithm is used as the global optimization method, and constructs the energy function of FFSP–PB to apply its asymptotically stable characteristic. Due to the limitations, such as small search range and high probability of falling into local extremum, this algorithm introduces the simulated annealing algorithm idea such that the algorithm is able to accept poor fitness solution and further expand its search scope during asymptotic convergence. In the process of local scheduling, considering the transferring time of workpieces moving into and out of public buffer and the manufacturing state of workpieces in the production process, this study designed serval local scheduling rules to control the moving process of the workpieces between the public buffer and the limited buffer between the stages. These local scheduling rules can also be used to reduce the production blockage and improve the efficiency of the workpiece transfer. Evaluated by the groups of simulation schemes with the actual production data of one bus manufacturing enterprise, the proposed method outperforms other methods in terms of searching efficiency and optimization target.

Keywords: flexible flow shop; limited buffer; public buffer; Hopfield neural network; local scheduling; simulated annealing algorithm

---

1. Introduction

With rapid development of information, advanced manufacturing, and artificial intelligence technology, Germany proposed "Industry 4.0" national strategy, which promotes progress in manufacturing industry and provides producing solution schemes for many complex industrial systems [1–3]. In the production of steel casting, assembly of heavy machinery, and other industries, a scheduling problem with non-deterministic polynomial hard (NP-hard) attributes [4–8], such as the flexible flow shop scheduling problem with the characteristics of customized production and parallel processing, was experienced [9,10]. During the actual production of the large equipment manufacturing workshop, only the buffer with limited capacity can be set in the workshop, owing to physical factors such as the workshop space and storage equipment capacity. When the required capacity of the production task fluctuates in the workshop or the takt time between stages is inconsistent, the buffer capacity might reach the upper limit. Consequently, the completed workpieces that are ready to enter the limited buffer cannot enter the buffer for temporary storage; they are stagnant on their processing workstation while waiting for available space in the limited buffer, which effectuates production blockage, thereby delaying the production process [11–14]. In the actual production of a large equipment manufacturer, it is uncommon to configure the limited buffer with redundant capacity or to temporarily adjust the capacity of limited buffer to avoid the production blockage. The manufacturer often sets up the public buffer in the workshop to dynamically receive workpieces that cannot enter the limited buffer between the stages. This is equivalent to dynamically expanding the capacity of the limited buffer, which can reduce the production blockage and ensure fluent production process. In the actual workshop, the public buffer is often located at a designated position not adjacent to the workstation due to physical factors such as production space. Thus, the transit time between the workstation and public buffer cannot be neglected. It creates the transfer scheduling problem of the workpiece between limited buffer and public buffer, which further increases the uncertainty of the scheduling result and the difficulty of resolving the scheduling problem [15]. Consequently, it is necessary to explore an effective scheduling method for the flexible flow shop scheduling problem with public buffer (FFSP–PB) due to its role in reducing the production blockage and improving the utilization of production resources [16,17]. Therefore, it is of great theoretical and engineering value to solve this problem.

The scheduling problem with the public buffer is derived from the limited buffer scheduling problem, which is closely related to actual engineering. The buffer between stages in the limited buffer scheduling problem is set to the upper limit of the capacity. Once the buffer capacity reaches the upper limit, the workpiece cannot enter this buffer [18]. Presently, the problem is systematically studied worldwide. Zhang et al. [19] proposed two rapidly generating heuristic algorithms with minimum makespan as the criterion. Rooeinfar et al. [20] proposed a novel optimization model and two types of solutions to resolve the uncertain flexible flow shop scheduling problem with limited buffers. Jiang et al. [21] developed an effective multi-objective optimization algorithm in the framework of a multi-objective evolutionary algorithm based on decomposition to solve the hybrid flow shop scheduling problem with limited buffer according to energy orientation. Zeng et al. [22] set forth a two-stage algorithm based on neighborhood search to resolve this issue in accordance with the job shop scheduling problem with limited output buffers.

Investigators have carried out relevant research on the production blockage caused by limited buffer, while studying the scheduling problem with limited buffers. Since the production blockage decreases the production efficiency of enterprises and delays the production process, solving the problem of production blockage of limited buffer in flexible flow shop has been under intensive focus in recent years. Ribas et al. [23] proposed an iterative greedy algorithm to solve the problem of parallel blocking flow shop and distributed blocking flow shop by minimizing the total delay time of the workpiece. Chang [24] established a multi-state manufacturing system (MMS) model to study the reliability of parallel production line manufacturing system with limited-capacity buffer stations to avoiding blockage and starvation. Johri [25] studied the blockage and starvation of limited buffer by linear programming, thereby proposing a method by increasing the selectivity of buffer space to resolve the problem of capacity loss caused by the small capacity buffer.

The above literature demonstrates that the current research on the limited buffer scheduling problem is mainly focused on the various types of workshops, with emphasis on the improvement of global optimization algorithm. However, only a few investigators have addressed the issue of solving the production blockage led by limited buffer stresses on adjusting the production plan and buffer space via alleviating the production blockage by setting the public buffer in the workshop. However, they have not explored the impact of transit time on the production process caused by the movement of the workpiece between public buffer and limited buffer among the stages. Herein, the flexible flow shop scheduling problem with public buffer was more complicated than the generally limited buffer scheduling problem. It considered not only the restricted capacity of limited buffer but also the transfer scheduling problem of workpieces among workstation, limited buffer, and public buffer.

As the research problems become more complicated, it is necessary to explore a global optimization algorithm that can effectively solve these complex problems. The majority of the swarm intelligence algorithms are based on a random search with slow optimization rate, which renders finding the global optimal value challenging. The Hopfield neural network (HNN) algorithm based on non-linear control theory has great advantages in global optimization speed and avoids the shortcomings of the random search of swarm intelligence algorithm [26]. However, the issues pertain small optimization range, easy to fall into, and difficult to break out of the local extremum. Therefore, the idea of a simulated annealing algorithm is introduced to HNN algorithm. During each generation training, neuron input adds random disturbance and Metropolis acceptance criteria controls whether the energy function value which is generated by disturbance input in the next generation of optimization. Thus, the HNN algorithm allows to accept the solution with poor fitness during asymptotic convergence, further changing the optimizing direction of HNN algorithm, expanding its optimizing range and enhancing the ability to jump out of local extremum. By comparison with analysis of groups of simulation schemes, combining the methods of the simulated annealing algorithm-based Hopfield neural network (SAA–HNN) algorithm and local scheduling rules to control the moving process of workpieces between public buffer and the limited buffer can be verified for the efficiency of solving the FFSP–PB. The SAA–HNN algorithm is applied to solve the flexible flow shop scheduling problem with public buffer by extending the application field of neural network algorithm.

With the rise of Industry 4.0, customized production mode has become more popular among manufacturing enterprises, and the varieties of products that cater to customers' needs have diversified. It is difficult for enterprises to control the takt, which leads to production blockage. This increases the importance of public buffer setting on the production line for relieving the production blockage and stabilizing the operation of the whole production line. Because intelligent transportation equipment such as automated guided vehicle (AGV) is widely used in the actual production workshop, it is more convenient for work in progress (WIP) to transport back and forth between the public buffer and the limited buffer, which plays the role of the public buffer. Therefore, the relevant scheduling optimization technology for automatic production lines with public buffer has an extensive application prospect, which would improve the intellectualization of the manufacturing automation technology.

2. FFSP–PB Mathematical Model

2.1. Problem Description

As shown in Figure 1, the FFSP–PB could be described as follows: *m* stages in the workshop and the processing queues of *n* workpieces need to be processed in order with *m* processing stages. At least one of the *m* stages consists of two or more parallel workstations, and the processing times of the workpiece are the same on different parallel workstations at one stage. A buffer with limited capacity is set up between stages; if the limited buffer capacity between stages reaches the upper limit, production blockage is likely to occur. In order to alleviate production blockage, a public buffer is set up in the production workshop, and this area provides services for all stages. If the capacity of the limited buffer between the stages reaches the upper limit, the workpiece can be moved into the public buffer for temporary storage. All workpieces are processed online from the first stage, completing all stages sequentially. If the capacity of the limited buffer between stages reaches the upper limit, the newly completed workpiece is transferred to the public buffer. Under certain conditions, the workpiece during transfer can also be returned to the limited buffer. If the workpiece in the limited buffer is moved to the workstation of the next stage for processing, the limited buffer will have an available space. Subsequently, the workpiece in the public buffer that should have accessed the limited buffer is transferred back to the limited buffer. The transit time between limited buffer and public buffer cannot be ignored. Under preconditions of the online sequence of the workpiece, the standard processing time for transferring the workpiece, the standard processing time of the workpiece at each stage and the online sequence is optimized by global optimization method, and the movement of the workpiece among the workstation, the limited buffer, and the public buffer is controlled by the local scheduling rules. Thus, the scheduling results of the processing workstations, the start time, and the completion time of all workpieces at each stage and the information of the transport process are obtained.

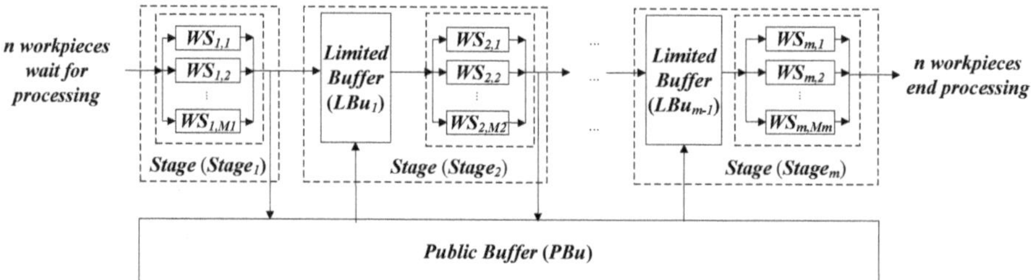

Figure 1. Mathematical model of flexible flow shop with limited buffer and public buffer.

2.2. Parameters in the Model

The parameters used in this study are as follows:

n: Total number of workpieces to be processed;

m: Total number of stages;

Wp_i: Workpiece i, $i \in \{1, \ldots, n\}$;

$Stage_j$: Stage j, $j \in \{1, \ldots, m\}$;

M_j: Total number of workstations in stage $Stage_j$, $j \in \{1, \ldots, m\}$;

$WS_{j,l}$: Workstation l of stage $Stage_j$, $j \in \{1, \ldots, m\}$, $l \in \{1, \ldots, M_j\}$;

LBu_j: Limited buffer of stage $Stage_j$, $j \in \{2, \ldots, m\}$;

Kl_j: Maximum buffer capacity in the limited buffer LBu_j of stage $Stage_j$, $j \in \{1, \ldots, m\}$;

$Bl_{j,k}$: Buffer position k of limited buffer LBu_j, $j \in \{2, \ldots, m\}$, $k \in \{1, \ldots, Kl_j\}$;

$WAl_j(t)$: At the t time, the workpieces in the limited buffer LBu_j;

PBu: Public buffer;

Kp: Maximum buffer capacity in the public buffer PBu, $1 \leq Kp \leq n - \min\{Kl_j\}$;

$Ts_{i,j}$: The start time of the workpiece Wp_i at stage $Stage_j$, $j \in \{1, \ldots, m\}$;

$Tc_{i,j}$: The completion time of the workpiece Wp_i at stage $Stage_j$, $j \in \{1, \ldots, m\}$;

$Tb_{i,j}$: The standard processing time of the workpiece Wp_i at stage $Stage_j$, $j \in \{1, \ldots, m\}$;

$Ti_{i,j}$: The entry time of the workpiece Wp_i into stage $Stage_j$, $j \in \{1, \ldots, m\}$;

$To_{i,j}$: The departure time of the workpiece Wp_i out of stage $Stage_j$, $j \in \{1, \ldots, m\}$;

$Tli_{i,j}$: The entry time of the workpiece Wp_i into limited buffer LBu_j, $j \in \{2, \ldots, m\}$;

$Tlo_{i,j}$: The departure time of the workpiece Wp_i out of limited buffer LBu_j, $j \in \{2, \ldots, m\}$;

$Tpi_{i,j}$: The entry time of the workpiece Wp_i into public buffer PBu at stage $Stage_j$, $j \in \{2, \ldots, m\}$;

$Tpo_{i,j}$: The departure time of the workpiece Wp_i out of public buffer PBu at stage $Stage_j$, $j \in \{2, \ldots, m\}$;

Tbt: The standard processing time for transferring the workpiece;

Rp_i: The workpiece Wp_i is on the way from the workstation to the public buffer;

Rl_i: The workpiece Wp_i is on the way from the public buffer to the limited buffer;

Rb_i: The workpiece Wp_i is on the way back to the limited buffer;

Cfp: The time for electric flat carriage to finish the current task and return to the public buffer;

$Ttw_{i,j}(t)$: At the t time, the transit time of the workpiece Wp_i from the workstation of stage $Stage_j$ to the public buffer.

2.3. Constraints

The variables used in this study and the constraints that exist between variables are as follows:

2.3.1. Assumptions

$$At_{i,j,l} = \begin{cases} 1 & \text{Workpiece } Wp_i \text{ is assigned to be processed on workstation } WS_{j,l} \\ 0 & \text{Workpiece } Wp_i \text{ is not assigned to be processed on workstation } WS_{j,l} \end{cases} \tag{1}$$

$$OAl_{i,j}(t) = \begin{cases} 1 & \text{At time } t,\ \text{workpiece } Wp_i \text{ is in limited buffer } LBu_j \\ 0 & \text{At time } t,\ \text{workpiece } Wp_i \text{ is not in limited buffer } LBu_j \end{cases} \tag{2}$$

$$OAp_{i,j}(t) = \begin{cases} 1 & \begin{array}{l} \text{At time } t,\ \text{workpiece } Wp_i \text{ that should have entered} \\ \text{limited buffer } LBu_j \text{ is in public buffer} \end{array} \\ 0 & \begin{array}{l} \text{At time } t,\ \text{workpiece } Wp_i \text{ that should have entered} \\ \text{limited buffer } LBu_j \text{ is not in public buffer} \end{array} \end{cases} \tag{3}$$

$$OAt_{i,j}(t) = \begin{cases} 1 & \begin{array}{l} \text{At time } t,\ \text{workpiece } Wp_i \text{ that should have} \\ \text{entered limited buffer } LBu_j \text{ is in transit} \end{array} \\ 0 & \begin{array}{l} \text{At time } t,\ \text{workpiece } Wp_i \text{ that should have} \\ \text{entered limited buffer } LBu_j \text{ is not in transit} \end{array} \end{cases} \tag{4}$$

$$Pan\_Car(t) = \begin{cases} 0 & \text{At time } t,\ \text{electric flat carriage is on workstation} \\ 1 & \text{At time } t,\ \text{electric flat carriage is in limited buffer} \\ 2 & \text{At time } t,\ \text{electric flat carriage is in public buffer} \\ 3 & \begin{array}{l} \text{At time } t,\ \text{electric flat carriage is in transit from} \\ \text{workstation to public buffer} \end{array} \\ 4 & \begin{array}{l} \text{At time } t,\ \text{electric flat carriage is in transit from} \\ \text{public buffer to limited buffer} \end{array} \end{cases} \tag{5}$$

2.3.2. General Constraint of Flexible Flow Shops Scheduling

$$\sum_{l=1}^{M_j} At_{i,j,l} = 1,\ i \in \{1,\ldots,n\},\ j \in \{1,\ldots,m\} \tag{6}$$

$$Tc_{i,j} = Ts_{i,j} + Tb_{i,j},\ i \in \{1,\ldots,n\},\ j \in \{1,\ldots,m\} \tag{7}$$

$$Tc_{i,j-1} \leq Ts_{i,j},\ i \in \{1,\ldots,n\},\ j \in \{2,\cdots,m\} \tag{8}$$

Equation (6) indicates that the constraint that the workpiece Wp_i can only be processed at one workstation of the stage $Stage_j$ during the processing. Equation (7) indicates the constraints that the completion time of the workpiece Wp_i in the stage $Stage_j$ is equal to the sum of its start time and standard processing time in the stage $Stage_j$, which guarantees close machining between the workpieces. Equation (8) indicates the constraint that the workpiece Wp_i needs to complete the processing task of the current stage before the processing task of the next stage. This constraint limits the processing sequence of each workpiece in all stages. Equations (6)–(8) ensure that the whole processing is in accordance with the processing characteristics of the flexible flow shop.

2.3.3. Constraints of the Limited Buffers

$$Tli_{i,j} \geq Tc_{i,j-1}, \; j \in \{2,\ldots,m\} \tag{9}$$

Equation (9) indicates the constraint that time $Tli_{i,j}$ for the workpiece to enter the limited buffer LBu_j cannot be less than the completion time $Tc_{i,j-1}$ of this workpiece processed at the previous stage $Stage_{j-1}$.

$$WAl_j(t) = \left\{ J_i \middle| OAl_{i,j}(t) = 1 \right\} \tag{10}$$

Equation (10) indicates the constraint and the workpieces in the limited buffer LBu_j at the t time.

$$card\left(WAl_j(t)\right) \leq Kl_j \tag{11}$$

Equation (11) indicates the constraint that at any time, the total number of workpieces in the collection WAl_j waiting to be processed cannot be greater than the maximum buffer capacity Kl_j in the limited buffer. This constraint guarantees that the characteristics of the limited buffer correspond to the actual processing.

$$Tlo_{i,j} \geq Tli_{i,j}, \; j \in \{2,\ldots,m\} \tag{12}$$

Equation (12) denotes that time for the workpiece to leave the limited buffer LBu_j cannot be less than the time for the workpiece to enter the limited buffer LBu_j.

2.3.4. Constraints of the Public Buffer

$$Tpo_{i,j} \geq Tpi_{i,j}, \; j \in \{2,3,\ldots,m\} \tag{13}$$

Equation (13) indicates the time for the workpiece in the public buffer PBu that should have entered the limited buffer LBu_j to leave the public buffer PBu should not be less than the time for it to enter the public buffer PBu. This constraint ensures that the moving in and out of the public buffer conforms to the actual processing.

$$WAp_j(t) = \left\{ Wp_i \middle| OAp_{i,j}(t) = 1 \right\} \tag{14}$$

Equation (14) represents the collection of all workpieces contained in the public buffer PBu at time t.

$$card\left(WAp_j(t)\right) \leq Kp \tag{15}$$

Equation (15) indicates the constraint that at any time, the sum of workpieces contained in the to-be-processed collection WAp_j is less than or equal to the maximum buffer capacity Kp of the public buffer. This constraint guarantees that the characteristics of the public buffer conform to the actual processing.

$$Ttw_{i,j}(t) = (t - To_{i,j}) \tag{16}$$

Equation (16) shows that at time t, the time for the workpiece Wp_i to be transferred from the workstation of the stage $Stage_j$ to the public buffer is equal to the difference between the time for the workpiece Wp_i to leave the workstation of stage $Stage_j$ at time t.

$$Ttw_{i,j}(t) \leq Tbt \tag{17}$$

Equation (17) indicates the constraint that at time t, the time for the workpiece Wp_i to be transferred from the workstation of the stage $Stage_j$ to the public buffer should be less than the standard processing time for transferring the workpiece.

2.3.5. Other Constraints

(1) Continuous processing constraint: If the workpiece has started the processing task of a certain stage, it cannot be interrupted until the task is completed.
(2) Workstation uniqueness constraint: A workstation can only process one workpiece simultaneously.
(3) Workstation availability constraint: All workstations are available at the scheduling time.
(4) Time simplification constraint: Irrespective of the transit time of the workpiece between the limited buffer and workstation, only the processing time of the workpiece at each stage and the transit time of the workpiece on the electric flat carriage are considered.

3. Research on FFSP–PB Local Scheduling Rules

During production, when the limited buffer capacity of the current stage reaches the upper limit, the finished workpiece of the previous stage is directly transferred to the public buffer. The available workstation at this stage allows the processing of the workpiece in limited buffer at the current stage. Strikingly, there is available space in the limited buffer at this stage. If the workpiece is directly transferred from the public buffer to the limited buffer of the current stage, the available space would not be occupied by the finished workpiece of the previous stage during the period when the workpiece is transported back to the limited buffer of the current stage; otherwise, the workpiece transferred from the public buffer will collide with the completed workpiece of the previous stage while entering the available space. In order to prevent this conflict and as long as the workpiece in the public buffer starts to be transported to the available space in the limited buffer of the current stage, this space cannot be occupied, which might block the finished workpiece of the previous stage at its processing workstation. According to the above analysis, after adding the public buffer to the workshop, if the corresponding local scheduling rules are not established, the production blockage and completion time of gross workpieces cannot be reduced effectively. In order to play the role of public buffer and further alleviate the production blockage, the reentrant rules of the electric flat carriage and the workpiece transfer rules in the public buffer are designed. These local heuristics rules can control the transit of the workpiece among limited buffer, public buffer, and workstation. These rules exert the role of public buffer to reduce the production blockage.

3.1. Reentrant Rules of Electric Flat Carriage

At time t, if there is available space in the limited buffer LBu_{j+1} and workpieces that have completed the processing task of stage $Stage_j$ and transferred to public buffer PBu, the already-spent transit time $Ttw_{i,j}(t)$ of the workpiece in transit will be compared to the estimated minimum completion time $\min\left\{\left(Tc_{i,j}-t\right)\middle|\left(Ts_{i,j}-t\right)\le 0,\left(Tc_{i,j}-t\right)>0\right\}$ of all workpieces at the current stage $Stage_j$. If the estimated minimum completion time is longer than the already-spent transit time of the workpiece, the electric flat carriage will begin to turn back, transporting the workpiece back to the limited buffer LBu_{j+1} of the next stage $Stage_{j+1}$.

At time t, if $card\left(OAl_{i,j+1}(t)\right) < Kl_{j+1}$ & $OAt_{i,j+1}(t) \cdot Ttw_{i,j}(t) < \min\left\{\left(Tc_{i,j}-t\right)\middle|\left(Ts_{i,j}-t\right)\le 0,\left(Tc_{i,j}-t\right)>0\right\}$ & $Pan\_Car(t) = 3$ & $OAt_{i,j+1}(t) \ne 0$, the workpiece begins to be transported, and the state of electric flat carriage becomes $Pan\_Car(t) = 4$.

3.2. Workpiece Transfer Rules in Public Buffer

At time t, there is available space in the limited buffer LBu_{j+1} and workpieces in the public buffer that should have entered the limited buffer LBu_{j+1}. If the electric flat carriage is in the public buffer

$Pan\_Car(t) = 2$, the standard processing time for transferring the workpiece Tbt will be compared to the estimated minimum completion time $\min\left\{\left(Tc_{i,j} - t\right)\middle|\left(Ts_{i,j} - t\right) \leq 0, \left(Tc_{i,j} - t\right) > 0\right\}$ of all workpieces at stage $Stage_j$. If the estimated minimum completion time is longer than the standard processing time for transferring the workpiece, the workpieces currently in the public buffer PBu that should have entered the limited buffer LBu_{j+1} are transported back to the limited buffer LBu_{j+1}; however, if the estimated minimum completion time is short, the available space in the limited buffer LBu_{j+1} will remain idle until the workpiece with the estimated minimum completion time at the current stage is accessed.

If the electric flat carriage is not in the public buffer, i.e., $Pan\_Car(t) \neq 2$, the sum of the standard processing time for transferring the workpiece Tbt and the time Cfp for electric flat carriage to the public buffer after completing the current task will be compared to the estimated minimum completion time $\min\left\{\left(Tc_{i,j} - t\right)\middle|\left(Ts_{i,j} - t\right) \leq 0, \left(Tc_{i,j} - t\right) > 0\right\}$ of all workpieces at stage $Stage_j$. If the estimated minimum completion time is longer than that described above, the workpieces currently in the public buffer PBu that should have entered the limited buffer LBu_{j+1} are transported back to the limited buffer LBu_{j+1}. If the estimated minimum completion time is shorter than that mentioned above, the available space in the limited buffer LBu_{j+1} will remain idle until the workpiece with the estimated minimum completion time at the current stage is processed and accessed.

At time t, if $card\left(OAl_{i,j+1}(t)\right) < Kl_{j+1}$ & $OAp_{i,j+1}(t) \cdot Tbt < \min\left\{\left(Tc_{i,j} - t\right)\middle|\left(Ts_{i,j} - t\right) \leq 0, \left(Tc_{i,j} - t\right) > 0\right\}$ & $Pan\_Car(t) = 2$ & $OAp_{i,j+1}(t) \neq 0$, the workpiece begins to be transported, and the state of electric flat carriage becomes $Pan\_Car(t) = 4$.

At time t, if $card\left(OAl_{i,j+1}(t)\right) < Kl_{j+1}$ & $OAp_{i,j+1}(t) \cdot Tbt + Cfp < \min\left\{\left(Tc_{i,j} - t\right)\middle|\left(Ts_{i,j} - t\right) \leq 0, \left(Tc_{i,j} - t\right) > 0\right\}$ & $Pan\_Car(t) \neq 2$ & $OAp_{i,j+1}(t) \neq 0$, the workpiece begins to be transported, and the state of electric flat carriage becomes $Pan\_Car(t) = 2$ after time Cfp.

4. Local Scheduling Rules for Multi-Queue Limited Buffers

In this study, HNN is primarily used for optimization. The energy function in the continuous HNN is monotonically decreasing, and the gradually decreasing process of the energy function is the process of neural network optimization. The algorithm employs this feature to solve the optimization problem and search for the optimal solution [27].

When the standard HNN algorithm solves the NP-hard problem, it establishes a non-linear correlation between the input and output of the network, since the activation function of the analog neurons in the neural network is a non-linear transfer function. Moreover, the output of the problem solution is a non-linear space, which might encompass multiple poles, renders the algorithm as an optimal local solution in the event of failure to obtain the optimal global solution. Also, the algorithm cannot break after falling into local extremum, which makes the overall evolutionary trend irreversible. Thus, the simulated annealing algorithm is introduced into the standard HNN algorithm to prevent its premature convergence, expand the search ability of the feasible solution, and improve the global optimization effect.

4.1. HNN Algorithm

4.1.1. Establishing the Permutation Matrix

The permutation matrix is a bridge connecting the buffer dynamic capacity-increase problem in a flexible flow shop with public buffer and the improved HNN algorithm. This study applied the workpiece processing sequence in the first stage to construct the matrix. For example, the FFSP permutation matrix of the six workpieces to be processed is shown in Table 1. The permutation matrix in Table 2 indicates the processing sequence of the six workpieces $\{Wp_2, Wp_1, Wp_5, Wp_3, Wp_4, Wp_6\}$.

Table 1. FFSP permutation matrix of six workpieces.

Workpiece	Processing Sequence					
	1	2	3	4	5	6
Wp_1	0	1	0	0	0	0
Wp_2	1	0	0	0	0	0
Wp_3	0	0	0	1	0	0
Wp_4	0	0	0	0	1	0
Wp_5	0	0	1	0	0	0
Wp_6	0	0	0	0	0	1

Table 2. Algorithm test result.

Example	LB	ICA			CGA			HNN			SAA–HNN		
		C_{max}	ARE	Time	C_{max}	ARE	Time	C_{max}	ARE	Time	C_{max}	ARE	Time
j15c5c1	85	91.3	7.4%	6.6 s	90.2	6.1%	6.3 s	97.3	14.5%	3.5 s	90.1	6.1%	4.1 s
j15c5c2	90	102.1	13.5%	6.7 s	99.9	11.0%	6.4 s	109.6	21.8%	3.7 s	96.7	7.4%	4.6 s
j15c5d1	167	206.4	23.6%	6.9 s	204.2	22.3%	6.6 s	242.1	45.0%	4.1 s	190.7	14.2%	4.8 s
j15c5d2	82	101.3	23.6%	6.7 s	96.9	18.2%	6.4 s	111.8	36.4%	3.8 s	92.1	12.3%	4.5 s
j80c4a1	—	1415.4	—	24.3 s	1389.5	—	29.5 s	1453.7	—	6.5 s	1375.9	—	7.7 s
j80c4a2	—	1434.7	—	21.1 s	1419.4	—	24.5 s	1474.2	—	6.1 s	1408.3	—	6.2 s
j80c8a1	—	2088.4	—	24.2 s	2027.2	—	26.4 s	2170.2	—	6.4 s	2011.2	—	7.5 s
j80c8a2	—	1856.4	—	24.5 s	1827.7	—	24.4 s	1956.6	—	5.8 s	1811.1	—	7.5 s
j120c20a1	—	3764.7	—	44.6 s	3699.1	—	48.3 s	3954.9	—	13.9 s	3442.5	—	14.8 s
j200c10a1	—	4953.2	—	76.3 s	4787.8	—	79.7 s	5211.5	—	18.6 s	4364.9	—	21.5 s

4.1.2. Establishing the Energy Function

As a major feedback of the network, the energy function can easily determine the stability of the system.

(1) Energy function row constraint

$$E_1 = \frac{A}{2}\sum_{x=1}^{n}\sum_{i=1}^{n-1}\sum_{j=i+1}^{n} V_{xi}V_{xj} = 0 \tag{18}$$

Equation (18) indicates that the sum $\sum_{x=1}^{n}\sum_{i=1}^{n-1}\sum_{j=i+1}^{n} V_{xi}V_{xj}$ of all elements of n rows multiplied by each other in order should be 0, i.e., each row in the FFSP permutation matrix only has one '1', indicating that a workpiece can only be processed once at each stage. V_{xi} represents the element of the i column of the x row of the FFSP permutation matrix, and A is the coefficient.

(2) Energy function column constraint

$$E_2 = \frac{B}{2}\sum_{i=1}^{n}\sum_{x=1}^{n-1}\sum_{y=x+1}^{n} V_{xi}V_{yi} = 0 \tag{19}$$

Equation (19) denotes that the sum $\sum_{i=1}^{n}\sum_{x=1}^{n-1}\sum_{y=x+1}^{n} V_{xi}V_{yi}$ of all elements of n columns multiplied by each other in a specific order should be 0, i.e., each column in the FFSP permutation matrix only has one '1', indicating that a workpiece can only be processed once at one stage, and B is the coefficient.

(3) Energy function overall constraint

$$E_3 = \frac{C}{2}\left(\sum_{i=1}^{n}\sum_{x=1}^{n} V_{xi} - n\right)^2 \tag{20}$$

Equation (20) indicates that the sum $\sum_{i=1}^{n}\sum_{x=1}^{n} V_{xi} - n$ of all elements should be 0, i.e., there are n '1' in the FFSP permutation matrix, indicating that all workpieces are to be processed at one stage. In the Equation, C is the coefficient, and the square value is used to conform to the expression of energy, as well as embody a punishment for not meeting the constraint.

(4) Energy function target item

Since the main optimization goal of the FFSP–PB is the makespan, the fourth item of the energy function needs to be expressed in combination with the makespan, as shown in Equation (21), and D is the coefficient.

$$E_4 = \frac{D}{2}C_{max} \tag{21}$$

Combining Equations (18)–(21), the energy function for constructing FFSP–PB is as follows:

$$E = \frac{A}{2}E_1 + \frac{B}{2}E_2 + \frac{C}{2}E_3 + \frac{D}{2}E_4 = \frac{A}{2}\sum_{x=1}^{n}\sum_{i=1}^{n-1}\sum_{j=i+1}^{n} V_{xi}V_{xj} + \frac{B}{2}\sum_{i=1}^{n}\sum_{x=1}^{n-1}\sum_{y=x+1}^{n} V_{xi}V_{yi} + \frac{C}{2}\left(\sum_{i=1}^{n}\sum_{x=1}^{n} V_{xi} - n\right)^2 + \frac{D}{2}C_{max} \tag{22}$$

According to a previous study [26], Equation (22) can be improved to:

$$E = \frac{A}{2}\sum_{x=1}^{n}\left(\sum_{i=1}^{n} V_{xi} - 1\right)^2 + \frac{B}{2}\sum_{i=1}^{n}\left(\sum_{x=1}^{n} V_{xi} - 1\right)^2 + DC_{max} \tag{23}$$

To ensure the symmetry for the solution of HNN algorithm, the value of A in Equation (23) needs to be equal to the value of B.

4.1.3. Establishing HNN Dynamic Differential Equation

According to another study [27], the connection weight coefficient is calculated as follows:

$$\frac{du_{xi}}{dt} = -\frac{\partial E}{\partial V_{xi}} \tag{24}$$

Equations (23) and (24) can derive HNN dynamic equation as follows:

$$\frac{du_{xi}}{dt} = -A\left(\sum_{i=1}^{n} V_{xi} - 1\right) - A\left(\sum_{y=1}^{n} V_{yi} - 1\right) \tag{25}$$

The correlation function between input u_{xi} and output V_{xi} of the simulating neuron in HNN algorithm is:

$$V_{xi}(t_0) = \varphi_{xi}(u_{xi}) = \frac{1}{2}\left[1 + \tanh\left(\frac{u_{xi}(t_0)}{u_0}\right)\right] \tag{26}$$

According to the HNN dynamic equation, the input bias Δu_{xi} is:

$$\Delta u_{xi}(t_0) = \frac{du_{xi}}{dt}V_{yj} \tag{27}$$

In the evolution process of HNN, the input is updated by the first-order Euler Equation, which is shown in Equation (28):

$$u_{xi}(t_0 + \Delta t) = u_{xi}(t_0) + \left.\frac{du_{xi}}{dt}\right|_{t=t_0}\Delta t \tag{28}$$

4.2. Improvement of the HNN Algorithm

Since the energy function of the standard HNN algorithm decreases monotonically, the optimization range of the algorithm is narrow with a fixed optimization direction, which ultimately leads the algorithm to easily fall into a local extremum and is difficult to jump out. The idea of the simulated annealing algorithm exists between the given solution and the new solution is generated in the local area by the given solution. The solution with the better fitness is accepted by Metropolis acceptance criteria [28], while the solution with the poorer fitness is selected to expand the searching range of the solution space. Therefore, in the process of optimizing the standard HNN algorithm, the idea of the simulated annealing algorithm is introduced to expand the optimization range of the standard HNN algorithm and further achieve a better solution.

After the standard HNN algorithm is introduced, the idea of the simulated annealing algorithm during each training process of the standard HNN algorithm causes the neuron input to increase the random disturbance in the HNN algorithm after energy function value, which is computed. The energy function value of the HNN after disturbance input is computed. Energy function value of the original input is compared to the energy function value of the disturbance input; if the energy function value is smaller, it indicates that the value is better. If the original energy function value is not smaller than the energy function value of the disturbance input, the energy function value of the disturbance input is selected to replace the original energy function value, and then the HNN algorithm starts the next generation of optimization. If the original energy function value is smaller than the energy function value of the disturbance input, according to the Metropolis acceptance criteria, the energy value function of the disturbance input is compared to the original energy function value. If the Metropolis acceptance criteria are fulfilled, the energy function value of the disturbance input is chosen to replace the original energy function value, and then the HNN algorithm starts the next generation optimization. Otherwise, the original energy function value does not change and enters the next generation of optimization. After the above process of algorithm optimization is repeated several times, a part of the energy function does not show monotone decreasing characteristic during convergence. However, the global convergence trend of the energy function still shows a decreasing tendency and converges to the optimal equilibrium point. Finally, the ability of the standard HNN algorithm to jump out of local extremum is improved.

The main steps for the HNN algorithm based on the simulated annealing algorithm for solving the scheduling problem are as follows:

Step 1: Set the initial parameters $A, D, u_0, t_0, \Delta t$ of the HNN algorithm.

Step 2: Set the maximum generation K_{max} and the evolutionary generation $K = 0$.

Step 3: Set the initial value of $u_{xi}(t_0)$, whose value is within the interval $[-1, 1]$.

Step 4: Set the initial value of each element V_{xi} in the network initial permutation matrix to 0.

Step 5: Calculate the output $V(t_0)$ of each neuron according to Equation (26), and judge whether the permutation matrix at this time point is valid. The validity of the permutation matrix needs to be judged strictly by the energy function constraint. If the permutation matrix is legal, the output is obtained, C_{max} is calculated, and continue to step 6; if not, return to step 2.

Step 6: Calculate the energy $E(t_0)$ of the network at this time point according to Equation (23).

Step 7: Calculate $\frac{du_{xi}}{dt}$ according to Equation (24).

Step 8: Increase the random disturbance Δt_0 to the input of neuron $u_{xi}(t_0)$, and the neuron input changes into $u_{xi}(t_0 + \Delta t_0)$ at this time. $u_{xi}(t_0 + \Delta t_0)$ is calculated according to Equation (28), and the permutation matrix $V(t_0 + \Delta t_0)$ is updated by $u_{xi}(t_0 + \Delta t_0)$.

Step 9: Calculate the value of energy function in network $E(t_0 + \Delta t)$ at this time.

Step 10: If $E(t_0 + \Delta t_0) \le E(t_0)$, then $E(t_0) = E(t_0 + \Delta t_0)$; if not, then judge whether the result satisfies the Metropolis acceptance criteria. If the criteria are fulfilled, then $E(t_0) = E(t_0 + \Delta t_0)$; if not, then $E(t_0) = E(t_0)$.

Step 11: If the permutation matrix output is legal at this time, then continue to step 12, otherwise return to step 5. At the same time point, evolutionary generation K is processed as $K = K + 1$.

Step 12: If the iteration K has reached the maximum generation K_{max} and the permutation matrix output is legal at this time point, then output the result, otherwise return to step 3.

The flowchart of the SAA–HNN algorithm is shown in Figure 2.

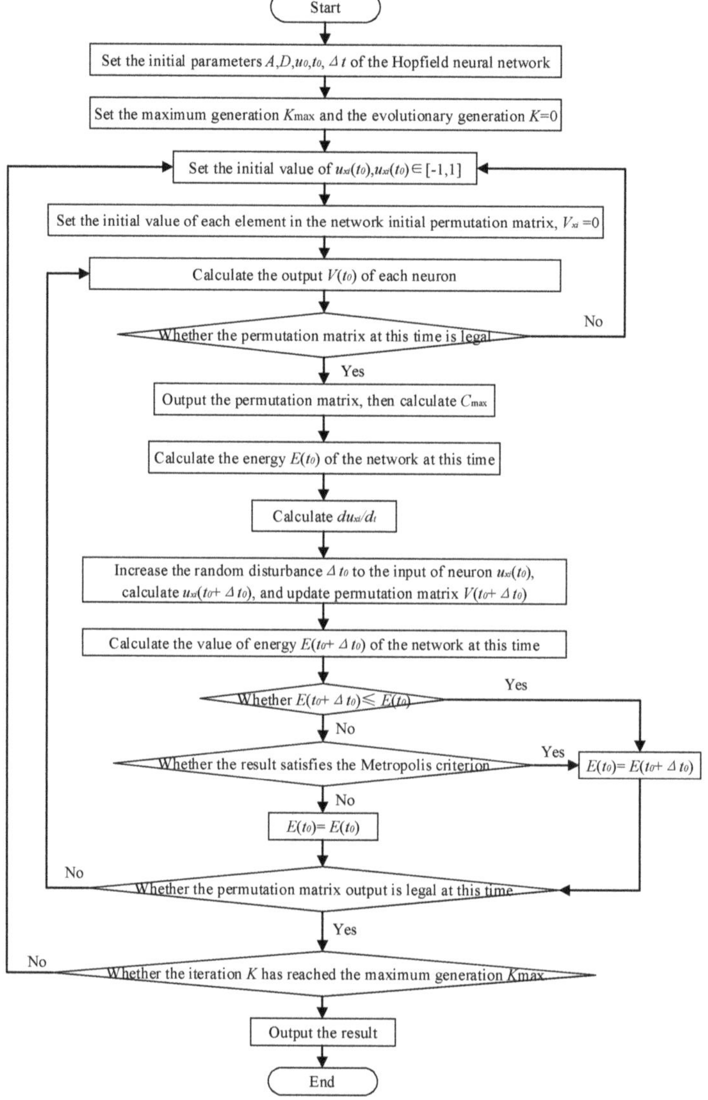

Figure 2. The flowchart of the SAA–HNN algorithm.

4.3. Optimization Performance Testing on the SAA–HNN Algorithm

In order to verify the effect of optimization performance of the HNN algorithm, which adds the idea of the simulated annealing algorithm in comparison to the optimization effect of the improved HNN algorithm and the typical algorithms of the swarm intelligence algorithm, four groups of small-scale FFSP standard example data and six groups of large-scale FFSP instance data were used to test and analyze the SAA–HNN algorithm and its comparison algorithms. The comparison algorithms included ICA, CGA, and HNN.

Four groups of small-scale FFSP standard example data and four groups of large-scale FFSP instance data were used to, respectively, test and analyze. Moreover, the SAA–HNN algorithm was compared to ICA, CGA, and HNN to verify the effect of optimization performance of HNN algorithm, which adds the idea of the simulated annealing algorithm.

The four groups of small-scale FFSP test data were from the standard examples [29,30] proposed by Neron and Taillard based on the standard FFSP, while six groups of large-scale FFSP test data adopted the actual production data. Each algorithm was run 30 times under each group of data, and the average makespan $\overline{C_{max}}$ was applied as the main evaluation index to obtain the result. In addition, the maximum evolution of the four algorithms was 500 generations. The test results are shown in Table 2.

In Table 2, the 'j15c5c1' standard example is taken as an example. 'j15' indicates that the total number of processed workpieces is 15. 'c5' indicates that the total number of processing stages is 5, and since there is only one machine per stage, the total number of machines is also 5. 'c1' represents the difficulty of the standard example. The easy-to-solve examples include six j10c10c* classes, i.e., the total examples of 'a' class and 'b' class. The hard-to-solve examples include the other 'c' classes, i.e., the total examples of 'd' class and 'e' class. *LB* is the lower bound of the makespan of the standard example whose value has been given in the literature [31,32], and *ARE* represents the average relative error of the solution obtained by this algorithm as compared to the lower bound of the makespan. The smaller the *ARE*, the better the optimization effect of the algorithm. $\overline{Time}$ means the average optimization time of the standard example; the smaller the $\overline{Time}$, the better the optimization speed of the algorithm.

According to the test results in Table 2, based on four groups of small-scale standard example data, in terms of optimization effect, *ARE* between the ICA algorithm and the CGA algorithm is 7.4% and 6.1%, respectively, under j15c5c1 standard example data. Therefore, its optimization effect is improved. The *ARE* of the HNN algorithm under j15c5c1 standard example data is 14.5%, which increases to 7.1% and 8.4% in comparison to ICA and CGA, respectively. So the optimization effect is relatively poor. The *ARE* of the SAA–HNN algorithm is 6.1% under j15c5c1 standard example data and is smaller than those of the other three algorithms, so the optimizing effect of the SAA–HNN algorithm is relatively better. The other small-scale standard example data also show the same status. In terms of optimization time, ICA is similar to CGA in average optimization time, which is 6–7 s time range. However, the average optimization time of HNN algorithm is 3–4 s time range and is shortened to about 50% as much as ICA and CGA with obvious improvement in the optimization speed. Although average optimization time of the SAA–HNN algorithm increases about 0.5 s as much as the HNN algorithm, it retains high optimization speed. In terms of four groups of large-scale data, on the basis of optimization effect, the SAA–HNN algorithm increases about 4% as compared to the other three algorithms. However, in terms of optimization time, the optimization speed of the SAA–HNN algorithm accelerates about 70% as much as ICA and CGA.

Based on the above tests using four groups of small-scale data and six groups of large-scale data for four algorithms, the SAA–HNN algorithm has faster optimization speed than the comparison algorithms, with significantly improved evaluation index under small-scale data. Also, under large-scale data, the SAA–HNN algorithm still maintains a high speed of optimization, and the evaluation index is relatively better. Therefore, the SAA–HNN algorithm is suitable for solving large-scale and complicated scheduling optimization problems, and hence, has a broad applicability.

During the test, four algorithms were considered based on 30 simulation experiments with large-scale data. The makespan C_{max} obtained at each time point exhibited obvious fluctuation. Since the HNN algorithm can easily fall into local extremum, its C_{max} may appear as a large outlier. In order to compare and evaluate the optimization effect of the four algorithms on large-scale data, j80c8a2 data were considered as the example, and the C_{max} values obtained by running four algorithms for 30 times are drawn into a box plot (Figure 3).

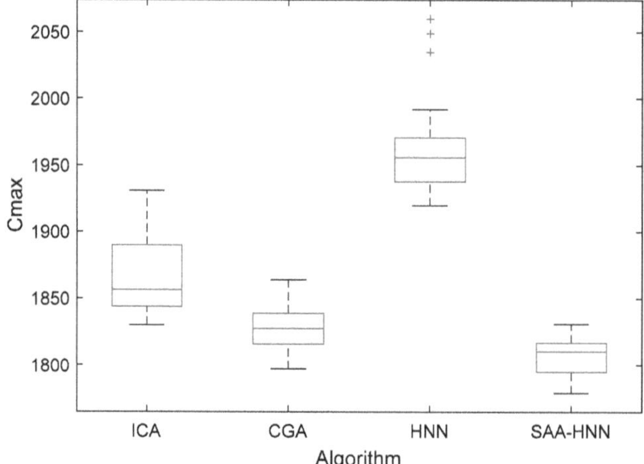

Figure 3. Box-plot of four algorithms for large-scale data j80c8a2.

A box-plot is a statistical graph for describing the discrete degree of a group of data. The stability of the optimization effect can be reflected by the box-plot. The interquartile range *IQR* was used to measure the discrete degree of the data in the box-plot.

As can be seen from Figure 3, the HNN algorithm generates three large outliers of 2033, 2047, and 2056, respectively, in the 30 running times of the algorithm, indicating that the HNN algorithm easily falls into the local extremum. The median of box-plot produced by the SAA–HNN algorithm is 1815, and other medians of box-plot produced by other algorithms are 1854, 1828, and 1956, respectively. Therefore, the box-plot of the SAA–HNN algorithm is in the lowest position, which indicates that the overall quality of the solution generated by the SAA–HNN algorithm is better than the other three algorithms. Besides, the *IQR* of the box-plot generated by the SAA–HNN algorithm is 23, and the *IQR* of the box-plot generated by the other three algorithms are 25, 46, and 38, respectively, indicating that the SAA–HNN algorithm produces the smallest discrete degree, and the stability of the scheduling results under large-scale data was optimal among the four algorithms.

Based on the above analysis, the SAA–HNN algorithm is better than the ICA, CGA, and HNN algorithms, and it also maintains fast optimization of the HNN algorithm in terms of optimization speed while solving the scheduling problem of small-scale and large-scale data. This indicates that the idea of the simulated annealing algorithm overcomes the HNN algorithm for falling into local extremum, and the SAA–HNN algorithm has a strong ability of continuous evolution.

5. FFSP–PB Instance Test

Taking a bus manufacturer in actual large-scale equipment manufacturing enterprises as an example, simulation data similar to the production operation of the body shop and paint shop for the bus manufacturer were constructed. The body shop of the bus manufacturer was a rigid flow shop with multiple production lines that could be simplified into one production stage. The paint shop could be simplified into three stages. Therefore, the simulation data included four stages: $\{Stage_1, Stage_2, Stage_3, Stage_4\}$. The parallel workstation of these four stages was $\{M_j\} = \{3, 3, 3, 3\}$. A limited buffer occurred between each stage, and the maximum capacity of each limited buffer was $\{LBu_2, LBu_3, LBu_4\} = \{2, 2, 2\}$. In addition, a public buffer was set on the production line, and the standard processing time for transferring the workpiece TBt was 5. The maximum buffer capacity in the public buffer was 3.

The simulation test first analyzed the impact of the relevant local scheduling rules for FFSP–PB on the scheduling results, and discussed the role of the public buffer and local scheduling rules for the public buffer in alleviating the production blockage and improving the scheduling results. Finally, the SAA–HNN algorithm and other global optimization algorithms were combined with the local scheduling rules, respectively, to solve the FFSP–PB problem under different data scales, which verified the optimization performance of the SAA–HNN algorithm with respect to the complex scheduling problems. Thus, the efficiency of the combination of the SAA–HNN algorithm and local scheduling rules for solving the FFSP–PB problem was assessed.

5.1. Evaluation Index of Scheduling Results

In order to better analyze and study the scheduling results during the scheduling process, we employed the makespan C_{max} as the optimization goal, establishing other evaluation indexes related to the actual production, including the total workstation idle time $TWIT$, the total plant factor TPF, and total workpiece blockage time $TWBT$. Except for TPF, the smaller the value, the better the other evaluation indexes:

(1) Makespan

$$C_{max} = \max\{Tc_{i,m}\}, \; i \in \{1, \ldots n\} \tag{29}$$

In Equation (29), makespan C_{max} indicates the maximum value of all workpieces that completed processing at the last stage.

(2) Total workstation idle time

$$TWIT = \sum_{j=1}^{m} \sum_{l=1}^{M_j} \left(\left(\max\{To_{i,j} \cdot At_{i,j,l}\} - \min\{S_{i,j} \cdot At_{i,j,l}\} \right) - \sum_{i=1}^{n} \left(Tb_{i,j} \cdot At_{i,j,l} \right) \right) \tag{30}$$

In Equation (30), TWT represents the sum of the idle time for the workstation between the start time of the first processed workpiece and the completion time of the last processed workpiece.

(3) Total plant factor

$$TPF = \frac{\sum\limits_{j=1}^{m} \sum\limits_{i=1}^{n} \left(Tb_{i,j} \right)}{\sum\limits_{j=1}^{m} \sum\limits_{i=1}^{n} \sum\limits_{l=1}^{M_j} \left(\max\{To_{i,j} \cdot At_{i,j,l}\} - \min\{S_{i,j} \cdot At_{i,j,l}\} \right)} \tag{31}$$

In Equation (31), TPF represents the total plant factor of all workstations, which is the ratio of the effective processing time of all workstations to the occupied period of all workstations. This period starts from the first workpiece processed on the workstation at each stage to the last workpiece that leaves the workstation after completion [33].

(4) Total workpiece blockage time

$$TWBT = \sum_{i=1}^{n} \sum_{j=2}^{m} \left(To_{i,j-1} - C_{i,j-1} \right) \tag{32}$$

In Equation (32), $TWBT$ indicates the sum of blockage time of all workpieces stuck on the workstations since the limited buffer is full and the electric flat carriage is in transit during the production.

5.2. Instance Test of FFSP–PB Local Scheduling Rules

5.2.1. Simulation Scheme

In order to verify the efficiency of the public buffer in reducing the production blockage during the process of buffer dynamic capacity-increase in flexible flow shop and analyze the role of the relevant local scheduling rules for the public buffer, three groups of simulation schemes were designed to solve the flexible flow shop scheduling problem with the limited buffer. Scheme 1: There is no public buffer and no reentrant rules of electric flat carriage or workpiece transfer rules in the public buffer. Scheme 2: There is a public buffer and no reentrant rules of electric flat carriage or workpiece transfer rules in the public buffer. Scheme 3: There is a public buffer, reentrant rules of electric flat carriage, and workpiece transfer rules in the public buffer.

5.2.2. Simulation Results and Analysis

To obtain a better evaluation of the scheduling results, 30 different online sequences were randomly generated, and the values of the evaluation index were obtained through simulation tests. Moreover, the average values of each evaluation index for each scheme based on 30 different online sequences were calculated (Table 3), and Equation (33) was established. $IR(A/B)$ indicates the improvement range of A with respect to the designated evaluation index of B, and the meanings of A and B would be modified according to the actual situation.

$$IR(A/B) = \left|[(A-B)/B]\right| \times 100\% \tag{33}$$

Table 3. Comparison of evaluation indexes for the three scheme scheduling results.

Scheme	Evaluation Index			
	C_{max}	$TWIT$	TPF	$TWBT$
1	182.6	40.4	0.924	26.3
2	168.5	37.1	0.951	18.4
3	159.3	22.5	0.978	5.9
$IR(2/1)$	7.72%	8.17%	2.92%	30.04%
$IR(3/1)$	12.76%	44.31%	5.84%	77.57%
$IR(3/1)$	5.46%	39.35%	2.84%	67.93%

As is shown in Table 3, the main evaluation index makespan C_{max} and the total workpiece blockage time $TWBT$ of Scheme 2 with the public buffer decreases to 14.1 and 7.9, respectively, as compared to that of Scheme 1 without the public buffer, and the improvement is 7.72% and 30.04%, respectively. If production line has a public buffer but does not establish corresponding local scheduling rules, it might lead to new production blockage. Thus, the reentrant rules of electric flat carriage and workpiece transfer rules in the public buffer are established. The main evaluation index makespan C_{max} and total workpiece blockage time $TWBT$ of the reentrant rules of electric flat carriage in Table 3 and workpiece transfer rules in the public buffer of Scheme 3 decreases to 9.2 and 12.5, respectively, as compared to that of Scheme 2 with the public buffer but without the corresponding local scheduling rules, and the improvement is 5.46% and 67.93%, respectively. Meanwhile, the other evaluation indexes are also improved. The above analysis indicates that the public buffer adds in the flexible flow shop and relevant local scheduling rules can be established in order to relieve the production blockage effectively.

5.2.3. Gantt Chart Analysis of the Scheduling Result

Figure 4 shows the Gantt chart of the scheduling result of Scheme 3. The abscissa is the time axis, while the ordinate indicates the workstation of each stage, the limited buffer, and the public buffer. The violet part in the figure represents the time for the workpiece that is temporarily stored in the limited buffer; the red part indicates that the electric flat carriage transfers the workpiece from the workstation to the public buffer; the blue part indicates the time for the electric flat carriage that transfers the workpiece back to the limited buffer when transporting the workpiece from the workstation to the public buffer; the yellow part denotes that the workpiece is stored temporarily in the public buffer; the green part denotes that the electric flat carriage transports the workpiece from the public buffer to the limited buffer; the orange part in the figure represents the blockage time of the workpiece in the stage. Figure 4 demonstrates that restricted by Equation (6), the workpiece in each stage is processed only once at one workstation, and hence, the processing route of the workpiece Wp_5 is $\{WS_{1,2}, Rp_5, PBu, Rl_5, Bl_{2,1}, WS_{2,1}, Bl_{3,2}, WS_{3,1}, Rp_5, Rb_5, Bl_{4,1}, WS_{4,1}\}$, i.e., the processing route is represented by the connecting lines between the blocks in the figure.

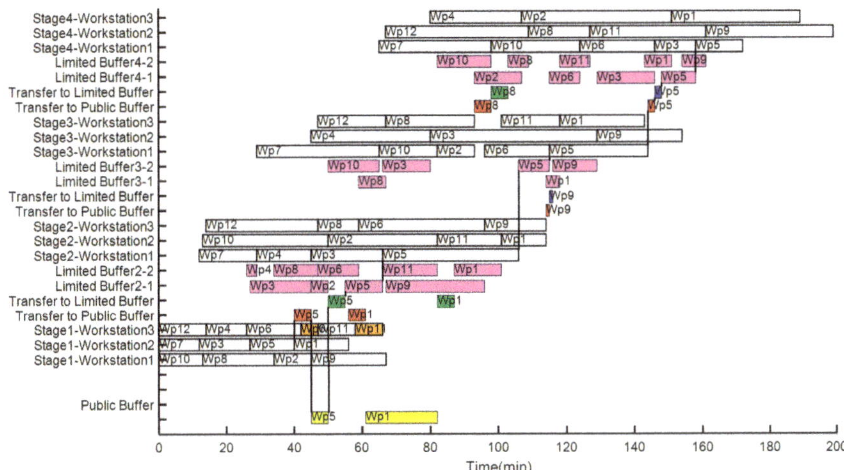

Figure 4. Gantt chart of scheme 3 scheduling result.

At time $t = 40$, the workpiece Wp_5 completed the manufacturing task at workstation $WS_{1,2}$ of stage $Stage_1$. Simultaneously, restricted by Equation (11), the limited buffer LBu_2 reached the upper limit of its capacity $WAl_2(40) = \{Wp_3, Wp_8\}$; thus, the workpiece Wp_5 started to be transferred to the public buffer. At time $t = 45$, the workpiece Wp_5 entered the public buffer. At time $t = 50$, the workpiece Wp_{10} completed the manufacturing task at workstation $WS_{2,2}$ of stage $Stage_2$, and then left the workstation. The workpiece Wp_2 entered the workstation $WS_{2,2}$ of stage $Stage_2$, the space $Bl_{2,1}$ in the limited buffer LBu_2 was available, and the workpiece Wp_5 in the public buffer that should enter the limited buffer LBu_2 existed. Therefore, the workpiece transfer rules in the public buffer took effect. Since the electric flat carriage was at the position of the public buffer $Pan\_Car(50) = 2$, the standard processing time for transferring the workpiece $Tbt = 5$ was compared to the estimated minimum completion time $\min\{(Tc_{i,1} - t) | (Ts_{i,1} - t) \leq 0, (Tc_{i,1} - t) > 0, i \in \{1,9,11\}\} = 6$ for all workpieces at stage $Stage_1$. Since the standard processing time for transferring the workpiece was less than the estimated minimum completion time for all workpieces of stage $Stage_1$, the electric flat carriage transfered the workpiece Wp_5 from the public buffer to the space $Bl_{2,1}$ in the limited buffer LBu_2. At time $t = 55$, the workpiece Wp_5 entered the space $Bl_{2,1}$ of the limited buffer LBu_2.

At time $t = 144$, the workpiece Wp_5 completed the manufacturing task at workstation $WS_{3,1}$ of stage $Stage_3$. In addition, restricted by Equation (11), the limited buffer LBu_4 reached the upper limit of its capacity $WAl_4(144) = \{Wp_1, Wp_3\}$, following which, the workpiece Wp_5 started to be transferred to the public buffer. At time $t = 146$, the workpiece Wp_6 completed the manufacturing task at workstation $WS_{4,1}$ of stage $Stage_4$ and left the workstation. When the workpiece Wp_3 entered the workstation $WS_{4,1}$ of stage $Stage_4$, the space $Bl_{4,1}$ in the limited buffer LBu_4 was available and the workpiece Wp_5

was still in transportation $Pan\_Car(146) = 3$, so the reentrant rules of the electric flat carriage took effect. Also, the already-spent transit time $Ttw_{5,3}(146) = 2$ for transporting the workpiece Wp_5 from the workstation of stage $Stage_3$ to the public buffer was compared to the estimated minimum completion time $\min\left\{(Tc_{i,4} - t)\middle|(Ts_{i,4} - t) \le 0, (Tc_{i,4} - t) > 0, i \in \{9\}\right\} = 8$ for all workpieces at stage $Stage_3$. Since the already-spent transit time of the workpiece Wp_5 was less than the estimated minimum completion time for all workpieces of stage $Stage_3$, the electric flat carriage returned the workpiece Wp_5 to space $Bl_{4,1}$ in the limited buffer LBu_4. Thus, at time $t = 148$, the workpiece Wp_5 entered the space $Bl_{4,1}$ of the limited buffer LBu_4. Based on the above analysis, restricted by Equation (17), the already-spent transit time $Ttw_{5,3}(146)$ for transporting the workpiece Wp_5 from the workstation of stage $Stage_3$ to the public buffer should be less than the standard processing time for transferring workpiece.

From the above specific analysis of the Gantt chart about scheduling results, the reentrant rules of the electric flat carriage designed for the public buffer state that, when the workpiece is transferred to the public buffer, it should enter the available space in the limited buffer, following which, the workpiece reenters and is stored in the limited buffer to reduce its storage time in the public buffer and the total transfer time. Based on the workpiece transfer rules in public buffer, and in the event of available space in the limited buffer and that the workpiece transfers from the public buffer to the limited buffer and the previous stage of the limited buffer does not have the the completed workpiece, the workpiece in the public buffer is transferred to the available space in the limited buffer. This avoids competition for the available space in the limited buffer, which in turn, effectively reduces the new production blockage caused by the increase in the public buffer.

5.3. Instance Test of FFSP–PB Global Optimization Algorithm

5.3.1. Parameter Settings of the Optimization Algorithm

ICA, CGA, HNN, and SAA–HNN were utilized as global optimization algorithms. Moreover, these algorithms were combined with the reentrant rules of electric flat carriage and workpiece transfer rules in the public buffer designed for the public buffer to solve the FFSP–PB, which are optimal for comparing and analyzing the optimization performance and verifying its efficiency with local scheduling rules to resolve FFSP–PB. The parameter settings of the four global optimization algorithms are shown in Table 4.

Table 4. Swarm evolutionary algorithm parameters.

Optimization Algorithm	Algorithm Parameter
ICA	Maximum evolutional generation $Gen = 500$; number of imperialist countries $N_d = 5$; number of colonial countries $N_z = 25$; colonial impact factor $K = 0.15$; similarity threshold $\alpha = 0.3$.
CGA	Maximum evolutional generation $Gen = 500$; number of populations $NP = 4$; adjustment override of the learning rate $\beta = 0.08$.
HNN	Maximum evolutional generations $Gen = 500$; $A = 1.5$; $D = 1$; $u_0 = 0.02$; $\Delta t = 0.1$
SAA–HNN	Maximum evolutional generations $Gen = 500$; $A = 1.5$; $D = 1$; $u_0 = 0.02$; $\Delta t = 0.1$; initial temperature T_{max}; end temperature T_{min}; cooling coefficient b.

5.3.2. Simulation Results and Analysis

(1) Evaluation Index of Scheduling Results

Each of the four algorithms were tested on the small-scale, medium-scale, and large-scale data. The total number n of the processed workpieces was set to 12 in small-scale data, 40 in medium-scale data, and 80 in large-scale data.

i   Small-scale data

The four algorithms were run 30 times for small-scale data, and the average values of the evaluation indexes obtained from 30 simulations are summarized in Table 5.

Table 5. Comparison of the evaluation indexes of four algorithms' scheduling results (small-scale data).

Algorithms	Evaluation Indexes			
	C_{max}	*TWIT*	*TPF*	*TWBT*
ICA	176.54	46.42	0.931	27.42
CGA	157.89	39.56	0.952	22.17
HNN	189.55	47.32	0.927	33.30
SAA–HNN	129.92	22.59	0.978	9.89
$IR(\frac{SAA\text{-}HNN}{ICA})$	26.41%	51.34%	5.04%	63.93%
$IR(\frac{SAA\text{-}HNN}{CGA})$	17.71%	42.90%	2.73%	55.39%
$IR(\frac{SAA\text{-}HNN}{HNN})$	31.46%	52.26%	5.50%	70.3%

From the simulation results of small-scale data in Table 5, during the solving of FFSP–PB, the main evaluation index makespan C_{max} of the SAA–HNN algorithm decreases to 46.62, 27.97, and 59.63 in comparison to the ICA, CGA, and HNN algorithms, respectively, and the improvement is 26.42%, 17.71%, and 31.46%, respectively. Moreover, the total workpiece blockage time *TWBT* of the SAA–HNN algorithm decreases 17.53, 12.28, and 23.41 in comparison to the other three algorithms, respectively, and the improvement is 63.93%, 55.39%, and 70.3%, respectively. The total plant factor *TPF* and total workstation idle time *TWBT* also improves, thereby indicating that using the SAA–HNN algorithm as the global optimization algorithm to solve the FFSP–PB problem improves each evaluation index and reduces the production blockage in small-scale data simulation.

ii   Medium-scale and large-scale data

The four algorithms were run 30 times for medium-scale and large-scale data, respectively, and the average values of the evaluation indexes obtained from the 30 simulations under the two data scales are listed in Tables 6 and 7.

Table 6. Comparison of the evaluation indexes of four algorithms' scheduling results (medium-scale data).

Algorithms	Evaluation Indexes			
	C_{max}	*TWIT*	*TPF*	*TWBT*
ICA	834.15	269.26	0.885	92.46
CGA	801.37	243.16	0.916	67.71
HNN	893.32	312.55	0.881	99.21
SAA–HNN	704.85	197.87	0.979	16.94
$IR(\frac{SAA\text{-}HNN}{ICA})$	15.50%	26.51%	10.62%	81.68%
$IR(\frac{SAA\text{-}HNN}{CGA})$	12.04%	18.63%	6.88%	74.24%
$IR(\frac{SAA\text{-}HNN}{HNN})$	21.10%	36.69%	11.12%	82.93%

Table 7. Comparison of the evaluation indexes of four algorithms' scheduling results (large-scale data).

Algorithms	Evaluation Indexes			
	$\overline{C_{max}}$	$\overline{TWIT}$	$\overline{TPF}$	$\overline{TWBT}$
ICA	1794.39	687.21	0.879	145.22
CGA	1650.60	582.57	0.899	106.06
HNN	1943.35	706.02	0.837	160.42
SAA–HNN	1379.16	462.18	0.982	25.15
$IR(\frac{SAA\text{-}HNN}{ICA})$	23.14%	32.75%	11.72%	82.68%
$IR(\frac{SAA\text{-}HNN}{CGA})$	16.44%	20.67%	9.23%	76.29%
$IR(\frac{SAA\text{-}HNN}{HNN})$	29.03%	34.54%	17.32%	84.32%

From the simulation results in Tables 6 and 7, under medium-scale data, the main evaluation index makespan C_{max} of the SAA–HNN algorithm decreases to 129.3, 96.52, and 188.47 in comparison to the ICA, CGA, and HNN algorithms, respectively, and the improvement is 15.50%, 12.04%, and 21.10%, respectively. Under large-scale data, the main evaluation index makespan C_{max} of the SAA–HNN algorithm decreases to 415.23, 271.44, and 564.19 in comparison to the ICA, CGA and HNN algorithms, respectively, and the improvement is 23.14%, 16.44%, and 29.03%, respectively. Under medium-scale data and large-scale data, other evaluation indexes of the SAA–HNN algorithm also improve. The SAA–HNN algorithm performs optimally and has perfect adaptability during the solving of the FFSP–PB of medium-scale and large-scale data.

Based on the comprehensive analysis of the simulated results in the FFSP–PB of different data scales, it can be concluded that the SAA–HNN algorithm combines the reentrant rules of the electric flat carriage and workpiece transfer rules in the public buffer which are designed for the public buffer, effectively solves the FFSP–PB, and reduces the production blockage.

(2) Scheduling Evolutionary Process Analysis

The correlation between the makespan C_{max} and the iterations of four algorithms under actual production data are shown in Figure 5.

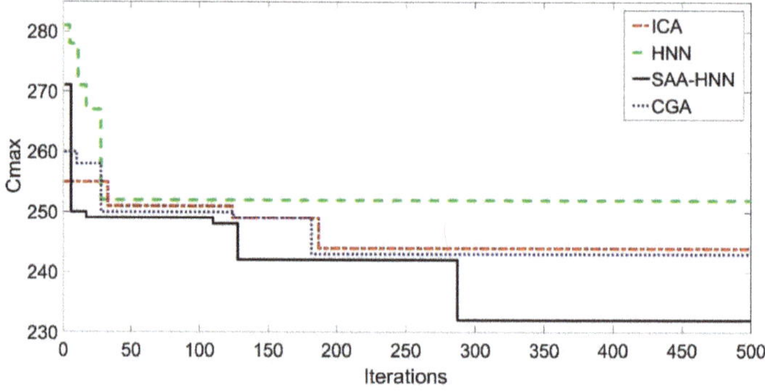

Figure 5. Correlation between makespan C_{max} and the iterations of four algorithms.

Figure 5 shows that the HNN algorithm converges rapidly in the initial stage of evolution but falls into local extremum prematurely due to the monotonous decrease in the energy function; it stagnates evolving in the 31st generation when C_{max} converges to 252 eventually. The ICA and CGA algorithms have the ability of rapid optimization and convergence in the initial stage; however, they display integral convergence after evolving for a specific number of generations, making them easy to fall into local extremum. These phenomena stagnate evolving in the 221st and 185th generations, respectively, and their C_{max} finally converge to 246 and 245, respectively. The SAA–HNN algorithm, which maintains the fast optimization feature of the HNN algorithm, converges rapidly in the initial stage of evolution, but falls into local extremum in the 127th generation. By introducing the idea of the simulated annealing algorithm, the ability of the SAA–HNN algorithm to jump out of local extremum is enhanced. Also, the SAA–HNN algorithm reactivates the evolutional process in the 278th generation, whose C_{max} converges to 233 eventually.

From the above analysis of the scheduling evolutionary process, the SAA–HNN algorithm not only maintains the fast optimization characteristics of the HNN algorithm, but also jumps out of the local extremum and continues evolution in the process of solving the FFSP–PB problem with the improved local scheduling rules. Also, the SAA–HNN algorithm has better optimization effect than the ICA and CGA algorithms.

6. Conclusions

The present study investigated the FFSP–PB problem. Since the standard HNN algorithm easily falls into local extremum and is difficult for continuous evolution, this study proposed the SAA–HNN algorithm for global optimization. It adopts the Metropolis acceptance mechanism of the simulated annealing algorithm, such that the HNN algorithm can accept the non-optimal solution. Thus, the evolutionary vitality of the HNN algorithm is enhanced, rendering it the ability to jump out of the local extremum. Consecutively, considering the influence of the public buffer on the scheduling process, the reentrant rules of the electric flat carriage and workpiece transfer rules in the public buffer for controlling the movement of the workpiece are designed according to the transit time-cost of the workpiece among the workstation, the limited buffer, and the public buffer, as well as the processing status of the workpiece during the production. This phenomenon reduces the production blockage and improves the utilization of production resources. Finally, the simulation experiment proves that the SAA–HNN algorithm, combined with the improved local scheduling rules, can solve the FFSP–PB problem.

The actual production process has some local scheduling rules, such as setup time rules, process specification rules, and customer specification rules. If these local scheduling rules coexist with the relevant local scheduling rules established for the public buffer in the present study, the complexity of the whole scheduling optimization process would increase, and certain conflicts would occur between some of the local scheduling rules. Since the problem of conflict resolution between local scheduling rules are beyond the scope of this study, the main directions of future work are divided into the following:

(1) Since the complex scheduling problem was investigated, the methods in the field of artificial intelligence should be explored to further improve the optimization effect and intellectualization of production scheduling algorithms.

(2) Various local scheduling rules conflict with each other in the actual production system, necessitating that the method of effectively eliminating the conflicts between various local scheduling rules is explored further.

Author Contributions: Conceptualization, Z.H. and C.H.; methodology, S.L.; validation, X.D. and H.S.; writing—original draft preparation, Z.H.; writing—review and editing, C.H.; supervision, H.S.; funding acquisition, Z.H.

Funding: This research was funded by the Liaoning Provincial Science Foundation, China (grant number: 2018106008), the Natural Science Foundation of China (grant number: 61873174), the Project of Liaoning Province Education Department, China (grant number: LJZ2017015), and the Shenyang Municipal Science and Technology Project, China (grant number: Z18-5-015).

Conflicts of Interest: The authors declare no conflicts of interest.

References

1. Gao, Z.W.; Saxen, H.; Gao, C.H. Guest Editorial Special Section on Data-Driven Approaches for Complex Industrial Systems. *IEEE Trans. Ind. Inform.* **2013**, *9*, 2210–2212. [CrossRef]
2. Gao, Z.W.; Kong, D.X.; Gao, C.H. Modeling and Control of Complex Dynamic Systems: Applied Mathematical Aspects. *J. Appl. Math.* **2012**, *2012*, 1–18. [CrossRef]
3. Zambon, I.; Egidi, G.; Rinaldi, F. Applied Research Towards Industry 4.0: Opportunities for SMEs. *Processes* **2019**, *7*, 344. [CrossRef]
4. Khamseh, A.; Jolai, F.; Babaei, M. Integrating sequence-dependent group scheduling problem and preventive maintenance in flexible flow shops. *Int. J. Adv. Manuf. Technol.* **2015**, *77*, 173–185. [CrossRef]
5. Gerstl, E.; Mosheiov, G.; Sarig, A. Batch scheduling in a two-stage flexible flow shop problem. *Found. Comput. Decis. Sci.* **2014**, *39*, 3–16. [CrossRef]
6. Gupta, J.N.D. Two-stage, Hybrid flow shop scheduling problem. *Oper. Res.* **1988**, *39*, 359–364. [CrossRef]
7. Han, Z.H.; Ma, X.F.; Yao, L.L.; Shi, H.B. Cost Optimization Problem of Hybrid Flow-Shop Based on PSO Algorithm. *Adv. Mater. Res.* **2012**, *532*, 1616–1620. [CrossRef]
8. Han, Z.H.; Sun, Y.; Ma, X.F.; Lv, Z. Hybrid flow shop scheduling with finite buffers. *Int. J. Simul. Process Model.* **2018**, *13*, 156–166. [CrossRef]
9. Lalami, I.; Frein, Y.; Gayon, J.P. Production planning in automotive powertrain plants: A case study. *Int. J. Prod. Res.* **2017**, *55*, 5378–5393. [CrossRef]
10. Han, Z.H.; Zhu, Y.H.; Ma, X.F.; Chen, Z.L. Multiple rules with game theoretic analysis for flexible flow shop scheduling problem with component altering times. *Int. J. Model. Identif. Control* **2016**, *26*, 1–18. [CrossRef]
11. Nahas, N. Buffer allocation and preventive maintenance optimization in unreliable production lines. *J. Intell. Manuf.* **2017**, *28*, 85–93. [CrossRef]
12. Hibino, H.; Yamamoto, M.; Yamaguchi, M.; Kobayashi, T. A study on Lot-Size dependence of energy consumption per unit of production throughput considering buffer capacity. *Int. J. Autom. Technol.* **2017**, *11*, 46–55. [CrossRef]
13. Xi, S.H.; Chen, Q.X.; Mao, N.; Li, X.; Yu, A.L.; Zhang, H.Y. Capacity optimal configuration method of large-scale finite buffer production line. *Comput. Integr. Manuf. Syst.* **2017**, *23*, 2200–2210.
14. Dolgui, A.B.; Eremeev, A.V.; Sigaev, V.S. Analysis of a multicriterial buffer capacity optimization problem for a production line. *Autom. Remote Control* **2017**, *78*, 1276–1289. [CrossRef]
15. Gao, Z.W.; Nguang, S.K.; Kong, D.X. Advances in Modelling, Monitoring, and Control for Complex Industrial Systems. *Complexity* **2019**. [CrossRef]
16. Wang, X.H.; Gu, X.W.; Liu, Z.B. Production Process Optimization of Metal Mines Considering Economic Benefit and Resource Efficiency Using an NSGA-II Model. *Processes* **2018**, *6*, 228. [CrossRef]
17. Georgiadis, G.P.; Elekidis, A.P.; Georgiadis, M.C. Optimization-Based Scheduling for the Process Industries: From Theory to Real-Life Industrial Applications. *Processes* **2019**, *7*, 438. [CrossRef]
18. Han, Z.H.; Zhang, Q.; Shi, H.B. An Improved Compact Genetic Algorithm for Scheduling Problems in a Flexible Flow Shop with a Multi-Queue Buffer. *Processes* **2019**, *7*, 302. [CrossRef]
19. Zhang, G.H.; Xing, K.Y. Differential evolution metaheuristics for distributed limited-buffer flowshop scheduling with makespan criterion. *Comput. Oper. Res.* **2019**, *108*, 33–43. [CrossRef]
20. Rooeinfar, R.; Raissi, S.; Ghezavati, V.R. Stochastic flexible flow shop scheduling problem with limited buffers and fixed interval preventive maintenance: A hybrid approach of simulation and metaheuristic algorithms. *Simulation* **2019**, *95*, 509–528. [CrossRef]
21. Jiang, S.L.; Zhang, L. Energy-Oriented Scheduling for Hybrid Flow Shop with Limited Buffers Through Efficient Multi-Objective Optimization. *IEEE Access* **2019**, *7*, 34477–34487. [CrossRef]
22. Zeng, C.K.; Liu, S.X. Job Shop Scheduling Problem with Limited Output Buffer. *Dongbei Daxue Xuebao J. Northeast. Univ.* **2018**, *39*, 1679–1684. [CrossRef]
23. Ribas, I.; Companys, R.; Tort-Martorell, X. An iterated greedy algorithm for solving the total tardiness parallel blocking flow shop scheduling problem. *Expert Syst. Appl.* **2019**, *121*, 347–361. [CrossRef]
24. Chang, P.C. Reliability with finite buffer size for a multistate manufacturing system with parallel production lines. *J. Chin. Inst. Eng.* **2017**, *40*, 275–283. [CrossRef]
25. Johri, P.K. A linear programming approach to capacity estimation of automated production lines with finite buffers. *Int. J. Prod. Res.* **1987**, *25*, 851–867. [CrossRef]

26. Sun, S.Y.; Zheng, J.L. A modified algorithm and theoretical analysis for hopfield network solving TSP. *Acta Electron. Sin.* **1995**, *23*, 73–78.
27. Yan, Y.L. A Solving Method to TSP Based on Improved Hopfield Neural Networks. *J. Minnan Norm. Univ.* **2014**, *27*, 37–43.
28. Metropolis, N.; Rosenbluth, A.W.; Rosenbluth, M.N.; Teller, A.H.; Teller, E. Equation of state calculations by fast computing machines. *J. Chem. Phys.* **1953**, *21*, 1087–1092. [CrossRef]
29. Neron, E.; Baptiste, P.; Gupta, J.N.D. Solving hybrid flow shop problem using energetic reasoning and global operations. *Omega* **2001**, *29*, 501–511. [CrossRef]
30. Taillard, E. Benchmarks for basic scheduling problems. *Eur. J. Oper. Res.* **1993**, *22*, 278–285. [CrossRef]
31. Carlier, J.; Neron, E. An exact met-hod for solving the multi-processor flow-shop. *RAIRO Oper. Res.* **2000**, *34*, 1–2. [CrossRef]
32. Santos, D.L.; Hunsucker, J.L.; Deal, D.E. Global lower bounds for flow shop with multiple processors. *Eur. J. Oper. Res.* **1995**, *80*, 112–120. [CrossRef]
33. Han, Z.H.; Dong, X.T.; Shi, H.B. Improved DE algorithm for hybrid flow shop load balancing scheduling problem. *Comput. Integr. Manuf. Syst.* **2016**, *22*, 548–557.

Article

An Improved Compact Genetic Algorithm for Scheduling Problems in a Flexible Flow Shop with a Multi-Queue Buffer

Zhonghua Han [1,2,3,4], **Quan Zhang** [2,*], **Haibo Shi** [1,3,4] and **Jingyuan Zhang** [2]

[1] Department of Digital Factory, Shenyang Institute of Automation, the Chinese Academy of Sciences (CAS), Shenyang 110016, China; xiaozhonghua1977@163.com (Z.H.); hbshi@sia.cn (H.S.)
[2] Faculty of Information and Control Engineering, Shenyang Jianzhu University, Shenyang 110168, China; Nlnlznl_0307@163.com
[3] Key Laboratory of Network Control System, Chinese Academy of Sciences, Shenyang 110016, China
[4] Institutes for Robotics and Intelligent Manufacturing, Chinese Academy of Sciences, Shenyang 110016, China
[*] Correspondence: zhangq0716@163.com; Tel.: +86-24-24690045

Received: 3 April 2019; Accepted: 15 May 2019; Published: 21 May 2019

Abstract: Flow shop scheduling optimization is one important topic of applying artificial intelligence to modern bus manufacture. The scheduling method is essential for the production efficiency and thus the economic profit. In this paper, we investigate the scheduling problems in a flexible flow shop with setup times. Particularly, the practical constraints of the multi-queue limited buffer are considered in the proposed model. To solve the complex optimization problem, we propose an improved compact genetic algorithm (ICGA) with local dispatching rules. The global optimization adopts the ICGA, and the capability of the algorithm evaluation is improved by mapping the probability model of the compact genetic algorithm to a new one through the probability density function of the Gaussian distribution. In addition, multiple heuristic rules are used to guide the assignment process. Specifically, the rules include max queue buffer capacity remaining (MQBCR) and shortest setup time (SST), which can improve the local dispatching process for the multi-queue limited buffer. We evaluate our method through the real data from a bus manufacture production line. The results show that the proposed ICGA with local dispatching rules and is very efficient and outperforms other existing methods.

Keywords: flexible flow shop scheduling; multi-queue limited buffers; improved compact genetic algorithm; probability density function of the Gaussian distribution

---

1. Introduction

The body shop and paint shop for the bus manufacturers are flexible flow shops. The processing flow is divided into multiple stages, and there are multiple parallel machines to process jobs in each stage. Due to the large volume of the buses and the long production cycle, only buffers with a limited number of spaces can be deployed in the production line. At the same time, the bus is not equipped with a chassis for starting the engine in the installation workshop, and they can only be carried on the skids. Therefore, there is usually a buffer between the body shop and paint shop, and the buffer is usually divided into multiple lanes for the ease of scheduling and operation. Each lane has an equal number of spaces for the bus bodies. The bus enters the lane from one side and exits the lane from the other side. The bus body which is waiting for processing will form a waiting queue in each lane. Thus, for multiple lanes, there could exist multiple waiting queues. These waiting queues together are called a multi-queue limited buffer in this paper. When the bus completes the processing in the body shop, it needs to select one of the lanes in the multi-queue limited buffer and enter the corresponding waiting queue, and the bus is carried by an electric flat carriage to move in the queue. When there is an idle machine in the paint shop, one bus in the multiple waiting queues can be moved out of the buffer to the paint shop for processing. In the actual production line, if the properties such as the model and color are different from those of the previous bus processed by the machine, the cleaning and adjustment of the equipment also have to be done on the machine before proceeding to the next, which results in an extra setup time in addition to the standard processing time. Therefore, these scheduling problems in a bus manufacturer can be characterized as the multi-queue limited buffer scheduling problems in a flexible flow shop with setup times.

The research status of the limited buffers scheduling problem and the research status of the scheduling problem considering the setup times are described below. The scheduling problem with limited buffers is of significant value for practical production scenarios, but is also very challenging in theory. In recent years, the scheduling problem with limited buffers has gained the immense attention of researchers. Zhao et al. [1] designed an improved particle swarm optimization (LDPSO) with a linearly decreasing disturbance term for flow shop scheduling with limited buffers. Ventura and Yoon [2] studied the lot-streaming flow shop scheduling problem with limited capacity buffers and proposed a new genetic algorithm (NGA) to solve the problem. Han et al. [3] studied the hybrid flow shop scheduling problem with limited buffers and used a novel self-adaptive differential evolution algorithm to effectively improve the production efficiency of the hybrid flow shop. Zeng et al. [4] proposed an adaptive cellular automata variation particles warm optimization algorithm with better optimization and robustness to solve the problems of flexible flow shop batch scheduling. Zhang et al. [5] presented a hybrid artificial bee colony algorithm combined with the weighted profile fitting based on Nawaz–Enscore–Ham (WPFE) heuristic algorithm which effectively solved the flow shop scheduling problem with limited buffers.

In recent years, due to the awareness of both the importance of production preparation and the necessity of separating the setup time and processing time, the scheduling problem considering the setup time has gained a lot of attention from the academic and industrial circles. Zhang et al. [6] employed an enhanced version of ant colony optimization (E-ACO) algorithm to solve flow shop scheduling problems with setup times. Shen et al. [7] presented a Tabu search algorithm with specific neighborhood functions and a diversification structure to solve the job shop scheduling problem with limited buffers. Tran et al. [8] studied the unrelated parallel machine scheduling problem with setup times and proposed a branch-and-check hybrid algorithm. Vallada and Ruiz [9] studied the unrelated parallel machine scheduling problem with sequence dependent setup times and proposed a genetic algorithm to solve the problem. Benkalai et al. [10] studied the problem of scheduling a set of jobs with setup times on a set of machines in a permutation flow shop environment, proposing an improved migrating bird optimization algorithm to solve the problem. An et al. [11] studied the two-machine scheduling problem with setup times and proposed a branch and bound algorithm to solve this problem.

The related work in recent years showed that the current research on the limited buffers scheduling problem mainly addressed the shop scheduling problems under a given buffer capacity and mainly highlighted global optimization algorithms. By improving algorithms or combining different algorithms, researchers proposed algorithms with more accuracy in terms of optimization. However, few scholars systematically studied one specific type of limited buffer. At present, the research on scheduling problems with setup times mainly focus on the impact of setup times under the single condition on the processing time. The current studies mainly focus on the improvement in traditional algorithms, while few scholars study the scheduling problems with setup times under the conditions of a dynamic combination of multiple properties (in actual production, setup times are affected by quite a few factors and the combination of multiple factors will produce multiple setup times). Thus, there is no further exploration of the scheduling problem with setup times under the production constraints. The multi-queue limited buffers scheduling problem studied in this paper is more complex than the general limited buffers scheduling problem. This problem requires not only the consideration of the capacity for limited buffers but also the investigation of (1) the distribution problem when the job enters the lane in the multi-queue limited buffers and (2) the problem of selecting a job from multiple lanes in the multi-queue limited buffers to enter the next stage. On this basis, the research in this paper also considers the influence of setup times on the scheduling process under the dynamic combination of multiple properties, which can greatly increase the complexity of scheduling problems and the uncertainty of scheduling results. The scheduling problems in a flexible flow shop have long proved to be a non-deterministic polynomial hard (NP-hard) problem [12], and thus the multi-queue limited buffers scheduling problem in a flexible flow shop with setup times studied in this paper is also of the NP-hard nature.

As the situations become complicated, more efficient optimization methods needed to be further explored. The compact genetic algorithm (CGA) is a distribution estimation algorithm proposed by Harik [13]. It has great advantages in regard to computational complexity and evolutionary speed, but still the search range is smaller, and it is easy to fall into the local extremum. The scholars overcame its deficiencies mainly by using multiple population and probability distributions [14]. These two approaches, however, could not significantly improve the algorithm's optimization and might lose the advantage of the CGA in the fast search. The probabilistic model and the process of generating new individuals in a probabilistic model are explored to prevent the CGA from being premature and improve the quality and diversity of new individuals. This study proposes an improved compact genetic algorithm (ICGA) that maps the original probabilistic model to a new one through the probability density function of the Gaussian distribution to enhance the algorithm's evolutionary vigor and its ability to jump out of the local extremum, which can better solve the multi-queue limited buffers scheduling problems in a flexible flow shop with setup times.

2. Mathematical Model

2.1. Problem Description

Figure 1 illustrates the scheduling problem in which n jobs are required to be processed in m stages [15]. At least one stage in the m stages includes multiple parallel machines. Each job is needed to be assigned one machine at each stage. Also, buffers with given capacity exist between stages. The completed job enters the buffer and waits for the availability of the next stage. If the buffer is full, the completed job of the previous stage will stay on its current machine. In this situation, the machine is unavailable and cannot process other jobs until the buffer has available spaces. The buffers contains multiple lanes, and each lane has a limited number of spaces. The job enters the lane from one side and exits the lane from the other side. The jobs that wait for processing form a waiting queue in each lane. When the job enters this buffer, it needs to select one of the lanes and enter its waiting queue. When there is an idle machine in the next stage, it is necessary to select one bus in multiple waiting queues to move out of the buffer for processing. When all the lanes of this buffer reach the upper limit of the capacity, there will also be cases where the job is blocked at the machine of the previous stage. When the job is assigned to the machine, the setup time should also be added besides the standard processing time if the property of the job did not match that of the previous job. The standard processing time of the job at each stage, the online sequence of the job in the production process and the job's setup times, start time, and completion time at each stage can be obtained through the scheduling, so as to achieve better scheduling results.

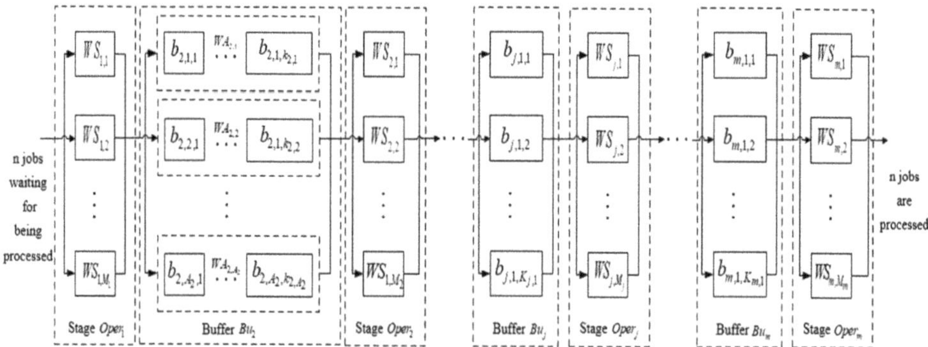

Figure 1. Model for the multi-queue limited buffers scheduling problems in a flexible flow shop.

2.2. Parameters in the Model

The parameters used in this paper are as follows:

n: number of jobs to be scheduled;

m: number of stages;

J_i: job i, $i \in \{1, \ldots, n\}$;

$Oper_j$: stage j, $j \in \{1, \ldots, m\}$;

M_j: number of machines in each stage, $j \in \{1, \ldots, m\}$;

$WS_{j,l}$: machine l of stage $Oper_j$, $j \in \{1, \ldots, m\}$, $l \in \left\{1, \ldots, M_j\right\}$;

Bu_j: the buffer of stage $Oper_j$, $j \in \{2, \ldots, m\}$;

A_j: number of waiting queue in the buffer of stage $Oper_j$, $j \in \{2, \ldots, m\}$;

$Bs_{j,a}$: the ath lane of buffer Bu_j at stage $Oper_j$, $j \in \{2, \ldots, m\}$, $a \in \left\{1, \ldots, A_j\right\}$;

$K_{j,a}$: number of spaces in lane $Bs_{j,a}$ of buffer Bu_j at stage $Oper_j$, $j \in \{2, \ldots, m\}$, $a \in \left\{1, \ldots, A_j\right\}$;

$b_{j,a,k}$: space k in lane $Bs_{j,a}$ of buffer Bu_j, $j \in \{2, \ldots, m\}$, $a \in \left\{1, \ldots, A_j\right\}$, $k \in \left\{1, \ldots, K_{j,a}\right\}$;

$WA_{j,a}(t)$: at time t, the waiting queue in the ath lane $Bs_{j,a}$ of buffer Bu_j, $j \in \{2, \ldots, m\}$, $a \in \left\{1, \ldots, A_j\right\}$;

$S_{i,j}$: the start time to process job J_i at stage $Oper_j$;

$C_{i,j}$: the completion time to process job J_i at stage $Oper_j$;

$Tb_{i,j}$: the standard processing time of job J_i at stage $Oper_j$, $j \in \{1, \ldots, m\}$;

$Te_{i,j}$: the entry time of job J_i into buffer Bu_j, $j \in \{2, \ldots, m\}$;

$Tl_{i,j}$: the departure time of job J_i out of buffer Bu_j, $j \in \{2, \ldots, m\}$;

$To_{i,j}$: the departure time of job J_i on its machine after job J_i is processed at stage $Oper_j$, $j \in \{1, \ldots, m\}$;

$Tw_{i,j}$: the waiting time of job J_i in buffer Bu_j, $j \in \{2, \ldots, m\}$;

$Ts_{i,j,l}$: the setup time of machine $WS_{j,l}$ when job J_i is processed on machine $WS_{j,l}$, $j \in \{2, \ldots, m\}$;

$Nsrv_{x,i',i''}^{l}$: the relationship between properties of job continuously processed on machine $WS_{j,l}$, $i', i'' \in \{1, \ldots n\}$ and $i' \neq i''$.

2.3. Constraints

2.3.1. Assumptions

The variables used in this paper are as follows:

$$At_{i,j,l} = \begin{cases} 1, & \begin{array}{l} \text{Job } J_i \text{ is assigned to be processed} \\ \text{on machine } WS_{j,l} \text{ at stage } Oper_j \end{array} \\ 0, & \begin{array}{l} \text{Job } J_i \text{ isn't assigned to be processed} \\ \text{on machine } WS_{j,l} \text{ at stage } Oper_j \end{array} \end{cases} \tag{1}$$

$$OA_{i,ja}(t) = \begin{cases} 1, & \begin{array}{l} \text{At time t, job } J_i \text{ is in waiting} \\ \text{processing queue } WA_{j,a}(t) \text{ at stage } Oper_j \end{array} \\ 0, & \begin{array}{l} \text{At time } t, \text{ job } J_i \text{ isn't in waiting} \\ \text{processing queue } WA_{j,a}(t) \text{ at stage } Oper_j \end{array} \end{cases} \tag{2}$$

$$QBC_{ja}(t) = \begin{cases} 1, & \begin{array}{l} \text{At time } t, \text{ the number of jobs in lane } Bs_{j,a} \text{ at stage} \\ Oper_j \text{ is greater than zero, namely, } card\big(WA_{j,a}(t)\big) > 0 \end{array} \\ 0, & \begin{array}{l} \text{At time } t, \text{ the number of jobs in lane } Bs_{j,a} \text{ at stage} \\ Oper_j \text{ is equal to zero, namely, } card\big(WA_{j,a}(t)\big) = 0 \end{array} \end{cases} \tag{3}$$

2.3.2. General Constraint of Flexible Flow Shops Scheduling

The general constraint of flexible flow shops scheduling are as follows:

$$\sum_{l=1}^{M_j} At_{i,j,l} = 1 \tag{4}$$

$$C_{i,j} = S_{i,j} + Tb_{i,j}, i \in \{1,2,\cdots,n\}, j \in \{1,2,\cdots,m\} \tag{5}$$

$$C_{i,j-1} \leq S_{i,j}, i \in \{1,2,\cdots,n\}, j \in \{1,2,\cdots,m\} \tag{6}$$

Equation (4) indicates job J_i at stage $Oper_j$ can only be processed on one machine. Equation (5) constrains the relationship between the start time and completion time for job J_i at stage $Oper_j$. Equation (6) represents that job J_i needs to complete the current stage before proceeding to the next stage. The general constraints of flexible flow shops are still valid for the multi-queue limited buffers scheduling problem in a flexible flow shop with setup times.

2.3.3. Constraints of Limited Buffers

The constraints of limited buffers are as follows:

$$To_{i,j} = \begin{cases} Te_{i,j+1} & j = \{1,\ldots,m-1\} \\ C_{i,j} & j = m \end{cases}. \tag{7}$$

In Equation (7), when the job is at stage $Oper_j$ ($i = \{1,\ldots,m-1\}$), the departure time of the job on the machine is equal to the entry time of the job into the buffer. When the job is at stage $Oper_m$, the departure time of the job on the machine equals the completion time.

$$Te_{i,j} \geq C_{i,j-1}, j \in \{2,\ldots,m\}. \tag{8}$$

Equation (8) indicates that the entry time of the job into the buffer is greater than or equal to the completion time of the job in the previous stage. If the limited buffers are blocked, the job will be retained on the machine after completing the previous stage.

$$WA_{j,a}(t) = \left\{ J_i \middle| OA_{i,j,a}(t) = 1 \right\}. \tag{9}$$

Equation (9) shows all jobs contained in the waiting queue of limited buffers at time t.

$$card\left(WA_{j,a}(t) \right) \leq K_{j,a}. \tag{10}$$

Equation (10) denotes that at any time, the number of jobs in the waiting queue $WA_{j,a}$ is less than or equal to the maximum number of spaces ($K_{j,a}$) in the lane where $WA_{j,a}$ is located.

2.3.4. Constraints of Multi-Queue Limited Buffers

Two new constraints of Equations (11) and (12) are added based on the constraints of Equations (8) and (10).

$$\sum_{a=1}^{A_j} \sum_{i=1}^{n} OA_{i,j,a}(t) \leq n. \tag{11}$$

In Equation (11), when $A_j > 1$, at any time, the number of jobs to be processed in the waiting queue of multi-queue limited buffers is less than or equal to the number of jobs n.

$$\sum_{a=1}^{A_j} QBC_{j,a}(t) \leq n. \tag{12}$$

In Equation (12), when $A_j > 1$, at any time, the number of jobs, which can proceed to the machine of stage $Oper_j$, in the waiting queue of multi-queue limited buffers is less than or equal to the number of jobs n.

If $\exists i', i'' \in \left\{ i \middle| OA_{i,j,a}(t) = 1 \right\}$, $Te_{i',j} \leq Te_{i'',j}$, $Tl_{i',j}$ and $Tl_{i'',j}$ will satisfy the following relationship

$$Tl_{i',j} \leq Tl_{i'',j}. \tag{13}$$

In Equation (13), when $A_j > 1$, the job of the ath lane must meet the requirement that the job entering the queue first should depart from the queue first.

2.3.5. Constraints of Setup Times

The constraints of setup times are as follows:

$$\sum_{l=1}^{M_j} \left(Ts_{i,j,l} \cdot At_{i,j,l} \right) + C_{i,j-1} \leq S_{i,j}, \ i \in \{1,\dots,n\}, \ l \in \left\{1,\dots,M_j\right\}, \ j \in \{2,\dots,m\}. \tag{14}$$

Equation (14) extends the basic constraint of Equation (6) of a flexible flow shop to obtain the relationship between setup times and start time as well as completion time. Equation (14) indicates that the start time of the current stage is greater than or equal to the completion time of the previous stage plus setup times.

X denotes the number of job's properties. $prop_{x,i}$ represents the property of job J_i. $Prop_i = \left\{ prop_{x,i} \right\}$ represents the collection of properties of job J_i, and $x \in \{1,\dots,X\}$.

$$Nsrv_{x,i',i''}^{l} = \begin{cases} 1 & prop_{x,i'} \neq prop_{x,i''} \\ 0 & prop_{x,i'} = prop_{x,i''} \end{cases}. \tag{15}$$

$J_{i'}$ and $J_{i''}$ ($i', i'' \in \{1,\ldots,n\}$) are jobs continuously processed on machine $WS_{j,l}$. When $Nsrv^l_{x,i',i''} = 0$, it means that when property $prop_{x,i'}$ of job $J_{i'}$ and property $prop_{x,i''}$ of job $J_{i''}$ are the same, no setup time is required to process the latter job on the machine. When $Nsrv^l_{x,i',i''} = 1$, it represents that when property $prop_{x,i'}$ of job $J_{i'}$ and property $prop_{x,i''}$ of job $J_{i''}$ are different, the setup time is required to process the latter job on the machine.

$$Ts_{i',j,l} = \sum_{x=1}^{X} Tsp_{j,x} \cdot Nsrv^l_{x,i',i''}, \quad i',i'' \in \{1,\ldots,n\}, \ j \in \{2,\ldots,m\}, \ l \in \{1,\ldots,M_j\}. \tag{16}$$

In Equation (16), at stage $Oper_j$, $Tsp_{j,x}$ represents the required setup times when there is a change in one property $prop_{x,i}$ of two consecutively processed jobs. $TSP_j = \{Tsp_{j,x}\}$ denotes the collection of setup times when there is a change in the property of two consecutively processed jobs at stage $Oper_j$, and $x \in \{1,\ldots,X\}$. $Ts_{i',j,l}$ is the required setup time when several properties of job $J_{i'}$ processed on machine $WS_{j,l}$ at stage $Oper_j$ are different from those of the previous job processed on the same machine.

Other constraints are as follows. Uninterruptible constraint: if the job has been started on the machine, it cannot be interrupted until the production process on the machine is completed. Machine availability constraint: all machines in the scheduling are available at production time. Time simplification constraint: not to consider the time of jobs transferred among spaces in the multi-queue limited buffers, and the time of jobs transferred between machines in front and back stages, that is to say, only consider the processing time of the job processed in each stage and setup times when the property of jobs processed successively on the machine is changed.

2.4. Evaluation Index of the Scheduling Result

- Makespan

$$C_{max} = \max\{C_{i,m}\}, \ i \in \{1,2,\ldots,n\}. \tag{17}$$

In Equation (17), C_{max} indicates the maximum completion time of all jobs processed at the last stage, which is also the time for all jobs to complete the process.

- Waiting processing time for total job

$$TWIP = \sum_{j=2}^{m} \sum_{i=1}^{n} \left(S_{ij} - C_{i,j-1}\right). \tag{18}$$

In Equation (18), the waiting time of the job is from the completion moment $C_{i,j-1}$ of job J_i at stage $Oper_{j-1}$ to the start moment S_{ij} at the next stage $Oper_j$. $TWIP$ represents the sum of the waiting times of all jobs processed in the entire production. In the production shop of limited buffers, the waiting time of each job between stages is equal to the sum of time for the job staying on the buffer ($Tl_{i,j} - Te_{i,j}$), blocking time of the job on the machine ($Te_{i,j} - C_{i,j-1}$), and setup times ($\sum_{l=1}^{M_j} (Ts_{i,j,l} \cdot At_{i,j,l})$).

- Idle time for total machine

$$TWT = \sum_{j=1}^{m} \sum_{l=1}^{M_j} \left(\left(\max\{To_{i,j} \cdot At_{i,j,l}\} - \min\{S_{i,j} \cdot At_{i,j,l}\}\right) - \sum_{i=1}^{n} \left(Tb_{i,j} \cdot At_{i,j,l}\right) \right). \tag{19}$$

The idle time for the machine in Equation (19) represents the time between the start time of the first job and completion time of the last job on each machine. TWT is the sum of idle time for all machines.

- Total device availability

$$FUR = \frac{\sum\limits_{j=1}^{m} \sum\limits_{i=1}^{n} \left(Tb_{i,j}\right)}{\sum\limits_{j=1}^{m} \sum\limits_{l=1}^{M_j} \left(\max\left\{To_{i,j} \cdot At_{i,j,l}\right\} - \min\left\{S_{i,j} \cdot At_{i,j,l}\right\}\right)}. \tag{20}$$

In Equation (20), FUR represents the total device availability of all machines in a flexible flow shop, which is the ratio of the effective processing time of all machines to the occupied time span of the machine. This time span starts from the first job processed on the machine to the last job that is finished and has left the machine.

- Total machine setup times

$$TS = \sum\limits_{i=1}^{n} \sum\limits_{j=1}^{m} \sum\limits_{l=1}^{M_j} \left(Ts_{i,j,l} \cdot At_{i,j,l}\right). \tag{21}$$

In Equation (21), TS is the sum of all stages' setup times in a flexible flow shop.

- Total job blocking time

$$TPB = \sum\limits_{i=1}^{n} \sum\limits_{j=2}^{m} \left(Te_{i,j} - C_{i,j-1}\right). \tag{22}$$

In Equation (22), TPB is the sum of blocking time of all jobs stuck on the machine due to the buffers being full after all jobs finish the processing in a flexible flow shop.

3. Improved Compact Genetic Algorithm

In the evolution of standard CGA, new individuals are generated based on probability distribution of the probabilistic model, and individuals conforming to the evolutionary trend are selected to update the distribution probability. The elements in the model, namely the probability values, represent the distribution of feasible solutions. The continuously accumulated optimization information in the evolution is reflected in the probability values of the probabilistic model. The CGA adopts the single individual to update the probabilistic model in each generation, and the new individual of each generation is also generated in the model. After several generations, if one of the probability values on the model's column (or row), which controlled the generation of individual gene fragments, is extremely large, it leads to similar genes appearing in the same position of new individuals generated in later evolutions, decreasing the diversity of new individuals. As the individuals generated from the probabilistic model will update the model conversely, the probability values will further increase. When elements (probability values) in the probabilistic model all become 0 or 1, the CGA will terminate the evolution process. Therefore, it is difficult for the CGA to jump out of the local extremum once it falls into this situation. Afterward, the overall trend for evolution is irreversible, leading to a premature convergence of the algorithm. As such, if only an expanding the number of new individuals generated by the model is considered, the optimization effect of the algorithm cannot be significantly improved, and the advantage of the CGA in the fast search will be lost. The probabilistic model and the process of generating new individuals from the model were explored to prevent the prematurity of the CGA and improve the quality and diversity of new individuals. First, the distribution of probability values in a column (or a row) that were related to gene fragments of new individuals in the probabilistic model were determined, and then the probability density function of the Gaussian distribution was introduced to map the probabilistic model from the original one to the new one. Under the premise of keeping the probability values' distribution of the original probabilistic model unchanged, the search ability of feasible solutions could be expanded, thereby developing the diversity of new individuals.

3.1. Establishing and Initializing the Probabilistic Model

The probabilistic model was responsible for counting and recording the distribution of genes in the individual after the evolution of the algorithm. According to the individual coding information, namely online sequence and machine assignment, the $n \times n$ matrix P^L was established as the online-sequence probabilistic model of the CGA to optimize the scheduling online sequence. In the probabilistic model, the 1st to nth rows corresponded to jobs J_1 to J_i, and the 1st to nth columns corresponded to individuals 1 to n. $P^1_{i,s}$ indicated the probability of job J_i appearing at position s of the online processing queue.

The probabilistic model was initialized by a uniform assignment. This method directly set the equal probability of each job appearing at each position, providing a more balanced optimization starting point for the algorithm. It used the uniform assignment to initialize the probabilistic model P, namely $\forall(i,s)$, $P^1_{i,s} = 1/n$, which could expand the search range of the algorithm's feasible solution. In addition, it was restricted by the constraint that $\forall(s)$, $\sum_{i=1}^{n} P^1_{i,s} = 1$.

3.2. Mapping the Original Probabilistic Model to the New Probabilistic Model

Some scholars used information entropy to evaluate the distribution of probability values in the probabilistic model [16]. This approach is not quite sensitive to the distribution of probability values. This study adopted the method of calculating the standard deviation of probability values in the probabilistic model to assess the probability distribution in the original probabilistic model so as to judge and evaluate the ability of the current model to search for feasible solutions.

The ICGA consisted of two models: the original probabilistic model P^L and the new probabilistic model F^L. The element $P^L_{i,s}$ in the probabilistic model referred to the probability of job J_i appearing at position s in online processing queue.

The probability value distribution in P^L was judged by calculating each column of standard deviation $\sigma_s{}^L$ in the original probability model P^L. Meanwhile, the standard deviation-based threshold value σ^T which started the mapping operation was set to judge whether the probability density function of the Gaussian distribution can be started to map P^L to the new probability model F^L. When $\sigma_s{}^L > \sigma^T$, mapping the P^L to the F^L using the probability density function of the Gaussian distribution shown in Equation (23). Then, a new feasible solution I^L (new individual) of the problem was generated in accordance with the F^L. The original probability model P^L was updated through this new feasible solution. After that, this new feasible solution was applied to update the P^L. The scope of searching a feasible solution was magnified through expanding the ability of the P^L to select new feasible solutions.

$$f(x|\mu_s^L, \sigma_s^L) = \frac{1}{\sigma_s^L \cdot \sqrt{2\pi}} e^{-\frac{(x-\mu_s^L)^2}{2\sigma_s^{L2}}} . \tag{23}$$

Equation (23) is the probability density function of the Gaussian distribution, where μ_s^L is the expectation value of the sth column in the probabilistic model P^L, and the calculation formula is shown in Equation (24). $\sigma_s{}^L$ is the standard deviation of the sth column in the probabilistic model P^L, and the calculation formula is shown in Equation (25). The size of the standard deviation determines the degree of steepness or flatness of the Gaussian distribution curve. The smaller the σ, the steeper the curve; and the larger the σ, the flatter the curve. With the evolution of the CGA, the distribution of probability values in the probabilistic model decreases, and the value of σ becomes larger (a probability value is dominant, while others are small and far from the average value). Further, the selection range after the mapping will become larger.

$$\mu_s^L = \sum_{i=1}^{n} P^L_{i,s} \cdot i \tag{24}$$

$$\sigma_s{}^L = \sqrt{\frac{1}{n} \cdot \sum_{s=1}^{n} \left(P_{i,s}^L - \frac{1}{n} \right)^2}.$$ (25)

At the beginning of evolution $P_{i,s}^L = 1/n$, and the initial value of $\sigma_s{}^L$ is 0. If the value of $\sigma_s{}^L$ is too small, the Gaussian distribution curve is too steep, the probability value after mapping is smaller. As such, the standard deviation adjustment parameter ξ is added to Equation (25) to obtain the standard deviation $\sigma_s{}^{\xi L}$ after adjustment in Equation (26).

$$\sigma_s{}^{\xi L} = \xi \cdot \sqrt{\frac{1}{n} \cdot \sum_{s=1}^{n} \left(P_{i,s}^L - \frac{1}{n} \right)^2}$$ (26)

$$\xi = \frac{n \cdot max\left\{ P_{i,s}^L \right\}}{\sum\limits_{i=1}^{n} P_{i,s}^L - max\left\{ P_{i,s}^L \right\}}.$$ (27)

In Equation (27), the standard deviation adjustment parameter ξ satisfies the constraints: when $max\left\{ P_{i,s}^L \right\} = 1$ or $\sum\limits_{i=1}^{n} P_{i,s}^L - max\left\{ P_{i,s}^L \right\} = 0$, $\xi = \sqrt{n}$.

After determining the probability density function of the Gaussian distribution corresponding to each column element in the P^L, this function was used to obtain the corresponding probability value $GP_{i,s}^L$ of each probability value $P_{i,s}^L$ in this column in the new probabilistic model F^L. The specific operations were: first, determine the range $[X_1, X_2]$ for each probability value $P_{i,s}^L$ on the X axis, where $X_1 = \sum\limits_{h=1}^{i-1} P_{h,s}^L$, $X_2 = \sum\limits_{h=1}^{i} P_{h,s}^L$. Then, determine the position of expectation value μ_s^L on the X axis. Finally, the corresponding probability value $GP_{i,s}^L$ of probability value $P_{i,s}^L$ in the F^L is obtained by using Equation (28)

$$GP_{i,s}^L = \begin{cases} |f(X_1) - f(X_2)| & \mu_s^L \notin [X_1, X_2] \\ 2 \cdot f(\mu_s^L) - f(X_1) - f(X_2) & \mu_s^L \in [X_1, X_2] \end{cases}.$$ (28)

After all the columns in the P^L have been mapped, the probability values in each column of the new probabilistic model F^L were normalized to ensure that the sum of the probability values in each column was 1. Then, the roulette was used to choose the positions of jobs which were arranged in the processing queue based on the F^L, so as to generate new individuals in more diversities. The superior individual among new individuals was chosen to update the original probabilistic model P^L through Equation (29).

$$P_{i,s}^{L+1} = \left(1 - \frac{\beta}{n} \right) \cdot P_{i,s}^L + \frac{\beta}{n} \cdot St, \quad St = \begin{cases} 1 & Gene_s = i \\ 0 & Gene_s \neq i \end{cases}$$ (29)

where St is the learning coefficient; n is the number of jobs to be processed; and β is the adjustment override of the learning rate. In each generation, the superior individual is selected to update the probabilistic model. The genetic value $Gene_s$ in the individual represents the sequential position of job J_i in online processing queue. Equation (29) shows that when the value of $Gene_s$ is i, the probability value $P_{i,s}^1$ in the sth column of the model which combines with learning coefficient St further increases the probability of job J_i being chosen on the position s in the online processing queue. In addition, other probability values of this column subtract $\frac{\beta}{n} \cdot P_{i,s}^L$.

3.3. Encoding and Decoding of New Individuals

Based on the probabilistic model, NP new individuals were generated. The process of generating each new individual was: in the probabilistic model P^L, based on the probability $P_{i,s}$, the genetic value of individual from the 1st to nth referred to the probability $P_{i,s}$ of the job J_i which appeared on the *sth* position. The job numbers were chosen on the basis of roulette in turn, that is, the job online order. In the process of individual decoding, the individual genetic value was decoded into the online processing sequence of n jobs in the first stage.

3.4. Procedure of the ICGA

Step 1: Initialize probabilistic model P^L. According to the principle of the maximum entropy, the probabilistic model P^L is initialized, where $\forall(i,s)P_{i,s}^L = 1/n$. Meanwhile, the evolutionary generation is set as $L = 0$.

Step 2: Map the probability value $P_{i,s}^L$ in the original probabilistic model P^L to the new probabilistic model F^L. Calculate the standard deviation $\sigma_s{}^L$ of the probability value of the *sth* column in the original probabilistic model P^L, and judge whether $\sigma_s{}^L$ is larger than the threshold value σ^T for initiating the mapping operation. If $\sigma_s{}^L \leq \sigma^T$, the probability value of the *sth* column in the P^L is taken as the *sth* column probability value in the new probabilistic model F^L. If $\sigma_s{}^L > \sigma^T$, execute Step 3.

Step 3: Calculate the expectation value μ_s^L of the probability value of the *sth* column in the original probabilistic model P^L. Determine its corresponding probability density function of the Gaussian distribution, and calculate the probability value $GP_{i,s}^L$ in the new probabilistic model F^L corresponding to each probability value $P_{i,s}^L$ in the *sth* column.

Step 4: Repeat Step 2 until all the columns in the original probabilistic model P^L are all mapped into the new probabilistic model F^L.

Step 5: Generate new individuals based on the new probabilistic model F^L. Through the roulette, the individual genetic sampling values are selected in turn according to the probability values of each column in the F^L, and these genetic values represent the positions of the jobs to be machined in the online queue. Once a job is selected, it no longer participates in the subsequent selection, while the unselected jobs go on to take part in the selection process until all jobs are arranged so as to generate a new individual. After that, NP new individuals $I_1, I_2, \ldots, I_{NP}$ are generated.

Step 6: NP new individuals are decoded and the fitness function value for each new individual can be calculated.

Step 7: Compare the fitness function values of NP new individuals, and select the optimal individual I_{better} among them, and its fitness function value is f_{better}.

Step 8: Judge whether f_{better} is better than the fitness function value f_{best} of the historically optimal individual I_{best}. If f_{better} is better than f_{best}, replace I_{best} with I_{better} and replace f_{best} with f_{better}.

Step 9: The historically optimal individual I_{best} is used to update the probabilistic model P^L, and the update operation is performed according to Equation (29). At the same time, evolutionary generation L is processed as $L = L + 1$.

Step 10: Judge whether the updated original probabilistic model P^L converges, that is, whether all probability values of model P^L are 1 or 0. If the convergence is met, the historically optimal individual I_{best} is output and the evolution process ends. Otherwise, execute Step 11.

Step 11: Judge whether the evolutionary generation L reaches the set maximum evolutionary generation $L_{\max}$. If $L = L_{\max}$, the historically optimal individual I_{best} is output and the evolution process ends. Otherwise, Step 2 is performed again.

The flowchart of the ICGA is shown in Figure 2.

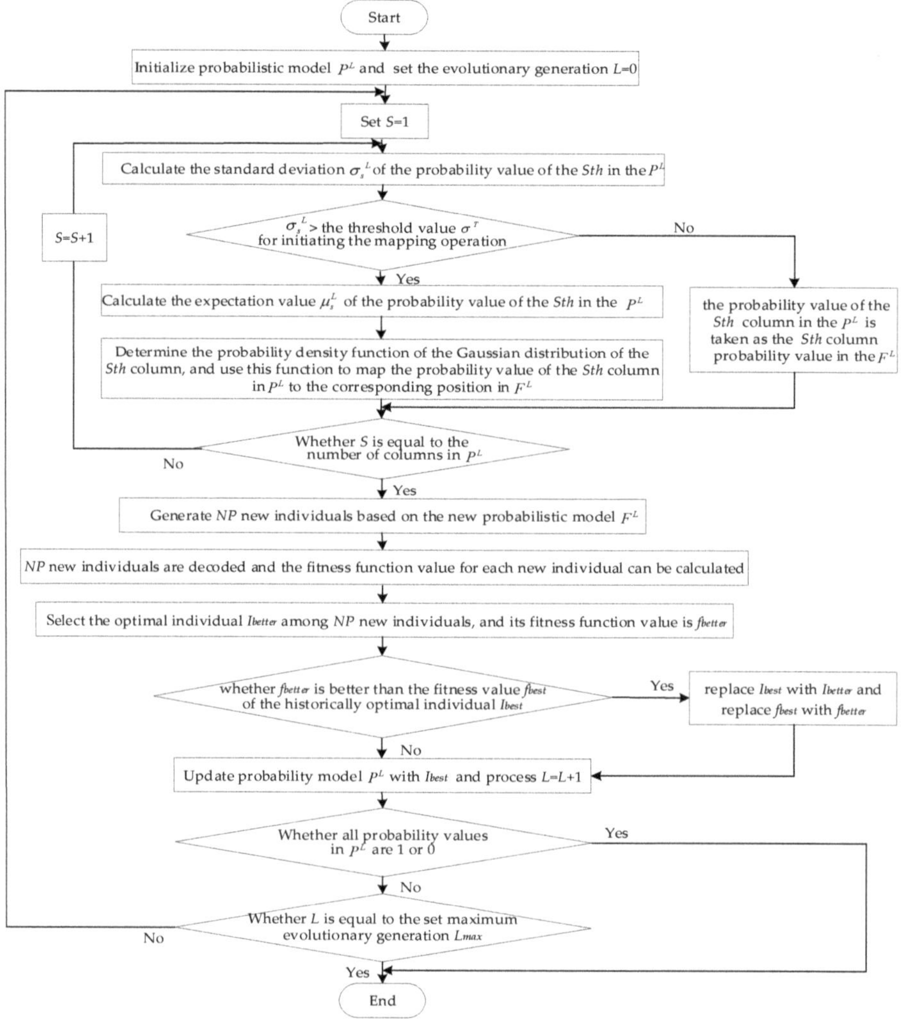

Figure 2. The flowchart of ICGA.

4. Local Scheduling Rules for Multi-Queue Limited Buffers

In order to reduce the setup times and reduce the impact of setup times on the scheduling process, this paper has developed a variety of local scheduling rules to guide the distribution of jobs for the process of entering and exiting multi-queue limited buffers. When the job enters the multi-queue limited buffer, the local scheduling rules employ the remaining capacity of max queue buffer (RCMQB) rule. When the job leaves the multi-queue limited buffer, the local scheduling rules use the shortest setup time (SST) rule, the first available machine (FAM) rule, and the first-come first-served (FCFS) rule, of which the SST rule is a priority [17].

4.1. Rules for Jobs Entering Multi-Queue Limited Buffers

- Max queue buffer capacity remaining rule
  When $\exists C_{i,j-1}$ and $t = C_{i,j-1}$,

$$Mca_{i,j}(t) = \left\{ Bs_{j,a} \middle| max\left\{ K_{j,a} - card\left(WA_{j,a}(t)\right) \middle| \left(K_{j,a} - card\left(WA_{j,a}(t)\right) > 0\right) \right\} \right\} \tag{30}$$

$$j \in \{2, \ldots, m\}$$

where $Mca_{i,j}(t)$ is the set of lane $Bs_{j,a}$ that the job in the previous stage $Oper_{j-1}$ can access. The difference between the maximum number of spaces ($K_{j,a}$) of each lane $Bs_{j,a}$ and the number of jobs in the waiting queue $WA_{j,a}$ in this lane is the remaining capacity (available space) of lane $Bs_{j,a}$ at the current stage. When the job finishes the previous stage, namely when $t = C_{i,j-1}$, the job enters the lane with the largest remaining capacity. When $card\left(Mca_j(t)\right) = 0$, it indicates there is no remaining space in the current lane. When $card\left(Mca_j(t)\right) > 1$, it illustrates that multiple lanes can be entered.

4.2. Rules for Jobs Leaving Multi-Queue Limited Buffers

When the job exits the buffer and is assigned to the machine, in the case of an available machine $WS_{j,l}$ and several jobs are waiting in the buffer, that is, the number of the selectable jobs to be processed in the multi-queue limited buffers is greater than 1 ($\sum_{a=1}^{A_j} QBC_{j,a}(t) > 1$). In the course of assigning jobs, the job with the minimum setup time ($\left\{ J_i \middle| min\left(Ts_{i',j,l}\right) \right\}$) is processed according to the SST rule. If the number of jobs with the minimum setup time is greater than 2, the job with the longest waiting time in the buffer ($\left\{ J_i \middle| max\left\{ (t - Te_{i,j}) \middle| \left\{ i \middle| OA_{i,j,a}(t) = 1 \right\} \right\} \right\}$) is selected for processing in accordance with the first in first out (FIFO) rule. In the case of available machines and only one job waiting to be processed, that is, the number of selectable jobs to be processed in the multi-queue limited buffers is equal to 1 ($\sum_{a=1}^{A_j} QBC_{j,a}(t) = 1$), the SST rule is used to select the machine with the minimum setup time in the course of assigning machines. If the number of selectable machines is greater than 2, the job is machined on the basis of the FAM rule. The multimachine-and-multi-job case is a combination of the aforementioned conditions.

5. Simulation Experiment

The ICGA was implemented using MATLAB 2016a simulation software, running on the PC with Windows 10 operating system, Core i5 processor, 2.30 GHz Central Processing Unit (CPU), and 6 GB memory. The multi-queue limited buffers scheduling problems in a flexible flow shop with setup times originate from the production practice of bus manufacturers. It is a complex scheduling problem. As standard examples are not available at present, the standard data of a flexible flow shop scheduling problem (FFSP) was employed to discuss and analyze the parameters in the improved compact genetic algorithm, so as to determine the optimal parameter [18]. In addition, multiple groups of large-scale and small-scale data were used to test the effect of the improved method on the optimization ability of the standard CGA. Furthermore, the instance data of multi-queue limited buffers scheduling in a flexible flow shop with setup times were used to verify the optimal performance of the ICGA for solving such problems.

5.1. Analysis of Algorithm Parameters

The parameter values of the algorithm had a significant effect on the algorithm's optimization performance. The ICGA has three key parameters: threshold value σ^T that started the mapping operation, adjustment override of the learning rate β, and thr number of new individuals generated in each generation NP. The FFSP standard examples of a d-class problem with five stages and 15 jobs in the 98 standard examples proposed by Carlier and Neron were applied to test each parameter. The example j15c5d3 was used to illustrate the orthogonal experiment. Three parameters, each with four levels (see Table 1), were taken for orthogonal experiments with the scale of L_{16} (4^3) [19]. The algorithm ran 20 times in each group of experiments. The average time of makespan ($\overline{C_{max}}$) was used as an evaluation index.

Table 1. Level of each parameter.

Parameter	Level			
	1	2	3	4
σ^T	5	10	15	20
β	1.0	1.5	2.0	2.5
NP	2	3	4	5

Through experiments, we can see that the influence of the change of each parameter on the performance of the algorithm is shown in Figure 3. From the figure we can see that β and NP had a greater impact on the performance of the algorithm, and σ^T had the least impact on the performance of the algorithm. The best combination of parameters was: $\sigma^T = 10$, $\beta = 1.5$, $NP = 4$.

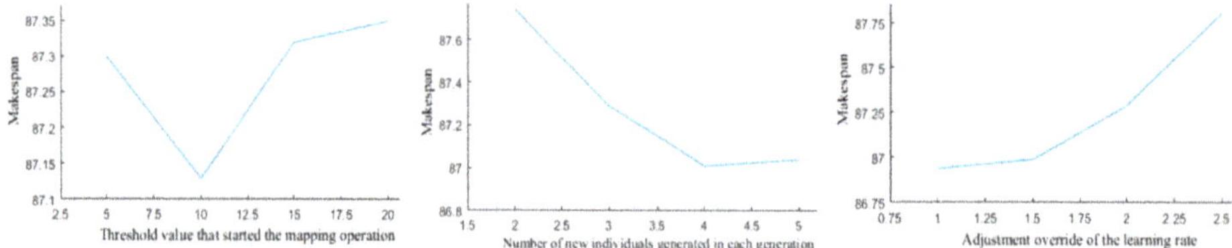

Figure 3. Trend chart of the algorithm's performance impacted by each parameter.

5.2. Optimization Performance Testing on the ICGA

In order to study the impact of the improved method (which is based on the probability density function of the Gaussian distribution mapping) on the optimization performance of the CGA, the ICGA was compared with CGA, bat algorithm (BA) [20] and whale optimization algorithm (WOA) [21]. These algorithms are the currently emerging intelligent optimization algorithms and are widely used in the field of optimization and scheduling [22–24]. In the BA, the number of individuals in the population $NP = 30$, pulse rate $\gamma = 0.9$, search pulse frequency range $[F_{min}, F_{max}] = [0, 2]$. In the WOA, the number of individuals in the population $NP = 30$. Each algorithm ran 30 times on each group of data. The maximum evolutionary generation of four algorithms was set to 500 generations. The average makespan $\overline{C_{max}}$ and the average running time $\overline{T_{cpu}}$ were used as the evaluation metrics.

5.2.1. Small-Scale Data Testing

The test data are from the 98 standard examples proposed by Carlier and Neron based on the standard FFSP. The set of examples was divided into five classes. According to the difficulty of solving the examples, the set of examples was divided into two categories by Neron et al. [25]: easy to solve and difficult to solve. Four groups of easy examples (j15c5a1, j15c5a2, j15c5b1, and j15c5b2) and four groups of difficult examples (j15c10c3, j15c10c4, j15c5d4, and j15c5d5) were chosen to test the ICGA so as to better evaluate the optimization performance of the ICGA for small-scale data. The test results are shown in Table 2.

Table 2. Small-scale data test results.

Standard Example	LB	BA		WOA		CGA		ICGA	
		C_{max}	$T_{cpu}(s)$	C_{max}	$T_{cpu}(s)$	C_{max}	$T_{cpu}(s)$	C_{max}	$T_{cpu}(s)$
j15c5a1	178	178	16.324	178	4.018	178	0.652	178	0.671
j15c5a2	165	165	16.474	165	4.029	165	0.738	165	0.754
j15c5b1	170	170	15.585	170	3.887	170	0.669	170	0.682
j15c5b2	152	152	15.767	152	3.863	152	0.636	152	0.710
j15c10c3	141	148.14	31.722	148.77	8.008	150.30	0.768	147.56	1.007
j15c10c4	124	132.86	48.289	133.28	8.056	135.65	0.764	132.26	1.121
j15c5d4	61	87.16	17.679	87.02	4.267	89.57	0.471	86.13	0.758
j15c5d5	67	82.03	17.368	82.28	4.296	84.65	0.435	81.72	0.756

In the table, *LB* represents the lower bound of makespan for the examples, whose optimal value was given by Santos and Neron [25,26]. It can be seen from Table 2 that even if the data size was small, the average running time of the CGA and ICGA under each group of data was significantly shorter than that of the BA and WOA. This indicates that CGA and ICGA have the advantages of the convergence speed of optimization. In terms of the optimization performance of small-scale data, the four algorithms did not have significant differences. But overall, the optimization effect of the ICGA on small-scale data was still the best among the four algorithms. The ICGA has achieved better solutions than the other three algorithms when solving all four groups of difficult examples (the four algorithms all reached the lower bound of makespan when solving the easy examples). Especially, compared with the CGA, when solving two groups of examples of the j15c5d class, the average relative error obtained by the ICGA was reduced by 5.64% and 4.37%, respectively. This shows that the optimization effect of the ICGA is greatly improved from the CGA when solving small-scale data.

5.2.2. Large-Scale Data Testing

Six groups of data, including 80 jobs with four stages, 80 jobs with eight stages, and 120 jobs with four stages, were used to test the optimization performance of the ICGA in solving large-scale complex problems. The test results are shown in Table 3.

Table 3. Large-scale data test results.

Instance Number	BA		WOA		CGA		ICGA	
	C_{max}	$T_{cpu}(s)$	C_{max}	$T_{cpu}(s)$	C_{max}	$T_{cpu}(s)$	C_{max}	$T_{cpu}(s)$
j80c4d1	1459.84	49.351	1465.12	17.485	1477.56	11.722	1455.44	13.731
j80c4d2	1367.44	49.735	1372.40	18.549	1385.16	12.595	1364.48	14.717
j80c8d1	1811.68	106.187	1813.88	51.967	1842.56	16.174	1814.28	19.078
j80c8d2	1822.56	104.954	1830.48	51.485	1851.92	16.079	1828.96	19.767
j120c4d1	2132.08	78.779	2138.04	38.171	2196.92	23.859	2139.88	28.942
j120c4d2	2243.52	79.821	2250.41	39.275	2306.52	24.221	2246.04	29.642

As can be seen from the table, when solving the problem of large-scale data, the advantages of the CGA and ICGA in the speed of optimization were more obvious. Especially when solving two groups of examples of j80c8d class, the average running time of CGA and ICGA was significantly smaller than the BA and WOA. It can be seen from the test results of the first four groups of examples that when the number of processes was increased from four to eight, the average running time of the BA and the WOA was greatly increased. On the other hand, although the average running time of the CGA and the ICGA was also increased, the rate of increase was small. This indicates that CGA and ICGA are suitable for solving the problem of scheduling optimization with large data scale with many processes. In general, when solving the problem of large-scale data, the optimization performance of the BA and

ICGA was better than that of the WOA and CGA. As for the aspect of optimization speed, the average running time of the BA was much larger than the other three algorithms under every group of data.

It can also be seen from the table that although the CGA had the fastest optimization speed under each group of large-scale data, its optimization performance was also the worst among the four algorithms. And as the size of the data increased, the gap between the CGA and the other three algorithms was getting larger. This is mainly because the CGA uses roulette to select gene fragments in the probabilistic model when generating new individuals. In the initial stage of the probabilistic model, the probability of each gene fragment being selected is $1/n$. When the size of the data is large ($n = 80$ and $n = 120$), the initial probability value of each gene fragment in the probability matrix becomes small (1/8 and 1/120). Further, in the process of evolution, the probability value of the unselected gene fragment will be reduced again, and the probability value of the selected gene fragment will be increased. This will lead to the probability values of certain gene fragments that are much larger than the probability values at other locations after several generations of evolution. Although this will make the evolution speed of the algorithm faster, it will also lead to a decline in the diversity of new individuals generated by the probabilistic model, which means the premature convergence of the algorithm. Compared with the CGA, when solving the problems of each group of large-scale data, the optimization effect of the ICGA was greatly improved. This indicates that the ICGA, to some extent, overcame the problem that the CGA was easy to prematurely converge.

By testing the four algorithms using large-scale and small-scale data, we can conclude that compared with the CGA, the ICGA has stronger capability to continuously evolve and jump out of the local extremum while maintaining the characteristic of fast convergence of the CGA. The ICGA, to some extent, overcomes the problem that the CGA is easy to prematurely converge. At the same time, compared with the BA and the WOA, for either small-scale data or large-scale data, the ICGA has an obvious advantage in optimization speed. Thus, the ICGA is suitable for solving the complicated problem of scheduling optimization with many processes.

5.3. Instance Test on Multi-Queue Limited Buffers in Flexible Flow Shops with Setup Times

5.3.1. Establishing Simulation Data

The simulation data of the production operations in the body shop and paint shop of the bus manufacturer were established as follows.

1. Parameters in the shop model

The body shop of the bus manufacturer is a rigid flow shop with multiple production lines, which can be simplified into one production stage. The production of the paint shop is simplified to three stages [27]. The simulation data for scheduling include four stages, namely $\{Oper_1, Oper_2, Oper_3, Oper_4\}$, whose parallel machine $\{M_j\}$ is {3, 2, 3, 2}. The buffer between the body shop and the paint shop is a multi-queue limited buffer. As such, the buffer of stage $Oper_2$ in the scheduling simulation data is set to the multi-queue limited buffer. The number of lanes in buffer Bu_2 of stage $Oper_2$ is equal to 2, namely $A_2 = 2$. The number of spaces in lane $Bs_{2,1}$ is 2, namely $K_{2,1} = 2$, and that of buffer $Bs_{2,2}$ is 2, namely $K_{2,2} = 2$. That is to say, the multi-queue limited buffers have two lanes and each lane has two spaces.

In the production of the paint shop, it is necessary to clean the machine and adjust production equipment if the model and color of the buses that are successively processed on the machine are different. Therefore, the simulation process uses the changes in the model and color as the basis for calculating setup times. Table 4 shows that the setup times parameters are set when the model and color of the buses that are processed successively on the machine changed. When the bus is assigned to the machine of the next stage from the buffer, the setup times of the machine is calculated using Equation (16).

Table 4. Model parameters.

	Parameter	Description	Value
Parameters of limited buffers	A_2	Number of lanes in buffer Bu_2 of stage $Oper_2$	2
	$K_{2,1}$	Capacity of lane $Bs_{2,1}$	2
	$K_{2,2}$	Capacity of lane $Bs_{2,2}$	2
	$K_{3,1}$	Capacity of limited buffer Bu_3 of stage $Oper_3$	1
	$K_{4,1}$	Capacity of limited buffer Bu_4 of stage $Oper_4$	1
Parameters of setups times	$Tsp_{2,1}$	Setup times when the model of buses processed successively on a parallel machine of stage $Oper_2$ changes	4
	$Tsp_{2,2}$	Setup times when the color of buses processed successively on a parallel machine of stage $Oper_2$ changes	4
	$Tsp_{3,1}$	Setup times when the model of buses processed successively on a parallel machine of stage $Oper_3$ changes	5
	$Tsp_{3,2}$	Setup times when the color of buses processed successively on a parallel machine of stage $Oper_3$ changes	5
	$Tsp_{4,1}$	Setup times when the model of buses processed successively on a parallel machine of stage $Oper_4$ changes	3
	$Tsp_{4,2}$	Setup times when the color of buses processed successively on a parallel machine of stage $Oper_4$ changes	3

2. Parameters of processing the object

The information of the bus model and color properties is shown in Table 5. The sum of bus properties is 2, namely $X = 2$. $Prop_1$ represented the model property of the bus, while $Prop_2$ denoted the color property of the bus. The value of model property ($PropValue_1$) is $\{BusType_1, BusType_2, BusType_3\}$, and the value of color property ($PropValue_2$) is $\{BusColor_1, BusColor_2, BusColor_3\}$. It is assumed that two successively processed buses on machine of stage $WS_{2,1}$ are buses J_1 and J_5. If model properties were as follows: $prop_{1,1} = BusType_1$ and $prop_{1,5} = BusType_1$, and color properties are as follows: $prop_{2,1} = BusColor_1$ and $prop_{2,5} = BusColor_3$, then $prop_{1,1} = prop_{1,5}, prop_{2,1} \neq prop_{2,5}$. Hence, $Nsrv^2_{2,1,5} = 1$. Using Equation (16), the setup time can be obtained as follows: $Ts_{5,2,2} = Tsp_{2,2} = 4$. And the Table 6 shows the standard processing time for bus production.

Table 5. Information of bus model and color properties.

Bus Property	J_1	J_2	J_3	J_4	J_5	J_6	J_7	J_8	J_9	J_{10}	J_{11}	J_{12}
Model $Prop_1$	$Type_1$	$Type_2$	$Type_3$	$Type_2$	$Type_1$	$Type_3$	$Type_1$	$Type_2$	$Type_3$	$Type_1$	$Type_2$	$Type_3$
Color $Prop_2$	$Color_1$	$Color_1$	$Color_2$	$Color_2$	$Color_3$	$Color_3$	$Color_1$	$Color_1$	$Color_2$	$Color_3$	$Color_2$	$Color_3$

Table 6. Standard processing time for bus production.

	J_1	J_2	J_3	J_4	J_5	J_6	J_7	J_8	J_9	J_{10}	J_{11}	J_{12}
$Oper_1$	8	11	15	19	10	16	12	21	22	13	20	14
$Oper_2$	30	38	28	25	26	36	20	24	22	32	35	34
$Oper_3$	34	38	44	42	52	40	46	48	35	36	45	50
$Oper_4$	42	36	26	24	34	30	28	32	38	40	44	22

5.3.2. Simulation Scheme

The scheduling problem of the bus manufacturer was investigated by using the ICGA, BA, WOA, and standard CGA as the global optimization algorithm, combined with local dispatching rules in the multi-queue limited buffers. This study further analyzed the optimization performance of the ICGA combined with local dispatching rules in solving the multi-queue limited buffers scheduling problems in a flexible flow shop with setup times. A total of eight groups of simulation schemes were designed, in which schemes 1–4 employed the FIFO rule as their local dispatching rules, and schemes 5–8 adopted the RCMQB rule when entering the buffer and used the SST rule, FAM rule, and FCFS rule as local dispatching rules when leaving the buffer. Among them, the SST rule was the priority. The information of the eight sets of simulation schemes is shown in Table 7.

Table 7. The eight groups of the simulation program.

Simulation Scheme	Global Optimization Algorithm	Local Dispatching Rules
Scheme 1	BA	FIFO
Scheme 2	WOA	FIFO
Scheme 3	CGA	FIFO
Scheme 4	ICGA	FIFO
Scheme 5	BA	RCMQB; SST, FAM, FCFS
Scheme 6	WOA	RCMQB; SST, FAM, FCFS
Scheme 7	CGA	RCMQB; SST, FAM, FCFS
Scheme 8	ICGA	RCMQB; SST, FAM, FCFS

5.3.3. Simulation Results and Analysis

1. Evaluation index of scheduling results

In the optimization process, makespan C_{max} was used as the fitness function value of the global optimization algorithm. Meanwhile, a number of evaluation indexes related to the actual production line were established, including T_{cpu}, *TWIP*, *TWT*, *FUR*, *TS* and *TPB*. Except *FUR*, the smaller the values, the better the remaining evaluation indexes.

Eight sets of simulation schemes were run 30 times. The average of 30 simulation results is presented in Table 8. As shown in the table, under the principle of adopting the same global optimization algorithm, compared with schemes 1–4, each metric of schemes 5–8 has been improved to some extent, except the T_{cpu}. Among the metrics, the optimization improvement of *TWIP*, *TS* and *TPB* is obvious. This is mainly because schemes 5–8 adopted RCMQB rules when entering multi-queue limited buffers, which can allocate resources of multi-queue limited buffers more reasonably and reduce the occurrence of blocking. The SST rule, the FAM rule and the FCFS rule were adopted when leaving the buffer, which can more effectively control the jobs to select the machine with the least change of properties among multiple machines for processing. This was beneficial to reduce the setup times. Further, *TWIP* is equal to the sum of the time for the job staying on the buffer, the blocking time of the job and setup times, so these three evaluation indexes were significantly improved. The reduction in blocking time and setup times will lead to the reduction in the makespan of the whole process and the reduction of the idle time for machines. Therefore, C_{max}, *TWT* and *FUR* of schemes 5–8 were also optimized to some extent. From the table we can also see that compared with the schemes 1–4, although the T_{cpu} of the schemes 5–8 had all increased, the rate of increase was small. This indicates that the complicated local rules adopted in schemes 5–8 can made full use of the capacity of the multi-queue limited buffers and effectively reduced the blocking time and setup times. At the same time, the running time costs were small and did not have much impact on the speed of the algorithm.

Table 8. Evaluation index comparison of scheduling results of eight schemes.

Evaluation Index		Scheme 1	Scheme 2	Scheme 3	Scheme 4	Scheme 5	Scheme 6	Scheme 7	Scheme 8
C_{max}	Optimum	296	295	297	295	288	288	290	286
	Worst	307	305	309	304	303	300	302	300
	Average	301.12	303.16	305.76	300.56	293.56	295.60	299.84	292.32
	variance	18.56	8.96	11.74	9.06	12.48	5.07	8.69	6.86
T_{cpu}	Optimum	529.15	103.35	14.30	15.29	503.83	113.27	14.264	15.64
	Worst	587.99	144.61	15.23	16.24	600.83	149.73	15.39	16.77
	Average	551.61	122.57	14.32	15.08	559.48	134.35	14.62	15.32
	variance	325.24	339.50	0.041	0.06	425.57	81.53	0.05	0.18
TWIP	Optimum	630	642	623	623	625	605	613	611
	Worst	768	730	745	750	734	707	718	713
	Average	691.52	689.72	698.32	681.96	660.88	656.48	672.64	647.44
	variance	824.88	550.12	958.45	764.27	781.56	421.86	795.19	635.52
TWT	Optimum	229	253	250	251	202	202	204	203
	Worst	358	321	347	343	282	278	309	302
	Average	282.81	282.96	287.56	285.60	240.36	235.48	255.40	248.04
	variance	673.44	428.41	514.16	564	543.59	276.83	493.68	517.95
FUR	Optimum	86.26%	85.03%	85.18%	85.43%	87.67%	87.62%	87.56%	87.59%
	Worst	80.36%	80.17%	80.54%	80.25%	83.59%	83.79%	82.91%	83.16%
	Average	83.58%	83.41%	83.14%	83.34%	85.68%	85.93%	84.94%	85.88%
	variance	0.05%	0.01%	0.02%	0.02%	0.02%	0.00%	0.01%	0.01%
TS	Optimum	110	121	126	115	105	108	108	98
	Worst	153	164	167	160	141	137	151	144
	Average	137.64	139.12	148.52	138.43	121.36	119.76	125.60	120.76
	variance	123.31	103.59	130.24	114.40	66.15	45.86	91.08	73.62
TPB	Optimum	73	73	77	74	70	72	75	72
	Worst	122	109	118	113	118	99	110	104
	Average	94.42	93.04	94.88	93.28	87.8	85.16	87.64	95.21
	variance	165.12	71.55	97.82	99.16	129.44	48.21	69.59	74.36

The comparison among schemes 1–4 or schemes 5–8 show that the running speed of the CGA and ICGA was significantly faster than that of the BA and WOA. The average running time of the BA reached 555.56 s, which is obviously not suitable for solving practical problems. In terms of optimization performance, under two different local dispatching rules, the optimization effect of ICGA for C_{max} was the best among the four algorithms. Specially, compared with the CGA, the optimization effect of the ICGA was significantly improved. The above results show that the ICGA, to some extent, overcame the problem that the CGA was easy to prematurely converge while maintaining the fast running speed of the CGA. This is mainly because in the early stage of evolution, there is no large probability value in the probabilistic model. Thus, the ICGA was still evolving as the procedure of the CGA. In this way, the ICGA maintained the advantage that the CGA can converge quickly in the early stage of evolution. In the later stage of evolution, when falling into the local extremum, the ICGA mapped the original probabilistic model to the new probabilistic model through the probability density function of the Gaussian distribution. In this way, while keeping probability values' distribution of the original probabilistic model unchanged, ICGA's searching ability of feasible solutions was expanded, and the diversity of population individuals was improved. Therefore, the algorithm could jump out of the local extremum and continued to evolve.

Through the above analysis, it can be concluded that compared with the other seven schemes, the combination of the ICGA and local dispatching rules adopted in scheme 8 was the best for reducing setup times of the machine, and it effectively decreased the blocking effect of the limited buffers, more reasonably arranged the jobs in and out of the multi-queue limited buffer, and orderly assigned the processing tasks. Various evaluation metrics were improved, and the problem of multi-queue limited buffers scheduling problems in a flexible flow shop with setup times was more effectively solved.

2. Gantt chart analysis of scheduling result

Figure 4 is the Gantt chart of the scheduling result of scheme 8, with time axis as the abscissa and machine of each stage as the ordinate. The green part indicates the residence time of the bus in the buffer. The red part indicates the setup times when the successively-processed buses are with different properties. The blue part denotes the blocking time of the bus on the machine after the completion of the process. The processing route of J_3 was $\{WS_{1,1}, b_{2,2,1}, b_{2,2,2}, WS_{2,2}, b_{3,1,1}, WS_{3,1}, b_{4,1,1}, WS_{4,1}\}$. In Figure 4, we can see that at time $t = 44$, J_3 completed the processing on machine $WS_{1,1}$ of stage $Oper_1$ (the competition time of J_3 was 44, namely $C3, 1 = 44$). At this time, the first lane $Bs_{2,1}$ of buffer Bu_2 had two jobs waiting to be processed (J_8 and J_4), that is to say, $WA_{2,1}(t) = \{J_8, J_4\}$, $card(WA_{2,1}(t)) = 2$. The second lane $Bs_{2,2}$ of buffer Bu_2 had one job waiting to be processed (J_9), that is to say, $WA_{2,2}(t) = \{J_9\}$, $card(WA_{2,2}(t)) = 1$. The capacity of lane $Bs_{2,1}$ was 2 ($K_{2,1} = 2$), $card(WA_{2,1}(t)) \le K_{2,1}$. And the capacity of lane $Bs_{2,2}$ was 2 ($K_{2,2} = 2$), $card(WA_{2,2}(t)) \le K_{2,2}$. Both of them satisfy the constraint of Equation (10) of the limited buffers. At this time, based on the RCMQB rule which regulates the job's access to the buffer, we can obtain $Mca_{3,2}(t) = \{Bs_{2,2}\}$ from Equation (26). The J_3 was assigned to space $b_{2,2,1}$ in lane $Bs_{2,2}$ of stage $Oper_2$, waiting to be processed at stage $Oper_2$. The entry time of the bus into the buffer ($Te_{3,2}$) was equal to $C3, 1$, which satisfied the constraint of Equation (8) of limited buffers.

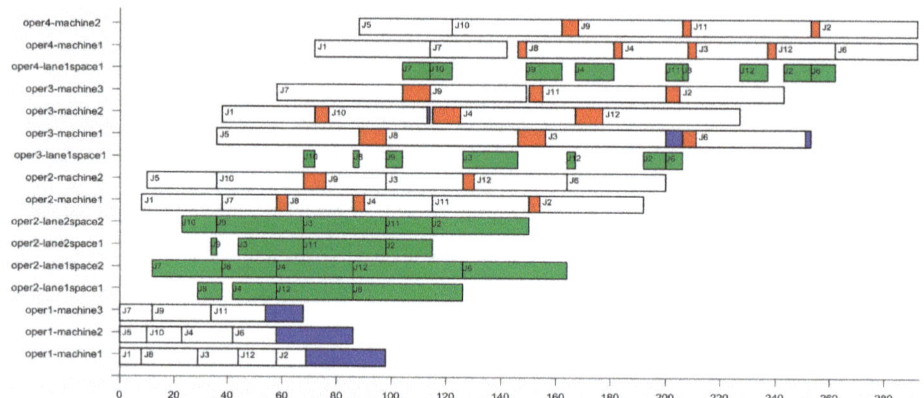

Figure 4. Gantt chart of the scheduling result of the ICGA.

At time $t = 68$, J_9 left the space $b_{2,2,2}$ in lane $Bs_{2,2}$, and the departure time out of the buffer ($Tl_{9,2}$) was 68. At the same time, J_3 was assigned to the space $b_{2,2,2}$, waiting to be processed on the machine of stage $Oper_2$. At time $t = 98$, $Tl_{3,2} = 98$, J_3 also departed from the buffer, and $Tl_{9,2} < Tl_{3,2}$, satisfying the constraint of Equation (13). From the Gantt chart drawn from scheduling results, it can be seen that the constraint of Equation (12) was always met during the use of multi-queue limited buffers.

Further analysis of the process that J_3 left the buffer is as follows. When $t = 98$, machine $WS_{2,2}$ completed the processing of J_9, so that machine $WS_{2,2}$ was available. At this time, J_{12} on space $b_{2,1,2}$ in lane $Bs_{2,2}$ was waiting for processing, and J_3 on space $b_{2,2,2}$ was also waiting for processing. According to the bus's property information in Table 3, in regard to machine $WS_{2,2}$, if we choose to process J_{12}, the setup time ($Ts_{12,2,2}$) will be equal to $Tsp_{2,2}$, namely $Ts_{12,2,2} = Tsp_{2,2} = 4$; if we choose to process J_3, the setup time ($Ts_{3,2,2}$) will be 0, namely $Ts_{3,2,2} = 0$. In accordance with the SST rule that controls the job leaving the buffer, J_3 was chosen to process.

3. Analysis of scheduling evolution

Figure 5 shows the relationship between fitness value and training iterations of schemes 5–8. As can be seen, the BA and WOA converged quickly at the initial stage of the evolution. This is mainly due to the fact that the initial numbers of population of the BA and WOAs ($NP = 30$) were much larger than that of the CGA and ICGAs ($NP = 4$). This means a relatively strong ability for searching the solution space. However, with the increase in evolutional iterations, the search ability of the BA and the WOA gradually deterioratesdand they fell into the local extremum in the 135th and 139th generations, respectively. Although the CGA should be good at fast search, the distribution of probability values in the probabilistic model all decreased with evolution, and thus and the search performance on the solution space decreased. The algorithm stopped evolving at the 87th generation, falling into the local extremum. On the other hand, the ICGA had the similar fast search performance as the CGA during the initial stage and encountered evolutionary stagnation at the 94th generation. But as evolution continues, the ICGA started the mapping operation to reactivate the algorithm's evolutionary ability and found for the optimal solution in the 263th generation.

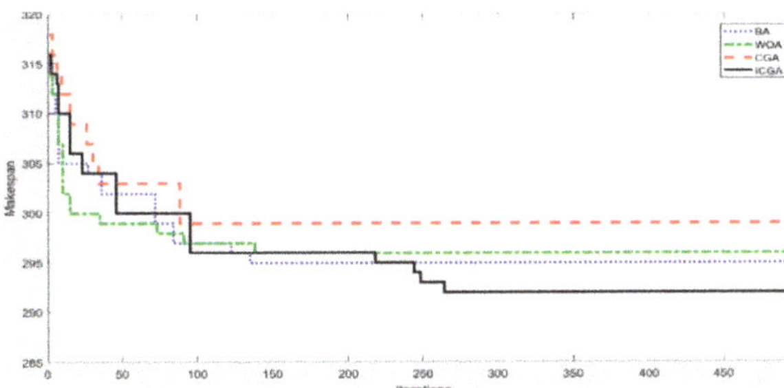

Figure 5. Relationship between fitness value and training iterations of schemes 5–8.

Figure 6 shows the relationship between fitness value and running time of schemes 5–8. As can be seen from the figure, the running speed of CGA and ICGA was very fast, but the CGA fell into the local extremum at 4 s. The optimization speed of the BA was the slowest among the four algorithms, and its optimal solution was 299 during the 35 s running time. From the figure, we can also see that the fitness value of the ICGA was the best among the four algorithms at the same CPU time. And the time used by the ICGA to get the same fitness value was the least among the four algorithms. The ICGA maintained the advantage that the CGA can converge quickly in the early stage of evolution. In the later stage of evolution, when falling into the local extremum, the ICGA improved the diversity of individuals in the population by mapping the original probabilistic model to a new probabilistic model, so that the algorithm could jump out of the local extremum and continue to evolve.

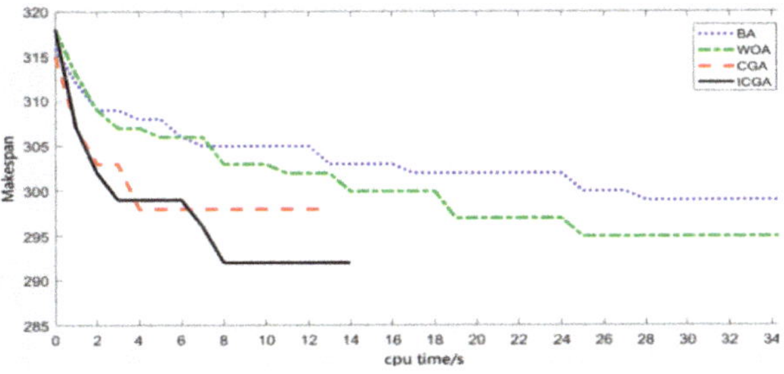

Figure 6. Relationship between fitness value and Running time of schemes 5–8.

4. Statistical analysis of scheduling results

Figure 7 shows the makespan C_{max} of schemes 5–8 running 10 times separately. For scheme 7, the average value of makespan C_{max} in 10 tests was 298.1, which was the worst among the four schemes. Although scheme 5 could sometimes obtain better results, the results obtained in 10 tests were not stable enough and were with large fluctuations, easily falling into the local extremum sometimes. Its average value of 10 tests was 295.2. Compared with the other three schemes, the results obtained by the scheme 6 were relatively stable. However, its average result was poor, at 296.3. The average value of the test results of scheme 8 was 292.3, which was the best of all the schemes. The fluctuations of the curve indicate that the magnitude of the change in the results of scheme 8 in 10 tests is small. This shows that when solving multi-queue limited buffers scheduling problems in a flexible flow shop with setup times, compared with the CGA, the ICGA has not only greatly improved the optimization performance, but also improved the stability of the optimization results.

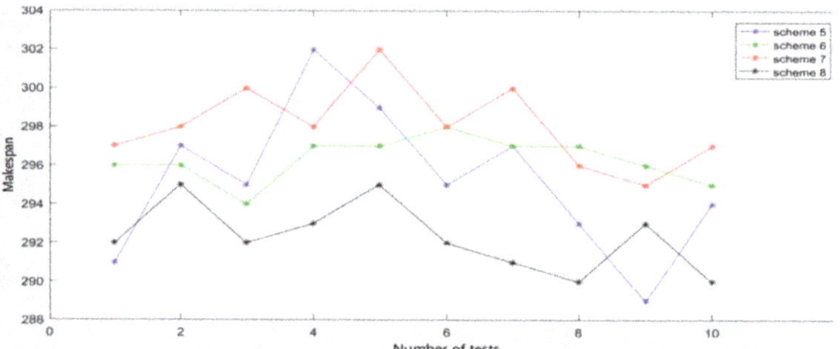

Figure 7. Results of 10 instance tests.

6. Conclusions

This study explored the multi-queue limited buffers scheduling problems in a flexible flow shop with setup times in a bus manufacturer. The study proposed an ICGA for global optimization to better improve the global optimization results, aiming at the shortcomings of the CGA that it is easy to fall into the local extremum and rapidly stops evolving. This algorithm employed the probability density function of the Gaussian distribution to map the original probabilistic model to a new probabilistic model so as to enhance the evolutionary vigor of the CGA. The job was processed on the specified online sequence in accordance with the individual decoding. Considering the impact of multi-queue limited buffers, the problems of the job into and out of the buffer during the subsequent scheduling were emphasized. When the job enters the buffer, the remaining capacity of buffers should be taken into account to reduce machining blocking and stagnation. When the job leaves the buffer and is assigned to the machine at the next stage, it is influenced by the setup times. In this study, the setup time of the machine was calculated based on the change in the properties of successively processed jobs. The SST rule was used to reduce the setup times. Finally, the findings of simulation experiments proved that combining the ICGA with local dispatching rules could better solve the multi-queue limited buffers scheduling problems in a flexible flow shop with setup times.

Author Contributions: Z.H. conceived and designed the research; Q.Z. performed the experiment and wrote the manuscript; H.S. and J.Z. checked the results of the whole manuscript.

Funding: This research was funded by the Liaoning Provincial Science Foundation, China (grant number: 2018106008), the Natural Science Foundation of China (grant number: 61873174), Project of Liaoning Province Education Department, China (grant number: LJZ2017015) and Shenyang Municipal Science and Technology Project, China (grant number: Z18-5-102).

Conflicts of Interest: The authors declare no conflict of interest.

References

1. Zhao, F.Q.; Tang, J.X.; Wang, J.B.; Jonrinaldi, J. An improved particle swarm optimization with a linearly decreasing disturbance term for flow shop scheduling with limited buffers. *Int. J. Comput. Integr. Manuf.* **2014**, *27*, 488–499. [CrossRef]
2. Ventura, J.A.; Yoon, S.H. A new genetic algorithm for lot-streaming flow shop scheduling with limited capacity buffers. *J. Intell. Manuf.* **2013**, *24*, 1185–1196.
3. Han, Z.H.; Sun, Y.; Ma, X.F.; Lv, Z. Hybrid flow shop scheduling with finite buffers. *Int. J. Simul. Process Model.* **2018**, *13*, 156–166.
4. Zeng, M.; Long, Q.Y.; Liu, Q.M. Cellular automata variation particles warm optimization algorithm for batch scheduling. In Proceedings of the 2012 Second International Conference on Intelligent System Design and Engineering Application, Sanya, Hainan, China, 6–7 January 2012; IEEE: New York, NY, USA, 2012.
5. Zhang, P.W.; Pan, Q.K.; Liu, J.Q.; Duan, J.H. Hybrid artificial bee colony algorithms for flowshop scheduling problem with limited buffers. *Comput. Integr. Manuf. Syst.* **2013**, *19*, 2510–2511.
6. Zhang, S.C.; Wong, T.N. Studying the impact of sequence-dependent set-up times in integrated process planning and scheduling with E-ACO heuristic. *Int. J. Prod. Res.* **2016**, *54*, 4815–4838. [CrossRef]
7. Shen, L.; Dsuzere, P.; Neufeld, J.S. Solving the flexible job shop scheduling problem with sequence-dependent setup times. *Eur. J. Oper. Res.* **2018**, *265*, 503–516. [CrossRef]
8. Tran, T.T.; Araujo, A.; Bech, J.C. Decomposition methods for the parallel machine scheduling problem with setups. *INFORMS J. Comput.* **2016**, *28*, 83–95. [CrossRef]
9. Vallada, E.; Ruiz, R. A genetic algorithm for the unrelated parallel machine scheduling problem with sequence dependent setup times. *Eur. J. Oper. Res.* **2011**, *211*, 612–622. [CrossRef]
10. Benkalai, I.; Rebaine, D.; Gagné, C.; Baptiste, P. Improving the migrating birds optimization metaheuristic for the permutation flow shop with sequence-dependent set-up times. *Int. J. Prod. Res.* **2017**, *55*, 6145–6157. [CrossRef]
11. An, Y.J.; Kim, Y.D.; Choi, S.W. Minimizing makespan in a two-machine flowshop with a limited waiting time constraint and sequence-dependent setup times. *Comput. Oper. Res.* **2016**, *71*, 127–136.
12. Lenstra, J.K.; Kan, A.; Brucker, P. Complexity of machine scheduling problems. *Stud. Integer Program.* **1977**, *1*, 343–362.
13. Harik, G.R.; Lobo, F.G.; Goldberg, D.E. The compact genetic algorithm. *IEEE Trans. Evol. Comput.* **1999**, *3*, 287–297. [CrossRef]
14. Rafael, D.S.; Carlos, E.L.; Heitor, L. Template matching in digital images using a compact genetic algorithm with elitism and mutation. *J. Circuits Syst. Comput.* **2010**, *19*, 91–106.
15. Sharifi, R.; Anvari-Moghaddam, A. A Flexible Responsive Load Economic Model for Industrial Demands. *Processes* **2019**, *7*, 147. [CrossRef]
16. Tran, V.; Ramkrishna, D. Simulating Stochastic Populations. Direct Averaging Methods. *Processes* **2019**, *7*, 132. [CrossRef]
17. Gao, Z.W.; Saxen, H.; Gao, C.H. Data-Driven Approaches for Complex Industrial Systems. *IEEE Trans. Ind. Inform.* **2013**, *9*, 2210–2212. [CrossRef]
18. Gao, Z.W.; Ding, S.X.; Cecati, C. Real-time Fault Diagnosis and Fault Tolerant. *IEEE Trans. Ind. Electron.* **2015**, *62*, 3752–3756. [CrossRef]
19. Han, Z.H.; Zhu, Y.H.; Ma, X.F.; Chen, Z.L. Multiple rules with game theoretic analysis for flexible flow shop scheduling problem with component altering times. *Int. J. Model. Identif. Control* **2016**, *26*, 1–18. [CrossRef]
20. Yang, X.S. A new metaheuristic bat-inspired algorithm. In *Nature Inspired Cooperative Strategies for Optimization*; Springer: Berlin, Germany, 2010; pp. 65–74.
21. Mirjalili, S.; Lewis, A. The whale optimization algorithm. *Adv. Eng. Softw.* **2016**, *95*, 51–67. [CrossRef]
22. Luo, Q.F.; Zhou, Y.Q.; Xie, J.; Ma, M.; Li, L.L. Discrete bat algorithm for optimal problem of permutation flow shop scheduling. *Sci. World J.* **2014**, *2014*. [CrossRef]
23. Zhang, J.J.; Li, Y.G. An improved bat algorithm and its application in permutation flow shop scheduling problem. *Adv. Mater. Res.* **2014**, *1049*, 1359–1362. [CrossRef]
24. Jiang, T.H.; Zhang, C.; Zhu, H.Q.; Gu, J.C.; Deng, G.L. Energy-efficient scheduling for a job shop using an improved whale optimization algorithm. *Mathematics* **2018**, *6*, 220. [CrossRef]

25. Neron, E.; Baptiste, P.; Gupta, J. Solving hybrid flow shop problem using energetic reasoning and global operations. *Omega* **2001**, *29*, 501–511. [CrossRef]
26. Santos, D.L.; Hunsucker, J.L.; Deal, D.E. Global lower bounds for flow shop with multiple processors. *Eur. J. Oper. Res.* **1995**, *80*, 112–120. [CrossRef]
27. Kim, M.K.; Narasimhan, R. Designing Supply Networks in Automobile and Electronics Manufacturing Industries: A Multiplex Analysis. *Processes* **2019**, *7*, 176. [CrossRef]

Article

Process Simulation of the Separation of Aqueous Acetonitrile Solution by Pressure Swing Distillation

Jing Li [1,*], Keliang Wang [1,*], Minglei Lian [1], Zhi Li [1] and Tingzhao Du [2]

[1]   College of Chemistry and Materials Engineering, Liupanshui Normal University, Liupanshui 553004, China
[2]   North China Company, China Petroleum Engineering Co., LTD., Renqiu 052100, China
*   Correspondence: woxinfeiyang1986@163.com (J.L.); wangkeliang84@163.com (K.W.)

Received: 24 May 2019; Accepted: 27 June 2019; Published: 1 July 2019

Abstract: The separation of aqueous acetonitrile solution by pressure swing distillation (PSD) was simulated and optimized through Aspen Plus software. The distillation sequence of the low pressure column (LPC) and high pressure column (HPC) was determined with a phase diagram. The pressures of the two columns were set to 1 and 4 atm, respectively. Total annual cost (TAC) was considered as the objective function, and design variables, such as the tray number, the reflux ratio, and the feeding position, were optimized. The optimum process parameters were obtained. For the reduction of energy consumption, the PSD with full-heat integration was designed. The TAC of this method is lower by 32.39% of that of the PSD without heat integration. Therefore, it is more economical to separate acetonitrile and water mixture by PSD with full-heat integration, which provides technical support for the separation design of such azeotropes.

Keywords: pressure swing distillation; full-heat integration; acetonitrile; water

1. Introduction

Acetonitrile is an organic solvent often used to purify butadiene and fatty acids. Given its high chemical activity, acetonitrile is also used as a raw material and synthetic intermediate for pharmaceutical chemicals [1–3]. A considerable amount of acetonitrile–water mixture is produced during the chemical production process. Acetonitrile is toxic and expensive, so its recovery from chemical production waste liquid is beneficial from environmental and economic points of view. However, owing to the azeotropic phenomenon, acetonitrile cannot be separated from its aqueous solution by a conventional rectification method. Thus, techniques, such as extractive distillation (ED) and pressure swing distillation (PSD), are often used in the industry [4–6].

ED involves the addition of an extractant to the azeotrope to increase the relative volatility of light and heavy components and achieve the purpose of separation [7–9]. Rodriguez-Donis et al. [10] used butyl acetate as extractant to separate acetonitrile and water mixture and analyzed the feasibility of seven configurations. Raeva et al. [11] proposed that the selection of extractants should not be limited to the relative volatility, but selective analysis is also an effective method. Sazonova et al. [12] found that when ED was used to separate acetonitrile and water mixture, the energy consumption of glycerol as an extractant was lower than that of dimethylsulfoxide and 1,2-ethandiol. You et al. [13] used a multi-objective genetic algorithm to simulate and optimize the separation process of acetonitrile and water mixture by ED with ethylene glycol as an entrainer. For multi-component mixtures, many scholars adopt the technology of the dividing wall column to reduce energy consumption and equipment investment [14,15].

The advantage of PSD over ED is that PSD does not introduce a third component and can efficiently separate pressure-sensitive azeotropes [16–18]. The separation of acetonitrile and water mixture by PSD has already been studied by some scholars [19,20]. Repke et al. [21] studied the separation of acetonitrile and water mixture by PSD and carried out dynamic control. It was found that when the feed concentration fluctuated greatly, it could also maintain stability. Kim and Huang [22,23] optimized the process variables by taking the reboiler heat duty and total energy consumption as objective functions, respectively. Huang and Matsuda [24,25] studied the rectifying/stripping section heat integration to reduce the energy consumption in PSD, i.e., heat transfer between the rectifying section of the high pressure column (HPC) and the stripping section of the low pressure column (LPC), with a remarkable energy-saving effect. However, there is another way of heat integration, that is, heat transfer between the condenser of the HPC and the reboiler of the LPC.

For increasingly complex industrial technologies, Gao and Dai [26–28] proposed the fault detection and diagnosis (FDD) and data-driven approaches in the modeling, control, and optimization of complex industrial systems. Many design variables are involved in the chemical distillation industry, so commercial software, such as Aspen Plus, can better design, optimize, and control the distillation process [29,30]. Aspen Plus (Aspen Tech, Bedford, MA, USA) is a large-scale general process simulation system for plant design, steady-state design, and dynamic simulation. It is recognized as a large scale process simulation software in the chemical industry, which has been widely used in industrial design and academic research [31,32]. In this study, the conceptual design and optimization of PSD for the separation of acetonitrile and water mixture was carried out, so Aspen Plus was selected for steady-state simulation and optimization design.

This work aims to compare PSD with full heat integration and traditional PSD for the separation of acetonitrile and water mixture. At present, there is no comparison between the two methods in the literature. Therefore, in this work, Aspen Plus software was used to simulate and optimize the PSD process. The entire optimization process uses the total annual cost (TAC) as the objective function to find the optimal values of the design variables through the sequential iteration method. At the same time, full-heat integration process design for PSD was optimized and compared with PSD without heat integration from an economic point of view. The results from this work can provide some technical support for such azeotrope separation designs.

2. Design of Separation Scheme for PSD

2.1. Feasibility Analysis

The calculation accuracy of the process simulation depends on the selection of the thermodynamic equation. Since the azeotropic concentration under different pressures predicted by the Wilson equation agrees well with the experimental data published in Azeotropic Data [33], the Wilson equation was used in the simulation process, and the binary interaction parameters involved in Wilson equation can be obtained in Aspen Plus software. Table 1 lists the variation of acetonitrile–water azeotrope with the pressure. The acetonitrile–water azeotrope composition is sensitive to the pressure changes. As the pressure increases from 1 to 4 atm, the mole concentration of acetonitrile decreases from 0.6971 to 0.5886, indicating that PSD is feasible for separating acetonitrile–water azeotrope.

Table 1. Effect of the pressure on acetonitrile–water azeotrope.

Operating Pressure/Atm	Boiling Point of Acetonitrile/°C	Boiling Point of Water/°C	Boiling Point of Azeotrope/°C	Azeotropic Concentration (Mole)	
				Acetonitrile	Water
1	81.48	100.02	76.40	0.6971	0.3029
2	105.33	120.69	98.06	0.6438	0.3562
3	120.91	134.05	112.01	0.6116	0.3884
4	132.79	144.16	122.58	0.5886	0.4114
5	142.52	152.41	131.18	0.5707	0.4293

2.2. Pressure Selection of Two Columns

Given the convenience of operation and the low cost of utility engineering, the pressure of the low pressure column (LPC) is selected to be 1 atm. In the pressure selection of the high pressure column (HPC), the cost of the heating medium used in the column reboiler should be considered. The bottom product of the HPC is acetonitrile, so the bottom temperature corresponds to the boiling point temperature of acetonitrile under this pressure. From Table 1, it can be seen that if the pressure of the HPC is 5 atm, the bottom temperature is 142.52 °C, and the low pressure (LP) steam of 160 °C cannot be used as the heating medium. Reduce the pressure of the HPC to 4 atm, there is no such situation. Here the bottom temperature is 132.79 °C, and the LP steam can meet the use requirements. In this way, the pressures for the LPC and HPC are 1 and 4 atm, respectively.

2.3. PSD Process Description

The acetonitrile aqueous mixture has 100 kmol/h flow rate at room temperature, 25 mol% acetonitrile, and 75 mol% water content. The final acetonitrile product has 99.9% mole purity, and the acetonitrile content of the wastewater does not exceed 0.1%. The *Txy* phase diagram for the acetonitrile–water mixture is plotted in Figure 1. The mixture is fed into the LPC, and the bottom stream B1 of the LPC is water with a very low acetonitrile content. The overhead azeotrope D1 is pressurized into the HPC, the bottom stream B2 of the HPC is high purity acetonitrile, and the top stream D2 of the HPC is the azeotrope. Through changing the pressure of the double columns, the azeotropic concentration under the corresponding pressure is crossed for the effective separation.

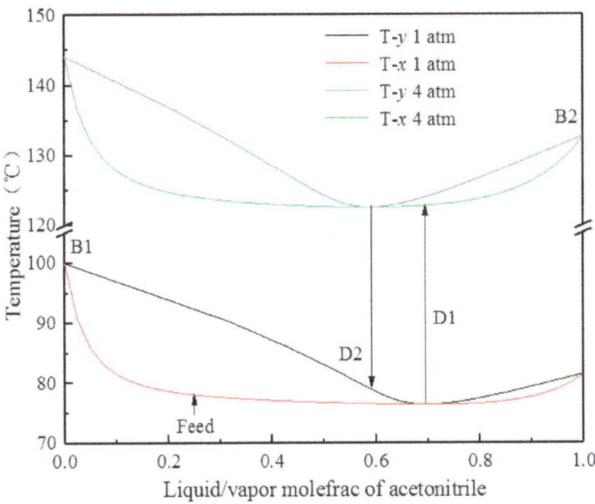

Figure 1. *Txy* phase diagram for the acetonitrile–water mixture.

2.4. The Objective Function of the Process Optimization

In the process of separating acetonitrile and water mixture by PSD, not only the separation requirements of product purity should be satisfied but also the economy of the process should be considered. TAC is a commonly used economic cost index for chemical process simulation and optimization, which is used in the evaluation of the economic rationality and feasibility of the process. The calculation of TAC is based on the following equations, which is taken from Luben and Chien [34]. Design variables, such as the tray numbers, the reflux ratios, and the feeding positions of the two columns, are optimized through the equations below, in which the minimum TAC and purity of acetonitrile are considered as the objective function and constraint condition, respectively.

Objective function:

$$\min \text{TAC} = C_{\text{operation}} + C_{\text{capital}}/3, \tag{1}$$

$$\text{TAC} = f(x), \tag{2}$$

$$C_{\text{capital}} = C_{\text{exchanger}} + C_{\text{column}}, \tag{3}$$

$$C_{\text{exchanger}} = 7296A^{0.65}, \tag{4}$$

$$C_{\text{column}} = 17640D^{1.066}L^{0.802}, \tag{5}$$

$$C_{\text{operation}} = C_{\text{steam}} + C_{\text{cooling water}}, \tag{6}$$

$$L = 1.2 \times 0.61 \times (N_T - 2), \tag{7}$$

$$A = Q/(K \times \Delta T_m). \tag{8}$$

Constraints:

$$w_{B1} < 0.001, \tag{9}$$

$$w_{B2} > 0.999. \tag{10}$$

where TAC is the total annual cost, $C_{\text{operation}}$ is the annual operation cost, and the annual operation time is 8000 h; C_{capital} is the equipment investment cost, and the payback period is 3 years; x is the whole process operation variable, including the feeding positions of the LPC and HPC (N_{F1}, N_{FR}, and N_{F2}), reflux ratios (RR1 and RR2), and tray numbers (N_{T1} and N_{T2}); $C_{\text{exchanger}}$ and C_{column} represent the heat exchanger (including condenser and reboiler) and the column equipment costs, respectively; A is the heat exchange area (m^2); L and D represent the height and diameter of the column (m), respectively; The diameter of the column is calculated by the tool of "Tray Sizing" in Aspen Plus software; 0.61 m is the typical distance between the trays; Q and ΔT_m are the heat duty (kW) and heat transfer temperature difference (K), respectively; K is the heat transfer coefficient, 0.852 kW/(K·m^2) for the condenser and 0.568 kW/(K·m^2) for the reboiler; C_{steam} and $C_{\text{cooling water}}$ represent the cost of heating medium and cooling water, respectively; The LP steam is used in this work with the price of \$7.78 /GJ, and the price of the cooling water is \$4.43/GJ. Additionally, w_{B1} and w_{B2} are the purity of acetonitrile at the bottom streams of the LPC and HPC, respectively.

3. Optimization of PSD

3.1. Optimization of PSD without Heat Integration

3.1.1. Process Optimization Sequence

The design variables of the PSD process include three feeding positions (N_{F1}, N_{FR}, and N_{F2}), the reflux ratios (RR1 and RR2), and the tray numbers (N_{T1} and N_{T2}) of the LPC and HPC. The variables are optimized through the sequential iteration method, which was commonly used to optimize the PSD process in the published literatures [17,35]. By editing the calculating formulas in Fortran, Aspen Plus software can automatically call the values of the corresponding variables to calculate TAC. The optimization sequence is illustrated in Figure 3. First, the feeding position is optimized as the innermost iteration, followed by the reflux ratio, and finally, the tray number is optimized. The objective function of the whole optimization process is to minimize TAC.

3.1.2. The Optimization Results of PSD

According to the optimization sequence in Figure 2, three feeding positions (N_{F1}, N_{FR}, and N_{F2}), the reflux ratios (RR1 and RR2), and the tray numbers (N_{T1} and N_{T2}) of the LPC and HPC were optimized iteratively, and the optimum process parameters were obtained. The results can be seen in Table 4. That is, N_{T1} of the LPC is 21, N_{F1} and N_{FR} are 19 and 16, respectively, and RR1 is 0.3; N_{T2} of the HPC is 25, N_{F2} is 16th, and RR2 is 0.5. The optimized flow diagram of PSD without heat integration for separating acetonitrile–water azeotrope is shown in Figure 3.

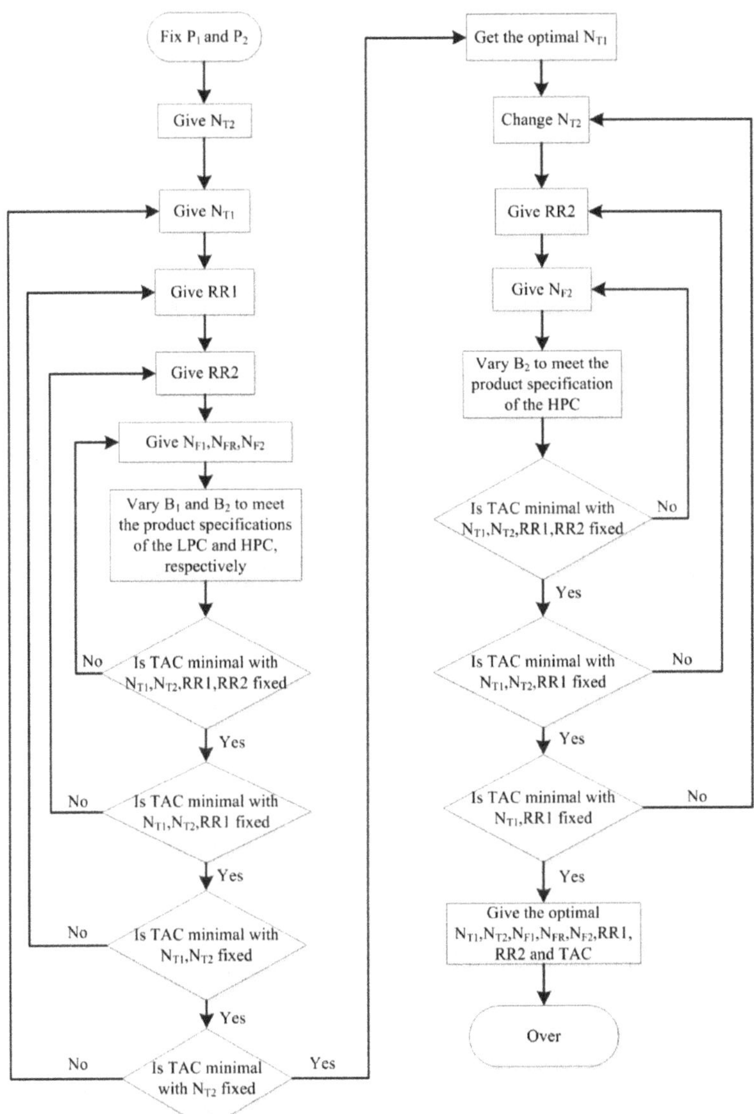

Figure 2. Optimization sequence of acetonitrile–water separation process by pressure swing distillation (PSD).

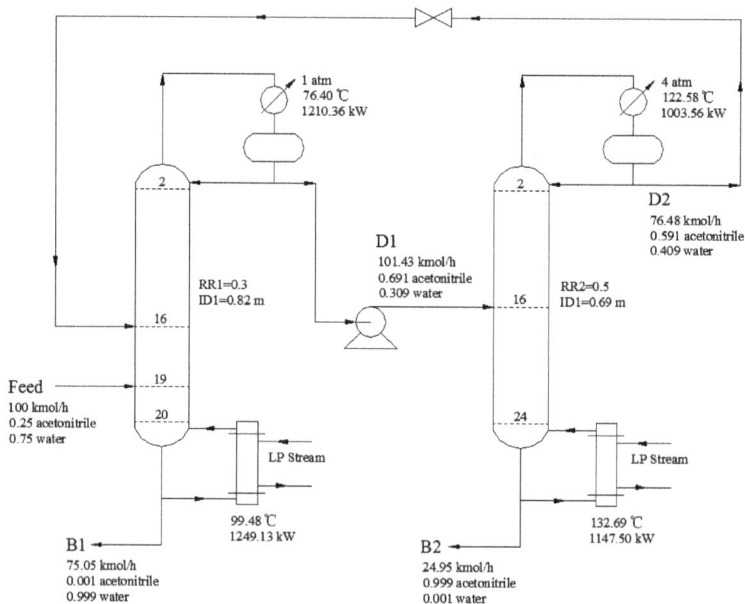

Figure 3. Optimized flow diagram of pressure swing distillation (PSD) without heat integration.

3.2. Optimization of PSD with Full-Heat Integration

PSD with full-heat integration is carried out by the huge temperature difference between the top stream of the HPC and the bottom stream of the LPC. That is, the top stream of the HPC can be used as the heating medium of the reboiler of the LPC. Full-heat integration can be carried out by adjusting the reflux ratios of the LPC and HPC.

3.2.1. Process Optimization Sequence of PSD with Full-Heat Integration

The design variables are optimized according to the sequential iteration method. The optimization process is shown in Figure 4. First, the reflux ratio is optimized as the innermost iteration, then the feeding position, and finally, the tray number is optimized. The objective function of the whole optimization process is to minimize TAC.

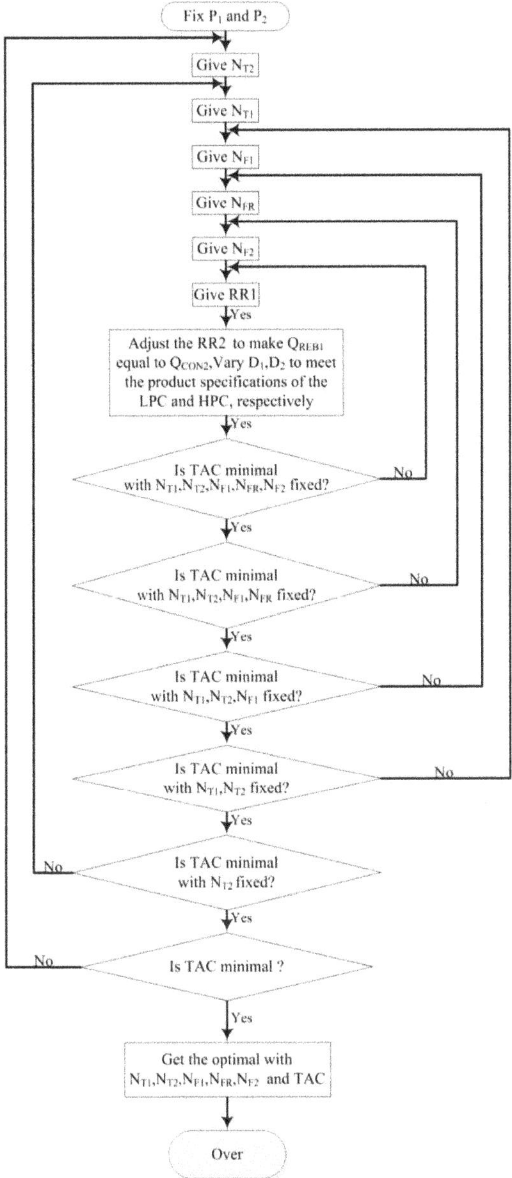

Figure 4. Optimization sequence of acetonitrile–water separation process by pressure swing distillation (PSD) with full-heat integration.

3.2.2. Optimization of the Reflux Ratios

In the whole process optimization, the reflux ratios of the two columns are used as the design variables, and the heat duty of the condenser of the HPC is equal to that of the reboiler of the LPC as the objective variable to achieve the full-heat integration design.

Figure 5 investigates the effect of RR1 in the LPC on RR2 in the HPC and heat duties of the condenser and reboiler (Q_{CON2} and Q_{REB2}) in HPC. It can be seen that with the increase of RR1, RR2 increases gradually, but Q_{CON2} and Q_{REB2} of the HPC appear to have a minimum. When RR1 is 0.2 and RR2 is 0.7, the values of Q_{REB2} and Q_{CON2} are minimum.

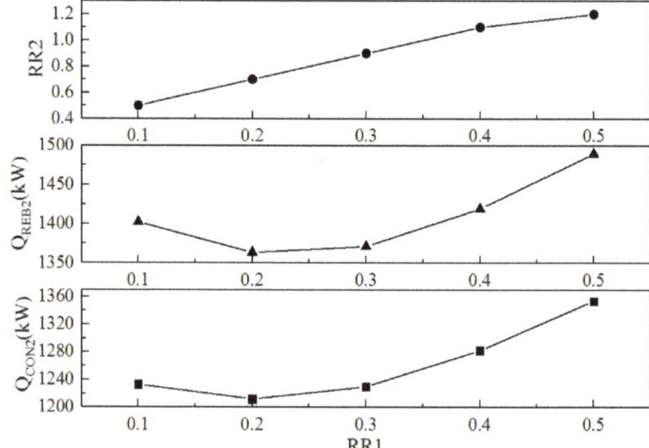

Figure 5. Effect of the reflux ratio RR1 in the low pressure column (LPC).

The effect of the reflux ratios of the two columns on the TAC for the whole process is shown in Figure 6. It can also be seen that TAC is the smallest when RR1 is 0.2 and RR2 is 0.7. This finding is consistent with the conclusion in Figure 5. The optimum reflux ratios of the LPC and HPC are 0.2 and 0.7, respectively.

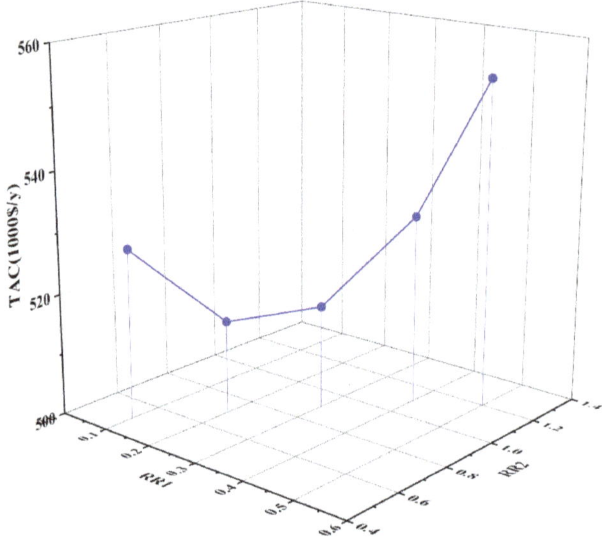

Figure 6. Effect of the reflux ratios of two columns on the total annual cost (TAC).

3.2.3. Optimization of the Feeding Positions

The feeding position of the acetonitrile–water mixture (N_{F1}), the feeding position of the recycle stream (N_{FR}), and the feeding position of the HPC (N_{F2}) affect the TAC to varying degrees, as shown in Figure 7. Taking the feeding position N_{F1} of acetonitrile–water mixture as an example, as the feeding position moves down, the TAC tends to decrease initially and then increase. The TAC is the smallest when feeding on the 20th tray. Similarly, the optimum feeding positions of N_{FR} and N_{F2} are determined to be the 16th and 9th trays, respectively.

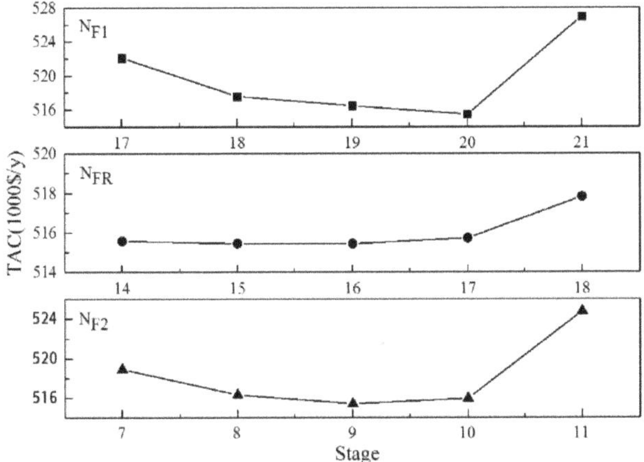

Figure 7. Effect of three feeding positions on the total annual cost (TAC).

3.2.4. Optimization of the Tray Numbers

The tray numbers directly affect the equipment investment cost and energy consumption, so a reasonable number of trays must be used for the minimization of the TAC. First, the number of the HPC trays is fixed, and that of the LPC trays is gradually changed. The corresponding TAC is calculated, and the specific results for each case are listed in Table 2. The flow rate of the recycle stream gradually decreases as the number of trays increases, and the smallest TAC is obtained when the number of the LPC trays is 23.

Table 2. Specific results for the tray number of the low pressure column (LPC).

Design Variables	Case 1	Case 2	Case 3	Case 4	Case 5
N_{T1}	21	22	23	24	25
N_{T2}	17	17	17	17	17
RR1	0.2	0.2	0.2	0.2	0.2
RR2	0.7	0.7	0.7	0.7	0.7
ID1/m	0.92	0.81	0.81	0.82	0.82
ID2/m	0.86	0.76	0.76	0.76	0.76
R_{REC}/kmol/h	116.27	83.05	82.97	82.96	82.96
Equipment investment cost/10^3 \$	738.52	625.32	622.82	631.88	640.12
Operation cost/10^3 \$/y	392.61	308.05	307.85	307.83	307.82
TAC/10^3 \$/y	638.78	516.49	515.45	518.45	521.20

The number of the LPC trays is fixed 23, the number of the HPC trays is changed, and the corresponding TAC is calculated. The specific results for each case can be seen in Table 3. The smallest TAC is obtained when the number of the HPC trays is 17.

Table 3. Specific results for the tray number of the high pressure column (HPC).

Design Variables	Case 6	Case 7	Case 8	Case 9	Case 10
N_{T1}	23	23	23	23	23
N_{T2}	15	16	17	18	19
RR1	0.2	0.2	0.2	0.2	0.2
RR2	0.7	0.7	0.7	0.7	0.7
ID1/m	0.83	0.82	0.81	0.81	0.81
ID2/m	0.77	0.76	0.76	0.76	0.76
R_{REC}/kmol/h	88.13	83.76	82.97	82.75	82.69
Equipment investment cost/10^3 \$	621.25	618.81	622.82	630.15	637.76
Operation cost/10^3 \$/y	320.56	309.79	307.85	307.31	307.14
TAC/10^3 \$/y	527.65	516.06	515.45	517.36	519.73

3.2.5. Process Optimization Results

The optimization results of the whole process are shown in Figure 8. The number of the LPC trays is 23, and the acetonitrile–water mixture and the recycle stream are fed at 20th and 16th, respectively, with a reflux ratio of 0.2. The number of the HPC trays is 17, the feeding position is 9th, and the reflux ratio is 0.7. Finally, the acetonitrile content in the wastewater at the bottom of the LPC is less than 0.1%, and the purity of acetonitrile product at the top of the HPC is 99.9%, which meets the separation requirements.

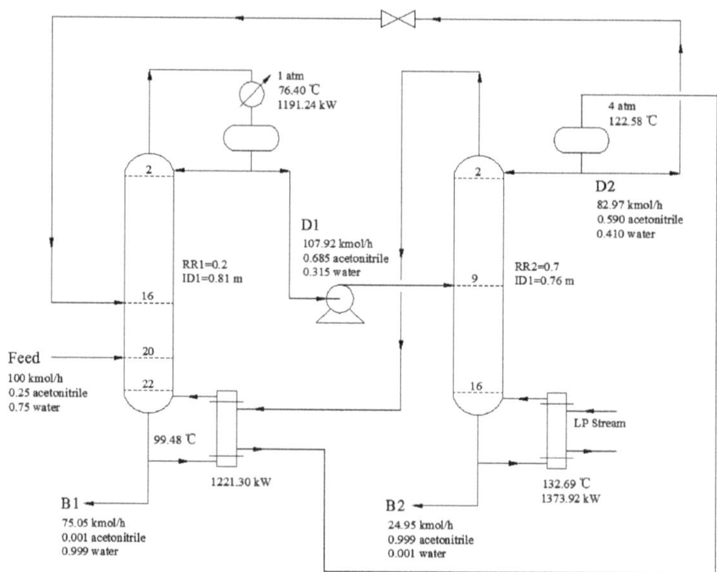

Figure 8. Optimized flow diagram of the pressure swing distillation (PSD) with full-heat integration.

3.3. Comparison of Two Technological Processes

After the optimization of the whole process, the two processes are compared. The results in Table 4 show that the TAC of the PSD without heat integration process is higher than that of the full-heat integration process. The latter can effectively reduce equipment investment and operation cost, thereby reducing the TAC to 515.45×10^3 \$/y. Thirty-two point three nine percent of the TAC of the PSD with full-heat integration is saved compared with that of the process without heat integration.

Table 4. Comparison of two technological processes.

Design Variables	Without Heat Integration	Full-Heat Integration
N_{T1}	21	23
N_{T2}	25	17
N_{F1}/N_{FR}	19/16	20/16
N_{F2}	16	9
RR1/RR2	0.3/0.5	0.20/0.70
Q_{CON1}/kW	1210.36	1191.24
Q_{REB1}/kW	1249.13	—
Q_{CON2}/kW	1003.56	—
Q_{REB2}/kW	1147.50	1373.92
Heat exchanger /kW	—	1221.30
ID1/m	0.82	0.81
ID2/m	0.69	0.76
R_{REC}/kmol/h	76.48	82.97
Equipment investment cost/10^3 $	676.05	622.82
Operation cost/10^3 $/y	537.00	307.85
TAC/10^3 $/y	762.35	515.45

4. Conclusions

PSD was used to separate acetonitrile and water mixture. The sensitivity of the acetonitrile–water azeotrope concentration to the pressure changes indicates the feasibility of PSD. The distillation sequence of the LPC and HPC was determined through a phase diagram. The pressures of the LPC and HPC were set to 1 and 4 atm, respectively. The final purity of the acetonitrile product reached 99.9%, and the content of acetonitrile in wastewater was less than 0.1%, which met the separation requirements.

Based on the principle of the minimum TAC, the design variables of the PSD without heat integration were optimized, and the optimum process parameters were obtained. Then, PSD with full-heat integration was optimized. The two processes were compared, and the results showed that 32.39% of the TAC of the PSD with full-heat integration is saved compared with that of the process without heat integration. Therefore, it is more economical to separate acetonitrile and water mixture by PSD with full-heat integration. This study provides technical support for the separation design of such azeotropes.

Author Contributions: J.L. and K.W. conceived and designed this case-study as well as wrote the paper; M.L. and Z.L. reviewed the paper; All authors interpreted the data; T.D. substantively revised the work and contributed the process simulation.

Funding: This work is financially supported by Guizhou Province United Fund (Qiankehe J zi LKLS[2013]27), Excellent engineers education training plan (LPSSY zyjypyjh201702), Guizhou Solid Waste Recycling Laboratory of Coal Utilization ([2011]278), Guizhou Province United Fund (Qiankehe LH zi [2015]7608) and Academician Workstation of Liupanshui Normal University (qiankehepingtairencai [2019]5604).

Conflicts of Interest: The authors declare no conflict of interest.

References

1. Wang, Y.L.; Bu, G.L.; Geng, X.L.; Zhu, Z.Y.; Cui, P.Z.; Liao, Z.W. Design optimization and operating pressure effects in the separation of acetonitrile/methanol/water mixture by ternary extractive distillation. *J. Clean. Prod.* **2019**, *218*, 212–224. [CrossRef]
2. Jhoany, A.E.; Ivonne, R.D.; Ulises, J.H.; Lauro, N.P.; Eladio, P.F. Recovery of acetonitrile from aqueous waste by a combined process: Solvent extraction and batch distillation. *Sep. Purif. Technol.* **2006**, *52*, 95–101.
3. Liang, K.; Li, W.S.; Luo, H.T.; Xia, M.; Xu, C.J. Energy-efficient extractive distillation process by combining preconcentration column and entrainer recovery column. *Ind. Eng. Chem. Res.* **2014**, *53*, 7121–7131. [CrossRef]
4. Luyben, W.L. Comparison of extractive distillation and pressure-swing distillation for acetone/chloroform separation. *Comp. Chem. Eng.* **2013**, *50*, 1–7. [CrossRef]

5. Liang, S.S.; Cao, Y.J.; Liu, X.Z.; Li, X.; Zhao, Y.T.; Wang, Y.K.; Wang, Y.L. Insight into pressure-swing distillation from azeotropic phenomenon to dynamic control. *Chem. Eng. Res. Des.* **2017**, *117*, 318–335. [CrossRef]

6. Yang, A.; Shen, W.F.; Wei, S.A.; Dong, L.C.; Li, J.; Gerbaud, V. Design and control of pressure-swing distillation for separating ternary systems with three binary minimum azeotropes. *AIChE J.* **2019**, *65*, 1281–1293. [CrossRef]

7. Shen, W.F.; Dong, L.C.; Wei, S.A.; Li, J.; Benyounes, H.; You, X.Q.; Gerbaud, V. Systematic design of extractive distillation for maximum-boiling azeotropes with heavy entrainers. *AIChE J.* **2015**, *61*, 3898–3910. [CrossRef]

8. Zhao, Y.T.; Ma, K.; Bai, W.T.; Du, D.Q.; Zhu, Z.Y.; Wang, Y.L.; Gao, J. Energy-saving thermally coupled ternary extractive distillation process by combining with mixed entrainer for separating ternary mixture containing bioethanol. *Energy* **2018**, *148*, 296–308. [CrossRef]

9. Rodriguez-Donis, I.; Acosta-Esquijarosa, J.; Gerbaud, V.; Joulia, X. Heterogeneous batch-extractive distillation of minimum boiling azeotropic mixtures. *AIChE J.* **2003**, *49*, 3074–3083. [CrossRef]

10. Rodriguez-Donis, I.; Papp, K.; Rev, E.; Lelkes, Z.; Gerbaud, V.; Joulia, X. Column configurations of continuous heterogeneous extractive distillation. *AIChE J.* **2007**, *53*, 1982–1993. [CrossRef]

11. Raeva, V.M.; Sazonova, A.Y. Separation of ternary mixtures by extractive distillation with 1, 2-ethandiol and glycerol. *Chem. Eng. Res. Des.* **2015**, *99*, 125–131. [CrossRef]

12. Sazonova, A.Y.; Raeva, V.M. Recovery of acetonitrile from aqueous solutions by extractive distillation–effect of entrainer. *Int. J. Chem. Mol. Nucl. Mater. Metall. Eng.* **2015**, *9*, 288–291.

13. You, X.; Gu, J.; Gerbaud, V.; Peng, C.; Liu, H. Optimization of pre-concentration, entrainer recycle and pressure selection for the extractive distillation of acetonitrile-water with ethylene glycol. *Chem. Eng. Sci.* **2018**, *177*, 354–368. [CrossRef]

14. Wang, X.; Du, Z.; Zhang, Y.; Wang, J.; Wang, J.; Sun, W. Optimization of distillation sequences with nonsharp separation columns. *Processes* **2019**, *7*, 323. [CrossRef]

15. Salvador, T.A.; Arturo, J.G.; Juergen, H. Analysis of multi-loop control structures of dividing-wall distillation columns using a fundamental model. *Processes* **2014**, *2*, 180–199.

16. Li, X.; Geng, X.L.; Cui, P.Z.; Yang, J.W.; Zhu, Z.Y.; Wang, Y.L.; Xu, D.M. Thermodynamic efficiency enhancement of pressure-swing distillation process via heat integration and heat pump technology. *Appl. Therm. Eng.* **2019**, *154*, 519–529. [CrossRef]

17. Wang, Y.L.; Ma, K.; Yu, M.X.; Dai, Y.; Yuan, R.J.; Zhu, Z.Y.; Gao, J. An improvement scheme for pressure-swing distillation with and without heat integration through an intermediate connection to achieve energy savings. *Comp. Chem. Eng.* **2018**, *119*, 439–449. [CrossRef]

18. Wang, K.L.; Li, J.; Liu, P.; Lian, M.L.; Du, T.Z. Pressure swing distillation for the separation of methyl acetate-methanol azeotrope. *Asia-Pac J. Chem. Eng.* **2019**, *14*, e2319. [CrossRef]

19. Varbanov, P.; Klein, A.; Repke, J.U.; Wozny, G. Minimising the startup duration for mass- and heat-integrated two-column distillation systems: A conceptual approach. *Chem. Eng. Process.* **2008**, *47*, 1456–1469. [CrossRef]

20. Rahman, I.; Sagar, S.A. Optimization of pressure-swing distillation by evolutionary techniques: Separation of ethanol-water and acetonitrile-water mixtures. *Chem. Prod. Process. Model.* **2018**, *13*, 20170007.

21. Repke, J.U.; Forner, F.; Klein, A. Separation of homogeneous azeotropic mixtures by pressure swing distillation-analysis of the operation performance. *Chem. Eng. Technol.* **2005**, *28*, 1151–1157. [CrossRef]

22. Kim, K.W.; Shin, J.S.; Kim, S.H.; Hong, S.K.; Cho, J.H.; Park, S.J. A computational study on the separation of acetonitrile and water azeotropic mixture using pressure swing distillation. *J. Chem. Eng. Jpn.* **2013**, *46*, 347–352. [CrossRef]

23. Huang, F.; Zheng, S.; Chen, Y.; Zhou, M.; Sun, X. Simulation and optimization of pressure-swing distillation system for high purity acetonitrile. In *ACSR-Advances in Computer Science Research, Proceedings of the 2016 International Conference on Modeling, Simulation and Optimization Technologies and Applications (MSOTA), Xiamen, China, 18–19 December 2016*; Atlantis Press: Paris, France, 2016; Volume 58, pp. 96–100.

24. Huang, K.; Shan, L.; Zhu, Q.; Qian, J. Adding rectifying/stripping section type heat integration to a pressure-swing distillation (PSD) process. *Appl. Therm. Eng.* **2008**, *28*, 923–932. [CrossRef]

25. Matsuda, K.; Huang, K.; Iwakabe, K.; Nakaiwa, M. Separation of binary azeotrope mixture via pressure-swing distillation with heat integration. *J. Chem. Eng. Jpn.* **2011**, *44*, 969–975. [CrossRef]

26. Gao, Z.; Saxen, H.; Gao, C. Guest editorial: Special section on data-driven approaches for complex industrial systems. *IEEE Trans. Ind. Informat.* **2013**, *9*, 2210–2212. [CrossRef]

27. Gao, Z.; Nguang, S.K.; Kong, D.X. Advances in modelling, monitoring, and control for complex industrial systems. *Complexity* **2019**, *2019*, 2975083. [CrossRef]

28. Dai, X.; Gao, Z. From model, signal to knowledge: A data-driven perspective of fault detection and diagnosis. *IEEE Trans. Ind. Informat.* **2013**, *9*, 2226–2238. [CrossRef]

29. Zhai, C.; Liu, Q.; Romagnoli, J.A.; Sun, W. Modeling/simulation of the dividing wall column by using the rigorous model. *Processes* **2019**, *7*, 26. [CrossRef]

30. Liang, T.; Guo, X.; Giwa, A.S.; Shi, J.; Li, Y.; Wei, Y.; Du, J. Design and optimization of a process for the production of methyl methacrylate via direct methylation. *Processes* **2019**, *7*, 377. [CrossRef]

31. Petrovic, B.A.; Masoudi, S.M. Optimization of post combustion CO_2 capture from a combined-cycle gas turbine power plant via taguchi design of experiment. *Processes* **2019**, *7*, 364. [CrossRef]

32. Sarda, P.; Hedrick, E.; Reynolds, K.; Bhattacharyya, D.; Zitney, S.E.; Omell, B. Development of a dynamic model and control system for load-following studies of supercritical pulverized coal power plants. *Processes* **2018**, *6*, 226. [CrossRef]

33. Gmehling, J.; Menke, J.; Fischer, K.; Krafczyk, J. *Azeotropic Data*; Wiley-VCH: Weinheim, Germany, 2004.

34. Luyben, W.L.; Chien, I.L. *Design and Control of Distillation Systems for Separating Azeotropes*; John Wiley & Sons: Hoboken, NJ, USA, 2011.

35. Luo, H.; Liang, K.; Li, W.; Li, Y.; Xia, M.; Xu, C. Comparison of pressure-swing distillation and extractive distillation methods for isopropyl alcohol/diisopropyl ether separation. *Ind. Eng. Chem. Res.* **2014**, *53*, 15167–15182. [CrossRef]

Article

Reactive Power Optimization of Large-Scale Power Systems: A Transfer Bees Optimizer Application

Huazhen Cao [1], Tao Yu [2], Xiaoshun Zhang [3], Bo Yang [4,*] and Yaxiong Wu [1]

[1] Power Grid Planning Research Center of Guangdong Power Grid Co., Ltd., Guangzhou 510062, China; zzw572@126.com (H.C.); jxpxlxwyx@163.com (Y.W.)

[2] College of Electric Power, South China University of Technology, Guangzhou 510640, China; taoyu1@scut.edu.cn

[3] College of Engineering, Shantou University, Shantou 515063, China; xszhang1990@sina.cn

[4] Faculty of Electric Power Engineering, Kunming University of Science and Technology, Kunming 650500, China

* Correspondence: yangbo_ac@outlook.com; Tel.: +86-183-1459-6103

Received: 6 May 2019; Accepted: 24 May 2019; Published: 31 May 2019

Abstract: A novel transfer bees optimizer for reactive power optimization in a high-power system was developed in this paper. Q-learning was adopted to construct the learning mode of bees, improving the intelligence of bees through task division and cooperation. Behavior transfer was introduced, and prior knowledge of the source task was used to process the new task according to its similarity to the source task, so as to accelerate the convergence of the transfer bees optimizer. Moreover, the solution space was decomposed into multiple low-dimensional solution spaces via associated state-action chains. The transfer bees optimizer performance of reactive power optimization was assessed, while simulation results showed that the convergence of the proposed algorithm was more stable and faster, and the algorithm was about 4 to 68 times faster than the traditional artificial intelligence algorithms.

Keywords: transfer bees optimizer; reinforcement learning; behavior transfer; state-action chains; reactive power optimization

---

1. Introduction

Nonlinear programming is a very common issue in the operation of power systems, including reactive power optimization (RPO) [1], unit commitment (UC) [2], economic dispatch (ED) [3]. In order to tackle this issue, several optimization approaches have been adopted, such as the Newton method [4], quadratic programming [5], interior-point method [6]. However, these methods are essentially gradient-based mathematic optimization methods, which highly depend on an accurate system model. When there is nonlinearity, there are discontinuous functions and constraints, and there usually exist many local minimum upon which the algorithm may be easily fall into one local optimum [7].

In the past decades, artificial intelligence (AI) [8–18] has been widely used as an effective alternative because of its high independence from an accurate system model and strong global optimization ability. Inspired by nectar gathering of bees in wild nature, the artificial bee colony (ABC) [19] has been applied to optimal distributed generation allocation [8], global maximum power point (GMPP) tracking [20], multi-objective UC [21], and so on, and has the merits of simple structure, high robustness, strong universality, and efficient local search.

However, the ABC mainly depends on a simple collective intelligence without self-learning or knowledge transfer, which is a common weakness of AI algorithms such as genetic algorithm (GA) [9], particle swarm optimization (PSO) [10], group search optimizer (GSO) [11], ant colony system (ACS) [12], interactive teaching–learning optimizer (ITLO) [13], grouped grey wolf optimizer

(GGWO) [14], memetic salp swarm algorithm (MSSA) [15], dynamic leader-based collective intelligence (DLCI) [16], and evolutionary algorithms (EA) [17]. Thus, a relatively low search efficiency may result, particularly while considering a new optimization task of a complex industrial system [22], e.g., the optimization of a large-scale power system with different complex new tasks. In fact, the computational time of these algorithms can be effectively reduced for RPO or optimal power flow (OPF) via the external equivalents in some areas (e.g., distribution networks) [23]. This is because the optimization scale and difficulty are significantly reduced as the number of optimization variables and constraints decreases. However, the optimization results are highly determined by the accuracy of the external equivalent model [24], which is usually worse than that obtained by global optimization. Hence, this paper aims to propose a fast and efficient AI algorithm for global optimization.

Previous studies [25] discovered that bees have evolved an instinct to memorize the beneficial weather conditions of their favorite flowers, e.g., temperature, air humidity, and illumination intensity, which may rapidly guide bees to find the best nectar source in a new environment with high probability, hence, the survival and prosperity of the whole species living in various environments can be guaranteed via the exploitation of such knowledge. The above natural phenomenon resulted from the struggle for existence in a harsh and unknown environment can be regarded as a knowledge transfer, which has been popularly investigated in machine learning and data mining [26]. In practice, prior knowledge is from the source tasks and then it is applied to a new but analogous assignment, such that fewer training data can be used to achieve a higher learning efficiency [27], e.g., learn to ride a bike before starting to ride a motorcycle. In fact, knowledge transfer-based optimization is essentially the knowledge-based history data-driven method [28], which can accelerate the optimization speed of a new task according to prior knowledge. Also, reinforcement learning (RL) can be accelerated by knowledge conversion [29], and agents learn new tasks faster and interact less with the environment. As a consequence, knowledge transfer reinforcement learning (KTRL) has been developed [30] through combining AI and behavior psychology [31] and is divided into behavior shift and information shift.

In this paper, behavior shift was used for Q-learning [32] to accelerate the learning of a new task, which was called *Q*-value transfer. The *Q*-value matrix was applied in knowledge learning, storage, and transfer. However, the practical application of conventional Q-learning was restricted to only a group of new tasks with small size due to the calculation burden. To deal with this obstacle, an associated state-action chain was introduced after the solution space was decomposed into several low-dimensional solution spaces. Therefore, this paper proposes a transfer bee optimizer (TBO) based on Q-learning and behavior transfer. The main novelties and contributions of this work are given as follows:

(1) Compared with the traditional mathematic optimization methods, the TBO has a stronger global search ability by employing the scouts and workers for exploitation and exploration. Besides, it can approximate the global optimum more closely via global optimization instead of external equivalent-based local optimization.
(2) Compared with the general AI algorithms, the TBO can effectively avoid a blind search in the initial optimization phase and implement a much faster optimization for a new task by utilizing prior knowledge from the source tasks.
(3) The optimization performance of the TBO was thoroughly tested by RPO of large-scale power systems. Because of its high optimization flexibility and superior optimization efficiency, it can be extended to other complex optimization problems.

The remaining of this paper is arranged as follows: Section 2 presents the basic principles of the TBO. Section 3 designs the TBO for RPO. Section 4 shows the simulation results, and Section 5 summarizes paper.

2. Transfer Bees Optimizer

2.1. State-Action Chain

Q-learning typically finds and learns different state-action pairs through a look-up table, i.e., $Q(s,a)$, but this is not enough to handle a complex task with multiple controllable variables because of the curse of dimension, as illustrated in Figure 1a. Suppose the optional operand of the controllable variable x_i is m_i, and $|A| = m_1 m_2 \cdots m_n$, n is the sum of the controlled variables, and A is the action set. If n is dramatically increased, the dimension of Q-value matrix will grow very fast, so the calculation convergence is slow and may even lead to failure. Hierarchical reinforcement learning (HRL) [33] is commonly used to avert this obstacle and decomposes the original complex task into several simpler subtasks, e.g., MAXQ [34]. However, it is easy to fall into a near-optimum for the overall task due to the fact that it is difficult to design and coordinate all the subtasks.

In contrast, the whole solution space is decomposed into several low-dimensional solution spaces by the associated state-action chain. In such frame, each controlled variable has a unique memory matrix Q^i, the size of Q-value matrix can be significantly reduced, it has the advantages of convenient storage and transfer, and the controlled variables are linked, as shown in Figure 1b.

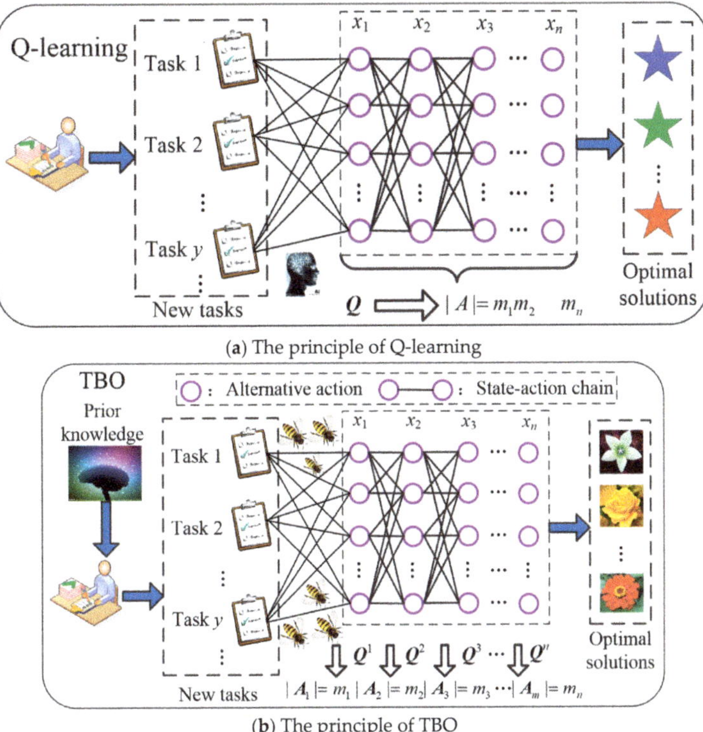

Figure 1. Comparison of Q-learning and transfer bee optimizer (TBO).

2.2. Knowledge Learning and Behavior Transfer

2.2.1. Knowledge learning

Figure 2 shows that all bees search a nectar source according to their prior knowledge (Q-value matrix); the obtained knowledge will then be updated after each interaction with the nectar source, therefore a cycle of knowledge learning and conscious exploration can be fully completed. As shown in

Figure 1a, a simple RL agent is usually adopted for traditional Q-learning to acquire knowledge [18,35], which is the cause of inefficient learning. In contrast, the TBO adopts the method of swarm collaborative exploration for knowledge learning, which can update multiple elements of the Q value matrix at the same time, thus greatly improving the learning efficiency, which can be described as [32]

$$Q_{k+1}^i(s_k^{ij}, a_k^{ij}) = Q_k^i(s_k^{ij}, a_k^{ij}) + \alpha[R^{ij}(s_k^{ij}, s_{k+1}^{ij}, a_k^{ij}) +$$

$$\gamma \max_{a^i \in A_i} Q_k^i(s_{k+1}^{ij}, a) - Q_k^i(s_k^{ij}, a_k^{ij})], \tag{1}$$

$$j = 1, 2, \cdots, J; i = 1, 2, \cdots, n$$

where α represents the factor of knowledge study; γ is the discount coefficient; the superscripts i and j signify the ith Q-value matrix (i.e., the ith controlled variable) and the jth bee, respectively; J is the population size of bees; (s_k, a_k) means a pair of state-action in the kth iteration; $R(s_k, s_{k+1}, a_k)$ is the reward function that transforms from state s_k to s_{k+1} used under an optional operation a_k; a^i means random optional action of the ith controlled variable x_i; and A_i represents the multiple active result sets of the ith controlled variable.

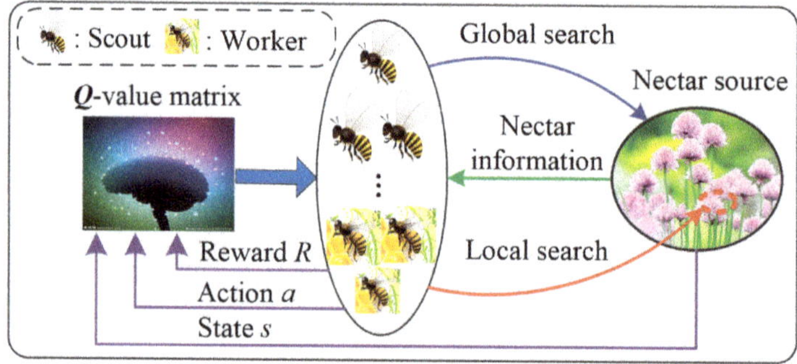

Figure 2. The principle of knowledge learning of the TBO inspired by the nectar gathering of bees.

2.2.2. Behavior transfer

In the initial process, the TBO needs to go through a series of source tasks to get the optimal Q-value matrix, so as to make use of and to prepare prior knowledge for similar new tasks in the future. The relevant prior knowledge of source task is shown in Figure 3. According to the similarity of the source task, the optimal Q-value matrix of the source task Q_S^* is shifted from the initial Q-value matrix to the new task Q_N^0, as

$$Q_N^0 = \begin{bmatrix} r_{11} & r_{12} & \cdots & r_{1h} & \cdots \\ r_{21} & r_{22} & \cdots & r_{2h} & \cdots \\ \vdots & \vdots & \ddots & \vdots & \cdots \\ r_{y1} & r_{y2} & \cdots & r_{yh} & \cdots \\ \vdots & \vdots & \cdots & \vdots & \ddots \end{bmatrix} Q_S^* \tag{2}$$

with

$$Q_N^0 = \begin{bmatrix} Q_{N1}^{10} & \cdots & Q_{N1}^{i0} & \cdots & Q_{N1}^{n0} \\ Q_{N2}^{10} & \cdots & Q_{N2}^{i0} & \cdots & Q_{N2}^{n0} \\ \vdots & \vdots & \vdots & \vdots & \vdots \\ Q_{Ny}^{10} & \cdots & Q_{Ny}^{i0} & \cdots & Q_{Ny}^{n0} \\ \vdots & \vdots & \vdots & \vdots & \vdots \end{bmatrix}, Q_S^* = \begin{bmatrix} Q_{S1}^{1*} & \cdots & Q_{S1}^{i*} & \cdots & Q_{S1}^{n*} \\ Q_{S2}^{1*} & \cdots & Q_{S2}^{i*} & \cdots & Q_{S2}^{n*} \\ \vdots & \vdots & \vdots & \vdots & \vdots \\ Q_{Sh}^{1*} & \cdots & Q_{Sh}^{i*} & \cdots & Q_{Sh}^{n*} \\ \vdots & \vdots & \vdots & \vdots & \vdots \end{bmatrix}$$

where Q^{i0}_{Ny} is the ith initial Q-value matrix in the yth new task; Q^{i*}_{Sh} is the ith optimized Q-value matrix in the hth source task; and r_{yh} represents the comparability between the hth source task and the yth new task; here, a large r_{yh} indicates that the yth new task can gain much knowledge from the hth source task, and $0 \leq r_{yh} \leq 1$.

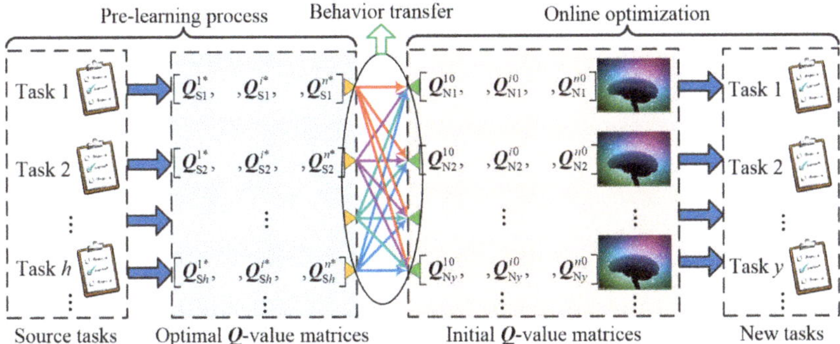

Figure 3. Behavior transfer of the TBO.

2.3. Exploration and Feedback

2.3.1. Action policy

There are two kinds of bees in Figure 2, e.g., scout and worker, determined by their nectar amounts (fitness values) [19], which are responsible for global searching and local searching. As a consequence, a bee colony can balance the exploration and exploitation through different action policies in a nectar source. In the TBO, 50% of bees with nectar amounts that rank in the half top of all bees are designated as worker, while the others are scout. On the basis of the ε-*Greedy* rule [31], the scouts' actions are based on the proportion of Q-value in the current status. As for the controlled variable x_i, one behavior of each scout is chosen as

$$a^{ij}_{k+1} = \begin{cases} \underset{a^i \in A_i}{\arg\max} Q^i_{k+1}(s^{ij}_{k+1}, a^i), & \text{if } \varepsilon \leq \varepsilon_0 \\ a_{\text{rg}}, & \text{otherwise} \end{cases} \tag{3}$$

where ε is any value, uniformly distributed between [0, 1]; ε_0 represents the exploration rate; and a_{rg} represents any global behavior ascertained by the distribution of the state-action probability matrix P^i, updated by

$$\begin{cases} P^i(s^i, a^i) = \dfrac{e^i(s^i, a^i)}{\sum\limits_{a' \in A_i} e^i(s^i, a')} \\ e^i(s^i, a^i) = \dfrac{1}{\max\limits_{a' \in A_i} Q^i(s^i, a') - \beta Q^i(s^i, a^i)} \end{cases} \tag{4}$$

where β is the discrepancy factor, and e^i is an evaluation matrix of the pairs of the state-action.

On the other hand, the workers keep exploring new nectar sources at nearby nectar sources, which can be written as [8]

$$\begin{cases} a^{ij}_{\text{new}} = a^{ij} + \text{Rnd}(0, 1)(a^{ij} - a^{ij}_{\text{rand}}) \\ a^{ij}_{\text{rand}} = a^i_{\text{min}} + \text{Rnd}(0, 1)(a^i_{\text{max}} - a^i_{\text{min}}) \end{cases} \tag{5}$$

where a^{ij}_{new}, a^{ij}, and a^{ij}_{rand} denote the new action, current action, and random action of the ith controlled variable selected by the jth worker; a^i_{max} and a^i_{min} are the maximum and minimum behavior, respectively, in the ith controlled variable.

2.3.2. Reward function

After each exploration, each bee will get an instant reward based on its fitness value. Because it is the goal of the TBO to maximize the expected long-term rewards for each state [28], the reward function is designed as

$$R^{ij}(s_k^{ij}, s_{k+1}^{ij}, a_k^{ij}) = \frac{C}{f_k^j} \tag{6}$$

where C is a positive multiplicator, and f_k^j represents the fitness function of the jth bee in the kth iteration. This is closely related to the target function.

After each bee obtains its new reward, the scouts and workers will swap their roles according to the obtained reward rank, precisely, 50% of bees who had a larger reward will become workers. As a result, a compromise is reached between exploration and development to ensure a global search for the TBO.

3. Design of the TBO for RPO

3.1. Mathematical Model of RPO

As the subproblem of OPF, the conventional RPO aims to lower the active power losses or other appropriate target functions via optimizing the different types of controlled variables (e.g., transformer tap ratio) under multiple equality constraints and inequality constraints [35]. In this article, the integrated target function of the active power loss and the deviation of supply voltage were used as follows [18]

$$\min f(x) = \mu P_{\text{loss}} + (1 - \mu) V_{\text{d}} \tag{7}$$

subject to

$$\begin{cases} P_{\text{G}i} - P_{\text{D}i} - V_i \sum_{j \in N_i} V_j(g_{ij} \cos \theta_{ij} + b_{ij} \sin \theta_{ij}) = 0, i \in N_0 \\ Q_{\text{G}i} - Q_{\text{D}i} - V_i \sum_{j \in N_i} V_j(g_{ij} \sin \theta_{ij} - b_{ij} \cos \theta_{ij}) = 0, i \in N_{\text{PQ}} \\ Q_{\text{G}i}^{\min} \le Q_{\text{G}i} \le Q_{\text{G}i}^{\max}, \qquad i \in N_{\text{G}} \\ V_i^{\min} \le V_i \le V_i^{\max}, \qquad i \in N_i \\ Q_{\text{C}i}^{\min} \le Q_{\text{C}i} \le Q_{\text{C}i}^{\max}, \qquad i \in N_{\text{C}} \\ T_k^{\min} \le T_k \le T_k^{\max}, \qquad k \in N_{\text{T}} \\ |S_l| \le S_l^{\max}, \qquad l \in N_{\text{L}} \end{cases} \tag{8}$$

where $x = [V_{\text{G}}, T_k, Q_{\text{C}}, V_{\text{L}}, Q_{\text{G}}]$ represents the variable vector, V_{G} represents the terminal voltage of generator; T_k means the transformer tapping ping ratio; Q_{C} is the reactive power of the shunt capacitor; V_{L} is the load-bus voltage; Q_{G} is the reactive power of the generator; P_{loss} is the power loss; V_{d} is the deviation of supply voltage; $0 \le \mu \le 1$ is the weight coefficient; $P_{\text{G}i}$ and $Q_{\text{G}i}$ are the generated active power; $P_{\text{D}i}$ and $Q_{\text{D}i}$ are the demanded active and reactive power, respectively; $Q_{\text{C}i}$ represents the reactive power compensation; V_i and V_j are the voltage magnitude of the ith and jth node, respectively; θ_{ij} is the phase difference of voltage; g_{ij} represents the conductance in the transmission line i-j; b_{ij} represents the susceptance of the transmission line i-j; S_l is the apparent power of the transmission line l; N_i is the node set; N_0 is the set of the slack bus; N_{PQ} is the set of active/reactive power (PQ) buses; N_{G} is the unit set; N_{C} is the compensation equipment; N_{T} is the set of transformer taps; and N_{L} is the branch set. In addition, the active power loss and the deviation of supply voltage can be computed by [18]

$$P_{\text{loss}} = \sum_{i,j \in N_L} g_{ij}[V_i^2 + V_j^2 - 2V_iV_j \cos \theta_{ij}] \tag{9}$$

$$V_{\mathrm{d}} = \sum_{i \in N_i} \left| \frac{2V_i - V_i^{\max} - V_i^{\min}}{V_i^{\max} - V_i^{\min}} \right| \tag{10}$$

3.2. Design of the TBO

3.2.1. Design of state and action

The terminal voltage of generator, transformer tapping ratio, and d shunt capacitor reactive power compensation were chosen as the controlled variables of RPO, in which each controlled variable had its own Q-value matrix Q^i and action set A_i, as shown in Figure 1. In addition, the operation sets for every controlled variable were the state sets for the next controlled variable, i.e., $S_{i+1} = A_i$, where the initial controlled variable depends on different tasks of RPO, thus each task can be considered as a specific state of x_1.

3.2.2. Design of the reward function

It can be found from (6) that the fitness function determines the reward function that represents the overall target function (7) and needs to satisfy all constraints (8). Hence, the fitness function is designed as

$$f^j = \mu P_{\mathrm{loss}}^j + (1 - \mu)V_{\mathrm{d}}^j + \eta q^j \tag{11}$$

where q represents the sum of inequalities that violate the constraints, and η represents the regularization factor.

3.2.3. Behavior transfer for RPO

Based on eq (2), the transfer efficiency of TBO mainly depends on getting the comparability between the source tasks and the new tasks [30]. In fact, the distribution of power flow principally determines the RPO in the power system, thus it is principally influenced by the power demand, because the topological structure of the power system cannot be changed much daily. Therefore, the active power demand was divided into several load intervals as follows

$$\left\{ \left[P_{\mathrm{D}}^1, P_{\mathrm{D}}^2 \right), \left[P_{\mathrm{D}}^2, P_{\mathrm{D}}^3 \right), \cdots, \left[P_{\mathrm{D}}^h, P_{\mathrm{D}}^{h+1} \right), \cdots, \left[P_{\mathrm{D}}^{H-1}, P_{\mathrm{D}}^H \right] \right\} \tag{12}$$

where P_{D}^h represents the demand of active power in the hth source task for RPO, $P_{\mathrm{D}}^1 < P_{\mathrm{D}}^2 < \cdots < P_{\mathrm{D}}^h < \cdots < P_{\mathrm{D}}^H$.

Suppose the active power required in the yth new task is P_{D}^y, $P_{\mathrm{D}}^1 < P_{\mathrm{D}}^y < P_{\mathrm{D}}^H$, then the comparability between two tasks will be computed by

$$r_{yh} = \frac{\left[W + \Delta P_{\mathrm{D}}^{\max} \right] - \left| P_{\mathrm{D}}^h - P_{\mathrm{D}}^y \right|}{\sum\limits_{h=1}^{H} \left\{ \left[W + \Delta P_{\mathrm{D}}^{\max} \right] - \left| P_{\mathrm{D}}^h - P_{\mathrm{D}}^y \right| \right\}} \tag{13}$$

$$\Delta P_{\mathrm{D}}^{\max} = \max_{h=1,2,\cdots,H} \left(\left| P_{\mathrm{D}}^h - P_{\mathrm{D}}^y \right| \right) \tag{14}$$

where W represents the transfer factor, and $\Delta P_{\mathrm{D}}^{\max}$ represents the ultimate error of active power demand, where $\sum\limits_{h=1}^{H} r_{yh} = 1$.

Note that a smaller deviation $\left| P_{\mathrm{D}}^h - P_{\mathrm{D}}^y \right|$ brings in a larger similarity r_{yh}, therefore the new yth task can develop more knowledge.

Therefore, the overall process of TBO behavior transfer is generalized as follows:

Step 1. Determine the source tasks according to a typical load curve in a day by Equation (12);
Step 2. Complete the source tasks in the initial study process and store their optimal Q-value matrices;

Step 3. Calculate the comparability between original tasks and new task according to the deviation of power demand according to Equations (13) and (14);

Step 4. Obtain the original Q-value matrices in the new task by Equation (2).

3.2.4. Parameters setting

For the TBO, eight parameters, α, γ, J, ε_0, β, C, η, and W are important and need to be set following the general guidelines below [18,26,32,35].

Of these, α represents the knowledge learning factor, with $0 < \alpha < 1$, α determines the rate of knowledge acquisition of the bees. A larger α will accelerate knowledge learning, which may bring about a local optimization, while a smaller value can improve the algorithm stability [35].

The discount factor is defined to exponentially discount the rewards obtained by the Q-value matrix in the future, and its value is $0 < \gamma < 1$. Since the future return on RPO is insignificant, it is required assume a value close to zero [18].

J is the number of bees, with $J \geq 1$; it determines the rate of convergence and the rate of solution. A large J makes the algorithm approximate a more accurate global optimization solution but results in a larger computational burden [26].

The exploration rate, $0 < \varepsilon_0 < 1$, balances the exploration and development of a nectar resource by the scouts. The scouts are encouraged to pick a voracious action instead of any action according to the state-action probability matrix by a larger ε_0.

The discrepancy factor, $0 < \beta < 1$, increases the differences among the elements of each row in Q-value matrices.

The positive multiplicator, $C > 0$, distributes the weight and fitness functions of the reward function. The bees are encouraged to get more rewards by the fitness function with a large C, while the difference in rewards is smaller among all bees.

The penalty factor, $\eta > 0$, makes sure to satisfy the restrain of inequality. Solution infeasibility may arise because of a smaller η [18]. Here, W is the shift factor, with $W \geq 0$, that identifies the comparability among two tasks.

The parameters were selected through trial and error, as shown in Table 1.

Table 1. Parameters used in TBO.

Parameter	Range	IEEE 118-Bus System		IEEE 300-Bus System	
		Pre-Learning	Online Optimization	Pre-Learning	Online Optimization
α	$0 < \alpha < 1$	0.95	0.99	0.9	0.95
γ	$0 < \gamma < 1$	0.2	0.2	0.3	0.3
J	$J \geq 1$	15	5	30	10
ε_0	$0 < \varepsilon_0 < 1$	0.9	0.98	0.95	0.98
β	$0 < \beta < 1$	0.95	0.95	0.98	0.98
C	$C > 0$	1	1	1	1
η	$\eta > 0$	10	10	50	50
W	$W \geq 0$	—	50	—	100

3.2.5. Execution Procedure of the TBO for RPO

At last, the overall implementation process of TBO is shown in Figure 4, and $\left\| Q_{k+1}^i - Q_k^i \right\|_2 < \sigma$ is the memory value difference of matrix 2-norm, with the precision factor $\sigma = 0.001$.

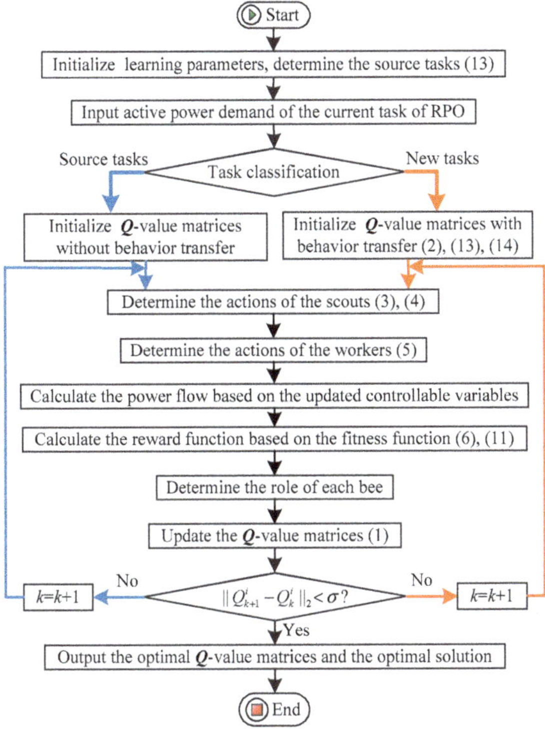

Figure 4. Flowchart of TBO for reactive power optimization (RPO).

4. Case Studies

The TBO for RPO was assessed by the IEEE 118-bus system and the IEEE 300-bus system and compared with those of ABC [8], GSO [11], ACS [12], PSO [10], GA [9], quantum genetic algorithm (QGA) [36], and ant colony based Q-learning (Ant-Q) [37]. Furthermore, the main parameters of other algorithms were obtained through trial and error and were set according to reference [38], while the weight coefficient μ applied to eq (7) assigned the active power loss and the deviation of the output voltage. The simulation is executed on Matlab 7.10 by a personal computer with Intel(R) Xeon (R) E5-2670 v3 CPU at 2.3 GHz with 64 GB of RAM.

The IEEE 118-bus system consists of 54 generators and 186 branches and is divided into three voltage levels, i.e., 138 kV, 161 kV, and 345 kV. The IEEE 300-bus system is constituted by 69 generators and 411 branches, with 13 voltage levels, i.e., 0.6 kV, 2.3 kV, 6.6 kV, 13.8 kV, 16.5 kV, 20 kV, 27 kV, 66 kV, 86 kV, 115 kV, 138 kV, 230 kV, and 345 kV [39–41]. The number of controlled variables in IEEE 118-bus system was 25, and the number of controlled variables in IEEE 300-bus system was 111. More specifically, reactive power compensation is divided into five levels from rated level [−40%, −20%, 0%, 20%, 40%], the transformer tapping is divided into three levels [0.98, 1.00, 1.02], and the terminal voltage of generator is uniformly discretized into [1.00, 1.01, 1.02, 1.03, 1.04, 1.05, 1.06].

According to the given typical daily load curves shown in Figure 5, the active power demand was discretized into 20 and 22 load intervals, respectively, where every interval was 125 MW and 500 MW, respectively, i.e., {[3500, 3625), [3625, 3750), . . . , [5875, 6000]} MW and {[19,000, 19,500), [19,500, 20,000), . . . , [28,500, 29,000]} MW. Moreover, the implementation time of RPO was set at 15 min. Hence, the number of new tasks per day was 96, while the source tasks of the IEEE 118-bus system was 21, and the original tasks of the IEEE 300-bus system was 23.

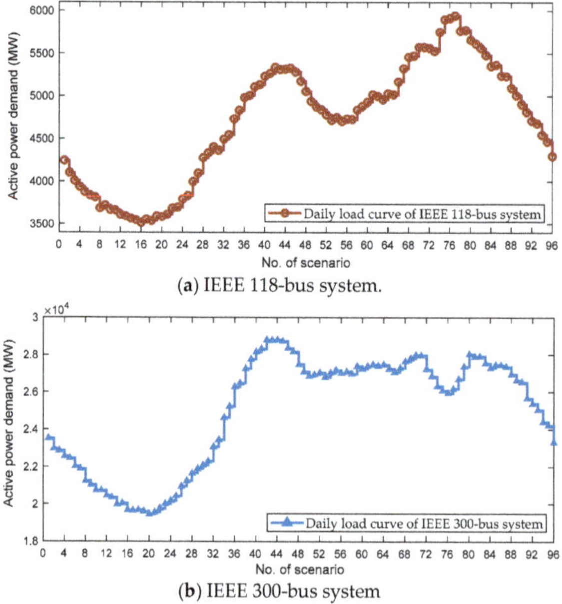

(**a**) IEEE 118-bus system.

(**b**) IEEE 300-bus system

Figure 5. Typical daily load curves of the IEEE 118-bus system and IEEE 300-bus system.

4.1. Study of the Pre-Learning Process

The TBO required a preliminary study to gain the optimal Q-value matrices for all source tasks, and then convert them to an initial Q-value. Figures 6 and 7 illustrate that the TBO will astringe to the optimal Q-value matrices of the source tasks while the optimal objective function can be obtained. Furthermore, the convergence of all Q-value matrices was consistent, as the same feedback rewards were used from the same bees to update the Q-value matrices at each iteration.

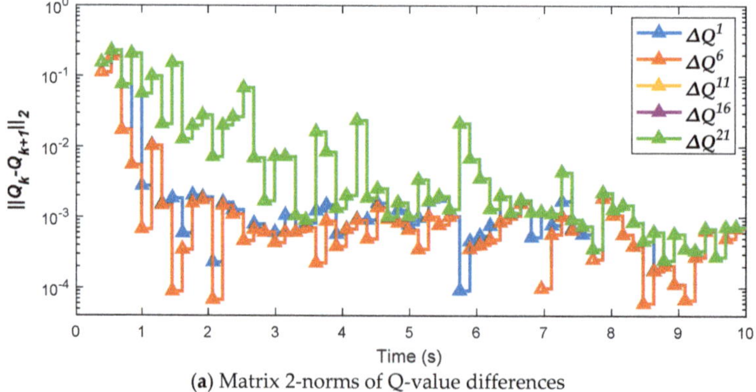

(**a**) Matrix 2-norms of Q-value differences

Figure 6. *Cont.*

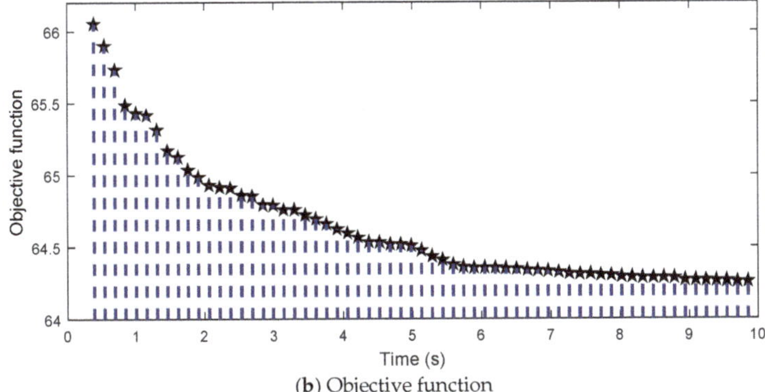

Figure 6. Convergence of the seventh source task of the IEEE 118-bus system obtained in the pre-learning process.

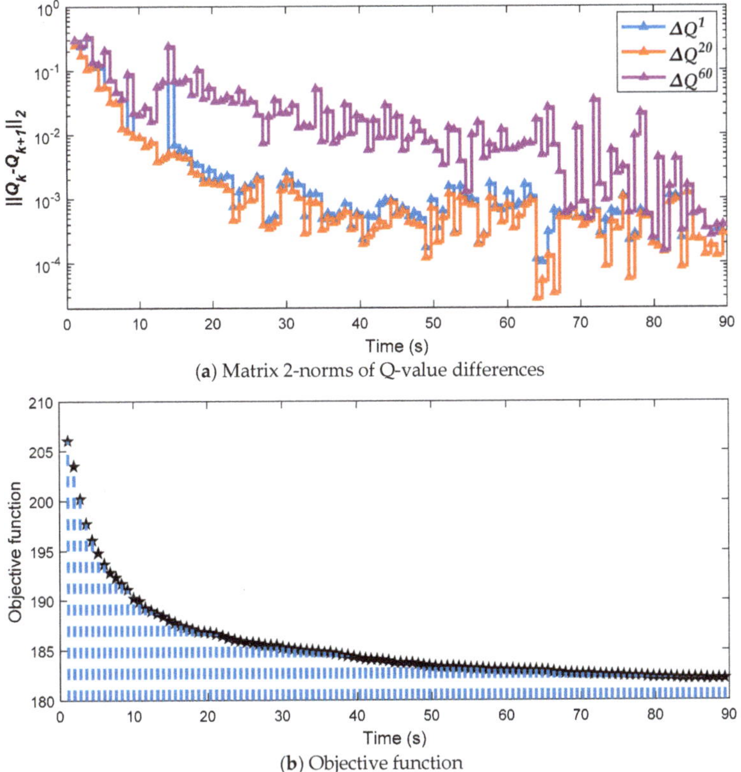

(a) Matrix 2-norms of Q-value differences

(b) Objective function

Figure 7. Convergence of the eighth source task of the IEEE 300-bus system obtained in the pre-learning process.

4.2. Study of Online Optimization

4.2.1. Study of behavior transfer

Through the preliminary study process, the TBO was ready for online optimization of RPO under different scenarios (different new tasks) with prior knowledge. The convergence of the target functions gained by different algorithms in online optimization is given in Figures 8 and 9. Compared to the preliminary study process, the convergence of the TBO was approximately 10 to 20 times that of other algorithms in online optimization, which verified the effectiveness of knowledge transfer. Furthermore, the convergence rate of the TBO algorithm was much faster than that of other algorithms due to transcendental knowledge exploitation.

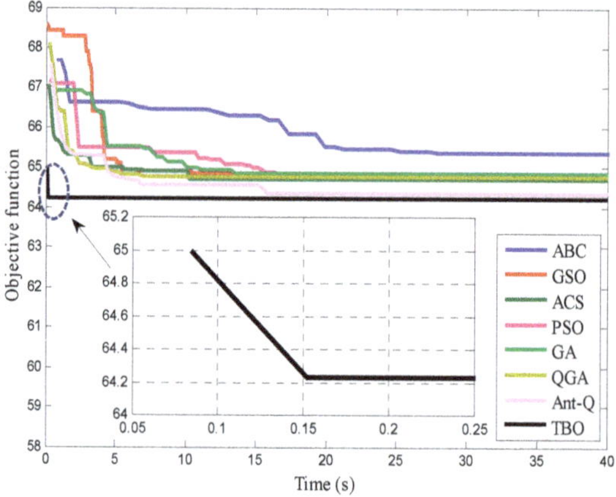

Figure 8. Convergence of the second new task of the IEEE 118-bus system obtained in the online optimization.

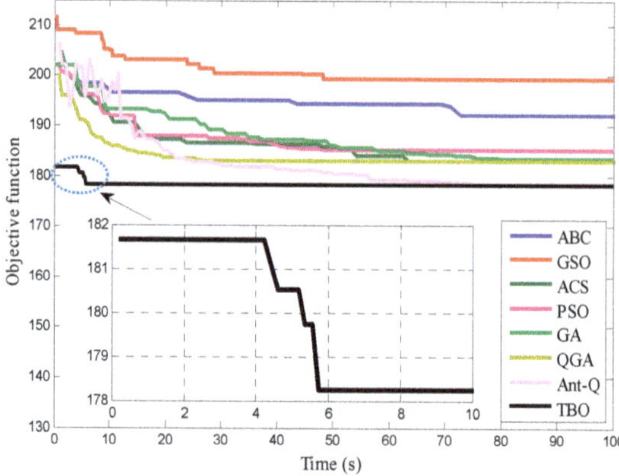

Figure 9. Convergence of the fourth new task of the IEEE 300-bus system obtained in the online optimization.

4.2.2. Comparative results and discussion

Tables 2 and 3 provide the performance and the statistical results gained by these algorithms in 10 runs, in which the convergence time was the average time of each scene, and the others were the average time of one day. The variance, standard deviation (Std. Dev.), and relative standard deviation (Rel. Std. Dev.) [42–44] were introduced in order to assess the stability. One can easily find that the convergence rate of the TBO was faster compared with that of other algorithms, as illustrated in Figure 10. Compared with that of other algorithms, the convergence rate of the TBO was about 4 to 68 times, indicating the benefit of cooperative exploration and action transfer. In addition, the optimization target function from the TBO was much smaller than that of other algorithms, which verified the advantageous effect of self-learning and global search. Note that the TBO would gain a better solution which is closer to the global optimum with respect to other algorithms with a smaller optimal objective function in most new tasks (72.92% of new tasks on the IEEE 118-bus system and 89.58% of new tasks on the IEEE 300-bus system), as shown in Figures 11 and 12.

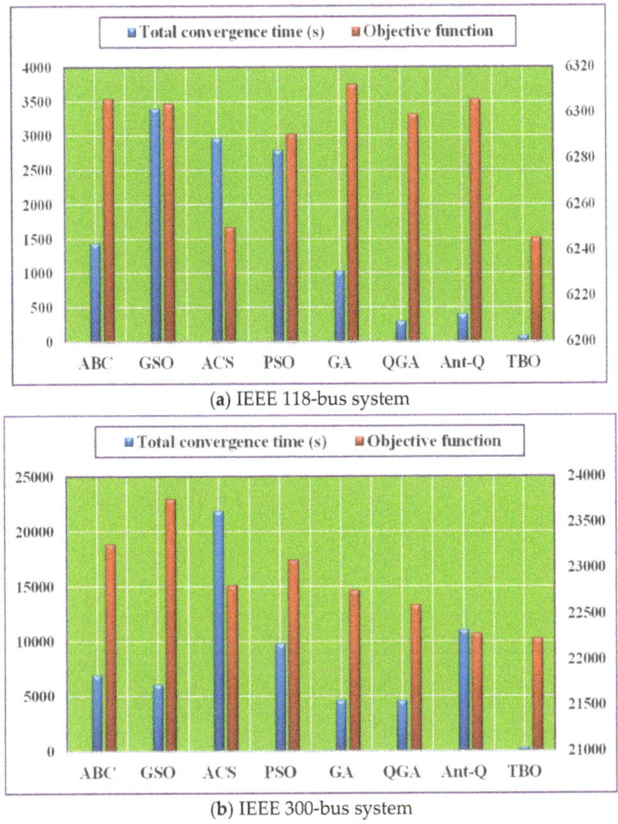

(**a**) IEEE 118-bus system

(**b**) IEEE 300-bus system

Figure 10. Comparison of performance of different algorithms obtained in 10 runs.

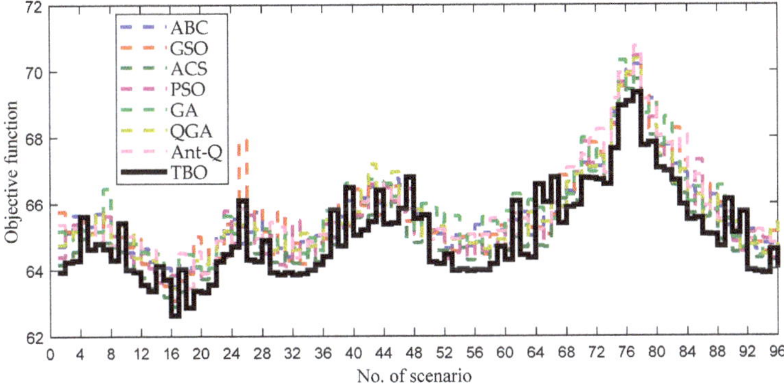

Figure 11. Optimal objective function of the IEEE 118-bus system obtained by different algorithms in 10 runs.

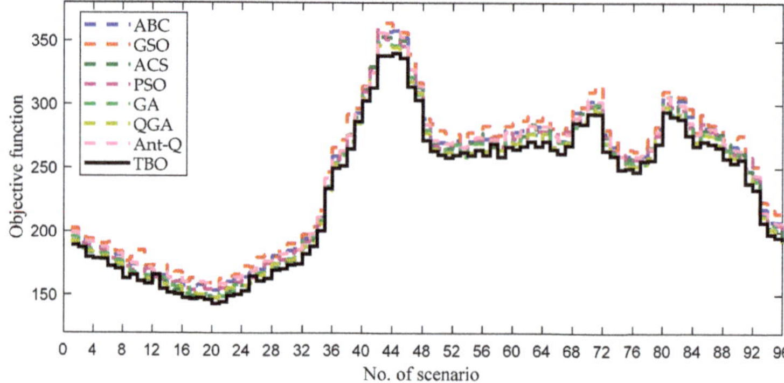

Figure 12. Optimal objective function of the IEEE 300-bus system obtained by different algorithms in 10 runs.

Table 2. Performance indices results of different algorithms on the IEEE 118-bus system obtained in 10 runs. ABC: artificial bee colony, GSO: group search optimizer, ACS: ant colony system, PSO: particle swarm optimization, GA: genetic algorithm, QGA: quantum genetic algorithm, Ant-Q: ant colony based Q-learning

Algorithm Index	ABC	GSO	ACS	PSO	GA	QGA	Ant-Q	TBO
Convergence time (s)	15	35.5	30.9	29.1	10.8	3.99	4.16	0.94
P_{loss} (MW)	1.11×10^4	1.11×10^4	1.11×10^4	1.11×10^4	1.11×10^4	1.11×10^4	1.11×10^4	1.10×10^4
V_{d} (%)	1.51×10^3	1.49×10^3	1.44×10^3	1.48×10^3	1.50×10^3	1.51×10^3	1.50×10^3	1.48×10^3
f	6.31×10^3	6.30×10^3	6.25×10^3	6.29×10^3	6.31×10^3	6.30×10^3	6.31×10^3	6.25×10^3
Best	6.30×10^3	6.30×10^3	6.24×10^3	6.28×10^3	6.31×10^3	6.30×10^3	6.30×10^3	6.24×10^3
Worst	6.31×10^3	6.31×10^3	6.25×10^3	6.30×10^3	6.32×10^3	6.30×10^3	6.31×10^3	6.25×10^3
Variance	4.02	19.3	6.43	16.3	12	6.37	8.57	2.27
Std. Dev.	2.01	4.39	2.54	4.03	3.46	2.52	2.93	1.51
Rel. Std. Dev	3.18×10^{-4}	6.96×10^{-4}	4.06×10^{-4}	6.41×10^{-4}	5.49×10^{-4}	4.01×10^{-4}	4.64×10^{-4}	2.41×10^{-4}

Table 3. Performance indices results of different algorithms on the IEEE 300-bus system obtained in 10 runs.

Algorithm Index	ABC	GSO	ACS	PSO	GA	QGA	Ant-Q	TBO
Convergence time (s)	72.3	63.4	228	102	48.3	47.8	115	3.37
P_{loss} (MW)	3.82×10^4	3.86×10^4	3.83×10^4	3.81×10^4	3.77×10^4	3.76×10^4	3.74×10^4	3.75×10^4
V_{d} (%)	8.34×10^3	8.87×10^3	7.36×10^3	8.07×10^3	7.78×10^3	7.56×10^3	7.14×10^3	6.94×10^3
f	2.33×10^4	2.38×10^4	2.28×10^4	2.31×10^4	2.28×10^4	2.26×10^4	2.23×10^4	2.22×10^4
Best	2.32×10^4	2.37×10^4	2.28×10^4	2.30×10^4	2.27×10^4	2.26×10^4	2.23×10^4	2.22×10^4
Worst	2.33×10^4	2.38×10^4	2.28×10^4	2.32×10^4	2.28×10^4	2.26×10^4	2.23×10^4	2.22×10^4
Variance	381	1.29×10^3	228	2.37×10^3	178	194	221	66.4
Std. Dev.	19.5	36	15.1	48.7	13.4	13.9	14.9	8.15
Rel. Std. Dev	8.39×10^{-4}	1.51×10^{-3}	6.61×10^{-4}	2.11×10^{-3}	5.87×10^{-4}	6.16×10^{-4}	6.67×10^{-4}	3.67×10^{-4}

On the other hand, Table 3 shows that the convergence stability of the TBO was the highest as the values of all of its indices were the lowest, and Rel. Std. Dev. of the TBO was only 17.39% with respect to that of PSO gotten from the IEEE 300-bus system and was up to 75.79% of that of ABC gotten from the IEEE 118-bus system. This was due to the exploitation of prior knowledge by scouts and workers, which beneficially avoids the randomness of global search, thus a higher search efficiency can be achieved.

5. Conclusions

In this article, a novel TBO incorporating behavior conversion was obtained for RPO. Like other AI algorithms, the TBO is highly independent from the accurate system model and has a much stronger global search ability and a higher application flexibility compared with the traditional mathematical optimization methods. Compared with network simplified model (e.g., external equivalent model)

-based methods, the TBO also can rapidly converge to an optimum for RPO but it can obtain a higher quality optimum via global optimization. By introducing the Q-learning-based optimization, the TBO can learn, store, and transfer knowledge between different optimization tasks, in which the state-action chain can significantly reduce the size of the Q-value matrix, and the cooperative exploration between scouts and workers can dramatically accelerate knowledge learning. Compared with the general AI algorithms, the greatest advantage of the TBO is that it can significantly accelerate the convergence rate for a new task via re-using prior knowledge from the source tasks. Through simulation comparisons on the IEEE 118-bus system and IEEE 300-bus system, the convergence rate of the TBO was 4 to 68 times faster than that of existing AI algorithms for RPO, while ensuring the quality and convergence stability of the optimal solutions. Thanks to its superior optimization performance, the TBO can be readily applied to other cases of nonlinear programming of large-scale power systems.

Author Contributions: Preparation of the manuscript was performed by H.C., T.Y., X.Z., B.Y., and Y.W.

Funding: This research was funded by [National Natural Science Foundation of China] grant number [51777078], and [Yunnan Provincial Basic Research Project-Youth Researcher Program] grant number [2018FD036], and The APC was funded by [National Natural Science Foundation of China].

Conflicts of Interest: The authors declare no conflict of interest.

References

1. Grudinin, N. Reactive power optimization using successive quadratic programming method. *IEEE Trans. Power Syst.* **1998**, *13*, 1219–1225. [CrossRef]
2. Li, C.; Johnson, R.B.; Svoboda, A.J. A new unit commitment method. *IEEE Trans. Power Syst.* **1997**, *12*, 113–119.
3. Zhou, B.; Chan, K.W.; Yu, T.; Chung, C.Y. Equilibrium-inspired multiple group search optimization with synergistic learning for multiobjective electric power dispatch. *IEEE Trans. Power Syst.* **2013**, *28*, 3534–3545. [CrossRef]
4. Sun, D.I.; Ashley, B.; Brewer, B. Optimal power flow by newton approach. *IEEE Trans. Power Appar. Syst.* **1984**, *103*, 2864–2880. [CrossRef]
5. Fan, J.Y.; Zhang, L. Real-time economic dispatch with line flow and emission constraints using quadratic programming. *IEEE Trans. Power Syst.* **1998**, *13*, 320–325. [CrossRef]
6. Wei, H.; Sasaki, H.; Kubokawa, J.; Yokoyama, R. An interior point nonlinear programming for optimal power flow problems with a novel data structure. *IEEE Trans. Power Syst.* **1998**, *13*, 870–877. [CrossRef]
7. Dai, C.; Chen, W.; Zhu, Y.; Zhang, X. Seeker optimization algorithm for optimal reactive power dispatch. *IEEE Trans. Power Syst.* **2009**, *24*, 1218–1231.
8. Abu-Mouti, F.S.; El-Hawary, M.E. Optimal distributed generation allocation and sizing in distribution systems via artificial bee colony algorithm. *IEEE Trans. Power Del.* **2011**, *26*, 2090–2101. [CrossRef]
9. Han, Z.; Zhang, Q.; Shi, H.; Zhang, J. An improved compact genetic algorithm for scheduling problems in a flexible flow shop with a multi-queue buffer. *Processes* **2019**, *7*, 302. [CrossRef]
10. Han, P.; Fan, G.; Sun, W.; Shi, B.; Zhang, X. Research on identification of LVRT characteristics of photovoltaic inverters based on data testing and PSO algorithm. *Processes* **2019**, *7*, 250. [CrossRef]
11. He, S.; Wu, Q.H.; Saunders, J.R. Group search optimizer: An optimization algorithm inspired by animal searching behavior. *IEEE Trans. Evol. Comput.* **2009**, *13*, 973–990. [CrossRef]
12. Gómez, J.F.; Khodr, H.M.; De Oliveira, P.M.; Ocque, L.; Yusta, J.M.; Urdaneta, A.J. Ant colony system algorithm for the planning of primary distribution circuits. *IEEE Trans. Power Syst.* **2004**, *19*, 996–1004. [CrossRef]
13. Yang, B.; Yu, T.; Zhang, X.S.; Huang, L.N.; Shu, H.C.; Jiang, L. Interactive teaching–learning optimiser for parameter tuning of VSC-HVDC systems with offshore wind farm integration. *IET Gener. Transm. Distrib.* **2017**, *12*, 678–687. [CrossRef]
14. Yang, B.; Zhang, X.S.; Yu, T.; Shu, H.C.; Fang, Z.H. Grouped grey wolf optimizer for maximum power point tracking of doubly-fed induction generator based wind turbine. *Energy Convers. Manag.* **2017**, *133*, 427–443. [CrossRef]

15. Yang, B.; Zhong, L.E.; Zhang, X.S.; Shu, H.C.; Li, H.F.; Jiang, L.; Sun, L.M. Novel bio-inspired memetic salp swarm algorithm and application to MPPT for PV systems considering partial shading condition. *J. Clean. Prod.* **2019**, *215*, 1203–1222. [CrossRef]

16. Yang, B.; Yu, T.; Zhang, X.S.; Li, H.F.; Shu, H.C.; Sang, Y.Y.; Jiang, L. Dynamic leader based collective intelligence for maximum power point tracking of PV systems affected by partial shading condition. *Energy Convers. Manag.* **2019**, *179*, 286–303. [CrossRef]

17. Montoya, F.G.; Alcayde, A.; Arrabal-Campos, F.M.; Raul, B. Quadrature current compensation in non-sinusoidal circuits using geometric algebra and evolutionary algorithms. *Energies* **2019**, *12*, 692. [CrossRef]

18. Yu, T.; Liu, J.; Chan, K.W.; Wang, J.J. Distributed multi-step Q (λ) learning for optimal power flow of large-scale power grids. *Int. J. Electr. Power Energy Syst.* **2012**, *42*, 614–620. [CrossRef]

19. Mühürcü, A. FFANN optimization by ABC for controlling a 2nd order SISO system's output with a desired settling time. *Processes* **2019**, *7*, 4. [CrossRef]

20. Sundareswaran, K.; Sankar, P.; Nayak, P.S.R.; Simon, S.P.; Palani, S. Enhanced energy output from a PV system under partial shaded conditions through artificial bee colony. *IEEE Trans. Sustain. Energy* **2014**, *6*, 198–209. [CrossRef]

21. Chandrasekaran, K.; Simon, S.P. Multi-objective unit commitment problem with reliability function using fuzzified binary real coded artificial bee colony algorithm. *IET Gener. Transm. Distrib.* **2012**, *6*, 1060–1073. [CrossRef]

22. Gao, Z. Advances in Modelling, monitoring, and control for complex industrial systems. *Complexity* **2019**, *2019*, 2975083. [CrossRef]

23. Tognete, A.L.; Nepomuceno, L.; Dos Santos, A. Framework for analysis and representation of external systems for online reactive-optimisation studies. *IEE Gener. Transm. Distrib.* **2005**, *152*, 755–762. [CrossRef]

24. Tognete, A.L.; Nepomuceno, L.; Dos Santos, A. Evaluation of economic impact of external equivalent models used in reactive OPF studies for interconnected systems. *IET Gener. Transm. Distrib.* **2007**, *1*, 140–145. [CrossRef]

25. Britannica Academic, S.V. The Life of Bee. Available online: http://academic.eb.com/EBchecked/topic/340282/The-Life-of-the-Bee (accessed on 24 April 2019).

26. Pan, S.J.; Yang, Q. A survey on transfer learning. *IEEE Trans. Knowl. Data Eng.* **2009**, *22*, 1345–1359. [CrossRef]

27. Ramon, J.; Driessens, K.; Croonenborghs, T. Transfer learning in reinforcement learning problems through partial policy recycling. In *Learn ECML*; Springer: Heidelberg/Berlin, Germany, 2007; pp. 699–707.

28. Gao, Z.; Saxen, H.; Gao, C. Data-driven approaches for complex industrial systems. *IEEE Trans. Ind. Inform.* **2013**, *9*, 2210–2212. [CrossRef]

29. Taylor, M.E.; Stone, P. Transfer learning for reinforcement learning domains: A survey. *J. Mach. Learn. Res.* **2009**, *10*, 1633–1685.

30. Pan, J.; Wang, X.; Cheng, Y.; Cao, G. Multi-source transfer ELM-based Q learning. *Neurocomputing* **2014**, *137*, 57–64. [CrossRef]

31. Bianchi, R.A.C.; Celiberto, L.A.; Santos, P.E.; Matsuura, J.P.; Mantaras, R.L.D. Transferring knowledge as heuristics in reinforcement learning: A case-based approach. *Artif. Intell.* **2015**, *226*, 102–121. [CrossRef]

32. Watkins, J.C.H.; Dayan, P. Q-learning. *Mach. Learn.* **1992**, *8*, 279–292. [CrossRef]

33. Ghavamzadeh, M.; Mahadevan, S. Hierarchical average reward reinforcement learning. *J. Mach. Learn. Res.* **2007**, *8*, 2629–2669.

34. Dietterich, T.G. Hierarchical reinforcement learning with the MAXQ value function decomposition. *J. Artif. Intel.* **2000**, *13*, 227–303. [CrossRef]

35. Yu, T.; Zhou, B.; Chan, K.W.; Chen, L.; Yang, B. Stochastic optimal relaxed automatic generation control in Non-Markov environment based on multi-step Q (λ) learning. *IEEE Trans. Power Syst.* **2011**, *26*, 1272–1282. [CrossRef]

36. Maloosini, A.; Blanzieri, E.; Calarco, T. Quantum genetic optimization. *IEEE Trans. Evol. Comput.* **2008**, *12*, 231–241. [CrossRef]

37. Dorigo, M.; Gambardella, L.M. A study of some properties of Ant-Q. In *International Conference on Parallel Problem Solving from Nature*; Springer-Verlag: Berlin, Germany, 1996; pp. 656–665.

38. Zhang, X.S.; Yu, T.; Yang, B.; Cheng, L. Accelerating bio-inspired optimizer with transfer reinforcement learning for reactive power optimization. *Knowl. Base. Syst.* **2017**, *116*, 26–38. [CrossRef]

39. Rajaraman, P.; Sundaravaradan, N.A.; Mallikarjuna, B.; Jaya, B.R.M.; Mohanta, D.K. Robust fault analysis in transmission lines using Synchrophasor measurements. *Prot. Control. Mod. Power Syst.* **2018**, *3*, 108–110.

40. Hou, K.; Shao, G.; Wang, H.; Zheng, L.; Zhang, Q.; Wu, S.; Hu, W. Research on practical power system stability analysis algorithm based on modified SVM. *Prot. Control Mod. Power Syst.* **2018**, *3*, 119–125. [CrossRef]

41. Ren, C.; Xu, Y.; Zhang, Y.C. Post-disturbance transient stability assessment of power systems towards optimal accuracy-speed tradeoff. *Prot. Control Mod. Power Syst.* **2018**, *3*, 194–203. [CrossRef]

42. Dabra, V.; Paliwal, K.K.; Sharma, P.; Kumar, N. Optimization of photovoltaic power system: A comparative study. *Prot. Control Mod. Power Syst.* **2017**, *2*, 29–39. [CrossRef]

43. Yang, B.; Yu, T.; Shu, H.C.; Dong, J.; Jiang, L. Robust sliding-mode control of wind energy conversion systems for optimal power extraction via nonlinear perturbation observers. *Appl. Energy* **2018**, *210*, 711–723. [CrossRef]

44. Yang, B.; Jiang, L.; Wang, L.; Yao, W.; Wu, Q.H. Nonlinear maximum power point tracking control and modal analysis of DFIG based wind turbine. *Int. J. Electr. Power Energy Syst.* **2016**, *74*, 429–436. [CrossRef]

Article

Grouping Method of Semiconductor Bonding Equipment Based on Clustering by Fast Search and Find of Density Peaks for Dynamic Matching According to Processing Tasks

Zhijun Gao [1], Wen Si [1,*], Zhonghua Han [1,2,3,4], Jiayu Peng [1] and Feng Qiao [1]

[1]   Faculty of Information and Control Engineering, Shenyang Jianzhu University, Shenyang 110168, China
[2]   Department of Digital Factory, Shenyang Institute of Automation, Chinese Academy of Sciences (CAS), Shenyang 110016, China
[3]   Key Laboratory of Network Control System, Chinese Academy of Sciences, Shenyang 110016, China
[4]   Institutes for Robotics and Intelligent Manufacturing, Chinese Academy of Sciences, Shenyang 110016, China
*   Correspondence: siwen950717@163.com; Tel.: +86-24-24690969

Received: 6 August 2019; Accepted: 20 August 2019; Published: 25 August 2019

Abstract: Given that the equipment for the semiconductor packaging line adopts the fixed grouping production method, thus failing to dynamically match the processing task demand capacity, in the present study, we proposed a semiconductor bonding equipment-grouping method based on processing task matching. This method sets the device group closed position constraint and the matching constraint between the device type and the processing type and uses the graph theory method to establish the device grouping model. The dynamic grouping of equipment under the capacity demand of different processing tasks was achieved by changing the relationship matrix between devices. The drawback of this grouping method is rather large grouping deviation, which we tried to solve with the clustering by fast search and find of density peaks (CFSFDP) that was added to cluster the sets of attribute information of the devices so as to obtain the maximum number of grouping groups obtained to reduce the grouping deviation. Simulation comparison experiments were carried out under different circumstances considering the size of the formation, the distribution of demand capacity, and the coefficient of difference in demand capacity. Compared with the standard device grouping method, the grouping method based on semiconductor bonding equipment and CFSFDP algorithm for dynamic matching according to processing tasks had better performance in solving the dynamic grouping problem.

Keywords: semiconductor bonding equipment-grouping method; graph theory; association matrix; CFSFDP algorithm

---

1. Introduction

Industry 4.0 has become the current trend of automation industries, which has great impacts on improving the reliability and operation performance of complex industrial systems. Therefore, it is paramount but challenging to develop effective techniques in modelling, monitoring, and control for complex industrial systems [1]. With the rapid development of artificial intelligence and the promotion of strategic emerging industries such as the Internet of Things, energy conservation and environmental protection, and new energy vehicles, the demand for semiconductors continues to increase. Semiconductor manufacturing mainly consists of three phases: chip design, front wafer fabrication, and back-end packaging testing. The process of semiconductor package testing is dicing, loading, bonding, plastic sealing, deflashing, electroplating, printing, cutting, molding, visual inspection, finished product testing, packaging, and shipping. The main production process of the package test is mainly to receive the wafer from the previous wafer fabrication, divide it into small wafers by using a dicing blade, and attach the adhesive to the corresponding frame substrate. After that, the pins of the wafer and the substrate are soldered by using ultra-fine metal wires to form the required circuit. Then, the molding compound is injected for protection, and finally a series of printing, cutting, forming, and the like are performed to form a complete product for test packaging output.

The semiconductor package line is a process of making a wafer, which after passing the test is made into a separate chip according to the product model and functional requirements [2]. The bonding process in the package test line includes a large number of devices with huge capacity differences, which represents a challenge for the production management and control of the semiconductor package line. Semiconductor bonding process of international large enterprises may include as many as one thousand devices, for which the equipment control should be in the form of equipment grouping. Generally, the equipment types or certain equipment qualification conditions are used as the basis for grouping. A device group is a processing unit or a station. The production workshop in the semiconductor industry is an open workshop, where the equipment can be dynamically adjusted, which usually happens once a week. The drawback of such a grouping method is that it cannot be dynamically changed as the required capacity of the processing list Lot is put into production in the semiconductor production line, and the obtained grouping result cannot be dynamically matched with the processing task. Since each equipment group produces only one Lot at a time, a batch of Lots is produced faster in some groups than in the others. Due to the load imbalance, some equipment is idle, which restricts the production efficiency as well as the potential of the existing resources. The bonding process is used as a bottleneck process in the semiconductor production [3]. The resources in the bottleneck process restrict the output of the production system, wherein the loss of resources imply the loss of the entire production system. Increasing the utilization rate of the bonding process equipment increases the working efficiency of the entire semiconductor production line. Therefore, it is of great engineering value and theoretical value to study the equipment dynamic grouping method of semiconductor packaging lines.

As the semiconductor industry shifts to a multi-variety small batch production model, the number of LOTs put into production increases. If the production potential of the equipment in the bonding process segment cannot be fully utilized under the existing production capacity, the development of the semiconductor industry will be restricted [4]. The traditional fixed group production mode can no longer meet the existing needs, thus exploration of the improved grouping methods that would meet production needs has become the focus of interest of many researchers. Wang et al. [5] has used the quantum genetic algorithm to implement the intelligent body grouping, but without achieving dynamic strategic adjustments to deal with uncertain events. Furthermore, Rong and colleagues [6] has achieved multi-group optimization model under different time periods based on Pareto search and two-stage algorithm information entropy; however, since the rail transit lines run independently, the dynamic grouping of vehicle sharing was not realized. Fang and his team [7] used a linear weighting method to transform the multi-objective group optimization model into a single-objective group optimization model, and solved the grouping results with a controlled random search algorithm. However, there are better optimization results for rail transit lines with large differences in section traffic. Jacob et al. [8] developed a new classification scheme code based on the train classification method simply characterized as a plan category. By applying this efficient coding, a simpler and more accurate classification was achieved; nonetheless, dynamic grouping under different requirements has not been resolved. The above listed studies have improved the fixed grouping by establishing mathematical models and by adding optimization algorithms. The generalized cost of urban rail transit is the optimization goal. The constraint conditions mainly refer to the passenger flow demand constraint and the grouping time constraint. The relationship between individuals or groups within a group has not been fully considered. While the above research has focused on the research of railway transportation, there are very few researches on the dynamic grouping of equipment under different production requirements in the semiconductor field. Hsieh et al. [9] have developed the proposed WPH management system for semiconductor wafer Fabs, which is capable of estimating and monitoring equipment WPH by recipes in a sustainable manner. It is undeniable that its unique contribution, however, it has not solved the problem from the perspective of maximizing the utilization of processing capacity. In order to better match the equipment group and processing tasks, this paper establishes a mathematical model of equipment grouping. Mathematical modeling offers ways to gain deeper understanding of the interdependence between variables of the processes [10].

Therefore, the current semiconductor package wire bonding equipment group cannot be dynamically matched with the processing task. At present, relevant papers have been used to model and control complex dynamic systems [11]. We proposed a grouping method of semiconductor bonding equipment based on clustering by fast search and find of density peaks (CFSFDP) for dynamic matching according to processing tasks. The method based on the processing task has foundations in graph theory. The equipment marshalling model implements equipment grouping under different processing tasks, and adds the CFSFDP algorithm to improve the equipment marshalling model, and to optimize the results of dynamic marshalling.

2. Semiconductor Bonding Device Grouping Method Based on Processing Task Matching

In actual production, the semiconductor equipment is produced according to a fixed grouping method, i.e., once the actual grouping is determined, the processing capacity of the grouping is also fixed. Although such a grouping method is easy to execute and manage, there are some problems [12]. For example, the production capacity of certain production equipment is wasted, leading to redundant production capacity. When several devices are organized into a single production unit, the group can only select one production unit to process the wafer given that the wafer is inseparable, which in turn wastes the production capacity of the other devices during the same processing time. If grouping was made according to the production capacity of the production task, it would increase the utilization rate of the production equipment, thereby improving the efficiency of the entire production line. Based on the equipment grouping method of processing task capability matching, the equipment can be grouped according to the production capacity requirements of the production tasks, effectively cooperating with the production beat to improve the equipment utilization rate, and adapt to the dynamic changes of production tasks and resource capabilities. Among them, the following restrictions are imposed on the equipment to be grouped:

(1) The process requirements for equipment processing in a group are the same.

(2) Equipment types within the group should be identical or equipment within the group should be able to process the same type of product.

(3) In order for workers to operate the control device, the grouping device must be an adjacent device, and the area constructed by the grouping device cannot enclose other grouped devices or ungrouped devices inside the region.

(4) The processing capacity generated by the equipment group should match the processing capacity required by the processing task.

2.1. Equipment Marshalling Process

The flow of the bonding device grouping method based on the processing task [13] matching semiconductor package line is as shown in Figure 1 below:

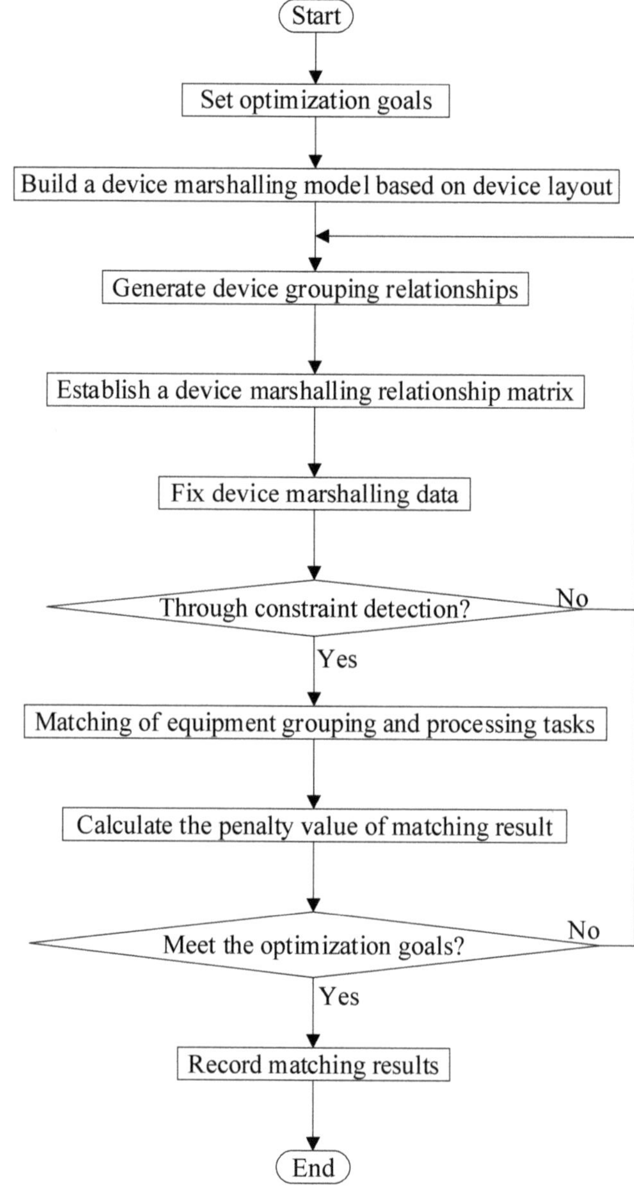

Figure 1. Flow chart of semiconductor bonding equipment grouping method based on processing task matching.

Step 1: The minimum penalty value obtained by equipment grouping to provide capacity and processing task demand capacity matching is the optimization goal of equipment grouping.

Semiconductor manufacturing mainly consists of three phases: chip design, front wafer fabrication, and back-end packaging testing. The process of semiconductor package testing is dicing, loading, bonding, plastic sealing, deflashing, electroplating, printing, cutting, molding, visual inspection, finished product testing, packaging, and shipping. Semiconductor manufacturing is a dynamic and continuous process, so the time-limited cost of the bonding process segment cannot be derived separately. The utilization rate of the equipment can only be measured by considering the difference in the capacity of the equipment provided by the difference between the capacity provided and the capacity required.

Insufficient capacity is provided by the equipment group matching the processing task W_P.

$$PC_p = \max\left(0, gw_p - gma_p\right) \tag{1}$$

Capacity redundancy is provided by the device group that matches the processing task W_P.

$$EC_p = \max\left(0, gma_p - gw_p\right) \tag{2}$$

gw_p indicates the equipment capacity required for the processing task p. gma_p indicates the processing capability of the equipment group matched by the processing task where p can provide, $p \in \{1, 2, L, PN\}$.

The punishment sum of equipment grouping and processing tasks.

$$f_o = \sum_{p=1}^{PN}\left(\alpha_p PC_p + \beta_p EC_p\right) \tag{3}$$

α_p indicates the penalty for the capacity of the processing task p β_p indicates the redundancy penalty for capacity of the processing task p.

Step 2: Using graph theory to simplify the representation of a device as a node on a two-dimensional plane [14], abstracting all device points of this process into n rows and m columns of device matrices. $m_{i,j}$ represents a device of the $i - th$ row and the $j - th$ column.

$$M = \begin{bmatrix} m_{1,1} & m_{1,2} & k & m_{1,m} \\ m_{2,1} & m_{2,2} & k & m_{2,m} \\ m & m & m & m \\ m_{n,1} & m_{n,2} & k & m_{n,m} \end{bmatrix} \tag{4}$$

Step 3: All devices are connected by wires to form a network structure with device as node [15]. As shown in Figure 2, the association coefficient between the devices is randomly generated and represented by 0 and 1. The "0" indicates no association between the devices, i.e., the two devices are not in one group, and the "1" indicates the existence of an association, i.e., the two devices are edited in a group [16,17].

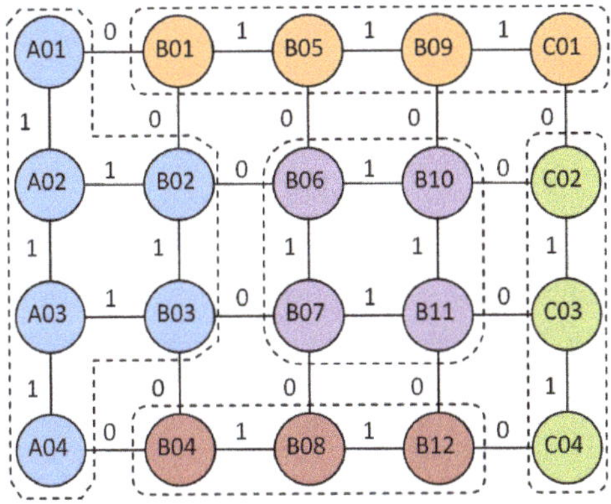

Figure 2. Schematic diagram of device grouping based on association relationship.

Step 4: According to the association coefficient between the devices, the device grouping row association matrix LR and the column association matrix LC are established by using the adjacency matrix, and the temporary device grouping G is established.

The correlation matrix LR between rows is as shown in Equation (5), and $l^r_{i,j}$ represents the row correlation coefficient between the $i-th$ row and the $i+1-th$ device of the $j-th$ column. When $l^r_{i,j}$ is "1", it means that two adjacent devices are associated and can be grouped together. On the contrary, $l^r_{i,j}$ takes "0" to indicate no association. The correlation matrix LC between the columns is as shown in Equation (6).

$$LR = \begin{bmatrix} l^r_{1,1} & l^r_{1,2} & 1 & l^r_{1,m} \\ l^r_{2,1} & l^r_{2,2} & 1 & l^r_{2,m} \\ m & m & l^r_{i,j} & m \\ l^r_{n-1,1} & l^r_{n-1,2} & 1 & l^r_{n-1,m} \end{bmatrix} \tag{5}$$

$$LC = \begin{bmatrix} l^c_{1,1} & l^c_{1,2} & 1 & l^c_{1,m-1} \\ l^c_{2,1} & l^c_{2,2} & 1 & l^c_{2,m-1} \\ m & m & l^c_{i,j} & m \\ l^c_{n,1} & l^c_{n,2} & 1 & l^c_{n,m-1} \end{bmatrix} \tag{6}$$

The associated device is grouped into a device group by the generated row and column association matrix. A temporary group number g_t is assigned to indicate the group number of the $t-th$ device group. GN represents the maximum group number for all groups.

Step 5: Correct the group association relationship. The temporary coding matrix of equipment grouping obtained in Step 4 may have a condition that conforms to the grouping rules but does not meet the actual grouping requirements. Therefore, it is necessary to correct the grouping relationship by employing the line scan and the column scan detection using the temporary group number g_t of each group identification in Step 4.

Line detection starts from the 1st line and detects the nth line stop. It is necessary to check whether the correlation coefficient of each line is correct. According to the Equation (7), if the device group number $g_{i,j}$ of the $i-th$ row and the $j-th$ column is equal to the device group number $g_{i,j+1}$ of the $i-th$ row and the $j+1-th$ column, the device row association relationship $l^c_{i,j}$ is "1". If it is "0", it should be set to "1".

$$l^c_{i,j} = \begin{cases} 0 & Mg_{i,j} \neq Mg_{i+1,j} \\ 1 & Mg_{i,j} = Mg_{i+1,j} \end{cases} \tag{7}$$

Column detection starts from the 1st column and detects the $n-th$ column stop. It is necessary to check whether the correlation coefficient of each column is correct. According to the Equation (8), if the device group number $g_{i,j}$ of the $j-th$ column and the $i-th$ row is equal to the device group number $g_{i+1,j}$ of the $j-th$ column and the $i+1-th$ row, the device column association relationship $l^r_{i,j}$ is "1". If it is "0", it should be set to "1".

$$l^r_{i,j} = \begin{cases} 0 & Mg_{i,j} \neq Mg_{i,j+1} \\ 1 & Mg_{i,j} = Mg_{i,j+1} \end{cases} \tag{8}$$

Step 6: Closed position constraint refers to a device group or other device group that does not belong to this group is included in a two-dimensional plane. Such a completely enclosed relationship can be understood as a closed position. Constraint detection of device grouping. Constraint detection is divided into device group closed position relationship constraint detection and device type and processing type matching constraint detection. If any constraint detection fails, the grouping result is invalid, and the grouping step returns to Step 3.

(1) Constraint detection of closed position relationship by equipment grouping

The enclosed area formed by the grouping device cannot contain devices that do not belong to the device group, nor can it contain another independent device group. If such a grouping situation exists, the jurisdiction of the two equipment group managers overlaps, causing confusion in the production management organization.

To determine whether the device is completely in the closed range of the grouping, it is necessary to perform row detection and column detection according to the grouping relationship in the device grouping matrix G. If it is detected that a closed area formed in a device group contains devices that are not in this group, it is considered that the device grouping is unreasonable, and the temporary device grouping needs to be re-established. This inspection process needs to determine whether the grouping result is reasonable based on the results of the row and column test.

When the grouping information of the group to which the device of the $i-th$ row and the $j-th$ column belong is the same as the grouping number of the $t-th$ group, $gz_{t,i,j}$ is equal to "1", otherwise it is equal to "0". $F_{t,i}$ indicates whether the $i-th$ row of the $t-th$ device group contains devices that do not belong to the $t-th$ group. When the value is "1", it means the device is not included, and when it is "0", it means that it contains equipment that does not belong to the $t-th$ group.

$$
F_{t,i} = \begin{cases} 1 & \sum_{j=m1_{t,i}}^{m2_{t,i}} gz_{t,i,j} = (m2_{t,i} - m1_{t,i}) \\ 0 & \sum_{j=m1_{t,i}}^{m2_{t,i}} gz_{t,i,j} < (m2_{t,i} - m1_{t,i}) \end{cases} \tag{9}
$$

F_t^r indicates the result of row detection by the $t-th$ device group. The value of F_t^r is "0", indicating that the $t-th$ device group has detected that there are $non-t-th$ grouped devices in a certain device after the line detection. F_t^c indicates the result of column detection by the $t-th$ device group. If F_t^c is "0", it indicates that the t device group is detected by column and there is a $non-t$ group device in a column device.

$$
F_t^r = \begin{cases} 1 & \left(\sum_{i=n1_t}^{n2_t} F_{t,j} = n2_t - n1_t \right) \cup \left(\left(\sum_{i=n1_t}^{n2_t} F_{t,j} < n2_t - n1_t \right) \cap \left((F_{t,n1_t} \neq 1) \cap (F_{t,n2_t} \neq 1) \right) \right) \\ 0 & \left(\sum_{i=n1_t}^{n2_t} F_{t,j} < n2_t - n1_t \right) \cap (F_{t,n1_t} = 1) \cap (F_{t,n2_t} = 1) \end{cases} \tag{10}
$$

F_t indicates whether the closed area of the $t-th$ equipment group contains equipment that is not in this group. If F_t is "0", it indicates that the F_t^r row detection result of the $t-th$ device group and the F_t^c column detection result simultaneously determine the device having the $non-t-th$ group in the group.

$$
F_t = \begin{cases} 1 & F_t^r \neq 0 \cup F_t^c \neq 0 \\ 0 & F_t^r = 0 \cap F_t^c = 0 \end{cases} \tag{11}
$$

F indicates that the device grouping result satisfies the device location constraint. When all device groupings constitute a closed area, none of the non-grouped devices are included, and F is equal to "1".

$$
F = \begin{cases} 1 & \sum_{t=1}^{GN} F_t = GN \\ 0 & \sum_{t=1}^{GN} F_t < GN \end{cases} \tag{12}
$$

(2)   Constraint detection matching device type and processing type

The processed product type is represented by k_x, a total of K_N product types. The device type is represented by K_N, which is a total of KN device types. The relationship matrix Kk whose product type matches the device type can be expressed as Equation (13). When Kkl_{K_y,k_x} is "1", the product type matches the device type. If it is "0", the two do not match.

$$Kk = \begin{bmatrix} Kkl_{1,1} & Kkl_{1,2} & L & Kkl_{1,kn} \\ Kkl_{2,1} & Kkl_{2,2} & L & Kkl_{2,kn} \\ M & M & Kkl_{i,j} & M \\ Kkl_{KN,1} & Kkl_{KN,2} & L & Kkl_{KN,kn} \end{bmatrix} \tag{13}$$

The matching relationship matrix MK between the device and the device type is as shown in Equation (14). $MKl_{i,j,y}$ indicates the matching relationship between device type and device. $MKl_{i,j,y}$ equals "1" to indicate that the device matches the device type. If it is "0", the device does not match the device type.

$$MK = \begin{bmatrix} Mkl_{1,1,1} & Mkl_{1,1,2} & L & Mkl_{1,1,KN} \\ Mkl_{1,2,1} & Mkl_{1,2,2} & L & Mkl_{1,2,KN} \\ M & M & Mkl_{i,j,y} & M \\ Mkl_{n,m,1} & Mkl_{n,m,2} & L & Mkl_{n,m,KN} \end{bmatrix} \tag{14}$$

Mk indicates that the device and product type matching relationship matrix is as shown in Equation (15), which is obtained by multiplying the product of MK and Kk.

$$Mk = \begin{bmatrix} Mkl_{1,1,1} & Mkl_{1,1,2} & L & Mkl_{1,1,kn} \\ Mkl_{1,2,1} & Mkl_{1,2,2} & L & Mkl_{1,2,kn} \\ M & M & Mkl_{i,j,x} & M \\ Mkl_{n,m,1} & Mkl_{n,m,2} & L & Mkl_{n,m,kn} \end{bmatrix} \tag{15}$$

For the device group g_t, the device set contained in the group is represented by $Mg_t = \{M_{i,j} | g_{i,j} = g_t\}$. If g_t and product type k_x are grouped for any device, and Mg_t is included in Mk_x and k_x belongs to k, the device grouping matches the device type and the processed product type matching constraint. If the constraint is not met, it is necessary to return to Step 3 to re-establish the correlation coefficient between the devices.

Step 7: Matching of equipment grouping and processing tasks

(1)   Capacity provided by equipment grouping

The processing speed matrix V is constructed according to the relationship matrix Mk where the device matches the product type. $v_{i,j,x}$ is defined as the processing speed of the k_x type product by the device in the $i-th$ row and the $j-th$ column.

$$V_{i,j,k_x} = \begin{cases} v_{i,j,k_x} & Mkl_{i,j,k_x} = 1 \\ 0 & Mkl_{i,j,k_x} = 0 \end{cases} \tag{16}$$

$$V = \begin{bmatrix} v_{1,1,k_1} & v_{1,2,k_2} & l & v_{1,m,k_{kn}} \\ v_{1,2,k_1} & v_{1,2,k_2} & l & v_{1,2,k_{kn}} \\ m & m & v_{i,j,k_x} & m \\ v_{n,m,k_1} & v_{n,m,k_2} & l & v_{n,m,k_{kn}} \end{bmatrix} \tag{17}$$

A matrix is constructed based on product type to provide capacity matrix Gga, where ga_{t,k_x} indicates that the equipment group g_t produces the processing capacity of the k_x type product,

$$ga_{t,k_x} = \left(\sum_{i=n1_t}^{n2_t} \sum_{j=m1_t}^{m2_t} V_{i,j,k_x} \right) gT, i \in \{n1_t, \ldots, n2_t\}, j \in \{m1_t, \ldots, m2_t\}.$$

$$Gga = \begin{bmatrix} ga_{1,k_1} & ga_{1,k_2} & l & ga_{1,k_{kn}} \\ ga_{2,k_2} & ga_{2,k_2} & l & ga_{2,k_{kn}} \\ m & m & ga_{t,k_x} & m \\ ga_{GN,k_1} & ga_{GN,k_2} & l & ga_{GN,k_{kn}} \end{bmatrix} \tag{18}$$

(2) Demand capacity of processing tasks

The processing task set is represented by $W_p, W_p = \{W_p | p \in \{1, 2, \ldots, PN\}\}$. W_p represents the $p - th$ processing task, the product type of the $p - th$ processing task is represented by K_p, and the product type set constituting the processing task is represented by KP, $KP = \{K_p | p \in \{1, 2, \ldots, PN\}\}$. The set of processing tasks for product type k_x is represented by $WP_x, WP_x = \{W_p | k_x = k_p, p \in \{1, 2, \ldots, PN\}\}$, and $WP_x \subseteq WP$, gw_p indicates the demand capacity of the $p - th$ processing task, and the processing task demand capacity set is represented by $GW, GW = \{gw_p | p \in \{1, 2, \ldots, PN\}\}$.

(3) Equipment grouping matching the processing task

The total number of production equipment marshalling GN should be greater than or equal to the total number of tasks to be assigned PN. The number of selectable device groups corresponding to the product type k_x should be greater than or equal to the number of tasks in the processing task set whose product type is k_x.

The matrix matching the processing task demand ability and the equipment group capability is represented by $Gamc$. $Gamc = \{\{\langle gw_p, gma_p \rangle | p \in \{1, 2, \ldots, PN\}\}\}$. gma_p indicates the processing capacity of the equipment group, gw_p indicates the demand capacity of the processing task. The PN processing tasks are assigned to the equipment group. It is necessary to sort the processing task W_p according to the descending order of the required task capacity GW. Then it is necessary to obtain the sorted device task collection queue $W_{p'}$ and the product type sequence $k_{p'}$ corresponding to $W_{p'}$.

According to the relationship matrix Mk matched by the device and the product type, the device group capable of processing the product is selected by the product type of the processing task, and the matching relationship matrix Gk of the device grouping and the product type are constructed.

$$Gk = \begin{bmatrix} gkl_{1,k_1} as_1 & gkl_{1,k_2} as_1 & L & gkl_{1,k_{kn}} as_1 \\ gkl_{2,k_1} as_2 & gkl_{2,k_2} as_2 & L & gkl_{2,k_{kn}} as_2 \\ M & M & gkl_{t,k_x} as_t & M \\ gkl_{GN,k_1} as_{GN} & gkl_{GN,k_2} as_{GN} & L & gkl_{GN,k_{kn}} as_{GN} \end{bmatrix} \tag{19}$$

The element $gkl_{t,x}$ in the matrix Gk indicates the matching relationship between product type and device grouping. Its value of "1" means that the device group g_t contains the device that satisfies the matching relationship between the device type and the product type.

When the device group is not assigned a task in the initial state, all as_t is 1. The processing product type is $k_{p'}$ and processing task is $W_{p'}$. Selected device grouping is set as $Gg_{k_{p'}}$, and each device group is calculated in $Gg_{k_{p'}}$. The processing capacity and the established $Gg_{k_{p'}}$ corresponds to the equipment group capacity set $Gga_{k_{p'}}$, $Gga_{k_{p'}} = \{ga_{t,k_x} | gkl_{t,k_x} = 1, as_t = 1, k_x = k_{p'}, t \in \{1, 2, \ldots, PN\}\}$. ga_{t,k_x} is the processing capability of equipment group g_t to produce k_x type products. When $Gg_{k_{p'}}$ is not an empty set, the processing task $W_{p'}$ should be assigned to $Gga_{k_{p'}}$. The device group with the largest processing capability in the set is represented by $gma_{p'}$, $gma_{p'} = \{ga_{q,k_{p'}} | q = \min\{t | \max(Gga_{k_{p'}})\}\}$. The task $W_{p'}$ is matched to the device group g_q, where $q = \min\left\{t \middle| \max\left(Gga_{k_{p'}}\right)\right\}$, and the row corresponding to the g_q in the matrix corresponds to $as_q = 0$, indicating that the device group was assigned tasks. When the next task matches, the device group no longer participates in the matching of devices and tasks. After all the processing tasks are matched, the device group capability matching set $Gamc'$ and the device group matching set Gmc' are matched with GW' and $W_{p'}$. Then, according to the order of the original processing task WP the equipment group matching set Gmc matching the order of the processing tasks is obtained, $Gmc = \left\{\left\langle W_{p'}, g_{q'}\right\rangle \middle| p \in \{12,\ldots,PN\}, q' \in \{1,2,\ldots,GN\}\right\}$. The processing task requirement capability and the device group capability matching set $Gamc$ corresponding to the device group matching set Gmc are obtained.

Step 8: The optimization goal of device grouping is the penalty value f_o minimum for device grouping and processing tasks. If the optimization range is satisfied, the temporary grouping result is saved. If the requirement is not met, the process returns to Step 3 until the maximum number of iterations is reached, and the grouping corresponding to the minimum penalty value is stopped as the final grouping result.

2.2. Analysis of Equipment Grouping Methods Based on Processing Task Matching

Compared with the prior art, the grouping method has the following advantages:

(1) The concept of directed graph in graph theory is used to abstract the device set into a matrix, and to establish a device marshalling model, and change the adjacency matrix to change the relationship between devices.

(2) The situation of the device fixed grouping of the bonding process segment is changed, and the device can be dynamically grouped according to the released production task. In the process of grouping equipment, the equipment in the group is in a complete unicom area, which is convenient for the staff of the production line.

(3) The evaluation index of the capacity matching between the capacity provided by the equipment group and the processing task is established to better meet the purpose of matching the equipment grouping and processing tasks, and to ensure the rationality of the dynamic grouping results.

However, the proposed bonding device grouping method based on processing task matching semiconductor packaging lines is not without defects. For example, when the correlation matrix is generated. Since the correlation coefficient is 0 and 1 randomly generated by the rand function, the association relationship between the devices is random, and the devices that can be associated together and grouped together are also randomly combined. The randomness generated by the correlation coefficient may cause the generated temporary grouping number to be too large or too small, thus generating a grouping deviation. The grouping deviation refers to the absolute value of the difference between the number of groups obtained by the equipment grouping and the number of processing tasks. Too many groups may lead to devices with no processing tasks to match, thus wasting processing power. If the number of groups is too small or even smaller than the number of processing tasks, the grouping cannot be matched with the processing and production tasks. Considering that the production equipment itself has many characteristics, it can be clustered according to certain dimensions. Therefore, the density peak-based clustering method (CFSFDP) with better effect in unsupervised learning has been improved. In the paper, I made a modification and made a detailed explanation of the device grouping method improved by adding the CFSFDP algorithm. Therefore, based on the proposed grouping method, the CFSFDP algorithm is added to the divergence of the number of groups GN, and the reasonable range of the number of device groupings is obtained according to the attribute information of the device clustering. Attribute information refers to the self-contained attributes such as processing speed, device type, device location, and device parameters for different types of products.

3. Grouping Method of Semiconductor Bonding Equipment Based on CFSFDP Algorithm for Dynamic Matching According to Processing Tasks

Clustering is called unsupervised learning in the field of machine learning and pattern recognition. In the absence of prior knowledge, clustering classifies data according to the similarity between data objects. Usually, the similarity between data objects is the distance between data objects is determined. The CFSFDP clustering algorithm is based on the idea of density clustering. The implementation process is simple. Only the data object distance is calculated in the algorithm processing process, which greatly reduces the computational cost of the algorithm. The algorithm only has one parameter and does not require iteration. This algorithm is used to improve the device grouping method. Starting from the grouping deviation that has the greatest impact on the equipment grouping result, the effective range of the number of groupings is used to reduce the invalid grouping result and reduce the grouping deviation.

3.1. Overview of CFSFDP Clustering Algorithm

The CFSFDP algorithm [18,19] is a clustering algorithm based on density peaks. The premise fora this clustering approach is mainly based on two assumptions: first, compared with other data points, the local density of the cluster center point is larger. Second, the distance between the cluster center and other points with higher local density is greater. The cluster center selected by the CFSFDP algorithm has a very large density, and the distance between different cluster centers is very far. This selection rule is consistent with the actual law. In the process of processing a large number of data sets, the superiority and simplicity of the algorithm are reflected.

For any data point x_i, the data point local density ρ_i and the point distance δ_i are defined. The local density ρ_i represents the number of data points in the data set S that is less than d_c from the data point x_i. The adjacent density point distance δ_i represents the minimum distance between the point where the local density is greater than the local density of the data point x_i and the x_i point in the data set $S = \{x_i\}_{i=1}^{N}$. For any point x_i in the data set $S = \{x_i\}_{i=1}^{N}$, there exists (ρ_i, δ_i), and a plane rectangular coordinate system is established, the abscissa is ρ, and the ordinate is δ. The (ρ_i, δ_i) corresponding to each data point x_i in the data set is drawn in the established plane rectangular coordinate system, and the clustering decision diagram of the data set is established. The description of the cluster center by the CFSFDP algorithm is available, and data points with large ρ and δ in the data set may become cluster centers.

3.2. Application of CFSFDP Clustering Algorithm in Equipment Marshalling Process

The attribute information of all devices are collected in a certain area of the bonding process segment, and the continuous attributes to create a new multi-dimensional data set are selected, and the CFSFDP algorithm is used to cluster the data sets. After clustering, several clusters are automatically generated. The number of classes can provide a constraint range for the number of *GN*s generated in the device grouping process, and reduce the grouping bias. Figure 3 shows improved device grouping flow chart.

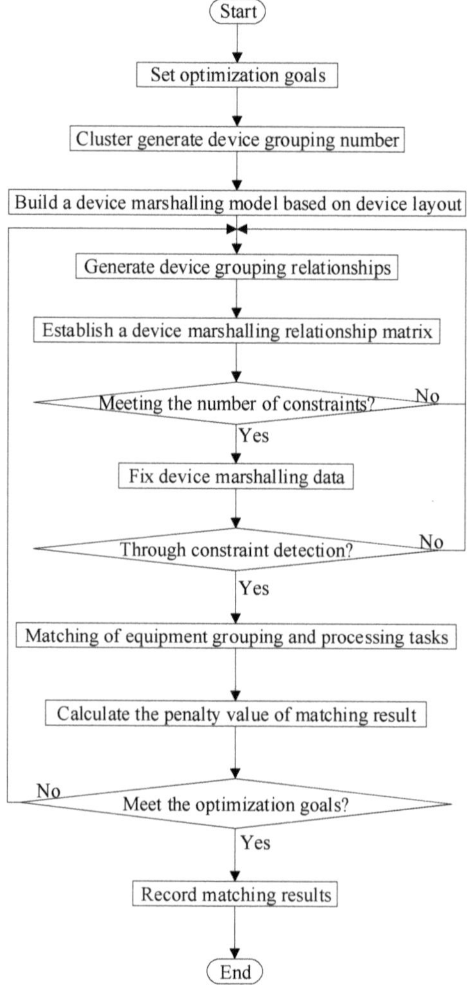

Figure 3. Improved device grouping flow chart.

Through the CFSFDP algorithm, the devices in the bonding process are automatically clustered, and the maximum number of clusters *GN* is generated by the number of clusters generated by the clustering result. Taking a small-scale device group as an example, a 4 × 5 device group is selected, and the collection device information is organized into an input data set. The CFSFDP algorithm reads the information of each device and numbers them according to the input order, and saves the attribute information of multiple dimensions. The Euclidean distance between the devices in all bonding process segments is then calculated and a similarity matrix is formed. The parameter d_c is determined according to the number of data points of the data set, and the local density ρ_i of each data point and the relative density point distance δ_i are calculated. γ_i is a value obtained by multiplying the local density ρ_i corresponding to all data points by its corresponding relative density point distance δ_i, and then grading the γ_i in descending order. Then, through the principle of decision graph, several data points with relatively large local density ρ and relative density point distance δ are selected as the cluster center of clusters [18]. A reasonable range of the number of groups of these devices is obtained, and used as a condition to judge whether the temporary grouping is reasonable and effective.

As shown in the decision diagram of Figure 4 below, the local density ρ is the coordinate axis x, and the relative density point distance δ is the coordinate axis y. The local density ρ_i and relative density point distance δ_i of these 20 devices are represented in the decision graph. The cluster center selected according to the CFSFDP algorithm is based on the principle that the density of the data points is very large and the distance between different cluster centers is very far. From 20 data points, a data point having a large local density compared with other data points and a large distance from other points with higher local density is found as the cluster center point of the 20-device data. According to Figure 4, the selected data points 6, 7, 10, 11, 12, 14, and 15 are points at which the local density ρ_i and the relative density point distance δ_i are simultaneously large. The product γ_i of the local density ρ_i corresponding to the 20 data points and the relative density point distance δ_i are arranged in descending order shown in Figure 5. In the screening of cluster centers, it mainly depends on two important parameters, local density ρ and distance δ of adjacent density points. The larger the product of the two, the more likely it is to become the center of the cluster. The product's product is obtained in descending order, and it can be seen that the falling slope has a distinct "inflection point". According to the cluster center map, the three points of 4, 6, and 11 have made significant changes in both the slope and the cluster center weight. For the cluster center, the weight of the cluster center has a very small number and the value is obviously large. As the weight of the cluster center decreases, the weight of the cluster center becomes slower and the value becomes smaller. The weight of the cluster center in the non-cluster has a large number, the value is obviously small and the difference between the points is small. Therefore, these three points may belong to the cluster center. Then, the average value α is obtained for the cluster center weight γ of all points, and the γ_i corresponding to each data point is compared with α. When γ_i is greater than α, the corresponding data points are used as the cluster center. The cluster center weight γ of point 4 is lower than the average value α, so it is excluded as the cluster center. The number of cluster clusters obtained is taken as the maximum number of device groups [20].

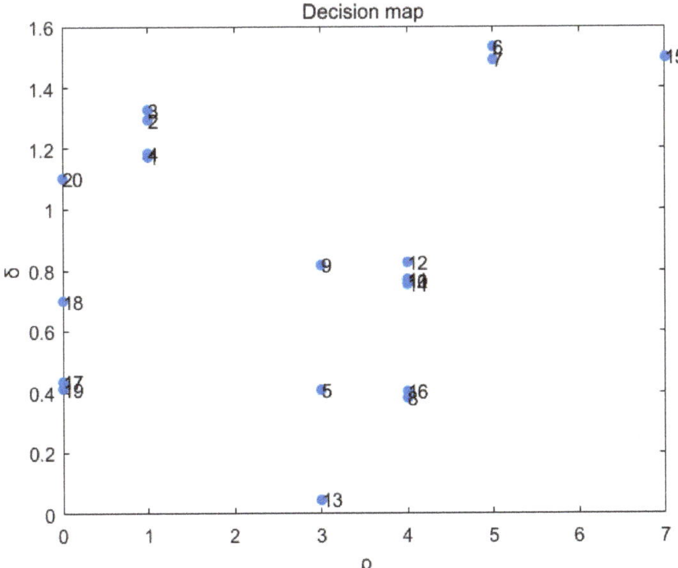

Figure 4. Cluster decision diagram of 4 × 5 device group.

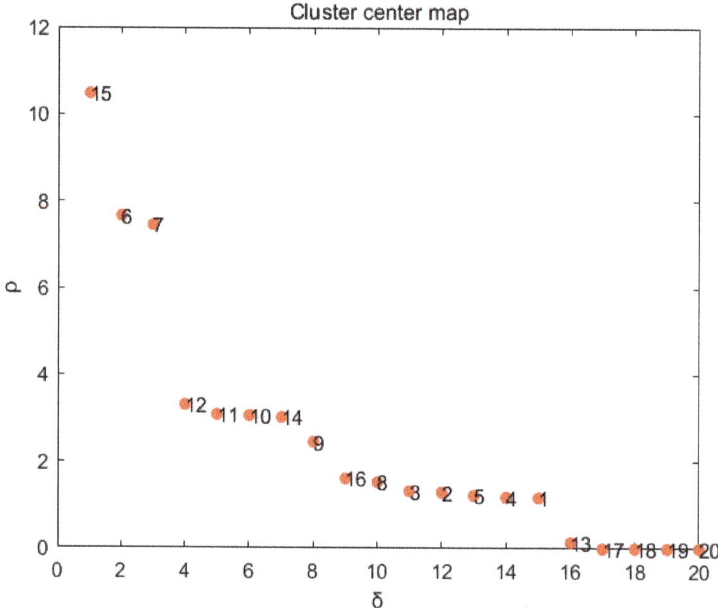

Figure 5. Cluster center map of 4 × 5 device group.

After the temporary device group G is created, it is necessary to determine whether the number of groups conforms to the maximum number of groups generated by the CFSFDP algorithm and the number of groups formed by the number of work tasks. If the grouping range requirements are met, the steps to modify the grouping data are performed. If they are not met, the relationship matrix is regenerated.

The simplified steps of the device grouping method improved by adding the CFSFDP algorithm are as follows.

Step 1: The minimum penalty value obtained by equipment grouping to provide capacity and processing task demand capacity matching is the optimization goal of equipment grouping.

The punishment sum of equipment grouping and processing tasks.

$$f_o = \sum_{p=1}^{PN} \left(\alpha_p PC_p + \beta_p EC_p \right) \tag{20}$$

α_p indicates the penalty for the capacity of the processing task p. β_p indicates the redundancy penalty for capacity of the processing task p.

Step 2: Using graph theory to simplify the representation of a device as a node on a two-dimensional plane [14], abstracting all device points of this process into n rows and m columns of device matrices. $m_{i,j}$ represents a device of the $i-th$ row and the $j-th$ column.

$$M = \begin{bmatrix} m_{1,1} & m_{1,2} & k & m_{1,m} \\ m_{2,1} & m_{2,2} & k & m_{2,m} \\ m & m & m & m \\ m_{n,1} & m_{n,2} & k & m_{n,m} \end{bmatrix} \tag{21}$$

Step 3: The collection device information is organized into an input data set, and the CFSFDP algorithm reads the information of each device and numbers them in the input order. The Euclidean distance between the devices in all bonding process segments is then calculated and a similarity matrix is formed. The parameter d_c is determined according to the number of data points of the data set, and the local density ρ_i of each data point and the relative density point distance δ_i are calculated, and the product γ_i of the local density ρ_i corresponding to all data points and the relative density point distance δ_i is descended. Then, through the principle of decision graph, several data points with relatively large local density ρ and relative density point distance δ are selected as the cluster center of the cluster, and then the reasonable number of groups of these devices is obtained.

Step 4: All devices are connected by wires to form a network structure with device as node. The association coefficient between the devices is randomly generated and represented by 0 and 1. The "0" indicates no association between the devices, i.e., the two devices are not in one group, and the "1" indicates the existence of an association, i.e., the two devices are edited in a group.

Step 5: According to the association coefficient between the devices, the device grouping row association matrix LR and the column association matrix LC are established by using the adjacency matrix, and the temporary device grouping G is established.

$$
LR = \begin{bmatrix}
l^r_{1,1} & l^r_{1,2} & l & l^r_{1,m} \\
l^r_{2,1} & l^r_{2,2} & l & l^r_{2,m} \\
m & m & l^r_{i,j} & m \\
l^r_{n-1,1} & l^r_{n-1,2} & l & l^r_{n-1,m}
\end{bmatrix}
\tag{22}
$$

$$
LC = \begin{bmatrix}
l^c_{1,1} & l^c_{1,2} & l & l^c_{1,m-1} \\
l^c_{2,1} & l^c_{2,2} & l & l^c_{2,m-1} \\
m & m & l^c_{i,j} & m \\
l^c_{n,1} & l^c_{n,2} & l & l^c_{n,m-1}
\end{bmatrix}
\tag{23}
$$

The associated device is grouped into a device group by the generated row and column association matrix. A temporary group number g_t is assigned to indicate the group number of the $t-th$ device group. GN represents the maximum group number for all groups.

Step 6: It is determined whether the temporary device grouping number GN is within the effective grouping number composed of the maximum number of groupings obtained by clustering and the number of processing tasks. If the thief continues to step 7 within this range, otherwise return to step 4 to re-establish the association.

Step 7: Correct the group association relationship. The temporary coding matrix of equipment grouping obtained in Step 5 may have a condition that conforms to the grouping rules but does not meet the actual grouping requirements. Therefore, it is necessary to correct the grouping relationship by employing the line scan and the column scan detection using the temporary group number g_t of each group identification in Step 5.

$$
l^c_{i,j} = \begin{cases} 0 & Mg_{i,j} \neq Mg_{i+1,j} \\ 1 & Mg_{i,j} = Mg_{i+1,j} \end{cases}
\tag{24}
$$

$$
l^r_{i,j} = \begin{cases} 0 & Mg_{i,j} \neq Mg_{i,j+1} \\ 1 & Mg_{i,j} = Mg_{i,j+1} \end{cases}
\tag{25}
$$

Step 8: Closed position constraint refers to a device group or other device group that does not belong to this group is included in a two-dimensional plane. Such a completely enclosed relationship can be understood as a closed position. Constraint detection of device grouping. Constraint detection is divided into device group closed position relationship constraint detection and device type and processing type matching constraint detection. If any constraint detection fails, the grouping result is invalid, and the grouping step returns to Step 4. Constraint detection of closed position relationship by equipment grouping.

Step 9: Matching of equipment grouping and processing tasks

(1)   Capacity provided by equipment grouping

A matrix is constructed based on product type to provide capacity matrix Gga, where ga_{t,k_x} indicates that the equipment group g_t produces the processing capacity of the k_x type product,

$$ga_{t,k_x} = \left(\sum_{i=n1_t}^{n2_t} \sum_{j=m1_t}^{m2_t} V_{i,j,k_x} \right) gT, i \in \{n1_t, \dots, n2_t\}, j \in \{m1_t, \dots, m2_t\}.$$

$$Gga = \begin{bmatrix} ga_{1,k_1} & ga_{1,k_2} & l & ga_{1,k_{kn}} \\ ga_{2,k_2} & ga_{2,k_2} & l & ga_{2,k_{kn}} \\ m & m & ga_{t,k_x} & m \\ ga_{GN,k_1} & ga_{GN,k_2} & l & ga_{GN,k_{kn}} \end{bmatrix} \tag{26}$$

(2)   Demand capacity of processing tasks

The processing task set is represented by W_p, $W_p = \{W_p | p \in \{1, 2, \dots, PN\}\}$. W_p represents the $p - th$ processing task, the product type of the $p - th$ processing task is represented by K_p, and the product type set constituting the processing task is represented by KP, $KP = \{K_p | p \in \{1, 2, \dots PN\}\}$. The set of processing tasks for product type k_x is represented by WP_x, $WP_x = \{W_p | k_x = k_p, p \in \{1, 2, \dots, PN\}\}$, and $WP_x \subseteq WP$, gw_p indicates the demand capacity of the $p - th$ processing task, and the processing task demand capacity set is represented by GW, $GW = \{gw_p | p \in \{1, 2, \dots PN\}\}$.

(3)   Equipment grouping matching the processing task

The total number of production equipment marshalling GN should be greater than or equal to the total number of tasks to be assigned PN. The number of selectable device groups corresponding to the product type k_x should be greater than or equal to the number of tasks in the processing task set whose product type is k_x.

The matrix matching the processing task demand ability and the equipment group capability is represented by $Gamc$. $Gamc = \{\{\langle gw_p, gma_p \rangle | p \in \{1, 2, \dots, PN\}\}\}$. gma_p indicates the processing capacity of the equipment group, gw_p indicates the demand capacity of the processing task.

Step 10: The optimization goal of device grouping is the penalty value f_o minimum for device grouping and processing tasks. If the optimization range is satisfied, the temporary grouping result is saved. If the requirement is not met, the process returns to Step 3 until the maximum number of iterations is reached, and the grouping corresponding to the minimum penalty value is stopped as the final grouping result.

4. Simulation Verification and Results Analysis

4.1. Simulation Experiment Design Related Information

The simulation experiment is divided into the comparison of the equipment grouping methods before and after the application of the processing task matching under different equipment scales, and the comparison of the equipment grouping methods before and after the application of the processing task matching with different discrete degree requirements. For the simulation experiment, we used MATLAB R2016a platform, with the following hardware parameters: CPU R 2.50 dual core and 4G memory.

The semiconductor industry is a dynamic production process, and there are many process segments. The bonding is only one process segment of the package test part, and the equipment of the bonding process segment is divided into multiple regions. This article only selects the equipment of a certain region. In the simulation experiment, the equipment was divided into three types: *A*, *B*, and *C*. The processing task product types were divided into two types: *a* and *b*. Type *a* products could only be processed by type *A*, and type *B* equipment, and type *b* products could only be processed by type *B* and type *C* equipment. The device grouping scale was divided into five types: 3×4 device group, 4×5 device group, 7×7 device group, 10×10 device group, and 15×15 device group.

Demand capacity was divided into dense and discrete according to the size of the difference coefficient [21]. The difference coefficient is the relative difference quantity, which can be used to compare the differences between data units, and can also be used to compare the differences of similar phenomena at different levels. The most commonly used coefficient of variation is proposed by Pearson (K.). The coefficient of variation removes the average standard deviation and multiplies it by 100% to form a percentage. The simulation experiment is carried out by selecting the demand capacity set of five different difference coefficients. The calculation formula is as follows, where S is the standard deviation and M is the average.

$$CV = \frac{S}{M} \times 100\% \tag{27}$$

4.2. Definition of Evaluation Indicators

The production grouping deviation d is the difference between the number of equipment groupings GN and the number of processing tasks PN.

$$d = GN - PN \tag{28}$$

Since the production line requires that the product type that can be processed in the equipment group is the same as that of the processing task, and the equipment constructed by the grouping equipment cannot contain equipment or equipment groups that are not grouped, the resulting grouping result may not be detected by the constraint. The proportion of detection by constraint m indicates the ratio of the generated grouping result by the constraint detection, GN indicates the number of generating device grouping groups, and F_g indicates the number of grouping groups that have not passed the constraint detection.

$$r = \frac{GN - F_g}{GN} \times 100\% \tag{29}$$

In actual production, production capacity redundancy can speed up production, but it can also lead to the waste of resources, and insufficient capacity allocation that can cause production delays. The optimal equipment grouping result is obtained when the penalty value for equipment grouping and processing tasks is the smallest. The device grouping and processing task penalty function is as follows [22].

$$f_o = \sum_{p=1}^{PN} \left(\alpha_p PC_p + \beta_p EC_p \right) \tag{30}$$

In the formula, α_p indicates the penalty for the capacity of the processing task p. β_p indicates the redundancy penalty for capacity of the processing task p.

4.3. Analysis of Experimental Results

In Tables 1–3, method one represents semiconductor bonding device grouping method based on dynamic matching of processing task, method two represents method for grouping semiconductor bonding equipment based on CFSFDP algorithm for dynamic matching of processing tasks.

Table 1. Comparison of pre-improvement equipment methods and improved equipment methods under different equipment scales when demand capacity distribution is intensive.

Equipment Scale	Distance Parameter d_c	Device Grouping Method Based on Processing Task Matching			Machining Task Matching Semiconductor Bonding Device Grouping Method Based on CFSFDP Algorithm		
		f_o	d	r	f_o	d	r
3×4	0.20	113.15	1.60	89.90	108.52	0	90.20
4×5	0.40	113.35	2.10	80.80	106.84	1.40	83.80
7×7	0.63	139.21	9.42	97.70	120.69	6.84	97.30
10×10	1.16	256.92	22.10	92.30	123.79	11.61	95.00
15×15	2.93	303.39	52.30	93.90	307.30	50.50	94.80

Table 2. Comparison of pre-improvement equipment methods and improved equipment methods under different equipment scales when demand capacity distribution is discrete.

Equipment Scale	Distance Parameter d_c	Device Grouping Method Based on Processing Task Matching			Machining Task Matching Semiconductor Bonding Device Grouping Method Based on CFSFDP Algorithm		
		f_o	d	r	f_o	d	r
3×4	0.20	134.82	1.50	77.00	133.92	0	78.70
4×5	0.40	89.17	1.30	88.50	87.03	1.02	91.10
7×7	0.63	154.31	9.59	96.40	106.20	6.80	98.40
10×10	1.16	258.46	22.3	92.50	110.502	11.52	94.80
15×15	2.93	282.82	52.23	95.90	273.852	44.80	94.20

Table 3. Comparison of pre-improvement equipment methods and improved equipment methods for different discrete-degree demand capacity.

Demand Capacity	Difference Coefficient CV	Device Grouping Method Based on Processing Task Matching			Machining Task Matching Semiconductor Bonding Device Grouping Method Based on CFSFDP Algorithm		
		f_o	d	r	f_o	d	r
GW1	0.09	102.24	0.93	74.00	101.59	0.64	76.50
GW2	0.11	115.04	1.03	83.30	108.94	0.63	86.82
GW3	0.28	114.06	0.86	84.51	112.77	0.55	87.40
GW4	0.39	112.04	0.95	83.10	108.13	0.71	85.70
GW5	0.47	114.14	0.97	80.35	112.79	0.65	85.00

Table 1 shows the experimental results of grouping equipment groups with different equipment scales based on the processing of job matching and the processing of the semiconductor bonding equipment based on the CFSFDP algorithm. Table 2 shows the experimental results of grouping device groups based on different device sizes when the internal data distribution of the demand capacity set is discrete. The experimental results show that under the demand capacity set with the same difference coefficient, the equipment scale remains unchanged, and the evaluation index of the semiconductor bonding equipment grouping method based on CFSFDP algorithm is better than the equipment grouping method before the improvement. Table 1 shows that after adding the CFSFDP algorithm, the grouping deviation d is reduced by about 24.4% on average, while from the Table 2 it can be seen that the grouping deviation d is reduced by about 35.6% on average. Before the device is grouped, the cluster of the device based on the density peak is obtained, as well as the cluster number of the device set. The maximum number of the device groups is obtained, thereby generating a reasonable range of device grouping. Therefore, the grouping deviation is reduced without changing the number of processing tasks. It can be seen from Table 1 that the coincidence rate r was increased by 1.3%, and the coincidence rate r in Table 2 was increased by 1.7% on average. The CFSDP algorithm aggregates similar devices into one class, and reduces unreasonable grouping results. This increases the proportion of devices that are constrained by the closed area constraints and the matching of the equipment to the processed product type. The optimization goal of equipment grouping is to obtain the lowest penalty value for the equipment grouping and processing tasks. As is shown in Table 1, the equipment grouping and processing task penalty and the value f_o were reduced by 24.6% on average, and the f_o in Table 2 was reduced by 22.6% on average. This shows that the difference between the capacity of the device group and the capacity required for the processing task is reduced after adding the CFSDP algorithm and improving the processing task sequence. Moreover, the value of PC_p and EC_p are reduced, and equipment marshalling and processing tasks are better matched.

Comprehensive analysis of the data shown in Tables 1 and 2, reveal that under the same device grouping method, the size of the device may have a certain impact on the device grouping results. As shown in Figures 6 and 7, following the increase of equipment scale, the grouping deviation, equipment grouping, processing task penalty and numerical value revealed a significant increase, and the coincidence rate had no obvious changes, indicating the dynamics of equipment grouping and processing tasks. The matching effect was worse, suggesting that the size of the device may affect the results of the device grouping.

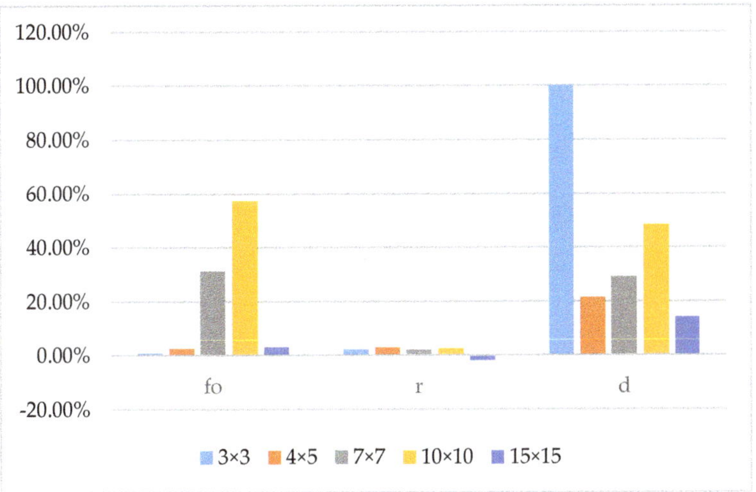

Figure 6. Schematic diagram of percentage changes of various evaluation indicators of the pre-improvement equipment grouping method under different equipment scales.

At present, the equipment in the workshop is equipped with wheels at the bottom, which can move the position according to production needs. Stationary grouping equipment usually adjusts the position every two days, which takes a certain amount of time. The equipment grouping method proposed in this paper can group equipment according to the processing task. The change of each grouping only changes the affiliation of the equipment, and does not need to change the position of the equipment. It only needs to be adjusted once a week, which greatly saves the time cost. Labor costs are basically unchanged here, because the managers in each area are a team. No matter how the devices in this area are grouped, the people who manage the area are not changed.

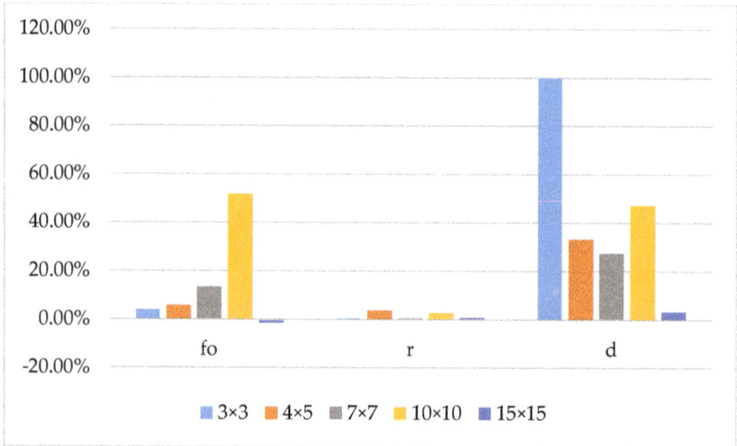

Figure 7. Schematic diagram of percentage changes in various evaluation indicators of the device grouping method with Please confirm that the added definition is correct algorithm under different device scales.

Table 3 shows the results of the semiconductor grouping method based on the processing task matching method and the processing task matching method based on the CFSFDP algorithm for the equipment of the same scale under the different discrete demand capacity. As shown in Figures 8–10, as the demand capacity difference coefficient increases from 0.09 to 0.47, the penalty and value obtained by the semiconductor bonding device grouping method based on the processing task matching increase by 11.64%, and the grouping deviation increases. At 17.8%, the coincidence rate was reduced by 12.43%. The penalty and value of the semiconductor bonding equipment grouping method based on CFSFDP algorithm increased by 11.02%, the grouping deviation increased by 12.7%, and the coincidence rate decreased by 12.47%. This suggests that the experimental results of the device grouping method improved by adding the CFSFDP algorithm. Under the premise of the same equipment scale, the capacity of the equipment group remains unchanged and as the coefficient of difference in demand capacity increases, the greater the dispersion of demand capacity. The capacity redundancy after the matching is PC_p or the capacity is insufficient EC_p increases, resulting in the equipment grouping and processing task penalty and value gradually increasing, the utilization rate of the equipment is decreasing. The coincidence rate gradually increases to a certain range, and as the fluctuation of demand capacity increases, the downward trend begins to occur. The degree of dispersion is the degree of difference between the values of the observed variables. It is an indicator used to measure the magnitude of the risk. Here is the difference coefficient. Therefore, the degree of dispersion of demand capacity will also affect the effect of dynamic grouping of equipment. The device grouping method combined with the CPSFDP algorithm can reduce such effects.

As shown in Figure 11, the penalty results obtained by the grouping of 4 × 5 device scales and the result of the smallest grouping are taken as an example. The figure shows that 20 devices are programmed into 5 production units, and the number of equipment groups is the same as the number of processing tasks. The grouping deviation is 0, and all the equipment is processed and produced. The device schematic reveals that the results of this grouping satisfy the closed constraint condition, and the equipment units do not cross each other. Also, the type *A* and type *B* equipment are programmed in one production unit to process the type *a* product, and the type *B* and type *a* equipment are programmed in one production unit to process the type *b* product, which conforms to the constraint of the matching relationship between the product type and the equipment type. Therefore, it is obvious that the semiconductor bonding equipment grouping method combined with

CFSFDP algorithm can meet the actual production requirements, reduce the grouping deviation and improve the equipment utilization.

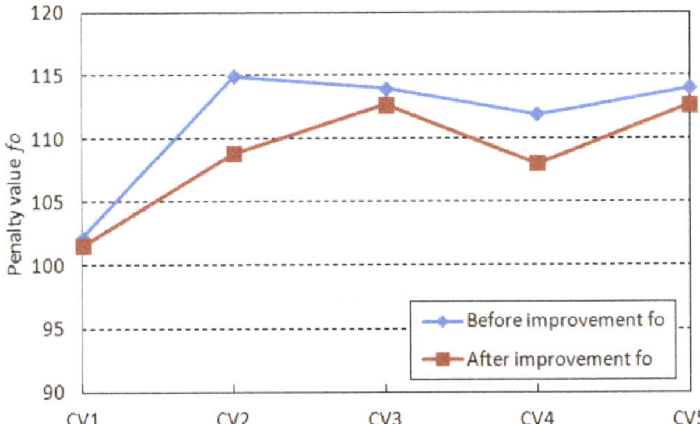

Figure 8. Comparison of the penalty value f_o obtained before the improvement of the different discrete degree demand capacity and the penalty value f_o obtained after the improvement.

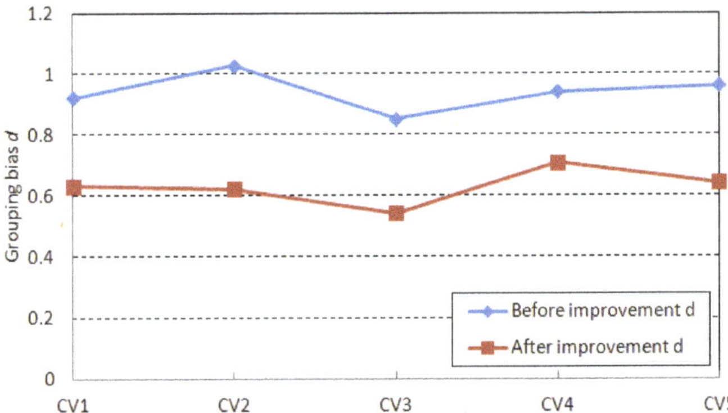

Figure 9. Comparison of the pre-improved marshalling deviation d and the improved marshalling deviation d under different discrete-degree demand capacities.

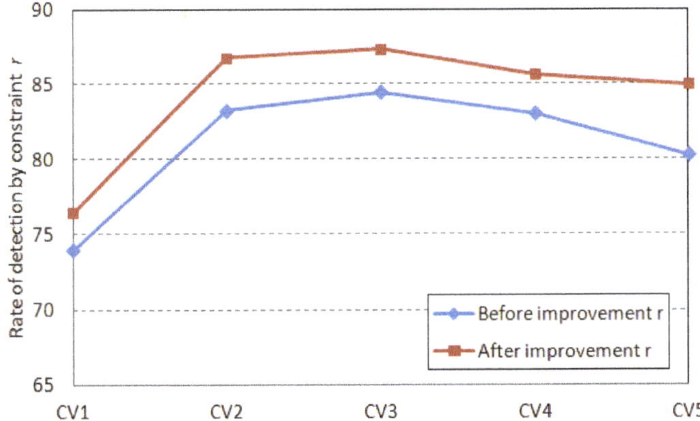

Figure 10. Comparison of the marshalling coincidence rate r of the pre-improvement and the improved grouping deviation rate r under different discrete-degree demand capacity.

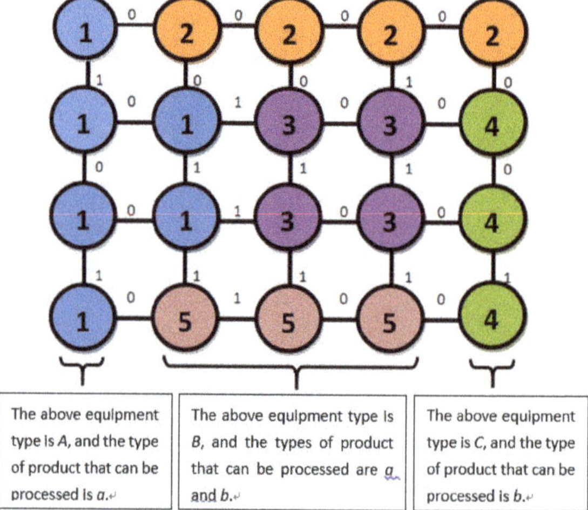

Figure 11. Take the grouping result of 4 × 5 device scale as an example.

Analysis from the perspective of time cost and labor cost, the equipment in the workshop is equipped with wheels at the bottom at present, which can move the position according to production needs. Stationary grouping equipment usually adjusts the position every two days, which takes a certain amount of time. The equipment grouping method proposed in this paper can group equipment according to the processing task. The change of each grouping only changes the affiliation of the equipment, and does not need to change the position of the equipment. It only needs to be adjusted once a week, which greatly saves the time cost. Labor costs are basically unchanged here, because the managers in each area are a team. No matter how the devices in this area are grouped, the people who manage the area are not changed. In summary, saving time costs means improving equipment utilization rate.

In summary, the device grouping method based on processing task matching achieves flexible matching of equipment grouping and processing tasks, and generates dynamic grouping. Based on the CFSFDP algorithm, the processing method of the semiconductor bonding equipment matching method is based on the dynamic matching of the equipment capacity and the demand capacity, further reducing the grouping deviation, saving time costs, improving the equipment utilization rate and achieving the dynamic grouping.

5. Conclusions

In this paper, we proposed a method for grouping semiconductor wire bonding equipment based on processing task matching. The basic process of building a semiconductor bonding equipment grouping model based on graph theory was discussed. We analyzed the number of groups in temporary groupings established following this method, given that a too large or too small of a number could lead to a defect causing grouping deviation. Aiming to solve this issue, before establishing the association relationship between devices in the device grouping method based on processing task matching, the CFSFDP algorithm was added to cluster the device attribute information sets and to obtain the maximum number of groups of device sets, so as to obtain the device group. Furthermore, the range constraint of the number of device groupings was obtained to reduce the grouping deviation.

Finally, the simulation experiments were carried out considering the size of the formation, the distribution of different demand capacities, and the difference coefficient of demand capacity. The experimental results showed that compared with the semiconductor bonding equipment grouping method based on processing task matching, the improved equipment grouping method with CFSFDP algorithm reduces the grouping deviation, improves the matching rate of the grouping result through production constraint detection, and reduces the equipment grouping and processing. The penalty of the task improves the equipment utilization of the bonding process segment and achieves a good dynamic grouping, thereby improving the efficiency of the entire semiconductor production line.

Author Contributions: Z.G. and Z.H. conceived and designed the experiments; W.S. performed the experiments; W.S. and J.P. analyzed the data; F.Q. contributed analysis tools; W.S. wrote the paper.

Funding: This research was funded by National Key R&D Program of China (grant number: 2018YFF0300304-04), Liaoning Provincial Science Foundation, China (grant number: 2018106008), Project of Liaoning Province Education Department, China (grant number: LJZ2017015).

Conflicts of Interest: The authors declare no conflict of interest.

References

1. Gao, Z.W.; Nguang, S.K.; Kong, D.X. Advances in Modelling, monitoring, and control for complex industrial systems. *Complexity* **2019**, *2019*, 2975083. [CrossRef]
2. Jia, P.D.; Wu, Q.D.; Li, L. Objective-driven dynamic dispatching rule for semiconductor wafer fabrication facilities. *Comput. Integr. Manuf. Syst.* **2014**, *20*, 2808–2813.
3. Zhang, G.H.; Liu, C.; Yao, L.L. A method of bottleneck detection of semiconductor assembly and test production line. *Chin. J. Electron Devices* **2015**, *38*, 44–48.
4. Wu, Q.D.; Ma, Y.M.; Li, L.; Qiao, F. Data-driven dynamic scheduling method for semiconductor production line. *Control Theory Appl.* **2015**, *32*, 1233–1239.
5. Wang, L.J.; Yao, P.Y. Cooperative task allocation methods in multiple groups using DLS-QGA. *Control Decis.* **2014**, *29*, 1562–1568.
6. Rong, Y.P.; Zhang, X.C. Optimization for train plan of urban rail transit based on hybrid train formation. *J. Transp. Syst. Eng. Inf. Technol.* **2016**, *16*, 117–122.
7. Wang, K.S.; Dong, Q.Z. An analysis on optimizing part and full routes of urban rail transit under different marshalling plans. *Railw. Transp. Econ.* **2018**, *40*, 94–99, 110.
8. Jacob, R.; Márton, P.; Maue, J.; Nunkesser, M. Multistage methods for freight train classification. *Networks* **2011**, *57*, 87–105. [CrossRef]
9. Hsieh, L.Y.; Hsieh, T.-J. A Throughput Management System for Semiconductor Wafer Fabrication Facilities: Design, Systems and Implementation. *Processes* **2018**, *6*, 16. [CrossRef]
10. Gao, Z.W.; Saxen, H.; Gao, C.H. Data-driven approaches for complex industrial systems. *IEEE Trans. Ind. Inform.* **2013**, *9*, 2210–2212. [CrossRef]
11. Gao, Z.W.; Kong, D.X.; Gao, C.H. Modeling and Control of Complex Dynamic Systems: Applied Mathematical Aspects. *J. Appl. Math.* **2012**, *2012*, 869792. [CrossRef]
12. Min, Y.; Wu, N.Q. Review of operations and control of cluster tools in semiconductor wafer fabrication. *Ind. Eng. J.* **2012**, *15*, 1–15.
13. Xiao, C.J.; Chen, H. Math model and scheduling method for the semiconductor assembling and testing line. *Trans. Beijing Inst. Technol.* **2013**, *33*, 1161–1164, 1170.
14. Ratkiewicz, A.; Truong, T.N. Application of chemical graph theory for automated mechanism generation. *J. Chem. Inf. Comput. Sci.* **2003**, *43*, 36–44. [CrossRef] [PubMed]
15. Fu, S.P.; Li, S.B.; Luo, N. Topological Transformation and Transmission Characteristics Analysis of Vehicle Auto Transmission Based on Graph Theory. *Shanghai Jiaotong Univ. Sci.* **2018**, *52*, 348–355.
16. Jayakrishna, K.; Vinodh, S.; Anish, S. A Graph Theory approach to measure the performance of sustainability enablers in a manufacturing organization. *Int. J. Sustain. Eng.* **2016**, *9*, 47–58. [CrossRef]
17. Chen, H.; Gan, M.X.; Song, M.Z. An improved recommendation algorithm based on graph model. *Appl. Mech. Mater.* **2013**, *380*, 1266–1269. [CrossRef]
18. Rodriguez, A.; Laio, A. Clustering by fast search and find of density peaks. *Science* **2014**, *344*, 1492–1496. [CrossRef] [PubMed]
19. Mehmood, R.; Zhang, G.; Bie, R.; Dawood, H.; Ahmad, H. Clustering by fast search and find of density peaks via heat diffusion. *Neurocomputing* **2016**, *208*, 210–217. [CrossRef]

20. Zhou, S.B.; Xu, W.X. A novel clustering algorithm based on relative density and decision graph. *Control Decis.* **2018**, *33*, 1921–1930.
21. Calif, R.; Soubdhan, T. On the use of the coefficient of variation to measure spatial and temporal correlation of global solar radiation. *Renew. Energy* **2016**, *88*, 192–199. [CrossRef]
22. Qiao, F.; Xu, X.H. Performance evaluation system for scheduling semiconductor wafer product line. *J. Tongji Univ. Nat. Sci.* **2007**, *4*, 537–542.

MDPI

St. Alban-Anlage 66

4052 Basel

Switzerland

Tel. +41 61 683 77 34

Fax +41 61 302 89 18

www.mdpi.com

Processes Editorial Office

E-mail: processes@mdpi.com

www.mdpi.com/journal/processes

www.ingramcontent.com/pod-product-compliance
Lightning Source LLC
LaVergne TN
LVHW071626170726

843515LV00009B/2494